T0297430

Applied and Numerical Harmonic Analysis

Series Editor
John J. Benedetto
University of Maryland
College Park, MD, USA

Editorial Advisory Board

Akram Aldroubi
Vanderbilt University
Nashville, TN, USA

Gitta Kutyniok
Technische Universität Berlin
Berlin, Germany

Douglas Cochran
Arizona State University
Phoenix, AZ, USA

Mauro Maggioni
Duke University
Durham, NC, USA

Hans G. Feichtinger
University of Vienna
Vienna, Austria

Zuowei Shen
National University of Singapore
Singapore, Singapore

Christopher Heil
Georgia Institute of Technology
Atlanta, GA, USA

Thomas Strohmer
University of California
Davis, CA, USA

Stéphane Jaffard
University of Paris XII
Paris, France

Yang Wang
Michigan State University
East Lansing, MI, USA

Jelena Kovačević
Carnegie Mellon University
Pittsburgh, PA, USA

More information about this series at http://www.springer.com/series/4968

Ole Christensen

An Introduction to Frames and Riesz Bases

Second Edition

 Birkhäuser

Ole Christensen
Department of Mathematics
    and Computer Science
Technical University of Denmark
Lyngby, Denmark

ISSN 2296-5009   ISSN 2296-5017   (electronic)
Applied and Numerical Harmonic Analysis
ISBN 978-3-319-25611-5   ISBN 978-3-319-25613-9   (eBook)
DOI 10.1007/978-3-319-25613-9

Library of Congress Control Number: 2015954170

Mathematics Subject Classification (2010): 41-01, 41-02, 42-01, 42-02, 42C15, 42C40

© Springer Science+Business Media New York 2003
© Springer International Publishing Switzerland 2016
This work is subject to copyright. All rights are reserved by the Publisher, whether the whole or part
of the material is concerned, specifically the rights of translation, reprinting, reuse of illustrations,
recitation, broadcasting, reproduction on microfilms or in any other physical way, and transmission or
information storage and retrieval, electronic adaptation, computer software, or by similar or dissimilar
methodology now known or hereafter developed.
The use of general descriptive names, registered names, trademarks, service marks, etc. in this publication does not imply, even in the absence of a specific statement, that such names are exempt from
the relevant protective laws and regulations and therefore free for general use.
The publisher, the authors and the editors are safe to assume that the advice and information in this
book are believed to be true and accurate at the date of publication. Neither the publisher nor the
authors or the editors give a warranty, express or implied, with respect to the material contained herein
or for any errors or omissions that may have been made.

Printed on acid-free paper

This Springer imprint is published by Springer Nature
The registered company is Springer International Publishing AG Switzerland

To Khadija, Jakob, Sara; and Karen

ANHA Series Preface

The *Applied and Numerical Harmonic Analysis (ANHA)* book series aims to provide the engineering, mathematical, and scientific communities with significant developments in harmonic analysis, ranging from abstract harmonic analysis to basic applications. The title of the series reflects the importance of applications and numerical implementation, but richness and relevance of applications and implementation depend fundamentally on the structure and depth of theoretical underpinnings. Thus, from our point of view, the interleaving of theory and applications and their creative symbiotic evolution is axiomatic.

Harmonic analysis is a wellspring of ideas and applicability that has flourished, developed, and deepened over time within many disciplines and by means of creative cross-fertilization with diverse areas. The intricate and fundamental relationship between harmonic analysis and fields such as signal processing, partial differential equations (PDEs), and image processing is reflected in our state-of-the-art *ANHA* series.

Our vision of modern harmonic analysis includes mathematical areas such as wavelet theory, Banach algebras, classical Fourier analysis, time-frequency analysis, and fractal geometry, as well as the diverse topics that impinge on them.

For example, wavelet theory can be considered an appropriate tool to deal with some basic problems in digital signal processing, speech and image processing, geophysics, pattern recognition, biomedical engineering, and turbulence. These areas implement the latest technology from sampling methods on surfaces to fast algorithms and computer vision methods. The underlying mathematics of wavelet theory depends not only on classical Fourier analysis but also on ideas from abstract harmonic analysis, including von Neumann algebras and the affine group. This leads to a study of the Heisenberg group and its relationship to Gabor systems and of the metaplectic group for a meaningful interaction of signal decomposition methods.

The unifying influence of wavelet theory in the aforementioned topics illustrates the justification for providing a means for centralizing and disseminating information from the broader, but still focused, area of harmonic analysis. This will be a key role of *ANHA*. We intend to publish with the scope and interaction that such a host of issues demands.

Along with our commitment to publish mathematically significant works at the frontiers of harmonic analysis, we have a comparably strong commitment to publish major advances in applicable topics such as the following, where harmonic analysis plays a substantial role:

Biomathematics, bioengineering, and biomedical signal processing;	*Machine learning;*
	Phaseless reconstruction;
Communications and RADAR;	*Quantum informatics;*
Compressive sensing (sampling) and sparse representations;	*Remote sensing;*
	Sampling theory;
Data science, data mining, and dimension reduction;	*Spectral estimation;*
Fast algorithms;	*Time-frequency and Time-scale analysis—Gabor theory and Wavelet theory*
Frame theory and noise reduction;	
Image processing and super-resolution;	

The above point of view for the *ANHA* book series is inspired by the history of Fourier analysis itself, whose tentacles reach into so many fields.

In the last two centuries Fourier analysis has had a major impact on the development of mathematics, on the understanding of many engineering and scientific phenomena, and on the solution of some of the most important problems in mathematics and the sciences. Historically, Fourier series were developed in the analysis of some of the classical PDEs of mathematical physics; these series were used to solve such equations. In order to understand Fourier series and the kinds of solutions they could represent, some of the most basic notions of analysis were defined, e.g., the concept of "function." Since the coefficients of Fourier series are integrals, it is no surprise that Riemann integrals were conceived to deal with uniqueness properties of trigonometric series. Cantor's set theory was also developed because of such uniqueness questions.

A basic problem in Fourier analysis is to show how complicated phenomena, such as sound waves, can be described in terms of elementary harmonics. There are two aspects of this problem: first, to find, or even define properly, the harmonics or spectrum of a given phenomenon, e.g., the spectroscopy problem in optics; second, to determine which phenomena can be constructed from given classes of harmonics, as done, for example, by the mechanical synthesizers in tidal analysis.

Fourier analysis is also the natural setting for many other problems in engineering, mathematics, and sciences. For example, Wiener's Tauberian theorem in Fourier analysis not only characterizes the behavior of the prime numbers but also provides the proper notion of spectrum for phenomena such as white light; this latter process leads to the Fourier analysis associated with correlation functions in filtering and prediction problems, and these problems, in turn, deal naturally with Hardy spaces in the theory of complex variables.

Nowadays, some of the theory of PDEs has given way to the study of Fourier integral operators. Problems in antenna theory are studied in terms of unimodular trigonometric polynomials. Applications of Fourier analysis abound in signal processing, whether with the fast Fourier transform (FFT), or filter design, or the adaptive modeling inherent in time-frequency-scale methods such as wavelet theory.

The coherent states of mathematical physics are translated and modulated Fourier transforms, and these are used, in conjunction with the uncertainty principle, for dealing with signal reconstruction in communications theory. We are back to the raison d'être of the *ANHA* series!

College Park, MD, USA   John J. Benedetto

Preface to the First Edition

Frames have fascinated me since day one. Every student in mathematics learns about bases in vector spaces, allowing one to represent each element in a convenient and unique way. One day in 1990, Henrik Stetkær, who was my master's thesis advisor, showed me the definition of a frame and told me that a frame is some kind of "overcomplete basis": one can also represent each element in the vector space via a frame, but the representation might not be unique. I was really surprised: How come that the question in, e.g., linear algebra always was how to *extract* a basis from an overcomplete set? Why did the idea that overcompleteness by itself could be useful never came up?

A search on Mathematical Reviews or Zentralblatt shows only a few titles of books or articles concerning frames published before 1991; among those we mention the original paper by Duffin and Schaeffer [262], the excellent book by Young [622], and the important papers by Daubechies, Grossmann, and Meyer [244], Daubechies [241], and Heil and Walnut [395]. Now, just ten years later, hundreds of papers have the word frame in the title, and perhaps a thousand discuss one or more results. Today, no single book can treat all the important and interesting results that have been published.

The aim of this book is to present parts of the modern theory for bases and frames in Hilbert spaces in a way that the material can be used in a graduate course, as well as by professional readers. For use in a graduate course, a number of exercises is included; they appear at the end of each chapter. The number of exercises gives a hint of the level of the chapter: there are many exercises in the introductory chapters, but only few in the advanced chapters. In the same spirit, almost all results in the introductory chapters appear with full proofs; in the advanced chapters several results are presented without proofs. We believe it is more useful to state a large number of results in a common framework than to see detailed proofs of significantly fewer statements; this feature also makes the book useful as a reference.

The content can be split naturally into three parts: Chapters 1–7 describe the theory on an abstract level, Chapters 9–18 describe explicit constructions in L^2-spaces, and Chapters 22–24 deal with selected research topics.

In Chapters 1–7 almost all results concern frames in general Hilbert spaces. The goal is an almost complete treatment of the known results for frames. For the explicit constructions in $L^2(-\pi, \pi)$ and $L^2(\mathbb{R})$, which appear in Chapters 9–18, the situation is different. For this part, I was forced to concentrate on selected parts of the theory. Since we are mainly interested in overcomplete systems, the theory presented in these chapters is part of a larger picture, and the reader will certainly benefit from knowledge of the background. Chapter 9 connects to the theory for nonharmonic Fourier series, cf. the book [622] by Young. Gabor frames arise naturally in the context of time-frequency analysis, and the book by Gröchenig [340] will clarify the role of Chapters 11–13 in time-frequency analysis. Finally, the role of wavelets is highlighted in the classic book [242] by Daubechies, which also gives the motivation for the study of frames arising from multiscale methods in Chapters 17–18. Technically, we do not rely on any of these books (only at a few places will we refer to results from them without proof), but they put the results of frame studies in the right perspective. Chapters 9–18 are also influenced by the fact that the material is used in several areas of applied mathematics; the reader will observe that although this is a book about mathematics, those chapters concentrate on applicable ways to construct frames rather than on abstract characterizations.

Let us describe the chapters in more detail. Chapter 1 presents basic results in finite-dimensional vector spaces with an inner product. This enables a reader with a basic knowledge of linear algebra to understand the idea behind frames without the technical complications in infinite-dimensional spaces. Many of the topics from the rest of the book are presented here, so Chapter 1 can also serve as an introduction to the later chapters.

Chapter 2 collects some definitions and conventions concerning infinite-dimensional vector spaces. Special attention is given to the Hilbert space $L^2(\mathbb{R})$ and operators hereon. We expect the reader to be familiar with this material; the chapter is too short to be considered as an introduction to the subject.

Chapter 3 describes the classical theory for bases in Hilbert spaces and Banach theory. The examples in this chapter are chosen so they lead naturally to the constructions in Chapters 9–18.

Chapter 4 highlights some of the limitations on the properties one can obtain from bases, and thus provides motivation for considering generalizations of bases.

Chapters 5–7 contain the core material about frames and Riesz bases. Chapter 5 is classical, but Chapter 7 contains several results published in the last five years. The interplay between frames and bases is discussed in detail in Chapter 7, and we also discuss frames that become bases when a certain number of elements are deleted.

Chapters 9–18 deal with frames having a special structure. A central part deals with various sufficient conditions for existence of those frames. The most fundamental frames, namely, frames of exponentials in $L^2(-\pi, \pi)$ and frames of translates in $L^2(\mathbb{R})$, are discussed in Chapter 9. If one wants to consider frames in $L^2(\mathbb{R})$, these frames easily lead to Gabor frames, which is the subject of Chapters 11–14. Wavelet frames are introduced in Chapter 15, and sufficient conditions to find them are given for arbitrary dilation parameter $a > 1$ and translation parameter $b > 0$. Some results concerning irregular wavelet frames are also presented there. Chapter 16 specializes to the important case $a = 2, b = 1$, which has attracted much attention during the past ten years. Constructions via multiscale methods are the focus in Chapters 17–18.

In Chapter 22, the question is stability of frames, i.e., whether a set of elements close to a frame is itself a frame. Since real-life measurements are never completely exact, this question is very important for applications.

Chapter 23 presents methods for the approximation of the inverse frame operator using finite subsets of the frame. Since every application of frame theory has to be performed in a finite-dimensional vector space, this topic is also of fundamental importance for applications.

Chapter 24 deals with extensions and generalizations of the material from the previous chapters. Expansions in Banach spaces and their relationship to frames in Hilbert spaces are discussed, as well as frames appearing via integrable group representations. The latter subject gives a unified description of the frames from Chapters 11–15.

Finally, an Appendix collects several basic results for easy reference. It also contains material on pseudo-inverse operators and splines which is not expected to be known in advance and therefore is treated in more detail.

For the purpose of a graduate course, we mention that if students have a good background in functional analysis, they can skip Chapter 1 and parts of Chapters 2–3. Chapter 4 is important as motivation, and Chapter 5 is also core material. But after covering these chapters, a course can continue in several ways. One possibility is to follow a theoretical track and consider the relationship between frames and bases in more detail; this could be followed by a study of one of the three final chapters. Another possibility is to continue with constructions of exponential frames and Gabor frames, or wavelet frames. If wavelets are chosen as the subject, it is worth noticing that the four wavelet chapters are almost independent of each other.

This book presents frames and Riesz bases from the functional analytic point of view, and we concentrate on their mathematical aspects. However, as demonstrated by several papers by Daubechies and others, frames are very useful in several areas of applied mathematics, including signal processing and image processing. But this part of the story should be told by the people who are directly involved in it, and we will only sketch a few applications.

It is a pleasure to thank the many colleagues and students who helped in the process of writing this book. The starting point was seventy pages of notes, which were written jointly with Torben Klint Jensen, who was at that time a master's student. My original idea was to write a book concentrating on frames in general Hilbert spaces; I am very happy that Thomas Strohmer and an anonymous reviewer suggested that I further go into detail with wavelet and Gabor systems. Their ideas added more than a hundred pages to the book and extended the scope considerably. Very useful suggestions for adding material were also given by Hans Feichtinger.

Alexander Lindner read a large part of the final manuscript and proposed many improvements. Elena Cordero, Niklas Grip, Per Christian Hansen, Reza Mahdavi, John Rassias, Henrik Stetkær, and Diana Stoeva read parts of the book and helped to spot mistakes; I am very grateful to all of them.

I am thankful to the Department of Mathematics at the Technical University of Denmark for providing me with the excellent working conditions that made it possible to concentrate on the book for two semesters. In addition, a large part of the book was written during a stay at the research group NuHAG at the University of Vienna. It is a pleasure to thank the group leader, Hans Feichtinger, and the members of NuHAG for their support.

I am thankful to John Benedetto for inviting me to write this book, and I thank the staff at Birkhäuser, especially Tom Grasso and Ann Kostant, for their assistance and support. Thanks are also given to Elizabeth Loew from Texniques, who helped with Latex problems.

Finally, I acknowledge support from the WAVE-program, sponsored by the Danish Science Foundation.

Ole Christensen
Kgs. Lyngby, Denmark
September 2002

Preface to the Second Edition

As I wrote the first edition of the book during 2001/2002, one of the goals was that at least the list of references should contain most of the frame literature. Now, 14 years later, even this very modest goal cannot be reached anymore. During the last 20 years frames have experienced an ever-increasing popularity, and they show up in many different contexts, explicitly or implicitly. Considering just four of the key topics, namely, (i) "Frames in finite-dimensional spaces," (ii) "Frames in general Hilbert spaces," (iii) "Frames in Gabor analysis," and (iv) "Frames and wavelet analysis," each of these topics could easily fill a book of the same size as the current book. Therefore one of the major decisions during the work on the second edition has been what to include – and at what level of details. My choice has been to follow the line from the first edition and present the core material (and frequently less known material that should belong to the core) in great detail, while other topics are treated as research topics with more focus on the connections between the results than the proofs. The fact that many recent and advanced results are presented without proofs made it possible still to give a quite broad picture of the frame theory; but clearly it also leaves a gap open for other authors who would like to give a detailed presentation focusing on one of the topics.

The new material mainly occurs in new chapters and sections, but of course the entire book has been updated with additional results and comments. On very compressed form the main additions can be described as follows:

- Sections 1.2 on tight frames and dual pairs of frames in finite-dimensional spaces. Section 1.8 on fusion frames. Section 1.9 on applications of frames. Finally, Section 1.10 which relates the properties of the harmonic frames to the ongoing research within finite frame theory.

- Extension and rearrangement of Chapter 2. Many results from the former appendix now appear here.

- Section 3.7, a new section on Riesz sequences; and Section 3.10 on sampling an analog-digital conversion.

- Several updates and additions in Chapter 4, which motivate the step from bases to frames.

- Section 5.2, a new section on frame sequences.

- Chapter 6, a new chapter that collects results about tight frames and dual pairs of frames in general Hilbert spaces.

- Section 7.2, a new section about relations between frames and their subsequences, with focus on the "strange" behavior of the lower frame bounds for finite subfamilies. And Section 7.7, a short section on the Feichtinger conjecture.

- Chapter 8, a new chapter on selected topics in general frame theory. It contains sections on G-frames, localization of frames, the R-dual sequences, a frame-like theory via unbounded operators, as well as a discussion of frames in the context of signal processing.

- Section 9.4, Section 9.5, and Section 9.7: new sections about the canonical dual of a frame of translates and oblique duals, as well as applications of frames of translates within sampling theory.

- Chapter 10, a new chapter on shift-invariant systems (parts of the material previously appeared within the presentation of Gabor frames).

- Extensions and updates in Section 11.6 on Gabor frames generated by special functions. Section 11.7, a new section collecting the known connections between B-splines and Gabor frames, as well as discussions about open problems.

- Chapter 12, a new chapter on dual pairs of Gabor frames and tight Gabor frames.

- Section 13.1, a new section about the duality principle. Section 13.5, a new section about localized Gabor frames. And Section 13.7, a new section on time-frequency localization.

- Chapter 14: new sections concerning duality of Gabor frames in $\ell^2(\mathbb{Z})$ and $\ell^2(\mathbb{Z}^d)$, and explicit construction of such frames based on dual pairs of Gabor frames for $L^2(\mathbb{R})$. Construction of periodic Gabor frames in $L^2(0, L)$, and description of the transition from a Gabor system in $L^2(\mathbb{R})$ to a finite-dimensional model in $\mathbb{C}^L$.

- Section 15.3, a new section on dual pairs of wavelet frames.

- Chapter 19, a new chapter on selected topics on wavelet frames. Some of the sections also appeared in the first edition of the book, but the sections on the extension problem and signal processing are new.

- Chapter 20, a new chapter on generalized shift-invariant systems.

- Chapter 21, a new chapter on frames on locally compact abelian groups.

- Section 23.3, a new section that yields convergence estimates in the context of finite-dimensional approximations of the inverse frame operator.

- Chapter 24 on Banach frames: the entire chapter has been updated with more recent results.

- Section A.5 and Section A.6, new sections stating the key properties of the modulation spaces and the Feichtinger algebra. Section A.9 and Section A.10, new sections on exponential B-splines and splines on LCA groups.

I would like to thank all the friends, colleagues, and students who have contributed to the current second edition. First and foremost I would like to thank my coauthors, who have definitely inspired me and shaped my view and understanding of frames over the years. Many of the papers with my coauthors were used as the starting point for various sections and chapters. For example, my papers with Hong Oh Kim and Rae Young Kim form the basis for Sections 6.4, 11.7, 12.5, 12.6, and 12.7; similarly, the paper [176] with Say Song Goh was the driving force behind most of the sections in Chapter 21.

I would also like to thank Hong Oh Kim and Rae Young Kim for organizing and supporting about 20 visits to Korea Advanced Institute for Science and Technology (KAIST) over the years, and for the many pleasant hours we spend working on joint problems; and Say Song Goh, whose many invitations to National University of Singapore (NUS) also gave me scientific inspiration and excellent working conditions, with direct influence on the current book.

It is a great pleasure to thank Henrik Stetkær, who used the first edition as textbook in several master courses at the University of Aarhus; this led to the discovery of several misprints and imprecisions, which I have tried to correct.

I thank Jakob Lemvig and Mads Sielemann Jakobsen for giving me access to a note providing a direct proof of the duality principle in Gabor analysis.

During the preparation of the manuscript, I got help from many colleagues and students to spot typing mistakes, bad formulations, etc.; I thank Say Song Goh, Marzieh Hasannasab, Christina Hildebrandt, Mads Sielemann Jakobsen, Jakob Lemvig, Diana Stoeva, and Jordy van Velthoven for their help, which clearly improved the manuscript.

Finally I want to thank John Benedetto for his never-ending support and positive attitude. I also thank the staff at Birkhäuser, especially Danielle Walker, for their support during the entire process.

Ole Christensen
Kgs. Lyngby, Denmark
August 2015

Contents

1

Frames in Finite-Dimensional Inner Product Spaces

In the study of vector spaces, one of the most important concepts is that of a basis. A basis provides us with an expansion of all vectors in terms of "elementary building blocks" and hereby helps us by reducing many questions concerning general vectors to similar questions concerning only the basis elements. However, the conditions to a basis are very restrictive – no linear dependence between the elements is possible, and sometimes we even want the elements to be orthogonal with respect to an inner product. This makes it hard or even impossible to find bases satisfying extra conditions, and this is the reason that one might look for a more flexible tool.

Frames are such tools. A frame for a vector space equipped with an inner product also allows each element in the space to be written as a linear combination of the elements in the frame, but linear independence between the frame elements is not required. Intuitively, one can think about a frame as a basis to which one has added more elements. In this chapter, we present frame theory in finite-dimensional vector spaces. While this restriction makes part of the theory trivial, it also makes the basic idea more transparent. Furthermore, our intention is to present the results in such a way as to give the right feeling about the infinite-dimensional setting as well. This also means that we sometimes use unusual words in the finite-dimensional setting. For example, we will frequently use the word "operator" for a linear map.

There are other reasons for starting with a chapter on finite-dimensional frames. Every "real-life" application of frames has to be performed in a finite-dimensional vector space, so even if we want to apply results from the infinite-dimensional setting, the frames will have to be confined to a

© Springer International Publishing Switzerland 2016
O. Christensen, *An Introduction to Frames and Riesz Bases*,
Applied and Numerical Harmonic Analysis,
DOI 10.1007/978-3-319-25613-9_1

finite-dimensional space at some point. Chapter 14 contains a detailed discussion of the transition from $L^2(\mathbb{R})$ to $\mathbb{C}^n$ in the context of Gabor frames.

Most of the chapter can be fully understood with an elementary knowledge of linear algebra. In order not to make the proofs too cumbersome, we will at a few places use some more advanced results; exact references to the literature will always be given in this case.

This chapter is organized as follows. Section 1.1 contains the basic properties of frames. For example, it is proved that every set of vectors $\{f_k\}_{k=1}^m$ in a vector space with an inner product is a frame for span$\{f_k\}_{k=1}^m$ and that every frame leads to expansions of the vectors in the underlying space. We prove the existence of coefficients minimizing the ℓ^2-norm in a frame expansion and show how a frame for a subspace leads to a formula for the orthogonal projection onto the subspace. Section 1.2 describes how frames in various ways can be extended to computationally more convenient frames. In Section 1.3 the role of the frame bounds is highlighted. Then, in Sections 1.4–1.5, we consider frames in $\mathbb{C}^n$; in particular, we prove how we can obtain an overcomplete frame by a projection of a basis for a larger space. In fact, the vectors $\{f_k\}_{k=1}^m$ in a frame for $\mathbb{C}^n$ can be considered as the first n coordinates of some vectors in $\mathbb{C}^m$ constituting a basis for $\mathbb{C}^m$. We also prove that the frame property for $\{f_k\}_{k=1}^m$ is equivalent to certain properties for the $m \times n$ matrix having the vectors f_k as rows. Furthermore, we show that the discrete Fourier transform basis leads to frame constructions with attractive properties. In Section 1.6 we prove that the canonical coefficients from the frame expansion arise naturally by considering the pseudo-inverse of the so-called synthesis operator, and we show how to compute it in terms of the singular value decomposition. Section 1.7 connects frames in finite-dimensional vector spaces with the infinite-dimensional constructions that will appear in later chapters, and Section 1.8 gives a short introduction to fusion frames. Finally, Section 1.9 discusses frames in the context of data transmission, and Section 1.10 relates the properties of the harmonic frames to the ongoing research within finite frame theory.

Much more can of course be said about frames in finite-dimensional vector spaces. Our main purpose is to give an elementary entrance to frame theory, streamlined toward the work on frames in infinite-dimensional spaces which will dominate the book. A reader who is mainly interested in the finite-dimensional case can consider the collection of papers [139] and the references therein for much more information. Also notice that a nice introduction to finite-dimensional frames directed toward students is given in the book [376] by Han et al.

1.1 Some Basic Facts About Frames

Let V be a finite-dimensional vector space, equipped with an inner product $\langle \cdot, \cdot \rangle$, which we choose to be linear in the first entry. Recall that a sequence $\{e_k\}_{k=1}^{m}$ in V is a *basis* for V if the following two conditions are satisfied:

(i) $V = \text{span}\{e_k\}_{k=1}^{m}$;

(ii) $\{e_k\}_{k=1}^{m}$ is *linearly independent*, i.e., if $\sum_{k=1}^{m} c_k e_k = 0$ for some scalar coefficients $\{c_k\}_{k=1}^{m}$, then $c_k = 0$ for all $k = 1, \ldots, m$.

As a consequence of this definition, every $f \in V$ has a unique representation in terms of the elements in the basis, i.e., there exist unique scalar coefficients $\{c_k\}_{k=1}^{m}$ such that

$$f = \sum_{k=1}^{m} c_k e_k. \tag{1.1}$$

If $\{e_k\}_{k=1}^{m}$ is an *orthonormal basis*, i.e., a basis for which

$$\langle e_k, e_j \rangle = \delta_{k,j} = \begin{cases} 1 & \text{if } k = j \\ 0 & \text{if } k \neq j, \end{cases}$$

then the coefficients $\{c_k\}_{k=1}^{m}$ are easy to find. In fact, taking the inner product of f in (1.1) with an arbitrary e_j gives

$$\langle f, e_j \rangle = \langle \sum_{k=1}^{m} c_k e_k, e_j \rangle = \sum_{k=1}^{m} c_k \langle e_k, e_j \rangle = c_j,$$

so

$$f = \sum_{k=1}^{m} \langle f, e_k \rangle e_k. \tag{1.2}$$

We now introduce frames; in Theorem 1.1.5 below, we prove that a frame $\{f_k\}_{k=1}^{m}$ also allows a representation of the form (1.1).

Definition 1.1.1 *A countable sequence of elements $\{f_k\}_{k \in I}$ in V is a frame for V if there exist constants $A, B > 0$ such that*

$$A \, ||f||^2 \leq \sum_{k \in I} |\langle f, f_k \rangle|^2 \leq B \, ||f||^2, \quad \forall f \in V. \tag{1.3}$$

The numbers A, B are called *frame bounds*. They are not unique. The *optimal lower frame bound* is the supremum over all lower frame bounds, and the *optimal upper frame bound* is the infimum over all upper frame bounds. Note that the optimal frame bounds are actually frame bounds. The frame is *normalized* if $||f_k|| = 1$, $\forall k \in I$.

In a finite-dimensional vector space, it is somehow artificial (though possible) to consider families $\{f_k\}_{k \in I}$ having infinitely many elements. In this chapter, we will only consider finite families $\{f_k\}_{k=1}^m$, $m \in \mathbb{N}$. With this restriction, Cauchy–Schwarz' inequality shows that

$$\sum_{k=1}^m |\langle f, f_k \rangle|^2 \le \sum_{k=1}^m ||f_k||^2 \, ||f||^2, \; \forall f \in V,$$

i.e., the upper frame condition is automatically satisfied. However, one can often find a better upper frame bound than $\sum_{k=1}^m ||f_k||^2$; it will become clear in Section 1.3 that good estimates for the frame bounds are indeed important.

In order for the lower condition in (1.3) to be satisfied, it is necessary that $\text{span}\{f_k\}_{k=1}^m = V$. This condition turns out to be sufficient. In fact, every finite sequence is a frame for the span of the elements:

Proposition 1.1.2 *Let $\{f_k\}_{k=1}^m$ be a sequence in V. Then $\{f_k\}_{k=1}^m$ is a frame for the vector space $W := \text{span}\{f_k\}_{k=1}^m$.*

Proof. We can assume that not all f_k are zero. As we have seen, the upper frame condition is satisfied with $B = \sum_{k=1}^m ||f_k||^2$. Now consider the continuous mapping

$$\phi : W \to \mathbb{R}, \; \phi(f) := \sum_{k=1}^m |\langle f, f_k \rangle|^2.$$

The unit sphere in W is compact, so we can find $g \in W$ with $||g|| = 1$ such that

$$A := \sum_{k=1}^m |\langle g, f_k \rangle|^2 = \inf \left\{ \sum_{k=1}^m |\langle f, f_k \rangle|^2 \mid f \in W, \; ||f|| = 1 \right\}.$$

It is clear that $A > 0$. Now given $f \in W, f \ne 0$, we have

$$\sum_{k=1}^m |\langle f, f_k \rangle|^2 = \sum_{k=1}^m |\langle \frac{f}{||f||}, f_k \rangle|^2 \, ||f||^2 \ge A \, ||f||^2. \qquad \square$$

Proposition 1.1.2 immediately leads to an important characterization of frames in a finite-dimensional space:

Corollary 1.1.3 *A family of elements $\{f_k\}_{k=1}^m$ in V is a frame for V if and only if $\text{span}\{f_k\}_{k=1}^m = V$.*

Corollary 1.1.3 shows that a frame might contain more elements than needed to be a basis. In particular, if $\{f_k\}_{k=1}^m$ is a frame for V and $\{g_k\}_{k=1}^n$ is an arbitrary finite collection of vectors in V, then $\{f_k\}_{k=1}^m \cup \{g_k\}_{k=1}^n$ is also a frame for V. A frame which is *not* a basis is said to be *overcomplete* or *redundant*.

Consider now a vector space V equipped with a frame $\{f_k\}_{k=1}^m$ and define a linear mapping

$$T : \mathbb{C}^m \to V, \ T\{c_k\}_{k=1}^m = \sum_{k=1}^m c_k f_k. \tag{1.4}$$

The operator T is usually called the *synthesis operator* or the *pre-frame operator*. The *adjoint* operator is given by (Exercise 1.1)

$$T^* : V \to \mathbb{C}^m, \ \ T^* f = \{\langle f, f_k \rangle\}_{k=1}^m, \tag{1.5}$$

and is called the *analysis operator*. Composing T with its adjoint T^*, we obtain the *frame operator*

$$S : V \to V, \ \ Sf = TT^* f = \sum_{k=1}^m \langle f, f_k \rangle f_k. \tag{1.6}$$

Note that in terms of the frame operator,

$$\langle Sf, f \rangle = \sum_{k=1}^m |\langle f, f_k \rangle|^2, \ f \in V; \tag{1.7}$$

the lower frame condition can thus be considered as some kind of "lower bound" on the frame operator.

A frame $\{f_k\}_{k=1}^m$ is *tight* if we can choose $A = B$ in the definition, i.e., if

$$\sum_{k=1}^m |\langle f, f_k \rangle|^2 = A \, ||f||^2, \ \forall f \in V. \tag{1.8}$$

For a tight frame, the exact value A in (1.8) is simply called the *frame bound*; in case $A = 1$, we call $\{f_k\}_{k=1}^m$ a *Parseval frame*. We note that for a tight frame (1.7) combined with Lemma 2.4.4 leads to a representation of $f \in V$ in terms of the frame elements (Exercise 1.2):

Proposition 1.1.4 *Assume that $\{f_k\}_{k=1}^m$ is a tight frame for V with frame bound A. Then $S = AI$ (here I is the identity operator on V), and*

$$f = \frac{1}{A} \sum_{k=1}^m \langle f, f_k \rangle f_k, \ \forall f \in V. \tag{1.9}$$

An interpretation of (1.9) is that if $\{f_k\}_{k=1}^m$ is a tight frame and we want to express $f \in V$ as a linear combination $f = \sum_{k=1}^m c_k f_k$, we can simply define $g_k = \frac{1}{A} f_k$ and take $c_k = \langle f, g_k \rangle$. Formula (1.9) is similar to the representation (1.2) via an orthonormal basis: the only difference is the factor $1/A$ in (1.9). For general frames we now prove that we still have a representation of each $f \in V$ of the form $f = \sum_{k=1}^m \langle f, g_k \rangle f_k$ for an appropriate choice of $\{g_k\}_{k=1}^m$. The obtained theorem is one of the most important results about frames, and (1.10) below is called the *frame decomposition*:

Theorem 1.1.5 *Let $\{f_k\}_{k=1}^m$ be a frame for V with frame operator S. Then the following hold:*

(i) *S is invertible and self-adjoint.*

(ii) *Every $f \in V$ can be represented as*

$$f = \sum_{k=1}^m \langle f, S^{-1} f_k \rangle f_k = \sum_{k=1}^m \langle f, f_k \rangle S^{-1} f_k. \tag{1.10}$$

(iii) *If $f \in V$ also has the representation $f = \sum_{k=1}^m c_k f_k$ for some scalar coefficients $\{c_k\}_{k=1}^m$, then*

$$\sum_{k=1}^m |c_k|^2 = \sum_{k=1}^m |\langle f, S^{-1} f_k \rangle|^2 + \sum_{k=1}^m |c_k - \langle f, S^{-1} f_k \rangle|^2.$$

Proof. Since $S = TT^*$, it is clear that S is self-adjoint. We now prove that S is injective. Let $f \in V$ and assume that $Sf = 0$. Then

$$0 = \langle Sf, f \rangle = \sum_{k=1}^m |\langle f, f_k \rangle|^2,$$

implying by the frame condition that $f = 0$. That S is injective actually implies that S is surjective, but let us give a direct proof. The frame condition implies by Corollary 1.1.3 that $\operatorname{span}\{f_k\}_{k=1}^m = V$, so the synthesis operator T is surjective. Given $f \in V$ we can therefore find $g \in \mathbb{C}^m$ such that $Tg = f$; we can choose $g \in \mathcal{N}_T^\perp = \mathcal{R}_{T^*}$, so it follows that $\mathcal{R}_S = \mathcal{R}_{TT^*} = V$. Thus, S is surjective, as claimed. Each $f \in V$ has the representation

$$f \;=\; SS^{-1} f = \sum_{k=1}^m \langle S^{-1} f, f_k \rangle f_k;$$

using that S is self-adjoint, we arrive at $f = \sum_{k=1}^m \langle f, S^{-1} f_k \rangle f_k$, as stated in (1.10). The second representation in (1.10) is obtained in the same way, using that $f = S^{-1} S f$. For the proof of (iii), suppose that $f = \sum_{k=1}^m c_k f_k$. We can write

$$\{c_k\}_{k=1}^m = \left(\{c_k\}_{k=1}^m - \{\langle f, S^{-1} f_k \rangle\}_{k=1}^m \right) + \{\langle f, S^{-1} f_k \rangle\}_{k=1}^m.$$

By the choice of $\{c_k\}_{k=1}^m$, we have

$$\sum_{k=1}^m \left(c_k - \langle f, S^{-1} f_k \rangle \right) f_k = 0,$$

i.e., $\{c_k\}_{k=1}^m - \{\langle f, S^{-1} f_k \rangle\}_{k=1}^m \in \mathcal{N}_T = \mathcal{R}_{T^*}^\perp$; since

$$\{\langle f, S^{-1} f_k \rangle\}_{k=1}^m = \{\langle S^{-1} f, f_k \rangle\}_{k=1}^m \in \mathcal{R}_{T^*},$$

we obtain (iii). $\qquad\square$

Every frame in a finite-dimensional space contains a subfamily which is a basis (Exercise 1.3). If $\{f_k\}_{k=1}^m$ is a frame but not a basis, there exist nonzero sequences $\{d_k\}_{k=1}^m$ such that $\sum_{k=1}^m d_k f_k = 0$. Therefore any given element $f \in V$ can be written

$$
\begin{aligned}
f &= \sum_{k=1}^m \langle f, S^{-1} f_k \rangle f_k + \sum_{k=1}^m d_k f_k \\
&= \sum_{k=1}^m \left(\langle f, S^{-1} f_k \rangle + d_k \right) f_k.
\end{aligned}
\tag{1.11}
$$

This demonstrates that in the redundant case f has many representations as superpositions of the frame elements. Theorem 1.1.5 shows that the coefficients $\{\langle f, S^{-1} f_k \rangle\}_{k=1}^m$ have minimal ℓ^2-norm among all sequences $\{c_k\}_{k=1}^m$ for which $f = \sum_{k=1}^m c_k f_k$. The numbers

$$
\langle f, S^{-1} f_k \rangle, \quad k = 1, \ldots, m
$$

are called *frame coefficients*. Note that because $S : V \to V$ is bijective, the sequence $\{S^{-1} f_k\}_{k=1}^m$ is also a frame by Corollary 1.1.3; it is called the *canonical dual frame* of $\{f_k\}_{k=1}^m$.

For frames consisting of only a few elements, the canonical dual frame and the corresponding frame decomposition can be found via elementary calculations:

Example 1.1.6 Let $\{e_k\}_{k=1}^2$ be an orthonormal basis for a two-dimensional vector space V with inner product. Let

$$
f_1 = e_1, \ f_2 = e_1 - e_2, \ f_3 = e_1 + e_2.
$$

Then $\{f_k\}_{k=1}^3$ is a frame for V. Using the definition of the frame operator,

$$
Sf = \sum_{k=1}^3 \langle f, f_k \rangle f_k,
$$

we obtain that

$$
Se_1 = e_1 + e_1 - e_2 + e_1 + e_2 = 3e_1
$$

and

$$
Se_2 = -(e_1 - e_2) + e_1 + e_2 = 2e_2.
$$

Thus,

$$
S^{-1} e_1 = \frac{1}{3} e_1, \ S^{-1} e_2 = \frac{1}{2} e_2.
$$

By linearity, the canonical dual frame is

$$
\begin{aligned}
\{S^{-1} f_k\}_{k=1}^3 &= \{S^{-1} e_1, S^{-1} e_1 - S^{-1} e_2, S^{-1} e_1 + S^{-1} e_2\} \\
&= \{\frac{1}{3} e_1, \frac{1}{3} e_1 - \frac{1}{2} e_2, \frac{1}{3} e_1 + \frac{1}{2} e_2\}.
\end{aligned}
$$

Via Theorem 1.1.5, the representation of $f \in V$ in terms of the frame is given by

$$f = \sum_{k=1}^{3} \langle f, S^{-1}f_k \rangle f_k$$

$$= \frac{1}{3}\langle f, e_1 \rangle e_1 + \langle f, \frac{1}{3}e_1 - \frac{1}{2}e_2 \rangle (e_1 - e_2) + \langle f, \frac{1}{3}e_1 + \frac{1}{2}e_2 \rangle (e_1 + e_2). \quad \square$$

Theorem 1.1.5 gives some special information in case $\{f_k\}_{k=1}^{m}$ is a basis:

Corollary 1.1.7 *Assume that $\{f_k\}_{k=1}^{m}$ is a basis for V. Then there exists a unique family $\{g_k\}_{k=1}^{m}$ in V such that*

$$f = \sum_{k=1}^{m} \langle f, g_k \rangle f_k, \ \forall f \in V. \tag{1.12}$$

In terms of the frame operator, $\{g_k\}_{k=1}^{m} = \{S^{-1}f_k\}_{k=1}^{m}$. Furthermore, $\langle f_j, g_k \rangle = \delta_{j,k}$.

Proof. The existence of a family $\{g_k\}_{k=1}^{m}$ satisfying (1.12) follows from Theorem 1.1.5; we leave the proof of the uniqueness to the reader. Applying (1.12) on a fixed element f_j and using that $\{f_k\}_{k=1}^{m}$ is a basis, we obtain that $\langle f_j, g_k \rangle = \delta_{j,k}$ for all $k = 1, 2, \ldots, m$. $\qquad \square$

We have already seen that, for given $f \in V$, the frame coefficients $\{\langle f, S^{-1}f_k \rangle\}_{k=1}^{m}$ have minimal ℓ^2-norm among all sequences $\{c_k\}_{k=1}^{m}$ for which $f = \sum_{k=1}^{m} c_k f_k$. We can also choose to minimize the norm in other spaces than ℓ^2; we now show the existence of coefficients minimizing the ℓ^1-norm.

Theorem 1.1.8 *Let $\{f_k\}_{k=1}^{m}$ be a frame for a finite-dimensional vector space V. Given $f \in V$, there exist coefficients $\{d_k\}_{k=1}^{m} \in \mathbb{C}^m$ such that $f = \sum_{k=1}^{m} d_k f_k$, and*

$$\sum_{k=1}^{m} |d_k| = \inf \left\{ \sum_{k=1}^{m} |c_k| \ \Big| \ f = \sum_{k=1}^{m} c_k f_k \right\}. \tag{1.13}$$

Proof. Fix $f \in V$. It is clear that we can choose a set of coefficients $\{c_k\}_{k=1}^{m}$ such that $f = \sum_{k=1}^{m} c_k f_k$; let $r := \sum_{k=1}^{m} |c_k|$. Since we want to minimize the ℓ^1-norm of the coefficients, it is also clear that we can now restrict our search for a minimizer to sequences $\{d_k\}_{k=1}^{m}$ belonging to the compact set

$$M := \left\{ \{d_k\}_{k=1}^{m} \in \mathbb{C}^m \ \big| \ |d_k| \leq r, \ k = 1, \ldots, m \right\}.$$

Now the result follows from the fact that the set

$$\left\{ \{d_k\}_{k=1}^m \in M \mid f = \sum_{k=1}^m d_k f_k \right\}$$

is compact and that the function $\phi : \mathbb{C}^m \to \mathbb{R}$, $\phi\{d_k\}_{k=1}^m := \sum_{k=1}^m |d_k|$ is continuous.   □

There are some important differences between Theorem 1.1.5 and Theorem 1.1.8. In Theorem 1.1.5 we find the sequence minimizing the ℓ^2-norm of the coefficients in the expansion of f explicitly; it is unique, and it depends linearly on f. On the other hand, Theorem 1.1.8 only gives the existence of an ℓ^1-minimizer, and it might not be unique (Exercise 1.9). Even if the minimizer is unique, it might not depend linearly on f (Exercise 1.10). An algorithm to find an ℓ^1-minimizer $\{d_k\}_{k=1}^m$ can be found in [151].

As we have seen in Proposition 1.1.2, every finite set of vectors $\{f_k\}_{k=1}^m$ is a frame for its span. In general, the frame decomposition associated with $\{f_k\}_{k=1}^m$ gives a convenient expression for the orthogonal projection onto span$\{f_k\}_{k=1}^m$:

Theorem 1.1.9 Let $\{f_k\}_{k=1}^m$ be a frame for a subspace W of the vector space V, with frame operator $S : W \to W$. Then the orthogonal projection of V onto W is given by

$$Pf = \sum_{k=1}^m \langle f, S^{-1} f_k \rangle f_k, \ f \in V. \tag{1.14}$$

Proof. It is enough to prove that if we define P by (1.14), then

$$Pf = f \text{ for } f \in W \text{ and } Pf = 0 \text{ for } f \in W^\perp.$$

The first equation follows by Theorem 1.1.5, and the second by the fact that the range of S^{-1} equals W because S is a bijection on W.   □

Let us connect to the topic of Chapter 22 and for a moment consider an orthonormal basis $\{e_k\}_{k=1}^n$ for V. It is clear that by adding a finite collection of vectors to $\{e_k\}_{k=1}^n$, we obtain a frame for V. Also, if we perturb the vectors $\{e_k\}_{k=1}^n$ slightly, we still have a basis, but in general not an orthonormal basis. More precisely, if $\{g_k\}_{k=1}^n$ is a family of vectors in V and

$$R := \left(\sum_{k=1}^n ||e_k - g_k||^2 \right)^{1/2} < 1,$$

then also $\{g_k\}_{k=1}^n$ is a basis for V. In fact, given a scalar sequence $\{c_k\}_{k=1}^n$, the opposite triangle inequality followed by Cauchy–Schwarz' inequality gives that

$$
\left\| \sum_{k=1}^{n} c_k g_k \right\| \geq \left\| \sum_{k=1}^{n} c_k e_k \right\| - \left\| \sum_{k=1}^{n} c_k (g_k - e_k) \right\|
$$

$$
\geq \left(\sum_{k=1}^{n} |c_k|^2 \right)^{1/2} - \left(\sum_{k=1}^{n} \|e_k - g_k\|^2 \right)^{1/2} \left(\sum_{k=1}^{n} |c_k|^2 \right)^{1/2}
$$

$$
= (1 - R) \left(\sum_{k=1}^{n} |c_k|^2 \right)^{1/2}.
$$

This shows that $\{g_k\}_{k=1}^{n}$ is linearly independent, and since $\dim V = n$, we conclude that $\{g_k\}_{k=1}^{n}$ is a basis. We return to more general perturbation results for frames in Chapter 22.   □

1.2   Extensions to Tight Frames and Dual Frames

The simplicity of the calculations in Example 1.1.6 is slightly misleading: for a general frame, calculation of the canonical dual frame might be very cumbersome and lengthy if the frame contains many elements. This explains the prominent role of tight frames, for which the representation (1.10) takes the much simpler form (1.9).

It is worth noticing that every frame can be extended to a tight frame by adding some vectors. This was first proved by Casazza and Leonhard [145]:

Proposition 1.2.1 Let $\{f_k\}_{k=1}^{m}$ be a frame for a vector space V with dimension n. Then there exist $n-1$ vectors $h_2, \ldots, h_n$ such that the collection $\{f_k\}_{k=1}^{m} \bigcup \{h_k\}_{k=2}^{n}$ forms a tight frame for V.

Proof.   Denote the frame operator for $\{f_k\}_{k=1}^{m}$ by $S : V \to V$. Since S is self-adjoint, Theorem A.1.1 shows that V has an orthonormal basis consisting of eigenvectors $\{e_k\}_{k=1}^{n}$ for S. Denote the corresponding eigenvalues by $\{\lambda_k\}_{k=1}^{n}$; they are real numbers because S is self-adjoint. We will assume that the eigenvectors and eigenvalues are ordered such that $\lambda_1 \geq \lambda_2 \geq \cdots \geq \lambda_n$. Now, for $k = 2, \ldots, n$, let $h_k := \sqrt{\lambda_1 - \lambda_k} e_k$. The frame operator $\widetilde{S}$ for the family $\{f_k\}_{k=1}^{m} \bigcup \{h_k\}_{k=2}^{n}$ is given by

$$
\widetilde{S} : V \to V, \quad \widetilde{S} f = Sf + \sum_{k=2}^{n} \langle f, h_k \rangle h_k. \tag{1.15}
$$

Now consider an arbitrary $f \in V$. Using that

$$
f = \sum_{k=1}^{n} \langle f, e_k \rangle e_k,
$$

we see that the action of the frame operator S on f is given by

$$Sf = \sum_{k=1}^{n} \langle f, e_k \rangle S e_k = \sum_{k=1}^{n} \lambda_k \langle f, e_k \rangle e_k.$$

Inserting this expression and the definition of h_k into (1.15) shows that

$$
\begin{aligned}
\widetilde{S} f &= \sum_{k=1}^{n} \lambda_k \langle f, e_k \rangle e_k + \sum_{k=2}^{n} (\lambda_1 - \lambda_k) \langle f, e_k \rangle e_k \\
&= \lambda_1 \langle f, e_1 \rangle e_1 + \sum_{k=2}^{n} \lambda_k \langle f, e_k \rangle e_k + \lambda_1 \sum_{k=2}^{n} \langle f, e_k \rangle e_k - \sum_{k=2}^{n} \lambda_k \langle f, e_k \rangle e_k \\
&= \lambda_1 \sum_{k=1}^{n} \langle f, e_k \rangle e_k \\
&= \lambda_1 f.
\end{aligned}
$$

This implies that for all $f \in V$,

$$\sum_{k=1}^{n} |\langle f, f_k \rangle|^2 + \sum_{k=2}^{n} |\langle f, h_k \rangle|^2 = \langle \widetilde{S} f, f \rangle = \lambda_1 \|f\|^2,$$

i.e., $\{f_k\}_{k=1}^{m} \bigcup \{h_k\}_{k=2}^{n}$ is a tight frame with frame bound λ_1.   □

Note that Proposition 1.2.1 implies a slightly stronger result: any finite sequence in a finite-dimensional space can be extended to a tight frame.

We also note that while it is possible to extend any finite sequence to a tight frame, the proof of Proposition 1.2.1 shows that it might not be convenient or fast to do so in practice.

Often there is an alternative way to obtain "simple" frame expansions. In fact, one can prove (see Lemma 6.3.1) that if $\{f_k\}_{k=1}^{m}$ is a frame but not a basis, there exist frames $\{g_k\}_{k=1}^{m} \neq \{S^{-1} f_k\}_{k=1}^{m}$ such that

$$f = \sum_{k=1}^{m} \langle f, g_k \rangle f_k. \tag{1.16}$$

Thus, rather than restricting the attention to tight frames, one could consider frames $\{f_k\}_{k=1}^{m}$, for which there is an easy way to find a frame $\{g_k\}_{k=1}^{m}$ satisfying (1.16). Any frame $\{g_k\}_{k=1}^{m}$ satisfying (1.16) is called a *dual frame* of $\{f_k\}_{k=1}^{m}$; in Exercise 1.6 all these frames are calculated explicitly for a given frame $\{f_k\}_{k=1}^{m}$. The following example shows that it might be an advantage to extend a given frame to a pair of dual frames rather than a tight frame. The exact meaning of what it means to extend to a dual frame pair will be discussed in Section 6.4; see in particular Theorem 6.4.1.

Example 1.2.2 Let $\{e_k\}_{k=1}^{10}$ be an orthonormal basis for $\mathbb{C}^{10}$, and consider the frame

$$\{f_k\}_{k=1}^{10} := \{2e_1\} \cup \{e_k\}_{k=2}^{10}.$$

Then $\{f_k\}_{k=1}^{10}$ is a frame for $\mathbb{C}^{10}$. By Proposition 1.2.1 there exist vectors $\{h_k\}_{k=1}^{9}$ such that $\{f_k\}_{k=1}^{10} \cup \{h_k\}_{k=1}^{9}$ is a tight frame. Furthermore, one can show (Exercise 1.7) that nine is the *smallest number* of vectors we can add if we want to extend $\{f_k\}_{k=1}^{10}$ to a tight frame.

On the other hand, it is clear that both of the systems

$$\{f_k\}_{k=1}^{10} \cup \{e_1\} \text{ and } \{f_k\}_{k=1}^{10} \cup \{-3e_1\}$$

form frames and that they are actually dual frames. Thus, the extension to a dual pair can be realized with the addition of just one vector.   □

Characterizations and explicit constructions of dual pairs of frames will be a key issue in our treatment of structured function systems (like Gabor systems and wavelet systems) in the later chapters.

1.3 Frame Bounds and Frame Algorithms

The speed of convergence in numerical procedures involving a strictly positive definite matrix depends heavily on the *condition number* of the matrix, which is defined as the ratio between the largest eigenvalue and the smallest eigenvalue. In the case of the frame operator, these eigenvalues correspond to the optimal frame bounds:

Theorem 1.3.1 *Let $\{f_k\}_{k=1}^{m}$ be a frame for V, with frame operator S. Then the following hold:*

(i) *The optimal lower frame bound is the smallest eigenvalue for S, and the optimal upper frame bound is the largest eigenvalue.*

(ii) *Let $\{\lambda_k\}_{k=1}^{n}$ denote the eigenvalues for S; each eigenvalue appears in the list corresponding to its algebraic multiplicity. Then*

$$\sum_{k=1}^{n} \lambda_k = \sum_{k=1}^{m} ||f_k||^2.$$

(iii) *Assume that V has dimension n. If $\{f_k\}_{k=1}^{m}$ is tight and $||f_k|| = 1$ for all k, then the frame bound is $A = m/n$.*

Proof. Assume that $\{f_k\}_{k=1}^{m}$ is a frame for V. Since the frame operator $S : V \to V$ is self-adjoint, Theorem A.1.1 shows that V has an orthonormal basis consisting of eigenvectors $\{e_k\}_{k=1}^{n}$ for S. Denote the corresponding eigenvalues by $\{\lambda_k\}_{k=1}^n$, since S is self-adjoint, they are real. Given $f \in V$, we can write $f = \sum_{k=1}^n \langle f, e_k \rangle e_k$. Then

$$Sf = \sum_{k=1}^n \langle f, e_k \rangle S e_k = \sum_{k=1}^n \lambda_k \langle f, e_k \rangle e_k,$$

and

$$\sum_{k=1}^m |\langle f, f_k \rangle|^2 = \langle Sf, f \rangle = \sum_{k=1}^n \lambda_k |\langle f, e_k \rangle|^2.$$

Therefore,

$$\lambda_{\min} ||f||^2 \le \sum_{k=1}^m |\langle f, f_k \rangle|^2 \le \lambda_{\max} ||f||^2.$$

So $\lambda_{\min}$ is a lower frame bound, and $\lambda_{\max}$ is an upper frame bound. That they are the optimal frame bounds follows by taking f to be an eigenvector corresponding to $\lambda_{\min}$ (respectively $\lambda_{\max}$).

For the proof of (ii), we have

$$\sum_{k=1}^n \lambda_k = \sum_{k=1}^n \lambda_k ||e_k||^2 = \sum_{k=1}^n \langle Se_k, e_k \rangle$$
$$= \sum_{k=1}^n \sum_{\ell=1}^m |\langle e_k, f_\ell \rangle|^2.$$

Interchanging the sums and using that $\{e_k\}_{k=1}^n$ is an orthonormal basis for V now gives (ii). For the proof of (iii), the assumptions imply that the set of eigenvalues $\{\lambda_k\}_{k=1}^n$ consists of the frame bound A repeated n times; thus, the result follows from (ii). $\qquad \square$

Corollary 1.3.2 *Let $\{f_k\}_{k=1}^m$ be a frame for V. Then the condition number for the frame operator is equal to the ratio between the optimal upper frame bound and the optimal lower frame bound.*

If we want to find an element $f \in V$ based on knowledge of the coefficients $\{\langle f, f_k \rangle\}_{k=1}^m$, we can use Theorem 1.1.5:

$$f = \sum_{k=1}^m \langle f, f_k \rangle S^{-1} f_k = S^{-1} T \{\langle f, f_k \rangle\}_{k=1}^m.$$

However, in order for this formula to be useful, we need to invert the frame operator, which can be complicated if the dimension of V is large. Another option is to use an algorithm to obtain approximations of f. A classical algorithm is known as the *frame algorithm*:

Lemma 1.3.3 *Let* $\{f_k\}_{k=1}^m$ *be a frame for* V *with frame bounds* A, B. *Given* $f \in V$, *define the sequence* $\{g_k\}_{k=0}^\infty$ *in* V *by*

$$g_0 = 0, \quad g_k = g_{k-1} + \frac{2}{A + B} S(f - g_{k-1}), \quad k \geq 1. \tag{1.17}$$

Then

$$\|f - g_k\| \leq \left(\frac{B - A}{B + A}\right)^k \|f\|.$$

Proof. Let I denote the identity operator on V. Using (1.7),

$$\langle (I - \frac{2}{A + B} S)f, f \rangle = \|f\|^2 - \frac{2}{A + B} \sum_{k=1}^m |\langle f, f_k \rangle|^2, \ \forall f \in V,$$

so via the frame condition,

$$\langle (I - \frac{2}{A + B} S)f, f \rangle \leq \|f\|^2 - \frac{2A}{A + B} \|f\|^2 = \frac{B - A}{B + A} \|f\|^2.$$

Similarly,

$$-\frac{B - A}{B + A} \|f\|^2 \leq \langle (I - \frac{2}{A + B} S)f, f \rangle.$$

The two inequalities and (2.8) together give that

$$\left\| I - \frac{2}{A + B} S \right\| \leq \frac{B - A}{B + A}.$$

Using the definition of $\{g_k\}_{k=0}^\infty$,

$$\begin{aligned}
f - g_k &= f - g_{k-1} - \frac{2}{A + B} S(f - g_{k-1}) \\
&= \left(I - \frac{2}{A + B} S \right)(f - g_{k-1}),
\end{aligned}$$

and by repeating the argument,

$$f - g_k = \left(I - \frac{2}{A + B} S \right)^k (f - g_0).$$

Thus, applying (2.4) and (2.5),

$$\begin{aligned}
\|f - g_k\| &= \left\| \left(I - \frac{2}{A + B} S \right)^k (f - g_0) \right\| \\
&\leq \left\| I - \frac{2}{A + B} S \right\|^k \|f - g_0\| \leq \left(\frac{B - A}{B + A} \right)^k \|f\|,
\end{aligned}$$

as desired. □

In particular, the vectors g_k in (1.17) converge to f as $k \to \infty$. The algorithm depends on the knowledge of some frame bounds, and the guaranteed speed of convergence also depends on them. If B is much larger than A (either because only bad estimates for the optimal bounds are known or because the frame is far from being tight), the convergence might be slow. It is natural to apply some of the known acceleration algorithms from linear algebra to obtain faster convergence. Gröchenig showed in [338] how to apply the *Chebyshev method* and the *conjugate gradient method*. For the sake of the numerically oriented reader, we will give a short presentation of these results but refer to the original paper for the details. We begin with the Chebyshev method:

Theorem 1.3.4 *Let $\{f_k\}_{k=1}^m$ be a frame for V with frame bounds A, B, and let*

$$\rho := \frac{B - A}{B + A}, \quad \sigma := \frac{\sqrt{B} - \sqrt{A}}{\sqrt{B} + \sqrt{A}}.$$

Given $f \in V$, define the sequence $\{g_k\}_{k=0}^\infty$ in V and corresponding numbers $\{\lambda_k\}_{k=1}^\infty$ by

$$g_0 = 0, \quad g_1 = \frac{2}{A + B} Sf, \quad \lambda_1 = 2,$$

and for $k \geq 2$,

$$\lambda_k = \frac{1}{1 - \frac{\rho^2}{4}\lambda_{k-1}}, \quad g_k = \lambda_k \left(g_{k-1} - g_{k-2} + \frac{2}{A + B} S(f - g_{k-1}) \right) + g_{k-2}.$$

Then

$$\|f - g_k\| \leq \frac{2\sigma^k}{1 + \sigma^{2k}} \|f\|.$$

The Chebyshev algorithm guarantees a faster convergence than the frame algorithm when B is much larger than A. Knowledge of some frame bounds is also needed in order to apply the Chebyshev algorithm. In contrast, the conjugate gradient algorithm described below works without knowledge of the frame bounds: only when we want to estimate the error $\|f - g_k\|$ do we need them. Following Gröchenig, we formulate the result using the norm

$$\||f\|| = \langle f, Sf \rangle^{1/2}, \quad f \in V.$$

We leave it to the reader to check that $\|| \cdot \||$ is in fact a norm on V. Remember also that all norms on a finite-dimensional vector space are equivalent; that is, there exist constants $C_1, C_2 > 0$ such that

$$C_1 \|f\| \leq \||f\|| \leq C_2 \|f\|, \quad \forall f \in V.$$

This means that an error estimate in the norm $\|| \cdot \||$ can be transferred into an error estimate in the usual norm.

Theorem 1.3.5 *Let $\{f_k\}_{k=1}^m$ be a frame for V. Let $f \in V \setminus \{0\}$ and define the vectors $\{g_k\}_{k=0}^\infty, \{r_k\}_{k=0}^\infty, \{p_k\}_{k=-1}^\infty$ and numbers $\{\lambda_k\}_{k=0}^\infty$ by*

$$g_0 = 0, \ r_0 = p_0 = Sf, \ p_{-1} = 0$$

and, for $k \geq 0$,

$$\lambda_k = \frac{\langle r_k, p_k \rangle}{\langle p_k, Sp_k \rangle},$$

$$g_{k+1} = g_k + \lambda_k p_k,$$

$$r_{k+1} = r_k - \lambda_k Sp_k,$$

$$p_{k+1} = Sp_k - \frac{\langle Sp_k, Sp_k \rangle}{\langle p_k, Sp_k \rangle}p_k - \frac{\langle Sp_k, Sp_{k-1} \rangle}{\langle p_{k-1}, Sp_{k-1} \rangle}p_{k-1}.$$

Then $g_k \to f$ as $k \to \infty$. If we let A denote the smallest eigenvalue for S and B the largest eigenvalue and let $\sigma = \frac{\sqrt{B}-\sqrt{A}}{\sqrt{B}+\sqrt{A}}$, the speed of convergence can be estimated by

$$|||f - g_k||| \leq \frac{2\sigma^k}{1 + \sigma^{2k}}|||f|||.$$

In the expression for p_{k+1}, the last term is interpreted as zero for $k = 0$.

1.4   Frames in $\mathbb{C}^n$

The natural examples of finite-dimensional vector spaces are

$$\mathbb{R}^n = \{(c_1, c_2, \ldots, c_n) \mid c_i \in \mathbb{R}, \ i = 1, \ldots, n\}$$

and

$$\mathbb{C}^n = \{(c_1, c_2, \ldots, c_n) \mid c_i \in \mathbb{C}, \ i = 1, \ldots, n\};$$

the latter is equipped with the inner product

$$\langle \{c_k\}_{k=1}^n, \{d_k\}_{k=1}^n \rangle = \sum_{k=1}^n c_k \overline{d_k}$$

and the associated norm

$$||\{c_k\}_{k=1}^n|| = \sqrt{\sum_{k=1}^n |c_k|^2}.$$

This corresponds to the definitions in $\mathbb{R}^n$, except that complex conjugation and modulus is not needed in the real case. We will describe the theory for bases and frames in $\mathbb{C}^n$, but the results have direct implications for frames in $\mathbb{R}^n$ as well. If, for example, $\{f_k\}_{k=1}^m$ is a frame for $\mathbb{C}^n$, then the $2m$ vectors consisting of the real parts, respectively the imaginary parts, of the frame vectors will be a frame for $\mathbb{R}^n$ (Exercise 1.11); in particular, if the vectors $\{f_k\}_{k=1}^m$ have real coordinates, they constitute a frame for $\mathbb{R}^n$. On the other hand, a frame for $\mathbb{R}^n$ is automatically a frame for $\mathbb{C}^n$ (Exercise 1.12).

The canonical basis for $\mathbb{C}^n$ consists of the vectors $\{\delta_k\}_{k=1}^n$, where δ_k is the vector in $\mathbb{C}^n$ having 1 at the kth entry and otherwise 0. We will consequently identify vectors in $\mathbb{C}^n$ with their representation in this basis.

From elementary linear algebra, we know many equivalent conditions for a set of vectors to constitute a basis for $\mathbb{C}^n$. Let us list the most important characterizations:

Theorem 1.4.1 *Consider n vectors in $\mathbb{C}^n$ and write them as columns in an $n \times n$ matrix*

$$\Lambda = \begin{pmatrix} \lambda_{11} & \lambda_{12} & \cdot & \cdot & \lambda_{1n} \\ \lambda_{21} & \lambda_{22} & \cdot & \cdot & \lambda_{2n} \\ \cdot & \cdot & \cdot & \cdot & \cdot \\ \cdot & \cdot & \cdot & \cdot & \cdot \\ \lambda_{n1} & \lambda_{n2} & \cdot & \cdot & \lambda_{nn} \end{pmatrix}.$$

Then the following are equivalent:

(i) The columns in Λ (i.e., the given vectors) constitute a basis for $\mathbb{C}^n$.

(ii) The rows in Λ constitute a basis for $\mathbb{C}^n$.

(iii) The determinant of Λ is nonzero.

(iv) Λ is invertible.

(v) Λ defines an injective mapping from $\mathbb{C}^n$ into $\mathbb{C}^n$.

(vi) Λ defines a surjective mapping from $\mathbb{C}^n$ onto $\mathbb{C}^n$.

(vii) The columns in Λ are linearly independent.

(viii) Λ has rank equal to n.

Recall that the *rank* of a matrix E is defined as the dimension of its range $\mathcal{R}_E$. We also remind the reader that any basis can be turned into an orthonormal basis by applying the Gram–Schmidt orthogonalization procedure.

We now turn to a discussion of frames for $\mathbb{C}^n$. Note that we *consequently identify operators $U : \mathbb{C}^n \to \mathbb{C}^m$ with their matrix representations with respect to the canonical bases in $\mathbb{C}^n$ and $\mathbb{C}^m$.* Letting $\{e_k\}_{k=1}^n$ denote the canonical orthonormal basis in $\mathbb{C}^n$ and $\{\tilde{e}_k\}_{k=1}^m$ the canonical orthonormal basis in $\mathbb{C}^m$, the matrix representation of U is the $m \times n$ matrix, where the kth column consists of the coordinates of the image under U of the kth basis vector in U, in terms of the given basis in W. The jkth entry in the matrix representation is $\langle Ue_k, \tilde{e}_j \rangle$.

In case $\{f_k\}_{k=1}^m$ is a frame for $\mathbb{C}^n$, the synthesis operator T defined in (1.4) maps $\mathbb{C}^m$ onto $\mathbb{C}^n$. Its matrix with respect to the canonical bases in $\mathbb{C}^n$ and $\mathbb{C}^m$ is

$$T = \begin{pmatrix} | & | & \cdot & \cdot & | \\ f_1 & f_2 & \cdot & \cdot & f_m \\ | & | & \cdot & \cdot & | \end{pmatrix} \tag{1.18}$$

i.e., the $n \times m$ matrix having the vectors f_k as columns.

Since m vectors can at most span an m-dimensional space, we necessarily have $m \geq n$ when $\{f_k\}_{k=1}^m$ is a frame for $\mathbb{C}^n$, i.e., the matrix T has at least as many columns as rows.

Theorem 1.4.2 *Let $\{f_k\}_{k=1}^m$ be a frame for $\mathbb{C}^n$. Then the following holds:*

(i) The vectors f_k can be considered as the first n coordinates of some vectors g_k in $\mathbb{C}^m$ constituting a basis for $\mathbb{C}^m$.

(ii) If $\{f_k\}_{k=1}^m$ is tight, then the vectors f_k are the first n coordinates of some vectors g_k in $\mathbb{C}^m$ constituting an orthogonal basis for $\mathbb{C}^m$.

Proof. Let $\{f_k\}_{k=1}^m$ be an arbitrary frame for $\mathbb{C}^n$. Then $m \geq n$. Consider the mapping

$$F : \mathbb{C}^n \to \mathbb{C}^m, \quad Fx = \{\langle x, f_k \rangle\}_{k=1}^m.$$

F is the adjoint of the synthesis operator T. The matrix for F with respect to the canonical bases is the $m \times n$ matrix where the kth row is the complex conjugate of f_k, i.e.,

$$F = \begin{pmatrix} - & \overline{f_1} & - \\ - & \overline{f_2} & - \\ \cdot & \cdot & \cdot \\ \cdot & \cdot & \cdot \\ - & \overline{f_m} & - \end{pmatrix}.$$

If $Fx = 0$, then $0 = ||Fx||^2 = \sum_{k=1}^m |\langle x, f_k \rangle|^2$. Since $\mathrm{span}\{f_k\}_{k=1}^m = \mathbb{C}^n$, it follows that $x = 0$, so F is an injective mapping. We can therefore extend F to a bijection $\widetilde{F}$ of $\mathbb{C}^m$ onto $\mathbb{C}^m$: for example, still letting $\{\delta_k\}_{k=1}^m$ be the canonical basis for $\mathbb{C}^m$, let $\{\phi_k\}_{k=n+1}^m$ be a basis for the orthogonal complement of $\mathcal{R}_F$ in $\mathbb{C}^m$ and extend F by the definition $\widetilde{F}\delta_k := \phi_k$, $k = n+1, n+2, \ldots, m$. The matrix for $\widetilde{F}$ is an $m \times m$ matrix, whose first n columns are the columns from F:

$$\widetilde{F} = \begin{pmatrix} - & \overline{f_1} & - & | & | & \cdot & | \\ \cdot & \cdot & \cdot & | & \phi_{n+1} & \cdot & \phi_m \\ - & \overline{f_m} & - & | & | & \cdot & | \end{pmatrix}.$$

Since $\widetilde{F}$ is surjective, the columns span $\mathbb{C}^m$. The rank of the rows equals the rank of the columns, so also the rows in $\widetilde{F}$ span $\mathbb{C}^m$, and they are linearly independent. Thus, they constitute a basis for $\mathbb{C}^m$.

If $\{f_k\}_{k=1}^m$ is a tight frame for $\mathbb{C}^n$ with frame bound A and $\{\delta_k\}_{k=1}^n$ still denotes the canonical basis for $\mathbb{C}^n$, Proposition 1.1.4 shows that

$$\langle TT^*\delta_\ell, \delta_j \rangle \;=\; A\delta_{j,\ell}, \;\; j,\ell = 1,\ldots,n.$$

$\langle TT^*\delta_\ell, \delta_j \rangle$ is the j,ℓth entry in the matrix representation for TT^*, so this calculation shows that the n rows in the matrix representation (1.18) for T are orthogonal, considered as vectors in $\mathbb{C}^m$. By adding $m-n$ rows we can extend the matrix for T to an $m \times m$ matrix in which the rows are orthogonal. Therefore the columns are orthogonal. $\qquad\square$

Geometrically, Theorem 1.4.2 means that if $\{f_k\}_{k=1}^m$ is a frame for $\mathbb{C}^n$, there exist vectors $\{h_k\}_{k=1}^m$ in $\mathbb{C}^{m-n}$ such that the columns in the matrix

$$\begin{pmatrix} | & | & \cdot & \cdot & | \\ f_1 & f_2 & \cdot & \cdot & f_m \\ | & | & \cdot & \cdot & | \\ h_1 & h_2 & \cdot & \cdot & h_m \\ | & | & \cdot & \cdot & | \end{pmatrix} \tag{1.19}$$

constitute a basis for $\mathbb{C}^m$.

So far, we have considered frames $\{f_k\}_{k=1}^m$ as columns in a matrix; see (1.18). The next result gives a condition for the rows in a matrix Λ to constitute a frame; this change of format is due to the nature of the condition that will be stated in (1.20). If we want to consider the frame property of the columns in the matrix, we can just apply the result on the transposed matrix Λ^T rather than on Λ itself.

Proposition 1.4.3 *For an $m \times n$ matrix*

$$\Lambda = \begin{pmatrix} \lambda_{11} & \cdot & \cdot & \lambda_{1n} \\ \cdot & \cdot & \cdot & \cdot \\ \cdot & \cdot & \cdot & \cdot \\ \lambda_{m1} & \cdot & \cdot & \lambda_{mn} \end{pmatrix},$$

the following are equivalent:

(i) There exists a constant $A > 0$ such that

$$A \sum_{k=1}^n |c_k|^2 \leq ||\Lambda\{c_k\}_{k=1}^n||^2, \;\; \forall \{c_k\}_{k=1}^n \in \mathbb{C}^n. \tag{1.20}$$

(ii) The columns in Λ constitute a basis for their span in $\mathbb{C}^m$.

(iii) The rows in Λ constitute a frame for $\mathbb{C}^n$.

Proof.   Denote the columns in Λ by $g_1, \ldots, g_n$; they are vectors in $\mathbb{C}^m$. By definition, (i) means that for all $\{c_k\}_{k=1}^n \in \mathbb{C}^n$,

$$A \sum_{k=1}^n |c_k|^2 \leq \left\| \sum_{k=1}^n c_k g_k \right\|^2, \qquad (1.21)$$

which is equivalent to $\{g_k\}_{k=1}^n$ being a basis for its span in $\mathbb{C}^m$ (use an argument such as in the proof of Proposition 1.1.2). On the other hand, denoting the rows in Λ by $f_1, \ldots, f_m$, (i) can also be written as

$$A \sum_{k=1}^n |c_k|^2 \leq \sum_{k=1}^n \left| \left\langle f_k, \begin{pmatrix} \overline{c_1} \\ \cdot \\ \overline{c_n} \end{pmatrix} \right\rangle \right|^2, \ \forall \{c_k\}_{k=1}^n \in \mathbb{C}^n,$$

which is equivalent to (iii). $\square$

Example 1.4.4 As an illustration of Proposition 1.4.3, consider the matrix

$$\Lambda = \begin{pmatrix} 1 & 0 \\ 0 & 1 \\ 1 & 0 \end{pmatrix};$$

it is clear that the rows $\begin{pmatrix} 1 \\ 0 \end{pmatrix}, \begin{pmatrix} 0 \\ 1 \end{pmatrix}, \begin{pmatrix} 1 \\ 0 \end{pmatrix}$ constitute a frame for $\mathbb{C}^2$.

The columns $\begin{pmatrix} 1 \\ 0 \\ 1 \end{pmatrix}, \begin{pmatrix} 0 \\ 1 \\ 0 \end{pmatrix}$ constitute a basis for their span in $\mathbb{C}^3$, but the span is only a two-dimensional subspace of $\mathbb{C}^3$. $\square$

As an immediate consequence of the proof of Proposition 1.4.3, we have the following useful fact:

Corollary 1.4.5 *Let Λ be an $m \times n$ matrix. Denote the columns by $g_1, \ldots, g_n$ and the rows by $f_1, \ldots, f_m$, i.e.,*

$$\Lambda = \begin{pmatrix} \lambda_{11} & \cdot & \cdot & \lambda_{1n} \\ \cdot & \cdot & & \cdot \\ \cdot & & \cdot & \cdot \\ \lambda_{m1} & \cdot & \cdot & \lambda_{mn} \end{pmatrix} = \begin{pmatrix} | & | & \cdot & \cdot & | \\ g_1 & g_2 & \cdot & \cdot & g_n \\ | & | & \cdot & \cdot & | \end{pmatrix} = \begin{pmatrix} - & f_1 & - \\ - & f_2 & - \\ \cdot & \cdot & \cdot \\ \cdot & \cdot & \cdot \\ - & f_m & - \end{pmatrix}.$$

Given $A, B > 0$, the vectors $\{f_k\}_{k=1}^m$ constitute a frame for $\mathbb{C}^n$ with bounds A, B if and only if

$$A \sum_{k=1}^n |c_k|^2 \leq \left\| \sum_{k=1}^n c_k g_k \right\|^2 \leq B \sum_{k=1}^n |c_k|^2, \ \forall \{c_k\}_{k=1}^n \in \mathbb{C}^n.$$

Example 1.4.6 Consider the vectors

$$
\begin{pmatrix} 0 \\ \sqrt{\frac{1}{3}} \\ \sqrt{\frac{2}{3}} \end{pmatrix}, \begin{pmatrix} 0 \\ -\sqrt{\frac{1}{3}} \\ \sqrt{\frac{2}{3}} \end{pmatrix}, \begin{pmatrix} 0 \\ 1 \\ 0 \end{pmatrix}, \begin{pmatrix} \sqrt{\frac{5}{6}} \\ 0 \\ \sqrt{\frac{1}{6}} \end{pmatrix}, \begin{pmatrix} -\sqrt{\frac{5}{6}} \\ 0 \\ \sqrt{\frac{1}{6}} \end{pmatrix} \tag{1.22}
$$

in $\mathbb{C}^3$. Corresponding to these vectors, we consider the matrix

$$
\Lambda = \begin{pmatrix} 0 & \sqrt{\frac{1}{3}} & \sqrt{\frac{2}{3}} \\ 0 & -\sqrt{\frac{1}{3}} & \sqrt{\frac{2}{3}} \\ 0 & 1 & 0 \\ \sqrt{\frac{5}{6}} & 0 & \sqrt{\frac{1}{6}} \\ -\sqrt{\frac{5}{6}} & 0 & \sqrt{\frac{1}{6}} \end{pmatrix}.
$$

The reader can check that the columns $\{g_k\}_{k=1}^3$ are orthogonal in $\mathbb{C}^5$ and all have length $\sqrt{\frac{5}{3}}$. Therefore,

$$
\left\| \sum_{k=1}^3 c_k g_k \right\|^2 = \frac{5}{3} \sum_{k=1}^3 |c_k|^2
$$

for all $c_1, c_2, c_3 \in \mathbb{C}$. By Corollary 1.4.5 we conclude that the vectors defined by (1.22) constitute a tight frame for $\mathbb{C}^3$ with frame bound $\frac{5}{3}$. The frame is normalized.   □

For later use we state a special case of Corollary 1.4.5; we ask the reader to provide the proof in Exercise 1.13.

Corollary 1.4.7 *Let Λ be an $m \times n$ matrix. Then the following are equivalent:*

(i) $\Lambda^* \Lambda = I$, the $n \times n$ identity matrix.

(ii) The columns $g_1, \ldots, g_n$ in Λ constitute an orthonormal system in $\mathbb{C}^m$.

(iii) The rows $f_1, \ldots, f_m$ in Λ constitute a Parseval frame for $\mathbb{C}^n$.

1.5 Frames and the Discrete Fourier Transform

When working with frames and bases in $\mathbb{C}^n$, one has to be particularly careful with the meaning of the notation. For example, we have used f_k and g_k to denote vectors in $\mathbb{C}^n$, whereas c_k in general is the kth coordinate of a sequence $\{c_k\}_{k=1}^n \in \mathbb{C}^n$, i.e., c_k is a scalar. In order to avoid confusion, we will change the notation slightly in this section. The key to the new notation is the observation that to have an element in $\mathbb{C}^n$ is equivalent to having a function

$$f : \{1,\ldots,n\} \to \mathbb{C};$$

the jth entry in the sequence corresponds to the jth function value $f(j)$. Thus, we will often denote a sequence in $\mathbb{C}^n$ by $\{f(j)\}_{j=1}^n$.

We begin this section by the construction of a special orthonormal basis for $\mathbb{C}^n$ and related frame constructions with attractive properties. At the end of the section, we connect the constructions to a problem in engineering.

Let $z := e^{2\pi i/n}$ and consider the $n \times n$ *discrete Fourier transform matrix* (DFT matrix) given by

$$\frac{1}{\sqrt{n}} \begin{pmatrix} 1 & 1 & 1 & \cdot & \cdot & 1 \\ 1 & z & z^2 & \cdot & \cdot & z^{n-1} \\ 1 & z^2 & z^4 & \cdot & \cdot & z^{2(n-1)} \\ 1 & \cdot & \cdot & \cdot & \cdot & \cdot \\ 1 & \cdot & \cdot & \cdot & \cdot & \cdot \\ 1 & z^{n-1} & z^{2(n-1)} & \cdot & \cdot & z^{(n-1)(n-1)} \end{pmatrix}. \tag{1.23}$$

We will now consider the columns $e_k, k = 1,\ldots,n$, in the matrix (1.23). That is, for $k = 1,\ldots,n$, we define the vectors $e_k \in \mathbb{C}^n$ by

$$e_k(j) = \frac{1}{\sqrt{n}} e^{2\pi i(j-1)\frac{k-1}{n}}, \ j = 1,\ldots,n; \tag{1.24}$$

or

$$e_k = \frac{1}{\sqrt{n}} \begin{pmatrix} 1 \\ e^{2\pi i\frac{k-1}{n}} \\ e^{4\pi i\frac{k-1}{n}} \\ \cdot \\ \cdot \\ e^{2\pi i(n-1)\frac{k-1}{n}} \end{pmatrix}, \ k = 1,\ldots n. \tag{1.25}$$

Theorem 1.5.1 *The vectors $\{e_k\}_{k=1}^n$ defined by (1.24) constitute an orthonormal basis for $\mathbb{C}^n$.*

Proof. Since $\{e_k\}_{k=1}^n$ are n vectors in an n-dimensional vector space, it is enough to prove that they constitute an orthonormal system. It is clear that $||e_k|| = 1$ for all k. Now, given $k \neq \ell$,

$$\langle e_k, e_\ell \rangle = \frac{1}{n} \sum_{j=1}^n e^{2\pi i(j-1)\frac{k-1}{n}} e^{-2\pi i(j-1)\frac{\ell-1}{n}} = \frac{1}{n} \sum_{j=0}^{n-1} e^{2\pi ij\frac{k-\ell}{n}}.$$

Using the formula $(1-x)(1+x+\cdots+x^{n-1}) = 1 - x^n$ with $x = e^{2\pi i\frac{k-\ell}{n}}$, we get

$$\langle e_k, e_\ell \rangle = \frac{1}{n} \frac{1 - (e^{2\pi i\frac{k-\ell}{n}})^n}{1 - e^{2\pi i\frac{k-\ell}{n}}} = 0. \qquad \square$$

The basis $\{e_k\}_{k=1}^{n}$ is called the *discrete Fourier transform basis*. Using this basis, every sequence $f \in \mathbb{C}^n$ has a representation

$$ f = \sum_{k=1}^{n} \langle f, e_k \rangle e_k = \frac{1}{\sqrt{n}} \sum_{k=1}^{n} \sum_{\ell=1}^{n} f(\ell) e^{-2\pi i (\ell-1)\frac{k-1}{n}} e_k. $$

Written out in coordinates, this means that

$$ f(j) = \frac{1}{n} \sum_{k=1}^{n} \sum_{\ell=1}^{n} f(\ell) e^{-2\pi i (\ell-1)\frac{k-1}{n}} e^{2\pi i (j-1)\frac{k-1}{n}} $$

$$ = \frac{1}{n} \sum_{k=1}^{n} \sum_{\ell=1}^{n} f(\ell) e^{2\pi i (j-\ell)\frac{k-1}{n}}, \ j = 1, \ldots, n. $$

Applications often ask for tight frames because the cumbersome inversion of the frame operator is avoided in this case; see (1.9). It was observed by Zimmermann [642] that overcomplete tight frames in $\mathbb{C}^n$ can be obtained by projecting the discrete Fourier transform basis in any $\mathbb{C}^m$, $m > n$, onto $\mathbb{C}^n$. In other words: if we consider the $m \times m$ discrete Fourier transform matrix and remove the last $m - n$ rows, the columns in the remaining matrix form a tight frame for $\mathbb{C}^n$:

Proposition 1.5.2 *Let $m > n$ and define the vectors $\{f_k\}_{k=1}^{m}$ in $\mathbb{C}^n$ by*

$$ f_k = \frac{1}{\sqrt{m}} \begin{pmatrix} 1 \\ e^{2\pi i \frac{k-1}{m}} \\ \cdot \\ \cdot \\ e^{2\pi i (n-1)\frac{k-1}{m}} \end{pmatrix}, \quad k = 1, 2, \ldots, m. $$

Then $\{f_k\}_{k=1}^{m}$ is an overcomplete Parseval frame for $\mathbb{C}^n$, and $\|f_k\| = \sqrt{\frac{n}{m}}$ for all $k = 1, \ldots, m$.

Proof. Let $\{\delta_j\}_{j=1}^{n}$ be the canonical basis for $\mathbb{C}^n$, and let $\{e_k\}_{k=1}^{m}$ be the discrete Fourier transform basis for $\mathbb{C}^m$, i.e.,

$$ e_k = \frac{1}{\sqrt{m}} \begin{pmatrix} 1 \\ e^{2\pi i \frac{k-1}{m}} \\ \cdot \\ e^{2\pi i (n-1)\frac{k-1}{m}} \\ \cdot \\ e^{2\pi i (m-1)\frac{k-1}{m}} \end{pmatrix}, \quad k = 1, \ldots, m. $$

Identifying $\mathbb{C}^n$ with a subspace of $\mathbb{C}^m$, the orthogonal projection of e_k onto $\mathbb{C}^n$ is $Pe_k = f_k$; now the result follows from Exercise 1.14.   □

The frames $\{f_k\}_{k=1}^{m}$ constructed in Proposition 1.5.2 are called *harmonic frames*. It is important to notice that all the vectors f_k in $\{f_k\}_{k=1}^{m}$ have

the same norm. If needed, we can therefore normalize them while keeping
a tight frame; we only have to adjust the frame bound accordingly. We
formulate the result as an existence result, but it is important to keep in
mind that we actually have an explicit construction:

Corollary 1.5.3 *For any $m \geq n$, there exists a tight frame in $\mathbb{C}^n$
consisting of m normalized vectors.*

Note that Zimmermann also constructed real-valued tight frames for $\mathbb{R}^n$,
based on the same idea; see [642] for details.

Example 1.5.4 The discrete Fourier transform basis in $\mathbb{C}^4$ consists of the
vectors

$$\frac{1}{2}\begin{pmatrix}1\\1\\1\\1\end{pmatrix}, \frac{1}{2}\begin{pmatrix}1\\i\\-1\\-i\end{pmatrix}, \frac{1}{2}\begin{pmatrix}1\\-1\\1\\-1\end{pmatrix}, \frac{1}{2}\begin{pmatrix}1\\-i\\-1\\i\end{pmatrix}.$$

Via Proposition 1.5.2, the vectors

$$\frac{1}{2}\begin{pmatrix}1\\1\end{pmatrix}, \frac{1}{2}\begin{pmatrix}1\\i\end{pmatrix}, \frac{1}{2}\begin{pmatrix}1\\-1\end{pmatrix}, \frac{1}{2}\begin{pmatrix}1\\-i\end{pmatrix}$$

constitute a Parseval frame in $\mathbb{C}^2$. The vectors have length $1/\sqrt{2}$. Changing
the length of the vectors, i.e., considering the vectors

$$\frac{1}{\sqrt{2}}\begin{pmatrix}1\\1\end{pmatrix}, \frac{1}{\sqrt{2}}\begin{pmatrix}1\\i\end{pmatrix}, \frac{1}{\sqrt{2}}\begin{pmatrix}1\\-1\end{pmatrix}, \frac{1}{\sqrt{2}}\begin{pmatrix}1\\-i\end{pmatrix},$$

we obtain a tight frame with frame bound 2, consisting of normalized
vectors. □

Recall that the key property of frames is the possibility of redundancy: if
$\{f_k\}_{k=1}^m$ is a frame for $\mathbb{C}^n$ and $m > n$, we know from (1.11) that each $f \in \mathbb{C}^n$
has several expansions in terms of $\{f_k\}_{k=1}^m$, so we have the possibility to
select the most convenient one in concrete cases. Also, in Section 1.9 we will
discuss a case where it is important that removal of one or more elements
from a frame does not destroy the frame property; this clearly requires that
the given frame is redundant.

The *excess* of a frame $\{f_k\}_{k=1}^m$ of a finite-dimensional space V is the
number of elements that has to be removed in order for the remaining set
to form a basis; if $V = \mathbb{C}^n$, the excess of a frame $\{f_k\}_{k=1}^m$ is simply $m - n$.
Note, however, that the excess alone does not provide sufficient information
about stability of the frame property against removal of selected elements:

Example 1.5.5 Let $\{c_k\}_{k=1}^2$ be an orthonormal basis for $\mathbb{C}^2$.

(i) The sequence $\{e_1, e_1, e_1, e_2\}$ is a frame for $\mathbb{C}^2$ with excess 2; however, removal of the single element e_2 leaves a set that is not a frame for $\mathbb{C}^2$.

(ii) The sequence $\{e_1, e_2, e_1+e_2, e_1-e_2\}$ is also a frame for $\mathbb{C}^2$ with excess 2. In this case removal of two *arbitrary* vectors leaves a set which is still a frame for $\mathbb{C}^2$. □

We note that Bodmann, Casazza, and Kutyniok have provided general definitions of *upper and lower redundancies* in order to describe the direction dependence of the redundancy that we encountered in Example 1.5.5; we refer to the paper [68] for details.

If we have information on the lower frame bound and the norm of the frame elements, we can provide a criterion for how many elements we can (at least) remove:

Proposition 1.5.6 *Let $\{f_k\}_{k=1}^m$ be a normalized frame for $\mathbb{C}^n$ with lower frame bound $A > 1$. Then, for any index set $I \subset \{1, \ldots, m\}$ with $|I| < A$, the family $\{f_k\}_{k \notin I}$ is a frame for $\mathbb{C}^n$ with lower bound $A - |I|$.*

Proof. Given $f \in \mathbb{C}^n$,

$$\sum_{k \in I} |\langle f, f_k \rangle|^2 \le \sum_{k \in I} \|f_k\|^2 \, \|f\|^2 = |I| \, \|f\|^2.$$

Thus,

$$\sum_{k \notin I} |\langle f, f_k \rangle|^2 \ge (A - |I|)\|f\|^2.$$ □

Theorem 1.3.1 shows that if $\{f_k\}_{k=1}^m$ is a tight normalized frame, then Proposition 1.5.6 applies if $|I| < \frac{m}{n}$. Considering an arbitrary frame $\{f_k\}_{k=1}^m$ for $\mathbb{C}^n$, the maximal number of elements one can hope to remove while keeping the frame property is $m - n$. If we want to be able to remove $m - n$ arbitrary elements, it is not enough to assume that $\{f_k\}_{k=1}^m$ is a normalized tight frame, as demonstrated by the frame in Example 1.4.6; in this example $m - n = 2$, but the three first vectors in (1.22) do not constitute a frame for $\mathbb{C}^3$. Concerning the stability against removal of vectors, the harmonic frame $\{f_k\}_{k=1}^m$ in Proposition 1.5.2 behaves well: $m - n$ arbitrary elements can be removed without destroying the frame property of the remaining set!

Proposition 1.5.7 *Consider the harmonic frame $\{f_k\}_{k=1}^m$ for $\mathbb{C}^n$ defined in Proposition 1.5.2. Any subset containing at least n elements of this frame forms a frame for $\mathbb{C}^n$.*

Proof. Consider an arbitrary subset $\{k_1, k_2, \ldots, k_n\} \subseteq \{1, 2, \ldots, m\}$. Placing the vectors $\{f_{k_i}\}_{i=1}^n$ as rows in an $n \times n$ matrix and letting $z := e^{\frac{2\pi i}{m}}$, we obtain

$$
\begin{pmatrix} -f_{k_1}- \\ -f_{k_2}- \\ \cdot \\ \cdot \\ -f_{k_n}- \end{pmatrix} = \frac{1}{\sqrt{m}} \begin{pmatrix} 1 & e^{2\pi i \frac{k_1-1}{m}} & \cdot & \cdot & e^{2\pi i \frac{(k_1-1)(n-1)}{m}} \\ 1 & e^{2\pi i \frac{k_2-1}{m}} & \cdot & \cdot & e^{2\pi i \frac{(k_2-1)(n-1)}{m}} \\ \cdot & \cdot & \cdot & & \cdot \\ \cdot & \cdot & & \cdot & \cdot \\ 1 & e^{2\pi i \frac{k_n-1}{m}} & \cdot & \cdot & e^{2\pi i \frac{(k_n-1)(n-1)}{m}} \end{pmatrix}
$$

$$
= \frac{1}{\sqrt{m}} \begin{pmatrix} 1 & z^{k_1-1} & \cdot & \cdot & z^{(k_1-1)(n-1)} \\ 1 & z^{k_2-1} & \cdot & \cdot & z^{(k_2-1)(n-1)} \\ \cdot & \cdot & \cdot & & \cdot \\ \cdot & \cdot & & \cdot & \cdot \\ 1 & z^{k_n-1} & \cdot & \cdot & z^{(k_n-1)(n-1)} \end{pmatrix};
$$

this is a Vandermonde matrix with determinant

$$
\frac{1}{m^{n/2}} \prod_{i,j=1, i \neq j}^{n} (z^{k_i-1} - z^{k_j-1}) \neq 0.
$$

Thus, $\{f_{k_i}\}_{i=1}^n$ is a basis for $\mathbb{C}^n$ by Theorem 1.4.1. $\square$

A frame $\{f_k\}_{k=1}^m$ for $\mathbb{C}^n$ is said to have *full spark* if every subset containing n elements is linearly independent; that is, if removal of $m - n$ arbitrary elements leaves a basis for $\mathbb{C}^n$. The harmonic frames are the standard examples of such frames.

Note that in the definition of the harmonic frames $\{f_k\}_{k=1}^m$ for $\mathbb{C}^n$ in Proposition 1.5.2, we defined the vectors f_k by taking the n *first* elements from the columns forming the $m \times m$ DFT matrix. One can prove more general ways of obtaining frames, even with full spark, by selecting other rows than just the n first rows in the DFT matrix; see the paper [10] by Alexeev, Cahill, and Mixon.

The redundancy of frames is the key to most of the frame applications in engineering. For example, the redundancy leads to a reduction in the inevitable *quantization error,* which appears in all applications of series representations. We will now describe this in more detail. Let $\{f_k\}_{k=1}^m$ be a frame for $\mathbb{C}^n$, with a dual frame $\{g_k\}_{k=1}^m$. General frame theory to be discussed later (see Lemma 6.3.2) yields that $\{f_k\}_{k=1}^m$ is also a dual frame of $\{g_k\}_{k=1}^m$, so any $f \in \mathbb{C}^n$ has the exact representation

$$
f = \sum_{k=1}^m \langle f, f_k \rangle g_k. \tag{1.26}
$$

Unfortunately, in real life we often have to give up the beauty of exact mathematics. In computer-based applications, we cannot work with arbitrary numbers, but only a finite collection of numbers, a so-called *alphabet.*

This means that the coefficients $c_k := \langle f, f_k \rangle$ have to be replaced by numbers $\widetilde{c}_k$ from the alphabet, and the exact representation of f in (1.26) will be replaced by an approximation

$$\widetilde{f} = \sum_{k=1}^{m} \widetilde{c}_k g_k. \tag{1.27}$$

The number $\|f - \widetilde{f}\|$ is called the *quantization error*. Note that if $\{f_k\}_{k=1}^{m}$ is an orthonormal basis, we know that the unique choice of the dual frame is $\{g_k\}_{k=1}^{m} = \{f_k\}_{k=1}^{m}$; in this case the quantization error is

$$\|f - \widetilde{f}\| = \sqrt{\sum_{k=1}^{m} |c_k - \widetilde{c}_k|^2}.$$

It is reasonable to expect that we can reduce the quantization error if we replace the orthonormal basis $\{f_k\}_{k=1}^{m}$ by an overcomplete frame: in this case, any dual frame $\{g_k\}_{k=1}^{m}$ is overcomplete as well, and the set of coefficients $\{c_k\}_{k=1}^{m}$ that yield an exact representation $f = \sum_{k=1}^{m} c_k g_k$ form an affine subspace of $\mathbb{C}^m$. At least in an intuitive sense, the flexibility in the choice of coefficients should make it possible to obtain good approximations using coefficients from the alphabet – with better approximations whenever the redundancy increases. This intuition has been confirmed in a series of papers by Powell, Yilmaz, and their collaborators [55, 64, 547]. For $m > n$, let $\{f_k\}_{k=1}^{m}$ denote the harmonic frame in $\mathbb{C}^n$, as constructed in Proposition 1.5.2. Then there exists for any $r \in \mathbb{N}$ a procedure called *rth-order sigma–delta quantization* that yields coefficients $\widetilde{c}_k$ and corresponding approximations $\widetilde{f}$ as in (1.27) such that

$$\|f - \widetilde{f}\| \le Cm^{-r}, \tag{1.28}$$

for a constant C that is independent of m. In other words: increasing the redundancy, i.e., the number m of elements in the frame, leads to a decay of the quantization error. The approach is in fact not restricted to harmonic frames, but works for certain other classes of frames as well.

The proof of the decay estimate (1.28) is based on a careful choice of the dual frame $\{g_k\}_{k=1}^{m}$. In fact, given any $r \in \mathbb{N}$, the procedure selects a particular dual frame $\{g_k\}_{k=1}^{m}$ such that (1.28) holds. This particular dual is called the *rth-order Sobolev dual;* explicit formulas are given in [65] and [547]. It is interesting to note that the choice of this particular dual frame instead of the canonical dual frame is essential. In fact, a concrete example in [473] yields a situation where the use of the canonical dual frame cannot lead to an approximation order better than $1/m^2$, regardless of the choice of r. Thus, it is necessary to use the full flexibility of the frame theory if we want to obtain a fast reduction in the quantization error by increasing the redundancy.

The analysis of sigma–delta quantization adds a new perspective to our discussion of tight frames versus dual frame pairs. So far, the point of view has been that the tight frames are theoretically perfect, because the canonical dual equals the frame itself (up to a constant) and thus yields a simple form of the frame decomposition. Now we know that even in the case of a tight frame, practical issues might prompt us to apply the general theory for dual frames.

1.6   Pseudo-inverses and the Singular Value Decomposition

For matrices that are not invertible, various types of generalized inverses exist in the literature. The right definition of a generalized inverse depends on the properties we are interested in, and we shall only define the so-called *pseudo-inverse*. Given an $m \times n$ matrix E, we consider it as a linear mapping of $\mathbb{C}^n$ into $\mathbb{C}^m$. E is not necessarily injective, but by restricting E to the orthogonal complement of the kernel $\mathcal{N}_E$, we obtain an injective linear mapping

$$\widetilde{E} : \mathcal{N}_E^\perp \to \mathbb{C}^m.$$

E and $\widetilde{E}$ have the same range, $\mathcal{R}_{\widetilde{E}} = \mathcal{R}_E$; thus, $\widetilde{E}$ considered as a mapping from $\mathcal{N}_E^\perp$ to $\mathcal{R}_E$ has an inverse,

$$(\widetilde{E})^{-1} : \mathcal{R}_E \to \mathcal{N}_E^\perp.$$

We can extend $(\widetilde{E})^{-1}$ to an operator $E^\dagger : \mathbb{C}^m \to \mathbb{C}^n$ by defining

$$E^\dagger(y + z) = (\widetilde{E})^{-1}y \text{ if } y \in \mathcal{R}_E, z \in \mathcal{R}_E^\perp. \tag{1.29}$$

With this definition,

$$EE^\dagger x = x, \ \forall x \in \mathcal{R}_E. \tag{1.30}$$

The operator $E^\dagger$ is called the *pseudo-inverse* of E. From the definition, we immediately have that

$$\mathcal{N}_{E^\dagger} = \mathcal{R}_E^\perp = \mathcal{N}_{E^*}, \quad \mathcal{R}_{E^\dagger} = \mathcal{N}_E^\perp = \mathcal{R}_{E^*}. \tag{1.31}$$

We state two characterizations of the pseudo-inverse:

Proposition 1.6.1 *Let E be an $m \times n$ matrix. Then*

(i) $E^\dagger$ is the unique $n \times m$ matrix for which $EE^\dagger$ is the orthogonal projection onto $\mathcal{R}_E$ and $E^\dagger E$ is the orthogonal projection onto $\mathcal{R}_{E^\dagger}$.

(ii) $E^\dagger$ is the unique $n \times m$ matrix for which $EE^\dagger$ and $E^\dagger E$ are self-adjoint and

$$EE^\dagger E = E, \ E^\dagger EE^\dagger = E^\dagger.$$

Proof. We first prove the equivalence between the conditions stated in (i) and (ii). If a matrix $E^\dagger$ satisfies (i), it immediately follows that (ii) is satisfied. On the other hand, if (ii) is satisfied, then

$$(EE^\dagger)^2 = EE^\dagger EE^\dagger = EE^\dagger.$$

Since $EE^\dagger$ is self-adjoint, it follows that $EE^\dagger$ is the orthogonal projection onto $\mathcal{R}_{EE^\dagger}$. Finally, the identity $EE^\dagger E = E$ shows that $\mathcal{R}_{EE^\dagger} = \mathcal{R}_E$. The proof that $E^\dagger E$ is the orthogonal projection onto $\mathcal{R}_{E^\dagger}$ is similar. Thus, (i) is satisfied.

We now prove the equivalence between the properties in Proposition 1.6.1 and the definition (1.29) of the pseudo-inverse. First we note that with our definition of the pseudo-inverse, the conditions in (i) are satisfied; the main ingredients in the following argument are the relations (1.30) and (1.31). In fact, if $y \in \mathcal{R}_E$, then $EE^\dagger y = y$; and if $y \in \mathcal{R}_E^\perp = \mathcal{N}_{E^\dagger}$, then $EE^\dagger y = 0$. This proves that $EE^\dagger$ is the orthogonal projection onto $\mathcal{R}_E$. Also, if $y \in \mathcal{R}_{E^\dagger}^\perp = \mathcal{N}_E$, then $E^\dagger E y = 0$; and if $y \in \mathcal{R}_{E^\dagger}$, $y = E^\dagger x$ for some x, then

$$E^\dagger E y = E^\dagger E E^\dagger x = E^\dagger x - E^\dagger (I - EE^\dagger)x = E^\dagger x = y.$$

Here we used that $I - EE^\dagger$ is the orthogonal projection onto $\mathcal{R}_E^\perp = \mathcal{N}_{E^\dagger}$. We have now proved that $E^\dagger E$ is the orthogonal projection onto $\mathcal{R}_{E^\dagger}$.

To conclude we only have to prove that if a matrix $E^\dagger$ satisfies (i) and (ii), then it fulfills the requirements in the definition of the pseudo-inverse, i.e., (1.29) is satisfied. First, we note that (ii) implies that

$$E^* = (EE^\dagger E)^* = (E^\dagger E)^* E^* = E^\dagger E E^*;$$

this shows that

$$\mathcal{N}_E^\perp = \mathcal{R}_{E^*} \subseteq \mathcal{R}_{E^\dagger}.$$

Now, if $y \in \mathcal{R}_E$, then we can find $x \in \mathcal{N}_E^\perp$ such that $y = Ex$; thus,

$$E^\dagger y = E^\dagger E x = x = (\widetilde{E})^{-1} Ex = (\widetilde{E})^{-1} y.$$

Finally, if $z \in \mathcal{R}_E^\perp = \mathcal{N}_{E^*}$, then by (i), $EE^\dagger z = 0$; using (ii),

$$E^\dagger z = E^\dagger E E^\dagger z = 0. \qquad \square$$

The pseudo-inverse gives the solution to an important minimization problem:

Theorem 1.6.2 *Let E be an $m \times n$ matrix. Given $y \in \mathcal{R}_E$, the equation $Ex = y$ has a unique solution of minimal norm, namely, $x = E^\dagger y$.*

Proof. By (1.30), we know that $x := E^\dagger y$ is a solution to the equation $Ex = y$. All solutions have the form $x = E^\dagger y + z$, where $z \in \mathcal{N}_E$. Since

$E^\dagger y \in \mathcal{N}_E^\perp$, the norm of the general solution is given by

$$||x||^2 = ||E^\dagger y + z||^2 = ||E^\dagger y||^2 + ||z||^2.$$

This expression is minimal when $z = 0$. $\sqcap$

Historically, (i) and (ii) were given as definitions of a "generalized inverse" by Moore and Penrose, respectively. For this reason, the pseudo-inverse is frequently called the *Moore–Penrose inverse*.

For computational purposes, it is important to notice that the pseudo-inverse can be found using the singular value decomposition of E. We begin with a lemma.

Lemma 1.6.3 *Let E be an $m \times n$ matrix with rank $r \geq 1$. Then there exist constants $\sigma_1, \ldots, \sigma_r > 0$ and orthonormal bases $\{u_k\}_{k=1}^r$ for $\mathcal{R}_E$ and $\{v_k\}_{k=1}^r$ for $\mathcal{R}_{E^*}$ such that*

$$Ev_k = \sigma_k u_k, \ k = 1, \ldots, r. \tag{1.32}$$

Proof. Observe that E^*E is a self-adjoint $n \times n$ matrix; by Theorem A.1.1, this implies that there exists an orthonormal basis $\{v_k\}_{k=1}^n$ for $\mathbb{C}^n$ consisting of eigenvectors for E^*E. Let $\{\lambda_k\}_{k=1}^n$ denote the corresponding eigenvalues. Note that for each k,

$$\lambda_k = \lambda_k ||v_k||^2 = \langle E^*Ev_k, v_k \rangle = ||Ev_k||^2 \geq 0.$$

The rank of E is given by

$$r = \dim \mathcal{R}_E = \dim \mathcal{R}_{E^*};$$

since $\mathcal{R}_E^\perp = \mathcal{N}_{E^*}$, we have

$$\mathcal{R}_{E^*} = \mathcal{R}_{E^*E} = \mathrm{span}\{E^*Ev_k\}_{k=1}^n = \mathrm{span}\{\lambda_k v_k\}_{k=1}^n. \tag{1.33}$$

Thus, the rank is equal to the number of nonzero eigenvalues, counted with multiplicity. We can assume that the eigenvectors $\{v_k\}_{k=1}^n$ are ordered such that $\{v_k\}_{k=1}^r$ corresponds to the nonzero eigenvalues. Then (1.33) shows that $\{v_k\}_{k=1}^r$ is an orthonormal basis for $\mathcal{R}_{E^*}$. Note that for $k > r$, we have $||Ev_k||^2 = \langle E^*Ev_k, v_k \rangle = 0$, i.e.,

$$Ev_k = 0, \ k > r. \tag{1.34}$$

Defining

$$u_k := \frac{1}{\sqrt{\lambda_k}} Ev_k, \ k = 1, \ldots, r,$$

we therefore obtain that $\{u_k\}_{k=1}^r$ spans $\mathcal{R}_E$, and it is an orthonormal basis for $\mathcal{R}_E$ because for all $k, \ell = 1, \ldots, r$, we have

$$
\begin{aligned}
\langle u_k, u_\ell \rangle &= \frac{1}{\sqrt{\lambda_k}} \frac{1}{\sqrt{\lambda_\ell}} \langle E v_k, E v_\ell \rangle \\
&= \frac{1}{\sqrt{\lambda_k \lambda_\ell}} \langle E^* E v_k, v_\ell \rangle \\
&= \sqrt{\frac{\lambda_k}{\lambda_\ell}} \langle v_k, v_\ell \rangle \\
&= \delta_{k,\ell}.
\end{aligned}
$$

Thus, the conditions in Lemma 1.6.3 are fulfilled with

$$
\sigma_k = \sqrt{\lambda_k}, \ k = 1, \ldots, r. \qquad \square
$$

Lemma 1.6.3 leads to the *singular value decomposition* of E:

Theorem 1.6.4 *Every $m \times n$ matrix E with rank $r \geq 1$ has a decomposition*

$$
E = U \begin{pmatrix} D & 0 \\ 0 & 0 \end{pmatrix} V^*, \tag{1.35}
$$

where U is a unitary $m \times m$ matrix, V is a unitary $n \times n$ matrix, and $\begin{pmatrix} D & 0 \\ 0 & 0 \end{pmatrix}$ is an $m \times n$ block matrix in which D is an $r \times r$ diagonal matrix with positive entries $\sigma_1, \ldots, \sigma_r$ in the diagonal.

Proof. We use the proof of Lemma 1.6.3. Let $\{v_k\}_{k=1}^n$ be the orthonormal basis for $\mathbb{C}^n$ considered there, ordered such that $\{v_k\}_{k=1}^r$ is an orthonormal basis for $\mathcal{R}_{E^*}$. Let V be the $n \times n$ matrix having the vectors $\{v_k\}_{k=1}^n$ as columns. Extend the orthonormal basis $\{u_k\}_{k=1}^r$ for $\mathcal{R}_E$ to an orthonormal basis $\{u_k\}_{k=1}^m$ for $\mathbb{C}^m$, and let U be the $m \times m$ matrix having these vectors as columns. Finally, let D be the $r \times r$ diagonal matrix having $\sigma_1, \ldots, \sigma_r$ in the diagonal. Via (1.32) and (1.34),

$$
\begin{aligned}
EV &= \begin{pmatrix} \sigma_1 u_1 & \cdot & \cdot & \sigma_r u_r & 0 & \cdot & \cdot & 0 \end{pmatrix} \\
&= U \begin{pmatrix} D & 0 \\ 0 & 0 \end{pmatrix}.
\end{aligned}
$$

Multiplying with V^* from the right gives the result. $\qquad \square$

The numbers $\sigma_1, \ldots, \sigma_r$ are called *singular values* for E; the proof of Lemma 1.6.3 shows that they are the square roots of the positive eigenvalues for $E^* E$.

Corollary 1.6.5 *With the notation in Theorem 1.6.4, the pseudo-inverse of E is given by*

$$E^\dagger = V \begin{pmatrix} D^{-1} & 0 \\ 0 & 0 \end{pmatrix} U^*, \tag{1.36}$$

where $\begin{pmatrix} D^{-1} & 0 \\ 0 & 0 \end{pmatrix}$ *is an* $n \times m$ *block matrix in which* D^{-1} *is the* $r \times r$ *matrix having* $1/\sigma_1, \ldots, 1/\sigma_r$ *in the diagonal.*

Proof.   We check that the matrix $E^\dagger$ defined by (1.36) satisfies the requirements in Proposition 1.6.1(ii). First, via (1.35),

$$EE^\dagger \;\; = \;\; U \begin{pmatrix} D & 0 \\ 0 & 0 \end{pmatrix} V^* V \begin{pmatrix} D^{-1} & 0 \\ 0 & 0 \end{pmatrix} U^* = U \begin{pmatrix} I & 0 \\ 0 & 0 \end{pmatrix} U^*,$$

which shows that $EE^\dagger$ is self-adjoint. The proof that $E^\dagger E$ is self-adjoint is similar. Furthermore, using the derived expression for $EE^\dagger$,

$$EE^\dagger E \;\; = \;\; U \begin{pmatrix} I & 0 \\ 0 & 0 \end{pmatrix} U^* U \begin{pmatrix} D & 0 \\ 0 & 0 \end{pmatrix} V^* = E.$$

Similarly, one can verify that $E^\dagger E E^\dagger = E^\dagger$.   □

Let us return to the setting where $\{f_k\}_{k=1}^m$ is a frame for $\mathbb{C}^n$ with synthesis operator $T : \mathbb{C}^m \to \mathbb{C}^n$. The calculation of the frame coefficients amounts to finding the pseudo-inverse $T^\dagger$:

Theorem 1.6.6 *Let* $\{f_k\}_{k=1}^m$ *be a frame for* $\mathbb{C}^n$, *with synthesis operator* T *and frame operator* S. *Then*

$$T^\dagger f = \{\langle f, S^{-1} f_k \rangle\}_{k=1}^m, \ \forall f \in \mathbb{C}^n. \tag{1.37}$$

Proof.   Let $f \in \mathbb{C}^n$. Expressed in terms of the synthesis operator T, the equation $f = \sum_{k=1}^m c_k f_k$ means that $T\{c_k\}_{k=1}^m = f$. The result now follows by combining Theorem 1.1.5 and Theorem 1.6.2.   □

One interpretation of Theorem 1.6.6 is that when $\{f_k\}_{k=1}^m$ is a frame for $\mathbb{C}^n$, the matrix for $T^\dagger$ is obtained by placing the complex conjugate of the vectors in the canonical dual frame $\{S^{-1} f_k\}_{k=1}^m$ as rows in an $m \times n$ matrix:

$$T^\dagger = \begin{pmatrix} -\overline{S^{-1} f_1}- \\ -\overline{S^{-1} f_2}- \\ \cdot \\ \cdot \\ -\overline{S^{-1} f_m}- \end{pmatrix}.$$

In operator terms, (1.37) means that

$$T^\dagger = T^* (TT^*)^{-1},$$

a formula that is known to hold generally for the pseudo-inverse of an arbitrary surjective operator T.

Given any $f \in \mathbb{C}^n$ and a frame $\{f_k\}_{k=1}^m$, the singular value decomposition gives a natural way to obtain coefficients $\{c_k\}_{k=1}^m$ such that $f = \sum_{k=1}^m c_k f_k$. Let $\{f_k\}_{k=1}^m$ be an overcomplete frame for $\mathbb{C}^n$ with synthesis operator $T : \mathbb{C}^m \to \mathbb{C}^n$. Considered as a matrix, T is an $n \times m$ matrix, and we know that $m > n$. Since T is surjective, its rank equals n, so according to Theorem 1.6.4, its singular value decomposition is

$$T = U \begin{pmatrix} D & 0 \end{pmatrix} V^*.$$

Note that D is now an $n \times n$ matrix; $\begin{pmatrix} D & 0 \end{pmatrix}$ is an $n \times m$ matrix, U is an $n \times n$ matrix, and V is an $m \times m$ matrix. Given any $(m - n) \times n$ matrix F and any $f \in \mathbb{C}^n$, we have that

$$\begin{aligned} TV \begin{pmatrix} D^{-1} \\ F \end{pmatrix} U^* f &= U \begin{pmatrix} D & 0 \end{pmatrix} V^* V \begin{pmatrix} D^{-1} \\ F \end{pmatrix} U^* f \\ &= UIU^* f = f. \end{aligned}$$

This means that we can use the coefficients

$$\{c_k\}_{k=1}^m = V \begin{pmatrix} D^{-1} \\ F \end{pmatrix} U^* f$$

for the reconstruction of f, regardless how the entries in the matrix F are chosen. By Corollary 1.6.5, the choice $F = 0$ leads to the pseudo-inverse, which, as noted already in Theorem 1.1.5, is optimal in the sense that the ℓ^2-norm of the coefficients is minimized. However, there are many cases where other properties than minimal ℓ^2-norm are more relevant. The matrix

$$V \begin{pmatrix} D^{-1} \\ F \end{pmatrix} U^*$$

is frequently called a *generalized inverse* of T.

1.7   Finite-Dimensional Function Spaces

The rest of the book will deal with frames in infinite-dimensional vector spaces, with concrete constructions in function spaces like $L^2(-\pi, \pi)$ and $L^2(\mathbb{R})$; the exact definition of these spaces will be given in Chapter 2, and for the moment we simply consider $L^2(I), I \subseteq \mathbb{R}$ as the set of functions for which

$$\int_I |f(x)|^2 dx < \infty.$$

It is important to notice that in every real-life application where these spaces appear, one will at some point have to confine to finite-dimensional subspaces. For this reason we conclude this chapter with a short description of frames in finite-dimensional function spaces.

Given $a, b \in \mathbb{R}$ with $a < b$, let $C[a, b]$ denote the set of continuous functions $f : [a, b] \to \mathbb{C}$. We equip $C[a, b]$ with the supremums-norm,

$$||f||_\infty = \sup_{x \in [a,b]} |f(x)| .$$

The *Weierstrass' approximation theorem* says that every $f \in C[a, b]$ can be approximated arbitrarily well by a polynomial:

Theorem 1.7.1 *Let $f \in C[a, b]$. Given $\epsilon > 0$, there exists a polynomial $P(x) = \sum_{k=0}^{n} c_k x^k$ such that*

$$||f - P||_\infty \leq \epsilon.$$

It is essential for the conclusion that $[a, b]$ is a finite and closed interval (Exercise 1.17). Also, we note that the degree of the approximating polynomial P depends as well on the chosen ϵ as the given function f. On the other hand, via an appropriate affine transformation $\varphi(x) := \alpha x + \beta$, any interval $[c, d]$ can be mapped onto the interval $[a, b]$; this implies that the degree of the approximating polynomials does not depend on the actual interval.

The polynomials $\{1, x, x^2, \dots\} = \{x^k\}_{k=0}^{\infty}$ are linearly independent and do not span a finite-dimensional subspace of $C[a, b]$. But for a given $n \in \mathbb{N}$, the vector space

$$V := \mathrm{span}\{1, x, \dots, x^n\}$$

is a finite-dimensional subspace of $C[a, b]$ with the polynomials $\{x^k\}_{k=0}^{n}$ as basis.

If we equip V with the $||\cdot||_\infty$-norm, we do not have the benefit of a norm arising from an inner product. But all norms on a finite-dimensional vector space are equivalent (see page 15), and V can also be equipped with the norm

$$||f|| = \left(\int_a^b |f(x)|^2 dx \right)^{1/2}$$

arising from the inner product

$$\langle f, g \rangle = \int_a^b f(x)\overline{g(x)}dx. \tag{1.38}$$

Via the Gram–Schmidt orthogonalization procedure, one can construct an orthonormal basis for $(V, ||\cdot||)$; see, e.g., Exercise 1.19.

In classical Fourier analysis, one expands functions in $L^2(0, 1)$ in terms of the complex exponential functions $\{e^{2\pi i k x}\}_{k \in \mathbb{Z}}$. In Chapter 9 we will obtain more general results with $\{e^{2\pi i k x}\}_{k \in \mathbb{Z}}$ replaced by $\{e^{i\lambda_k x}\}_{k \in \mathbb{Z}}$ for some real sequence $\{\lambda_k\}_{k \in \mathbb{Z}}$ satisfying certain density conditions. Let us for the moment consider a finite collection of exponential functions $\{e^{i\lambda_k x}\}_{k=1}^n$, where $\{\lambda_k\}_{k=1}^n$ is a sequence of real numbers. Unless $\{\lambda_k\}_{k=1}^n$ contains repetitions, such a family of exponentials is always linearly independent:

Lemma 1.7.2 *Let $\{\lambda_k\}_{k=1}^n$ be a sequence of real numbers, and assume that $\lambda_k \neq \lambda_j$ for $k \neq j$. Let $I \subseteq \mathbb{R}$ be an arbitrary nonempty interval, and consider the complex exponentials $\{e^{i\lambda_k x}\}_{k=1}^n$ as functions on I. Then the functions $\{e^{i\lambda_k x}\}_{k=1}^n$ are linearly independent.*

Proof. It is enough to prove that the functions $\{e^{i\lambda_k x}\}_{k\in\mathbb{Z}}$ are linearly independent as functions on any bounded interval $]a, b[$, where $a, b \in \mathbb{R}$, $a < b$. Assume that for some coefficients $\{c_k\}_{k=1}^n$,

$$\sum_{k=1}^n c_k e^{i\lambda_k x} = 0, \ \forall x \in]a, b[.$$

When x runs through the interval $]\frac{a-b}{2}, \frac{b-a}{2}[$, the variable $x + \frac{a+b}{2}$ runs through $]a, b[$; it follows that

$$\sum_{k=1}^n c_k e^{i\lambda_k (x+\frac{a+b}{2})} = 0, \ \forall x \in]\frac{a-b}{2}, \frac{b-a}{2}[.$$

Writing $d_k := c_k e^{i\lambda_k \frac{a+b}{2}}$, this leads to

$$\sum_{k=1}^n d_k e^{i\lambda_k x} = 0, \ \forall x \in]\frac{a-b}{2}, \frac{b-a}{2}[.$$

By differentiating this equation j times, $j = 0, 1, \cdots$, we obtain that

$$\sum_{k=1}^n d_k (i\lambda_k)^j e^{i\lambda_k x} = 0, \ \forall x \in]\frac{a-b}{2}, \frac{b-a}{2}[, \ j = 0, 1, \cdots.$$

Putting $x = 0$ and writing the corresponding equations for $j = 0, \ldots, n-1$ as a matrix equation gives

$$\begin{pmatrix} 1 & 1 & \cdot & \cdot & 1 \\ \lambda_1 & \lambda_2 & \cdot & \cdot & \lambda_n \\ \cdot & \cdot & \cdot & \cdot & \cdot \\ \cdot & \cdot & \cdot & \cdot & \cdot \\ \lambda_1^{n-1} & \lambda_2^{n-1} & \cdot & \cdot & \lambda_n^{n-1} \end{pmatrix} \begin{pmatrix} d_1 \\ d_2 \\ \cdot \\ \cdot \\ d_n \end{pmatrix} = \begin{pmatrix} 0 \\ 0 \\ \cdot \\ \cdot \\ 0 \end{pmatrix}.$$

The system matrix is a Vandermonde matrix with determinant

$$\Delta = \prod_{k,j=1, k\neq j}^n (\lambda_k - \lambda_j) \neq 0;$$

therefore, the unique solution is $d_1 = d_2 = \cdots = d_n = 0$, which implies that $c_1 = \cdots = c_n = 0$. Thus, $\{e^{i\lambda_k x}\}_{k=1}^n$ are linearly independent. $\square$

In words, Lemma 1.7.2 means that complex exponentials do not give natural examples of frames in finite-dimensional spaces: if $\lambda_k \neq \lambda_j$ for $k \neq j$, then the complex exponentials $\{e^{i\lambda_k x}\}_{k=1}^n$ form a basis for their span in $L^2(I)$ for any interval I of finite length, but never an overcomplete system. We cannot obtain overcompleteness by adding extra exponentials (except by repeating some of the λ-values) – this will just enlarge the space but keep the independence. In Exercise 1.20 the similar problem for sines and cosines is considered.

The complex exponentials do not belong to $L^2(\mathbb{R})$, but by multiplying them with a function $g \in L^2(\mathbb{R})$, we obtain a class of functions in $L^2(\mathbb{R})$. In Chapters 11–14 we will work with systems of functions in $L^2(\mathbb{R})$ of the form

$$\{E_{mb}T_{na}g\}_{m,n\in\mathbb{Z}} := \{e^{2\pi i mbx}g(x-na)\}_{m,n\in\mathbb{Z}};$$

here, g is a given function in $L^2(\mathbb{R})$, and the parameters a, b are positive real numbers. Such a family of functions is called a *Gabor system*. It was proved by Linnell [499] in 1997 that if $g \neq 0$, then an arbitrary finite subfamily $\{e^{2\pi i mbx}g(x-na)\}_{(m,n)\in\mathcal{F}}$, $\mathcal{F} \subset \mathbb{Z}^2$, is linearly independent. This yields a partial answer to a conjecture formulated by Heil, Ramanathan, and Topiwala [393] in 1995, where the points $\{(na, mb)\}_{m,n\in\mathbb{Z}}$ are replaced by arbitrary distinct points in $\mathbb{R}^2$.

The HRT Conjecture: Given any finite collection of distinct points $\{(\mu_k, \lambda_k)\}_{k\in\mathcal{F}}$ in $\mathbb{R}^2$ and a function $g \neq 0$, the Gabor system

$$\{e^{2\pi i \lambda_k x}g(x-\mu_k)\}_{k\in\mathcal{F}}$$

is linearly independent.

Considerable effort has been invested in the conjecture, but despite the fact that it "just" deals with finite collections of functions, it is still open. We return to this conjecture in its right context on page 343. Also, in Section 14.6, we will construct frames in $\mathbb{C}^n$ having the Gabor structure.

Wavelets is another important class of functions in $L^2(\mathbb{R})$; we consider them in detail in Chapters 15–19. A wavelet system consists of functions of the form

$$\psi_{j,k}(x) = 2^{j/2}\psi(2^j x - k), \ j, k \in \mathbb{Z},$$

where $\psi \in L^2(\mathbb{R})$ is a given function. Linearly dependent wavelet systems exist. For example, by letting $\psi := \chi_{[0,1[}$, one has

$$\psi_{0,0} = \frac{1}{\sqrt{2}}(\psi_{1,0} + \psi_{1,1}).$$

If a finite wavelet system, $\{\psi_{j,k}\}_{|j|,|k|\leq N}$ for some $N \in \mathbb{N}$, happens to be linearly independent, one could ask for the *minimal* number m_N of independent sets it can be split into. One could expect m_N to grow with N; however, in case ψ has compact support and $|\psi| > 0$ on some interval of positive length, it is proved in [195] that one can find a number $m \in \mathbb{N}$ such that $\{\psi_{j,k}\}_{|j|,|k| \leq N}$ can be split into m linearly independent sets, regardless of how large N is. It is not known whether a similar result holds if ψ is not assumed to have compact support. For results about linear independence of special wavelet systems, we refer to the paper [87] by Bownik and Speegle.

1.8 Fusion Frames

Fusion frames were introduced by Casazza and Kutyniok in [140] and further developed in their joint paper [141] with Li. The theory for fusion frames is available in arbitrary separable Hilbert spaces (finite-dimensional or not), but since the concept is motivated by its applications, we will restrict the description to the finite-dimensional case.

The motivation behind fusion frames comes from signal processing, more precisely, the desire to process and analyze large data sets efficiently. A natural idea is to split such data sets into suitable smaller "blocks" which can be treated independently. In more mathematical terms, this could correspond to a splitting of a signal belonging to a high-dimensional vector space into its components in lower-dimensional subspaces.

From a pure mathematical point of view, fusion frames are special cases of the g-frames discussed in detail in Section 8.1. However, the connection to concrete applications is less apparent from the more abstract definition of g-frames.

The key ingredients in a fusion frame for $\mathbb{C}^n$ are formed by a collection of subspaces $\{\mathcal{V}_k\}_{k=1}^m$ of $\mathbb{C}^n$ and a corresponding collection of strictly positive numbers $w_k,\ k = 1, \ldots, m$. For $k = 1, \ldots, m$, let P_k denote the orthogonal projection of $\mathbb{C}^n$ onto $\mathcal{V}_k$. Following [140], we say that the pair $(\{\mathcal{V}_k\}_{k=1}^m, \{w_k\}_{k=1}^m)$ is a *fusion frame* for $\mathbb{C}^n$ if there exist constants $A, B > 0$ such that

$$A \, ||f||^2 \leq \sum_{k=1}^m w_k ||P_k f||^2 \leq B \, ||f||^2, \ \forall f \in \mathbb{C}^n. \tag{1.39}$$

The numbers A, B are called *bounds* for the fusion frame. The connection to frame theory is clear: if $\{f_k\}_{k=1}^m$ is a frame for $\mathbb{C}^n$ and we let

$$\mathcal{V}_k := \{cf_k \big| \ c \in \mathbb{C}\}, \ k = 1, \ldots, m,$$

then $(\{\mathcal{V}_k\}_{k=1}^m, \{||f_k||^2\}_{k=1}^m)$ is a fusion frame with the same bounds. In this particular situation, the spaces $\mathcal{V}_k$ are one dimensional; the additional freedom in fusion frames compared to frame theory arises because we can choose to decompose $\mathbb{C}^n$ by considering projections onto higher-dimensional subspaces.

Given a fusion frame $(\{\mathcal{V}_k\}_{k=1}^m, \{w_k\}_{k=1}^m)$ for $\mathbb{C}^n$, the associated *fusion frame operator* is defined by

$$S : \mathbb{C}^n \to \mathbb{C}^n, \; Sf = \sum_{k=1}^m w_k P_k f.$$

Following the lines of the proof of Theorem 1.1.5, it is easy to see that S is invertible on $\mathbb{C}^n$, which leads to the *fusion frame decomposition*

$$f = S^{-1} S f = S^{-1} \sum_{k=1}^m w_k P_k f = \sum_{k=1}^m w_k S^{-1} P_k f. \qquad (1.40)$$

Mathematically, there is a clear link between frames and fusion frames and clear ways to transfer results from one setting to the other. We leave the proof of the following result to the reader (Exercise 1.21).

Proposition 1.8.1 *Let $\{\mathcal{V}_k\}_{k=1}^m$ be a collection of subspaces of $\mathbb{C}^n$ and $\{w_k\}_{k=1}^m$ a collection of positive scalars. For each $k = 1, \ldots, m$, let $\{e_{j,k}\}_{j \in J_k}$ denote an orthonormal basis for $\mathcal{V}_k$. Then the following are equivalent:*

(i) $(\{\mathcal{V}_k\}_{k=1}^m, \{w_k\}_{k=1}^m)$ is a fusion frame for $\mathbb{C}^n$, with bounds A, B;

(ii) $\{\sqrt{w_k} u_{j,k}\}_{k=1,\ldots,m, j \in J_k}$ is a frame for $\mathbb{C}^n$, with bounds A, B.

For a much more detailed discussion of fusion frames, we refer to Chapter 13 of the book [139]; see also the discussion of the more general g-frames in Section 8.1.

1.9   Applications of Finite Frames

The option of having overcompleteness in a frame makes the concept more flexible than that of a basis. In this section we will show that overcompleteness is also useful in the context of signal transmission.

Modern communication networks act by transporting packets of data. Each packet contains the "essential information," i.e., the data we want to transmit, as well as a collection of "control parameters." The purpose of these extra parameters is to check that the data is delivered correctly: in case an error occurs, no packet will be delivered at all. It is clear that if there are no relationships between the various packets, the data belonging to a lost packet cannot be recovered. However, if there is some redundancy built into the system, i.e., a relationship between the information in the packets, there is some hope that at least parts of the missing data can be recovered.

Mathematically, one can model the packets to transmit as frame coefficients. Thus, a packet that is not delivered amounts to removal of an element from the frame. If the frame is a basis, it is no longer a basis after removal of an element; however, if it is overcomplete, it is possible that it remains a frame after deletion of an element.

In practice one might lose more than one packet, i.e., more than one frame element. Thus, we are facing the question of how to construct frames that are stable toward removal of more than one element. We have already in Proposition 1.5.7 seen that the harmonic frame $\{f_k\}_{k=1}^m$ for $\mathbb{C}^n$ performs optimally with regard to erasure of elements: any subset containing at least n elements of this frame forms a frame for $\mathbb{C}^n$, i.e., the frame has full spark.

Not all frames behave as well as the one in Proposition 1.5.2: regardless how many elements a frame has, the idea from Example 1.5.5 shows that removal of a single particular element might destroy the frame property. If we have information on the lower frame bound and the norm of the frame elements, Proposition 1.5.6 provides a criterion for how many elements we can (at least) remove.

In the context of signal transmission, the overcompleteness of frames has a very useful noise-suppressing effect. We will first give an intuitive explanation and return to a more detailed statistical argument afterward. Let us assume that we want to transmit the signal f belonging to a vector space V from a transmitter $\mathcal{A}$ to a receiver $\mathcal{R}$. If both $\mathcal{A}$ and $\mathcal{R}$ have knowledge of a frame $\{f_k\}_{k=1}^m$ for V, this can be done if $\mathcal{A}$ transmits the coefficients $\{\langle f, f_k \rangle\}_{k=1}^m$; based on knowledge of these numbers, the receiver $\mathcal{R}$ can reconstruct the signal f using the frame decomposition

$$f = \sum_{k=1}^m \langle f, f_k \rangle S^{-1} f_k.$$

Now assume that $\mathcal{R}$ receives a noisy signal, i.e., a perturbation

$$\{\langle f, f_k \rangle + c_k\}_{k=1}^m$$

of the correct coefficients; this might, for example, happen due to quantization, as explained in Section 1.5. Based on the received coefficients, $\mathcal{R}$ will expect that the transmitted signal was

$$
\begin{aligned}
\sum_{k=1}^m \left(\langle f, f_k \rangle + c_k \right) S^{-1} f_k &= \sum_{k=1}^m \langle f, f_k \rangle S^{-1} f_k + S^{-1} \sum_{k=1}^m c_k f_k \\
&= f + S^{-1} \sum_{k=1}^m c_k f_k;
\end{aligned}
$$

this differs from the correct signal f by the term $S^{-1} \sum_{k=1}^m c_k f_k$. If $\{f_k\}_{k=1}^m$ is overcomplete, the synthesis operator $T\{c_k\}_{k=1}^m = \sum_{k=1}^m c_k f_k$ has a non-trivial kernel, implying that parts of the noise contribution might add up to zero and cancel. This will *never* happen if $\{f_k\}_{k=1}^m$ is an orthonormal basis! In that case $S = I$ and

$$\|S^{-1} \sum_{k=1}^{m} c_k f_k\| = \sqrt{\sum_{k=1}^{m} |c_k|^2},$$

so (at least intuitively) each noise contribution will make the reconstruction worse.

The above arguments can be refined using statistical models for noise. Following [328], we will use this to analyze how one should choose the frame $\{f_k\}_{k=1}^{m}$ in order to obtain the maximal noise-suppressing effect. Let us again assume that $\mathcal{A}$ transmits the coefficients $\{\langle f, f_k \rangle\}_{k=1}^{m}$ to the receiver $\mathcal{R}$ and that $\mathcal{R}$ receives a noisy signal $\{\langle f, f_k \rangle + \eta_k\}_{k=1}^{m}$. In contrast with the simplified setting above, we now consider each noise component η_k as a random variable; we will assume that each η_k has mean zero and variance σ^2 and that η_k and η_ℓ are uncorrelated for $k \neq \ell$. Letting E denote the mean, these assumptions can be expressed as

$$E[\eta_k] = 0, \quad E[\eta_k \eta_\ell] = \sigma^2 \delta_{k,\ell}, \ k, \ell = 1, \ldots, m. \tag{1.41}$$

As above, based on the coefficients $\{\langle f, f_k \rangle + \eta_k\}_{k=1}^{m}$, the receiver will reconstruct the signal as

$$\widetilde{f} = \sum_{k=1}^{m} \left(\langle f, f_k \rangle + \eta_k \right) S^{-1} f_k = f + \sum_{k=1}^{m} \eta_k S^{-1} f_k.$$

Thus, the difference between the reconstructed signal $\widetilde{f}$ and the original signal f is

$$\widetilde{f} - f = \sum_{k=1}^{m} \eta_k S^{-1} f_k.$$

Remember that $\widetilde{f} - f$ is a vector with n coordinates, which depend on the random variables η_k. The associated *mean-square error* (MSE) is defined by

$$MSE := \frac{1}{n} E \|\widetilde{f} - f\|^2.$$

Now, inserting the expression for $\widetilde{f} - f$ shows that

$$MSE = \frac{1}{n} E(\langle \widetilde{f} - f, \widetilde{f} - f \rangle)$$

$$= \frac{1}{n} E \left[\sum_{k=1}^{m} \sum_{\ell=1}^{m} \eta_k \eta_\ell \langle S^{-1} f_k, S^{-1} f_\ell \rangle \right]$$

$$= \frac{1}{n} \sum_{k=1}^{m} \sum_{\ell=1}^{m} E[\eta_k \eta_\ell] \langle S^{-1} f_k, S^{-1} f_\ell \rangle.$$

Via the assumptions (1.41), this implies that

$$MSE = \frac{1}{n}\sum_{k=1}^{m}\sum_{\ell=1}^{m}\sigma^2\delta_{k,\ell}\langle S^{-1}f_k, S^{-1}f_\ell\rangle = \frac{1}{n}\sigma^2\sum_{k=1}^{m}||S^{-1}f_k||^2. \quad (1.42)$$

We now show that among all normalized frames containing a fixed number of elements, this expression is minimized for tight frames. We will use the following well-known lemma.

Lemma 1.9.1 *Let $\{a_k\}_{k=1}^{n}$ be a sequence of positive numbers. Then the harmonic mean of the sequence is smaller than or equal to the arithmetic mean, i.e.,*

$$\frac{n}{\sum_{k=1}^{n}\frac{1}{a_k}} \leq \frac{1}{n}\sum_{k=1}^{n}a_k.$$

The inequality is an equality if and only if all the a_k are equal.

Theorem 1.9.2 *Consider normalized frames $\{f_k\}_{k=1}^{m}$ for $\mathbb{R}^n$, where $n, m \in \mathbb{N}$ are fixed. Among all such frames $\{f_k\}_{k=1}^{m}$, the MSE is minimal if and only if the frame is tight. The attained minimal value is*

$$MSE = \frac{n}{m}\sigma^2. \quad (1.43)$$

Proof. Let $\lambda_1, \ldots, \lambda_n$ denote the eigenvalues for the frame operator S associated with $\{f_k\}_{k=1}^{m}$. By Theorem 1.3.1,

$$\sum_{k=1}^{n}\lambda_k = \sum_{k=1}^{m}||f_k||^2 = m. \quad (1.44)$$

The frame $\{S^{-1}f_k\}_{k=1}^{m}$ has S^{-1} as frame operator, and this operator has the eigenvalues $\lambda_1^{-1}, \ldots, \lambda_n^{-1}$. Now, (1.42) together with Theorem 1.3.1 imply that

$$MSE = \frac{1}{n}\sigma^2\sum_{k=1}^{m}||S^{-1}f_k||^2 = \frac{1}{n}\sigma^2\sum_{k=1}^{n}\frac{1}{\lambda_k}. \quad (1.45)$$

Our goal is now to minimize the expression in (1.45) under the constraint (1.44); equivalently, we want to *maximize* the expression

$$\frac{1}{\sum_{k=1}^{n}\frac{1}{\lambda_k}}$$

under the condition that $\sum_{k=1}^{n}\lambda_k = m$. According to Lemma 1.9.1, this happens if and only if all eigenvalues λ_k are equal, i.e., for

$$\lambda_k = \frac{m}{n}, \ k = 1, \ldots, m.$$

This implies that $\{f_k\}_{k=1}^m$ is a tight frame with frame bound m/n. The attained minimal value of the mean-square error is

$$MSE = \frac{1}{n}\sigma^2 \sum_{k=1}^n \frac{1}{\lambda_k} = \frac{1}{n}\sigma^2 n \frac{n}{m} = \frac{n}{m}\sigma^2. \qquad \Box$$

The expression (1.43) shows that for a fixed dimension, i.e., a fixed value of n, the MSE decreases when the number of elements in the frame increases, i.e., for higher redundancy. In this sense, the redundancy in a frame helps to reduce the mean-square error.

Let us end this section with a few words about another topic with clear relation to engineering, namely, *phaseless reconstruction*. Given a frame $\{f_k\}_{k=1}^m$ for a finite-dimensional vector space V, the question is when and how one can reconstruct a vector $f \in V$ based on knowledge of the *magnitudes* of the inner products between f and the frame vectors, i.e., the numbers $|\langle f, f_k \rangle|$. Clearly this is an impossible task if $\{f_k\}_{k=1}^m$ is a basis for V. On the other hand, there exist frames for which the reconstruction is indeed possible. A class of tight frames with this property is constructed in the paper [30] by Balan, Casazza, and Edidin; more results can be found in [29, 38] and the references therein.

1.10   Remarks on Recent Frame Constructions

The development of frame theory in finite-dimensional spaces started relatively late, around the millennium. The paper by Zimmermann [642] can be seen as the starting point: it is based on an answer to a question by Feichtinger, posed at a conference in Haus Bommerholz in 2000. Relatively soon hereafter, in 2003, Benedetto (who was also at the conference) published the paper [46] with Fickus. Since then the number of contributions to the theory for finite frames has exploded. A large part of the literature deals with frame constructions with attractive properties from the numerical point of view or with respect to signal processing and thus goes in a somewhat different direction than the more functional analytic approach that will dominate the rest of this book. For this reason we will not go in detail with the concrete constructions and algorithms, but we will highlight some of the central issues. Along the way suggestions for further research will be discussed.

Some of the work by Zimmermann [642] was already presented in Proposition 1.5.2. The paper also consider corresponding real-valued frames for $\mathbb{R}^n$. The paper [46] by Benedetto and Fickus characterizes all finite normalized tight frames $\{f_k\}_{k=1}^m$ in $\mathbb{R}^n$ and $\mathbb{C}^n$ as minimizers of the *frame potential*, i.e., the mapping that to any sequence $\{f_k\}_{k=1}^m$ in $\mathbb{C}^n$ associates the number $\sum_{k=1}^m \sum_{\ell=1}^m |\langle f_k, f_\ell \rangle|^2$. Already in that paper, the short name

FNTF was introduced for the finite normalized tight frames; later, the abbreviation *FUNTF* has also been used. A systematic method to construct unit-norm tight frames $\{f_k\}_{k=1}^m$ in $\mathbb{R}^n$, provided that $m \geq n$, was introduced under the name *spectral tetris* in the paper [135] by Casazza et al. and further analyzed in [133]; in particular, the authors use the method to construct tight fusion frames.

Let us state some of the properties that are relevant for applications of finite frames:

- That the elements in the frame have the same norm; such frames are usually called *equal-norm frames*. This property is satisfied for the harmonic frames.

- That the condition number for the frame operator can be controlled. This is clearly satisfied if the given frame is tight and hence for the harmonic frames. Recall also from Theorem 1.9.2 that for equal-norm frames, the MSE is minimized whenever the frame is tight.

- Maximal stability against erasures, i.e., that $\{f_k\}_{k=1}^m$ has full spark. We have already noticed that the harmonic frames have full spark; see Proposition 1.5.7.

A further relevant property is that the angle between the elements in the frame is constant; such frames are said to be *equiangular*. If the elements in $\{f_k\}_{k=1}^m$ have the same length, the condition of being equiangular amounts to the existence of a constant C such that

$$|\langle f_k, f_j \rangle| = C, \ \forall k \neq j.$$

In particular, any orthonormal basis $\{e_k\}_{k=1}^n$ for $\mathbb{C}^n$ is equiangular. The question of existence of redundant equiangular frames is much more subtle, and many questions remain open. The following elegant result appeared in the paper [591] by Strohmer and Heath:

Theorem 1.10.1 *Consider a unit-norm frame $\{f_k\}_{k=1}^m$ for either $\mathbb{C}^n$ or $\mathbb{R}^n$; then*

$$\max_{k \neq j} |\langle f_k, f_j \rangle| \geq \sqrt{\frac{m-n}{n(m-1)}}. \tag{1.46}$$

Equality holds in (1.46) if and only if $\{f_k\}_{k=1}^m$ is an equiangular tight frame.

(i) In the case of $\mathbb{C}^n$, equality in (1.46) can only occur if $m \leq n(n+1)/2$;

(ii) In the case of $\mathbb{R}^n$, equality in (1.46) can only occur if $m \leq n^2$.

Already Theorem 1.10.1 demonstrates that equiangular unit-norm frames $\{f_k\}_{k=1}^m$ in $\mathbb{C}^n$ or $\mathbb{R}^n$ only exist for certain choices of the parameters m, n, but the existence is much more restricted than that; see the papers [606] by Sustik et al. and [411] by Holmes and Paulsen. For a closer analysis of the special case $m = 2n$, we refer to the paper [590] by Strohmer.

A frame that for fixed parameters $m, n \in \mathbb{N}$ minimizes the expression $\max_{k \neq j} |\langle f_k, f_j \rangle|$ over all unit-norm frames $\{f_k\}_{k=1}^m$ for $\mathbb{C}^n$ is called a *Grassmannian frame;* see [591]. Thus, in case the parameters m, n are chosen such that there exist equiangular tight frames $\{f_k\}_{k=1}^m$ in $\mathbb{C}^n$ or $\mathbb{R}^n$, the Grassmannian frames correspond to exactly these. For more information on Grassmannian frames, we refer to the original paper [591] and [617] by Tropp et al.

The paper [591] contains several examples of equiangular tight frames, e.g., certain versions of the harmonic frames where the columns are generated by different roots of unity. More results on equiangular frames and explicit examples can be found in the mentioned papers as well as [618] by Xia et al.

The large number of recent frame constructions is highly motivated by applications. The reader can consult the papers [591, 66] and [437] for applications of equiangular tight frames to coding theory, and [521] for an application to digital fingerprinting.

Despite the huge information that is available about finite frames, it is still possible to pose seemingly innocent questions, where no answer is known. Here is an interesting question that was posed by Thomas Strohmer at the SAMPTA conference in Washington, 2015:

Question: Let $\{f_k\}_{k=1}^m$ be a frame for $\mathbb{C}^n$, for which we only know the direction of the vectors f_k but not the norms $\|f_k\|$. Assume that we for an unknown vector $f \in \mathbb{C}^n$ know the inner products

$$\langle f, f_k \rangle, \; k = 1, \ldots, m. \tag{1.47}$$

How – and under which conditions – can we recover the direction of the vector f based on the information (1.47)?

Note that the information in (1.47) will not allow us to determine the length of f if we do not know the norm of the vectors f_k : in fact, if (1.47) holds for a vector f and a frame $\{f_k\}_{k=1}^m$, it also holds for the vector cf and the frame $\{c^{-1}f_k\}_{k=1}^m$ whenever c is an arbitrary nonzero real number.

We also observe that the question has an easy answer if we actually know the norms $\|f_k\|$. In that case we know the frame $\{f_k\}_{k=1}^m$ completely, and knowledge of the numbers in (1.47) allow us to compute the frame operator and apply the frame decomposition in Theorem 1.1.5 (ii) directly.

1.11   Exercises

1.1 Prove that the adjoint of the synthesis operator T in (1.4) is given by the expression in (1.5).

1.2 Prove Proposition 1.1.4. (Hint: use that if U is a linear self-adjoint map on V for which $\langle Ux, x \rangle = 0$ for all $x \in V$, then $U = 0$; see Lemma 2.4.4.)

1.3 Show that every frame $\{f_k\}_{k=1}^m$ for a finite dimensional vector space V contains a subset which is a basis for V.

1.4 Can a frame in a finite-dimensional space contain infinitely many elements?

1.5 Let $\{f_k\}_{k\in I}$ be a frame for a finite-dimensional vector space V and assume that $||f_k||$ is bounded below. Prove that I is finite. (w.l.o.g. you may assume that $V = \mathbb{R}^n$ and that $||f_k|| = 1$, $\forall k$; explain why if you want to use this fact!)

1.6 Find a noncanonical dual frame associated with the frame considered in Example 1.1.6.

1.7 Show that it is necessary to add at least 9 vectors in order to extend the frame $\{f_k\}_{k=1}^{10}$ in Example 1.2.2 to a tight frame.

1.8 Show that the vectors

$$\left(\begin{matrix} 0 \\ \sqrt{2/3} \end{matrix} \right), \left(\begin{matrix} -1/\sqrt{2} \\ -1/\sqrt{6} \end{matrix} \right), \left(\begin{matrix} 1/\sqrt{2} \\ -1/\sqrt{6} \end{matrix} \right)$$

constitute a tight frame for $\mathbb{C}^2$ with frame bound $A = 1$. Make a draft of the vectors; for obvious reasons the frame is called the *Mercedes-Benz frame.*

1.9 Construct a frame $\{f_k\}_{k=1}^m$ for $\mathbb{C}^2$ for which there exists $f \in \mathbb{C}^2$ such that the coefficients $\{d_k\}_{k=1}^m$ in Theorem 1.1.8 are not unique.

1.10 Let $\{e_1, e_2\}$ be the canonical orthonormal basis for $\mathbb{C}^2$ and consider the frame $\{f_k\}_{k=1}^3 = \{e_1, e_2, e_1 + e_2\}$.

(i) Find the coefficients with minimal ℓ^2-norm among all sequences $\{c_k\}_{k=1}^3$ for which $e_1 = \sum_{k=1}^3 c_k f_k$.

(ii) Find the coefficients $\{c_k^{(1)}\}_{k=1}^3$ and $\{c_k^{(2)}\}_{k=1}^3$ which minimize the ℓ^1-norm in the representation of e_1 and e_2, respectively.

(iii) Clearly, $e_1 + e_2 = \sum_{k=1}^3 (c_k^{(1)} + c_k^{(2)}) f_k$, but is $\{c_k^{(1)} + c_k^{(2)}\}_{k=1}^3$ minimizing the ℓ^1-norm among all sequences representing $e_1 + e_2$?

1.11 Assume that $\{f_k\}_{k=1}^m$ is a frame for $\mathbb{C}^n$. Prove that the $2m$ vectors consisting of the real parts, respectively the imaginary parts, of the frame vectors constitute a frame for $\mathbb{R}^n$.

1.12 Show that a frame for $\mathbb{R}^n$ is also a frame for $\mathbb{C}^n$.

1.13 Prove Corollary 1.4.7.

1.14 Let $\{f_k\}_{k=1}^m$ be a frame for V with bounds A, B, and let P denote the orthogonal projection of V onto a subspace W. Prove that $\{Pf_k\}_{k=1}^m$ is a frame for W with frame bounds A, B.

1.15 Let $\{f_k\}_{k=1}^m$ be a normalized tight frame. Prove that the frame bound A is at least 1 and that $A = 1$ if and only if $\{f_k\}_{k=1}^m$ is an orthonormal basis.

1.16 Let $\{f_k\}_{k=1}^m$ be a frame for an n-dimensional vector space V, and let B denote the optimal upper bound. Prove that

$$B \leq \sum_{k=1}^m \|f_k\|^2 \leq nB.$$

1.17 Prove that Theorem 1.7.1 fails if the closed and bounded interval $[a, b]$ is replaced by an open interval or an unbounded interval.

1.18 Prove that for any $n \in \mathbb{N}$, the polynomials $\{1, x, \ldots, x^n\}$ are linearly independent in $C(0, 1)$.

1.19 Consider the polynomials $\{1, x, x^2\}$ as functions on the interval $[0, 1]$, and let $V = \text{span}\{1, x, x^2\}$. Equip V with the inner product (1.38) and find an orthonormal basis for V.

1.20 Let $\{\lambda_k\}_{k=1}^n$ be a sequence of real numbers.

(i) Prove that $\{\cos \lambda_k x\}_{k=1}^n$ are linearly independent in $C(-1, 1)$ if and only if $|\lambda_k| \neq |\lambda_j|$ for $k \neq j$.

(ii) Prove that $\{\sin \lambda_k x\}_{k=1}^n$ are linearly independent in $C(-1, 1)$ if and only if all λ_k are nonzero and $|\lambda_k| \neq |\lambda_j|$ for $k \neq j$.

(iii) Under which conditions on sequences $\{\lambda_k\}_{k=1}^n, \{\mu_k\}_{k=1}^m$ are the functions

$$\{\cos \lambda_k x\}_{k=1}^n \cup \{\sin \mu_k x\}_{k=1}^m$$

linearly independent in $C(-1, 1)$?

(iv) Replace the interval $]-1, 1[$ by an arbitrary nonempty interval and generalize (i),(ii), and (iii).

1.21 Prove Proposition 1.8.1.

2

Infinite-Dimensional Vector Spaces and Sequences

After the introduction to frames in finite-dimensional vector spaces in Chapter 1, the rest of the book will deal with expansions in infinite-dimensional vector spaces. Here great care is needed: we need to replace finite sequences $\{f_k\}_{k=1}^n$ by infinite sequences $\{f_k\}_{k=1}^\infty$, and suddenly the question of convergence properties becomes a central issue. The vector space itself might also cause problems, e.g., in the sense that Cauchy sequences might not be convergent. We expect the reader to have a basic knowledge about these problems and the way to circumvent them, but for completeness we repeat the central definitions and results concerning Banach spaces and operators hereon in Sections 2.1–2.2. In Sections 2.3–2.4 we specialize to Hilbert spaces and their operators. Section 2.5 deals with pseudo-inverse operators; this subject is not expected to be known and is treated in more detail. Section 2.6 introduces the so-called moment problems in Hilbert spaces. In Sections 2.7–2.9, we discuss the Hilbert space $L^2(\mathbb{R})$ consisting of the square integrable functions on $\mathbb{R}$ and three classes of operators hereon, as well as the Fourier transform. The material in those sections is not needed for the study of frames and bases on abstract Hilbert spaces, but it forms the basis for all the constructions in Chapters 9–20.

2.1 Banach Spaces and Sequences

A central theme in this book is to find conditions on a sequence $\{f_k\}$ in a vector space X such that every $f \in X$ has a representation as a

© Springer International Publishing Switzerland 2016
O. Christensen, *An Introduction to Frames and Riesz Bases*,
Applied and Numerical Harmonic Analysis,
DOI 10.1007/978-3-319-25613-9_2

superposition of the vectors f_k. In most spaces appearing in functional analysis, this cannot be done with a finite sequence $\{f_k\}$. We are therefore forced to work with infinite sequences, say, $\{f_k\}_{k=1}^\infty$, and the representation of f in terms of $\{f_k\}_{k=1}^\infty$ will be via an infinite series. For this reason, the starting point must be a discussion of convergence of infinite series. We collect the basic definitions here together with some conventions.

Throughout the section, we let X denote a complex vector space. A *norm* on X is a function $||\cdot|| : X \to [0, \infty[$ satisfying the following three conditions:

(i) $||x|| = 0 \Leftrightarrow x = 0$;

(ii) $||\alpha x|| = |\alpha|\, ||x||, \ \forall x \in X, \ \alpha \in \mathbb{C}$;

(iii) $||x + y|| \le ||x|| + ||y||, \ \forall x, y \in X$.

In situations where more than one vector space appear, we will frequently denote the norm on X by $||\cdot||_X$. If X is equipped with a norm, we say that X is a *normed vector space*. The *opposite triangle inequality* is satisfied in any normed vector space:

$$||x - y|| \ge |\, ||x|| - ||y||\, |, \ x, y \in X. \tag{2.1}$$

We say that a sequence $\{x_k\}_{k=1}^\infty$ in X

(i) converges to $x \in X$ if

$$||x - x_k|| \to 0 \text{ for } k \to \infty;$$

(ii) is a *Cauchy sequence* if for each $\epsilon > 0$ there exists $N \in \mathbb{N}$ such that

$$||x_k - x_\ell|| \le \epsilon \text{ whenever } k, \ell \ge N.$$

A convergent sequence is automatically a Cauchy sequence, but the opposite is not true in general. There are, however, normed vector spaces in which a sequence is convergent if and only if it is a Cauchy sequence; a space X with this property is called a *Banach space*.

Imitating the finite-dimensional setting described in Chapter 1, we want to study sequences $\{f_k\}_{k=1}^\infty$ in X with the property that each $f \in X$ has a representation $f = \sum_{k=1}^\infty c_k f_k$ for some coefficients $c_k \in \mathbb{C}$. In order to do so, we have to explain exactly what we mean by convergence of an infinite series. There are, in fact, at least three different options; we will now discuss these options.

First, the notation $\{f_k\}_{k=1}^\infty$ indicates that we have chosen some ordering of the vectors f_k,

$$f_1, f_2, f_3, \dots, f_k, f_{k+1}, \dots \ .$$

We say that an *infinite series* $\sum_{k=1}^\infty c_k f_k$ is convergent with sum $f \in X$ if

$$\left|\left| f - \sum_{k=1}^n c_k f_k \right|\right| \to 0 \text{ as } n \to \infty.$$

If this condition is satisfied, we write

$$f = \sum_{k=1}^{\infty} c_k f_k. \tag{2.2}$$

Thus, the definition of a convergent infinite series corresponds exactly to our definition of a convergent sequence with $x_n = \sum_{k=1}^{n} c_k f_k$.

Above we insisted on a fixed ordering of the sequence $\{f_k\}_{k=1}^{\infty}$. It is very important to notice that convergence properties of $\sum_{k=1}^{\infty} c_k f_k$ not only depend on the sequence $\{f_k\}_{k=1}^{\infty}$ and the coefficients $\{c_k\}_{k=1}^{\infty}$ but also on the ordering. Even if we consider a sequence in the simplest possible Banach space, i.e., a sequence $\{a_k\}_{k=1}^{\infty}$ in $\mathbb{C}$, it can happen that $\sum_{k=1}^{\infty} a_k$ is convergent but that $\sum_{k=1}^{\infty} a_{\sigma(k)}$ is divergent for a certain permutation σ of the natural numbers (Exercise 2.1). This observation leads to the second definition of convergence. If $\{f_k\}_{k=1}^{\infty}$ is a sequence in X and $\sum_{k=1}^{\infty} f_{\sigma(k)}$ is convergent for all permutations σ, we say that $\sum_{k=1}^{\infty} f_k$ is *unconditionally convergent*. In that case, the limit is the same regardless of the order of summation.

As soon as we have defined frames and Riesz bases in Hilbert spaces, it will become clear that they automatically lead to unconditionally convergent expansions. For this reason, we never need to prove by hand that a given series converges unconditionally. For the sake of completeness, we refer to [495] and [577] for a more detailed analysis of the different types of convergence and the proof of the following lemma.

Lemma 2.1.1 *Let $\{f_k\}_{k=1}^{\infty}$ be a sequence in a Banach space X, and let $f \in X$. Then the following are equivalent:*

(i) $\sum_{k=1}^{\infty} f_k$ converges unconditionally to $f \in X$.

(ii) For every $\epsilon > 0$ there exists a finite set F such that

$$\left\| f - \sum_{k \in I} f_k \right\| \leq \epsilon$$

for all finite sets $I \subset \mathbb{N}$ containing F.

Finally, an infinite series $\sum_{k=1}^{\infty} f_k$ is said to be *absolutely convergent* if

$$\sum_{k=1}^{\infty} \|f_k\| < \infty.$$

In any Banach space, absolute convergence of $\sum_{k=1}^{\infty} f_k$ implies that the series converges unconditionally (Exercise 2.2), but the opposite does not hold in infinite-dimensional spaces (see page 51 in [401] or page 68 in [464] and the references therein). In finite-dimensional spaces, the two types of convergence are identical.

A subset $Z \subseteq X$ (countable or not) is said to be *dense* in X if for each $f \in X$ and each $\epsilon > 0$ there exists $g \in Z$ such that

$$||f - g|| \leq \epsilon.$$

In words, this means that elements in X can be approximated arbitrarily well by elements in Z.

For a given sequence $\{f_k\}_{k=1}^{\infty}$ in X, we let span$\{f_k\}_{k=1}^{\infty}$ denote the vector space consisting of all *finite* linear combinations of vectors f_k, i.e.,

$$\text{span}\,\{f_k\}_{k=1}^{\infty} = \{\alpha_1 f_1 + \alpha_2 f_2 + \cdots + \alpha_N f_N \mid N \in \mathbb{N}, \alpha_1, \alpha_2, \ldots, \alpha_N \in \mathbb{C}\}.$$

The definition of convergence shows that if each $f \in X$ has a representation of the type (2.2), then each $f \in X$ can be approximated arbitrarily well in norm by elements in span$\{f_k\}_{k=1}^{\infty}$, i.e.,

$$\overline{\text{span}}\{f_k\}_{k=1}^{\infty} = X. \tag{2.3}$$

A sequence $\{f_k\}_{k=1}^{\infty}$ having the property (2.3) is said to be *complete* or *total*. We note that there exist normed spaces where no sequence $\{f_k\}_{k=1}^{\infty}$ is complete. A normed vector space, in which a countable and dense family exists, is said to be *separable*.

When considering expansions of the form (2.2), the coefficients c_k are real or complex numbers. In case at most finitely many entries c_k are nonzero, we say that $\{c_k\}_{k=1}^{\infty}$ is a *finite sequence*.

2.2   Operators on Banach Spaces

Let X and Y denote Banach spaces. A linear map $U : X \to Y$ is called an *operator,* and U is *bounded* or *continuous* if there exists a constant $K > 0$ such that

$$||Ux||_Y \leq K\,||x||_X, \ \forall x \in X. \tag{2.4}$$

Usually, it will be clear from the context which norm we use, so we will write $||\cdot||$ for both $||\cdot||_X$ and $||\cdot||_Y$. The *norm* of the operator U, denoted by $||U||$, is the smallest constant K that can be used in (2.4). Alternatively,

$$||U|| = \sup\{||Ux|| \mid x \in X, ||x|| = 1\}.$$

If U_1 and U_2 are operators for which the range of U_2 is contained in the domain of U_1, we can consider the composed operator $U_1 U_2$; if U_1 and U_2 are bounded, then also $U_1 U_2$ is bounded, and

$$||U_1 U_2|| \leq ||U_1||\,||U_2||. \tag{2.5}$$

Now consider a sequence of operators $U_n : X \to Y$, $n \in \mathbb{N}$, which converges pointwise to a mapping $U : X \to Y$, i.e.,

$$U_n x \to Ux, \ \text{as } n \to \infty, \ \forall x \in X.$$

We say that U_n converges to U in the *strong operator topology*. The *Banach–Steinhaus theorem*, also known as the *uniform boundedness principle*, states the following (see page 69 in [621]) or page 14 in [401]):

Theorem 2.2.1 *Let $U_n : X \to Y$, $n \in \mathbb{N}$, be a sequence of bounded operators, which converges pointwise to a mapping $U : X \to Y$. Then U is linear and bounded. Furthermore, the sequence of norms $||U_n||$ is bounded, and $||U|| \leq \liminf ||U_n||$.*

An operator $U : X \to Y$ is *invertible* if U is surjective and injective. For a bounded, invertible operator, the inverse operator is bounded; see, e.g., page 286 in [464]:

Theorem 2.2.2 *A bounded bijective operator between Banach spaces has a bounded inverse.*

In case $X = Y$, it makes sense to speak about the identity operator I on X. The *Neumann theorem* states that an operator $U : X \to X$ is invertible if it is close enough to the identity operator; a proof can be found on page 48 in [401].

Theorem 2.2.3 *If $U : X \to X$ is bounded and $||I - U|| < 1$, then U is invertible, and*

$$U^{-1} = \sum_{k=0}^{\infty} (I - U)^k. \tag{2.6}$$

Furthermore,

$$||U^{-1}|| \leq \frac{1}{1 - ||I - U||}.$$

Note that (2.6) should be interpreted in the sense of the operator norm, i.e., as

$$\left\| U^{-1} - \sum_{k=0}^{N} (I - U)^k \right\| \to 0 \text{ as } N \to \infty.$$

A special role is played by the continuous linear operators $\Phi : X \to \mathbb{C}$; they are called *functionals*, and the collection of all functionals is the *dual space* X^* of X. The dual X^* is itself a Banach space with respect to the norm

$$||\Phi|| = \sup\{|\Phi(x)| \,|\, x \in X, ||x|| = 1\}.$$

It is well known that X is isometrically isomorphic to a subspace of the double dual $X^{**} := (X^*)^*$ and thus can be identified with a subspace of X^{**}; in case $X = X^{**}$, we say that X is *reflexive*.

2.3 Hilbert Spaces

A special class of normed vector spaces is formed by *inner product spaces.*
Recall that an inner product on a complex vector space X is a mapping
$\langle \cdot, \cdot \rangle : X \times X \to \mathbb{C}$ for which

  (i) $\langle \alpha x + \beta y, z \rangle = \alpha \langle x, z \rangle + \beta \langle y, z \rangle$, $\forall x, y, z \in X$, $\alpha, \beta \in \mathbb{C}$;

  (ii) $\langle x, y \rangle = \overline{\langle y, x \rangle}$, $\forall x, y \in X$;

  (iii) $\langle x, x \rangle \geq 0$, $\forall x \in X$, and $\langle x, x \rangle = 0 \Leftrightarrow x = 0$.

Note that we have chosen to let the inner product be linear in the first
entry. It implies that the inner product is conjugated linear in the second
entry. Frequently, the opposite convention is used in the literature.

A vector space X with an inner product $\langle \cdot, \cdot \rangle$ can be equipped with the
norm

$$||x|| := \sqrt{\langle x, x \rangle}, \ x \in X.$$

If X is a Banach space with respect to this norm, then X is called a *Hilbert
space.* We reserve the letter $\mathcal{H}$ for these spaces. We will always assume that
$\mathcal{H}$ is *nontrivial,* i.e., that $\mathcal{H} \neq \{0\}$. The standard examples are the spaces
$L^2(\mathbb{R})$ and $\ell^2(\mathbb{N})$ discussed in Section 2.7.

Two elements $x, y \in \mathcal{H}$ are *orthogonal* if $\langle x, y \rangle = 0$; and the *orthogonal
complement* of a subspace U of $\mathcal{H}$ is

$$U^{\perp} = \{x \in \mathcal{H} : \ \langle x, y \rangle = 0, \ \forall y \in U\}.$$

The above definitions apply whether $\mathcal{H}$ is finite-dimensional or infinite-
dimensional. Also note that norms and inner products are defined in a
similar way on real vector spaces (just replace the scalars $\mathbb{C}$ by the real
scalars $\mathbb{R}$).

We will now collect a few elementary results concerning Hilbert spaces
that will be used repeatedly during the book.

Lemma 2.3.1 *Let $\mathcal{H}$ denote a Hilbert space. Then the following hold:*

  (i) For any $x, y \in \mathcal{H}$,

$$|\langle x, y \rangle| \leq ||x|| \ ||y||.$$

  (ii) For any $x \in \mathcal{H}$,

$$||x|| = \sup_{||y||=1} |\langle x, y \rangle|.$$

  (iii) If $\mathcal{H}$ is a complex Hilbert space, then for any $x, y \in \mathcal{H}$,

$$\langle x, y \rangle \ = \ \frac{1}{4} \left(||x + y||^2 - ||x - y||^2 + i \ ||x + iy||^2 - i \ ||x - iy||^2 \right); (2.7)$$

in case $\mathcal{H}$ is a real Hilbert space,

$$\langle x, y \rangle = \frac{1}{4} \left(||x + y||^2 - ||x - y||^2 \right).$$

(iv) Assume that $x, y \in \mathcal{H}$ satisfy that

$$\langle x, z \rangle = \langle y, z \rangle, \ \forall z \in \mathcal{H}.$$

Then $x = y$.

(v) For any sequence $\{x_k\}_{k=1}^{\infty}$ in $\mathcal{H}$, the following are equivalent:

   (a) $\{x_k\}_{k=1}^{\infty}$ is complete.
   (b) If $\langle x, x_k \rangle = 0$ for all $k \in \mathbb{N}$, then $x = 0$.

Note that Lemma 2.3.1 (i) is the *Cauchy–Schwarz inequality;* the classical proofs in $\mathbb{R}^n$ carries over to the Hilbert space setting. The result in (ii) shows that the norm in the Hilbert space can be recovered with knowledge of the inner product; we ask the reader to prove this in Exercise 2.3. On the other hand, the result in (iii) shows that we can also recover the inner product in $\mathcal{H}$ from the norm (Exercise 2.4); (iii) is known in the literature under the name *the polarization identity.* The proof of the results in (iv) and (v) are also left to the reader (Exercise 2.5 and Exercise 2.6).

Among the linear operators on a Hilbert space, a special role is played by the functionals, i.e., the continuous linear operators $\Phi : \mathcal{H} \to \mathbb{C}$. They are characterized in the well-known *Riesz' representation theorem* (see, e.g., page 81 in [565] for a proof):

Theorem 2.3.2 *Let $\Phi : \mathcal{H} \to \mathbb{C}$ be a continuous linear mapping. Then there exists a unique $y \in \mathcal{H}$ such that $\Phi x = \langle x, y \rangle$ for all $x \in \mathcal{H}$.*

Note that the uniqueness of the element $y \in \mathcal{H}$ associated with a given functional is a consequence of Lemma 2.3.1 (iv).

Corollary 2.3.3 *The dual of a Hilbert space $\mathcal{H}$ can be identified with $\mathcal{H}$.*

2.4   Operators on Hilbert Spaces

Let U be a bounded operator from the Hilbert space $(\mathcal{K}, \langle \cdot, \cdot \rangle_{\mathcal{K}})$ into the Hilbert space $(\mathcal{H}, \langle \cdot, \cdot \rangle_{\mathcal{H}})$. The *adjoint* operator is defined as the unique operator $U^* : \mathcal{H} \to \mathcal{K}$ satisfying that

$$\langle x, Uy \rangle_{\mathcal{H}} = \langle U^* x, y \rangle_{\mathcal{K}}, \ \forall x \in \mathcal{H}, y \in \mathcal{K}.$$

Usually, we will write $\langle \cdot, \cdot \rangle$ for both inner products; it will always be clear from the context in which space the inner product is taken.

We collect some relationships between U and U^*; the proofs can be found in, e.g., Theorem 4.14 and Theorem 4.15 in [566].

Lemma 2.4.1 *Let $U : \mathcal{K} \to \mathcal{H}$ be a bounded operator. Then the following hold:*

(i) $||U|| = ||U^*||$, and $||UU^*|| = ||U||^2$.

(ii) $\mathcal{R}_U$ is closed in $\mathcal{H}$ if and only if $\mathcal{R}_{U^*}$ is closed in $\mathcal{K}$.

(iii) U is surjective if and only if there exists a constant $C > 0$ such that

$$||U^*y|| \geq C \, ||y||, \quad \forall y \in \mathcal{H}.$$

An operator $U : \mathcal{K} \to \mathcal{H}$ is *compact* if $V := \overline{\{Ux \; : \; ||x|| \leq 1\}}$ is compact, i.e., if every sequence from V has a convergent subsequence. A compact operator is bounded. Among the compact operators, we find all operators having *finite rank*, i.e., a finite-dimensional range. We collect some of the most important properties of compact operators; the proofs are in [566].

Lemma 2.4.2 *Let $U : \mathcal{K} \to \mathcal{H}$ be a compact operator. Then*

(i) *The composition of U and a bounded operator (from left or right) is a compact operator.*

(ii) *The adjoint operator U^* is compact.*

(iii) *If $\mathcal{K} = \mathcal{H}$ and $\lambda \neq 0$, then $U - \lambda I$ has closed range; here I denotes the identity operator on $\mathcal{H}$.*

In the rest of this section, we consider the case $\mathcal{K} = \mathcal{H}$. A bounded operator $U : \mathcal{H} \to \mathcal{H}$ is *unitary* if $UU^* = U^*U = I$. If U is unitary, then

$$\langle Ux, Uy \rangle = \langle x, y \rangle, \quad \forall x, y \in \mathcal{H}.$$

A bounded operator $U : \mathcal{H} \to \mathcal{H}$ is *self-adjoint* if $U = U^*$. When U is self-adjoint,

$$||U|| = \sup_{||x||=1} |\langle Ux, x \rangle|. \tag{2.8}$$

For a self-adjoint operator U, the inner product $\langle Ux, x \rangle$ is real for all $x \in \mathcal{H}$. One can introduce a partial order on the set of self-adjoint operators by

$$U_1 \leq U_2 \Leftrightarrow \langle U_1 x, x \rangle \leq \langle U_2 x, x \rangle, \quad \forall x \in \mathcal{H}. \tag{2.9}$$

Using this order, one can work with self-adjoint operators almost as with real numbers. For example, under certain conditions, it is possible to "multiply" an operator inequality with a bounded operator. The precise statement below can be found in [401].

Theorem 2.4.3 *Let U_1, U_2, U_3 be self-adjoint operators. If $U_1 \leq U_2$, $U_3 \geq 0$, and U_3 commutes with U_1 and U_2, then $U_1 U_3 \leq U_2 U_3$.*

An important class of self-adjoint operators consists of the *orthogonal projections*. Given a closed subspace V of $\mathcal{H}$, the orthogonal projection of $\mathcal{H}$ onto V is the operator $P : \mathcal{H} \to \mathcal{H}$ for which

$$Px = x, \ x \in V, \ Px = 0, \ x \in V^\perp.$$

If $\{e_k\}_{k=1}^\infty$ is an orthonormal basis for V (see Definition 3.4.1), the operator P is given explicitly by

$$Px = \sum_{k=1}^\infty \langle x, e_k \rangle e_k, \ x \in \mathcal{H}.$$

Lemma 2.4.4 *Let $U : \mathcal{H} \to \mathcal{H}$ be a bounded operator, and assume that $\langle Ux, x \rangle = 0$ for all $x \in \mathcal{H}$. Then the following hold:*

(i) If $\mathcal{H}$ is a complex Hilbert space, then $U = 0$.

(ii) If $\mathcal{H}$ is a real Hilbert space and U is self-adjoint, then $U = 0$.

Proof. If $\mathcal{H}$ is a complex Hilbert space, a direct calculation shows that

$$
\begin{aligned}
4\langle Ux, y \rangle \ &= \ \langle U(x + y), x + y \rangle - \langle U(x - y), x - y \rangle \\
&\quad + i\langle U(x + iy), x + iy \rangle - i\langle U(x - iy), x - iy \rangle, \ \forall x, y \in \mathcal{H}.
\end{aligned}
$$

Thus, if $\langle Ux, x \rangle = 0$ for all $x \in \mathcal{H}$, then $\langle Ux, y \rangle = 0$ for all $x, y \in \mathcal{H}$, and therefore $U = 0$.

In case $\mathcal{H}$ is a real Hilbert space, we must use a different approach. Let $\{e_k\}_{k=1}^\infty$ be an orthonormal basis for $\mathcal{H}$. Then, for arbitrary $j, k \in \mathbb{N}$,

$$
\begin{aligned}
0 \ &= \ \langle U(e_k + e_j), e_k + e_j \rangle \\
&= \ \langle Ue_k, e_k \rangle + \langle Ue_j, e_j \rangle + \langle Ue_k, e_j \rangle + \langle Ue_j, e_k \rangle \\
&= \ \langle Ue_k, e_j \rangle + \langle e_j, Ue_k \rangle = 2\langle Ue_j, e_k \rangle;
\end{aligned}
$$

therefore, $U = 0$. $\qquad\square$

Note that without the assumption $U = U^*$, Lemma 2.4.4 (ii) would fail; consider, e.g., the case where U is a rotation of $90°$ in $\mathbb{R}^2$.

A bounded operator $U : \mathcal{H} \to \mathcal{H}$ is *positive* if $\langle Ux, x \rangle \geq 0$, $\forall x \in \mathcal{H}$. On a complex Hilbert space, every bounded positive operator is self-adjoint. For a positive operator U, we will often use the following result about the existence of a *square root*, i.e., a bounded operator W such that $W^2 = U$; a proof can be found, e.g., at page 476 in [464].

Lemma 2.4.5 *Every bounded and positive operator $U : \mathcal{H} \to \mathcal{H}$ has a unique bounded and positive square root W. The operator W has the following properties:*

(i) If U is self-adjoint, then W is self-adjoint.

(ii) If U is invertible, then W is also invertible.

(iii) W can be expressed as a limit (in the strong operator topology) of a sequence of polynomials in U and commutes with U.

Frequently, the study of an operator is easier if it can be represented as a sum or product of "simple" operators. We mention a few examples of such representations:

Lemma 2.4.6 *Let $\mathcal{H}$ be a complex Hilbert space. Then the following hold:*

(i) Every bounded and invertible operator $U : \mathcal{H} \to \mathcal{H}$ has a unique representation $U = WP$, where W is unitary and P is positive.

(ii) Every positive operator P on $\mathcal{H}$ with $\|P\| \leq 1$ can be written as an average of unitary operators, namely,

$$P = \frac{1}{2}(W + W^*) \text{ with } W = P + i\sqrt{I - P^2}.$$

The representation $U = WP$ in (i) is called the *polar decomposition;* a proof of this result can be found on page 315 in [566]. The representation in (ii) is probably less known, but it is proved by direct verification. That $W = P + i\sqrt{I - P^2}$ is unitary follows by calculating WW^* and WW^* using that the square root of $I - P^2$ can be considered as a limit of polynomials in $I - P^2$ and therefore commutes with P. Note that (ii) applies if P is an orthogonal projection.

2.5   The Pseudo-inverse Operator

For operators that are not invertible, various types of generalized inverses exist in the literature; see, e.g., the book [43]. Among these generalized inverses, we will focus on a particular one, which will be called the *pseudo-inverse.* We first prove that if an operator U from one Hilbert space to another has closed range, there exists a " right-inverse operator" $U^\dagger$ in the following sense:

Lemma 2.5.1 *Let $\mathcal{H}, \mathcal{K}$ be Hilbert spaces, and suppose that $U : \mathcal{K} \to \mathcal{H}$ is a bounded operator with closed range $\mathcal{R}_U$. Then there exists a bounded operator $U^\dagger : \mathcal{H} \to \mathcal{K}$ for which*

$$UU^\dagger x = x, \ \forall x \in \mathcal{R}_U. \tag{2.10}$$

Proof.   Consider the restriction of U to an operator on the orthogonal complement of the kernel of U, i.e., let

$$\widetilde{U} := U_{|\mathcal{N}_U^\perp} : \mathcal{N}_U^\perp \to \mathcal{H}.$$

Clearly, $\widetilde{U}$ is linear and bounded. $\widetilde{U}$ is also injective: if $\widetilde{U}x = 0$, it follows that $x \in \mathcal{N}_U^\perp \cap \mathcal{N}_U = \{0\}$. We now prove that the range of $\widetilde{U}$ equals the range of U. Given $y \in \mathcal{R}_U$, there exists $x \in \mathcal{K}$ such that $Ux = y$. By writing $x = x_1 + x_2$, where $x_1 \in \mathcal{N}_U^\perp$, $x_2 \in \mathcal{N}_U$, we obtain that

$$\widetilde{U}x_1 = Ux_1 = U(x_1 + x_2) = Ux = y.$$

It follows from Theorem 2.2.2 that $\widetilde{U}$ has a bounded inverse

$$\widetilde{U}^{-1} : \mathcal{R}_U \to \mathcal{N}_U^\perp.$$

Extending $\widetilde{U}^{-1}$ by zero on the orthogonal complement of $\mathcal{R}_U$, we obtain a bounded operator $U^\dagger : \mathcal{H} \to \mathcal{K}$ for which $UU^\dagger x = x$ for all $x \in \mathcal{R}_U$.   □

The operator $U^\dagger$ constructed in the proof of Lemma 2.5.1 is called the *pseudo-inverse* of U. In the literature, one will often see the pseudo-inverse of an operator U with closed range defined as the unique operator $U^\dagger$ satisfying that

$$\mathcal{N}_{U^\dagger} = \mathcal{R}_U^\perp, \quad \mathcal{R}_{U^\dagger} = \mathcal{N}_U^\perp, \text{ and } UU^\dagger x = x, x \in \mathcal{R}_U; \qquad (2.11)$$

this definition is equivalent to the above construction (Exercise 2.7). We collect some properties of $U^\dagger$ and its relationship to U.

Lemma 2.5.2 *Let $U : \mathcal{K} \to \mathcal{H}$ be a bounded operator with closed range. Then the following hold:*

(i) *The orthogonal projection of $\mathcal{H}$ onto $\mathcal{R}_U$ is given by $UU^\dagger$.*

(ii) *The orthogonal projection of $\mathcal{K}$ onto $\mathcal{R}_{U^\dagger}$ is given by $U^\dagger U$.*

(iii) *U^* has closed range, and $(U^*)^\dagger = (U^\dagger)^*$.*

(iv) *On $\mathcal{R}_U$, the operator $U^\dagger$ is given explicitly by*

$$U^\dagger = U^*(UU^*)^{-1}. \qquad (2.12)$$

Proof. All statements follow from the characterization of $U^\dagger$ in (2.11). For example, it shows that

$$UU^\dagger = I \text{ on } \mathcal{R}_U \text{ and that } UU^\dagger = 0 \text{ on } \mathcal{N}_{U^\dagger} = \mathcal{R}_U^\perp;$$

this gives (i) by the definition of an orthogonal projection. The proof of (ii) is similar. That $\mathcal{R}_{U^*}$ is closed was stated already in Lemma 2.4.1; thus, $(U^*)^\dagger$ is well defined. That $(U^*)^\dagger$ equals $(U^\dagger)^*$ follows by verifying that $(U^\dagger)^*$ satisfies (2.11) with U replaced by U^*. Finally, UU^* is invertible as an operator on $\mathcal{R}_U$, and the operator given by

$$U^*(UU^*)^{-1} \text{ on } \mathcal{R}_U \text{ and } 0 \text{ on } \mathcal{R}_U^\perp$$

satisfies the conditions (2.11) characterizing $U^\dagger$.   □

The pseudo-inverse gives the solution to an important optimization problem:

Theorem 2.5.3 *Let* $U : \mathcal{K} \to \mathcal{H}$ *be a bounded surjective operator. Given* $y \in \mathcal{H}$, *the equation* $Ux = y$ *has a unique solution of minimal norm, namely,* $x = U^\dagger y$.

Proof.   The proof is identical with the proof of Theorem 1.6.2. Alternatively, if x is a solution to the equation $Ux = y$, then

$$x = (x - U^\dagger y) + U^\dagger y \in \mathcal{N}_U + \mathcal{N}_U^\perp;$$

thus, the norm of x is minimal precisely when $x = U^\dagger y$.   □

2.6   A Moment Problem

Before we leave the discussion of abstract Hilbert spaces, we mention a special class of equations, known as moment problems. The general version of a *moment problem* is as follows: given a collection of elements $\{x_k\}_{k=1}^\infty$ in a Hilbert space $\mathcal{H}$ and a sequence $\{a_k\}_{k=1}^\infty$ of complex numbers, can we find an element $x \in \mathcal{H}$ such that

$$\langle x, x_k \rangle = a_k, \text{ for all } k \in \mathbb{N}?$$

Many of the equations that will appear throughout the book will be formulated in terms of moment problems. We will need a special moment problem in Section 9.5:

Lemma 2.6.1 *Let* $\{x_k\}_{k=1}^N$ *be a collection of vectors in* $\mathcal{H}$ *and consider the moment problem*

$$\langle x, x_k \rangle = \begin{cases} 1 & \text{if } k = 1, \\ 0 & \text{if } k = 2, \dots, N. \end{cases} \qquad (2.13)$$

Then the following are equivalent:

(i) The moment problem (2.13) has a solution x.

(ii) If $\sum_{k=1}^N c_k x_k = 0$ *for some scalar coefficients* c_k, *then* $c_1 = 0$.

(iii) $x_1 \notin span\{x_k\}_{k=2}^N$.

In case the moment problem (2.13) has a solution, it can be chosen of the form $x = \sum_{k=1}^N d_k x_k$ *for some scalar coefficients* d_k.

Proof. (i)⇒ (ii). Assume first that (i) is satisfied, i.e., (2.13) has a solution x. Then, if $\sum_{k=1}^N c_k x_k = 0$ for some coefficients $\{c_k\}_{k=1}^N$, we have that

$$0 = \langle x, \sum_{k=1}^N c_k x_k \rangle = \sum_{k=1}^N \overline{c_k} \langle x, x_k \rangle = \overline{c_1},$$

i.e., (ii) holds.

(ii) $\Rightarrow$(iii). This implication is clear.

(iii) $\Rightarrow$(i). Let P denote the orthogonal projection of $\mathcal{H}$ onto span$\{x_k\}_{k=2}^N$, and put $\varphi = x_1 - Px_1$. Then

$$\langle \varphi, x_1 \rangle = \langle x_1 - Px_1, x_1 - Px_1 \rangle + \langle x_1 - Px_1, Px_1 \rangle = ||x_1 - Px_1||^2 \neq 0,$$

and $\langle \varphi, x_k \rangle = 0$ for $k = 2, \ldots, N$. Thus, the element

$$x := \frac{\varphi}{||x_1 - Px_1||^2} \tag{2.14}$$

solves the moment problem (2.13), i.e., (i) is satisfied.

In case the equivalent conditions are satisfied, the construction of x in (2.14) shows that $x \in$ span$\{x_k\}_{k=1}^N$. $\square$

2.7 The Spaces $L^p(\mathbb{R})$, $L^2(\mathbb{R})$, $\ell^p(\mathbb{N})$, and $\ell^2(\mathbb{N})$

The most important class of Banach spaces is formed by the L^p-spaces, $1 \leq p \leq \infty$. We expect these spaces and their basic properties to be known, so we only provide a quick overview; proofs and further results can be found in any standard book on the subject, e.g., [565].

First, $L^\infty(\mathbb{R})$ is the space of essentially bounded (Lebesgue) measurable functions $f : \mathbb{R} \to \mathbb{C}$, equipped with the essential supremums-norm. For $1 \leq p < \infty$, $L^p(\mathbb{R})$ is the space of functions f for which $|f|^p$ is integrable with respect to the Lebesgue measure:

$$L^p(\mathbb{R}) := \left\{ f : \mathbb{R} \to \mathbb{C} \mid f \text{ is measurable and } \int_{-\infty}^{\infty} |f(x)|^p \, dx < \infty \right\}.$$

The norm on $L^p(\mathbb{R})$ is

$$||f|| = \left(\int_{-\infty}^{\infty} |f(x)|^p \, dx \right)^{1/p}.$$

To be more precise, $L^p(\mathbb{R})$ consists of equivalence classes of functions that are equal almost everywhere and for which a representative (and hence all) for the equivalence class satisfies the integrability condition. In order not to be too tedious, we adopt the standard terminology and speak about functions in $L^p(\mathbb{R})$ rather than equivalence classes.

The case $p = 2$ plays a special role: in fact, the space

$$L^2(\mathbb{R}) = \left\{ f : \mathbb{R} \to \mathbb{C} \mid f \text{ is measurable and } \int_{-\infty}^{\infty} |f(x)|^2 \, dx < \infty \right\}$$

is the only one of the $L^p(\mathbb{R})$-spaces that can be equipped with an inner product. Actually, $L^2(\mathbb{R})$ is a Hilbert space with respect to the inner product

$$\langle f, g \rangle = \int_{-\infty}^{\infty} f(x)\overline{g(x)} \, dx, \ f, g \in L^2(\mathbb{R}).$$

Thus, we can apply all the results in Section 2.3 to the space $L^2(\mathbb{R})$. In particular, Cauchy–Schwarz inequality states that for all $f, g \in L^2(\mathbb{R})$,

$$\left| \int_{-\infty}^{\infty} f(x)g(x) \, dx \right| \le \left(\int_{-\infty}^{\infty} |f(x)|^2 \, dx \right)^{1/2} \left(\int_{-\infty}^{\infty} |g(x)|^2 \, dx \right)^{1/2}. \quad (2.15)$$

The spaces $L^2(\Omega)$, where Ω is an open subset of $\mathbb{R}$, are defined similarly. According to the general definition, a sequence of functions $\{g_k\}_{k=1}^{\infty}$ in $L^2(\Omega)$ converges to $g \in L^2(\Omega)$ if

$$||g - g_k|| = \left(\int_{\Omega} |g(x) - g_k(x)|^2 \, dx \right)^{1/2} \to 0 \text{ as } k \to \infty.$$

It is essential to be aware that the concept of convergence in L^2-sense is different from pointwise convergence. However, there is a relationship that will play an important role in several proofs: convergence in L^2-sense implies the existence of a subsequence that converges pointwise almost everywhere. This result is known as *Riesz' subsequence theorem;* we state it formally here and refer to page 68 in [565] for a proof.

Theorem 2.7.1 *Let $\Omega \subseteq \mathbb{R}$ be an open set, and let $\{g_k\}$ be a sequence in $L^2(\Omega)$ that converges to $g \in L^2(\Omega)$. Then $\{g_k\}$ has a subsequence $\{g_{n_k}\}_{k=1}^{\infty}$ such that*

$$g(x) = \lim_{k \to \infty} g_{n_k}(x)$$

for a.e. $x \in \Omega$.

The result holds no matter how we choose the representatives for the equivalence classes. This is typical for this book, where we rarely deal with a specific representative for a given class. There are, however, a few important exceptions. When we speak about a continuous function, it is clear that we have chosen a specific representative, and the same is the case when we discuss *Lebesgue points*. By definition, a point $y \in \mathbb{R}$ is a Lebesgue point for a measurable function $f : \mathbb{R} \to \mathbb{C}$ if

$$\lim_{\epsilon \to 0} \frac{1}{\epsilon} \int_{y-\frac{1}{2}\epsilon}^{y+\frac{1}{2}\epsilon} |f(y) - f(x)| \, dx = 0.$$

If f is continuous in y, then y is a Lebesgue point (Exercise 2.8). More generally, one can prove that if $f \in L^1(\mathbb{R})$, then almost every $y \in \mathbb{R}$ is a Lebesgue point; see page 138 in [565] for a proof of this fact, as well as a more detailed discussion.

It is clear from the definition that different representatives for the same equivalence class will have different Lebesgue points. For example, every $y \in \mathbb{R}$ is a Lebesgue point for the function $f = 0$; changing the definition of f in a single point y will not change the equivalence class, but y will no longer be a Lebesgue point. See Exercise 2.8 for some related observations.

The discrete analogue of $L^p(\mathbb{R})$ is $\ell^p(I)$, the space of p-summable scalar-valued sequences with a countable index set I. For $1 \le p < \infty$, let

$$\ell^p(I) := \left\{ \{x_k\}_{k \in I} \mid x_k \in \mathbb{C}, \ \sum_{k \in I} |x_k|^p < \infty \right\}.$$

For $1 \le p < \infty$ the space $\ell^p(\mathbb{R})$ is a Banach space with respect to the norm

$$\|\{x_k\}_{k \in I}\|_p = \left(\sum_{k \in I} |x_k|^p \right)^{1/p}.$$

In particular, $\ell^2(I)$ is a Hilbert space with respect to the inner product

$$\langle \{x_k\}_{k \in I}, \{y_k\}_{k \in I} \rangle = \sum_{k \in I} x_k \overline{y_k};$$

and Cauchy–Schwarz inequality states that

$$\left| \sum_{k \in I} x_k \overline{y_k} \right|^2 \le \sum_{k \in I} |x_k|^2 \sum_{k \in I} |y_k|^2, \quad \{x_k\}_{k \in I}, \{y_k\}_{k \in I} \in \ell^2(I). \qquad (2.16)$$

The Banach space $\ell^\infty(I)$ is the set of bounded scalar-valued sequences $\{x_k\}_{k \in I}$, equipped with the norm

$$\|\{x_k\}_{k \in I}\|_\infty = \sup_{k \in I} |x_k|.$$

We will frequently use the following discrete version of Fatou's lemma (the general version is stated in Lemma A.2.3):

Lemma 2.7.2 *Let I be a countable index set and $f_n : I \to [0, \infty]$, $n \in \mathbb{N}$, a sequence of functions. Then*

$$\sum_{k \in I} \liminf_{n \to \infty} f_n(k) \le \liminf_{n \to \infty} \sum_{k \in I} f_n(k).$$

2.8   The Fourier Transform and Convolution

The Fourier transform will be one of the central ingredients in our analysis of structured function systems in Chapters 9–21. In this section, we give a short introduction to the Fourier transforms on $L^2(\mathbb{R})$ and $\ell^2(\mathbb{Z})$. As for most of the other topics in this section, we expect the reader to have a basic knowledge about the subject, so we only collect the main definitions and results here. For more information, we refer to any of the standard texts, e.g., [465, 22], or [629].

For $f \in L^1(\mathbb{R})$, the *Fourier transform* $\widehat{f} : \mathbb{R} \to \mathbb{C}$ is defined by

$$\widehat{f}(\gamma) := \int_{-\infty}^{\infty} f(x) e^{-2\pi i x \gamma} \, dx, \ \gamma \in \mathbb{R}. \tag{2.17}$$

We will also denote the Fourier transform of f by $\mathcal{F}f$. This notation indicates that we will also consider the Fourier transform as an operator; the *Riemann–Lebesgue lemma* says that $\mathcal{F}$ maps $L^1(\mathbb{R})$ into $C_0(\mathbb{R})$, the vector space consisting of continuous functions vanishing at infinity.

The Fourier transform has an extension to a unitary operator on $L^2(\mathbb{R})$. One can show that if $(L^1 \cap L^2)(\mathbb{R})$ is equipped with the $L^2(\mathbb{R})$-norm, the Fourier transform is an isometry from $(L^1 \cap L^2)(\mathbb{R})$ into $L^2(\mathbb{R})$. If $f \in L^2(\mathbb{R})$ and $\{f_k\}_{k=1}^{\infty}$ is a sequence of functions in $(L^1 \cap L^2)(\mathbb{R})$ that converges to f in L^2-sense, then the sequence $\{\widehat{f_k}\}_{k=1}^{\infty}$ is also convergent in $L^2(\mathbb{R})$, with a limit that is independent of the choice of $\{f_k\}_{k=1}^{\infty}$. Defining

$$\widehat{f} := \lim_{k \to \infty} \widehat{f_k}$$

then extends the Fourier transform to a unitary mapping of $L^2(\mathbb{R})$ onto $L^2(\mathbb{R})$. We will use the same notation to denote this extension.

The above construction and the polarization identity immediately yields *Plancherel's equation*,

$$\langle \widehat{f}, \widehat{g} \rangle = \langle f, g \rangle, \ \forall f, g \in L^2(\mathbb{R}), \ \text{and} \ ||\widehat{f}|| = ||f||. \tag{2.18}$$

If $f \in L^1(\mathbb{R})$, then $\widehat{f}$ is continuous. If the function f as well as $\widehat{f}$ belong to $L^1(\mathbb{R})$, the *inversion formula* describes how to come back to f from the function values $\widehat{f}(\gamma)$, see [22].

Theorem 2.8.1 *Assume that $f, \widehat{f} \in L^1(\mathbb{R})$. Then*

$$f(x) = \int_{-\infty}^{\infty} \widehat{f}(\gamma) e^{2\pi i x \gamma} d\gamma, \ a.e. \ x \in \mathbb{R}. \tag{2.19}$$

If f is continuous, the pointwise formula (2.19) holds for all $x \in \mathbb{R}$. In general, it holds at least for all Lebesgue points for f.

We note that (2.19) also holds if $f \in L^2(\mathbb{R})$ and $\widehat{f} \in L^1(\mathbb{R})$.

Given two functions $f, g \in L^1(\mathbb{R})$, the *convolution* $f * g : \mathbb{R} \to \mathbb{C}$ is defined by

$$f * g(y) = \int_{-\infty}^{\infty} f(y - x) g(x) \, dx, \ y \in \mathbb{R}.$$

The function $f * g$ is well defined for all $y \in \mathbb{R}$ and belongs to $L^1(\mathbb{R})$. If $f \in L^1(\mathbb{R})$ and $g \in L^p(\mathbb{R})$ for some $p \in [1, \infty[$, the convolution $f * g(y)$ is well-defined for a.e. $y \in \mathbb{R}$ and defines a function in $L^p(\mathbb{R})$.

The Fourier transform and convolution are related by the following important result:

Theorem 2.8.2 *If $f, g \in L^1(\mathbb{R})$, then $\widehat{f * g}(\gamma) = \widehat{f}(\gamma)\widehat{g}(\gamma)$ for all $\gamma \in \mathbb{R}$; if $f \in L^1(\mathbb{R})$ and $g \in L^p(\mathbb{R})$, the formula holds for a.e. $\gamma \in \mathbb{R}$.*

We will also need the Fourier transform on $\ell^2(\mathbb{Z})$. Given a sequence $h = \{h_k\}_{k\in\mathbb{Z}} \in \ell^2(\mathbb{Z})$, we define its *Fourier transform* as the Fourier series

$$\widehat{h}(\nu) = \sum_{j\in\mathbb{Z}} h_k e^{-2\pi i k\nu}, \ a.e. \ \nu \in \mathbb{R}.$$

Given two scalar-valued sequences $g = \{g_k\}_{k\in\mathbb{Z}}$ and $h = \{h_k\}_{k\in\mathbb{Z}}$, their *convolution* is formally defined as the sequence $g * h$ whose jth coordinate is

$$(g * h)_j = \sum_{k\in\mathbb{Z}} g_k h_{j-k}.$$

If $g \in \ell^1(\mathbb{Z})$ and $h \in \ell^p(\mathbb{Z})$ for some $p \in [1, \infty[$, then the convolution $g * h$ is well-defined and belongs to $\ell^p(\mathbb{Z})$; *Young's inequality* states that

$$||g * h||_p \le ||g||_1 ||h||_p. \tag{2.20}$$

2.9 Operators on $L^2(\mathbb{R})$

In this section, we consider three classes of operators on $L^2(\mathbb{R})$ that will play a key role in our analysis of Gabor frames and wavelets. Their definitions are as follows:

Definition 2.9.1 *Consider the following classes of linear operators:*

(i) For $a \in \mathbb{R}$, the operator T_a, called translation by a, is defined by

$$T_a : L^2(\mathbb{R}) \to L^2(\mathbb{R}), \quad (T_a f)(x) := f(x - a), \ x \in \mathbb{R}. \tag{2.21}$$

(ii) For $b \in \mathbb{R}$, the operator E_b, called modulation by b, is defined by

$$E_b : L^2(\mathbb{R}) \to L^2(\mathbb{R}), \quad (E_b f)(x) := e^{2\pi i b x} f(x), \ x \in \mathbb{R}. \tag{2.22}$$

(iii) For $a \ne 0$, the operator D_a, called scaling by a, is defined by

$$D_a : L^2(\mathbb{R}) \to L^2(\mathbb{R}), \quad (D_a f)(x) := \frac{1}{\sqrt{a}} f(\frac{x}{a}), \ x \in \mathbb{R}. \tag{2.23}$$

The operator D_a is also called a dilation operator.

(iv) The dyadic scaling operator is

$$D : L^2(\mathbb{R}) \to L^2(\mathbb{R}), \quad (Df)(x) := D_{1/2} f(x) = 2^{1/2} f(2x). \tag{2.24}$$

A comment about notation: we will usually skip the parentheses and simply write $T_a f(x)$ and similarly for the other operators. Frequently, we will also let E_b denote the function $x \mapsto e^{2\pi i b x}$.

All the operators in Definition 2.9.1 indeed map $L^2(\mathbb{R})$ onto $L^2(\mathbb{R})$ as stated and are bounded (Exercise 2.9). They are even unitary:

Lemma 2.9.2 *The translation operators satisfy the following:*

(i) T_a *is unitary for all* $a \in \mathbb{R}$.

(ii) *For each* $f \in L^2(\mathbb{R})$, *the mapping* $y \mapsto T_y f$ *is continuous from* $\mathbb{R}$ *to* $L^2(\mathbb{R})$.

Similar statements hold for $E_b, b \in \mathbb{R}$, *and* D_a, $a \neq 0$.

Proof. Let us prove that the operators T_a are unitary. Since

$$\langle T_a f, g \rangle = \int_{-\infty}^{\infty} f(x-a)\overline{g(x)}\, dx \ = \ \int_{-\infty}^{\infty} f(x)\overline{g(x+a)}\, dx$$

$$= \ \langle f, T_{-a} g \rangle, \ \forall f, g \in L^2(\mathbb{R}),$$

we see that $T_a^* = T_{-a}$. On the other hand, T_a is clearly an invertible operator with $T_a^{-1} = T_{-a}$, so we conclude that $T_a^{-1} = T_a^*$.

To prove the continuity of the mapping $y \mapsto T_y f$, we first assume that the function f is continuous and has compact support, say, contained in the bounded interval $[c, d]$. For notational convenience, we prove the continuity in $y_0 = 0$. First, for $y \in\ ]-\frac{1}{2}, \frac{1}{2}[$ the function

$$\phi(x) = T_y f(x) - T_{y_0} f(x) = f(x-y) - f(x)$$

has support in the interval $[-\frac{1}{2}+c, d+\frac{1}{2}]$. Since f is uniformly continuous, we can for any given $\epsilon > 0$ find $\delta > 0$ such that

$$|f(x-y) - f(x)| \leq \epsilon \ \text{ for all } x \in \mathbb{R} \text{ whenever } |y| \leq \delta.$$

With this choice of δ, we thus obtain that

$$\|T_y f - T_{y_0} f\| \ = \ \left(\int_{-\frac{1}{2}+c}^{\frac{1}{2}+d} |f(x-y) - f(x)|^2\, dx \right)^{1/2}$$

$$\leq \ \epsilon \sqrt{d-c+1}.$$

This proves the continuity in the considered special case. The case of an arbitrary function $f \in L^2(\mathbb{R})$ follows by an approximation argument, using that the continuous functions with compact support are dense in $L^2(\mathbb{R})$ (Exercise 2.10). The proofs of the statements for E_b and D_a are left to the reader (Exercise 2.11). □

Chapters 11–20 will deal with Gabor systems, wavelet systems, and generalized shift-invariant systems in $L^2(\mathbb{R})$; all classes consist of functions that are defined by compositions of some of the operators $T_a, E_b,$

and D_a. For this reason, the following *commutator relations* are important (Exercise 2.12) :

$$T_a E_b f(x) \;=\; e^{-2\pi i b a} E_b T_a f(x) = e^{2\pi i b(x-a)} f(x-a), \qquad (2.25)$$

$$T_b D_a f(x) \;=\; D_a T_{b/a} f(x) = \frac{1}{\sqrt{|a|}} f\!\left(\frac{x}{a} - \frac{b}{a}\right), \qquad (2.26)$$

$$D_a E_b f(x) \;=\; E_{\frac{b}{a}} D_a f(x) = \frac{1}{\sqrt{|a|}} e^{2\pi i x b/a} f\!\left(\frac{x}{a}\right). \qquad (2.27)$$

With this notation, the commutator relation (2.26) in particular implies that

$$T_k D^j = D^j T_{2^j k} \text{ and } D^j T_k = T_{2^{-j}k} D^j, \; j,k \in \mathbb{Z}. \qquad (2.28)$$

We will often use the Fourier transformation in connection with Gabor systems and wavelet systems. In this context, we need the commutator relations (Exercise 2.13)

$$\mathcal{F} T_a = E_{-a}\mathcal{F}, \;\; \mathcal{F} E_a = T_a \mathcal{F}, \;\; \mathcal{F} D_a = D_{1/a}\mathcal{F}, \;\; \mathcal{F} D = D^{-1}\mathcal{F}. \qquad (2.29)$$

2.10   Exercises

2.1 Find a sequence $\{a_k\}_{k=1}^{\infty}$ of real numbers for which $\sum_{k=1}^{\infty} a_k$ is convergent but not unconditionally convergent.

2.2 Let $\{f_k\}_{k=1}^{\infty}$ be a sequence in a Banach space. Prove that absolute convergence of $\sum_{k=1}^{\infty} f_k$ implies unconditional convergence.

2.3 Prove Lemma 2.3.1(ii).

2.4 Prove Lemma 2.3.1(iii).

2.5 Prove Lemma 2.3.1 (iv).

2.6 Prove Lemma 2.3.1 (v).

2.7 Prove that the conditions in (2.11) are equivalent to the construction of the pseudo-inverse in Lemma 2.5.1.

2.8 Here we ask the reader to prove some results concerning Lebesgue points.

(i) Assume that $f : \mathbb{R} \to \mathbb{C}$ is continuous. Prove that every $y \in \mathbb{R}$ is a Lebesgue point.

(ii) Prove that $x = 0$ is not a Lebesgue point for the function $\chi_{[0,1]}$.

(iii) Let $f = \chi_{\mathbb{Q}}$. Prove that every $y \notin \mathbb{Q}$ is a Lebesgue point and that the rational numbers are not Lebesgue points.

2.9 Show that the operators in Definition 2.9.1 map $L^2(\mathbb{R})$ into $L^2(\mathbb{R})$ and are bounded.

2.10 Complete the proof of Lemma 2.9.2 by showing the continuity of the mapping $y \mapsto T_y f$ for $f \in L^2(\mathbb{R})$.

2.11 Prove the statements about E_b and D_a in Lemma 2.9.2.

2.12 Prove the commutator relations (2.25)–(2.27).

2.13 Prove the commutator relations (2.29).

3

Bases

Bases play a prominent role in the analysis of vector spaces, as well in the finite-dimensional as in the infinite-dimensional case. The idea is the same in both cases, namely, to consider a family of elements such that all vectors in the considered space can be expressed in a unique way as superpositions of these elements. In the infinite-dimensional case, the situation is complicated: we are forced to work with infinite series, and different concepts of a basis are possible, depending on how we want the series to converge. For example, are we asking for the series to converge with respect to a fixed order of the elements (conditional convergence) or do we want it to converge regardless of how the elements are ordered (unconditional convergence)? We define the relevant types of bases in general Banach spaces in Section 3.1; the case of a basis in a Hilbert space is considered in Section 3.3. Sequences satisfying the Bessel inequality are considered in Section 3.2 and characterized in terms of an associated operator, the synthesis operator. In Section 3.4 we discuss the most important properties of orthonormal bases in Hilbert spaces; we expect the reader to have some basic knowledge about this subject. Section 3.5 deals with the Gram matrix and its relationship with Bessel sequences. In Section 3.6, one of the key subjects of the current book, namely, Riesz bases, is introduced and treated in detail; a subspace version of these is discussed in Section 3.7. Several characterizations of Riesz bases and Riesz sequences are provided. Orthonormal bases and Riesz bases both satisfy the Bessel inequality, which is the key to the observation that they deliver unconditionally convergent expansions and can be ordered in an arbitrary way.

© Springer International Publishing Switzerland 2016   67
O. Christensen, *An Introduction to Frames and Riesz Bases*,
Applied and Numerical Harmonic Analysis,
DOI 10.1007/978-3-319-25613-9_3

Concrete examples of bases in function spaces are given in Section 3.8, where the basic theory for Fourier series is revisited (again this subject is expected to be known). Section 3.8 also introduce Gabor bases for $L^2(\mathbb{R})$. In Section 3.9 we consider wavelet bases for $L^2(\mathbb{R})$ and explain how to construct them using a multiresolution analysis. Sections 3.8–3.9 form the background for Chapters 9–20. Finally, Section 3.10 introduces the concept of sampling and relates Shannon's sampling theorem to engineering.

In the entire chapter, X denotes a Banach space, and $\mathcal{H}$ is a Hilbert space with the inner product $\langle \cdot, \cdot \rangle$ linear in the first entry. We will assume that the spaces are separable and infinite-dimensional, and we leave the modifications in the finite-dimensional case to the reader.

3.1 Bases in Banach Spaces

The most fundamental concept of a basis was introduced by Schauder [569] in 1927. It takes place in a Banach space X and captures the basic idea of having a family of vectors with the property that each $f \in X$ has a unique expansion in terms of the given vectors. All bases considered in this book are Schauder bases. Much more information about bases can be found in the monographs [389, 495, 577].

Before giving the formal definition, we emphasize once more that a *sequence* $\{e_k\}_{k=1}^\infty$ in X is an *ordered* set, i.e.,

$$\{e_k\}_{k=1}^\infty = \{e_1, e_2, \dots\}.$$

Definition 3.1.1 *Let X be a Banach space. A sequence of vectors $\{e_k\}_{k=1}^\infty$ belonging to X is a (Schauder) basis for X if, for each $f \in X$, there exist unique scalar coefficients $\{c_k(f)\}_{k=1}^\infty$ such that*

$$f = \sum_{k=1}^\infty c_k(f)e_k. \tag{3.1}$$

We refer to (3.1) as the *expansion* of f in the basis $\{e_k\}_{k=1}^\infty$. Equation (3.1) merely means that the series $f = \sum_{k=1}^\infty c_k(f)e_k$ converges with respect to the chosen order of the elements. If the series (3.1) converges unconditionally for each $f \in X$, we say that $\{e_k\}_{k=1}^\infty$ is an *unconditional basis*. One can prove that $\{e_k\}_{k=1}^\infty$ is an unconditional basis if and only if $\{e_{\sigma(k)}\}_{k=1}^\infty$ is a basis for every permutation σ of $\mathbb{N}$ (cf. [577]). In other words, if $\{e_k\}_{k=1}^\infty$ is a basis which is not unconditional, there exists a permutation σ for which $\{e_{\sigma(k)}\}_{k=1}^\infty$ is not a basis. It is known that every Banach space which has a basis also has a conditional basis (cf. [536]).

Besides the existence of an expansion of each $f \in X$, Definition 3.1.1 asks for *uniqueness*. This is usually obtained by requiring $\{e_k\}_{k=1}^\infty$ to be independent in an appropriate sense. In infinite-dimensional Banach spaces, different concepts of independence exist:

Definition 3.1.2 *Let $\{f_k\}_{k=1}^{\infty}$ be a sequence in X. We say that*

(i) *$\{f_k\}_{k=1}^{\infty}$ is linearly independent if every finite subsequence of $\{f_k\}_{k=1}^{\infty}$ is linearly independent;*

(ii) *$\{f_k\}_{k=1}^{\infty}$ is ω-independent if whenever the series $\sum_{k=1}^{\infty} c_k f_k$ is convergent and equal to zero for some scalar coefficients $\{c_k\}_{k=1}^{\infty}$, then necessarily $c_k = 0$ for all $k \in \mathbb{N}$.*

(iii) *$\{f_k\}_{k=1}^{\infty}$ is minimal if $f_j \notin \overline{span}\{f_k\}_{k\neq j}, \ \forall j \in \mathbb{N}$.*

The relationship between the definitions is as follows:

Lemma 3.1.3 *Let $\{f_k\}_{k=1}^{\infty}$ be a sequence in X. Then the following holds:*

(i) *If $\{f_k\}_{k=1}^{\infty}$ is minimal, then $\{f_k\}_{k=1}^{\infty}$ is ω-independent.*

(ii) *If $\{f_k\}_{k=1}^{\infty}$ is ω-independent, then $\{f_k\}_{k=1}^{\infty}$ is linearly independent.*

The opposite implications in (i) and (ii) are not valid.

Proof. For the proof of (i), assume that $\{f_k\}_{k=1}^{\infty}$ is not ω-independent. Choose scalar coefficients $\{c_k\}_{k=1}^{\infty}$ with $c_j \neq 0$ for some j, such that $\sum_{k=1}^{\infty} c_k f_k = 0$; then $f_j = \sum_{k\neq j} \frac{-c_k}{c_j} f_k$, implying that $f_j \in \overline{span}\{f_k\}_{k\neq j}$. That is, $\{f_k\}_{k=1}^{\infty}$ is not minimal. The statement (ii) is obvious, and the fact that the opposite implications are not valid is demonstrated by examples in Exercise 3.4. $\qquad\square$

A Banach space having a basis is necessarily separable. Most of the known separable Banach spaces have a basis; the first example of a separable Banach space not having a basis was constructed by Enflo [269] in 1972.

It is clear that a basis for X is complete and consists of nonzero vectors. Adding an extra condition leads to a characterization of bases:

Theorem 3.1.4 *A complete family of nonzero vectors $\{e_k\}_{k=1}^{\infty}$ in X is a basis for X if and only if there exists a constant K such that for all $m, n \in \mathbb{N}$ with $m \leq n$,*

$$\left\| \sum_{k=1}^{m} c_k e_k \right\| \leq K \left\| \sum_{k=1}^{n} c_k e_k \right\| \tag{3.2}$$

for all scalar-valued sequences $\{c_k\}_{k=1}^{\infty}$.

Proof. Suppose that $\{e_k\}_{k=1}^{\infty}$ is a basis. Then each $f \in X$ has a unique expansion $f = \sum_{k=1}^{\infty} c_k e_k$, and

$$|||f||| := \sup_{m \in \mathbb{N}} \left\| \sum_{k=1}^{m} c_k e_k \right\| < \infty. \tag{3.3}$$

Note that if $|||f||| = 0$, then $||\sum_{k=1}^{m} c_k e_k|| = 0$ for all $m \in \mathbb{N}$; it follows that $c_k = 0$ for all $k \in \mathbb{N}$, and $f = 0$. One can check (Exercise 3.1) that $||| \cdot |||$ satisfies the other conditions for a norm on X and that X is a Banach space with respect to this norm. By definition of $||| \cdot |||$, we have $||f|| \leq |||f|||$, $\forall f \in X$, meaning that the identity operator is a continuous and injective mapping of $(X, ||| \cdot |||)$ onto $(X, || \cdot ||)$. By Theorem 2.2.2, it follows that this operator has a continuous inverse, i.e., that there exists a constant $K > 0$ such that $|||f||| \leq K \, ||f||$ for all $f \in X$. In particular, fixing an arbitrary $n \in \mathbb{N}$ and considering $f = \sum_{k=1}^{n} c_k e_k$, we obtain (3.2).

For the implication (ii)$\Rightarrow$(i), assume that a complete family $\{e_k\}_{k=1}^{\infty}$ of nonzero vectors satisfies (3.2). We begin by showing an inequality that will be used in the proof. Consider any $f \in X$ with an expansion $f = \sum_{k=1}^{\infty} c_k e_k$; then, for any choice of $i \in \mathbb{N}$ and $m \geq i$, (3.2) shows that

$$
\begin{aligned}
|c_i| \, ||e_i|| = \left\| \sum_{k=1}^{i} c_k e_k - \sum_{k=1}^{i-1} c_k e_k \right\| &\leq \left\| \sum_{k=1}^{i} c_k e_k \right\| + \left\| \sum_{k=1}^{i-1} c_k e_k \right\| \\
&\leq K \left\| \sum_{k=1}^{m} c_k e_k \right\| + K \left\| \sum_{k=1}^{m} c_k e_k \right\| \\
&= 2K \left\| \sum_{k=1}^{m} c_k e_k \right\|.
\end{aligned}
\tag{3.4}
$$

Now let $\mathcal{A}$ denote the vector space consisting of all $f \in X$, which can be expanded as $f = \sum_{k=1}^{\infty} c_k e_k$ for some coefficients $\{c_k\}_{k=1}^{\infty}$. We will prove that $\mathcal{A} = X$; because $\{e_k\}_{k=1}^{\infty}$ is assumed to be complete, we know that $\mathcal{A}$ is dense in X, so it is enough to prove that $\mathcal{A}$ is closed. Let $f \in X$, and choose a sequence $\{f_j\}_{j=1}^{\infty} \subset \mathcal{A}$ such that $f_j \to f$ as $j \to \infty$. Write $f_j = \sum_{k=1}^{\infty} c_k^{(j)} e_k$ for appropriate coefficients $\{c_k^{(j)}\}_{k=1}^{\infty}$. By (3.4), for each $i \in \mathbb{N}$ and all $n \geq m \geq i$, we have for all $j, \ell \in \mathbb{N}$ that

$$
\begin{aligned}
|c_i^{(j)} - c_i^{(\ell)}| \, ||e_i|| &\leq 2K \left\| \sum_{k=1}^{m} \left(c_k^{(j)} - c_k^{(\ell)} \right) e_k \right\| \tag{3.5} \\
&\leq 2K^2 \left\| \sum_{k=1}^{n} \left(c_k^{(j)} - c_k^{(\ell)} \right) e_k \right\| \\
&\leq 2K^2 \left(\left\| \sum_{k=1}^{n} c_k^{(j)} e_k - f_j \right\| + ||f_j - f|| \right) \\
&\quad + 2K^2 \left(||f - f_\ell|| + \left\| f_\ell - \sum_{k=1}^{n} c_k^{(\ell)} e_k \right\| \right).
\end{aligned}
$$

Given $\epsilon > 0$, choose $N \in \mathbb{N}$ such that

$$
||f - f_j|| \leq \frac{\epsilon}{4K^2} \text{ for } j \geq N.
$$

Letting $n \to \infty$, it follows from the above estimate that

$$|c_i^{(j)} - c_i^{(\ell)}| \, ||e_i|| \le \epsilon \text{ for all } i \in \mathbb{N}, \; j, \ell \ge N, \tag{3.6}$$

and, via the intermediate step (3.5),

$$2K \left\| \sum_{k=1}^{m} \left(c_k^{(j)} - c_k^{(\ell)} \right) e_k \right\| \le \epsilon \text{ for all } m \in \mathbb{N}, \; j, \ell \ge N. \tag{3.7}$$

For each $i \in \mathbb{N}$, the sequence $\{c_i^{(\ell)}\}_{\ell=1}^{\infty}$ is convergent by (3.6), say, $c_i^{(\ell)} \to c_i$ as $\ell \to \infty$. Letting $\ell \to \infty$ in (3.6) and (3.7), we obtain that

$$|c_i^{(j)} - c_i| \, ||e_i|| \le \epsilon \text{ for all } i \in \mathbb{N}, \; j \ge N, \tag{3.8}$$

and

$$2K \left\| \sum_{k=1}^{m} \left(c_k^{(j)} - c_k \right) e_k \right\| \le \epsilon \text{ for all } m \in \mathbb{N}, \; j \ge N. \tag{3.9}$$

Now, for given $m \in \mathbb{N}$ and all $j \in \mathbb{N}$,

$$\left\| f - \sum_{k=1}^{m} c_k e_k \right\| \le \; ||f - f_j|| $$
$$+ \; \left\| f_j - \sum_{k=1}^{m} c_k^{(j)} e_k \right\| + \left\| \sum_{k=1}^{m} \left(c_k^{(j)} - c_k \right) e_k \right\|.$$

We will now show that $\sum_{k=1}^{\infty} c_k e_k$ converges to f. In fact, for a given $\epsilon > 0$, we can choose $N \in \mathbb{N}$ so that (3.9) holds. By fixing a sufficiently large value for $j > N$, we obtain that $||f - f_j|| \le \epsilon$; after that, we can obtain that $\left\| f_j - \sum_{k=1}^{m} c_k^{(j)} e_k \right\| \le \epsilon$ by choosing $m \in \mathbb{N}$ sufficiently large. Thus,

$$\left\| f - \sum_{k=1}^{m} c_k e_k \right\| \le 2\epsilon + \frac{\epsilon}{2K} = \epsilon \left(2 + \frac{1}{2K} \right)$$

for m sufficiently large. We conclude that $f = \sum_{k=1}^{\infty} c_k e_k$, i.e., $f \in \mathcal{A}$ as desired. To prove that $\{e_k\}_{k=1}^{\infty}$ is a basis, we only need to show that if $\sum_{k=1}^{\infty} c_k e_k = 0$, then $c_k = 0$ for all $k \in \mathbb{N}$. This again follows from (3.4). In fact, if $\sum_{k=1}^{\infty} c_k e_k = 0$, then for each $i \in \mathbb{N}$ and all $n \ge i$,

$$|c_i| \, ||e_i|| \le 2K \left\| \sum_{k=1}^{n} c_k e_k \right\|;$$

from here we obtain that $c_i = 0$ by letting $n \to \infty$. □

Theorem 3.1.4 is often formulated using the *basis constant,* which for an arbitrary sequence $\{e_k\}_{k=1}^{\infty}$ is defined by

$$K := \sup\left\{\left\|\sum_{k=1}^{m} c_k e_k\right\| \mid m \le n, \left\|\sum_{k=1}^{n} c_k e_k\right\| = 1\right\}. \qquad (3.10)$$

If $\{e_k\}_{k=1}^{\infty}$ is a basis, this is clearly the smallest constant that can be used in (3.2). On the other hand, if the basis constant is infinite, then $\{e_k\}_{k=1}^{\infty}$ is not a basis. For a finite sequence $\{e_k\}_{k=1}^{N}$ the basis constant is defined as above, with the addition that we consider $n \le N$.

The basis constant K tells whether the sequence $\{e_k\}_{k=1}^{\infty}$ can be a basis with respect to the chosen order of the elements. We note that a similar characterization of unconditional bases exists (cf. [577]): a complete sequence $\{e_k\}_{k=1}^{\infty}$ consisting of nonzero elements is an unconditional basis if and only if its *unconditional basis constant*

$$\sup\left\{\left\|\sum \sigma_k c_k e_k\right\| \mid \left\|\sum c_k e_k\right\| = 1 \text{ and } \sigma_k = \pm 1, \forall k\right\}$$

is finite.

Given a basis $\{e_k\}_{k=1}^{\infty}$, it is clear that the coefficients $\{c_k(f)\}_{k=1}^{\infty}$ in (3.1) depend linearly on f. The mappings $f \mapsto c_k(f)$ are called *coefficient functionals.* As a consequence of Theorem 3.1.4, they are continuous:

Corollary 3.1.5 *The coefficient functionals* $\{c_k\}_{k=1}^{\infty}$ *associated to a basis* $\{e_k\}_{k=1}^{\infty}$ *for X are continuous and are thus elements in the dual X^*. If there exists a constant $C > 0$ such that $\|e_k\| \ge C$ for all $k \in \mathbb{N}$, then the norms of* $\{c_k\}_{k=1}^{\infty}$ *are uniformly bounded.*

Proof. We use Theorem 3.1.4 and the notation introduced there. Given $f \in X$, write $f = \sum_{k=1}^{\infty} c_k(f)e_k$. Then, for any $j \in \mathbb{N}$ and all $n \ge j$,

$$|c_j(f)| \; \|e_j\| \le 2K \left\|\sum_{k=1}^{n} c_k(f)e_k\right\|.$$

Letting $n \to \infty$ we obtain that

$$|c_j(f)| \le \frac{2K}{\|e_j\|} \; \|f\|. \qquad \square$$

Exercise 3.2 exhibits a case where $\{e_k\}_{k=1}^{\infty}$ is not norm bounded below and examines some of the properties of the coefficient functionals.

A sequence $\{f_k\}_{k=1}^{\infty}$ in X and a sequence $\{g_k\}_{k=1}^{\infty}$ in X^* are said to be *biorthogonal* if

$$g_k(f_j) = \delta_{k,j} := \begin{cases} 1 & \text{if} \quad k = j, \\ 0 & \text{if} \quad k \ne j. \end{cases} \qquad (3.11)$$

Corollary 3.1.6 *Suppose that $\{e_k\}_{k=1}^\infty$ is a basis for X. Then $\{c_k\}_{k=1}^\infty$ and the coefficient functionals $\{c_k\}_{k=1}^\infty$ constitute a biorthogonal system.*

We leave the proof to the reader (Exercise 3.3). For completeness, we mention the following results about the coefficient functionals; they are proved in, e.g., [495, 622].

Theorem 3.1.7 *Let $\{e_k\}_{k=1}^\infty$ be a basis for X and let $\{c_k\}_{k=1}^\infty$ be the associated coefficient functionals. Then*

(i) $\{c_k\}_{k=1}^\infty$ *is a basis for its closed span in X^*, and its associated biorthogonal system is $\{e_k\}_{k=1}^\infty$ (considered as elements in X^{**}).*

(ii) *If X is reflexive, then $\{c_k\}_{k=1}^\infty$ is a basis for X^*.*

3.2 Bessel Sequences in Hilbert Spaces

The rest of this chapter concerns sequences in Hilbert spaces. For convenience, we index all sequences by the natural numbers in this section. We shall soon see that all results actually hold with arbitrary countable index sets.

Lemma 3.2.1 *Let $\{f_k\}_{k=1}^\infty$ be a sequence in $\mathcal{H}$, and suppose that $\sum_{k=1}^\infty c_k f_k$ is convergent for all $\{c_k\}_{k=1}^\infty \in \ell^2(\mathbb{N})$. Then*

$$T : \ell^2(\mathbb{N}) \to \mathcal{H}, \ T\{c_k\}_{k=1}^\infty := \sum_{k=1}^\infty c_k f_k \tag{3.12}$$

defines a bounded linear operator. The adjoint operator is given by

$$T^* : \mathcal{H} \to \ell^2(\mathbb{N}), \quad T^* f = \{\langle f, f_k\rangle\}_{k=1}^\infty. \tag{3.13}$$

Furthermore,

$$\sum_{k=1}^\infty |\langle f, f_k\rangle|^2 \le ||T||^2 \, ||f||^2, \ \forall f \in \mathcal{H}. \tag{3.14}$$

Proof. Consider the sequence of bounded linear operators

$$T_n : \ell^2(\mathbb{N}) \to \mathcal{H}, \ T_n\{c_k\}_{k=1}^\infty := \sum_{k=1}^n c_k f_k.$$

Clearly $T_n \to T$ pointwise as $n \to \infty$, so T is bounded by Theorem 2.2.1. In order to find the expression for T^*, let $f \in \mathcal{H}, \{c_k\}_{k=1}^\infty \in \ell^2(\mathbb{N})$. Then

$$\langle f, T\{c_k\}_{k=1}^\infty\rangle_\mathcal{H} = \langle f, \sum_{k=1}^\infty c_k f_k\rangle_\mathcal{H} = \sum_{k=1}^\infty \langle f, f_k\rangle \overline{c_k}. \tag{3.15}$$

We mention two ways to find T^*f from here:

1) The convergence of the series $\sum_{k=1}^{\infty}\langle f, f_k\rangle\overline{c_k}$ for all $\{c_k\}_{k=1}^{\infty} \in \ell^2(\mathbb{N})$ implies that $\{\langle f, f_k\rangle\}_{k=1}^{\infty} \in \ell^2(\mathbb{N})$; see for example [401], page 145. Thus, we can write

$$\langle f, T\{c_k\}_{k=1}^{\infty}\rangle_{\mathcal{H}} = \langle\{\langle f, f_k\rangle\}, \{c_k\}\rangle_{\ell^2(\mathbb{N})}$$

and conclude that

$$T^*f = \{\langle f, f_k\rangle\}_{k=1}^{\infty}.$$

2) Alternatively, when $T : \ell^2(\mathbb{N}) \to \mathcal{H}$ is bounded, we already know that T^* is a bounded operator from $\mathcal{H}$ to $\ell^2(\mathbb{N})$. Therefore, the kth coordinate function is bounded from $\mathcal{H}$ to $\mathbb{C}$; by Riesz representation theorem, T^* therefore has the form

$$T^*f = \{\langle f, g_k\rangle\}_{k=1}^{\infty}$$

for some $\{g_k\}_{k=1}^{\infty}$ in $\mathcal{H}$. By definition of T^*, (3.15) now shows that

$$\sum_{k=1}^{\infty}\langle f, g_k\rangle\overline{c_k} = \sum_{k=1}^{\infty}\langle f, f_k\rangle\overline{c_k}, \ \forall\{c_k\}_{k=1}^{\infty} \in \ell^2(\mathbb{N}), \ f \in \mathcal{H}.$$

It follows from here that $g_k = f_k$.

The adjoint of a bounded operator T is itself bounded, and $||T|| = ||T^*||$. Under the assumption in Lemma 3.2.1, we therefore have

$$||T^*f||^2 \leq ||T||^2 \, ||f||^2, \ \forall f \in \mathcal{H},$$

which leads to (3.14). □

Sequences $\{f_k\}_{k=1}^{\infty}$ for which an inequality of the type (3.14) holds will play a crucial role in the entire book.

Definition 3.2.2 *A sequence $\{f_k\}_{k=1}^{\infty}$ in $\mathcal{H}$ is called a Bessel sequence if there exists a constant $B > 0$ such that*

$$\sum_{k=1}^{\infty}|\langle f, f_k\rangle|^2 \leq B \, ||f||^2, \ \forall f \in \mathcal{H}. \tag{3.16}$$

Any number B satisfying (3.16) is called a *Bessel bound* for $\{f_k\}_{k=1}^{\infty}$. The *optimal bound* for a given Bessel sequence $\{f_k\}_{k=1}^{\infty}$ is the smallest possible value of $B > 0$ satisfying (3.16). Except for the case $f_k = 0$, $\forall k \in \mathbb{N}$, the optimal bound always exists.

We will now show that the Bessel condition can be expressed in terms of the operator T in (3.12). The operator is called the *synthesis operator* or *pre-frame operator*. We will use the wording that "T is well-defined on $\ell^2(\mathbb{N})$" if the infinite series in (3.12) converges for all $\{c_k\}_{k=1}^{\infty} \in \ell^2(\mathbb{N})$.

Theorem 3.2.3 *Let $\{f_k\}_{k=1}^{\infty}$ be a sequence in $\mathcal{H}$ and $B > 0$ be given. Then $\{f_k\}_{k=1}^{\infty}$ is a Bessel sequence with Bessel bound B if and only if*

$$T : \{c_k\}_{k=1}^{\infty} \to \sum_{k=1}^{\infty} c_k f_k$$

is a well-defined bounded operator from $\ell^2(\mathbb{N})$ into $\mathcal{H}$ and $\|T\| \leq \sqrt{B}$.

Proof. First assume that $\{f_k\}_{k=1}^{\infty}$ is a Bessel sequence with Bessel bound B. Let $\{c_k\}_{k=1}^{\infty} \in \ell^2(\mathbb{N})$. First we want to show that $T\{c_k\}_{k=1}^{\infty}$ is well-defined, i.e., that $\sum_{k=1}^{\infty} c_k f_k$ is convergent. Consider $n, m \in \mathbb{N}, n > m$. Then

$$\left\| \sum_{k=1}^{n} c_k f_k - \sum_{k=1}^{m} c_k f_k \right\| = \left\| \sum_{k=m+1}^{n} c_k f_k \right\|$$

$$= \sup_{\|g\|=1} \left| \left\langle \sum_{k=m+1}^{n} c_k f_k, g \right\rangle \right|$$

$$\leq \sup_{\|g\|=1} \sum_{k=m+1}^{n} |c_k \langle f_k, g \rangle|$$

$$\leq \left(\sum_{k=m+1}^{n} |c_k|^2 \right)^{1/2} \sup_{\|g\|=1} \left(\sum_{k=m+1}^{n} |\langle f_k, g \rangle|^2 \right)^{1/2}$$

$$\leq \sqrt{B} \left(\sum_{k=m+1}^{n} |c_k|^2 \right)^{1/2}.$$

Since $\{c_k\}_{k=1}^{\infty} \in \ell^2(\mathbb{N})$, we know that $\{\sum_{k=1}^{n} |c_k|^2\}_{n=1}^{\infty}$ is a Cauchy sequence in $\mathbb{C}$. The above calculation now shows that $\{\sum_{k=1}^{n} c_k f_k\}_{n=1}^{\infty}$ is a Cauchy sequence in $\mathcal{H}$ and therefore convergent. Thus, $T\{c_k\}_{k=1}^{\infty}$ is well-defined. Clearly, T is linear; since $\|T\{c_k\}_{k=1}^{\infty}\| = \sup_{\|g\|=1} |\langle T\{c_k\}_{k=1}^{\infty}, g \rangle|$, a calculation as above shows that T is bounded and that $\|T\| \leq \sqrt{B}$. For the opposite implication, suppose that T is well-defined and that $\|T\| \leq \sqrt{B}$. Then (3.14) shows that $\{f_k\}_{k=1}^{\infty}$ is a Bessel sequence with bound B. $\qquad\square$

Lemma 3.2.1 shows that if we only need to know that $\{f_k\}_{k=1}^{\infty}$ is a Bessel sequence and the value for the Bessel bound is irrelevant, we can just check that the operator T is well defined:

Corollary 3.2.4 *If $\{f_k\}_{k=1}^{\infty}$ is a sequence in $\mathcal{H}$ and $\sum_{k=1}^{\infty} c_k f_k$ is convergent for all $\{c_k\}_{k=1}^{\infty} \in \ell^2(\mathbb{N})$, then $\{f_k\}_{k=1}^{\infty}$ is a Bessel sequence.*

The Bessel condition (3.16) remains the same, regardless of how the elements $\{f_k\}_{k=1}^{\infty}$ are numbered. This leads to a very important consequence of Theorem 3.2.3:

Corollary 3.2.5 *If $\{f_k\}_{k=1}^{\infty}$ is a Bessel sequence in $\mathcal{H}$, then $\sum_{k=1}^{\infty} c_k f_k$ converges unconditionally for all $\{c_k\}_{k=1}^{\infty} \in \ell^2(\mathbb{N})$.*

Thus, a reordering of the elements in $\{f_k\}_{k=1}^{\infty}$ will not affect the series $\sum_{k=1}^{\infty} c_k f_k$ when $\{c_k\}_{k=1}^{\infty}$ is reordered the same way: the series will converge toward the same element as before. For this reason, we can choose an arbitrary indexing of the elements in the Bessel sequence; in particular, it is not a restriction that we present all results with the natural numbers as index set. As we will see in the sequel, all orthonormal bases, Riesz bases, and frames are Bessel sequences.

It is enough to check the Bessel condition (3.16) on a dense subset of $\mathcal{H}$:

Lemma 3.2.6 *Suppose that $\{f_k\}_{k=1}^{\infty}$ is a sequence of elements in $\mathcal{H}$ and that there exists a constant $B > 0$ such that*

$$\sum_{k=1}^{\infty} |\langle f, f_k \rangle|^2 \leq B \, \|f\|^2$$

for all f in a dense subset V of $\mathcal{H}$. Then $\{f_k\}_{k=1}^{\infty}$ is a Bessel sequence with bound B.

Proof. We have to prove that the Bessel condition is satisfied for all elements in $\mathcal{H}$. Let $g \in \mathcal{H}$, and suppose by contradiction that

$$\sum_{k=1}^{\infty} |\langle g, f_k \rangle|^2 > B \, \|g\|^2.$$

Then there exists a finite set $F \subset \mathbb{N}$ such that $\sum_{k \in F} |\langle g, f_k \rangle|^2 > B \, \|g\|^2$. Since V is dense in $\mathcal{H}$, this implies that there exists $h \in V$ such that

$$\sum_{k \in F} |\langle h, f_k \rangle|^2 > B \, \|h\|^2,$$

but this is a contradiction. We conclude that (3.16) indeed holds for all $f \in \mathcal{H}$, as claimed. $\square$

See Exercise 3.11 for another sufficient condition for $\{f_k\}_{k=1}^{\infty}$ to be a Bessel sequence.

3.3 Bases and Biorthogonal Systems in $\mathcal{H}$

We now return to some of the concepts defined in Section 3.1. The first lemma actually holds in Banach spaces, but for our purpose, it suffices to consider a Hilbert space $\mathcal{H}$. Recall from Corollary 2.3.3 that $\mathcal{H}^* = \mathcal{H}$; thus, according to the general definition in (3.11), two sequences $\{f_k\}_{k=1}^{\infty}$ and $\{g_k\}_{k=1}^{\infty}$ in $\mathcal{H}$ are biorthogonal if

$$\langle g_k, f_j \rangle = \delta_{k,j}, \, \forall j, k \in \mathbb{N}.$$

Lemma 3.3.1 *Let $\{f_k\}_{k=1}^{\infty}$ be a sequence in $\mathcal{H}$. Then the following holds:*

(i) $\{f_k\}_{k=1}^{\infty}$ has a biorthogonal sequence $\{g_k\}_{k=1}^{\infty}$ if and only if $\{f_k\}_{k=1}^{\infty}$ is minimal.

(ii) If a biorthogonal sequence for $\{f_k\}_{k=1}^{\infty}$ exists, it is uniquely determined if and only if $\{f_k\}_{k=1}^{\infty}$ is complete in $\mathcal{H}$.

Proof. For the proof of (i), suppose first that $\{f_k\}_{k=1}^{\infty}$ has a biorthogonal system $\{g_k\}_{k=1}^{\infty}$. Then, for any given $j \in \mathbb{N}$,

$$\langle f_j, g_j \rangle = 1 \text{ and } \langle f_k, g_j \rangle = 0 \text{ for } k \neq j.$$

Therefore $f_j \notin \overline{\text{span}}\{f_k\}_{k\neq j}$, i.e., $\{f_k\}_{k=1}^{\infty}$ is minimal. For the other implication in (i), assume that $\{f_k\}_{k=1}^{\infty}$ is minimal. Given $j \in \mathbb{N}$, let P_j denote the orthogonal projection of $\mathcal{H}$ onto $\overline{\text{span}}\{f_k\}_{k\neq j}$. Then it follows that $(I - P_j)f_j \neq 0$, and

$$\langle f_j, (I - P_j)f_j \rangle = \langle P_j f_j + (I - P_j)f_j, (I - P_j)f_j \rangle = ||(I - P_j)f_j||^2 \neq 0.$$

For $k \neq j$, clearly $\langle f_k, (I - P_j)f_j \rangle = 0$. Defining

$$g_j := \frac{(I - P_j)f_j}{||(I - P_j)f_j||^2}, \; j \in \mathbb{N},$$

we obtain that $\{g_k\}_{k=1}^{\infty}$ is a biorthogonal system for $\{f_k\}_{k=1}^{\infty}$.

For the proof of (ii), assume that $\{f_k\}_{k=1}^{\infty}$ has a biorthogonal system $\{g_k\}_{k=1}^{\infty}$. If $\{f_k\}_{k=1}^{\infty}$ is not complete, then it has several biorthogonal systems. In fact, letting

$$\mathcal{H}_0 := \overline{\text{span}}\{f_k\}_{k=1}^{\infty},$$

we can replace $\{g_k\}_{k=1}^{\infty}$ by $\{g_k + h_k\}_{k=1}^{\infty}$ for some $h_k \in \mathcal{H}_0^{\perp} \setminus \{0\}$ and hereby obtain a new biorthogonal system for $\{f_k\}_{k=1}^{\infty}$. We leave it to the reader to verify that if $\{f_k\}_{k=1}^{\infty}$ is complete, then the biorthogonality condition can at most be satisfied for one family $\{g_k\}_{k=1}^{\infty}$. $\square$

We will now prove that every basis $\{e_k\}_{k=1}^{\infty}$ for $\mathcal{H}$ leads to an expansion of arbitrary elements $f \in \mathcal{H}$, with coefficients given as inner products between f and the elements in an appropriately chosen sequence $\{g_k\}_{k=1}^{\infty}$. Expansions of exactly that type will be a key topic throughout the entire book; in fact, we will later obtain similar expansions for frames.

Theorem 3.3.2 *Assume that $\{e_k\}_{k=1}^{\infty}$ is a basis for the Hilbert space $\mathcal{H}$. Then there exists a unique family $\{g_k\}_{k=1}^{\infty}$ in $\mathcal{H}$ for which*

$$f = \sum_{k=1}^{\infty} \langle f, g_k \rangle e_k, \; \forall f \in \mathcal{H}. \tag{3.17}$$

$\{g_k\}_{k=1}^{\infty}$ *is a basis for $\mathcal{H}$, and $\{e_k\}_{k=1}^{\infty}$ and $\{g_k\}_{k=1}^{\infty}$ are biorthogonal.*

Proof. By Corollary 3.1.5, the coefficient functionals $\{c_k\}_{k=1}^\infty$ associated to $\{e_k\}_{k=1}^\infty$ are continuous; using Riesz' representation theorem (Theorem 2.3.2), there exists a unique family $\{g_k\}_{k=1}^\infty$ in $\mathcal{H}$ such that

$$c_k(f) = \langle f, g_k \rangle, \ \forall f \in \mathcal{H};$$

that is,

$$f = \sum_{k=1}^\infty \langle f, g_k \rangle e_k, \ \forall f \in \mathcal{H}.$$

We leave it to the reader to verify that no other family $\{g_k\}_{k=1}^\infty$ can satisfy (3.17) and that $\{e_k\}_{k=1}^\infty$ and $\{g_k\}_{k=1}^\infty$ are biorthogonal. The fact that $\{g_k\}_{k=1}^\infty$ is a basis for $\mathcal{H}$ follows from Theorem 3.1.7. □

The basis $\{g_k\}_{k=1}^\infty$ satisfying (3.17) is called the *dual basis*, or the *biorthogonal basis*, associated to $\{e_k\}_{k=1}^\infty$. It is interesting to observe that the Bessel condition on $\{e_k\}_{k=1}^\infty$ implies some kind of "opposite inequalities" for $\{g_k\}_{k=1}^\infty$; inequalities of this type will play an important role as soon as we have defined frames in Chapter 5.

Lemma 3.3.3 *Let $\{e_k\}_{k=1}^\infty$ be a basis for $\mathcal{H}$ and $\{g_k\}_{k=1}^\infty$ the associated biorthogonal system. If $\{e_k\}_{k=1}^\infty$ is a Bessel sequence with bound B, then*

(i) $\frac{1}{B}\|f\|^2 \le \sum_{k=1}^\infty |\langle f, g_k \rangle|^2, \ \forall f \in \mathcal{H}.$

(ii) $\frac{1}{B}\sum_{k=1}^\infty |c_k|^2 \le \|\sum_{k=1}^\infty c_k g_k\|^2$ *for all finite sequences* $\{c_k\}_{k=1}^\infty.$

Proof. Let $f \in \mathcal{H}$. Using $f = \sum_{k=1}^\infty \langle f, g_k \rangle e_k$ and Cauchy–Schwarz inequality, we obtain that

$$\|f\|^4 = \left|\sum_{k=1}^\infty \langle f, g_k \rangle \langle e_k, f \rangle\right|^2 \le \sum_{k=1}^\infty |\langle f, g_k \rangle|^2 \sum_{k=1}^\infty |\langle e_k, f \rangle|^2$$

$$\le B\|f\|^2 \sum_{k=1}^\infty |\langle f, g_k \rangle|^2.$$

(i) follows from this. For the proof of (ii), let $\{c_k\}_{k=1}^\infty$ be a finite sequence. Using the biorthogonal system $\{e_k\}_{k=1}^\infty, \{g_k\}_{k=1}^\infty$, we can write

$$\{c_k\}_{k=1}^\infty = \{\langle \sum_{j=1}^\infty c_j g_j, e_k \rangle\}_{k=1}^\infty$$

and

$$\sum_{k=1}^\infty |c_k|^2 = \sum_{k=1}^\infty \left|\langle \sum_{j=1}^\infty c_j g_j, e_k \rangle\right|^2 \le B\left\|\sum_{j=1}^\infty c_j g_j\right\|^2. \quad □$$

Note that it is essential that $\{c_k\}_{k=1}^\infty$ is finite in (ii); for general sequences $\{c_k\}_{k=1}^\infty \in \ell^2(\mathbb{N})$, the series $\sum_{k=1}^\infty c_k g_k$ might not converge (Exercises 3.2).

3.4 Orthonormal Bases

We are now ready to introduce one of the central themes, namely, orthonormal bases in Hilbert spaces. They are the abstract (infinite-dimensional) counterparts of the canonical bases in $\mathbb{C}^n$ and have many similar properties. Orthonormal bases are widely used in mathematics as well as physics, signal processing, and many other areas where expansions in terms of "convenient building blocks" are needed.

Definition 3.4.1 *A sequence* $\{e_k\}_{k=1}^{\infty}$ *in* $\mathcal{H}$ *is an orthonormal system if*

$$\langle e_k, e_j \rangle = \delta_{k,j}.$$

An orthonormal basis is an orthonormal system $\{e_k\}_{k=1}^{\infty}$ *which is a basis for* $\mathcal{H}$.

Note that an orthonormal system $\{e_k\}_{k=1}^{\infty}$ is a Bessel sequence. In fact, if $\{c_k\}_{k=1}^{\infty} \in \ell^2(\mathbb{N})$ and $m, n \in \mathbb{N}, n > m$, then

$$\left\| \sum_{k=1}^{n} c_k e_k - \sum_{k=1}^{m} c_k e_k \right\|^2 = \left\| \sum_{k=m+1}^{n} c_k e_k \right\|^2 = \sum_{k=m+1}^{n} |c_k|^2;$$

as in the proof of Theorem 3.2.3, this implies that $\sum_{k=1}^{\infty} c_k e_k$ is convergent and that

$$\left\| \sum_{k=1}^{\infty} c_k e_k \right\|^2 = \sum_{k=1}^{\infty} |c_k|^2.$$

The next theorem gives equivalent conditions for an orthonormal system $\{e_k\}_{k=1}^{\infty}$ to be an orthonormal basis.

Theorem 3.4.2 *For an orthonormal system* $\{e_k\}_{k=1}^{\infty}$ *in a Hilbert space* $\mathcal{H}$, *the following are equivalent:*

(i) $\{e_k\}_{k=1}^{\infty}$ *is an orthonormal basis.*

(ii) $f = \sum_{k=1}^{\infty} \langle f, e_k \rangle e_k, \ \forall f \in \mathcal{H}$.

(iii) $\langle f, g \rangle = \sum_{k=1}^{\infty} \langle f, e_k \rangle \langle e_k, g \rangle, \ \forall f, g \in \mathcal{H}$.

(iv) $\sum_{k=1}^{\infty} |\langle f, e_k \rangle|^2 = ||f||^2, \ \forall f \in \mathcal{H}$.

(v) $\overline{span}\{e_k\}_{k=1}^{\infty} = \mathcal{H}$.

(vi) *If* $\langle f, e_k \rangle = 0, \ \forall k \in \mathbb{N}$, *then* $f = 0$.

Proof. For the proof of (i) $\Rightarrow$ (ii), let $f \in \mathcal{H}$. If $\{e_k\}_{k=1}^{\infty}$ is an orthonormal basis, there exist coefficients $\{c_k\}_{k=1}^{\infty}$ such that $f = \sum_{k=1}^{\infty} c_k e_k$. Given any $j \in \mathbb{N}$, we have $\langle f, e_j \rangle = \sum_{k=1}^{\infty} c_k \delta_{k,j} = c_j$, and (ii) follows. (iii) is an obvious consequence of (ii), and (iv) is a special case of (iii). The implications (iv) $\Rightarrow$ (v) $\Rightarrow$ (vi) are clear. For the proof of (vi) $\Rightarrow$ (i), let $f \in \mathcal{H}$.

Since $\{e_k\}_{k=1}^{\infty}$ is a Bessel sequence, we know that $g := \sum_{k=1}^{\infty}\langle f, e_k\rangle e_k$ is well defined; furthermore, $\langle f - g, e_j\rangle = 0$ for all $j \in \mathbb{N}$, so by (vi), $f = g = \sum_{k=1}^{\infty}\langle f, e_k\rangle e_k$. To prove that $\{e_k\}_{k=1}^{\infty}$ is a basis, we only need to show that no other linear combination of $\{e_k\}_{k=1}^{\infty}$ can be equal to f, and this follows by the argument we used to prove that (ii) follows from (i). $\square$

The equality in (iv) is called *Parseval's equation*. Via Corollary 3.2.5, we obtain the following important consequence of Theorem 3.4.2:

Corollary 3.4.3 *If $\{e_k\}_{k=1}^{\infty}$ is an orthonormal basis, then each $f \in \mathcal{H}$ has an unconditionally convergent expansion*

$$f = \sum_{k=1}^{\infty}\langle f, e_k\rangle e_k. \tag{3.18}$$

In particular, the dual basis equals the basis itself.

Theorem 3.4.4 *Every separable Hilbert space $\mathcal{H}$ has an orthonormal basis.*

Proof. Since $\mathcal{H}$ is assumed separable, we can choose a sequence $\{f_k\}_{k=1}^{\infty}$ in $\mathcal{H}$ such that $\overline{\text{span}}\{f_k\}_{k=1}^{\infty} = \mathcal{H}$. By extracting a subsequence if necessary, we can assume that for each $n \in \mathbb{N}, f_{n+1} \notin \text{span}\{f_k\}_{k=1}^{n}$. By applying the Gram–Schmidt process to $\{f_k\}_{k=1}^{\infty}$, we obtain an orthonormal system $\{e_k\}_{k=1}^{\infty}$ in $\mathcal{H}$ for which $\overline{\text{span}}\{e_k\}_{k=1}^{\infty} = \overline{\text{span}}\{f_k\}_{k=1}^{\infty} = \mathcal{H}$. $\square$

Often we want to have a concrete orthonormal basis for a given Hilbert space, rather than just its existence. The simplest case is $\ell^2(\mathbb{N})$:

Example 3.4.5 Let e_k be the sequence in $\ell^2(\mathbb{N})$ whose k-th entry is 1, and all other entries are zero. Then $\{e_k\}_{k=1}^{\infty}$ is an orthonormal basis for $\ell^2(\mathbb{N})$; it is called the *canonical orthonormal basis*. We will often denote this special basis by $\{\delta_k\}_{k=1}^{\infty}$. $\square$

We will later construct orthonormal bases for other Hilbert spaces, e.g., $L^2(-\pi, \pi)$ and $L^2(\mathbb{R})$.

Orthonormal bases are certainly the most convenient bases to use because the biorthogonal basis equals the basis itself. That is, the representation (3.18) is directly available, while the representation (3.17) via a general basis requires that we find the biorthogonal sequence $\{g_k\}_{k=1}^{\infty}$. Unfortunately, the conditions for $\{e_k\}_{k=1}^{\infty}$ being an orthonormal basis are strong, and often it is impossible to construct orthonormal bases satisfying extra conditions. We discuss this in more detail in Chapter 4. Note also that it is not always a good idea to use the Gram–Schmidt orthonormalization procedure to construct an orthonormal basis from a given basis: it might destroy special properties of the basis at hand. For example, the special structure of Gabor bases and wavelet bases (to be discussed in Sections 3.8–3.9) will get lost:

Based on Theorem 3.4.4, we can prove that every separable Hilbert space can be identified with $\ell^2(\mathbb{N})$:

Theorem 3.4.6 *Every separable infinite-dimensional Hilbert space $\mathcal{H}$ is isometrically isomorphic to $\ell^2(\mathbb{N})$:*

Proof. Let $\{e_k\}_{k=1}^{\infty}$ be an orthonormal basis for $\mathcal{H}$. We have already observed that $\sum_{k=1}^{\infty} c_k e_k$ is convergent for all $\{c_k\}_{k=1}^{\infty} \in \ell^2(\mathbb{N})$. Furthermore each $f \in \mathcal{H}$ has a unique expansion with ℓ^2-coefficients, namely, $f = \sum_{k=1}^{\infty} \langle f, e_k \rangle e_k$. By letting $\{\delta_k\}_{k=1}^{\infty}$ be the canonical orthonormal basis for $\ell^2(\mathbb{N})$, we can thus define the operator

$$U : \mathcal{H} \to \ell^2(\mathbb{N}), \ U\left(\sum_{k=1}^{\infty} c_k e_k\right) = \sum_{k=1}^{\infty} c_k \delta_k, \ \{c_k\}_{k=1}^{\infty} \in \ell^2(\mathbb{N}).$$

Then U maps $\mathcal{H}$ bijectively onto $\ell^2(\mathbb{N})$. For $f \in \mathcal{H}, f = \sum_{k=1}^{\infty} \langle f, e_k \rangle e_k$, we have

$$||Uf||^2 \ = \ \left\|\sum_{k=1}^{\infty} \langle f, e_k \rangle \delta_k\right\|^2 = \sum_{k=1}^{\infty} |\langle f, e_k \rangle|^2 = ||f||^2;$$

thus, U is an isometry. □

The following theorem characterizes all orthonormal bases for $\mathcal{H}$ starting with one orthonormal basis.

Theorem 3.4.7 *Let $\{e_k\}_{k=1}^{\infty}$ be an orthonormal basis for $\mathcal{H}$. Then the orthonormal bases for $\mathcal{H}$ are precisely the sets $\{Ue_k\}_{k=1}^{\infty}$, where $U : \mathcal{H} \to \mathcal{H}$ is a unitary operator.*

Proof. Let $\{f_k\}_{k=1}^{\infty}$ be an orthonormal basis for $\mathcal{H}$. Define the operator

$$U : \mathcal{H} \to \mathcal{H}, \ U\left(\sum_{k=1}^{\infty} c_k e_k\right) = \sum_{k=1}^{\infty} c_k f_k, \ \{c_k\}_{k=1}^{\infty} \in \ell^2(\mathbb{N}).$$

Then U maps $\mathcal{H}$ boundedly and bijectively onto $\mathcal{H}$. For $f, g \in \mathcal{H}$, write $f = \sum_{k=1}^{\infty} \langle f, e_k \rangle e_k$ and $g = \sum_{k=1}^{\infty} \langle g, e_k \rangle e_k$; then, via the definition of U and Theorem 3.4.2,

$$\begin{aligned}
\langle U^*Uf, g \rangle \ &= \ \langle Uf, Ug \rangle \\
&= \ \left\langle \sum_{k=1}^{\infty} \langle f, e_k \rangle f_k, \sum_{k=1}^{\infty} \langle g, e_k \rangle f_k \right\rangle \\
&= \ \sum_{k=1}^{\infty} \langle f, e_k \rangle \overline{\langle g, e_k \rangle} = \langle f, g \rangle;
\end{aligned}$$

thus, $U^*U = I$. Since U is surjective, it follows that U is unitary. On the other hand, if U is a given unitary operator, then

$$\langle Ue_k, Ue_j \rangle = \langle U^*Ue_k, e_j \rangle = \langle e_k, e_j \rangle = \delta_{k,j},$$

i.e., $\{Ue_k\}_{k=1}^\infty$ is an orthonormal system. That it is a basis follows from the fact that U is surjective. □

Condition (iv) in Theorem 3.4.2 has an interpretation in terms of frames; see Definition 5.1.2. Without assuming that $\{e_k\}_{k=1}^\infty$ is an orthonormal system, it implies that $\{e_k\}_{k=1}^\infty$ is an orthonormal basis if the vectors are normalized:

Proposition 3.4.8 *Assume that $\{e_k\}_{k=1}^\infty$ is a sequence of normalized vectors in $\mathcal{H}$ and that*

$$\sum_{k=1}^\infty |\langle f, e_k \rangle|^2 = ||f||^2, \ \forall f \in \mathcal{H}.$$

Then $\{e_k\}_{k=1}^\infty$ is an orthonormal basis for $\mathcal{H}$.

Proof. By Theorem 3.4.2 we only have to prove that $\{e_k\}_{k=1}^\infty$ is an orthonormal system. For each $j \in \mathbb{N}$, we have

$$1 = ||e_j||^2 = \sum_{k=1}^\infty |\langle e_j, e_k \rangle|^2 = 1 + \sum_{k \neq j} |\langle e_j, e_k \rangle|^2,$$

which shows that $\langle e_j, e_k \rangle = 0$ for $k \neq j$. □

3.5 The Gram Matrix

If $\{f_k\}_{k=1}^\infty$ is a Bessel sequence in a Hilbert space $\mathcal{H}$, we can compose the bounded operators T^* and T; hereby we obtain the bounded operator

$$T^*T : \ell^2(\mathbb{N}) \to \ell^2(\mathbb{N}), \ T^*T\{c_k\}_{k=1}^\infty = \left\{ \left\langle \sum_{\ell=1}^\infty c_\ell f_\ell, f_k \right\rangle \right\}_{k=1}^\infty.$$

Letting $\{e_k\}_{k=1}^\infty$ be the canonical orthonormal basis for $\ell^2(\mathbb{N})$, the jk-th entry in the matrix representation for T^*T is

$$\langle T^*Te_k, e_j \rangle = \langle Te_k, Te_j \rangle = \langle f_k, f_j \rangle.$$

Identifying T^*T with its matrix representation, we write

$$T^*T = \{\langle f_k, f_j \rangle\}_{j,k=1}^\infty.$$

The matrix $\{\langle f_k, f_j \rangle\}_{j,k=1}^\infty$ is called the *Gram matrix* associated with $\{f_k\}_{k=1}^\infty$, and the above argument shows that it defines a bounded operator on $\ell^2(\mathbb{N})$ when $\{f_k\}_{k=1}^\infty$ is a Bessel sequence. One can in principle consider the Gram matrix associated to any sequence $\{f_k\}_{k=1}^\infty$ in $\mathcal{H}$, but if we want it to define a bounded operator on $\ell^2(\mathbb{N})$, we cannot avoid the Bessel condition:

Lemma 3.5.1 *For a sequence $\{f_k\}_{k=1}^\infty$ in $\mathcal{H}$, the following are equivalent:*

(i) $\{f_k\}_{k=1}^\infty$ *is a Bessel sequence with bound B.*

(ii) *The Gram matrix associated to $\{f_k\}_{k=1}^\infty$ defines a bounded operator on $\ell^2(\mathbb{N})$, with norm at most B.*

Proof. The implication (i) $\Rightarrow$ (ii) follows from the arguments above together with the norm estimate $\|T\| \le \sqrt{B}$ in Theorem 3.2.3. Now assume that (ii) is satisfied, and let $\{c_k\}_{k=1}^\infty \in \ell^2(\mathbb{N})$. Then

$$\sum_{j=1}^\infty \left| \sum_{k=1}^\infty \langle f_k, f_j \rangle c_k \right|^2 \le B^2 \sum_{k=1}^\infty |c_k|^2. \tag{3.19}$$

Given arbitrary $n, m \in \mathbb{N}, n > m$,

$$\left\| \sum_{k=1}^n c_k f_k - \sum_{k=1}^m c_k f_k \right\|^4 = \left\| \sum_{k=m+1}^n c_k f_k \right\|^4$$

$$= \left| \left\langle \sum_{k=m+1}^n c_k f_k, \sum_{j=m+1}^n c_j f_j \right\rangle \right|^2$$

$$= \left| \sum_{j=m+1}^n \overline{c_j} \sum_{k=m+1}^n c_k \langle f_k, f_j \rangle \right|^2$$

$$\le \left(\sum_{j=m+1}^n |c_j|^2 \right) \left(\sum_{j=m+1}^n \left| \sum_{k=m+1}^n c_k \langle f_k, f_j \rangle \right|^2 \right),$$

where Cauchy–Schwarz' inequality was used on the sum over j in the last step. Via (3.19) applied to the finite sequence

$$(\cdots, 0, 0, c_{m+1}, c_{m+2}, \cdots, c_n, 0, 0, \cdots),$$

$$\sum_{j=m+1}^n \left| \sum_{k=m+1}^n c_k \langle f_k, f_j \rangle \right|^2 \le \sum_{j=1}^\infty \left| \sum_{k=m+1}^n c_k \langle f_k, f_j \rangle \right|^2$$

$$\le B^2 \sum_{j=m+1}^n |c_j|^2.$$

Altogether we arrive at

$$\left\|\sum_{k=1}^{n} c_k f_k - \sum_{k=1}^{m} c_k f_k\right\|^4 \leq B^2 \left(\sum_{j=m+1}^{\infty} |c_j|^2\right)^2.$$

It follows that $\sum_{k=1}^{\infty} c_k f_k$ is convergent and, by repeating the argument,

$$\left\|\sum_{k=1}^{\infty} c_k f_k\right\| \leq \sqrt{B} \left(\sum_{j=1}^{\infty} |c_j|^2\right)^{1/2}.$$

By Theorem 3.2.3 we conclude that $\{f_k\}_{k=1}^{\infty}$ is a Bessel sequence with bound B. □

Lemma 3.5.2 *Assume that $\{f_k\}_{k=1}^{\infty}$ is a Bessel sequence in $\mathcal{H}$ with synthesis operator T. Then the Gram matrix defines an injective operator from $\mathcal{R}_{T^*}$ into $\mathcal{R}_{T^*}$. Its range is dense in $\mathcal{R}_{T^*}$.*

Proof. It is clear that T^*T maps $\mathcal{R}_{T^*}$ into itself. This restriction of T^*T is injective: if $\{c_k\}_{k=1}^{\infty} \in \mathcal{R}_{T^*}$ and $T^*T\{c_k\}_{k=1}^{\infty} = 0$, then

$$\|T\{c_k\}_{k=1}^{\infty}\|^2 = \langle T^*T\{c_k\}_{k=1}^{\infty}, \{c_k\}_{k=1}^{\infty}\rangle = 0,$$

i.e., $\{c_k\}_{k=1}^{\infty} \in \mathcal{R}_{T^*} \cap \mathcal{N}_T = \{0\}$. Using that $\mathcal{H} = \overline{\mathcal{R}_T} + \mathcal{N}_{T^*}$, we see that

$$\mathcal{R}_{T^*} = T^*\mathcal{H} = T^*\overline{\mathcal{R}_T},$$

so $\mathcal{R}_{T^*T}$ is dense in $\mathcal{R}_{T^*}$ by continuity of T^*. □

Proposition 3.5.4 will give a sufficient condition for $\{f_k\}_{k=1}^{\infty}$ being a Bessel sequence. The proof uses *Schur's lemma*:

Lemma 3.5.3 *Let $M = \{M_{j,k}\}_{j,k=1}^{\infty}$ be a matrix for which $M_{j,k} = \overline{M_{k,j}}$ for all $j, k \in \mathbb{N}$, and for which there exists a constant $B > 0$ such that*

$$\sum_{k=1}^{\infty} |M_{j,k}| \leq B, \ \forall j \in \mathbb{N}.$$

Then M defines a bounded operator on $\ell^2(\mathbb{N})$ of norm at most B.

Proof. Let $\{c_k\}_{k=1}^{\infty} \in \ell^2(\mathbb{N})$. The assumptions imply that $M\{c_k\}_{k=1}^{\infty}$ is well defined as a sequence indexed by $\mathbb{N}$, whose j-th coordinate is $\sum_{k=1}^{\infty} M_{j,k}c_k$. It is, however, not immediately clear that this sequence belongs to $\ell^2(\mathbb{N})$. Abusing the notation, it is enough to show that the map

$$\{d_k\}_{k=1}^{\infty} \mapsto \langle \{d_k\}_{k=1}^{\infty}, M\{c_k\}_{k=1}^{\infty}\rangle_{\ell^2(\mathbb{N})} \tag{3.20}$$

is a continuous linear functional on $\ell^2(\mathbb{N})$. In fact, this implies that $M\{c_k\}_{k=1}^{\infty}$ belongs to the dual of $\ell^2(\mathbb{N})$, which is $\ell^2(\mathbb{N})$ itself. Now, for

$\{d_k\}_{k=1}^{\infty} \in \ell^2(\mathbb{N})$,

$$\sum_{j=1}^{\infty}\left|\sum_{k=1}^{\infty}\overline{M_{j,k}c_k}d_j\right| \leq \sum_{j=1}^{\infty}\sum_{k=1}^{\infty}|M_{j,k}c_kd_j|$$

$$= \sum_{j=1}^{\infty}\sum_{k=1}^{\infty}\left(|M_{j,k}|^{1/2}|c_k|\right)\left(|M_{j,k}|^{1/2}|d_j|\right) = (*).$$

Using Cauchy–Schwarz inequality,

$$(*) \leq \left(\sum_{j=1}^{\infty}\sum_{k=1}^{\infty}|M_{j,k}|\,|c_k|^2\right)^{1/2}\left(\sum_{j=1}^{\infty}\sum_{k=1}^{\infty}|M_{j,k}|\,|d_j|^2\right)^{1/2}$$

$$\leq B\left(\sum_{k=1}^{\infty}|c_k|^2\right)^{1/2}\left(\sum_{j=1}^{\infty}|d_j|^2\right)^{1/2}.$$

This shows that (3.20) indeed defines a continuous linear functional on $\ell^2(\mathbb{N})$, so M maps $\ell^2(\mathbb{N})$ into $\ell^2(\mathbb{N})$. Also,

$$||M\{c_k\}_{k=1}^{\infty}|| = \sup_{||\{d_k\}||=1}\left|\langle\{d_k\}_{k=1}^{\infty}, M\{c_k\}_{k=1}^{\infty}\rangle_{\ell^2(\mathbb{N})}\right|$$

$$\leq B\left(\sum_{k=1}^{\infty}|c_k|^2\right)^{1/2},$$

which completes the proof (see Exercise 3.15 for a question about the proof). □

An application of Schur's lemma gives a sufficient condition for the Gram matrix defining a bounded operator on $\ell^2(\mathbb{N})$ and thus for $\{f_k\}_{k=1}^{\infty}$ being a Bessel sequence. For the proof, we just have to refer to Lemma 3.5.1.

Proposition 3.5.4 *Let $\{f_k\}_{k=1}^{\infty}$ be a sequence in $\mathcal{H}$ and assume that there exists a constant $B > 0$ such that*

$$\sum_{k=1}^{\infty}|\langle f_j, f_k\rangle| \leq B, \ \forall j \in \mathbb{N}. \tag{3.21}$$

Then $\{f_k\}_{k=1}^{\infty}$ is a Bessel sequence with bound B.

Compared with the Bessel condition (3.16), Proposition 3.5.4 has the advantage that it only involves inner products between the elements in $\{f_k\}_{k=1}^{\infty}$; that is, only a countable number of conditions must be verified, while the Bessel condition (3.16) has to be checked for all $f \in \mathcal{H}$. A natural way to obtain (3.21) is to impose decay conditions on $|\langle f_j, f_k\rangle|$, i.e., conditions that force these numbers to decrease whenever $|j - k|$ increases:

Corollary 3.5.5 *Assume that $\{f_k\}_{k=1}^{\infty}$ is a sequence in $\mathcal{H}$ and that either*

(i) there exist $s > 1$ and a constant $C > 0$ such that

$$|\langle f_j, f_k \rangle| \le \frac{C}{(1 + |j - k|)^s}, \ \forall j, k \in \mathbb{N} \tag{3.22}$$

or

(ii) there exist $\alpha > 0$ and a constant $C > 0$ such that

$$|\langle f_j, f_k \rangle| \le C e^{-\alpha |j-k|}, \ \forall j, k \in \mathbb{N}. \tag{3.23}$$

Then $\{f_k\}_{k=1}^{\infty}$ is a Bessel sequence.

We leave the proof of Corollary 3.5.5 to the reader (Exercise 3.8). The conditions (3.22) and (3.23) will later play the key role in the context of localization of frames; see Section 8.2.

3.6 Riesz Bases

In Theorem 3.4.7 we characterized all orthonormal bases in terms of unitary operators acting on a single orthonormal basis. Formally, the definition of a *Riesz basis* appears by weakening the condition on the operator:

Definition 3.6.1 *A Riesz basis for $\mathcal{H}$ is a family of the form $\{Ue_k\}_{k=1}^{\infty}$, where $\{e_k\}_{k=1}^{\infty}$ is an orthonormal basis for $\mathcal{H}$ and $U : \mathcal{H} \to \mathcal{H}$ is a bounded bijective operator.*

A Riesz basis is actually a basis (Exercise 3.6.3). The dual basis associated to a Riesz basis is also a Riesz basis:

Theorem 3.6.2 *If $\{f_k\}_{k=1}^{\infty}$ is a Riesz basis for $\mathcal{H}$, there exists a unique sequence $\{g_k\}_{k=1}^{\infty}$ in $\mathcal{H}$ such that*

$$f = \sum_{k=1}^{\infty} \langle f, g_k \rangle f_k, \ \forall f \in \mathcal{H}. \tag{3.24}$$

$\{g_k\}_{k=1}^{\infty}$ is also a Riesz basis, and $\{f_k\}_{k=1}^{\infty}$ and $\{g_k\}_{k=1}^{\infty}$ are biorthogonal. Moreover, the series (3.24) converges unconditionally for all $f \in \mathcal{H}$.

Proof. According to the definition we can write $\{f_k\}_{k=1}^{\infty} = \{Ue_k\}_{k=1}^{\infty}$, where U is a bounded bijective operator and $\{e_k\}_{k=1}^{\infty}$ is an orthonormal basis. Let now $f \in \mathcal{H}$. By expanding $U^{-1}f$ in the orthonormal basis $\{e_k\}_{k=1}^{\infty}$, we have

$$U^{-1}f = \sum_{k=1}^{\infty} \langle U^{-1}f, e_k \rangle e_k = \sum_{k=1}^{\infty} \langle f, (U^{-1})^* e_k \rangle e_k.$$

Therefore, with $g_k := (U^{-1})^* e_k$,

$$f = UU^{-1}f \;=\; \sum_{k=1}^{\infty}\langle f, (U^{-1})^* e_k\rangle U e_k = \sum_{k=1}^{\infty}\langle f, g_k\rangle f_k.$$

Since $(U^{-1})^*$ is bounded and bijective, $\{g_k\}_{k=1}^{\infty}$ is a Riesz basis by definition. For $f \in \mathcal{H}$,

$$
\begin{aligned}
\sum_{k=1}^{\infty}|\langle f, f_k\rangle|^2 = \sum_{k=1}^{\infty}|\langle f, U e_k\rangle|^2 &= \;||U^* f||^2 \\
&\leq \;||U^*||^2\,||f||^2 \\
&= \;||U||^2\,||f||^2, \qquad (3.25)
\end{aligned}
$$

this proves that a Riesz basis is a Bessel sequence. Thus, the series (3.24) converges unconditionally by Corollary 3.2.5. The rest follows from Theorem 3.3.2 (or direct verification). $\qquad\square$

The unique sequence $\{g_k\}_{k=1}^{\infty}$ satisfying (3.24) is called the *dual Riesz basis* of $\{f_k\}_{k=1}^{\infty}$. Let us find the dual of $\{g_k\}_{k=1}^{\infty}$. In the notation used in the proof of Theorem 3.6.2, we have that the dual of $\{f_k\}_{k=1}^{\infty} = \{U e_k\}_{k=1}^{\infty}$ is given by $\{g_k\}_{k=1}^{\infty} = \{(U^{-1})^* e_k\}_{k=1}^{\infty}$; thus, the dual of $\{g_k\}_{k=1}^{\infty}$ is

$$\left\{\left(\left((U^{-1})^*\right)^{-1}\right)^* e_k\right\}_{k=1}^{\infty} = \{U e_k\}_{k=1}^{\infty} = \{f_k\}_{k=1}^{\infty}.$$

That is, $\{f_k\}_{k=1}^{\infty}$ and $\{g_k\}_{k=1}^{\infty}$ are duals of each other. For this reason, we frequently speak about a *pair of dual Riesz bases*. In particular, this implies a symmetric version of (3.24):

Corollary 3.6.3 *Let $\{f_k\}_{k=1}^{\infty}$ and $\{g_k\}_{k=1}^{\infty}$ be a pair of dual Riesz bases. Then*

$$f = \sum_{k=1}^{\infty}\langle f, g_k\rangle f_k = \sum_{k=1}^{\infty}\langle f, f_k\rangle g_k, \; \forall f \in \mathcal{H}. \qquad (3.26)$$

For later use, we note that a Riesz basis not only satisfies the Bessel inequality: it also satisfies some kind of "opposite inequality."

Proposition 3.6.4 *If $\{f_k\}_{k=1}^{\infty} = \{U e_k\}_{k=1}^{\infty}$ is a Riesz basis for $\mathcal{H}$, there exist constants $A, B > 0$ such that*

$$A\,||f||^2 \leq \sum_{k=1}^{\infty}|\langle f, f_k\rangle|^2 \leq B\,||f||^2, \; \forall f \in \mathcal{H}. \qquad (3.27)$$

The largest possible value for the constant A is $\frac{1}{||U^{-1}||^2}$, and the smallest possible value for B is $||U||^2$.

Proof. That a Riesz basis $\{Ue_k\}_{k=1}^{\infty}$ is a Bessel sequence with optimal upper bound $||U||$ follows already from the estimate in (3.25). The result about the lower bound follows from

$$||f|| = ||(U^*)^{-1}U^*f|| \leq ||(U^*)^{-1}|| \; ||U^*f|| = ||U^{-1}|| \; ||U^*f||. \qquad \square$$

Our next aim is to characterize Riesz bases. For this purpose we need a technical result about operators. A standard way of constructing an operator is to define it on a basis and then extend by linearity; the following lemma gives some conditions for this being possible.

Lemma 3.6.5 *Let $\mathcal{H}, \mathcal{K}$ be Hilbert spaces, and let $\{h_k\}_{k=1}^{\infty}$ be a sequence in $\mathcal{H}, \{g_k\}_{k=1}^{\infty}$ a sequence in $\mathcal{K}$. Assume that $\{g_k\}_{k=1}^{\infty}$ is a Bessel sequence with bound B, that $\{h_k\}_{k=1}^{\infty}$ is complete in $\mathcal{H}$, and that there exists a constant $A > 0$ such that*

$$A \sum_{k=1}^{\infty} |c_k|^2 \leq \left\| \sum_{k=1}^{\infty} c_k h_k \right\|^2 \tag{3.28}$$

for all finite scalar sequences $\{c_k\}_{k=1}^{\infty}$. Then

$$U \left(\sum_{k=1}^{\infty} c_k h_k \right) := \sum_{k=1}^{\infty} c_k g_k$$

defines a linear bounded operator from $\mathrm{span}\{h_k\}_{k=1}^{\infty}$ into $\mathrm{span}\{g_k\}_{k=1}^{\infty}$, and U has a unique extension to a bounded operator from $\mathcal{H}$ into $\mathcal{K}$; the norm of U as well as its extension is at most $\sqrt{\frac{B}{A}}$.

Proof. By the assumption (3.28), every $h \in \mathrm{span}\{h_k\}_{k=1}^{\infty}$ has a unique representation $h = \sum_{k=1}^{\infty} c_k h_k$ with $\{c_k\}_{k=1}^{\infty}$ finite; it follows that U is well defined and linear. Given a finite sequence $\{c_k\}_{k=1}^{\infty}$,

$$\left\| U \left(\sum_{k=1}^{\infty} c_k h_k \right) \right\|^2 = \left\| \sum_{k=1}^{\infty} c_k g_k \right\|^2$$

$$\leq B \sum_{k=1}^{\infty} |c_k|^2 \leq \frac{B}{A} \left\| \sum_{k=1}^{\infty} c_k h_k \right\|^2.$$

Thus, U is bounded. Since $\{h_k\}_{k=1}^{\infty}$ is assumed to be complete, U has an extension to a bounded operator on $\mathcal{H}$. The rest is standard. $\qquad \square$

The next theorem gives equivalent conditions for $\{f_k\}_{k=1}^{\infty}$ being a Riesz basis. Note in particular condition (ii), which will be used throughout the book and, in fact, by several authors is used as the definition of a Riesz basis.

Theorem 3.6.6 *For a sequence $\{f_k\}_{k=1}^{\infty}$ in $\mathcal{H}$, the following conditions are equivalent:*

(i) $\{f_k\}_{k=1}^{\infty}$ *is a Riesz basis for $\mathcal{H}$.*

(ii) $\{f_k\}_{k=1}^{\infty}$ *is complete in $\mathcal{H}$, and there exist constants $A, B > 0$ such that for every finite scalar sequence $\{c_k\}$*

$$A \sum_{k=1}^{\infty} |c_k|^2 \leq \left\| \sum_{k=1}^{\infty} c_k f_k \right\|^2 \leq B \sum_{k=1}^{\infty} |c_k|^2. \tag{3.29}$$

(iii) $\{f_k\}_{k=1}^{\infty}$ *is complete, and its Gram matrix $\{\langle f_k, f_j \rangle\}_{j,k=1}^{\infty}$ defines a bounded, invertible operator on $\ell^2(\mathbb{N})$.*

(iv) $\{f_k\}_{k=1}^{\infty}$ *is a complete Bessel sequence, and it has a complete biorthogonal sequence $\{g_k\}_{k=1}^{\infty}$ which is also a Bessel sequence.*

Proof. (i)$\Rightarrow$(ii). Assume that $\{f_k\}_{k=1}^{\infty}$ is a Riesz basis, and write it in the form $\{Ue_k\}_{k=1}^{\infty}$ as in the definition. Note that $\{f_k\}_{k=1}^{\infty}$ is complete. Given any finite scalar sequence $\{c_k\}_{k=1}^{\infty}$,

$$\left\| \sum_{k=1}^{\infty} c_k f_k \right\|^2 = \left\| U \left(\sum_{k=1}^{\infty} c_k e_k \right) \right\|^2 \leq \|U\|^2 \left\| \sum_{k=1}^{\infty} c_k e_k \right\|^2 = \|U\|^2 \sum_{k=1}^{\infty} |c_k|^2$$

and

$$\left\| \sum_{k=1}^{\infty} c_k e_k \right\|^2 = \left\| U^{-1} U \left(\sum_{k=1}^{\infty} c_k e_k \right) \right\|^2 \leq \|U^{-1}\|^2 \left\| \sum_{k=1}^{\infty} c_k f_k \right\|^2.$$

From this we deduce that

$$\frac{1}{\|U^{-1}\|^2} \sum_{k=1}^{\infty} |c_k|^2 \leq \left\| \sum_{k=1}^{\infty} c_k f_k \right\|^2 \leq \|U\|^2 \sum_{k=1}^{\infty} |c_k|^2.$$

(ii)$\Rightarrow$(i). The right-hand inequality in (3.29) implies that $\{f_k\}_{k=1}^{\infty}$ is a Bessel sequence with bound B (Exercise 3.13). Choose an orthonormal basis $\{e_k\}_{k=1}^{\infty}$ for $\mathcal{H}$, and extend by Lemma 3.6.5 the mapping $Ue_k := f_k$ to a bounded operator on $\mathcal{H}$. In the same way, extend $Vf_k := e_k$ to a bounded operator on $\mathcal{H}$. Then $VU = UV = I$, so U is invertible; thus, $\{f_k\}_{k=1}^{\infty}$ is a Riesz basis.

(i)$\Rightarrow$(iii). Write again $\{f_k\}_{k=1}^{\infty} = \{Ue_k\}_{k=1}^{\infty}$. For any $k, j \in \mathbb{N}$,

$$\langle f_k, f_j \rangle = \langle Ue_k, Ue_j \rangle = \langle U^*Ue_k, e_j \rangle$$

i.e., the Gram matrix is the matrix representing the bounded invertible operator U^*U in the basis $\{e_k\}_{k=1}^{\infty}$.

(iii)$\Rightarrow$(ii). Assume that (iii) is satisfied. Then Lemma 3.5.1 together with Theorem 3.2.3 shows that the upper condition in (3.29) is satisfied. Let G denote the operator on $\ell^2(\mathbb{N})$ given by the Gram matrix $\{\langle f_k, f_j \rangle\}_{j,k=1}^{\infty}$.

Given a sequence $\{c_k\}_{k=1}^\infty \in \ell^2(\mathbb{N})$, the jth element in the image sequence $G\{c_k\}_{k=1}^\infty$ is $\sum_{k=1}^\infty \langle f_k, f_j \rangle c_k$. Thus,

$$\langle G\{c_k\}_{k=1}^\infty, \{c_k\}_{k=1}^\infty \rangle = \sum_{j=1}^\infty \sum_{k=1}^\infty \langle f_k, f_j \rangle c_k \overline{c_j} = \left\| \sum_{k=1}^\infty c_k f_k \right\|^2.$$

Thus, G is positive, and a similar calculation shows that G is self-adjoint. Let V denote the square root of G (cf. Lemma 2.4.5). Then the above calculation gives that

$$\left\| \sum_{k=1}^\infty c_k f_k \right\|^2 = \|V\{c_k\}_{k=1}^\infty\|^2 \geq \frac{1}{\|V^{-1}\|^2} \sum_{k=1}^\infty |c_k|^2.$$

(i) $\Rightarrow$ (iv). Follows from Theorem 3.6.2 and Proposition 3.6.4.

(iv) $\Rightarrow$ (i). Every $f \in \operatorname{span}\{f_k\}_{k=1}^\infty$ has a representation $f = \sum_{k=1}^\infty c_k f_k$ for a finite sequence $\{c_k\}_{k=1}^\infty$, and under the assumptions in (iv) it is unique: if $f = \sum_{k=1}^\infty c_k f_k$, then $c_k = \langle f, g_k \rangle$. Letting $\{e_k\}_{k=1}^\infty$ be an orthonormal basis for $\mathcal{H}$, we can therefore define an operator

$$V : \operatorname{span}\{f_k\}_{k=1}^\infty \to \mathcal{H}, \ V f_k = e_k.$$

Writing $f \in \operatorname{span}\{f_k\}_{k=1}^\infty$ as $f = \sum_{k=1}^\infty \langle f, g_k \rangle f_k$, and letting C denote a Bessel bound for $\{g_k\}_{k=1}^\infty$, we have

$$\|Vf\|^2 = \left\| \sum_{k=1}^\infty \langle f, g_k \rangle e_k \right\|^2 = \sum_{k=1}^\infty |\langle f, g_k \rangle|^2 \leq C\|f\|^2.$$

By completeness of $\{f_k\}_{k=1}^\infty$, V has an extension to a bounded operator on $\mathcal{H}$. Since the assumptions in (iv) are symmetric in f_k and g_k, we can also extend $Tg_k := e_k$ to a bounded operator on $\mathcal{H}$.

Consider finite linear combinations of $\{f_k\}_{k=1}^\infty$ and $\{g_k\}_{k=1}^\infty$, say,

$$f = \sum_{k=1}^\infty c_k f_k, \ \ g = \sum_{k=1}^\infty d_k g_k.$$

Because $\{f_k\}_{k=1}^\infty$ and $\{g_k\}_{k=1}^\infty$ are biorthogonal, we have

$$\langle Vf, Tg \rangle = \left\langle \sum_{k=1}^\infty c_k e_k, \sum_{k=1}^\infty d_k e_k \right\rangle = \sum_{k=1}^\infty c_k \overline{d_k} = \langle f, g \rangle;$$

by continuity and completeness, we therefore have $\langle Vf, Tg \rangle = \langle f, g \rangle$ for all $f, g \in \mathcal{H}$. Thus, for any $h \in \mathcal{H}$,

$$\|h\|^2 = \langle h, h \rangle = \langle Vh, Th \rangle \leq \|Vh\| \ \|T\| \ \|h\|.$$

It follows that V is injective. The operator V is also surjective: given $g \in \mathcal{H}$, write $g = \sum_{k=1}^\infty \langle g, e_k \rangle e_k = V\left(\sum_{k=1}^\infty \langle g, e_k \rangle f_k \right)$. Since $f_k = V^{-1} e_k$, we conclude that $\{f_k\}_{k=1}^\infty$ is a Riesz basis. $\qquad\square$

If (3.29) holds for all finite scalar sequences $\{c_k\}_{k=1}^{\infty}$, then it automatically holds for all $\{c_k\}_{k=1}^{\infty} \in \ell^2(\mathbb{N})$ (Exercise 3.13). If $\{f_k\}_{k=1}^{\infty}$ is a Riesz basis, numbers $A, B > 0$ which satisfy (3.29) are called *lower Riesz bounds* and *upper Riesz bounds*, respectively. They are clearly not unique, and we define the *optimal Riesz bounds* as the largest possible value for A and the smallest possible value for B. The optimal Riesz bounds can be characterized in terms of the operators appearing in the proof of Theorem 3.6.6:

Proposition 3.6.7 *Let* $\{f_k\}_{k=1}^{\infty} = \{Ue_k\}_{k=1}^{\infty}$ *be a Riesz basis for* $\mathcal{H}$*, and let* $G : \ell^2(\mathbb{N}) \to \ell^2(\mathbb{N})$ *be the Gram matrix. Then the optimal Riesz bounds are*

$$A = \frac{1}{||U^{-1}||^2} = \frac{1}{||G^{-1}||} \quad and \quad B = ||U||^2 = ||G||.$$

Proof.   The bounds involving U follow directly from the proof of Theorem 3.6.6. Also, by Lemma 2.4.1,

$$||G|| = ||U^*U|| = ||U||^2 \text{ and } ||G^{-1}|| = ||(U^*U)^{-1}|| = ||U^{-1}||^2.$$

That the optimal upper Riesz bound equals $||G||$ was also proved in Lemma 3.5.1.   □

Note that the same optimal bounds involving U were obtained in the inequalities in Proposition 3.6.4.

If (3.29) holds with $A = B = 1$, the sequence $\{f_k\}_{k=1}^{\infty}$ is orthonormal:

Proposition 3.6.8 *Assume that* $\overline{span}\{f_k\}_{k=1}^{\infty} = \mathcal{H}$ *and that*

$$\left|\left| \sum_{k=1}^{\infty} c_k f_k \right|\right|^2 = \sum_{k=1}^{\infty} |c_k|^2$$

for all finite scalar sequences $\{c_k\}_{k=1}^{\infty}$*. Then* $\{f_k\}_{k=1}^{\infty}$ *is an orthonormal basis for* $\mathcal{H}$*.*

Proof.   The assumptions imply by Theorem 3.6.6 that $\{f_k\}_{k=1}^{\infty}$ is a Riesz basis for $\mathcal{H}$, so by letting $\{e_k\}_{k=1}^{\infty}$ be an orthonormal basis for $\mathcal{H}$, we can write $\{f_k\}_{k=1}^{\infty} = \{Ue_k\}_{k=1}^{\infty}$ for an appropriate bounded invertible operator U. Then, for all $\{c_k\}_{k=1}^{\infty} \in \ell^2(\mathbb{N})$,

$$\sum_{k=1}^{\infty} |c_k|^2 = \left|\left| \sum_{k=1}^{\infty} c_k f_k \right|\right|^2 = \left|\left| U\left(\sum_{k=1}^{\infty} c_k e_k \right) \right|\right|^2.$$

It follows from here that $||U|| = ||U^{-1}|| = 1$; by Proposition 3.6.4 we conclude that $\sum_{k=1}^{\infty} |\langle f, f_k \rangle|^2 = ||f||^2$, $\forall f \in \mathcal{H}$. Since $||f_k|| = 1, \forall k \in \mathbb{N}$, we now obtain the result via Proposition 3.4.8.   □

Let us finally observe that one can characterize Riesz bases in terms of bases satisfying extra conditions:

Lemma 3.6.9 *A sequence* $\{f_k\}_{k=1}^{\infty}$ *is a Riesz basis for* $\mathcal{H}$ *if and only if it is an unconditional basis for* $\mathcal{H}$ *and*

$$0 < \inf_k ||f_k|| \le \sup_k ||f_k|| < \infty.$$

Lemma 3.6.9 was proved by Köthe and Lorch and has been rediscovered/reproved many times (see the discussion in [622]). We refer to, e.g., [325] or [495] for a proof.

3.7   Riesz Sequences

We will often encounter sequences $\{f_k\}_{k=1}^{\infty}$ in a Hilbert space $\mathcal{H}$, which satisfy (3.29) but not necessarily span the entire Hilbert space. This motivates the following definition:

Definition 3.7.1 *A sequence* $\{f_k\}_{k=1}^{\infty}$ *satisfying* (3.29) *for all finite sequences* $\{c_k\}_{k=1}^{\infty}$ *is called a Riesz sequence.*

By Theorem 3.6.6 a Riesz sequence $\{f_k\}_{k=1}^{\infty}$ is a Riesz basis for the Hilbert space $\overline{\text{span}}\{f_k\}_{k=1}^{\infty}$, which might just be a subspace of $\mathcal{H}$. Note that if the condition (3.29) is satisfied for a family $\{f_k\}_{k=1}^{\infty}$, then it is clearly satisfied for any subsequence of $\{f_k\}_{k=1}^{\infty}$. This leads to the following important consequence of Theorem 3.6.6.

Corollary 3.7.2 *Every subfamily of a Riesz basis is a Riesz sequence.*

Theorem 3.6.6 also characterizes Riesz sequences, simply by applying the conditions on the Hilbert space $\overline{\text{span}}\{f_k\}_{k=1}^{\infty}$ rather than on $\mathcal{H}$. As a minor modification, we also obtain the following result, which will be useful later:

Proposition 3.7.3 *Let* $\{f_k\}_{k=1}^{\infty}$ *be a Bessel sequence in* $\mathcal{H}$. *Then the following are equivalent:*

(i) $\{f_k\}_{k=1}^{\infty}$ *is a Riesz sequence with lower bound* A;

(ii) $\{f_k\}_{k=1}^{\infty}$ *has a biorthogonal system* $\{g_k\}_{k=1}^{\infty}$, *which is a Bessel sequence with bound* A^{-1}.

Proof. Assume first that (i) holds. Taking any orthonormal basis $\{e_k\}_{k=1}^{\infty}$ for $\overline{\text{span}}\{f_k\}_{k=1}^{\infty}$, there is a bounded bijective operator

$$U : \overline{\text{span}}\{f_k\}_{k=1}^{\infty} \to \overline{\text{span}}\{f_k\}_{k=1}^{\infty}$$

such that $f_k = Ue_k$. By Proposition 3.6.7 we know that $A \le ||U^{-1}||^{-2}$. Now, let $\{g_k\}_{k=1}^{\infty}$ denote the dual Riesz basis of $\{f_k\}_{k=1}^{\infty}$ within the space $\overline{\text{span}}\{f_k\}_{k=1}^{\infty}$; then $\{f_k\}_{k=1}^{\infty}$ and $\{g_k\}_{k=1}^{\infty}$ are biorthogonal by Theorem 3.6.2. Furthermore, the proof of Theorem 3.6.2 shows that

$g_k = (U^{-1})^* e_k$, which again by Proposition 3.6.7 implies that the optimal Bessel bound for $\{g_k\}_{k=1}^\infty$ is

$$\|(U^{-1})^*\|^2 = \|U^{-1}\|^2 \le A^{-1},$$

as desired.

Now assume that (ii) holds. Applying the assumption that $\{g_k\}_{k=1}^\infty$ is a Bessel sequence with bound A^{-1} on any finite sum $f = \sum_{j=1}^\infty c_j f_j$ yields that

$$\sum_{k=1}^\infty \left| \langle \sum_{j=1}^\infty c_j f_j, g_k \rangle \right|^2 \le A^{-1} \left\| \sum_{j=1}^\infty c_j f_j \right\|^2,$$

or

$$\left\| \sum_{j=1}^\infty c_j f_j \right\|^2 \ge A \sum_{k=1}^\infty |c_k|^2;$$

thus, $\{f_k\}_{k=1}^\infty$ is a Riesz sequence with lower bound A.   □

By definition, $\{f_k\}_{k=1}^\infty$ is a Riesz sequence if the inequalities (3.29) are satisfied for all finite scalar sequences $\{c_k\}_{k=1}^\infty$. We will now simplify this (Exercise 3.10):

Corollary 3.7.4 *There exists a countable collection of finite normalized sequences $\{c_k^\ell\}_{k=1}^\infty$, $\ell \in \mathbb{N}$ with the following property: a given sequence $\{f_k\}_{k=1}^\infty$ in a Hilbert space $\mathcal{H}$ is a Riesz sequence with bounds $A, B > 0$ if and only if*

$$A \le \left\| \sum_{k=1}^\infty c_k^\ell f_k \right\|^2 \le B, \ \forall \ell \in \mathbb{N}. \tag{3.30}$$

Corollary 3.7.4 reduces the verification that $\{f_k\}_{k=1}^\infty$ is a Riesz sequence to a calculation of a countable collection of numbers. This is a significant reduction compared with verification using the definition and can in principle be implemented via a computer program [which of course will not finish in finite time]. We will return to this comment in connection with our discussion of Riesz sequences versus frames in Section 8.3.

We have in Sections 3.4–3.7 concentrated on theoretical properties of orthonormal bases and Riesz bases in general Hilbert spaces. In the following sections we will connect such bases with the structured function systems that will dominate the book from Chapter 9 and onward. For now we just mention one important class of Riesz sequences, based on integer-translates of the B-splines B_n defined in Section A.8.

Lemma 3.7.5 *Let $n \in \mathbb{N}$ and consider the B-spline B_n. Then $\{T_k B_n\}_{k \in \mathbb{Z}}$ is a Riesz sequence in $L^2(\mathbb{R})$.*

The result is clear for $n = 1$ because $\{B_1(\cdot - k)\}_{k \in \mathbb{Z}}$ is an orthonormal system. For $n > 1$ it is an easy consequence of Theorem 9.2.5, so we postpone the proof till page 208.

3.8   Fourier Series and Gabor Bases

We will now take the first steps toward the analysis of bases in function spaces like $L^2(I)$ and $L^2(\mathbb{R})$. Here we will use other index sets than the natural numbers; as we have seen in Corollary 3.2.5, Bessel sequences can be ordered any way we want without affecting the convergence of the relevant series expansions, so we can apply all results presented so far without problems.

The starting point is *Fourier series*. We expect the reader to be familiar with the basic theory, so we only give a short repetition.

Fourier series can be associated to functions in any space $L^2(I)$, where I is a bounded interval in $\mathbb{R}$. For our purpose it will be convenient to consider functions in $L^2(0, 1/b)$, where $b > 0$; recall that $L^2(0, 1/b)$ is a Hilbert space with respect to the inner product

$$\langle f, g \rangle = \int_0^{1/b} f(x)\overline{g(x)}\, dx, \quad f, g \in L^2(0, 1/b).$$

We will consider functions $f \in L^2(0, 1/b)$ as periodic functions on $\mathbb{R}$, with period $1/b$. Since the functions

$$e_k(x) := b^{1/2} E_{kb}(x) = b^{1/2} e^{2\pi i k b x}, \quad k \in \mathbb{Z} \tag{3.31}$$

constitute an orthonormal basis for $L^2(0, 1/b)$, every $f \in L^2(0, 1/b)$ has an expansion

$$f = \sum_{k \in \mathbb{Z}} \langle f, e_k \rangle e_k. \tag{3.32}$$

We will usually expand the functions f directly in terms of the functions $\{e^{2\pi i k b x}\}_{k \in \mathbb{Z}}$ rather than $\{e_k\}_{k \in \mathbb{Z}}$. Thus, we arrive at

$$f(\cdot) = \sum_{k \in \mathbb{Z}} c_k e^{2\pi i k b(\cdot)}, \tag{3.33}$$

where

$$c_k = b^{1/2} \langle f, e_k \rangle = b \int_0^{1/b} f(x) e^{-2\pi i k b x}\, dx. \tag{3.34}$$

The expansion (3.33) is called the *Fourier series* of f, and the numbers $\{c_k\}_{k \in \mathbb{Z}}$ in (3.34) are the *Fourier coefficients*.

The exact meaning of the Fourier expansion (3.33) is that

$$\left\| f - \sum_{k=-n}^{n} c_k e^{2\pi i k b(\cdot)} \right\|_{L^2(0,1/b)} = \left(\int_0^{1/b} \left| f(x) - \sum_{k=-n}^{n} c_k e^{2\pi i k b x} \right|^2 dx \right)^{1/2}$$
$$\to 0 \text{ as } n \to \infty.$$

Convergence in $L^2(0, 1/b)$-sense is different from pointwise convergence, so we cannot claim that (3.33) holds for a given $x \in [0, 1/b]$ without extra assumptions. For an arbitrary function in $L^2(0, 1/b)$, the Fourier series converges pointwise almost everywhere; conditions implying convergence for all x are stated in the following well-known result.

Theorem 3.8.1 *Assume that $f \in L^2(0, 1/b)$ is continuous, periodic with period $1/b$, and that the Fourier coefficients $\{c_k\}_{k\in\mathbb{Z}} \in \ell^1(\mathbb{Z})$. Then*

$$f(x) = \sum_{k\in\mathbb{Z}} c_k e^{2\pi i k b x},$$

pointwise for all $x \in \mathbb{R}$.

Parseval's equation (see Theorem 3.4.2) gives us an important relationship between a given function $f \in L^2(0, 1/b)$ and its Fourier coefficients $\{c_k\}_{k\in\mathbb{Z}}$:

$$b \int_0^{1/b} |f(x)|^2 dx = \sum_{k\in\mathbb{Z}} |c_k|^2. \tag{3.35}$$

We now state a lemma, which is an immediate consequence of the functions $\{e_k\}_{k=1}^{\infty}$ in (3.31) being an orthonormal basis for $L^2(0, 1/b)$.

Lemma 3.8.2 *Let $f, g \in L^2(0, 1/b)$ for some $b > 0$, and consider two series expansions*

$$f = \sum_{k\in\mathbb{Z}} a_k e_k, \quad g = \sum_{k\in\mathbb{Z}} b_k e_k,$$

with e_k given by (3.31) and $\{a_k\}_{k\in\mathbb{Z}}, \{b_k\}_{k\in\mathbb{Z}} \in \ell^2(\mathbb{Z})$. Then

$$\langle f, g \rangle = \sum_{k\in\mathbb{Z}} a_k \overline{b_k}.$$

A $\frac{1}{b}$-periodic function $f : \mathbb{R} \to \mathbb{C}$ can equally well be considered as a function in $L^2(0, 1/b)$ as in $L^2(-\frac{1}{2b}, \frac{1}{2b})$; the latter choice will sometimes be more convenient, e.g., in our discussion of sampling problems in Section 3.10. Whenever we consider our functions as members in $L^2(-\frac{1}{2b}, \frac{1}{2b})$, it is often convenient to exchange the integrals over $]0, 1/b[$ with integrals over $]-\frac{1}{2b}, \frac{1}{2b}[$; for example, the expression for the Fourier coefficients in (3.34) takes the form

$$c_k = b \int_{-\frac{1}{2b}}^{\frac{1}{2b}} f(x) e^{-2\pi i k b x} dx. \tag{3.36}$$

In the following example, we show how to construct an orthonormal basis for $L^2(\mathbb{R})$ based on the orthonormal basis $\{e^{2\pi i k x}\}_{k \in \mathbb{Z}}$ for $L^2(0,1)$. The example gives the first introduction to Gabor systems in $L^2(\mathbb{R})$.

Example 3.8.3 Let $\chi_{[0,1]}$ denote the characteristic function for the interval $[0,1]$. Then $\{e^{2\pi i k x} \chi_{[0,1]}(x)\}_{k \in \mathbb{Z}}$ is an orthonormal basis for $L^2(0,1)$; by translation, we see that for each $n \in \mathbb{Z}$ the space $L^2(n, n+1)$ has the orthonormal basis

$$\{e^{2\pi i k (x-n)} \chi_{[0,1]}(x-n)\}_{k \in \mathbb{Z}} = \{e^{2\pi i k x} \chi_{[0,1]}(x-n)\}_{k \in \mathbb{Z}}.$$

Putting these bases together, we obtain that $L^2(\mathbb{R})$ has the orthonormal basis

$$\{e^{2\pi i k x} \chi_{[0,1]}(x-n)\}_{k,n \in \mathbb{Z}}.$$

Note that all elements in the basis consist of translated versions of $\chi_{[0,1]}$ which have been *modulated*, i.e., multiplied with a complex exponential function. Using the operators introduced in Section 2.9, we can write the basis as $\{E_k T_n g\}_{k,n \in \mathbb{Z}}$, where $g = \chi_{[0,1]}$. Bases of the form $\{E_k T_n g\}_{k,n \in \mathbb{Z}}$ are called *Gabor bases*. Calculations with Gabor bases are convenient because of their *coherent structure:* all the elements in the basis appear by the action of a family of operators, namely, $E_k T_n, k, n \in \mathbb{Z}$, on the single function g. We will consider some of the limitations of such bases in Chapter 4 and extensions to frames in Chapters 11–13. □

In concrete applications, a Fourier expansion will always need to be truncated to a finite sum. A function f that is a *finite* linear combination of the type

$$f(x) = \sum_{k=N_1}^{N_2} c_k e^{2\pi i k x} \quad \text{for some } c_k \in \mathbb{C}, \ N_1, N_2 \in \mathbb{Z}, N_2 \geq N_1 \tag{3.37}$$

is called a *trigonometric polynomial*. A trigonometric polynomial f can also be written as a linear combination of functions $\sin(2\pi k x), \cos(2\pi k x)$, in general with complex coefficients. It will be useful later to note that if the function f in (3.37) is real-valued and the coefficients c_k are real, then f is a linear combination of functions $\cos(2\pi k x)$ alone:

Lemma 3.8.4 *Assume that the trigonometric polynomial f in (3.37) is real-valued and that the coefficients $c_k \in \mathbb{R}$. Then*

$$f(x) = \sum_{k=N_1}^{N_2} c_k \cos(2\pi k x). \tag{3.38}$$

We leave the short proof to the reader. Note that we need the assumption that $c_k \in \mathbb{R}$: for example, the function

$$f(x) = \frac{1}{2i}e^{2\pi i x} - \frac{1}{2i}e^{-2\pi i x} = \sin(2\pi x),$$

is real-valued but does not have the form (3.38).

For later use, we also mention that a positive-valued trigonometric polynomial with real coefficients has a square root (in the sense of (3.41) below), which again is a trigonometric polynomial. For convenience, we formulate the result for a slight rewriting of the series (3.38):

Lemma 3.8.5 *Let f be a positive-valued trigonometric polynomial of the form*

$$f(x) = \sum_{k=0}^{N} c_k \cos(2\pi k x), \quad c_k \in \mathbb{R}. \tag{3.39}$$

Then there exists a trigonometric polynomial

$$g(x) = \sum_{k=0}^{N} d_k e^{2\pi i k x} \text{ with } d_k \in \mathbb{R}, \tag{3.40}$$

such that

$$|g(x)|^2 = f(x), \quad \forall x \in \mathbb{R}. \tag{3.41}$$

The procedure of finding the trigonometric polynomial g in (3.40) is called *spectral factorization*; a constructive proof can be found in [242]. Note that by definition, the function g in (3.40) is complex-valued, unless f is constant; actually, despite the fact that f is assumed to be positive, there might not exist a *positive* trigonometric polynomial g satisfying (3.41). See Exercise 3.16.

3.9 Wavelet Bases

Wavelet bases constitute another important class of bases. Given a function $\psi \in L^2(\mathbb{R})$ and $j, k \in \mathbb{Z}$, let

$$\psi_{j,k}(x) := 2^{j/2}\psi(2^j x - k), \quad x \in \mathbb{R}. \tag{3.42}$$

In terms of the translation operators T_k and the dilation operator D introduced in Section 2.9,

$$\psi_{j,k} = D^j T_k \psi, \quad j, k \in \mathbb{Z}.$$

If $\{\psi_{j,k}\}_{j,k \in \mathbb{Z}}$ is an orthonormal basis for $L^2(\mathbb{R})$, the function ψ is called a *wavelet*. The first example of such a function appeared long time before the systematic study of wavelet bases began:

Example 3.9.1 The *Haar function* is defined by

$$\psi(x) = \begin{cases} 1 & \text{if } 0 \le x < \frac{1}{2}, \\ -1 & \text{if } \frac{1}{2} \le x < 1, \\ 0 & \text{otherwise.} \end{cases} \tag{3.43}$$

Already in 1910 it was proved by Haar [361] that the functions $\{\psi_{j,k}\}_{j,k\in\mathbb{Z}}$ constitute an orthonormal basis for $L^2(\mathbb{R})$ for this choice of ψ. For the orthonormality, one can argue as follows. If we first consider $\psi_{j,k}$ and $\psi_{j,k'}$, i.e., elements with the same dilation parameter, then

$$\langle \psi_{j,k}, \psi_{j,k'} \rangle = \langle D^j T_k \psi, D^j T_{k'} \psi \rangle = \langle T_k \psi, T_{k'} \psi \rangle = \delta_{k,k'}.$$

Now assume that $j' \neq j$, say, $j' > j$. Using the commutator relations (2.28),

$$\langle \psi_{j,k}, \psi_{j',k'} \rangle \;\; = \;\; \langle D^j T_k \psi, D^{j'} T_{k'} \psi \rangle = \langle D^{j-j'} T_{-k'2^{j-j'}+k} \psi, \psi \rangle.$$

The function $D^{j-j'} T_{-k'2^{j-j'}+k} \psi$ has support in the interval

$$I : \;\; = \;\; [2^{j'-j}(-k'2^{j-j'}+k), 2^{j'-j}(-k'2^{j-j'}+k+1)[$$
$$= \;\; [-k'+2^{j'-j}k, -k'+2^{j'-j}(k+1)[.$$

The length of I is $2^{j'-j}$, which can take the values $2, 4, 8, \ldots\ldots$ Now, the support of ψ has length 1 and is contained in an interval on which $D^{j-j'} T_{-k'2^{j-j'}+k} \psi$ is constant (make a picture!); it follows that

$$\langle \psi_{j',k'}, \psi_{j,k} \rangle = \int_{-\infty}^{\infty} \left(D^{j-j'} T_{-k'2^{j-j'}+k} \psi \right)(x) \psi(x) dx = 0.$$

For the proof of the basis property, we refer to [242, 400], or [637].   □

Strömberg [592] constructed in 1982 wavelet orthonormal bases $\{\psi_{j,k}\}_{j,k\in\mathbb{Z}}$ for which ψ has exponential decay and $\psi \in C^\ell(\mathbb{R})$; here $\ell \in \mathbb{N}$ is arbitrary. Meyer [519, 482] found in 1985 wavelet bases for which $\psi \in C^\infty(\mathbb{R})$ and $\widehat{\psi} \in C^\ell(\mathbb{R})$, $\ell \in \mathbb{N}$. In 1986, Mallat and Meyer introduced *multiresolution analysis* as a general tool to construct wavelet orthonormal bases [510]:

Definition 3.9.2 *A multiresolution analysis for $L^2(\mathbb{R})$ consists of a sequence of closed subspaces $\{V_j\}_{j\in\mathbb{Z}}$ of $L^2(\mathbb{R})$ and a function $\phi \in V_0$, such that*

(i) $\cdots V_{-1} \subset V_0 \subset V_1 \cdots$, *i.e., the spaces V_j are nested.*

(ii) $\overline{\cup_{j\in\mathbb{Z}} V_j} = L^2(\mathbb{R})$ *and* $\cap_{j\in\mathbb{Z}} V_j = \{0\}$.

(iii) $f \in V_j \Leftrightarrow [x \to f(2x)] \in V_{j+1}$.

(iv) $f \in V_0 \Rightarrow T_k f \in V_0, \; \forall k \in \mathbb{Z}$.

(v) $\{T_k \phi\}_{k\in\mathbb{Z}}$ *is an orthonormal basis for V_0.*

Definition 3.9.2 leads to a general method for construction of wavelets and can be seen as the beginning of modern wavelet analysis. The topic is already well covered with many books (see, e.g., [308, 633] and [165] for elementary treatments, or [242, 520, 637] for more advanced presentations). We will only explain some of the key steps in the construction of a wavelet based on a multiresolution analysis; the reader will observe that these steps serve as motivation and guideline for the more advanced frame constructions that will appear in Chapters 17–18.

Assume that the conditions in Definition 3.9.2 are satisfied. For $j \in \mathbb{Z}$, we let W_j denote the orthogonal complement of V_j in V_{j+1}. By letting Q_j denote the orthogonal projection onto W_j, it follows from Definition 3.9.2 (i) and (ii) that each $f \in L^2(\mathbb{R})$ has a representation $f = \sum_{j \in \mathbb{Z}} Q_j f$, where $Q_j f \perp Q_{j'} f$ for $j \neq j'$; that is,

$$L^2(\mathbb{R}) = \bigoplus_{j \in \mathbb{Z}} W_j. \tag{3.44}$$

The spaces W_j satisfy the same dilation relationship as V_j, i.e.,

$$\psi \in W_0 \Leftrightarrow [x \to \psi(2^j x)] \in W_j. \tag{3.45}$$

In order to obtain an orthonormal basis $\{\psi_{j,k}\}_{j,k \in \mathbb{Z}}$ for $L^2(\mathbb{R})$, it is now enough to find $\psi \in W_0$ such that $\{\psi(\cdot - k)\}_{k \in \mathbb{Z}}$ is an orthonormal basis for W_0; via the dilation property (3.45) and (3.44), this implies that $\{\psi_{j,k}\}_{j,k \in \mathbb{Z}}$ is an orthonormal basis for $L^2(\mathbb{R})$. One way of choosing ψ is as follows. First, the condition $\phi \in V_0 \subset V_1$ implies by Definition 3.9.2 (iii) that $\frac{1}{\sqrt{2}} D^{-1} \phi \in V_0$. Since $\{T_k \phi\}_{k \in \mathbb{Z}}$ is an orthonormal basis for V_0, there exist coefficients $\{c_k\}_{k \in \mathbb{Z}} \in \ell^2(\mathbb{Z})$ such that

$$\frac{1}{\sqrt{2}} D^{-1} \phi = \sum_{k \in \mathbb{Z}} c_k T_k \phi.$$

Using the Fourier transform and the commutator relations in (2.29), it follows that $\frac{1}{\sqrt{2}} D\widehat{\phi} = \sum_{k \in \mathbb{Z}} c_k E_{-k} \widehat{\phi}$; defining the 1-periodic function $H_0 := \sum_{k \in \mathbb{Z}} c_k E_{-k}$, this can be written as

$$\widehat{\phi}(2\gamma) = H_0(\gamma)\widehat{\phi}(\gamma), \quad a.e.\, \gamma \in \mathbb{R}. \tag{3.46}$$

The equation (3.46) is called a *scaling equation* or *refinement equation*. Now, it turns out that with a certain choice of a 1-periodic function H_1, the function ψ defined via

$$\widehat{\psi}(2\gamma) = H_1(\gamma)\widehat{\phi}(\gamma) \tag{3.47}$$

generates a wavelet orthonormal basis $\{D^j T_k \psi\}_{j,k \in \mathbb{Z}}$. One choice of H_1 is to take

$$H_1(\gamma) = \overline{H_0(\gamma + \tfrac{1}{2})} e^{-2\pi i \gamma}. \tag{3.48}$$

Note that (3.47) leads to an explicit expression of the function ψ in terms of the given function ϕ:

Lemma 3.9.3 *Assume that (3.47) holds for a 1-periodic and bounded function H_1 with Fourier expansion $H_1 = \sum_{k\in\mathbb{Z}} c_k E_k$. Then*

$$\psi(x) = \sqrt{2}\sum_{k\in\mathbb{Z}} c_k DT_{-k}\phi(x) = 2\sum_{k\in\mathbb{Z}} c_k\phi(2x+k), \ \ a.e. \ x \in \mathbb{R}. \quad (3.49)$$

In particular, if H_1 is a trigonometric polynomial, $H_1(x) = \sum_{k=N_1}^{N_2} c_k e^{2\pi ikx}$, then

$$\psi(x) = \sqrt{2}\sum_{k=N_1}^{N_2} c_k DT_{-k}\phi(x) = 2\sum_{k=N_1}^{N_2} c_k\phi(2x+k), \ \ \forall x \in \mathbb{R}. \quad (3.50)$$

Proof.   We can rewrite (3.47) as $\widehat{\psi}(\gamma) = H_1(\gamma/2)\widehat{\phi}(\gamma/2)$; formulated in terms of the Fourier series for H_1 and the dilation operator D, this means that

$$\mathcal{F}\psi = \sqrt{2}\sum_{k\in\mathbb{Z}} c_k E_{k/2} D^{-1}\mathcal{F}\phi = \sqrt{2}\sum_{k\in\mathbb{Z}} c_k E_{k/2}\mathcal{F}D\phi.$$

Now, using the commutator relations in Section 2.9,

$$\mathcal{F}\psi = \sqrt{2}\mathcal{F}\sum_{k\in\mathbb{Z}} c_k T_{-k/2} D\phi = \sqrt{2}\mathcal{F}\sum_{k\in\mathbb{Z}} c_k DT_{-k}\phi.$$

Applying the inverse Fourier transform now yields the result.   □

The Haar basis can be constructed via the multiresolution analysis defined by $\phi = \chi_{[0,1[}$, and

$$V_j \ = \ \{f \in L^2(\mathbb{R}) \mid f \text{ is constant on } [2^{-j}k, 2^{-j}(k+1)[, \ \forall k \in \mathbb{Z}\}.$$

In terms of the function ϕ, the Haar function in (3.43) is

$$\psi = \frac{1}{\sqrt{2}}\phi_{1,0} - \frac{1}{\sqrt{2}}\phi_{1,1}. \quad (3.51)$$

The Haar function is a special case of a *spline wavelet*. In fact, one can consider higher-order splines $\widetilde{B_n}$ (see (A.18) for the definition) and define associated multiresolution analyses, which leads to wavelets of the type

$$\psi(x) = \sum_{k\in\mathbb{Z}} c_k \widetilde{B_n}(2x-k). \quad (3.52)$$

We ask the reader to verify the instrumental scaling equation (3.46) directly; see Exercise 3.17. The resulting wavelets are called *Battle–Lemarié wavelets*. The coefficients $\{c_k\}_{k\in\mathbb{Z}}$ are calculated in, e.g., [242]; except for the case $n = 1$, all coefficients c_k are nonzero, which implies that ψ has support equal to $\mathbb{R}$. However, the wavelets have exponential decay.

Wavelets are characterized in, e.g., [400]; we will provide a proof of the following result on page 412.

Lemma 3.9.4 *A function $\psi \in L^2(\mathbb{R})$ is a wavelet if and only if $||\psi|| = 1$ and the equations*

$$\sum_{j\in\mathbb{Z}} |\widehat{\psi}(2^j\gamma)|^2 = 1, \tag{3.53}$$

$$\sum_{j=0}^{\infty} \widehat{\psi}(2^j\gamma)\overline{\widehat{\psi}(2^j(\gamma+q))} = 0 \ \ \text{for all odd integers } q \tag{3.54}$$

hold for almost all $\gamma \in \mathbb{R}$.

Most of the important wavelet bases for $L^2(\mathbb{R})$ are constructed via the approach sketched above, e.g., the bases by Daubechies [242]. However, not all wavelets can be constructed via multiresolution analysis. Among all wavelets, the wavelets generated from a multiresolution analysis are characterized by the equation

$$\sum_{j=1}^{\infty}\sum_{k\in\mathbb{Z}} |\widehat{\psi}(2^j(\gamma+k)|^2 = 1,$$

a result which is also proved in [400].

Let us now return to the multiresolution analysis setup. As we have seen, the conditions in Definition 3.9.2 determine the spaces V_j uniquely; in fact, $V_0 = \overline{\text{span}}\{T_k\phi\}_{k\in\mathbb{Z}}$, and via the condition (iii),

$$V_j = \overline{\text{span}}\{D^j T_k\phi\}_{k\in\mathbb{Z}}. \tag{3.55}$$

On the other hand, assuming that ϕ is a given function such that $\{T_k\phi\}_{k\in\mathbb{Z}}$ forms an orthonormal basis for its closed linear span, we only have to verify the conditions in Definition 3.9.2 (i) and (ii) in order to show that ϕ and the spaces V_j in (3.55) form a multiresolution analysis. It turns out that these conditions are satisfied under very weak assumptions. Let us state a general result obtained by de Boor, DeVore, and Ron [71]:

Lemma 3.9.5 *Let $\phi \in L^2(\mathbb{R})$ and define the spaces V_j by (3.55). Then the following holds:*

(i) $\cap_{j\in\mathbb{Z}} V_j = \{0\}$.

(ii) Assume that the spaces V_j in (3.55) are nested. If

$$|\widehat{\phi}| > 0 \tag{3.56}$$

on a neighborhood of 0, then $\cup_{j\in\mathbb{Z}} V_j$ is dense in $L^2(\mathbb{R})$.

Thus, if (3.56) is satisfied, all that we need is a condition ensuring that the spaces V_j are nested. But under a weak condition, this also follows from the assumption that $\{T_k\phi\}_{k\in\mathbb{Z}}$ forms an orthonormal basis for its closed linear span. In fact, it is enough to assume that $\{T_k\phi\}_{k\in\mathbb{Z}}$ forms a Bessel sequence; this extension will play a role in Chapter 18.

Lemma 3.9.6 *Assume that $\phi \in L^2(\mathbb{R})$ and that $\{T_k\phi\}_{k\in\mathbb{Z}}$ is a Bessel sequence. Define the spaces V_j by (3.55). Then the following holds:*

(i) *If $\psi \in L^2(\mathbb{R})$ and there exists a bounded 1-periodic function H_1 such that $\widehat{\psi}(2\gamma) = H_1(\gamma)\widehat{\phi}(\gamma)$, then $\psi \in V_1$.*

(ii) *If there exists a bounded 1-periodic function H_0 such that*

$$\widehat{\phi}(2\gamma) = H_0(\gamma)\widehat{\phi}(\gamma), \tag{3.57}$$

then $V_j \subseteq V_{j+1}$ for all $j \in \mathbb{Z}$.

Proof. If the conditions in (i) are satisfied, the expression for the function ψ in Lemma 3.9.3 shows that $\psi \in V_1$. This proves (i). For the proof of (ii), we note that, via (i), $\phi \in V_1$; since V_1 is closed and invariant under integer translations, it follows that $V_0 \subseteq V_1$. A scaling now implies that $V_j \subseteq V_{j+1}$ for all $j \in \mathbb{Z}$. □

Via Lemma 3.9.5 and Lemma 3.9.6, we obtain the following:

Theorem 3.9.7 *Let $\phi \in L^2(\mathbb{R})$, and assume that $|\widehat{\phi}| > 0$ on a neighborhood of 0. Assume further that (3.57) is satisfied for a bounded 1-periodic function H_0. Define the spaces V_j by (3.55). Then the following holds:*

(i) *If $\{T_k\phi\}_{k\in\mathbb{Z}}$ is an orthonormal system, then ϕ and the spaces V_j form a multiresolution analysis.*

(i) *If $\{T_k\phi\}_{k\in\mathbb{Z}}$ is a Bessel sequence, then the spaces V_j satisfy the conditions (i)–(iv) in Definition 3.9.2.*

As already mentioned the conditions to an orthonormal basis are very strong, and there are indeed a number of limitations on the properties one can obtain for a wavelet. Some of these limitations will be discussed in Chapter 4. In Chapters 15–19, we will discuss frames having the wavelet structure; in particular, Chapters 17–18 will provide constructions based on a multiresolution setup, and we will see that frame theory allows to eliminate some of the restrictions.

3.10   Sampling and Analog–Digital Conversion

A short and not yet precise formulation of the *sampling problem* is: How can we recover a function $f : \mathbb{R} \to \mathbb{C}$ if we only know a countable set of function values $\{f(\lambda_k)\}_{k\in I}$? Formulated this way the problem is ill-posed:

there are infinitely many functions that take the same prescribed values on a given countable set, so we need to impose some condition on the function f for the problem to make sense. Traditionally, this is done by requiring f to belong to a certain function space. A classical example is to consider a space of *band-limited* functions, i.e., functions for which the Fourier transform has compact support. Let us consider the *Paley–Wiener space PW*, defined by

$$PW := \left\{ f \in L^2(\mathbb{R}) \mid \operatorname{supp} \widehat{f} \subseteq [-\tfrac{1}{2}, \tfrac{1}{2}] \right\}. \qquad (3.58)$$

As always when dealing with L^2-functions, the Paley–Wiener space really consists of equivalence classes of functions; however, due to the fact that the Fourier transform of a function $f \in PW$ has compact support, each of these equivalence classes contains a continuous function. We will always select the continuous representation for the equivalence classes in the Paley–Wiener space.

We will now show that the Paley–Wiener space has an orthonormal basis consisting of translates of a single function. Define the *sinc function* by

$$\operatorname{sinc}(x) = \begin{cases} \frac{\sin(\pi x)}{\pi x} & \text{if } x \neq 0, \\ 1 & \text{if } x = 0. \end{cases}$$

Shannon's sampling theorem states that any continuous function in the Paley–Wiener space can be fully recovered from its samples at the integers.

Theorem 3.10.1 *The functions $\{\operatorname{sinc}(\cdot - k)\}_{k \in \mathbb{Z}}$ form an orthonormal basis for the Paley–Wiener space PW. If $f \in PW$ is continuous, then*

$$f(x) = \sum_{k \in \mathbb{Z}} f(k)\operatorname{sinc}(x - k),$$

with convergence of the symmetric partial sums in $L^2(\mathbb{R})$ and pointwise for all $x \in \mathbb{R}$.

Proof.   The proof is based on classical Fourier analysis as described in Section 3.8. Because of our definition of the Paley–Wiener space, it will be convenient to work with Fourier series in the space $L^2(-1/2, 1/2)$ rather than $L^2(0, 1)$.

We first show that the functions $\{\operatorname{sinc}(\cdot - k)\}_{k \in \mathbb{Z}}$ form an orthonormal sequence in $L^2(\mathbb{R})$. We know that the functions $\{e^{2\pi i k(\cdot)}\chi_{]-1/2,1/2[}(\cdot)\}_{k \in \mathbb{Z}}$ form an orthonormal sequence in $L^2(\mathbb{R})$; taking the Fourier transform of these functions, we arrive at

$$\mathcal{F}\left(e^{2\pi i k(\cdot)}\chi_{]-1/2,1/2[}(\cdot)\right)(\gamma) = \int_{-1/2}^{1/2} e^{2\pi i k x} e^{-2\pi i x \gamma} dx = \operatorname{sinc}(\gamma - k).$$

Because the Fourier transform is unitary, this implies that the functions $\{\operatorname{sinc}(\cdot - k)\}_{k \in \mathbb{Z}}$ are orthonormal as well.

Now let $f \in PW$ be the continuous representative for a given equivalence class. On the interval $]-1/2, 1/2[$ we can expand $\widehat{f}$ in a Fourier series,

$$\widehat{f}(\cdot) = \sum_{k \subset \mathbb{Z}} c_k e^{2\pi i k(\cdot)},$$

where

$$c_k = \int_{-1/2}^{1/2} \widehat{f}(\gamma) e^{-2\pi i k \gamma} d\gamma. \tag{3.59}$$

Recall that the partial sums of the Fourier series converge in the norm of $L^2(-1/2, 1/2)$, i.e.,

$$\int_{-1/2}^{1/2} \left| \widehat{f}(\gamma) - \sum_{k=-N}^{N} c_k e^{2\pi i k \gamma} \right|^2 d\gamma \to 0 \text{ as } N \to \infty.$$

Note that because we are dealing with a finite interval, convergence in $L^2(-1/2, 1/2)$ implies convergence in $L^1(-1/2, 1/2)$, so

$$\int_{-1/2}^{1/2} \left| \widehat{f}(\gamma) - \sum_{k=-N}^{N} c_k e^{2\pi i k \gamma} \right| d\gamma \to 0 \text{ as } N \to \infty. \tag{3.60}$$

Because $\operatorname{supp} \widehat{f} \subseteq [-\frac{1}{2}, \frac{1}{2}]$, the expression for c_k in (3.59) implies by Theorem 2.8.1 that $c_k = f(-k)$. Using Theorem 2.8.1 once more, we arrive at the following formula, valid pointwise for all $x \in \mathbb{R}$:

$$f(x) = \int_{-\infty}^{\infty} \widehat{f}(\gamma) e^{2\pi i x \gamma} d\gamma = \int_{-1/2}^{1/2} \left(\sum_{k \in \mathbb{Z}} f(-k) e^{2\pi i k \gamma} \right) e^{2\pi i x \gamma} d\gamma.$$

Because of (3.60), we can interchange the order of summation and integration; thus, for all $x \in \mathbb{R}$,

$$f(x) = \sum_{k \in \mathbb{Z}} f(-k) \int_{-1/2}^{1/2} e^{2\pi i (x+k) \gamma} d\gamma = \sum_{k \in \mathbb{Z}} f(-k) \operatorname{sinc}(x+k)$$

$$= \sum_{k \in \mathbb{Z}} f(k) \operatorname{sinc}(x-k).$$

The series converges in $L^2(\mathbb{R})$ as well: in fact, since $\{\operatorname{sinc}(\cdot - k)\}_{k \in \mathbb{Z}}$ is an orthonormal system,

$$\left\| f - \sum_{k=-N}^{N} f(k) \operatorname{sinc}(\cdot - k) \right\| = \left\| \sum_{|k|>N} f(k) \operatorname{sinc}(\cdot - k) \right\| = \sqrt{\sum_{|k|>N} |f(k)|^2},$$

which converges to 0 as $N \to \infty$ because $\{f(k)\}_{k \in \mathbb{Z}} \in \ell^2(\mathbb{Z})$ (we just saw that they are Fourier coefficients). Finally, that $\{\operatorname{sinc}(\cdot - k)\}_{k \in \mathbb{Z}}$ forms an orthonormal basis for PW follows from the fact that all equivalence classes in PW contain a continuous function. $\square$

Note that, via an appropriate scaling, the result in Theorem 2.8.1 can be extended to functions whose Fourier transform has support in an arbitrary fixed interval (Exercise 3.18).

Shannon's sampling theorem dates back to 1950, but it was actually discovered even earlier, independently by Whittaker and Kotelnikov. It marks the beginning of sampling theory, which is still a very active field of research. We refer to the books [268] by Eldar and [627] by Vetterli et al. for excellent introductions to sampling theory, which also highlight the connection to mathematics. We will return to sampling in Section 9.7 and Chapter 14.

The principle in Shannon's sampling theorem is the basis for all modern communication. Most signals appearing in practice depend on a continuous variable (very often, the time). Processing of such a signal is facilitated greatly if it can be stored and handled in terms of a sequence of samples. As a concrete case, consider a piece of music, modeled as the function f that measures the current running through the cable to the speaker when the music is played. In principle, all frequencies might appear in the signal, but the human ear can only hear frequencies belonging to a certain range (at most up to 20.000 Hz). Thus, we can remove the high frequencies and consider the resulting signal as band-limited. Via an appropriate scaling (see Exercise 3.18), Theorem 3.10.1 shows that this signal f can be recovered from its samples $\{f(k/\alpha)\}_{k\in\mathbb{Z}}$ at sufficiently dense equidistant time intervals. This principle forms the cornerstone in conversion of an analog signal to a digital signal and thus for the modern communication technology.

In concrete applications of Shannon's sampling theorem, we might think about the samples $\{f(k)\}_{k\in\mathbb{Z}}$ as measurements of an unknown function f, realized at equidistant points $k \in \mathbb{Z}$. In practice a physical device is never able to measure an exact function value: for example, a measurement of a current at time k will rather give an average of the current over a very small time interval around k. Thus, a mathematical exact modeling will have to replace the exact values $\{f(k)\}_{k\in\mathbb{Z}}$ by a sequence of averages of f on intervals around the points $k \in \mathbb{Z}$. In the literature this problem has been considered under the name *local average sampling*; and it has been proved that in many cases it is still possible to obtain exact reconstruction of the signal f. We refer to the paper [603] and the references therein for details.

3.11 Exercises

3.1 Prove that $||| \cdot |||$ (introduced in the proof of Theorem 3.1.4) defines a norm on X and that X is a Banach space with respect to this norm.

3.2 Let $\{e_k\}_{k=1}^{\infty}$ be an orthonormal basis for a Hilbert space $\mathcal{H}$, and define $\{f_k\}_{k=1}^{\infty}$ by $f_k = \frac{1}{k}e_k, k \in \mathbb{N}$.

(i) Prove that $\{f_k\}_{k=1}^{\infty}$ is a basis for $\mathcal{H}$, and find the biorthogonal system $\{g_k\}_{k=1}^{\infty}$.

(ii) Prove that the coefficient functionals associated to $\{f_k\}_{k=1}^{\infty}$ are not uniformly bounded.

(iii) Show that there exists $\{d_k\}_{k=1}^{\infty} \in \ell^2(\mathbb{N})$ for which $\sum_{k=1}^{\infty} d_k g_k$ is divergent.

3.3 Prove Corollary 3.1.6.

3.4 Let $\{e_k\}_{k=1}^{\infty}$ be an orthonormal basis for a Hilbert space $\mathcal{H}$.

(i) Prove that $\{\sum_{j=1}^{\infty} \frac{1}{j}e_j\} \cup \{e_k\}_{k=1}^{\infty}$ is linearly independent, but not ω-independent.

(ii) Prove that $\{e_1\} \cup \{e_k + e_{k+1}\}_{k=1}^{\infty}$ is ω-independent, but not minimal. (*Hint:* in Example 5.4.6, we prove that $\{e_k + e_{k+1}\}_{k=1}^{\infty}$ is complete.)

3.5 Let δ_k denote the sequence in $\ell^2(\mathbb{N})$ for which the kth entry is 1 and all other entries are 0. Prove that $\{\delta_k\}_{k=1}^{\infty}$ forms an orthonormal basis for $\ell^2(\mathbb{N})$.

3.6 Assume that $\{f_k\}_{k=1}^{\infty}$ is a Bessel sequence with bound B. Prove that

(i) $\|f_k\|^2 \le B$ for all $k \in \mathbb{N}$;

(ii) If $\|f_k\|^2 = B$ for some $k \in \mathbb{N}$, then $f_k \perp f_j$ for all $j \in \mathbb{N} \setminus \{k\}$.

3.7 Assume that $\{f_k\}_{k=1}^{\infty}$ is a Bessel sequence, and let $\{c_k\}_{k=1}^{\infty} \in \ell^2(\mathbb{N})$. The purpose of this exercise is to give a direct proof of the fact that $\sum_{k=1}^{\infty} c_k f_k$ is independent of the indexing of the sequences.

(i) Show that for any $f \in \mathcal{H}$, the series $\sum_{k=1}^{\infty} c_k \langle f_k, f \rangle$ is absolutely convergent.

(ii) Show that for any permutation σ of the natural numbers,

$$\langle \sum_{k=1}^{\infty} c_k f_k, f \rangle = \langle \sum_{k=1}^{\infty} c_{\sigma(k)} f_{\sigma(k)}, f \rangle.$$

(Hint: use that absolute convergence in $\mathbb{C}$ implies unconditional convergence.)

(iii) Conclude that for any permutation σ of the natural numbers,

$$\sum_{k=1}^{\infty} c_k f_k = \sum_{k=1}^{\infty} c_{\sigma(k)} f_{\sigma(k)}.$$

3.8 Prove Corollary 3.5.5.

3.9 Prove directly via the definition that a Riesz basis is a basis.

3.10 Let $\{f_k\}_{k=1}^\infty$ be a sequence in a Hilbert space $\mathcal{H}$, and let $A, B > 0$. Consider the set of finite sequences in the unit sphere of $\ell^2(\mathbb{N})$, i.e., let

$$\mathcal{S} := \left\{ \{c_k\}_{k=1}^\infty \in \ell^2(\mathbb{N}) \,\big|\, \{c_k\}_{k=1}^\infty \text{ is finite and } \sum_{k=1}^\infty |c_k|^2 = 1 \right\}.$$

(i) Show that $\{f_k\}_{k=1}^\infty$ is a Riesz sequence with bounds A, B if

$$A \le \left\| \sum_{k=1}^\infty c_k f_k \right\|^2 \le B, \ \forall \{c_k\}_{k=1}^\infty \in \mathcal{S}. \tag{3.61}$$

(ii) Show that there exists a countable collection of sequences $\{c_j^\ell\}_{j=1}^\infty \in \mathcal{S}$, $\ell \in \mathbb{N}$, such that (3.61) holds if

$$A \le \left\| \sum_{k=1}^\infty c_k^\ell f_k \right\|^2 \le B, \ \forall \ell \in \mathbb{N}. \tag{3.62}$$

Hint: for each $n \in \mathbb{N}$, there is a countable and dense subset of the unit sphere in $\mathbb{C}^n$.

(iii) Complete the proof of Corollary 3.7.4.

3.11 Prove that if $\{f_k\}_{k=1}^\infty$ is a sequence in a Hilbert space $\mathcal{H}$ and

$$\sum_{k=1}^\infty |\langle f, f_k \rangle|^2 < \infty, \ \forall f \in \mathcal{H},$$

then $\{f_k\}_{k=1}^\infty$ is a Bessel sequence.

3.12 Prove that the upper and lower conditions in (3.29) are unrelated: there exists a sequence $\{f_k\}_{k=1}^\infty$ satisfying the upper condition for all finite sequences $\{c_k\}_{k=1}^\infty$, but not the lower condition, and vice versa.

3.13 Let $\{f_k\}_{k=1}^\infty$ be a sequence in a Hilbert space $\mathcal{H}$. Prove that

(i) If there exists $B > 0$ such that

$$\left\| \sum_{k=1}^\infty c_k f_k \right\|^2 \le B \sum_{k=1}^\infty |c_k|^2$$

for all finite sequences $\{c_k\}$, then $\sum_{k=1}^{\infty} c_k f_k$ converges for all $\{c_k\}_{k=1}^{\infty} \in \ell^2(\mathbb{N})$ and $\{f_k\}_{k=1}^{\infty}$ is a Bessel sequence with bound B.

(ii) If (3.29) holds for all finite scalar sequences $\{c_k\}$, then it holds for all $\{c_k\}_{k=1}^{\infty} \in \ell^2(\mathbb{N})$.

(iii) If $\{f_k\}_{k=1}^{\infty}$ is a Riesz basis, then

$$\sum_{k=1}^{\infty} c_k f_k \text{ is convergent} \Leftrightarrow \{c_k\}_{k=1}^{\infty} \in \ell^2(\mathbb{N}).$$

3.14 Prove that a basis in a Hilbert space is minimal.

3.15 Consider the proof of Lemma 3.5.3. Where is the assumption

$$M_{j,k} = \overline{M_{k,j}}$$

used?

3.16 Consider the positive trigonometric polynomial

$$f(x) = 1 + \cos(2\pi x).$$

Find by direct calculation all trigonometric polynomials

$$g(x) = d_0 + d_1 e^{2\pi i x}, \ d_0, d_1 \in \mathbb{R},$$

for which $|g(x)|^2 = f(x)$.

3.17 Consider the B-spline $\widetilde{B_n}$, $n \in \mathbb{N}$.

(i) Show that the scaling equation

$$\widetilde{B_n}(2\gamma) = H_0(\gamma)\widetilde{B_n}(\gamma), \ \forall \gamma \in \mathbb{R}$$

is satisfied with

$$H_0(\gamma) = \left(\frac{1 + e^{-2\pi i \gamma}}{2}\right)^n.$$

(ii) Show that H_0 is periodic with period 1.

3.18 Let $f \in L^2(\mathbb{R})$ be a continuous function for which

$$\operatorname{supp} \widehat{f} \subseteq [-\alpha/2, \alpha/2]$$

for some $\alpha > 0$. Show that f can be recovered from its samples $\{f(k/\alpha)\}_{k \in \mathbb{Z}}$ via

$$f(x) = \sum_{k \in \mathbb{Z}} f(\frac{k}{\alpha}) \operatorname{sinc}(\alpha x - k), \ x \in \mathbb{R}.$$

4

Bases and Their Limitations

The next chapters will deal with generalizations of the basis concept, so it is natural to ask why they are needed. Bases exist in all separable Hilbert spaces and in practically all Banach spaces of interest, so why do we have to search for generalizations?

In this chapter, we will give some answers to this question. As we will see, the main point is the missing *flexibility*: the conditions for being a basis are so strong that

- It is often impossible to construct bases with special properties;

- Even a slight modification of a basis might destroy the basis property.

In Section 4.1, we will consider simple modifications of bases that destroy the basis property but keep the essential expansion property. Section 4.2 and Section 4.3 will consider a number of limitations on the properties one can expect from bases having Gabor structure or wavelet structure.

4.1 Bases and the Expansion Property

The starting point for a more detailed discussion must be to clarify why we are at all interested in bases! One reason is that a basis $\{e_k\}$ for a normed vector space X allows us to represent every $f \in X$ as a (maybe infinite) linear combination of the basis elements,

$$f = \sum c_k e_k, \tag{4.1}$$

© Springer International Publishing Switzerland 2016
O. Christensen, *An Introduction to Frames and Riesz Bases*,
Applied and Numerical Harmonic Analysis,
DOI 10.1007/978-3-319-25613-9_4

with coefficients $\{c_k\}$ which depend linearly on f. We will refer to this by saying that $\{e_k\}$ has the *expansion property*. We have already seen concrete cases where the expansion property is very useful. For example, the discussion following Theorem 3.10.1 showed how to represent band limited signals f on the form (4.1), with coefficients c_k obtained as equidistant samples $\{f(k/\alpha)\}_{k\in\mathbb{Z}}$; as explained in Section 3.10, this lays the foundation for modern communication technology.

The expansion property also makes it possible to reduce many questions about elements in X to the elements $\{e_k\}$ in the basis. For example, the action of a bounded operator U on a vector f can be found if we know the representation (4.1) and the action of U on the basis $\{e_k\}$:

$$ Uf = U\left(\sum c_k e_k\right) = \sum c_k U e_k. $$

Bases are characterized by the expansion property (4.1) with *unique* coefficients $\{c_k\}$ associated to each $f \in X$. One might ask whether uniqueness is really needed. Our answer is no: it is usually enough to know the *existence* of some usable coefficients, together with a recipe for finding them. This turns out to be the key in the transition from bases to frames: we will revise the conditions in such a way that we keep the expansion property, but we gain flexibility by giving up the requirement of uniqueness of the expansion coefficients.

In this chapter we discuss some cases where (4.1) holds without $\{e_k\}$ being a basis. We begin with the simple observation that if $\{e_k\}$ is a basis for X and ϕ is an arbitrary element in X, then $\{e_k\} \cup \phi$ is not a basis, despite the fact that each $f \in X$ has representations of the form

$$ f = \sum c_k e_k + d\phi. \tag{4.2} $$

In fact, the sequence $\{e_k\} \cup \phi$ is not linearly independent, i.e., several choices for the coefficients $\{c_k\}$ and d are possible. One choice is to take $d = 0$ and let $\{c_k\}$ be the coefficients representing f in the basis $\{e_k\}$; another choice is to take $\{c_k\}$ such that $f - \phi = \sum c_k e_k$ and $d = 1$.

By this argument, the basis property is destroyed when an arbitrary nonempty collection of vectors is added to $\{e_k\}$, but the expansion property is preserved.

At first glance, the above construction might appear artificial: why would one like to add elements to a basis? One reason is that we gain some freedom: the coefficients in (4.1) are unique, but in (4.2), we can *choose* between several options. We will encounter several scenarios where this is useful, e.g., in Section 16.1 where the freedom is used to find coefficients of a particularly convenient form in the wavelet case. Also, Section 8.5 will show that having more elements than needed for a basis has a certain *noise-suppressing* effect.

The following example shows that non-bases with the expansion property actually appear in a natural fashion in function spaces.

Example 4.1.1 Let us return to the orthonormal basis $\{e_k\}_{k\in\mathbb{Z}}$ for $L^2(0,1)$ considered in Section 3.8, i.e., the functions $e_k(x) = e^{2\pi ikx}$. We will now consider these functions on an open subinterval $I \subset]0,1[$ with $|I| < 1$. We can identify $L^2(I)$ with the subspace of $L^2(0,1)$ consisting of the functions which are zero on $]0,1[\backslash I$. Hereby a function $f \in L^2(I)$ is identified with a function (which we still denote f) in $L^2(0,1)$, which has the expansion

$$f = \sum_{k\in\mathbb{Z}}\langle f, e_k\rangle e_k \text{ in } L^2(0,1). \tag{4.3}$$

Since

$$\left\|f - \sum_{|k|\leq n}\langle f, e_k\rangle e_k\right\|_{L^2(I)} = \left(\int_I \left|f(x) - \sum_{k=-n}^{n}\langle f, e_k\rangle e^{2\pi ikx}\right|^2 dx\right)^{1/2}$$

$$\leq \left(\int_0^1 \left|f(x) - \sum_{k=-n}^{n}\langle f, e_k\rangle e^{2\pi ikx}\right|^2 dx\right)^{1/2}$$

$$\to 0 \text{ as } n \to \infty,$$

we also have

$$f = \sum_{k\in\mathbb{Z}}\langle f, e_k\rangle e_k \text{ in } L^2(I). \tag{4.4}$$

That is, the functions $\{e_k\}_{k\in\mathbb{Z}}$ also have the expansion property in $L^2(I)$. However, they are not a basis for $L^2(I)$! To see this, define the function

$$\widetilde{f}(x) = \begin{cases} f(x) & \text{if } x \in I, \\ 1 & \text{if } x \notin I. \end{cases}$$

Then $\widetilde{f} \in L^2(0,1)$ and we have the representation

$$\widetilde{f} = \sum_{k\in\mathbb{Z}}\langle \widetilde{f}, e_k\rangle e_k \text{ in } L^2(0,1). \tag{4.5}$$

By restricting to I, the expansion (4.5) is also valid in $L^2(I)$; since $f = \widetilde{f}$ on I, this shows that

$$f = \sum_{k\in\mathbb{Z}}\langle \widetilde{f}, e_k\rangle e_k \text{ in } L^2(I). \tag{4.6}$$

Thus, (4.4) and (4.6) are both expansions of f in $L^2(I)$, and they are non-identical; the argument is that since $f \neq \widetilde{f}$ in $L^2(0,1)$, the expansions (4.3) and (4.5) show that

$$\{\langle f, e_k\rangle\}_{k\in\mathbb{Z}} \neq \{\langle \widetilde{f}, e_k\rangle\}_{k\in\mathbb{Z}}.$$

The conclusion is that the restriction of the functions $\{e_k\}_{k\in\mathbb{Z}}$ to I is not a basis for $L^2(I)$, but the expansion property is preserved. In the terminology used in Section 5, the sequence $\{e_k\}_{k\in\mathbb{Z}}$ is a *tight frame* for $L^2(I)$; see Example 5.4.5.   □

Example 4.1.1 is actually just a concrete manifestation of the following general result. It shows that the orthogonal projection of an orthonormal basis onto a nontrivial subspace V always yields a sequence with the expansion property (on the subspace) but with expansion coefficients that are not unique. We ask the reader to provide the proof in Exercise 4.1.

Proposition 4.1.2 *Let $\{e_k\}_{k=1}^{\infty}$ denote an orthonormal basis for a Hilbert space $\mathcal{H}$, let P denote the orthogonal projection of $\mathcal{H}$ onto a closed nontrivial subspace V, and put $f_k := Pe_k$. Fix any $g \in V^{\perp}$. Then each $f \in V$ has the expansions*

$$f = \sum_{k=1}^{\infty} \langle f, e_k \rangle f_k = \sum_{k=1}^{\infty} \langle f + g, e_k \rangle f_k.$$

Furthermore, for any $f \in V$ and any choice of $g \in V^{\perp} \setminus \{0\}$,

$$\{\langle f, e_k \rangle\}_{k=1}^{\infty} \neq \{\langle f + g, e_k \rangle\}_{k=1}^{\infty}.$$

Again, Proposition 4.1.2 shows that expansions with nonunique coefficients appear naturally. In the terminology used in Section 5, the sequence $\{Pe_k\}_{k=1}^{\infty}$ is a *tight frame* for V.

In a finite-dimensional vector space X, we know that every family of vectors which spans X contains a basis (Exercise 1.3). In an infinite-dimensional Hilbert space, the situation is dramatically different: there exists a family of vectors $\{f_k\}_{k=1}^{\infty}$ such that

- Each $f \in \mathcal{H}$ has an unconditionally convergent expansion

$$f = \sum_{k=1}^{\infty} c_k f_k \ \text{ with } \{c_k\}_{k=1}^{\infty} \in \ell^2(\mathbb{N});$$

- No subsequence of $\{f_k\}_{k=1}^{\infty}$ is a basis for $\mathcal{H}$.

We present an explicit construction of such a sequence $\{f_k\}_{k=1}^{\infty}$ in Section 7.5. Intuitively, this kind of example is difficult to understand: it shows that we might have the expansion property for families which have no relationship to a basis. The existence of such examples is a strong argument for considering generalizations of bases.

4.2 Gabor Systems and the Balian–Low Theorem

We already encountered Gabor systems in Section 3.8; in particular we saw in Example 3.8.3 that the Gabor system

$$\{e^{2\pi imx}\chi_{[0,1]}(x-n)\}_{m,n\in\mathbb{Z}} = \{E_m T_n \chi_{[0,1]}(x)\}_{m,n\in\mathbb{Z}}$$

forms an orthonormal basis for $L^2(\mathbb{R})$. Exactly this example touches one of the limitations on the properties we can expect from a Gabor basis, as we will see now. Observe that

$$\widehat{\chi}_{[0,1]}(\gamma) = \int_0^1 e^{-2\pi ix\gamma}dx = e^{-\pi i\gamma}\frac{\sin \pi\gamma}{\pi\gamma}.$$

The fact that $\chi_{[0,1]}$ is discontinuous, and the oscillations and slow decay of $\widehat{\chi}_{[0,1]}$, makes the characteristic function unattractive from the point of view of, e.g., time–frequency analysis. A natural idea is to examine whether we can replace the function $\chi_{[0,1]}$ by a continuous (or even differentiable) function g and still obtain an orthonormal basis or Riesz basis $\{E_m T_n g\}_{m,n\in\mathbb{Z}}$. Unfortunately, the *Balian–Low theorem* shows that there are limitations on the properties such a function g can have:

Theorem 4.2.1 *If $\{E_m T_n g\}_{m,n\in\mathbb{Z}}$ is a Riesz basis for $L^2(\mathbb{R})$, then*

$$\left(\int_{-\infty}^{\infty}|xg(x)|^2 dx\right)\left(\int_{-\infty}^{\infty}|\gamma\widehat{g}(\gamma)|^2 d\gamma\right) = \infty. \tag{4.7}$$

For proofs of the Balian–Low theorem we refer to [241, 48], or [400]. In words, the Balian–Low theorem means that a function g generating a Gabor Riesz basis cannot be well localized in both time and frequency. For example, it is not possible that g and $\widehat{g}$ satisfy estimates like

$$|g(x)| \le \frac{C}{1+x^2}, \quad |\widehat{g}(\gamma)| \le \frac{C}{1+\gamma^2} \tag{4.8}$$

simultaneously. We note in passing that the Balian–Low theorem is close to describe the limit case of what can be obtained with Gabor bases. In fact, it has been proved by Benedetto et al. [45] that for any $\epsilon > 0$, we can construct orthonormal bases $\{E_m T_n g\}_{m,n\in\mathbb{Z}}$, where

$$\left(\int_{-\infty}^{\infty}|g(x)|^2\frac{1+|x|^2}{\log^{1+\epsilon}(2+|x|)}dx\right)\left(\int_{-\infty}^{\infty}|\widehat{g}(\gamma)|^2\frac{1+|\gamma|^2}{\log^{2+\epsilon}(2+|\gamma|)}d\gamma\right) < \infty.$$

If faster decay of g and $\widehat{g}$ than allowed by the Balian–Low theorem is needed, we have to ask whether we need all the properties characterizing a Riesz basis or whether we can relax some of them. The property we want to keep is that every $f \in L^2(\mathbb{R})$ has an unconditionally convergent expansion in terms of modulated and translated versions of the function g; together with Lemma 3.6.9, this shows that we do not gain anything by asking for $\{E_m T_n g\}_{m,n\in\mathbb{Z}}$ being merely a basis instead of a Riesz basis. However, it turns out that the (unconditionally convergent) expansion property actually can be combined with g and $\widehat{g}$ having fast decay, even exponential decay: the part of the definition of a basis which has to be given up is the *uniqueness* of such an expansion. This will bring us from bases to frames. The exact definition will be given in the next chapter, and the above description rather tries to state the difference between frames and bases than to give the key to the right definition. Frames in $L^2(\mathbb{R})$ having the Gabor structure will be the subject of Chapters 11–13.

In the analysis of Gabor systems in the forthcoming chapters, we will be more general than here and consider systems of the form $\{E_{mb}T_{na}g\}_{m,n\in\mathbb{Z}}$ for some parameters $a, b > 0$. One of the essential issues is how to obtain expansions of functions $f \in L^2(\mathbb{R})$ in terms of the countable family $\{E_{mb}T_{na}g\}_{m,n\in\mathbb{Z}}$, with a continuous and compactly supported "generating function" g. As the next result shows, these requirements cannot be combined with $\{E_{mb}T_{na}g\}_{m,n\in\mathbb{Z}}$ being a Riesz basis for $L^2(\mathbb{R})$; on the other hand, frames $\{E_{mb}T_{na}g\}_{m,n\in\mathbb{Z}}$ with these properties exist. We state the result here, although the formal frame definition is given later, in Definition 5.1.1:

Proposition 4.2.2 *Let g be a continuous function with compact support. Then the following hold:*

(i) $\{E_{mb}T_{na}g\}_{m,n\in\mathbb{Z}}$ *cannot be an orthonormal basis for $L^2(\mathbb{R})$.*

(ii) $\{E_{mb}T_{na}g\}_{m,n\in\mathbb{Z}}$ *cannot be a Riesz basis for $L^2(\mathbb{R})$.*

(iii) $\{E_{mb}T_{na}g\}_{m,n\in\mathbb{Z}}$ *can be a frame for $L^2(\mathbb{R})$ if $0 < ab < 1$.*

A more precise version of Proposition 4.2.2 (iii) says that for any $a, b > 0$ with $ab < 1$, there exists a continuous function g with compact support such that $\{E_{mb}T_{na}g\}_{m,n\in\mathbb{Z}}$ is a frame for $L^2(\mathbb{R})$; we will present such a construction in Example 12.3.3. The proof of the results stated in (i) and (ii) uses the Zak transform and will be given on page 334.

Having a frame of the type $\{E_{mb}T_{na}g\}_{m,n\in\mathbb{Z}}$, one could ask whether there exists a subfamily which is a basis. We will not do so: on page 295, we argue that even if the answer is yes, it will in general not be an advantage to remove elements from $\{E_{mb}T_{na}g\}_{m,n\in\mathbb{Z}}$ because the computational benefits from the points $\{(na, mb)\}_{m,n\in\mathbb{Z}}$ forming a lattice in $\mathbb{R}^2$ will be lost.

Let us summarize some of the key properties we would like to obtain for a Gabor frame $\{E_{mb}T_{na}g\}_{m,n\in\mathbb{Z}}$ for $L^2(\mathbb{R})$:

- For computational reasons we would like the window function g to be explicitly given, with short support and high regularity;

- We would like $\{E_{mb}T_{na}g\}_{m,n\in\mathbb{Z}}$ to have "low redundancy," at the moment in the intuitive sense as "the frame does not contain too many extra elements compared to a basis." A formal definition will be given in Definition 11.3.3.

The analysis of Gabor frames in Chapters 11–13 will show that these properties actually can be realized; see in particular Section 12.6.

4.3 Bases and Wavelets

Wavelet orthonormal bases $\{\psi_{j,k}\}_{j,k \in \mathbb{Z}} = \{2^{j/2}\psi(2^j x - k)\}_{j,k \in \mathbb{Z}}$ form another important class of bases for $L^2(\mathbb{R})$. Also for these bases there are limitations on the properties which can be satisfied simultaneously:

Theorem 4.3.1 *Let $\psi \in L^2(\mathbb{R})$. Assume that ψ decays exponentially and that $\{\psi_{j,k}\}_{j,k \in \mathbb{Z}}$ is an orthonormal basis. Then ψ cannot be infinitely often differentiable with bounded derivatives.*

For a proof we refer to [242]. We will see in Example 15.2.7 that the properties in Theorem 4.3.1 can be combined if we allow ψ to generate a frame instead of a basis.

Some of the relevant properties for a wavelet basis $\{\psi_{j,k}\}_{j,k \in \mathbb{Z}}$ are:

- That ψ has a computationally convenient form, for example, that ψ is a piecewise polynomial (a spline);

- Regularity of ψ;

- Symmetry (or antisymmetry) of ψ, i.e., that

$$\psi(x) = \psi(-x) \text{ or } \psi(x) = -\psi(-x);$$

- Compact support of ψ or at least fast decay;

- That ψ has *vanishing moments*, i.e., that for a certain $m \in \mathbb{N}$,

$$\int_{-\infty}^{\infty} x^\ell \psi(x)dx = 0 \text{ for } \ell = 0, 1, \ldots, m.$$

We discuss the role played by these properties and how they motivated the development of the wavelet theory below. First, the following proposition shows that a large number of vanishing moments is important if we want to obtain smooth wavelets. For the proof we refer to [242].

Proposition 4.3.2 *Assume that $\psi \in L^2(\mathbb{R})$ is m times continuously differentiable with bounded derivatives, that $\{\psi_{j,k}\}_{j,k \in \mathbb{Z}}$ is an orthonormal system, and that there exist constants $C, \epsilon > 0$ such that*

$$|\psi(x)| \leq \frac{C}{(1 + |x|)^{1+m+\epsilon}}, \ \forall x \in \mathbb{R}. \tag{4.9}$$

Then

$$\int_{-\infty}^{\infty} x^\ell \psi(x)dx = 0 \text{ for all } \ell = 0, 1, \ldots, m. \tag{4.10}$$

In particular, a differentiable and compactly supported wavelet will automatically have a certain number of vanishing moments. For the Daubechies wavelets there is a connection between the regularity of the wavelet and the size of the support: the N-th Daubechies wavelet has a support with Lebesgue measure equal to $2N - 1$, and asymptotically as $N \to \infty$, it belongs to $C^{\mu N}$ with $\mu \sim 0.19$. We refer to [242] for a proof.

Vanishing moments are essential in the context of *compression*. Assuming that $\{\psi_{j,k}\}_{j,k \in \mathbb{Z}}$ is an orthonormal basis for $L^2(\mathbb{R})$, every $f \in L^2(\mathbb{R})$ has the representation

$$f = \sum_{j,k \in \mathbb{Z}} \langle f, \psi_{j,k} \rangle \psi_{j,k}. \tag{4.11}$$

If the function ψ is generated by a multiresolution analysis, classical wavelet analysis shows that with $\phi \in L^2(\mathbb{R})$ taken as in Definition 3.9.2,

$$f = \sum_{k \in \mathbb{Z}} \langle f, T_k \phi \rangle T_k \phi + \sum_{j=1}^{\infty} \sum_{k \in \mathbb{Z}} \langle f, \psi_{j,k} \rangle \psi_{j,k}, \ \forall f \in L^2(\mathbb{R}). \tag{4.12}$$

Let us assume that ϕ is chosen to be compactly supported. All information about f is stored in the coefficients

$$\{\langle f, T_k \phi \rangle\}_{k \in \mathbb{Z}} \cup \{\langle f, \psi_{j,k} \rangle\}_{j \in \mathbb{N}, k \in \mathbb{Z}},$$

and (4.12) tells us how to reconstruct f based on knowledge of the coefficients. However, in practice, we cannot store an infinite sequence of nonzero numbers, so we have to select a finite number of the coefficients to keep. If the function f is compactly supported, the sequence $\{\langle f, T_k \phi \rangle\}_{k \in \mathbb{Z}}$ is finite; thus, we will focus on the sequence $\{\langle f, \psi_{j,k} \rangle\}_{j \in \mathbb{N}, k \in \mathbb{Z}}$ in the rest of the argument. Usually this sequence is treated by *thresholding*: one chooses a certain $\epsilon > 0$ and keeps only the coefficients $\langle f, \psi_{j,k} \rangle$ for which $|\langle f, \psi_{j,k} \rangle| \geq \epsilon$. Here the vanishing moments come in: one can prove that if ψ has a large number of vanishing moments, then only relatively few coefficients $\langle f, \psi_{j,k} \rangle$ will be large for "natural signals" f. By keeping these coefficients and throwing the rest away, we have obtained an efficient compression of the signal f. We refer to the paper by Beylkin, Coifman, and Rokhlin [60] for more details.

Compact support (or at least fast decay) of ψ is essential for the use of computer-based methods, where a function with unbounded support always has to be truncated. For the same reason, we often want the support to be small. The condition of ψ being symmetric is relevant in image processing, where a nonsymmetric wavelet will generate nonsymmetric errors, which are more disturbing to the human eye than symmetric errors. The next result, which is also proved in [242], shows clear limitations on the properties we can obtain via the classical multiresolution analysis:

Proposition 4.3.3 *Assume that $\phi \in L^2(\mathbb{R})$ is real-valued and compactly supported, and let*

$$V_j = \overline{span}\{D^j T_k \phi\}_{k\in\mathbb{Z}}, \ j \in \mathbb{Z}.$$

Assume that $(\phi, \{V_j\})$ constitute a multiresolution analysis. Then, if the associated wavelet ψ in (3.47) is real-valued and compactly supported and has either a symmetry axis or an antisymmetry axis, ψ is necessarily the Haar wavelet.

The Haar wavelet only has one vanishing moment, and it is fair to say that the entire wavelet theory is developed in an attempt to avoid this wavelet; thus, the result in Proposition 4.3.3 is a serious shortcoming. Proposition 4.3.3 was one of the reasons for Cohen, Daubechies, and Feauveau to introduce *biorthogonal multiresolution analysis* [223], where one constructs a Riesz basis $\{\psi_{j,k}\}_{j,k\in\mathbb{Z}}$ for $L^2(\mathbb{R})$ instead of an orthonormal basis. As we have seen in Theorem 3.6.2, the coefficients in the expansion of a function in terms of a Riesz basis are given by inner products between the function and the elements in the dual Riesz basis. This is the reason for the name biorthogonal multiresolution analysis: one actually constructs two coupled multiresolution analyses, which deliver the wavelet Riesz basis $\{\psi_{j,k}\}_{j,k\in\mathbb{Z}}$ and a dual wavelet Riesz basis $\{\tilde{\psi}_{j,k}\}_{j,k\in\mathbb{Z}}$ for some function $\tilde{\psi} \in L^2(\mathbb{R})$. Note that for general wavelet Riesz bases, the dual Riesz basis might not have wavelet structure; see Section 16.1).

The Daubechies wavelets are not given by an explicit formula. This is not a problem for the typical applications which barely need the wavelets themselves but only certain associated algorithms. On a more general level, it is clearly relevant that a wavelet has a computationally convenient form. The Battle–Lemarié wavelets in (3.52) are splines, but for $n > 1$, they do not have compact support. The construction of biorthogonal wavelets in [223] allows the functions $\psi, \tilde{\psi}$ to be symmetric and compactly supported, but only one of them can be a spline. A related result by Chui and Wang [222] allows ψ as well as $\tilde{\psi}$ to be symmetric splines, but only one can have compact support. The difficulty of getting compactly supported spline wavelets is real, and in Section 18.5, we will indeed show that two compactly supported functions ψ and $\tilde{\psi}$ of the form (3.52) cannot generate dual wavelet Riesz bases.

This is only a glimpse of the intense activity in the area, which took place around 1990–1993. For our purpose we only mention one more step, which is important for the presentation in Chapters 15–19. This is the idea of using *multiwavelets*. Here we give up the basic requirement that a wavelet system is generated by translated and scaled versions of *one* function. In fact, we begin instead with a finite collection of functions $\psi_1, \ldots, \psi_n \in L^2(\mathbb{R})$ and consider the system of functions which we obtain by translation and dilation of all these functions. In [256], Donovan, Geronimo, and Hardin proved that one can construct orthonormal bases of multiwavelets, where the functions $\psi_1, \ldots, \psi_n$ are symmetric splines with compact support. This is close in spirit to the frame constructions that will be constructed in Chapter 18. Indeed we will show that dual multiwavelet frames generated by compactly supported splines on the form (3.52) exist and are relatively easy to construct.

From this short description, it is clear that the purpose of the different extensions of the first multiresolution scheme is to gain more flexibility. This is also the key reason for extending the theory to frames, as we will do in Chapters 17–19.

4.4 General Shortcomings

Another annoying fact about bases is their lack of stability against applications of operators. If, for example, $\{e_k\}_{k=1}^\infty$ is an orthonormal basis, then only very special operators (the unitary ones) will make $\{Ue_k\}_{k=1}^\infty$ an orthonormal basis. Even though we have not given all the relevant definitions yet, let us give an overview that shows how the generalizations in the subsequent chapters stepwise weaken the conditions on the operator U. Assuming that $\{e_k\}_{k=1}^\infty$ is an orthonormal basis for $\mathcal{H}$,

- The orthonormal bases are the sequences of the form $\{Ue_k\}_{k=1}^\infty$ where U is a unitary operator on $\mathcal{H}$ (Theorem 3.4.7);

- The Riesz bases are the sequences of the form $\{Ue_k\}_{k=1}^\infty$ where U is a bounded bijective operator on $\mathcal{H}$ (Definition 3.6.1);

- The frames are the sequences of the form $\{Ue_k\}_{k=1}^\infty$ where U is a bounded surjective operator on $\mathcal{H}$ (Theorem 5.5.4);

- $\{Ue_k\}_{k=1}^\infty$ leads to a frame-like expansion if U is closed and surjective (Theorem 8.4.1).

The condition that U is surjective appears in all the statements; it can be replaced by the assumption that U has closed range if we only need a frame expansion on a subspace. Note also that the Bessel sequences are the sequences of the form $\{Ue_k\}_{k=1}^\infty$ where U is just a bounded operator on $\mathcal{H}$ (Theorem 3.2.3), but Bessel sequences do not lead to any series expansion by themselves.

The limitations on the possible constructions of bases give theoretical reasons to consider frames. We will in Sections 8.5 and 13.8 describe cases where bases actually exist but where frames simply perform better.

4.5 Exercises

4.1 Prove Proposition 4.1.2.

5

Frames in Hilbert Spaces

The main feature of a basis $\{f_k\}_{k=1}^{\infty}$ in a Hilbert space $\mathcal{H}$ is that every $f \in \mathcal{H}$ can be represented as a superposition of the elements f_k in the basis:

$$f = \sum_{k=1}^{\infty} c_k(f) f_k. \tag{5.1}$$

The coefficients $c_k(f)$ are unique. We now introduce the concept of *frames*. A frame is also a sequence of elements $\{f_k\}_{k=1}^{\infty}$ in $\mathcal{H}$, which allows every $f \in \mathcal{H}$ to be written as in (5.1). However, the corresponding coefficients are not necessarily unique. Thus a frame might not be a basis; arguments for generalizing the basis concept were given in Chapter 4.

The history of frames is a nice example of the development of mathematics. Frames were introduced already in 1952 by Duffin and Schaeffer in their fundamental paper [262]; they used frames as a tool in the study of *nonharmonic Fourier series*, i.e., sequences of the type $\{e^{i\lambda_n x}\}_{n \in \mathbb{Z}}$, where $\{\lambda_n\}_{n \in \mathbb{Z}}$ is a family of real or complex numbers. Apparently, the importance of the concept was not realized by the mathematical community; at least it took almost 30 years before the next treatment appeared in print. In 1980, Young wrote his book [622], which contains the basic facts about frames. Frames were presented in the abstract setting and again used in the context of nonharmonic Fourier series. Then, in 1985, as the wavelet era began, Daubechies, Grossmann, and Meyer [244] observed that frames can be used to find series expansions of functions in $L^2(\mathbb{R})$ which are very similar to the expansions using orthonormal bases. This was probably the

© Springer International Publishing Switzerland 2016   119
O. Christensen, *An Introduction to Frames and Riesz Bases*,
Applied and Numerical Harmonic Analysis,
DOI 10.1007/978-3-319-25613-9_5

time when many mathematicians started to see the potential of the topic; this point became more clear via Daubechies' important paper [241], her book [242], and the combined survey/research paper by Heil and Walnut [395]. Since then, the number of papers concerning frames has increased drastically, and a single book cannot present all the important results. Our aim is, however, to give a comprehensive presentation of the fundamental results which hold for frames in general Hilbert spaces. The limitations will mainly appear in the later chapters, where we are only able to present some of the many results about structured function systems like Gabor frames and wavelet frames.

A subject like frames can be approached in different ways. One way is to look at frame theory as a branch of functional analysis and ask what we can prove for general frames in general Hilbert spaces. Another approach is to consider a class of frames, which is used in, e.g., signal processing (this could, e.g., be frames having the wavelet structure), and examine the properties for this special class of frames. Most papers concentrate on one of these two aspects (frame theory would actually benefit from a closer coordination) and we will treat them separately here, too. In this chapter and Chapters 6–8 we present the general theory, while the subsequent chapters will go into details with specific constructions.

Section 5.1 is instrumental for a good understanding of frames; here, their basic properties are presented, and the important frame decompositions are discussed. A reader who is mainly interested in Gabor frames or wavelet frames might go directly to the relevant later chapters in the book after reading this section; in fact, the theory for these frames is to a large extent independent of the results for general frames. Section 5.2 discusses sequences that might only form frames for certain subspaces of the given Hilbert space, and Section 5.3 deals with preservation of the frame property under the action of various operators. Section 5.4 takes the first steps toward the analysis of the relationship between frames and bases; much more will be said about this in Chapter 7. Section 5.5 characterizes the frame property, e.g., in terms of operators. Section 5.6 gives a short introduction to continuous frames.

5.1 Frames and Their Properties

We are now ready to give the central definition. In the entire chapter, $\mathcal{H}$ will denote a separable Hilbert space with inner product $\langle \cdot, \cdot \rangle$.

Definition 5.1.1 *A sequence $\{f_k\}_{k=1}^{\infty}$ of elements in $\mathcal{H}$ is a frame for $\mathcal{H}$ if there exist constants $A, B > 0$ such that*

$$A\,||f||^2 \le \sum_{k=1}^{\infty} |\langle f, f_k \rangle|^2 \le B\,||f||^2, \quad \forall f \in \mathcal{H}. \tag{5.2}$$

The numbers A, B are called *frame bounds*. They are not unique. The *optimal upper frame bound* is the infimum over all upper frame bounds, and the *optimal lower frame bound* is the supremum over all lower frame bounds. Note that the optimal bounds are actually frame bounds. We collect a few more definitions:

Definition 5.1.2

  (i) *A frame is tight if we can choose $A = B$ as frame bounds; a tight frame with bound $A = B = 1$ is called a Parseval frame.*

  (ii) *If a frame ceases to be a frame when an arbitrary element is removed, it is called an exact frame.*

When we speak about the *frame bound* for a tight frame, we mean the exact value A which is at the same time an upper and lower frame bound. Note that this is slightly different from the terminology for general frames, where, e.g., an upper frame bound is just *some* number for which the Bessel condition is satisfied.

It follows from the definition that if $\{f_k\}_{k=1}^{\infty}$ is a frame for $\mathcal{H}$, then

$$\overline{\text{span}}\{f_k\}_{k=1}^{\infty} = \mathcal{H}.$$

We often need to consider sequences which are not complete in $\mathcal{H}$; they cannot form frames for $\mathcal{H}$, but they can very well form frames for the closed linear span of their elements:

Definition 5.1.3 *Let $\{f_k\}_{k=1}^{\infty}$ be a sequence in $\mathcal{H}$. We say that $\{f_k\}_{k=1}^{\infty}$ is a frame sequence if it is a frame for $\overline{\text{span}}\{f_k\}_{k=1}^{\infty}$.*

Before we develop the theory for frames, we mention a few examples of frames. They might appear quite "constructed," but they are useful for the theoretical understanding of frames. In Chapters 9–20 we consider frames which are more interesting by themselves, for example, frames in $L^2(\mathbb{R})$ having Gabor structure or wavelet structure.

Example 5.1.4 *Let $\{e_k\}_{k=1}^{\infty}$ be an orthonormal basis for $\mathcal{H}$.*

  (i) By repeating each element in $\{e_k\}_{k=1}^{\infty}$ twice we obtain

$$\{f_k\}_{k=1}^{\infty} = \{e_1, e_1, e_2, e_2, ..\},$$

which is a tight frame with frame bound $A = 2$. If only e_1 is repeated we obtain

$$\{f_k\}_{k=1}^{\infty} = \{e_1, e_1, e_2, e_3, ..\},$$

which is a frame with bounds $A = 1, B = 2$.

(ii) Let

$$\{f_k\}_{k=1}^{\infty} := \left\{ e_1, \frac{1}{\sqrt{2}}e_2, \frac{1}{\sqrt{2}}e_2, \frac{1}{\sqrt{3}}e_3, \frac{1}{\sqrt{3}}e_3, \frac{1}{\sqrt{3}}e_3, \cdots \right\};$$

that is, $\{f_k\}_{k=1}^{\infty}$ is the sequence where each vector $\frac{1}{\sqrt{k}}e_k$ is repeated k times. Then, for each $f \in \mathcal{H}$,

$$
\begin{aligned}
\sum_{k=1}^{\infty} |\langle f, f_k \rangle|^2 &= \sum_{k=1}^{\infty} k \, |\langle f, \frac{1}{\sqrt{k}}e_k \rangle|^2 \\
&= ||f||^2.
\end{aligned}
$$

So $\{f_k\}_{k=1}^{\infty}$ is a tight frame for $\mathcal{H}$ with frame bound $A = 1$.

(iii) If $I \subset \mathbb{N}$ is a proper subset, then $\{e_k\}_{k \in I}$ is not complete in $\mathcal{H}$ and cannot be a frame for $\mathcal{H}$. However, $\{e_k\}_{k \in I}$ is a frame for $\overline{\text{span}}\{e_k\}_{k \in I}$, i.e., it is a frame sequence.   □

Since a frame $\{f_k\}_{k=1}^{\infty}$ is a Bessel sequence, the operator

$$T : \ell^2(\mathbb{N}) \to \mathcal{H}, \quad T\{c_k\}_{k=1}^{\infty} = \sum_{k=1}^{\infty} c_k f_k \tag{5.3}$$

is bounded by Theorem 3.2.3; T is called the *synthesis operator* or the *pre-frame operator*. By Lemma 3.2.1, the adjoint operator is given by

$$T^* : \mathcal{H} \to \ell^2(\mathbb{N}), \quad T^*f = \{\langle f, f_k \rangle\}_{k=1}^{\infty}. \tag{5.4}$$

The operator T^* is called the *analysis operator*. By composing T and T^*, we obtain the *frame operator*

$$S : \mathcal{H} \to \mathcal{H}, \quad Sf = TT^*f = \sum_{k=1}^{\infty} \langle f, f_k \rangle f_k. \tag{5.5}$$

Note that since $\{f_k\}_{k=1}^{\infty}$ is a Bessel sequence, the series defining S converges unconditionally for all $f \in \mathcal{H}$ by Corollary 3.2.5. We state some of the important properties of S:

Lemma 5.1.5 *Let $\{f_k\}_{k=1}^{\infty}$ be a frame with frame operator S and frame bounds A, B. Then the following hold:*

(i) *S is bounded, invertible, self-adjoint, and positive.*

(ii) *$\{S^{-1}f_k\}_{k=1}^{\infty}$ is a frame with bounds B^{-1}, A^{-1}; if A, B are the optimal bounds for $\{f_k\}_{k=1}^{\infty}$, then the bounds B^{-1}, A^{-1} are optimal for $\{S^{-1}f_k\}_{k=1}^{\infty}$. The frame operator for $\{S^{-1}f_k\}_{k=1}^{\infty}$ is S^{-1}.*

Proof. (i): S is bounded as a composition of two bounded operators. By Theorem 3.2.3,

$$||S|| = ||TT^*|| = ||T|| \; ||T^*|| = ||T||^2 \le B.$$

Since $S^* = (TT^*)^* = TT^* = S$, the operator S is self-adjoint. The inequality (5.2) means that $A||f||^2 \le \langle Sf, f \rangle \le B||f||^2$ for all $f \in \mathcal{H}$, or, in the notation introduced in (2.9), $AI \le S \le BI$; thus S is positive. Furthermore, $0 \le I - B^{-1}S \le \frac{B-A}{B}I$, and consequently

$$||I - B^{-1}S|| = \sup_{||f||=1} |\langle (I - B^{-1}S)f, f \rangle| \le \frac{B-A}{B} < 1,$$

which by Theorem 2.2.3 shows that S is invertible.

(ii): Note that for $f \in \mathcal{H}$,

$$\sum_{k=1}^{\infty} |\langle f, S^{-1}f_k \rangle|^2 = \sum_{k=1}^{\infty} |\langle S^{-1}f, f_k \rangle|^2 \le B \, ||S^{-1}f||^2$$

$$\le B \, ||S^{-1}||^2 \, ||f||^2.$$

That is, $\{S^{-1}f_k\}_{k=1}^{\infty}$ is a Bessel sequence. It follows that the frame operator for $\{S^{-1}f_k\}_{k=1}^{\infty}$ is well defined. By definition, it acts on $f \in \mathcal{H}$ by

$$\sum_{k=1}^{\infty} \langle f, S^{-1}f_k \rangle S^{-1}f_k = S^{-1}\sum_{k=1}^{\infty} \langle S^{-1}f, f_k \rangle f_k = S^{-1}SS^{-1}f$$

$$= S^{-1}f; \qquad (5.6)$$

this shows that the frame operator for $\{S^{-1}f_k\}_{k=1}^{\infty}$ equals S^{-1}. The operator S^{-1} commutes with both S and I, so using Theorem 2.4.3 we can "multiply the inequality" $AI \le S \le BI$ with S^{-1}; this gives

$$B^{-1}I \le S^{-1} \le A^{-1}I,$$

i.e.,

$$B^{-1}||f||^2 \le \langle S^{-1}f, f \rangle \le A^{-1}||f||^2, \; \forall f \in \mathcal{H}.$$

Via (5.6),

$$B^{-1}||f||^2 \le \sum_{k=1}^{\infty} |\langle f, S^{-1}f_k \rangle|^2 \le A^{-1}||f||^2, \; \forall f \in \mathcal{H};$$

thus $\{S^{-1}f_k\}_{k=1}^{\infty}$ is a frame with frame bounds B^{-1}, A^{-1}. To prove the optimality of the bounds (in case A, B are optimal for $\{f_k\}_{k=1}^{\infty}$), let A be the optimal lower bound for $\{f_k\}_{k=1}^{\infty}$, and assume that the optimal upper bound for $\{S^{-1}f_k\}_{k=1}^{\infty}$ is $C < \frac{1}{A}$. By applying what we already proved to the frame $\{S^{-1}f_k\}_{k=1}^{\infty}$ having frame operator S^{-1}, we obtain that $\{f_k\}_{k=1}^{\infty} = \{(S^{-1})^{-1}S^{-1}f_k\}_{k=1}^{\infty}$ has the lower bound $\frac{1}{C} > A$, but this is a contradiction. Thus $\{S^{-1}f_k\}_{k=1}^{\infty}$ has the optimal upper bound $\frac{1}{A}$. The argument for the optimal lower bound is similar. □

The frame $\{S^{-1}f_k\}_{k=1}^\infty$ is called the *canonical dual frame* of $\{f_k\}_{k=1}^\infty$ because it plays the same role in frame theory as the dual of a basis; see Theorem 5.1.6 and Theorem 5.4.1.

The *frame decomposition*, stated below, is the most important frame result. It shows that if $\{f_k\}_{k=1}^\infty$ is a frame for $\mathcal{H}$, then every element in $\mathcal{H}$ has a representation as a superposition of the frame elements. Thus it is natural to view a frame as some kind of "generalized basis."

Theorem 5.1.6 *Let $\{f_k\}_{k=1}^\infty$ be a frame with frame operator S. Then*

$$f = \sum_{k=1}^\infty \langle f, S^{-1}f_k\rangle f_k, \quad \forall f \in \mathcal{H}, \tag{5.7}$$

and

$$f = \sum_{k=1}^\infty \langle f, f_k\rangle S^{-1}f_k, \quad \forall f \in \mathcal{H}. \tag{5.8}$$

Both series converge unconditionally for all $f \in \mathcal{H}$.

Proof.   Let $f \in \mathcal{H}$. Using the properties of the frame operator in Lemma 5.1.5,

$$f = SS^{-1}f = \sum_{k=1}^\infty \langle S^{-1}f, f_k\rangle f_k = \sum_{k=1}^\infty \langle f, S^{-1}f_k\rangle f_k.$$

Since $\{f_k\}_{k=1}^\infty$ is a Bessel sequence and $\{\langle f, S^{-1}f_k\rangle\}_{k=1}^\infty \in \ell^2(\mathbb{N})$, the fact that the series converges unconditionally follows from Corollary 3.2.5. The expansion (5.8) is proved similarly, using that $f = S^{-1}Sf$.   □

Theorem 5.1.6 shows that all information about the given $f \in \mathcal{H}$ is contained in the sequence $\{\langle f, S^{-1}f_k\rangle\}_{k=1}^\infty$. The numbers $\langle f, S^{-1}f_k\rangle$ are called *frame coefficients*.

Theorem 5.1.6 also immediately reveals one of the main difficulties in frame theory. In fact, in order for the expansions (5.7) and (5.8) to be applicable in practice, we need to be able to find the operator S^{-1} or at least to calculate its action on all f_k, $k \in \mathbb{N}$. In general, this is a major problem. One way of circumventing the problem is to consider tight frames:

Corollary 5.1.7 *If $\{f_k\}_{k=1}^\infty$ is a tight frame with frame bound A, then the canonical dual frame is $\{A^{-1}f_k\}_{k=1}^\infty$ and*

$$f = \frac{1}{A}\sum_{k=1}^\infty \langle f, f_k\rangle f_k, \quad \forall f \in \mathcal{H}. \tag{5.9}$$

Proof. If $\{f_k\}_{k=1}^{\infty}$ is a tight frame with frame bound A and frame operator S, the definition shows that

$$\langle Sf, f \rangle = \sum_{k=1}^{\infty} |\langle f, f_k \rangle|^2 = A \, ||f||^2 = \langle Af, f \rangle, \ \forall f \in \mathcal{H}.$$

By Lemma 2.4.4, this implies that $S = AI$; thus, S^{-1} acts by multiplication by A^{-1}, and the result follows from (5.7). □

Later we will discuss another way to avoid the problem of inverting the frame operator S. In fact, for frames $\{f_k\}_{k=1}^{\infty}$ that are *not* bases, we prove in Theorem 6.3.1 that one can find other frames $\{g_k\}_{k=1}^{\infty}$ than $\{S^{-1}f_k\}_{k=1}^{\infty}$, for which

$$f = \sum_{k=1}^{\infty} \langle f, g_k \rangle f_k, \ \forall f \in \mathcal{H}. \tag{5.10}$$

Such a frame $\{g_k\}_{k=1}^{\infty}$ is called a *dual frame* of $\{f_k\}_{k=1}^{\infty}$. Now, there is a chance that even if the canonical dual frame is difficult to find, there exist other duals that are (comparably) easy to find or have better properties. We will see several such examples in the analysis of concrete frames consisting of functions in $L^2(\mathbb{R})$, e.g., in the Gabor case in Section 12.5 or for wavelet frames in Section 18.8. For general frames in Hilbert spaces, all duals are characterized in Section 6.3.

Example 5.1.8 Let $\{e_k\}_{k=1}^{\infty}$ be an orthonormal basis for $\mathcal{H}$ and consider the frame

$$\{f_k\}_{k=1}^{\infty} = \{e_1, e_1, e_2, e_3, ..\};$$

see Example 5.1.4(i). The canonical dual frame is given by

$$\{S^{-1}f_k\}_{k=1}^{\infty} = \left\{ \frac{1}{2}e_1, \frac{1}{2}e_1, e_2, e_3, .. \right\}.$$

As examples of noncanonical dual frames, we mention

$$\{g_k\}_{k=1}^{\infty} = \{0, e_1, e_2, e_3, ..\}$$

and

$$\{g_k\}_{k=1}^{\infty} = \left\{ \frac{1}{3}e_1, \frac{2}{3}e_1, e_2, e_3, .. \right\}.$$

We leave the verifications to the reader (Exercise 5.6). □

The following lemma shows that it is enough to check the frame condition on a dense set.

Lemma 5.1.9 *Suppose that* $\{f_k\}_{k=1}^\infty$ *is a sequence of elements in* $\mathcal{H}$ *and that there exist constants* $A, B > 0$ *such that*

$$A \, ||f||^2 \leq \sum_{k=1}^\infty |\langle f, f_k \rangle|^2 \leq B \, ||f||^2 \tag{5.11}$$

for all f *in a dense subset* V *of* $\mathcal{H}$. *Then* $\{f_k\}_{k=1}^\infty$ *is a frame for* $\mathcal{H}$ *with bounds* A, B.

Proof. We proved already in Lemma 3.2.6 that $\{f_k\}_{k=1}^\infty$ is a Bessel sequence with bound B if (5.11) is satisfied. We now prove that (5.11) implies that the lower frame condition is satisfied on $\mathcal{H}$. Expressed in terms of the synthesis operator T, our assumption means that

$$A \, ||f||^2 \leq ||T^* f||^2, \ \forall f \in V. \tag{5.12}$$

Since T^* is bounded and V is dense in $\mathcal{H}$, it follows that (5.12) holds for all $f \in \mathcal{H}$. $\square$

A note: the proof that the lower frame condition extends from a dense set to $\mathcal{H}$ uses the assumption about the upper frame condition being satisfied.

Returning to the definition of a frame, it is clear that it is complicated to deal with the condition (5.2) in general: how can we possibly check a condition involving arbitrary vectors in an abstract Hilbert space? Let us end this section with an observation by Stoeva, showing that in case we know an orthonormal basis for the Hilbert space, we can turn the condition into a condition involving finite sequences of scalars (Exercise 5.4).

Proposition 5.1.10 *Let* $\{f_k\}_{k=1}^\infty$ *be a sequence in a Hilbert space* $\mathcal{H}$, *having an orthonormal basis* $\{e_k\}_{k=1}^\infty$. *Furthermore, let* $A, B > 0$ *be given. Finally, let*

$$\mathcal{S} := \left\{ \{c_k\}_{k=1}^\infty \in \ell^2(\mathbb{N}) \mid \{c_k\}_{k=1}^\infty \text{ is finite and } \sum_{k=1}^\infty |c_k|^2 = 1 \right\}.$$

Then the following hold:

(i) *If* $A \leq \sum_{k=1}^\infty \left| \sum_{j=1}^\infty c_j \langle e_j, f_k \rangle \right|^2 \leq B$ *for all* $\{c_k\}_{k=1}^\infty \in \mathcal{S}$, *then* $\{f_k\}_{k=1}^\infty$ *is a frame for* $\mathcal{H}$ *with bounds* A, B.

(ii) *There exists a countable collection of sequences* $\{c_j^{(\ell)}\}_{j=1}^\infty \in \mathcal{S}$, $\ell \in \mathbb{N}$, *such that* $\{f_k\}_{k=1}^\infty$ *is a frame for* $\mathcal{H}$ *with bounds* A, B *if*

$$A \leq \sum_{k=1}^\infty \left| \sum_{j=1}^\infty c_j^{(\ell)} \langle e_j, f_k \rangle \right|^2 \leq B, \ \forall \ell \in \mathbb{N}.$$

Proposition 5.1.10 is motivated by the characterization of Riesz sequences in Corollary 3.7.4. We note that for infinite-dimensional Hilbert spaces, the case of Riesz sequences is easier: the condition in Corollary 3.7.4 does not involve knowledge of an appropriate orthonormal basis, and no infinite sum appears in the condition. We return to a further comparison of these results in Section 8.3.

5.2 Frame Sequences

Frame sequences and Riesz sequences are useful concepts in cases where we only obtain (or are interested in) expansions in subspaces. As a mathematical example, consider $L^2(-\pi, \pi)$ versus $L^2(\mathbb{R})$: restricting the functions in a frame for $L^2(\mathbb{R})$ to the interval $]-\pi, \pi[$ gives a frame for $L^2(-\pi, \pi)$. On the other hand, if we extend the functions in a frame for $L^2(-\pi, \pi)$ to functions in $L^2(\mathbb{R})$, by defining them to be zero on $\mathbb{R}\setminus]-\pi, \pi[$, we obtain a frame sequence in $L^2(\mathbb{R})$. Concrete examples of frame sequences appear in Chapter 9, where we study frames of translates.

The terminology is also useful in signal processing, where it might be known that the class of relevant signals for a concrete application belongs to a certain subspace of $L^2(\mathbb{R})$ (e.g., the Paley–Wiener space of functions whose Fourier transform has support in $[-\pi, \pi]$).

We state a criteria for a frame sequence being a frame:

Lemma 5.2.1 *Let $\{f_k\}_{k=1}^{\infty}$ be a frame sequence in $\mathcal{H}$, with synthesis operator $T : \ell^2(\mathbb{N}) \to \mathcal{H}$. Then $\{f_k\}_{k=1}^{\infty}$ is a frame for $\mathcal{H}$ if and only if T^* is injective.*

Proof. Since $\{f_k\}_{k=1}^{\infty}$ is a frame sequence, $\mathcal{R}_T = \overline{\operatorname{span}}\{f_k\}_{k=1}^{\infty}$. Because $\mathcal{N}_{T^*} = \mathcal{R}_T^{\perp}$, the operator T^* is injective if and only if the range of T is dense in $\mathcal{H}$. $\qquad\square$

If $\{f_k\}_{k=1}^{\infty}$ is a frame sequence, we can extend Lemma 3.5.2 concerning the Gram matrix:

Proposition 5.2.2 *If $\{f_k\}_{k=1}^{\infty}$ is a frame sequence in $\mathcal{H}$, then the associated Gram matrix defines a bounded invertible operator from the Banach space $\mathcal{R}_{T^*}$ onto $\mathcal{R}_{T^*}$, with a bounded inverse.*

Proof. If $\{f_k\}_{k=1}^{\infty}$ is a frame sequence, then $\mathcal{R}_T = \overline{\operatorname{span}}\{f_k\}_{k=1}^{\infty}$. Since $\mathcal{H} = \mathcal{R}_T \oplus \mathcal{R}_T^{\perp} = \mathcal{R}_T \oplus \mathcal{N}_{T^*}$, we can write any $f \in \mathcal{H}$ as $f = T\{c_k\}_{k=1}^{\infty} + z$ for some $\{c_k\}_{k=1}^{\infty} \in \ell^2(\mathbb{N})$, $z \in \mathcal{N}_{T^*}$. Thus,

$$T^*f = T^*T\{c_k\}_{k=1}^{\infty}.$$

Therefore $\mathcal{R}_{T^*T} = \mathcal{R}_{T^*}$. The rest follows from Lemma 3.5.2 and Theorem 2.2.2. $\qquad\square$

If $\{f_k\}_{k=1}^\infty$ is a frame sequence, we can consider the frame operator as a bijective bounded operator $S : \overline{\mathrm{span}}\{f_k\}_{k=1}^\infty \to \overline{\mathrm{span}}\{f_k\}_{k=1}^\infty$ and write the frame decomposition on the usual form,

$$f = \sum_{k=1}^\infty \langle f, S^{-1}f_k\rangle f_k, \ f \in \overline{\mathrm{span}}\{f_k\}_{k=1}^\infty. \tag{5.13}$$

Let us follow up on the short discussion of dual frames on page 125. Also for a frame sequence $\{f_k\}_{k=1}^\infty$, we can look for sequences $\{g_k\}_{k=1}^\infty \ne \{S^{-1}f_k\}_{k=1}^\infty$ such that

$$f = \sum_{k=1}^\infty \langle f, g_k\rangle f_k, \ \forall f \in \overline{\mathrm{span}}\{f_k\}_{k=1}^\infty. \tag{5.14}$$

For frame sequences, the gained flexibility is even larger than for frames, because the sequence $\{g_k\}_{k=1}^\infty$ is not a priori required to belong to $\overline{\mathrm{span}}\{f_k\}_{k=1}^\infty$. Depending on the exact constraints put on $\{g_k\}_{k=1}^\infty$, it is called a *pseudodual* of $\{f_k\}_{k=1}^\infty$ (by Li and Ogawa in [492]) or an *oblique dual* (see [169]). We also note that Li and Ogawa [493] have defined the general concept of a *pseudoframe* for a subspace; in this case, neither the sequence itself nor its "dual" is required to belong to the subspace where the "frame-like" expansion takes place. We will not go into the theory of these types of duals but just refer to the above papers and the references therein. A concrete case will be considered in Section 9.5.

Let us end this section with a result that connects frames, frame sequences, and orthogonal projections.

Proposition 5.2.3 *Let V denote a closed subspace of a Hilbert space $\mathcal{H}$. Then the following holds:*

(i) *If $\{f_k\}_{k=1}^\infty$ is a frame for $\mathcal{H}$ with frame bounds A, B and P denotes the orthogonal projection of $\mathcal{H}$ onto V, then $\{Pf_k\}_{k=1}^\infty$ is a frame for V with frame bounds A, B.*

(ii) *If $\{f_k\}_{k=1}^\infty$ is a frame for V with frame operator $S : V \to V$, then the orthogonal projection of $\mathcal{H}$ onto V is given by*

$$Pf = \sum_{k=1}^\infty \langle f, S^{-1}f_k\rangle f_k, \ f \in \mathcal{H}. \tag{5.15}$$

The proof of (i) is left to the reader (Exercise 5.10), and the proof of (ii) is identical to the proof of Theorem 1.1.9.

5.3   Frames and Operators

Lemma 5.1.5 shows that if $\{f_k\}_{k=1}^\infty$ is a frame, then the canonical dual $\{S^{-1}f_k\}_{k=1}^\infty$ is also a frame. This is a special case of a much more general result: $\{Uf_k\}_{k=1}^{\infty}$ is actually a frame for a large class of operators U. For later reference, we state some general versions of this result, where we assume that U is a bounded operator with closed range $\mathcal{R}_U$. We denote the *pseudo-inverse* (see Lemma 2.5.1) of such an operator U by $U^{\dagger}$.

Proposition 5.3.1 *Let* $\{f_k\}_{k=1}^{\infty}$ *be a frame for* $\mathcal{H}$ *with bounds* A, B, *and let* $U \neq 0$ *be a bounded operator on* $\mathcal{H}$ *with closed range. Then* $\{Uf_k\}_{k=1}^{\infty}$ *is a frame sequence with frame bounds* $A\,||U^{\dagger}||^{-2}$, $B\,||U||^2$.

Proof. If $f \in \mathcal{H}$, then

$$\sum_{k=1}^{\infty} |\langle f, Uf_k\rangle|^2 \leq B\,||U^*f||^2 \leq B\,||U||^2\,||f||^2,$$

which proves that $\{Uf_k\}_{k=1}^{\infty}$ is a Bessel sequence. For the lower frame condition, let $g \in \mathcal{R}_U$; we can write $g = Uf$ for some $f \in \mathcal{H}$. By Lemma 2.5.2, the operator $UU^{\dagger}$ is the orthogonal projection onto $\mathcal{R}_U$ and therefore self-adjoint. Therefore

$$g = Uf = (UU^{\dagger})^* Uf = (U^{\dagger})^* U^* Uf.$$

It follows that

$$
\begin{aligned}
||g||^2 &\leq ||(U^{\dagger})^*||^2\,||U^*Uf||^2 \\
&\leq \frac{||(U^{\dagger})^*||^2}{A} \sum_{k=1}^{\infty} |\langle U^*Uf, f_k\rangle|^2 \\
&= \frac{||U^{\dagger}||^2}{A} \sum_{k=1}^{\infty} |\langle g, Uf_k\rangle|^2.
\end{aligned}
$$

Thus the lower frame condition is satisfied for all $g \in \mathcal{R}_U$.   □

Exercise 5.11 shows that the conclusion in Proposition 5.3.1 might fail if U does not have closed range. And even if U has closed range, it is not enough to assume that $\{f_k\}_{k=1}^{\infty}$ is a frame sequence (Exercise 5.12).

Corollary 5.3.2 *Assume that* $\{f_k\}_{k=1}^{\infty}$ *is a frame for* $\mathcal{H}$ *with bounds* A, B *and that* $U : \mathcal{H} \to \mathcal{H}$ *is a bounded surjective operator. Then* $\{Uf_k\}_{k=1}^{\infty}$ *is a frame for* $\mathcal{H}$ *with frame bounds* $A\,||U^{\dagger}||^{-2}$, $B\,||U||^2$.

In the next result it is enough to assume that $\{f_k\}_{k=1}^{\infty}$ is a frame sequence. We leave the proof to the reader (Exercise 5.13).

Lemma 5.3.3 *If* $\{f_k\}_{k=1}^{\infty}$ *is a frame sequence with frame bounds* A, B *and* $U : \mathcal{H} \to \mathcal{H}$ *is a unitary operator, then* $\{Uf_k\}_{k=1}^{\infty}$ *is a frame sequence with frame bounds* A, B.

Corollary 5.3.4 *If $\{f_k\}_{k=1}^{\infty}$ is a frame for $\mathcal{H}$ with frame bounds A, B and $U : \mathcal{H} \rightarrow \mathcal{H}$ is a unitary operator, then $\{Uf_k\}_{k=1}^{\infty}$ is also a frame for $\mathcal{H}$ with frame bounds A, B.*

More connections between frames and operators will appear later, e.g., in Theorem 5.5.4.

5.4   Frames and Bases

Let us now mention some important relationships between frames and Riesz bases. Much more information is contained in Chapter 7.

Theorem 5.4.1 *A Riesz basis $\{f_k\}_{k=1}^{\infty}$ for $\mathcal{H}$ is a frame for $\mathcal{H}$, and the Riesz basis bounds coincide with the frame bounds. The dual Riesz basis is $\{S^{-1}f_k\}_{k=1}^{\infty}$.*

Proof. By Proposition 3.6.4, a Riesz basis $\{f_k\}_{k=1}^{\infty}$ for $\mathcal{H}$ is also a frame for $\mathcal{H}$; if we also involve Proposition 3.6.7, we obtain the statement about the bounds. The rest follows from the frame decomposition combined with the uniqueness part of Theorem 3.6.2.   □

A frame which is *not* a Riesz basis is said to be *overcomplete;* in the literature, the terms *nonexact frame* and *redundant frame* are also used. Theorem 7.1.1 (vii) will explain why the word "overcomplete" is used: in fact, if $\{f_k\}_{k=1}^{\infty}$ is a frame which is not a Riesz basis, there exist coefficients $\{c_k\}_{k=1}^{\infty} \in \ell^2(\mathbb{N}) \setminus \{0\}$ for which

$$\sum_{k=1}^{\infty} c_k f_k = 0.$$

We have already seen in Theorem 5.1.6 that the frame coefficients $\{\langle f, S^{-1}f_k \rangle\}_{k=1}^{\infty}$ lead to a representation of the given $f \in \mathcal{H}$. As in the finite-dimensional case, the frame coefficients $\{\langle f, S^{-1}f_k \rangle\}_{k=1}^{\infty}$ have minimal ℓ^2-norm among all sequences representing f:

Lemma 5.4.2 *Let $\{f_k\}_{k=1}^{\infty}$ be a frame for $\mathcal{H}$ and let $f \in \mathcal{H}$. If f has a representation $f = \sum_{k=1}^{\infty} c_k f_k$ for some coefficients $\{c_k\}_{k=1}^{\infty}$, then*

$$\sum_{k=1}^{\infty} |c_k|^2 = \sum_{k=1}^{\infty} |\langle f, S^{-1}f_k \rangle|^2 + \sum_{k=1}^{\infty} |c_k - \langle f, S^{-1}f_k \rangle|^2.$$

The proof is identical to the proof of Theorem 1.1.5(iii). As a consequence of Lemma 5.4.2 and Theorem 2.5.3, we obtain an explicit expression for the pseudo-inverse of the synthesis operator:

Theorem 5.4.3 *Let* $\{f_k\}_{k=1}^{\infty}$ *be a frame for* $\mathcal{H}$ *with synthesis operator* T *and frame operator* S. *Then*

$$T^{\dagger}f = \{\langle f, S^{-1}f_k\rangle\}_{k=1}^{\infty}, \forall f \in \mathcal{H}.$$

The optimal frame bounds can be expressed in terms of the operators T, S and their inverses/pseudo-inverses:

Proposition 5.4.4 *The optimal frame bounds* A, B *for a frame* $\{f_k\}_{k=1}^{\infty}$ *are given by*

$$A = ||S^{-1}||^{-1} = ||T^{\dagger}||^{-2}, \quad B = ||S|| = ||T||^2.$$

Proof. By definition and using (2.8),

$$B = \sup_{||f||=1} \sum_{k=1}^{\infty} |\langle f, f_k\rangle|^2 = \sup_{||f||=1} \langle Sf, f\rangle = ||S||.$$

Applying this on the dual frame $\{S^{-1}f_k\}_{k=1}^{\infty}$ (which has frame operator S^{-1} and the optimal upper bound $\frac{1}{A}$ by Lemma 5.1.5), we obtain $\frac{1}{A} = ||S^{-1}||$. For the rest, $S = TT^*$ implies that $||S|| = ||TT^*|| = ||T||^2$. Finally, via Theorem 5.4.3 and Lemma 5.1.5,

$$||T^{\dagger}f||^2 = \sum_{k=1}^{\infty} |\langle f, S^{-1}f_k\rangle|^2 \leq \frac{1}{A}||f||^2,$$

where $\frac{1}{A}$ is the smallest possible constant; thus $||T^{\dagger}||^2 = \frac{1}{A}$. □

Chapter 9–20 will give us several examples of overcomplete frames for $L^2(0,1)$ and $L^2(\mathbb{R})$. For the moment we do not go into those constructions, but we just note that already Example 4.1.1 illustrates what is meant by a frame being overcomplete:

Example 5.4.5 Let us return to Example 4.1.1, where we considered the orthonormal basis $\{e_k\}_{k\in\mathbb{Z}} = \{e^{2\pi ikx}\}_{k\in\mathbb{Z}}$ for $L^2(0,1)$. Let $I \subset [0,1]$ be a proper subinterval, $|I| < 1$. Since

$$\sum_{k\in\mathbb{Z}} |\langle f, e_k\rangle|^2 = ||f||^2, \forall f \in L^2(0,1),$$

the equality in particular holds for all $f \in L^2(I)$. So $\{e_k\}_{k\in\mathbb{Z}}$ is a tight frame for $L^2(I)$. We have already proved that it is overcomplete. However, recall from Lemma 1.7.2 that $\{e_k\}_{k\in\mathbb{Z}}$ is linearly independent. □

Example 5.4.5 points at a central property of frames: they can be overcomplete and linearly independent at the same time. The reason for this is the difference between linear independence (i.e., independence of all finite subsets) and ω-independence, as discussed in Section 3.1. For frames, the correct notion of independence is ω-independence; we return to this point in Section 7.1.

Misled by the situation in the finite-dimensional setting, one could expect that if $\overline{\operatorname{span}}\{f_k\}_{k=1}^\infty = \mathcal{H}$, then every $f \in \mathcal{H}$ would have an expansion $f = \sum_{k=1}^\infty c_k f_k$ for certain scalar coefficients $\{c_k\}_{k=1}^\infty$. However, in an infinite-dimensional Hilbert space, this does not necessarily hold. We give an example below; the example will be used several times in the sequel.

Example 5.4.6 Let $\{e_k\}_{k=1}^\infty$ be an orthonormal basis for $\mathcal{H}$ and define

$$f_k := e_k + e_{k+1}, \ k \in \mathbb{N}.$$

We will show that:

(i) $\overline{\operatorname{span}}\{f_k\}_{k=1}^\infty = \mathcal{H}$;

(ii) $\{f_k\}_{k=1}^\infty$ is a Bessel sequence but not a frame;

(iii) There exists $f \in \mathcal{H}$ that cannot be written on the form $\sum_{k=1}^\infty c_k f_k$ for any choice of the coefficients c_k.

(iv) $\{f_k\}_{k=1}^\infty$ is minimal and that its unique biorthogonal sequence $\{g_k\}_{k=1}^\infty$ is given by

$$g_k = (-1)^k \sum_{j=1}^k (-1)^j e_j, \ k \in \mathbb{N}.$$

To prove (i), i.e., that $\{f_k\}_{k=1}^\infty$ is complete, assume that $f \in \mathcal{H}$ and that

$$\langle f, f_k \rangle = 0 \text{ for all } k \in \mathbb{N}.$$

Then $\langle f, e_k \rangle = -\langle f, e_{k+1} \rangle$ for all $k \in \mathbb{N}$, implying that $|\langle f, e_k \rangle|$ is a constant. Since

$$\sum_{k=1}^\infty |\langle f, e_k \rangle|^2 = ||f||^2 < \infty,$$

we conclude that $\langle f, e_k \rangle = 0$, $\forall k$, so $f = 0$. Thus $\{f_k\}_{k=1}^\infty$ is complete.

We now prove (ii); we first show that $\{f_k\}_{k=1}^\infty$ is a Bessel sequence. For that purpose, we will use that for any $a, b \in \mathbb{R}$, the inequality $(a+b)^2 \leq 2(a^2 + b^2)$ holds. Now, for any $f \in \mathcal{H}$,

$$
\begin{aligned}
\sum_{k=1}^\infty |\langle f, e_k + e_{k+1} \rangle|^2 &= \sum_{k=1}^\infty |\langle f, e_k \rangle + \langle f, e_{k+1} \rangle|^2 \\
&\leq \sum_{k=1}^\infty (|\langle f, e_k \rangle| + |\langle f, e_{k+1} \rangle|)^2 \\
&\leq 2 \sum_{k=1}^\infty |\langle f, e_k \rangle|^2 + 2 \sum_{k=1}^\infty |\langle f, e_{k+1} \rangle|^2 \\
&\leq 4 ||f||^2.
\end{aligned}
$$

This proves that $\{f_k\}_{k=1}^\infty$ is a Bessel sequence. However, $\{f_k\}_{k=1}^\infty$ does not satisfy the lower frame condition. To see this, consider the vectors

$$g_j := \sum_{n=1}^j (-1)^{n+1} e_n, \ j \in \mathbb{N}.$$

We note that $\|g_j\|^2 = j$ for all $j \in \mathbb{N}$. Let us now calculate the inner products $\langle g_j, f_k \rangle$. Considering a fixed $j \in \mathbb{N}$, we see that

$$\langle g_j, f_k \rangle = \langle e_1 - e_2 + \cdots + (-1)^{j+1} e_j, e_k + e_{k+1} \rangle = \begin{cases} 0 & \text{if } k > j; \\ (-1)^{j+1} & \text{if } k = j; \\ 0 & \text{if } k < j. \end{cases}$$

Therefore

$$\sum_{k=1}^\infty |\langle g_j, f_k \rangle|^2 = 1 = \frac{1}{j}\, \|g_j\|^2.$$

Since this holds for all $j \in \mathbb{N}$, we see that $\{f_k\}_{k=1}^\infty$ does not satisfy the lower frame condition; this completes the proof of (ii).

Concerning (iii), despite the fact that $\{f_k\}_{k=1}^\infty$ is complete, there exists $f \in \mathcal{H}$ that cannot be written as $f = \sum_{k=1}^\infty c_k f_k$ for *any* choice of the coefficients $\{c_k\}_{k=1}^\infty$. As a concrete example, take $f = e_1$.

We now prove the first part of (iv), namely, that $\{f_k\}_{k=1}^\infty$ is minimal. Assume the opposite, i.e., that for some $j \in \mathbb{N}$,

$$\begin{aligned} f_j \ &\in \ \overline{\text{span}}\{f_k\}_{k \neq j} \\ &= \ \overline{\text{span}}\{e_1 + e_2, e_2 + e_3, \dots, e_{j-1} + e_j, e_{j+1} + e_{j+2}, \dots\}. \end{aligned}$$

$$(5.16)$$

Note that the space in (5.16) is an orthogonal sum of the subspaces $\text{span}\{e_1 + e_2, e_2 + e_3, \dots, e_{j-1} + e_j\}$ and $\overline{\text{span}}\{e_{j+1} + e_{j+2}, \dots\}$. Letting P denote the orthogonal projection of $\mathcal{H}$ onto $\text{span}\{e_1 + e_2, e_2 + e_3, \dots, e_{j-1} + e_j\}$ it follows that

$$e_j = P f_j \in \text{span}\{e_1 + e_2, e_2 + e_3, \dots, e_{j-1} + e_j\},$$

a conclusion that certainly does not hold. Thus $f_j \notin \overline{\text{span}}\{f_k\}_{k \neq j}$, i.e., $\{f_k\}_{k=1}^\infty$ is minimal. By Lemma 3.3.1, $\{f_k\}_{k=1}^\infty$ has a unique biorthogonal sequence $\{g_k\}_{k=1}^\infty$, which is determined by the conditions

$$\langle g_k, e_k + e_{k+1} \rangle = 1, \quad \langle g_k, e_j + e_{j+1} \rangle = 0 \text{ for } j \neq k. \quad (5.17)$$

In order to find $\{g_k\}_{k=1}^\infty$, fix $k \in \mathbb{N}$, and let $C := \langle g_k, e_k \rangle$. Then the first condition in (5.17) implies that $\langle g_k, e_{k+1} \rangle = 1 - C$; now the second condition implies that, in general for $j > k$, $|\langle g_k, e_j \rangle| = |1 - C|$. Because $\{e_j\}_{j=1}^\infty$ is a Bessel sequence, we know that

$$\sum_{j=k+1}^\infty |\langle g_k, e_j \rangle|^2 \leq \sum_{j=1}^\infty |\langle g_k, e_j \rangle|^2 < \infty,$$

so it follows that $C = 1$, i.e.,

$$\langle g_k, e_k \rangle = 1 \text{ and } \langle g_k, e_j \rangle = 0 \text{ for } j > k.$$

Now apply the second condition in (5.17) for $j = k - 1, k - 2, \ldots, 1$; this shows that $\langle g_k, e_j \rangle = (-1)^{k-j}$, $j = 1, \ldots, k$. We now put all the information together and conclude that the biorthogonal sequence is given by

$$g_k = \sum_{j=1}^{\infty} \langle g_k, e_j \rangle e_j = \sum_{j=1}^{k} (-1)^{k-j} e_j = (-1)^k \sum_{j=1}^{k} (-1)^j e_j.$$

This concludes the proof of (iv).

Note that the example demonstrates that the union of bases for subspaces might not be a basis or a frame. In fact, $\{\frac{1}{\sqrt{2}}(e_{2k-1} + e_{2k})\}_{k=1}^{\infty}$ is an orthonormal basis for $\overline{\text{span}}\{e_{2k-1} + e_{2k}\}_{k=1}^{\infty}$, and $\{\frac{1}{\sqrt{2}}(e_{2k} + e_{2k+1})\}_{k=1}^{\infty}$ is an orthonormal basis for $\overline{\text{span}}\{e_{2k} + e_{2k+1}\}_{k=1}^{\infty}$. But the union is the family

$$\left\{ \frac{1}{\sqrt{2}}(e_k + e_{k+1}) \right\}_{k=1}^{\infty},$$

which is not a basis or a frame for

$$\overline{\text{span}} \left\{ \frac{1}{\sqrt{2}}(e_k + e_{k+1}) \right\}_{k=1}^{\infty} = \mathcal{H}.$$

There does not exist a basis for $\mathcal{H}$ containing $\{f_k\}_{k=1}^{\infty}$ as a subset (Exercise 5.15). Furthermore, in Example 22.2.3, we prove that $\{f_k\}_{k=1}^{\infty}$ cannot be extended to a frame for $\mathcal{H}$ by adding a finite number of elements. But it follows from (ii) that if a sequence $\{h_k\}_{k=1}^{\infty}$ is a frame, then also $\{h_k\}_{k=1}^{\infty} \cup \{f_k\}_{k=1}^{\infty}$ is a frame.

Note that an extension is stated in Exercise 5.14: even if $\{f_k\}_{k=1}^{\infty}$ is assumed to be a Riesz basis, $\{f_k + f_{k+1}\}_{k=1}^{\infty}$ cannot be a frame. We will see in Proposition 9.2.8 that the situation is different if $\{f_k\}_{k=1}^{\infty}$ is allowed to be a frame. □

Because a frame $\{f_k\}_{k=1}^{\infty}$ might be overcomplete, it is possible that removal of an element f_j leaves us with a sequence $\{f_k\}_{k \neq j}$ that is still a frame. It turns out that whether this happens or not can be determined based on the value of the frame coefficient $\langle f_j, S^{-1} f_j \rangle$:

Theorem 5.4.7 *The removal of a vector f_j from a frame $\{f_k\}_{k=1}^{\infty}$ for $\mathcal{H}$ leaves either a frame or an incomplete set. More precisely:*

(i) If $\langle f_j, S^{-1} f_j \rangle \neq 1$, then $\{f_k\}_{k \neq j}$ is a frame for $\mathcal{H}$;

(ii) If $\langle f_j, S^{-1} f_j \rangle = 1$, then $\{f_k\}_{k \neq j}$ is incomplete.

Proof. Choose $j \in \mathbb{N}$ arbitrarily. By the frame decomposition,

$$f_j = \sum_{k=1}^{\infty} \langle f_j, S^{-1} f_k \rangle f_k.$$

Define, for notational convenience, $a_k = \langle f_j, S^{-1} f_k \rangle$, so $f_j = \sum_{k=1}^{\infty} a_k f_k$. Clearly, we also have $f_j = \sum_{k=1}^{\infty} \delta_{j,k} f_k$, so Lemma 5.4.2 yields the following relation between $\delta_{j,k}$ and a_k:

$$
\begin{aligned}
1 = \sum_{k=1}^{\infty} |\delta_{j,k}|^2 &= \sum_{k=1}^{\infty} |a_k|^2 + \sum_{k=1}^{\infty} |a_k - \delta_{j,k}|^2 \\
&= |a_j|^2 + \sum_{k \neq j} |a_k|^2 + |a_j - 1|^2 + \sum_{k \neq j} |a_k|^2.
\end{aligned}
$$

We consider the cases $a_j = 1$ and $a_j \neq 1$ separately. First, suppose that $a_j = 1$. From the above formula, $\sum_{k \neq j} |a_k|^2 = 0$, so that

$$a_k = \langle S^{-1} f_j, f_k \rangle = 0 \text{ for all } k \neq j.$$

Since $a_j = \langle S^{-1} f_j, f_j \rangle = 1$, we know that $S^{-1} f_j \neq 0$. Thus we have found a nonzero element $S^{-1} f_j$ which is orthogonal to $\{f_k\}_{k \neq j}$, so $\{f_k\}_{k \neq j}$ is incomplete. This proves (ii).

Suppose now that $a_j \neq 1$; then $f_j = \frac{1}{1-a_j} \sum_{k \neq j} a_k f_k$. For any $f \in \mathcal{H}$, Cauchy–Schwarz inequality gives

$$
\begin{aligned}
|\langle f_j, f \rangle|^2 &= \left| \frac{1}{1-a_j} \sum_{k \neq j} a_k \langle f_k, f \rangle \right|^2 \\
&\leq \frac{1}{|1-a_j|^2} \sum_{k \neq j} |a_k|^2 \sum_{k \neq j} |\langle f_k, f \rangle|^2 \\
&= C \sum_{k \neq j} |\langle f, f_k \rangle|^2,
\end{aligned}
$$

where $C = \frac{1}{|1-a_j|^2} \sum_{k \neq j} |a_k|^2$. Let A denote a lower frame bound for $\{f_k\}_{k=1}^{\infty}$. Then

$$
\begin{aligned}
A \|f\|^2 \leq \sum_{k=1}^{\infty} |\langle f, f_k \rangle|^2 &= \sum_{k \neq j} |\langle f, f_k \rangle|^2 + |\langle f, f_j \rangle|^2 \\
&\leq (1+C) \sum_{k \neq j} |\langle f, f_k \rangle|^2,
\end{aligned}
$$

showing that $\{f_k\}_{k \neq j}$ satisfies the lower frame condition with lower bound $\frac{A}{1+C}$. Clearly $\{f_k\}_{k \neq j}$ also satisfies the upper frame condition. Thus $\{f_k\}_{k \neq j}$ is a frame; this proves (i). $\qquad \square$

We return to a related result in Corollary 22.2.2. The proof of Theorem 5.4.7 shows an interesting property of $\{f_k\}_{k=1}^\infty$ and the dual frame $\{S^{-1}f_k\}_{k=1}^\infty$, which will be used later.

Proposition 5.4.8 *If* $\{f_k\}_{k=1}^\infty$ *is an exact frame, then* $\{f_k\}_{k=1}^\infty$ *and* $\{S^{-1}f_k\}_{k=1}^\infty$ *are biorthogonal and* $\{f_k\}_{k=1}^\infty$ *is a basis for* $\mathcal{H}$.

Proof.   Assume that $\{f_k\}_{k=1}^\infty$ is an exact frame and fix $j \in \mathbb{N}$. Then $\{f_k\}_{k \neq j}$ is not a frame, implying by Theorem 5.4.7 that $\langle f_j, S^{-1}f_j \rangle = 1$. The proof of Theorem 5.4.7 now shows that $\langle f_j, S^{-1}f_k \rangle = \delta_{j,k}$, i.e., that $\{f_k\}_{k=1}^\infty$ and $\{S^{-1}f_k\}_{k=1}^\infty$ are biorthogonal. By the frame decomposition, we have that every $f \in \mathcal{H}$ can be expressed as $f = \sum_{k=1}^\infty \langle f, S^{-1}f_k \rangle f_k$. In order to show that $\{f_k\}_{k=1}^\infty$ is a basis, it is enough to show that this representation is unique. But if $f = \sum_{k=1}^\infty b_k f_k$ for some coefficients b_k, then

$$\langle f, S^{-1}f_k \rangle = \left\langle \sum_{j=1}^\infty b_j f_j, S^{-1}f_k \right\rangle = \sum_{j=1}^\infty b_j \langle f_j, S^{-1}f_k \rangle = b_k. \qquad \square$$

In Theorem 7.1.1 we will prove more: an exact frame is actually a Riesz basis.

5.5 Characterization of Frames

Let us for a moment go back to the definition of a frame. In order to check that a sequence $\{f_k\}_{k=1}^\infty$ is a frame, we have to verify the existence of a positive lower frame bound A and a finite upper frame bound B. Intuitively, the lower frame condition is the most critical to verify: bad upper estimates on $\sum_{k=1}^\infty |\langle f, f_k \rangle|^2$ will sometimes force us to take a larger value for B than necessary, but bad lower estimates can easily make it impossible to find a value for A which can be used for all $f \in \mathcal{H}$. Later we will see more exact statements, which support this observation. For example, Proposition 11.5.2 will show that relatively weak decay conditions on a function $g \in L^2(\mathbb{R})$ imply that a Gabor system $\{E_{mb}T_{na}g\}_{m,n\in\mathbb{Z}}$ satisfies the upper frame condition, but no similar statement about the lower frame condition exists (unless we are allowed to vary the parameter b; see Proposition 11.5.3).

We now give a characterization of frames in terms of the synthesis operator, which was proved by Christensen [153]. It does not involve any knowledge of the frame bounds.

Theorem 5.5.1 *A sequence $\{f_k\}_{k=1}^{\infty}$ in $\mathcal{H}$ is a frame for $\mathcal{H}$ if and only if*

$$T : \{c_k\}_{k=1}^{\infty} \to \sum_{k=1}^{\infty} c_k f_k$$

is a well-defined mapping of $\ell^2(\mathbb{N})$ onto $\mathcal{H}$.

Proof. First, suppose that $\{f_k\}_{k=1}^{\infty}$ is a frame. By Theorem 3.2.3, T is a well-defined bounded operator from $\ell^2(\mathbb{N})$ into $\mathcal{H}$, and by Lemma 5.1.5(i), the frame operator $S = TT^*$ is surjective. Thus T is surjective. For the opposite implication, suppose that T is a well-defined operator from $\ell^2(\mathbb{N})$ onto $\mathcal{H}$. Then Lemma 3.2.1 shows that T is bounded and that $\{f_k\}_{k=1}^{\infty}$ is a Bessel sequence. Let $T^{\dagger} : \mathcal{H} \to \ell^2(\mathbb{N})$ denote the pseudo-inverse of T, as defined in Section 2.5. For $f \in \mathcal{H}$, we have

$$f = TT^{\dagger}f = \sum_{k=1}^{\infty}(T^{\dagger}f)_k f_k,$$

where $(T^{\dagger}f)_k$ denotes the k-th coordinate of $T^{\dagger}f$. Thus

$$
\begin{aligned}
||f||^4 &= |\langle f, f\rangle|^2 = |\langle \sum_{k=1}^{\infty}(T^{\dagger}f)_k f_k, f\rangle|^2 \\
&\leq \sum_{k=1}^{\infty}|(T^{\dagger}f)_k|^2 \sum_{k=1}^{\infty}|\langle f, f_k\rangle|^2 \leq ||T^{\dagger}||^2 \, ||f||^2 \sum_{k=1}^{\infty}|\langle f, f_k\rangle|^2;
\end{aligned}
$$

we conclude that

$$\sum_{k=1}^{\infty}|\langle f, f_k\rangle|^2 \geq \frac{1}{||T^{\dagger}||^2}||f||^2. \qquad \square$$

For an arbitrary sequence $\{f_k\}_{k=1}^{\infty}$ in a Hilbert space, $\overline{\text{span}}\{f_k\}_{k=1}^{\infty}$ is itself a Hilbert space, and Theorem 5.5.1 leads to a statement about frame sequences:

Corollary 5.5.2 *A sequence $\{f_k\}_{k=1}^{\infty}$ in $\mathcal{H}$ is a frame sequence if and only if the synthesis operator is well-defined on $\ell^2(\mathbb{N})$ and has closed range.*

In terms of the adjoint of the synthesis operator, we have:

Corollary 5.5.3 *For a sequence $\{f_k\}_{k=1}^{\infty}$ in $\mathcal{H}$, the following hold:*

(i) $\{f_k\}_{k=1}^{\infty}$ is a frame sequence if and only if

$$f \mapsto \{\langle f, f_k\rangle\}_{k=1}^{\infty} \tag{5.18}$$

is a well-defined map from $\mathcal{H}$ onto a closed subspace of $\ell^2(\mathbb{N})$.

(ii) If $\{f_k\}_{k=1}^{\infty}$ is a frame sequence, it is a frame for $\mathcal{H}$ if and only if the map (5.18) is injective.

Proof. The proof of (i) uses that a bounded operator has closed range if and only if its adjoint operator has closed range. First, assume that $\{f_k\}_{k=1}^\infty$ is a frame sequence. Then the synthesis operator T is well-defined and bounded, and the range is closed. Therefore T^* is well-defined and has closed range. For the opposite implication, if (5.18) maps $\mathcal{H}$ into $\ell^2(\mathbb{N})$, then $\{f_k\}_{k=1}^\infty$ is a Bessel sequence (Exercise 3.11). Thus the synthesis operator T is well defined and bounded; furthermore, if the range of the map in (5.18) is closed, the same is true for T. This implies by Corollary 5.5.2 that $\{f_k\}_{k=1}^\infty$ is a frame sequence.

For the proof of (ii), we note that $\overline{\mathcal{R}_T} = (\mathcal{N}_{T^*})^\perp$. Thus, if $\{f_k\}_{k=1}^\infty$ is a frame for $\mathcal{H}$, then T^* is injective. On the other hand, if (5.18) defines an injective mapping, then $\{f_k\}_{k=1}^\infty$ is complete in $\mathcal{H}$; thus, if $\{f_k\}_{k=1}^\infty$ is a frame sequence, it is a frame for $\mathcal{H}$. $\qquad\square$

Recall that Riesz bases for $\mathcal{H}$ are characterized as the families $\{Ue_k\}_{k=1}^\infty$, where $\{e_k\}_{k=1}^\infty$ is an orthonormal basis for $\mathcal{H}$ and $U : \mathcal{H} \to \mathcal{H}$ is bounded and invertible. We can now give a similar characterization of frames:

Theorem 5.5.4 *Let $\{e_k\}_{k=1}^\infty$ be an arbitrary orthonormal basis for $\mathcal{H}$. The frames for $\mathcal{H}$ are precisely the families $\{Ue_k\}_{k=1}^\infty$, where $U : \mathcal{H} \to \mathcal{H}$ is a bounded and surjective operator.*

Proof. Let $\{\delta_k\}_{k=1}^\infty$ be the canonical basis for $\ell^2(\mathbb{N})$ and $\{e_k\}_{k=1}^\infty$ an orthonormal basis for $\mathcal{H}$. Let $\Phi : \mathcal{H} \to \ell^2(\mathbb{N})$ be the isometric isomorphism defined by $\Phi e_k = \delta_k$. If $\{f_k\}_{k=1}^\infty$ is a frame, then the synthesis operator T is bounded and surjective, and $T\delta_k = f_k$. With $U := T\Phi$, we have $\{f_k\}_{k=1}^\infty = \{Ue_k\}_{k=1}^\infty$, and U is bounded and surjective. That every family $\{Ue_k\}_{k=1}^\infty$ of the described type is a frame follows from Theorem 5.5.1 (see Exercise 5.17). Alternatively, we can observe that $\sum_{k=1}^\infty |\langle f, Ue_k\rangle|^2 = ||U^*f||^2$, and refer to Lemma 2.4.1. $\qquad\square$

Via Theorem 5.5.1, the question of existence of an upper and a lower frame bound is replaced by an investigation of infinite series: we have to check that $\sum_{k=1}^\infty c_k f_k$ converges for all $\{c_k\}_{k=1}^\infty \in \ell^2(\mathbb{N})$ and that each $f \in \mathcal{H}$ can be represented via such an infinite series. The other results mentioned here do not involve the frame bounds either. We now state a characterization of frames which keeps the information about the frame bounds.

Lemma 5.5.5 *A sequence $\{f_k\}_{k=1}^\infty$ in $\mathcal{H}$ is a frame for $\mathcal{H}$ with bounds A, B if and only if the following conditions are satisfied:*

(i) $\{f_k\}_{k=1}^\infty$ is complete in $\mathcal{H}$.

(ii) The synthesis operator T is well-defined on $\ell^2(\mathbb{N})$ and

$$A \sum_{k=1}^\infty |c_k|^2 \leq ||T\{c_k\}_{k=1}^\infty||^2 \leq B \sum_{k=1}^\infty |c_k|^2, \ \forall\{c_k\}_{k=1}^\infty \in \mathcal{N}_T^\perp. \quad (5.19)$$

Proof. Theorem 3.2.3 gives the first part. the upper frame condition with bound B is equivalent to the right-hand inequality in (5.19) (it is clear that it is enough to check the condition for $\{c_k\}_{k=1}^\infty \in \mathcal{N}_T^\perp$). We therefore assume that $\{f_k\}_{k=1}^\infty$ is a Bessel sequence and prove the equivalence of the lower frame condition and the left-hand inequality in (5.19) together with completeness.

First, assume that $\{f_k\}_{k=1}^\infty$ satisfies the lower frame condition with bound A. Then (i) is satisfied. Note that $\mathcal{R}_{T^*}$ is closed because $\mathcal{R}_T$ is closed (the latter is equal to $\mathcal{H}$ because $\{f_k\}_{k=1}^\infty$ is a frame). Therefore

$$\mathcal{N}_T^\perp = \overline{\mathcal{R}_{T^*}} = \mathcal{R}_{T^*},$$

i.e., $\mathcal{N}_T^\perp$ consists of all sequences of the form $\{\langle f, f_k \rangle\}_{k=1}^\infty$, $f \in \mathcal{H}$. Now, given $f \in \mathcal{H}$,

$$\left(\sum_{k=1}^\infty |\langle f, f_k \rangle|^2 \right)^2 = |\langle Sf, f \rangle|^2 \leq \|Sf\|^2 \|f\|^2$$

$$\leq \|Sf\|^2 \frac{1}{A} \sum_{k=1}^\infty |\langle f, f_k \rangle|^2.$$

This implies that

$$A \sum_{k=1}^\infty |\langle f, f_k \rangle|^2 \leq \|Sf\|^2 = \|T\{\langle f, f_k \rangle\}_{k=1}^\infty\|^2,$$

as desired. For the other implication, assume that $\{f_k\}_{k=1}^\infty$ is complete and that the left-hand inequality in (5.19) is satisfied. We first prove that $\mathcal{R}_T = \mathcal{H}$. Since $\text{span}\{f_k\}_{k=1}^\infty \subset \mathcal{R}_T$, it is enough to prove that $\mathcal{R}_T$ is closed. Now, if $\{y_n\}$ is a sequence in $\mathcal{R}_T$, we can find a sequence $\{x_n\}$ in $\mathcal{N}_T^\perp$ such that $y_n = T x_n$; if y_n converges to some $y \in \mathcal{H}$, then (5.19) implies that $\{x_n\}$ is a Cauchy sequence. Therefore $\{x_n\}$ converges to some x, which by continuity of T satisfies $Tx = y$. Thus $\mathcal{R}_T$ is closed and hence $\mathcal{R}_T = \mathcal{H}$. Let $T^\dagger$ denote the pseudo-inverse of T. By Lemma 2.5.2 and (2.11), we know that the operator $T^\dagger T$ is the orthogonal projection onto $\mathcal{N}_T^\perp$ and that $TT^\dagger$ is the orthogonal projection onto $\mathcal{R}_T = \mathcal{H}$. Thus, for any $\{c_k\}_{k=1}^\infty \in \ell^2(\mathbb{N})$, the inequality (5.19) implies that

$$A \|T^\dagger T\{c_k\}_{k=1}^\infty\|^2 \leq \|TT^\dagger T\{c_k\}_{k=1}^\infty\|^2 = \|T\{c_k\}_{k=1}^\infty\|^2. \tag{5.20}$$

Again by (2.11), we have $\mathcal{N}_{T^\dagger} = \mathcal{R}_T^\perp$, so (5.20) gives that $\|T^\dagger\|^2 \leq \frac{1}{A}$. Using Lemma 2.5.2, we also have $\|(T^*)^\dagger\|^2 \leq \frac{1}{A}$. But $(T^*)^\dagger T^*$ is the orthogonal projection onto

$$\mathcal{R}_{(T^*)^\dagger} = \mathcal{R}_{(T^\dagger)^*} = \mathcal{N}_{T^\dagger}^\perp = \mathcal{R}_T = \mathcal{H},$$

so for all $f \in \mathcal{H}$,

$$\|f\|^2 = \|(T^*)^\dagger T^* f\|^2 \le \frac{1}{A} \|T^* f\|^2 = \frac{1}{A} \sum_{k=1}^\infty |\langle f, f_k \rangle|^2.$$

This shows that $\{f_k\}_{k=1}^\infty$ satisfies the lower frame condition as desired. $\square$

Lemma 5.5.5 is probably most useful for theoretical considerations; see the proof of Theorem 9.2.5 for an application.

A different approach to determine all frames for $\mathcal{H}$ was given by Aldroubi [1]. Assuming that $\{f_k\}_{k=1}^\infty$ is a frame, he considered the questions:

(i) Which conditions on the numbers $\{u_{lk}\}_{l,k\in\mathbb{N}}$ will imply that the vectors

$$\phi_l = \sum_{k=1}^\infty u_{lk} f_k, \ l \in \mathbb{N} \tag{5.21}$$

are well-defined and constitute a frame for $\mathcal{H}$?

(ii) Can all frames for $\mathcal{H}$ be constructed this way?

It is immediately clear that the answer to the second question is yes: the frame decomposition associated with $\{f_k\}_{k=1}^\infty$ says that for *any* sequence $\{\phi_l\}_{l=1}^\infty$ in $\mathcal{H}$, we can write the elements as $\phi_l = \sum_{k=1}^\infty u_{lk} f_k$ with

$$u_{lk} = \langle \phi_l, S^{-1} f_k \rangle.$$

In case $\{\phi_l\}_{l=1}^\infty$ is a frame, the operator defined by $\{u_{lk}\}_{l,k\in\mathbb{N}}$ is bounded:

Proposition 5.5.6 *Let $\{f_k\}_{k=1}^\infty$ and $\{\phi_l\}_{l=1}^\infty$ be frames for $\mathcal{H}$. Then the bi-infinite matrix U, where the lk-th entry is $u_{lk} = \langle \phi_l, S^{-1} f_k \rangle$, defines a bounded operator on $\ell^2(\mathbb{N})$.*

Proof. Let b denote an upper frame bound for $\{\phi_l\}_{l=1}^\infty$ and A a lower frame bound for $\{f_k\}_{k=1}^\infty$. The proof will use several frame results. First, we know from Lemma 5.1.5 that $\{S^{-1} f_k\}_{k=1}^\infty$ is a frame with upper bound $1/A$. It follows that $\{\langle \phi_l, S^{-1} f_k \rangle\}_{k=1}^\infty \in \ell^2(\mathbb{N})$ for all $l \in \mathbb{N}$ and that $\sum_{k=1}^\infty \overline{c_k} S^{-1} f_k$ is convergent for all $\{c_k\}_{k=1}^\infty \in \ell^2(\mathbb{N})$. If we also invoke Theorem 3.2.3, we see that for all $\{c_k\}_{k=1}^\infty \in \ell^2(\mathbb{N})$,

$$\|U\{c_k\}_{k=1}^\infty\|^2 = \sum_{l=1}^\infty \left| \sum_{k=1}^\infty \langle \phi_l, S^{-1} f_k \rangle c_k \right|^2 = \sum_{l=1}^\infty \left| \langle \phi_l, \sum_{k=1}^\infty \overline{c_k} S^{-1} f_k \rangle \right|^2$$

$$\le b \left\| \sum_{k=1}^\infty \overline{c_k} S^{-1} f_k \right\|^2 \le \frac{b}{A} \sum_{k=1}^\infty |c_k|^2.$$

$\square$

Concerning the first question, Aldroubi proves

Proposition 5.5.7 *Let $\{f_k\}_{k=1}^{\infty}$ be a frame and assume that a bi-infinite matrix $U = \{u_{lk}\}_{l,k\in\mathbb{N}}$ defines a bounded operator on $\ell^2(\mathbb{N})$. Then the vectors $\{\phi_l\}_{l=1}^{\infty}$ in (5.21) are well defined; they constitute a frame for $\mathcal{H}$ if and only if there exists a constant $C > 0$ such that*

$$\sum_{l=1}^{\infty} |\langle \phi_l, f\rangle|^2 \geq C \sum_{k=1}^{\infty} |\langle f_k, f\rangle|^2, \quad \forall f \in \mathcal{H}. \tag{5.22}$$

Proof. Let B denote an upper frame bound for $\{f_k\}_{k=1}^{\infty}$. If U is bounded on $\ell^2(\mathbb{N})$, then

$$\sum_{l=1}^{\infty} \left| \sum_{k=1}^{\infty} u_{lk} c_k \right|^2 \leq \|U\|^2 \sum_{k=1}^{\infty} |c_k|^2, \quad \forall \{c_k\}_{k=1}^{\infty} \in \ell^2(\mathbb{N}).$$

This implies that for any fixed $l \in \mathbb{N}$, the map $\{c_k\}_{k=1}^{\infty} \to \sum_{k=1}^{\infty} u_{lk} c_k$ is a continuous linear functional on $\ell^2(\mathbb{N})$; since the dual of $\ell^2(\mathbb{N})$ equals $\ell^2(\mathbb{N})$, we conclude that the rows in the matrix U are square summable, $\{u_{lk}\}_{k=1}^{\infty} \in \ell^2(\mathbb{N})$. Thus the vectors ϕ_l are well defined. By construction,

$$\{\langle \phi_l, f\rangle\}_{l=1}^{\infty} = U\{\langle f_k, f\rangle\}_{k=1}^{\infty}, \quad \forall f \in \mathcal{H};$$

from here, it follows that

$$\begin{aligned}
\sum_{l=1}^{\infty} |\langle \phi_l, f\rangle|^2 &= \|U\{\langle f_k, f\rangle\}_{k=1}^{\infty}\|^2 \leq \|U\|^2 \, \|\{\langle f_k, f\rangle\}_{k=1}^{\infty}\|^2 \\
&\leq \|U\|^2 \sum_{k=1}^{\infty} |\langle f_k, f\rangle|^2 \leq B \, \|U\|^2 \, \|f\|^2.
\end{aligned}$$

So $\{\phi_l\}_{l=1}^{\infty}$ is even a Bessel sequence. Now, if (5.22) is satisfied, it is clear that $\{\phi_l\}_{l=1}^{\infty}$ is a frame; on the other hand, if $\{\phi_l\}_{l=1}^{\infty}$ is a frame with lower bound a, then

$$\sum_{l=1}^{\infty} |\langle \phi_l, f\rangle|^2 \geq a \, \|f\|^2 \geq \frac{a}{B} \sum_{k=1}^{\infty} |\langle f_k, f\rangle|^2, \quad \forall f \in \mathcal{H}. \qquad \square$$

The condition (5.22) can also be written as

$$\sum_{l=1}^{\infty} \left| \sum_{k=1}^{\infty} \langle f_k, f\rangle u_{lk} \right|^2 \geq C \sum_{k=1}^{\infty} |\langle f_k, f\rangle|^2, \quad \forall f \in \mathcal{H}.$$

However, it is not clear from here which coefficients $\{u_{lk}\}_{k,l\in\mathbb{N}}$ will lead to a frame $\{\phi_l\}_{l=1}^{\infty}$. A sufficient condition on $\{u_{lk}\}_{k,l\in\mathbb{N}}$ is given by

Proposition 5.5.8 *Let* $\{f_k\}_{k=1}^{\infty}$ *be a frame with bounds* A, B. *If the numbers* $\{u_{lk}\}_{k,l \in \mathbb{N}}$ *satisfy the two conditions*

$$b := \sup_{k \in \mathbb{N}} \sum_{j=1}^{\infty} \left| \sum_{l=1}^{\infty} u_{lk} \overline{u_{lj}} \right| < \infty,$$

$$a := \inf_{k \in \mathbb{N}} \left(\sum_{l=1}^{\infty} |u_{lk}|^2 - \sum_{j \neq k} \left| \sum_{l=1}^{\infty} u_{lk} \overline{u_{lj}} \right| \right) > 0,$$

then $\{\phi_l\}_{l=1}^{\infty}$ *defined by (5.21) is a frame with bounds* aA, bB.

Proof. Let $f \in \mathcal{H}$. Then

$$\sum_{l=1}^{\infty} |\langle \phi_l, f \rangle|^2 = \sum_{l=1}^{\infty} \left| \langle \sum_{k=1}^{\infty} u_{lk} f_k, f \rangle \right|^2 = \sum_{l=1}^{\infty} \left| \sum_{k=1}^{\infty} u_{lk} \langle f_k, f \rangle \right|^2$$

$$= \sum_{l=1}^{\infty} \sum_{k=1}^{\infty} |u_{lk}|^2 \, |\langle f_k, f \rangle|^2$$

$$+ \sum_{l=1}^{\infty} \sum_{k=1}^{\infty} \sum_{j \neq k} u_{lk} \overline{u_{lj}} \langle f_k, f \rangle \langle f, f_j \rangle$$

$$= (*) + (**).$$

By Cauchy–Schwarz' inequality, we get

$$|(**)| \leq \sum_{k=1}^{\infty} \sum_{j \neq k} |\langle f_k, f \rangle \langle f, f_j \rangle| \left| \sum_{l=1}^{\infty} u_{lk} \overline{u_{lj}} \right|$$

$$\leq \left(\sum_{k=1}^{\infty} \sum_{j \neq k} |\langle f_k, f \rangle|^2 \left| \sum_{l=1}^{\infty} u_{lk} \overline{u_{lj}} \right| \right)^{1/2}$$

$$\times \left(\sum_{k=1}^{\infty} \sum_{j \neq k} |\langle f, f_j \rangle|^2 \left| \sum_{l=1}^{\infty} u_{lk} \overline{u_{lj}} \right| \right)^{1/2}.$$

The two terms in the last product are actually identical. In fact, by switching the order of summation and renaming the indices,

$$\sum_{k=1}^{\infty} \sum_{j \neq k} |\langle f, f_j \rangle|^2 \left| \sum_{l=1}^{\infty} u_{lk} \overline{u_{lj}} \right| = \sum_{j=1}^{\infty} \sum_{k \neq j} |\langle f, f_j \rangle|^2 \left| \sum_{l=1}^{\infty} u_{lk} \overline{u_{lj}} \right|$$

$$= \sum_{k=1}^{\infty} \sum_{j \neq k} |\langle f_k, f \rangle|^2 \left| \sum_{l=1}^{\infty} u_{lk} \overline{u_{lj}} \right|.$$

Thus

$$(**) \le \sum_{k=1}^{\infty} \sum_{j \ne k} |\langle f_k, f \rangle|^2 \left| \sum_{l=1}^{\infty} u_{lk} \overline{u_{lj}} \right|,$$

and by the calculation at the beginning of the proof,

$$\sum_{l=1}^{\infty} |\langle \phi_l, f \rangle|^2 \ge \sum_{l=1}^{\infty} \sum_{k=1}^{\infty} |u_{lk}|^2 |\langle f_k, f \rangle|^2$$

$$- \sum_{k=1}^{\infty} \sum_{j \ne k} |\langle f_k, f \rangle|^2 \left| \sum_{l=1}^{\infty} u_{lk} \overline{u_{lj}} \right|$$

$$= \sum_{k=1}^{\infty} |\langle f_k, f \rangle|^2 \left(\sum_{l=1}^{\infty} |u_{lk}|^2 - \sum_{j \ne k} \left| \sum_{l=1}^{\infty} u_{lk} \overline{u_{lj}} \right| \right)$$

$$\ge a \sum_{k=1}^{\infty} |\langle f_k, f \rangle|^2.$$

The upper frame condition is proved similarly. $\qquad\square$

Note that $\sum_{k=1}^{\infty} u_{lk} \overline{u_{jk}}$ is the inner product between the l-th and the j-th row in U. This gives a geometric interpretation of the conditions in Proposition 5.5.8: the lower condition, e.g., means that the inner product of any row with itself is larger (uniformly over all rows) than the sum of the absolute values of inner products between this row and all other rows.

Casazza proved in [113] that every frame in a complex Hilbert space is a multiple of a sum of three orthonormal bases:

Theorem 5.5.9 *Assume that $\mathcal{H}$ is a complex Hilbert space and that $\{f_k\}_{k=1}^{\infty}$ is a frame for $\mathcal{H}$ with synthesis operator T. Then, for every $\epsilon \in]0, 1[$, there exist three orthonormal bases $\{e_k^1\}_{k=1}^{\infty}, \{e_k^2\}_{k=1}^{\infty}$ and $\{e_k^3\}_{k=1}^{\infty}$ for $\mathcal{H}$ such that*

$$f_k = \frac{||T||}{1 - \epsilon} (e_k^1 + e_k^2 + e_k^3), \ \forall k \in \mathbb{N}.$$

Proof. Let $\{e_k\}_{k=1}^{\infty}$ denote an orthonormal basis for $\mathcal{H}$, and let $\{\delta_k\}_{k=1}^{\infty}$ be the canonical orthonormal basis for $\ell^2(\mathbb{N})$. Composing the synthesis operator for $\{f_k\}_{k=1}^{\infty}$ with the isometric isomorphism from $\mathcal{H}$ to $\ell^2(\mathbb{N})$ which maps e_k to δ_k, we obtain a bounded linear operator of $\mathcal{H}$ onto $\mathcal{H}$, which maps e_k to f_k; by a slight abuse of our standard notation, we denote it by T. Given $\epsilon \in]0, 1[$, consider the operator

$$U : \mathcal{H} \to \mathcal{H}, \ U := \frac{1}{2} I + \frac{1 - \epsilon}{2} \frac{T}{||T||}. \tag{5.23}$$

Since

$$\|I - U\| = \left\|\frac{1}{2}I - \frac{1-\epsilon}{2}\frac{T}{\|T\|}\right\| \le \frac{1}{2} + \frac{1-\epsilon}{2} < 1,$$

we see that U is invertible. Using the polar decomposition in Lemma 2.4.6, we can write $U = VP$, where V is unitary and P is a positive operator. Observe that

$$\|P\| = \|V^{-1}U\| \le \|U\| \le \frac{1}{2} + \frac{1-\epsilon}{2} < 1.$$

Now we apply the second part of Lemma 2.4.6 to write $P = \frac{1}{2}(W + W^*)$, where W, W^* are unitary. We can use these decompositions and (5.23) to obtain an expression for the operator T as

$$T = \frac{2\|T\|}{1-\epsilon}\left(U - \frac{1}{2}I\right) = \frac{2\|T\|}{1-\epsilon}\left(VP - \frac{1}{2}I\right) = \frac{\|T\|}{1-\epsilon}\left(VW + VW^* - I\right).$$

It follows from here that

$$\begin{aligned}
\{f_k\}_{k=1}^{\infty} &= \{Te_k\}_{k=1}^{\infty} \\
&= \frac{\|T\|}{1-\epsilon}\left(VW\{e_k\}_{k=1}^{\infty} + VW^*\{e_k\}_{k=1}^{\infty} - \{e_k\}_{k=1}^{\infty}\right).
\end{aligned}$$

Since VW and VW^* are unitary, the result now follows from Theorem 3.4.7. □

Note that on the other hand we cannot conclude that every sum of three orthonormal bases is a frame. Consider, for example, the one-dimensional Hilbert space $\mathbb{C}$; each of the complex numbers 1, $e^{\frac{2\pi i}{3}}$, $e^{\frac{4\pi i}{3}}$ constitutes an orthonormal basis for $\mathbb{C}$, but $1 + e^{\frac{2\pi i}{3}} + e^{\frac{4\pi i}{3}} = 0$.

In case $\mathcal{H}$ is a real Hilbert space, the representation of a positive operator P with $\|P\| \le 1$ as an average of two unitary operators is no longer valid. However, a representation as a sum of 16 unitary operators holds. Inserting this in the above proof, we obtain a representation of a frame as a sum of 17 orthonormal bases!

Theorem 5.5.9 is optimal in the sense that two orthonormal bases are not enough to represent all frames:

Example 5.5.10 Let $\{e_k\}_{k=1}^{\infty}$ be an orthonormal basis for $\mathcal{H}$ (a real or complex Hilbert space) and define the frame $\{f_k\}_{k=1}^{\infty}$ by

$$f_1 = 0, \quad f_k = e_{k-1}, \ k \ge 2.$$

Assume that we could find orthonormal systems $\{e_k^1\}_{k=1}^{\infty}, \{e_k^2\}_{k=1}^{\infty}$ and nonzero constants a, b such that $f_k = ae_k^1 + be_k^2$ for all k. Then in particular $f_1 = 0 = ae_1^1 + be_1^2$, which implies that span $e_1^1 =$ span e_1^2. By orthonormality we conclude that span $e_1^1 \subseteq \overline{\text{span}}\left(\{e_k^1\}_{k=2}^{\infty} \cup \{e_k^2\}_{k=2}^{\infty}\right)^{\perp}$. On the other hand $\mathcal{H} = \overline{\text{span}}\{f_k\}_{k=2}^{\infty} = \overline{\text{span}}\{ae_k^1 + be_k^2\}_{k=2}^{\infty}$, which is a contradiction. □

Using more advanced tools from operator theory one can prove that the class of frames which can be written as a linear combination of two orthonormal bases is exactly the class of Riesz bases. We refer to [113].

5.6   Continuous Frames

The frames discussed so far all lead to expansions of elements in Hilbert spaces in terms of infinite sums. One can consider these frames as manifestations of a broader theory, which in general leads to integral representations in Hilbert spaces. The following generalization of frames was proposed by Kaiser [444] and independently by Ali, Antoine, and Gazeau [11] in 1993:

Definition 5.6.1 *Let $\mathcal{H}$ be a complex Hilbert space and M a measure space with a positive measure μ. A continuous frame for $\mathcal{H}$ is a family of vectors $\{f_k\}_{k \in M}$ for which:*

(i) For all $f \in \mathcal{H}$, $k \mapsto \langle f, f_k \rangle$ is a measurable function on M;

(ii) There exist constants $A, B > 0$ such that

$$A\,||f||^2 \leq \int_M |\langle f, f_k \rangle|^2 d\mu(k) \leq B\,||f||^2, \ \forall f \in \mathcal{H}. \qquad (5.24)$$

$\{f_k\}_{k \in M}$ *is called a Bessel family if at least the upper condition in (5.24) is satisfied.*

Note that Kaiser used the name *generalized frames* for the continuous frames. In our treatment we will borrow words from our treatment of (discrete) frames without further comments (e.g., frame bounds, tight frames, Parseval frames).

The discrete frames considered in Sections 5.1–5.5 correspond to the case where $M = \mathbb{N}$, equipped with the counting measure. Whether a continuous frame $\{f_k\}_{k \in M}$ is in fact a discrete frame or not, depends on the measure space; thus, in certain cases it is necessary to be more careful and speak about a continuous frame for $\mathcal{H}$ *with respect to the measure space* (M, μ). Since $\{f_k\}_{k \in M}$ in general is not a sequence, we have chosen the wording "Bessel family" in Definition 5.6.1.

An important feature of continuous frames is that they unify several aspects of the theory for continuous/discrete Gabor systems and wavelet systems. We come back to this point in Chapters 11 and 15.

Let us derive the basic results for a continuous frame $\{f_k\}_{k \in M}$. First, Cauchy–Schwarz' inequality shows that the integral $\int_M \langle f, f_k \rangle \langle f_k, g \rangle d\mu(k)$ is well defined for all $f, g \in \mathcal{H}$. For a fixed $f \in \mathcal{H}$, the mapping

$$g \mapsto \int_M \langle f, f_k \rangle \langle f_k, g \rangle d\mu(k)$$

is clearly conjugated linear and bounded because

$$\left| \int_M \langle f, f_k \rangle \langle f_k, g \rangle d\mu(k) \right|^2 \leq \int_M |\langle f, f_k \rangle|^2 d\mu(k) \int_M |\langle f_k, g \rangle|^2 d\mu(k)$$
$$\leq B^2 \, ||f||^2 \, ||g||^2. \tag{5.25}$$

By Riesz' representation theorem (Theorem 2.3.2), there exists a unique element in $\mathcal{H}$ – we call it $\int_M \langle f, f_k \rangle f_k d\mu(k)$ – such that

$$\left\langle \int_M \langle f, f_k \rangle f_k d\mu(k), g \right\rangle = \int_M \langle f, f_k \rangle \langle f_k, g \rangle d\mu(k)$$

for all $g \in \mathcal{H}$. By this procedure, we have defined a mapping

$$S : \mathcal{H} \to \mathcal{H}, \quad Sf = \int_M \langle f, f_k \rangle f_k d\mu(k).$$

It is easy to check that S is linear; using that

$$||Sf|| = \sup_{||g||=1} |\langle Sf, g \rangle|,$$

it follows by (5.25) that S is bounded and that $||S|| \leq B$. Note that by definition S is positive,

$$A \, ||f||^2 \leq \langle Sf, f \rangle \leq B \, ||f||^2, \ \forall f \in \mathcal{H}.$$

Exactly as in the proof of Lemma 5.1.5, one can now prove that S is invertible. Thus, every $f \in \mathcal{H}$ has the representations

$$f = S^{-1}Sf = \int_M \langle f, f_k \rangle S^{-1} f_k d\mu(k),$$
$$f = SS^{-1}f = \int_M \langle f, S^{-1} f_k \rangle f_k d\mu(k).$$

Remember that these representations have to be interpreted in the weak sense. Sometimes stronger results (like pointwise convergence if $\mathcal{H} = L^2(\mathbb{R})$) can be obtained in concrete cases.

Similar to the terminology for discrete frames, two Bessel families $\{f_k\}_{k \in M}$ and $\{g_k\}_{k \in M}$ are called *dual continuous frames* if

$$f = \int_M \langle f, f_k \rangle g_k d\mu(k), \ \forall f \in \mathcal{H}. \tag{5.26}$$

Note that we above have introduced the frame operator directly, without using the synthesis operator as we did for discrete frames. This is just a matter of choice – in fact we could have followed exactly the same procedure as in Section 5.1, see (5.3) and (5.4). To be more precise, if $\{f_k\}_{k \in M}$ is a Bessel family, we can define the synthesis operator $T : L^2(M, \mu) \to \mathcal{H}$ in the weak sense by

$$T\{c_k\}_{k \in M} = \int_{k \in M} c_k f_k \, d\mu(k). \tag{5.27}$$

This is a bounded linear operator, with adjoint operator $T^* : \mathcal{H} \rightarrow$ $L^2(M, \mu)$ given by (Exercise 5.22)

$$T^* f = \{\langle f, f_k \rangle\}_{k \in M}. \tag{5.28}$$

The frame operator $S : \mathcal{H} \rightarrow \mathcal{H}$ is then $S = TT^*$, as in the case of discrete frames.

Continuous frames appear naturally in the setting of *reproducing kernel Hilbert spaces*:

Definition 5.6.2 *Let M denote a measure space with a positive measure μ. A Hilbert space $\mathcal{H}$ consisting of functions $f : M \rightarrow \mathbb{C}$, for which all the linear functionals Λ_x, $x \in M$, given by*

$$\Lambda_x : \mathcal{H} \rightarrow \mathbb{C}, \ \Lambda_x f := f(x),$$

are continuous, is called a reproducing kernel Hilbert space (RKHS).

Let us now consider an RKHS $\mathcal{H}$, which is a subset of $L^2(M, \mu)$ for some measure space M with a positive measure μ. We will assume that $\mathcal{H}$ is equipped with the norm inherited from $L^2(M, \mu)$. Riesz representation theorem shows that there for any $x \in M$ exists a unique element $K_x \in \mathcal{H}$ such that

$$f(x) = \langle f, K_x \rangle, \ \forall f \in \mathcal{H}.$$

Thus,

$$\int_M |\langle f, K_x \rangle|^2 \, d\mu(x) = \int_M |f(x)|^2 \, d\mu(x) = ||f||^2, \ \forall f \in \mathcal{H},$$

i.e., the functions $\{K_x\}_{x \in M}$ form a continuous Parseval frame for $L^2(M, \mu)$. The function

$$K : M \times M \rightarrow \mathbb{C}, \ K(x, y) := \langle K_y, K_x \rangle = K_y(x) \tag{5.29}$$

is called the *reproducing kernel*.

A concrete example of a RKHS is provided by the Paley–Wiener space; see (3.58). As always, we consider the continuous representatives for the equivalence classes in PW.

Lemma 5.6.3 *The Paley–Wiener space PW is a RKHS with reproducing kernel*

$$K(x, y) = sinc(y - x).$$

The functions $\{T_x sinc\}_{x \in \mathbb{R}}$ form a continuous Parseval frame for PW.

Proof.   Fix $x \in \mathbb{R}$, and consider $f \in PW$. Using the inverse Fourier transform and that $\mathrm{supp}\hat{f} \subseteq [-1/2, 1/2]$,

$$|f(x)| = \left| \int_{-\infty}^{\infty} \hat{f}(\gamma) e^{2\pi i x \gamma} d\gamma \right| = \left| \int_{-1/2}^{1/2} \hat{f}(\gamma) e^{2\pi i x \gamma} d\gamma \right|$$

$$\leq \left(\int_{-1/2}^{1/2} |\hat{f}(\gamma)|^2 \, d\gamma \right)^{1/2} = ||\hat{f}|| = ||f||.$$

Thus, for each $x \in \mathbb{R}$ the mapping $f \mapsto f(x)$ is indeed continuous on the Paley–Wiener space. Also, for $x \in \mathbb{R}$,

$$f(x) = \int_{-1/2}^{1/2} \hat{f}(\gamma) e^{2\pi i x \gamma} \, d\gamma = \int_{-\infty}^{\infty} \hat{f}(\gamma) \chi_{-1/2,1/2]}(\gamma) e^{2\pi i x \gamma} \, d\gamma$$

$$= \langle f, \mathcal{F}^{-1}[\chi_{-1/2,1/2]}(\cdot) e^{-2\pi i x \cdot}] \rangle$$

$$= \langle f, T_x \mathrm{sinc} \rangle.$$

This completes the proof.   □

Continuous frames will appear at several occasions throughout the book; see Section 11.1, Section 15.1, Section 21.7, and Section 24.1.

5.7   Exercises

5.1 Find an example of a sequence in a Hilbert space that is a basis but not a frame.

5.2 Prove that the upper and lower frame conditions are unrelated: in an arbitrary Hilbert space $\mathcal{H}$, there exists a sequence $\{f_k\}_{k=1}^{\infty}$ satisfying the upper condition for all $f \in \mathcal{H}$ but not the lower condition and vice versa.

5.3 Let $\{e_k\}_{k=1}^{\infty}$ be an orthonormal basis and consider the family $\{f_k\}_{k=1}^{\infty} := \{e_1 + \frac{1}{k}e_k, e_k\}_{k=2}^{\infty}$.

  (i) Prove that $\{f_k\}_{k=1}^{\infty}$ is not a Bessel sequence.

  (ii) Find all possible representations of e_1 as (infinite) linear combinations of $\{f_k\}_{k=1}^{\infty}$.

  (iii) Prove that there exists no set of coefficients having minimal ℓ^1-norm among all sequences representing e_1.

5.4 The purpose of this exercise is to prove Proposition 5.1.10. Let $A, B > 0$ be given, and consider a sequence $\{f_k\}_{k=1}^{\infty}$ in $\mathcal{H}$.

(i) Show that $\{f_k\}_{k=1}^{\infty}$ is a frame with bounds A, B if (5.11) holds for a dense set in

$$\Omega := \{f \in \mathcal{H} \mid ||f|| = 1\}.$$

(ii) Let $\{e_k\}_{k=1}^{\infty}$ denote an orthonormal basis for $\mathcal{H}$ and show that the set

$$\left\{ \sum_{k=1}^{N} c_k e_k \,\Big|\, N \in \mathbb{N},\ |\mathrm{Re}(c_k)|^2,\ |\mathrm{Im}(c_k)|^2 \in \mathbb{Q},\ \sum_{k=1}^{N} |c_k|^2 = 1 \right\}$$

is a countable and dense set in Ω.

5.5 Give an example of a frame $\{f_k\}_{k=1}^{\infty}$, for which $\sum_{k=1}^{\infty} c_k f_k$ converges for some $\{c_k\}_{k=1}^{\infty} \notin \ell^2(\mathbb{N})$ (compare with Exercise 3.13!).

5.6 Verify the statements in Example 5.1.8.

5.7 Let $\{f_k\}_{k=1}^{\infty}$ be a frame sequence in $\mathcal{H}$, with synthesis operator $T : \ell^2(\mathbb{N}) \to \mathcal{H}$. Prove that $\{f_k\}_{k=1}^{\infty}$ is a frame for $\mathcal{H}$ if and only if T^* is injective.

5.8 Let $\widetilde{\mathcal{H}}$ be the complexification of a real Hilbert space $\mathcal{H}$. Prove that a frame for $\mathcal{H}$ also is a frame for $\widetilde{\mathcal{H}}$.

5.9 Let $\{f_k\}_{k=1}^{\infty}$ be a Riesz basis with bounds A, B. Prove that

$$A \leq ||f_k||^2 \leq B \text{ for all } k \in \mathbb{N},$$

and that the elements in the dual Riesz basis $\{g_k\}_{k=1}^{\infty}$ satisfy

$$\frac{1}{B} \leq ||g_k||^2 \leq \frac{1}{A} \text{ for all } k \in \mathbb{N}.$$

5.10 Prove Proposition 5.2.3.

5.11 Prove that the conclusion in Proposition 5.3.1 might fail if U is not assumed to have closed range. (Hint: let $\{e_k\}_{k=1}^{\infty}$ be an orthonormal basis and define U by $Ue_k = e_k + e_{k+1}$.)

5.12 Let $\{e_k\}_{k=1}^{\infty}$ be an orthonormal basis for $\mathcal{H}$, and define an operator U on $\mathcal{H}$ by

$$Ue_{2k} = e_{2k}, \quad Ue_{2k-1} = \frac{1}{k}e_{2k}, \ k \in \mathbb{N}.$$

Prove that:

(i) U is a well-defined bounded operator on $\mathcal{H}$ and $\mathcal{R}_U$ is closed.

(ii) $\{e_{2k-1}\}_{k=1}^{\infty}$ is a frame sequence but $\{Ue_{2k-1}\}_{k=1}^{\infty}$ is not. Thus Proposition 5.3.1 does not extend to frame sequences.

5.13 Prove Lemma 5.3.3.

5.14 Assume that $\{f_k\}_{k=1}^{\infty}$ is a Riesz basis. Prove that $\{f_k + f_{k+1}\}_{k=1}^{\infty}$ cannot be a frame.

5.15 Show that the family $\{e_k + e_{k+1}\}_{k=1}^{\infty}$ in Example 5.4.6 cannot be extended to a basis for $\mathcal{H}$.

5.16 Let $\{f_k\}_{k\in\mathbb{Z}}$ be a Riesz basis. Our purpose is to show that

$$\{f_k + f_{k+1}\}_{k\in\mathbb{Z}}$$

cannot be a frame; compare with Exercise 5.14. Let $\{g_k\}_{k\in\mathbb{Z}}$ be the biorthogonal basis associated with $\{f_k\}_{k\in\mathbb{Z}}$. Let

$$h_j = \sum_{k=1}^{j}(-1)^k g_k.$$

(i) Prove that $\sum_{k\in\mathbb{Z}}|\langle h_j, f_k + f_{k+1}\rangle|^2 = 2$.

(ii) Prove that $||h_j||^2 \geq j/B$, where B is an upper frame bound for $\{f_k\}_{k\in\mathbb{Z}}$.

(iii) Conclude that $\{f_k + f_{k+1}\}_{k\in\mathbb{Z}}$ is not a frame.

5.17 Prove via Theorem 5.5.1 that $\{Ue_k\}_{k=1}^{\infty}$ is a frame whenever $\{e_k\}_{k=1}^{\infty}$ is an orthonormal basis and U is a bounded surjective operator.

5.18 Let $\{f_k\}_{k=1}^{\infty}$ be a frame sequence in $\mathcal{H}$. Show that the orthogonal projection Q of a sequence $\{c_k\}_{k=1}^{\infty} \in \ell^2(\mathbb{N})$ onto the range of T^* is given by

$$Q\{c_k\}_{k=1}^{\infty} = \left\{\left\langle \sum_{j=1}^{\infty} c_j S^{-1} f_j, f_k \right\rangle\right\}_{k=1}^{\infty}. \tag{5.30}$$

5.19 Let U be a bounded operator between Hilbert spaces. Prove that if at least one of the spaces $\mathcal{R}_U$ and $\mathcal{R}_{UU^*}$ is closed, then

$$\mathcal{R}_U = \mathcal{R}_{UU^*}.$$

5.20 Let $\{f_k\}_{k=1}^{\infty} = \{Ue_k\}_{k=1}^{\infty}$ be a Riesz basis for $\mathcal{H}$ as in Definition 3.6.1. Prove that the frame operator for $\{f_k\}_{k=1}^{\infty}$ is given by

$$S = UU^*.$$

5.21 Let $\{f_k\}_{k=1}^{\infty}$ be a frame with frame bounds A, B. Show that the frame operator S satisfies the inequalities

$$A \, ||f|| \leq ||Sf|| \leq B \, ||f||, \ \forall f \in \mathcal{H}.$$

5.22 Let $\{f_k\}_{k \in M}$ be a Bessel family in the sense of Definition 5.6.1. Show that the synthesis operator T in (5.27) is well defined in the weak sense as a bounded linear operator and that its adjoint operator T^* is given by (5.28).

6

Tight Frames and Dual Frame Pairs

We have already highlighted the frame decomposition, which shows that a frame $\{f_k\}_{k=1}^{\infty}$ for a Hilbert space $\mathcal{H}$ leads to the decomposition

$$f = \sum_{k=1}^{\infty} \langle f, S^{-1} f_k \rangle f_k, \quad \forall f \in \mathcal{H}; \tag{6.1}$$

here $S : \mathcal{H} \to \mathcal{H}$ denotes the frame operator. In practice, it is difficult to apply the general frame decomposition, due to the fact that we need to invert the frame operator. We have mentioned two ways to circumvent the problem. The first one is to restrict our attention to tight frames: as we have seen in Corollary 5.1.7, for a tight frame $\{f_k\}_{k=1}^{\infty}$ with frame bound A, the frame decomposition takes the much simpler form

$$f = \frac{1}{A} \sum_{k=1}^{\infty} \langle f, f_k \rangle f_k, \quad \forall f \in \mathcal{H}. \tag{6.2}$$

The second way to avoid the problem is to use the flexibility in the frame setup: if $\{f_k\}_{k=1}^{\infty}$ is a frame which is not a Riesz basis, we prove in Theorem 6.3.1 that one can find frames $\{g_k\}_{k=1}^{\infty} \neq \{S^{-1} f_k\}_{k=1}^{\infty}$, for which

$$f = \sum_{k=1}^{\infty} \langle f, g_k \rangle f_k, \quad \forall f \in \mathcal{H}. \tag{6.3}$$

As mentioned on page 125, such a frame $\{g_k\}_{k=1}^{\infty}$ is called a *dual frame* of $\{f_k\}_{k=1}^{\infty}$. In the literature the terminology *alternate dual frame* is also used. The hope is that one can find dual frames that are either easier to calculate

© Springer International Publishing Switzerland 2016
O. Christensen, *An Introduction to Frames and Riesz Bases*,
Applied and Numerical Harmonic Analysis,
DOI 10.1007/978-3-319-25613-9_6

than the canonical dual frame $\{S^{-1}f_k\}_{k=1}^{\infty}$ or have better properties. We will see several cases where this happens, e.g., in the context of Gabor frames in Section 12.5 and for wavelet frames in Section 18.8.

Tight frames and the concept of dual frames were both introduced in order to obtain what could be called *convenient frame expansions*. The purpose of this chapter is to go more into details with these concepts. We begin with a discussion of tight frames in Section 6.1. This is followed by a treatment of an important extension problem in Section 6.2; the main result states that any Bessel sequence in a separable Hilbert space $\mathcal{H}$ can be extended to a tight frame. In Section 6.3, we give a detailed analysis of frames and their various duals; the main results characterize all dual frames associated with a given frame in various ways. In Section 6.4, the extension problem in Section 6.2 is generalized to dual pairs of frames; finally, a concept of approximately dual frames is discussed in Section 6.5.

6.1 Tight Frames

Assume that the vectors $\{f_k\}_{k=1}^{\infty}$ in a Hilbert space $\mathcal{H}$ form a tight frame. By a scaling of the vectors, we can always obtain that the frame bound is $A = 1$; in that case, the frame decomposition (6.2) takes exactly the same form as the representation via an orthonormal basis; see (3.18). Thus, such frames can be used without any additional computational effort compared with the use of orthonormal bases. Recall that tight frames with frame bound $A = 1$ are referred to as Parseval frames.

Tight frames have other advantages. For the design of frames with prescribed properties, it is essential to be able to control the behavior of the canonical dual frame, but the complicated structure of the frame operator and its inverse makes this difficult. Assume, e.g., that we want to construct a frame consisting of functions in $L^2(\mathbb{R})$, in such a way that the frame $\{f_k\}_{k=1}^{\infty}$ and its canonical dual $\{S^{-1}f_k\}_{k=1}^{\infty}$ consist of functions with exponential decay. This is clearly a complicated task: even if we manage to construct the frame $\{f_k\}_{k=1}^{\infty}$ such that the functions f_k decay exponentially, how can we guarantee that the same is the case for $S^{-1}f_k$? For tight frames, questions of this type trivially have satisfying answers. Also, for a tight frame, the canonical dual frame automatically has the same structure as the frame itself: if the frame has wavelet structure or Gabor structure as described in Section 3.8 and Section 3.9, the same is the case for the canonical dual frame. In contrast, the canonical dual frame of a non-tight wavelet frame might not have the wavelet structure; a concrete example is given later; see Example 16.1.1.

It is important to notice that to every frame we can associate a *canonical tight frame* with frame bound $A = 1$:

Theorem 6.1.1 *Let $\{f_k\}_{k=1}^\infty$ be a frame for $\mathcal{H}$ with frame operator S. Then $\{S^{-1/2}f_k\}_{k=1}^\infty$ is a Parseval frame, and*

$$f = \sum_{k=1}^\infty \langle f, S^{-1/2}f_k\rangle S^{-1/2}f_k, \ \forall f \in \mathcal{H}.$$

Proof. The existence of a unique positive square root of S^{-1} follows from Lemma 2.4.5. Since $S^{-1/2}$ is a limit of a sequence of polynomials in S^{-1}, it commutes with S^{-1} and therefore with S. Therefore, for $f \in \mathcal{H}$,

$$f = S^{-1/2}SS^{-1/2}f = \sum_{k=1}^\infty \langle S^{-1/2}f, f_k\rangle S^{-1/2}f_k = \sum_{k=1}^\infty \langle f, S^{-1/2}f_k\rangle S^{-1/2}f_k.$$

By taking the inner product with f, we obtain that $\{S^{-1/2}f_k\}_{k=1}^\infty$ is a tight frame with frame bound $A = 1$. $\square$

Note that Theorem 6.1.1 does not solve the computational problems related to general frames: while $\{S^{-1/2}f_k\}_{k=1}^\infty$ forms a tight frame and in that sense leads to a convenient frame expansion, the calculation of the elements in the frame is in general prohibitively cumbersome.

6.2 Extension of Bessel Sequences to Tight Frames

The Bessel condition on a sequence $\{f_k\}_{k=1}^\infty$ in a Hilbert space $\mathcal{H}$ is a purely technical condition ensuring that the synthesis operator is well-defined from $\ell^2(\mathbb{N})$ to $\mathcal{H}$. The condition does not lead to any expansion property, as documented by the fact that a zero vector itself forms a Bessel sequence. Li and Sun proved in [488] that every Bessel sequence in a separable Hilbert space $\mathcal{H}$ can be extended to a tight frame for $\mathcal{H}$. The original proof uses g-frames (see Section 8.1) but we will give a direct proof.

Theorem 6.2.1 *Let $\{f_k\}_{k=1}^\infty$ be a Bessel sequence in a separable Hilbert space $\mathcal{H}$, with bound B. Then there exists a sequence $\{p_j\}_{j\in J}$ in $\mathcal{H}$ such that $\{f_k\}_{k=1}^\infty \cup \{p_j\}_{j\in J}$ is a tight frame for $\mathcal{H}$ with bound B.*

Proof. Let S denote the frame operator for $\{f_k\}_{k=1}^\infty$. Then $BI - S$ is a self-adjoint and positive operator. Considering its square root $(BI - S)^{1/2}$, we can write $Bf = Sf + (BI - S)^{1/2}(BI - S)^{1/2}f, \ \forall f \in \mathcal{H}$. Let $\{e_j\}_{j\in J}$ denote an orthonormal basis for $\mathcal{H}$; then, expanding $(BI - S)^{1/2}$ in terms of $\{e_j\}_{j\in J}$, we arrive at

$$Bf = \sum_{k=1}^\infty \langle f, f_k\rangle f_k + (BI-S)^{1/2}\sum_{j\in J}\langle(BI-S)^{1/2}f, e_j\rangle e_j$$

$$= \sum_{k=1}^\infty \langle f, f_k\rangle f_k + \sum_{j\in J}\langle f, (BI-S)^{1/2}e_j\rangle(BI-S)^{1/2}e_j, \ \forall f \in \mathcal{H}.$$

Taking inner product with f now yields that the sequence

$$\{f_k\}_{k=1}^\infty \cup \{(BI - S)^{1/2} e_j\}_{j \in J}$$

is a tight frame for $\mathcal{H}$ with bound B.   □

6.3 The Dual Frames

In order to motivate the concept of dual frames, let us consider again an arbitrary frame $\{f_k\}_{k=1}^\infty$ for a Hilbert space $\mathcal{H}$ and the associated frame decomposition (6.1). In Lemma 5.4.2, we have seen that the frame coefficients $\{\langle f, S^{-1} f_k \rangle\}_{k=1}^\infty$ have minimal ℓ^2-norm among all sequences which represent the vector f in terms of the frame $\{f_k\}_{k=1}^\infty$. However, minimality of the ℓ^2-norm of the coefficients $\{c_k(f)\}_{k=1}^\infty$ in the expansion

$$f = \sum_{k=1}^\infty c_k(f) f_k$$

is not always an important issue; there are cases where other criteria are more relevant. A serious issue is the computability: we know that in general it is difficult to calculate the frame coefficients, so it is natural to exploit the freedom in the choice of the coefficients $\{c_k(f)\}_{k=1}^\infty$ and search for coefficients that are easier to calculate than the frame coefficients.

Usually we want to work with coefficients which depend continuously and linearly on f; by Riesz' representation theorem (Theorem 2.3.2), this implies that the k-th coefficient in the expansion of f should have the form $c_k(f) = \langle f, g_k \rangle$ for some $g_k \in \mathcal{H}$. If $\{f_k\}_{k=1}^\infty$ is an overcomplete frame, there always exist several choices for $\{g_k\}_{k=1}^\infty$:

Lemma 6.3.1 *Assume that $\{f_k\}_{k=1}^\infty$ is an overcomplete frame. Then there exist frames $\{g_k\}_{k=1}^\infty \neq \{S^{-1} f_k\}_{k=1}^\infty$ for which*

$$f = \sum_{k=1}^\infty \langle f, g_k \rangle f_k, \ \forall f \in \mathcal{H}. \tag{6.4}$$

Proof. We split the proof in two cases and assume first that $f_\ell = 0$ for some $\ell \in \mathbb{N}$; in this case, $S^{-1} f_\ell = 0$. By letting $g_k := S^{-1} f_k$ for $k \neq \ell$ and choosing g_ℓ to be an arbitrary nonzero vector, the frame decomposition (6.1) shows that (6.4) holds and $\{g_k\}_{k=1}^\infty \neq \{S^{-1} f_k\}_{k=1}^\infty$.

Now we consider the case where $f_k \neq 0$ for all $k \in \mathbb{N}$. We will use a result proved later, namely, in Theorem 7.1.1: it says that since $\{f_k\}_{k=1}^\infty$ is overcomplete, there exists a sequence $\{c_k\}_{k=1}^\infty \in \ell^2(\mathbb{N}) \setminus \{0\}$ such that

$$0 = \sum_{k=1}^\infty c_k f_k.$$

For a certain $\ell \in \mathbb{N}$, we have $c_\ell \neq 0$, and we can write

$$f_\ell = \sum_{k \neq \ell} \frac{-c_k}{c_\ell} f_k.$$

Thus $\{f_k\}_{k \neq \ell}$ is complete in $\mathcal{H}$ and therefore a frame by Theorem 5.4.7. Denoting its canonical dual frame by $\{g_k\}_{k \neq \ell}$ and defining $g_\ell = 0$, we have found a frame $\{g_k\}_{k=1}^\infty$ for which (6.4) holds; it is different from the canonical dual of $\{f_k\}_{k=1}^\infty$ because $S^{-1} f_\ell \neq 0$. $\qquad \square$

Recall that a frame $\{g_k\}_{k=1}^\infty$ satisfying (6.4) is called a *dual frame* of $\{f_k\}_{k=1}^\infty$. Since $\{f_k\}_{k=1}^\infty$ and $\{g_k\}_{k=1}^\infty$ are assumed to be Bessel sequences, we can consider the synthesis operators; we denote the synthesis operator for $\{f_k\}_{k=1}^\infty$ by T and the synthesis operator for $\{g_k\}_{k=1}^\infty$ by U, i.e.,

$$T, U : \ell^2(\mathbb{N}) \to \mathcal{H}, \ T\{c_k\}_{k=1}^\infty = \sum_{k=1}^\infty c_k f_k, \ U\{c_k\}_{k=1}^\infty = \sum_{k=1}^\infty c_k g_k. \quad (6.5)$$

Composing T with the adjoint of U, we obtain the operator

$$TU^* : \mathcal{H} \to \mathcal{H}, \ TU^* f = \sum_{k=1}^\infty \langle f, g_k \rangle f_k. \quad (6.6)$$

The operator TU^* is called the *mixed frame operator* associated with $\{f_k\}_{k=1}^\infty$ and $\{g_k\}_{k=1}^\infty$. We note that in terms of this operator, (6.4) holds if and only if

$$TU^* = I.$$

We first prove a lemma, which shows that the roles of $\{f_k\}_{k=1}^\infty$ and $\{g_k\}_{k=1}^\infty$ can be interchanged and that the lower frame condition automatically is satisfied for Bessel sequences $\{f_k\}_{k=1}^\infty, \{g_k\}_{k=1}^\infty$ if (6.4) holds.

Lemma 6.3.2 *Assume that $\{f_k\}_{k=1}^\infty$ and $\{g_k\}_{k=1}^\infty$ are Bessel sequences in $\mathcal{H}$. Then the following are equivalent:*

(i) $f = \sum_{k=1}^\infty \langle f, g_k \rangle f_k, \ \forall f \in \mathcal{H}.$

(ii) $f = \sum_{k=1}^\infty \langle f, f_k \rangle g_k, \ \forall f \in \mathcal{H}.$

(iii) $\langle f, g \rangle = \sum_{k=1}^\infty \langle f, f_k \rangle \langle g_k, g \rangle, \ \forall f, g \in \mathcal{H}.$

In case the equivalent conditions are satisfied, $\{f_k\}_{k=1}^\infty$ and $\{g_k\}_{k=1}^\infty$ are dual frames for $\mathcal{H}$.

Proof. In terms of the synthesis operators, (i) means that $TU^* = I$; this is equivalent to

$$UT^* = I, \quad (6.7)$$

which is identical to the statement in (ii). It is also clear that (ii) implies (iii). To prove that (iii) implies (ii), we fix $f \in \mathcal{H}$ and note that $\sum_{k=1}^{\infty} \langle f, f_k \rangle g_k$ is well defined as an element in $\mathcal{H}$ because $\{f_k\}_{k=1}^{\infty}$ and $\{g_k\}_{k=1}^{\infty}$ are Bessel sequences. Now the assumption in (iii) shows that

$$\left\langle f - \sum_{k=1}^{\infty} \langle f, f_k \rangle g_k, g \right\rangle = 0, \ \forall g \in \mathcal{H},$$

and (ii) follows.

In case the equivalent conditions are satisfied, we now have to show that $\{f_k\}_{k=1}^{\infty}$ and $\{g_k\}_{k=1}^{\infty}$ indeed are frames, i.e., that the lower frame condition holds. We can write

$$\|f\|^2 = \langle f, f \rangle = \sum_{k=1}^{\infty} \langle f, g_k \rangle \langle f_k, f \rangle, \ \forall f \in \mathcal{H}.$$

Using Cauchy–Schwarz inequality and that *one* of the families $\{f_k\}_{k=1}^{\infty}$, $\{g_k\}_{k=1}^{\infty}$ is a Bessel sequence, we obtain that the *other* family satisfies the lower frame condition. This concludes the proof.   □

We now state the "dual frame version" of Corollary 5.3.4. It shows that if we apply a unitary operator to a pair of dual frames, we again obtain a pair of dual frames. We leave the proof to the reader as Exercise 6.2.

Lemma 6.3.3 *Let $\{f_k\}_{k=1}^{\infty}$ and $\{g_k\}_{k=1}^{\infty}$ be dual frames for a Hilbert space $\mathcal{H}$ and $U : \mathcal{H} \to \mathcal{H}$ a unitary operator. Then $\{U f_k\}_{k=1}^{\infty}$ and $\{U g_k\}_{k=1}^{\infty}$ also form a pair of dual frames for $\mathcal{H}$.*

The following result provides a convenient criterion to check that two Bessel sequences are dual frames; it originally appeared as a technical tool in the paper [398] by Hernandez, Labate, and Weiss. We ask the reader to provide the proof in Exercise 6.4.

Lemma 6.3.4 *Two Bessel sequences $\{f_k\}_{k=1}^{\infty}$ and $\{g_k\}_{k=1}^{\infty}$ are dual frames if the identity*

$$\|f\|^2 = \sum_{k=1}^{\infty} \langle f, f_k \rangle \langle g_k, f \rangle \tag{6.8}$$

holds for all f belonging to a dense subspace of $\mathcal{H}$.

A note on terminology is in order. By Lemma 6.3.2, we know that if $\{g_k\}_{k=1}^{\infty}$ is a dual frame of $\{f_k\}_{k=1}^{\infty}$, then $\{f_k\}_{k=1}^{\infty}$ is also a dual of $\{g_k\}_{k=1}^{\infty}$. For this reason, we will usually call $\{f_k\}_{k=1}^{\infty}$ and $\{g_k\}_{k=1}^{\infty}$ a *pair of dual frames* or a *dual frame pair* when (6.4) holds.

When (6.7) is satisfied, we say that U is a *left-inverse* of T^*. Following Li [489], our next goal is to characterize all dual frames $\{f_k\}_{k=1}^{\infty}$ associated with a given frame $\{f_k\}_{k=1}^\infty$. The first step is to characterize all the left inverses of the analysis operator associated with $\{f_k\}_{k=1}^\infty$:

Lemma 6.3.5 *Let* $\{f_k\}_{k=1}^\infty$ *be a frame for* $\mathcal{H}$ *and* $\{\delta_k\}_{k=1}^\infty$ *be the canonical orthonormal basis for* $\ell^2(\mathbb{N})$. *The dual frames for* $\{f_k\}_{k=1}^\infty$ *are precisely the families* $\{g_k\}_{k=1}^\infty = \{V\delta_k\}_{k=1}^\infty$, *where* $V : \ell^2(\mathbb{N}) \to \mathcal{H}$ *is a bounded left-inverse of* T^*.

Proof.   If V is a bounded left-inverse of T^*, then V is surjective; by Theorem 5.5.1, it follows that $\{g_k\}_{k=1}^\infty := \{V\delta_k\}_{k=1}^\infty$ is a frame. Note that in terms of $\{\delta_k\}_{k=1}^\infty$,

$$T^*f = \{\langle f, f_k \rangle\}_{k=1}^\infty = \sum_{k=1}^\infty \langle f, f_k \rangle \delta_k;$$

thus, for all $f \in \mathcal{H}$,

$$f = VT^*f = \sum_{k=1}^\infty \langle f, f_k \rangle g_k,$$

i.e., $\{g_k\}_{k=1}^\infty$ is a dual of $\{f_k\}_{k=1}^\infty$. For the other implication, assume that $\{g_k\}_{k=1}^\infty$ is a dual frame of $\{f_k\}_{k=1}^\infty$. Then the synthesis operator U for $\{g_k\}_{k=1}^\infty$ satisfies the conditions: in fact, $\{g_k\}_{k=1}^\infty = \{U\delta_k\}_{k=1}^\infty$, and by Lemma 6.3.2, $UT^* = I$. $\square$

Lemma 6.3.6 *Let* $\{f_k\}_{k=1}^\infty$ *be a frame with synthesis operator* T. *The bounded left-inverses of* T^* *are precisely the operators having the form* $S^{-1}T + W(I - T^*S^{-1}T)$, *where* $W : \ell^2(\mathbb{N}) \to \mathcal{H}$ *is a bounded operator and* I *denotes the identity operator on* $\ell^2(\mathbb{N})$.

Proof.   Straightforward calculation gives that an operator of the given form is a left-inverse of T^*. On the other hand, if U is a given left inverse of T^*, then by taking $W = U$,

$$S^{-1}T + W(I - T^*S^{-1}T) = S^{-1}T + U - UT^*S^{-1}T = U.   \square$$

We are now ready for the announced characterization of all dual frames associated to a given frame.

Theorem 6.3.7 *Let* $\{f_k\}_{k=1}^\infty$ *be a frame for* $\mathcal{H}$. *The dual frames of* $\{f_k\}_{k=1}^\infty$ *are precisely the families*

$$\{g_k\}_{k=1}^\infty = \left\{ S^{-1}f_k + h_k - \sum_{j=1}^\infty \langle S^{-1}f_k, f_j \rangle h_j \right\}_{k=1}^\infty, \tag{6.9}$$

where $\{h_k\}_{k=1}^\infty$ *is a Bessel sequence in* $\mathcal{H}$.

Proof. By Lemma 6.3.5 and Lemma 6.3.6, we can characterize the dual frames as all families of the form

$$\{g_k\}_{k=1}^\infty = \{S^{-1}T\delta_k + W(I - T^*S^{-1}T)\delta_k\}_{k=1}^\infty, \tag{6.10}$$

where $W : \ell^2(\mathbb{N}) \to \mathcal{H}$ is a bounded operator or, equivalently, an operator of the form $W\{c_j\}_{j=1}^\infty = \sum_{j=1}^\infty c_j h_j$ where $\{h_k\}_{k=1}^\infty$ is a Bessel sequence. Inserting this expression for W in (6.10), we get

$$
\begin{aligned}
\{g_k\}_{k=1}^\infty &= \{S^{-1}f_k + W\delta_k - WT^*S^{-1}T\delta_k\}_{k=1}^\infty \\
&= \left\{ S^{-1}f_k + h_k - \sum_{j=1}^\infty \langle S^{-1}f_k, f_j\rangle h_j \right\}_{k=1}^\infty,
\end{aligned}
$$

as desired. $\qquad\square$

Note that if $\{f_k\}_{k=1}^\infty$ is a Riesz basis, then $\{f_k\}_{k=1}^\infty$ and $\{S^{-1}f_k\}_{k=1}^\infty$ are biorthogonal by Theorem 5.4.1. Thus, independently of the choice of $\{h_k\}_{k=1}^\infty$, we have $\sum_{j=1}^\infty \langle S^{-1}f_k, f_j\rangle h_j = h_k$; that is, Theorem 6.3.7 gives that the unique dual is $\{S^{-1}f_k\}_{k=1}^\infty$, in accordance with Theorem 3.6.2.

Given a frame $\{f_k\}_{k=1}^\infty$, one could also ask for a characterization of all families $\{g_k\}_{k=1}^\infty$ (frames or not) such that (6.4) holds. We shall not go into this subject, but ask the reader to think about the different possibilities (Exercise 6.1). We also mention the paper [492], where a non-frame $\{g_k\}_{k=1}^\infty$ satisfying (6.4) is found for a frame $\{f_k\}_{k=1}^\infty$ in the context of wavelets.

In Chapters 15–18 we will see an important reason for considering other duals than the canonical. In fact, we will study frames which have a convenient special structure (namely, wavelet structure), and unfortunately, it turns out that the canonical duals might not have the same structure. However, often one can find other duals having the same structure as the frame itself. See in particular Section 16.1 and Section 18.8.

Let us relate tight frames and the question of finding dual frames.

Lemma 6.3.8 *Let $\{f_k\}_{k=1}^\infty$ be a frame. Then the following are equivalent:*

(i) $\{f_k\}_{k=1}^\infty$ *is tight.*

(ii) $\{f_k\}_{k=1}^\infty$ *has a dual of the form $g_k = Cf_k$ for some constant $C > 0$.*

In case (ii) holds, the frame bound is $1/C$.

Proof. (i)$\Rightarrow$(ii) follows by letting $\{g_k\}_{k=1}^\infty$ be the canonical dual of $\{f_k\}_{k=1}^\infty$; (ii)$\Rightarrow$(i) follows from Lemma 6.3.2 by taking $f = g$. Furthermore, if (ii) holds, then $f = C\sum_{k=1}^\infty \langle f, f_k\rangle f_k$ for all $f \in \mathcal{H}$; thus

$$\|f\|^2 = \langle f, f\rangle = C \sum_{k=1}^\infty |\langle f, f_k\rangle|^2, \forall f \in \mathcal{H},$$

implying that the frame is tight with frame bound $1/C$. $\qquad\square$

6.4   Extension Problems for Bessel Sequences

We have already seen that any Bessel sequence in a separable Hilbert space $\mathcal{H}$ can be extended to a tight frame for $\mathcal{H}$. As a natural generalization, we will now show that any pair of Bessel sequences has an extension to a dual pair of frames. The following result first appeared in [183].

Theorem 6.4.1 *Let $\{f_k\}_{k=1}^{\infty}$ and $\{g_k\}_{k=1}^{\infty}$ be Bessel sequences in a separable Hilbert space $\mathcal{H}$. Then there exist Bessel sequences $\{p_j\}_{j\in J}$ and $\{q_j\}_{j\in J}$ in $\mathcal{H}$ such that*

$$\{f_k\}_{k=1}^{\infty} \cup \{p_j\}_{j\in J} \text{ and } \{g_k\}_{k=1}^{\infty} \cup \{q_j\}_{j\in J}$$

form a pair of dual frames for $\mathcal{H}$.

Proof.   Let T and U denote the synthesis operators for $\{f_k\}_{k=1}^{\infty}$ and $\{g_k\}_{k=1}^{\infty}$; see (6.5). Let $\{a_j\}_{j\in J}, \{b_j\}_{j\in J}$ denote any pair of dual frames for $\mathcal{H}$. Then

$$
\begin{aligned}
f = UT^*f + (I - UT^*)f &= \sum_{k=1}^{\infty}\langle f, f_k\rangle g_k + \sum_{j\in J}\langle (I - UT^*)f, a_j\rangle b_j \\
&= \sum_{k=1}^{\infty}\langle f, f_k\rangle g_k + \sum_{j\in J}\langle f, (I - TU^*)a_j\rangle b_j.
\end{aligned}
$$

The sequences $\{f_k\}_{k=1}^{\infty}, \{g_k\}_{k=1}^{\infty}, \{a_j\}_{j\in J}$, and $\{b_j\}_{j\in J}$ are Bessel sequences by definition, and $\{(I - TU^*)a_j\}_{j\in J}$ is a Bessel sequence because the operator $I - TU^*$ is bounded. Thus, by Lemma 6.3.2, we can conclude that

$$\{f_k\}_{k=1}^{\infty} \cup \{(I - TU^*)a_j\}_{j\in J} \text{ and } \{g_k\}_{k=1}^{\infty} \cup \{b_j\}_{j\in J}$$

form a dual pair of frames for $\mathcal{H}$, ad desired. $\square$

With Theorem 6.2.1 and Theorem 6.4.1 at hand, there are two natural ways to obtain frame decompositions using a Bessel sequence $\{f_k\}_{k=1}^{\infty}$: we can extend $\{f_k\}_{k=1}^{\infty}$ to a tight frame or find sequences $\{p_j\}_{j\in J}$ and $\{q_j\}_{j\in J}$ in $\mathcal{H}$ such that

$$\{f_k\}_{k=1}^{\infty} \cup \{p_j\}_{j\in J} \text{ and } \{f_k\}_{k=1}^{\infty} \cup \{q_j\}_{j\in J}$$

form a pair of dual frames for $\mathcal{H}$. In Section 12.7, we will demonstrate that the seemingly more complicated extension to a pair of dual frames might allow properties that cannot be obtained via the extension to a tight frame.

While the extension problems in Theorem 6.2.1 and Theorem 6.4.1 always have solutions, additional issues turn up whenever concrete cases are considered. For example, if a certain application asks for the Gabor structure, it is not enough to know that a given Gabor system can be extended to a frame: we will also need that the extension can be done with functions forming a Gabor system themselves. In Section 12.7 we will see that this is indeed possible. Section 19.3 will discuss the wavelet case, where the answer to the analogue question is unknown.

6.5   Approximately Dual Frames

Consider two Bessel sequences $\{f_k\}_{k=1}^{\infty}$ and $\{g_k\}_{k=1}^{\infty}$ in a Hilbert space $\mathcal{H}$, with synthesis operators T and U, respectively. A reasonable measure for how far $\{f_k\}_{k=1}^{\infty}$ and $\{g_k\}_{k=1}^{\infty}$ are from being dual frames would be to consider the number

$$\|I - TU^*\| = \sup_{\|f\|=1} \left\| f - \sum_{k=1}^{\infty} \langle f, g_k \rangle f_k \right\|. \tag{6.11}$$

The Bessel sequences $\{f_k\}_{k=1}^{\infty}$ and $\{g_k\}_{k=1}^{\infty}$ are said to be *approximately dual frames* if $\|I - TU^*\| < 1$; see [192]. Considering a given $f \in \mathcal{H}$, this definition does not imply that the vector

$$TU^* f = \sum_{k=1}^{\infty} \langle f, g_k \rangle f_k$$

is close to f, but it yields an exact reconstruction formula in terms of the sequences $\{f_k\}_{k=1}^{\infty}, \{g_k\}_{k=1}^{\infty}$, and the operator TU^* (Exercise 6.5):

Proposition 6.5.1 *Assume that the Bessel sequences $\{f_k\}_{k=1}^{\infty}$ and $\{g_k\}_{k=1}^{\infty}$ are approximately dual frames, with synthesis operators T and U, respectively. Then the operator TU^* is invertible; furthermore, the sequences $\{g_k\}_{k=1}^{\infty}$ and $\{(TU^*)^{-1} f_k\}_{k=1}^{\infty}$ are dual frames.*

Due to the necessary inversion of the operator TU^*, the result in Proposition 6.5.1 is more theoretical interesting than of practical importance.

For concrete applications, the condition $\|I - TU^*\| < 1$ is usually too weak: it does not guarantee that $TU^* f$ is close to f for a given $f \in \mathcal{H}$. It is more reasonable to assume that $\|I - TU^*\| \leq \epsilon$, for an $\epsilon > 0$ that reflects the error margin that is acceptable in a given situation. In [192], it is shown that this can actually be obtained via any pair of approximately dual frames $\{f_k\}_{k=1}^{\infty}, \{g_k\}_{k=1}^{\infty}$. In fact, consider the Neumann series (see (2.6)),

$$(TU^*)^{-1} = \sum_{n=0}^{\infty} (I - TU^*)^n.$$

It is shown in [192] that for any $N \in \mathbb{N}$ the sequences $\{g_k\}_{k=1}^{\infty}$ and $\{\gamma_k^{(N)}\}_{k=1}^{\infty}$ given by

$$\gamma_k^{(N)} = \sum_{n=0}^{N}(I - TU^*)^n f_k = f_k + \sum_{n=1}^{N}(I - TU^*)^n f_k$$

also form approximately dual frames. Furthermore, letting Z_N denote the synthesis operator associated with $\{\gamma_k^{(N)}\}_{k=1}^{\infty}$, it is proved that

$$\|I - Z_N U^*\| \leq \|I - TU^*\|^{N+1}.$$

Thus, for any $\epsilon > 0$, we can obtain that $\|I - Z_N U^*\| \leq \epsilon$ by choosing $N \in \mathbb{N}$ sufficiently large.

We will later return to approximately dual frames in the context of Gabor systems; see Section 12.8. Other applications in the literature include [289] by Feichtinger, Onchis, and Wiesmeyr (dealing with wavelet systems), [279] by Feichtinger, Grybos, and Onchis (Gabor systems), and [258] by Dörfler and Matusiak (nonstationary Gabor systems).

6.6   Exercises

6.1 This exercise concerns the question of finding *generalized duals* which are not frames.

(i) Find an overcomplete frame $\{f_k\}_{k=1}^{\infty}$, for which a family $\{g_k\}_{k=1}^{\infty}$ satisfying (6.4) automatically is a frame.

(ii) Find an overcomplete frame $\{f_k\}_{k=1}^{\infty}$ and a non-Bessel sequence $\{g_k\}_{k=1}^{\infty}$ such that (6.4) is satisfied.

6.2 Prove Lemma 6.3.3

6.3 Find a tight frame $\{f_k\}_{k=1}^{\infty}$ for which dual frames $\{g_k\}_{k=1}^{\infty}$ with arbitrary large optimal Bessel bound exist. (This shows that for noncanonical dual frames, no expression for the upper frame bound in terms of the frame bounds for $\{f_k\}_{k=1}^{\infty}$ exists.)

6.4 The purpose of this exercise is to prove Lemma 6.3.4. Let $\mathcal{H}$ denote a complex Hilbert space with inner product $\langle \cdot, \cdot \rangle$.

(i) Show that

$$\langle f, g \rangle = \frac{1}{4}\sum_{n=0}^{3} i^n \langle f + i^n g, f + i^n g \rangle, \; \forall f, g \in \mathcal{H}.$$

(ii) Let $K(\cdot, \cdot) : \mathcal{H} \times \mathcal{H} \to \mathbb{C}$ denote a sesquilinear form, i.e., K is linear in the first entry and conjugated linear in the second. Show that then

$$K(f, g) = \frac{1}{4} \sum_{n=0}^{3} i^n K(f + i^n g, f + i^n g), \forall f, g \in \mathcal{H}. \qquad (6.12)$$

Now let $\{f_k\}_{k=1}^{\infty}$ and $\{g_k\}_{k=1}^{\infty}$ denote Bessel sequences in $\mathcal{H}$.

(iii) Show that $\{f_k\}_{k=1}^{\infty}$ and $\{g_k\}_{k=1}^{\infty}$ are dual frames if the identity

$$\langle f, g \rangle = \sum_{k=1}^{\infty} \langle f, f_k \rangle \langle g_k, g \rangle \qquad (6.13)$$

holds for all f and g belonging to a dense subspace of $\mathcal{H}$.

(iv) Show that $\{f_k\}_{k=1}^{\infty}$ and $\{g_k\}_{k=1}^{\infty}$ are dual frames if the identity

$$||f||^2 = \sum_{k=1}^{\infty} \langle f, f_k \rangle \langle g_k, f \rangle \qquad (6.14)$$

holds for all f belonging to a dense subspace of $\mathcal{H}$.
Hint: consider the sesquilinear functional

$$K(f, g) := \sum_{k=1}^{\infty} \langle f, f_k \rangle \langle g_k, g \rangle, \ f, g \in \mathcal{H}. \qquad (6.15)$$

6.5 Prove Proposition 6.5.1.

7

Frames Versus Riesz Bases

We have already seen that Riesz bases are frames. In this chapter we exploit the relationship between these two concepts further. In particular, we give a number of equivalent conditions for a frame to be a Riesz basis.

We have often spoken about a frame in an intuitive sense as some kind of "overcomplete basis." It turns out that, in the technical sense, one has to be careful with such statements. In fact, we will prove the existence of a frame which has no relation to a basis: no subfamily of the frame forms a basis. On the other hand, sufficient conditions for a frame to contain a Riesz basis as a subfamily are also given.

We begin with a detailed discussion of the relationship between frames and Riesz bases in Section 7.1. In particular, it is shown that a frame is a Riesz basis if and only if it is ω-independent. This is followed by a discussion of lower frame bounds for subsequences, in Section 7.2. Section 7.3 introduces a special class of frames, the so-called Riesz frames. These frames also appear in Section 7.4, where the main question is whether a frame can be reduced to a Riesz basis by deletion of selected elements; the general answer is no, but for Riesz frames it is yes. Section 7.5 gives the details of a surprising construction of a tight frame, where no subset is a basis. Section 7.6 yields another characterization of Riesz bases and relates it to a certain minimization problem; finally, Section 7.7 gives a short description of Feichtinger's conjecture, asserting that each frame that is norm-bounded below can be split into a union of a finite number of Riesz sequences (the conjecture was confirmed in 2013).

© Springer International Publishing Switzerland 2016
O. Christensen, *An Introduction to Frames and Riesz Bases*,
Applied and Numerical Harmonic Analysis,
DOI 10.1007/978-3-319-25613-9_7

7.1 Conditions for a Frame Being a Riesz Basis

In the entire chapter, we let $\mathcal{H}$ denote a separable Hilbert space with inner product $\langle \cdot, \cdot \rangle$. Our first purpose is to give a detailed analysis of the relation between frames and Riesz bases.

Recall from Lemma 3.1.3 that ω-independence and minimality are two different concepts. We now give some equivalent conditions for a frame to be a Riesz basis; in particular, we prove that for a frame, ω-independence is equivalent to minimality.

Theorem 7.1.1 *Let* $\{f_k\}_{k=1}^\infty$ *be a frame for* $\mathcal{H}$. *Then the following are equivalent:*

(i) $\{f_k\}_{k=1}^\infty$ *is a Riesz basis for* $\mathcal{H}$.

(ii) $\{f_k\}_{k=1}^\infty$ *is an exact frame.*

(iii) $\{f_k\}_{k=1}^\infty$ *is minimal.*

(iv) $\{f_k\}_{k=1}^\infty$ *has a biorthogonal sequence.*

(v) $\{f_k\}_{k=1}^\infty$ *and* $\{S^{-1}f_k\}_{k=1}^\infty$ *are biorthogonal.*

(vi) $\{f_k\}_{k=1}^\infty$ *is* ω-*independent.*

(vii) *If* $\sum_{k=1}^\infty c_k f_k = 0$ *for some* $\{c_k\}_{k=1}^\infty \in \ell^2(\mathbb{N})$, *then* $c_k = 0$, $\forall k \in \mathbb{N}$.

(viii) $\{f_k\}_{k=1}^\infty$ *is a basis.*

Proof. We proceed with the following steps:

(a) (i) $\Rightarrow$ (ii) $\Rightarrow$ (v) $\Rightarrow$ (iv)$\Rightarrow$ (iii)$\Rightarrow$ (ii)$\Rightarrow$ (i)

(b) (i) $\Rightarrow$ (vi) $\Rightarrow$ (vii)$\Rightarrow$ (i)

(c) (i)$\Rightarrow$(viii)$\Rightarrow$(iv)

Step (a):
(i)$\Rightarrow$(ii) A Riesz basis $\{f_k\}_{k=1}^\infty$ is an exact frame: if an arbitrary element is removed, the remaining family is not complete and therefore not a frame.
(ii)$\Rightarrow$(v). This is Proposition 5.4.8.
(v) $\Rightarrow$ (iv) is clear, and (iv)$\Rightarrow$(iii) is proved in Lemma 3.3.1 (i).
(iii)$\Rightarrow$(ii). Assume that $\{f_k\}_{k=1}^\infty$ is minimal. Then, for an arbitrary $j \in \mathbb{N}$, the family $\{f_k\}_{k\neq j}$ is incomplete in $\mathcal{H}$ and therefore not a frame for $\mathcal{H}$.
(ii)$\Rightarrow$(i). If $\{f_k\}_{k=1}^\infty$ is an exact frame, it is a basis by Proposition 5.4.8. Let as usual $\{\delta_k\}_{k=1}^\infty$ be the canonical basis for $\ell^2(\mathbb{N})$. The synthesis operator $T : \ell^2(\mathbb{N}) \to \mathcal{H}$ associated with $\{f_k\}_{k=1}^\infty$ is bounded and surjective by Theorem 5.5.1, and $T\delta_k = f_k$; in order to show that $\{f_k\}_{k=1}^\infty$ is a Riesz basis, it is enough to prove that T is invertible, and this follows from $\{f_k\}_{k=1}^\infty$ being a basis.

Step (b):

(i)⇒(vi). Assume that $\{f_k\}_{k=1}^{\infty}$ is a Riesz basis and that $\sum_{k=1}^{\infty} c_k f_k = 0$ for a given sequence $\{c_k\}_{k=1}^{\infty}$ of scalars. Then, by the result in Exercise 3.13, $\{c_k\}_{k=1}^{\infty} \in \ell^2(\mathbb{N})$. Denoting a lower Riesz bound by A, Exercise 3.13 also shows that

$$A \sum_{k=1}^{\infty} |c_k|^2 \leq \left\| \sum_{k=1}^{\infty} c_k f_k \right\|^2 = 0;$$

thus $c_k = 0$ for all k.

(vi)⇒(vii). Clear.

(vii)⇒(i). Let again $\{\delta_k\}_{k=1}^{\infty}$ be the canonical orthonormal basis for $\ell^2(\mathbb{N})$. The assumption (vii) assures that the synthesis operator T is injective, and T is also surjective because $\{f_k\}_{k=1}^{\infty}$ is a frame. Since $T\delta_k = f_k$, $\forall k$, the result follows from the definition of a Riesz basis.

Step (c):

(i)⇒(viii). Clear.

(viii)⇒(iv). This is Theorem 3.3.2.   □

Note the slight difference between the conditions (vi) and (vii) in Theorem 7.1.1. As seen in Exercise 5.5, one difference between frames and Riesz bases is that for a Riesz basis $\{f_k\}_{k=1}^{\infty}$, the series $\sum_{k=1}^{\infty} c_k f_k$ is only convergent for $\{c_k\}_{k=1}^{\infty} \in \ell^2(\mathbb{N})$; for general frames, the series might converge for coefficients $\{c_k\}_{k=1}^{\infty} \notin \ell^2(\mathbb{N})$. However, Theorem 7.1.1 shows that to check whether a frame is ω-independent, it is enough to consider ℓ^2-sequences.

The condition of ω-independence is a stronger condition than just linear independence. To illustrate that point, we prove two more characterization of Riesz bases. The result should be compared to the equivalence of (i) and (vi) in Theorem 7.1.1. The equivalence (i) ⇔ (ii) below first appeared in [450], and the equivalence (ii) ⇔ (iii) is from [193]. The proof is based on the knowledge that for a Riesz basis, the "Riesz basis bounds" and "frame bounds" coincide by Theorem 5.4.1

Proposition 7.1.2 *Let $\{f_k\}_{k=1}^{\infty}$ be a frame for $\mathcal{H}$. For $n \in \mathbb{N}$, let A_n denote the optimal lower frame bound for the frame sequence $\{f_k\}_{k=1}^{n}$. Then the following are equivalent:*

(i) $\{f_k\}_{k=1}^{\infty}$ *is a Riesz basis for $\mathcal{H}$.*

(ii) $\{f_k\}_{k=1}^{\infty}$ *is linearly independent and $\inf_{n \in \mathbb{N}} A_n > 0$.*

(iii) $\{f_k\}_{k=1}^{\infty}$ *is linearly independent and $\lim_{n \to \infty} A_n$ exists and is positive.*

Proof. For the proof of (i)⇒(ii), we note that any basis is linearly independent. Assume that $\{f_k\}_{k=1}^{\infty}$ is a Riesz basis, with lower Riesz bound A. Then each subfamily $\{f_k\}_{k=1}^{n}$ is a Riesz sequence, and A_n is also the optimal lower Riesz bound by Theorem 5.4.1; thus $A \leq A_n$ for all $n \in \mathbb{N}$ and

(ii) follows. For the proof of (ii) $\Rightarrow$ (i), assume that (ii) is satisfied. Then $A := \inf_{n\in\mathbb{N}} A_n$ is a lower frame bound for each of the Riesz sequences $\{f_k\}_{k=1}^n$, $n \in \mathbb{N}$. Letting B denote an upper frame bound for $\{f_k\}_{k=1}^\infty$, Theorem 3.2.3 shows that

$$A \sum_{k=1}^n |c_k|^2 \le \left\| \sum_{k=1}^n c_k f_k \right\|^2 \le B \sum_{k=1}^n |c_k|^2 \tag{7.1}$$

for all scalar sequences $\{c_k\}_{k=1}^n$. By Theorem 3.6.6, we conclude that $\{f_k\}_{k=1}^\infty$ is a Riesz basis for $\mathcal{H}$. That (ii) $\Leftrightarrow$ (iii) follows from the fact that when $\{f_k\}_{k=1}^\infty$ is linearly independent, the sequence of optimal frame bounds A_n, $n \in \mathbb{N}$, is equal to the sequence of optimal lower (Riesz) basis bounds, which is decreasing by definition. $\qquad\square$

Note that Example 5.4.5 exhibits an example of an overcomplete frame which is linearly independent. Theorem 11.3.1 combined with the discussion of the HRT conjecture in Section 13.4 will show us that the same happens for any Gabor frame $\{E_{mb}T_{na}g\}_{m,n\in\mathbb{Z}}$ with $ab < 1$.

7.2   Frames and Their Subsequences

The interplay between a frame and its finite subsequences is of great practical importance. In fact, even if a theoretical description is done via a frame $\{f_k\}_{k=1}^\infty$, a concrete realization always has to limit the attention to a finite subsequence, say $\{f_k\}_{k=1}^n$, for some $n \in \mathbb{N}$; of course the finite subsequence does not need to consist of the first n elements in the frame, but for the discussion in the current section, it suffices to assume this.

For the frames that are popular in applications, e.g., the Gabor frames, the finite subsequences have certain properties that are radically different from the properties of the frame itself. Thus the restriction to finite subsequences and the analysis hereof have to be performed with great care.

In order to explain this, consider an overcomplete frame $\{f_k\}_{k=1}^\infty$ for an infinite-dimensional Hilbert space $\mathcal{H}$. From Proposition 1.1.2, we know that for each $n \in \mathbb{N}$, the sequence $\{f_k\}_{k=1}^n$ is a frame for the finite-dimensional space

$$\mathcal{H}_n := \text{span}\{f_k\}_{k=1}^n.$$

By Theorem 7.1.1 the overcompleteness implies that there exist coefficients $\{c_k\}_{k=1}^\infty \in \ell^2(\mathbb{N})$, where not all c_k are zero but such that $\sum_{k=1}^\infty c_k f_k = 0$. Thus, $\{f_k\}_{k=1}^\infty$ is ω-dependent. Despite of this property, most frames that appear in applications have the property that finite subsequences are linearly independent! That is, if $\{f_k\}_{k\in\mathcal{F}}$ is any finite subsequences of $\{f_k\}_{k=1}^\infty$ and $\sum_{k\in\mathcal{F}} c_k f_k = 0$ for some coefficients c_k, then necessarily $c_k = 0$ for all $k \in \mathcal{F}$.

For overcomplete frames $\{f_k\}_{k=1}^{\infty}$ with the property that finite subsequences are linearly independent, the lower frame bounds of the finite subsequences also behave radically different than the frame bounds for the frame. In fact, Proposition 7.1.2 implies that the optimal lower frame bound A_n for $\{f_k\}_{k=1}^{n}$ tends to zero for $n \to \infty$. Let us state a more quantitative statement; it is a general version of a result by Gröchenig, stated in the setting of Gabor frames in [344].

Lemma 7.2.1 *Assume that*

(i) $\{f_k\}_{k=1}^{\infty}$ *is an overcomplete frame with upper bound B;*

(ii) *Each finite subsequence $\{f_k\}_{k=1}^{n}$, $n \in \mathbb{N}$, is linearly independent.*

Let A_n denote a lower frame bound for $\{f_k\}_{k=1}^{n}$ [as frame for $\mathcal{H}_n$] and take any sequence $\{c_k\}_{k=1}^{\infty} \in \ell^2(\mathbb{N}) \setminus \{0\}$ such that $\sum_{k=1}^{\infty} c_k f_k = 0$, normalized such that $\sum_{k=1}^{\infty} |c_k|^2 = 1$. Then, for n sufficiently large,

$$A_n \le 2B \sum_{k=n+1}^{\infty} |c_k|^2. \tag{7.2}$$

Proof. Take $\{c_k\}_{k=1}^{\infty} \in \ell^2(\mathbb{N}) \setminus \{0\}$ such that $\sum_{k=1}^{\infty} c_k f_k = 0$, normalized such that $\sum |c_k|^2 = 1$. Then, for any $n \in \mathbb{N}$, we have

$$\sum_{k=1}^{n} c_k f_k = - \sum_{k=n+1}^{\infty} c_k f_k,$$

and therefore

$$\left\| \sum_{k=1}^{n} c_k f_k \right\|^2 = \left\| \sum_{k=n+1}^{\infty} c_k f_k \right\|^2 \le B \sum_{k=n+1}^{\infty} |c_k|^2. \tag{7.3}$$

Since $\{f_k\}_{k=1}^{n}$ is assumed to be linearly independent, it is actually a Riesz basis for $\mathcal{H}_n$. Thus, the lower frame bound A_n for $\{f_k\}_{k=1}^{n}$ satisfies that

$$A_n \sum_{k=1}^{n} |c_k|^2 \le \left\| \sum_{k=1}^{n} c_k f_k \right\|^2. \tag{7.4}$$

Now, for $n \in \mathbb{N}$ sufficiently large,

$$1 = \sum_{k=1}^{\infty} |c_k|^2 \le 2 \sum_{k=1}^{n} |c_k|^2.$$

Then (7.3) implies that

$$\left\| \sum_{k=1}^{n} c_k f_k \right\|^2 \le B \sum_{k=n+1}^{\infty} |c_k|^2 \le \left(2B \sum_{k=n+1}^{\infty} |c_k|^2 \right) \sum_{k=1}^{n} |c_k|^2.$$

Using (7.4) this leads to (7.2). $\qquad\qquad\square$

Letting $S_n : \mathcal{H}_n \to \mathcal{H}_n$ denote the frame operator associated with $\{f_k\}_{k=1}^n$, the assumptions in Lemma 7.2.1 imply by Theorem 1.3.1 that the condition number of S_n tends to infinity as $n \to \infty$.

The problematic behavior of the lower frame bounds for finite subsequences encountered in Lemma 7.2.1 motivates the definition of Riesz frames in Section 7.3.

7.3   Riesz Frames and Near-Riesz Bases

Several variations on the definition of frames and Riesz bases are possible:

Definition 7.3.1 *Let $\{f_k\}_{k=1}^\infty$ be a sequence in $\mathcal{H}$. We say that a frame $\{f_k\}_{k=1}^\infty$ is*

(i) *A near-Riesz basis if it consists of a Riesz basis and a finite number of extra elements;*

(ii) *A Riesz frame if every subsequence of $\{f_k\}_{k=1}^\infty$ is a frame sequence, with uniform frame bounds A, B.*

Near-Riesz bases are relevant because they share many properties with Riesz bases. For an arbitrary sequence $\{f_k\}_{k=1}^\infty$ in $\mathcal{H}$, the *excess* is defined as

$$e(\{f_k\}_{k=1}^\infty) := \sup\{|J| \ : \ J \subseteq \mathbb{N} \text{ and } \overline{span}\{f_k\}_{k\in\mathbb{N}\setminus J} = \overline{span}\{f_k\}_{k=1}^\infty\}. \tag{7.5}$$

For a near-Riesz basis, the excess is equal to the number of elements which have to be removed in order to obtain a Riesz basis. Holub proved in [413] that for a near-Riesz basis $\{f_k\}_{k=1}^\infty$ with synthesis operator T,

$$e(\{f_k\}_{k=1}^\infty) = \dim(\mathcal{N}_T).$$

Furthermore, it was proved in [32] that if we consider a frame $\{f_k\}_{k=1}^\infty$ and denote the frame operator for by S, then

$$e(\{f_k\}_{k=1}^\infty) = \sum_{k=1}^\infty \left(1 - \langle f_k, S^{-1} f_k \rangle\right).$$

Note that it was proved by Bakić and Berić [25] that all dual frames associated with a given frame $\{f_k\}_{k=1}^\infty$ have the same excess.

To motivate the definition of Riesz frames, we recall Corollary 3.7.2: a subfamily of a Riesz sequence is a Riesz sequence. For frames, the situation is different. Consider, for example, the frame in Example 5.1.4 (ii); it contains the subfamily $\{\frac{1}{\sqrt{k}} e_k\}_{k=1}^\infty$, which is not a frame sequence. Riesz frames were introduced in [157] in order to avoid this situation. In the sequel, we will see that Riesz frames behave like Riesz bases in many contexts, despite the fact that a Riesz frame can be overcomplete. Unfortunately, at least so tar, Riesz frames are only relevant in abstract Hilbert spaces; for example, Gabor frames are in general not Riesz frames, and no connection between Riesz frames and wavelet frames is available in the literature.

Note that the concepts of Riesz frames and near-Riesz bases are unrelated (Exercise 7.1 and Exercise 23.1). For more information about excess, we refer to the paper [32].

7.4 Frames Containing a Riesz Basis

Intuitively, we think about a frame as some kind of "overcomplete basis," so a natural question is the following: given a frame $\{f_k\}_{k=1}^\infty$, is it possible to extract a basis $\{f_k\}_{k\in J}$, $J \subseteq \mathbb{N}$, from $\{f_k\}_{k=1}^\infty$, i.e., does $\{f_k\}_{k=1}^\infty$ contain a basis as a subset?

Clearly, the answer depends on which kind of basis we are interested in. In this section, we shall find sufficient conditions for a frame to contain a Riesz basis. In the next section, we construct a frame consisting of vectors having norm one, which does not even contain a Schauder basis. We begin with a technical lemma.

Lemma 7.4.1 Let $\{c_k\}_{k=1}^\infty$ be a sequence of nonnegative numbers for which $\sum_{k=1}^\infty c_k < \infty$. Suppose that $\{J_n\}_{n\in\mathbb{N}}$ is a family of subsets of $\mathbb{N}$ such that:

(i) $J_1 \supseteq J_2 \supseteq J_3 \supseteq \cdots$

(ii) There exists $c > 0$ such that $c \le \sum_{k\in J_n} c_k$, $\forall n \in \mathbb{N}$.

Then $c \le \sum_{k\in\cap J_n} c_k$.

Proof. Define a positive measure μ on the σ-algebra of subsets of $\mathbb{N}$ by

$$\mu(S) = \sum_{k\in S} c_k.$$

By Lemma A.2.3,

$$\mu(J_n) = \sum_{k\in J_n} c_k \to \mu(\cap J_n) = \sum_{k\in\cap J_n} c_k \text{ as } n \to \infty,$$

and the result follows. $\square$

Our purpose is to show that every Riesz frame contains a Riesz basis, a result which appeared in [157]. Our proof is based on the *axiom of choice*, also known as *Zorn's lemma*:

Lemma 7.4.2 Let $\mathcal{M}$ be a nonempty ordered set. If every totally ordered subset of $\mathcal{M}$ has an upper bound in $\mathcal{M}$, then $\mathcal{M}$ has at least one maximal element.

Theorem 7.4.3 *Every Riesz frame contains a Riesz basis.*

Proof. For convenience, we use the index set $\mathbb{N}$. Let $\{f_k\}_{k=1}^{\infty}$ be a Riesz frame, and let A be a common lower bound for all its subframes. Consider the set

$$\mathcal{M} := \left\{ \{f_k\}_{k \in J} \mid J \subseteq \mathbb{N}, \text{ and } A\|f\|^2 \leq \sum_{k \in J} |\langle f, f_k \rangle|^2, \ \forall f \in \mathcal{H} \right\}. \quad (7.6)$$

Clearly $\mathcal{M}$ is nonempty. Define an order on $\mathcal{M}$:

$$\{f_k\}_{k \in J} \prec \{f_k\}_{k \in K} \Leftrightarrow K \subseteq J.$$

Now consider a totally ordered family of elements $\{f_k\}_{k \in J_n} \in \mathcal{M}, n$ in some index set I. Such a family has an upper bound $\{f_k\}_{k \in \cap J_n}$, which is still an element from $\mathcal{M}$; to see this, we have to prove that

$$A\,\|f\|^2 \leq \sum_{k \in \cap J_n} |\langle f, f_k \rangle|^2, \ \forall f \in \mathcal{H}. \quad (7.7)$$

In case the index set I is countable, (7.7) follows from Lemma 7.4.1; for an uncountable index set, the result is still true, but a more general measure-theoretic argument is needed (we skip the argument; see, e.g., [523] for arguments of this type). So, by Zorn's lemma 7.4.2, $\mathcal{M}$ has a maximal element $\{f_k\}_{k \in J}$. Now we show that $\{f_k\}_{k \in J}$ is a Riesz basis. Clearly $\{f_k\}_{k \in J}$ is a frame, so by Theorem 7.1.1 it is enough to show that $\{f_k\}_{k \in J}$ is ω-independent. But if not, we could find an element $f_n, n \in J$ such that $\{f_k\}_{k \in J - \{n\}}$ was still complete and therefore a frame for $\mathcal{H}$ by Theorem 5.4.7. That is, $\{f_k\}_{k \in J - \{n\}} \in \mathcal{M}$, which contradicts the maximality of $\{f_k\}_{k \in J}$. $\square$

Via iterated application of Theorem 7.4.3, one can show that every Riesz frame is the union of a *finite* number of Riesz sequences. We refer to [195] for details.

The result in Theorem 7.4.3 can be slightly generalized: using a more complicated argument, it is shown in [118] that the conclusion holds for any frame $\{f_k\}_{k=1}^{\infty}$ with the property that each subfamily $\{f_k\}_{k \in J}, J \subseteq \mathbb{N}$ is a frame sequence. Such a frame is said to have the *subframe property*.

Lemma 7.4.4 *Assume that a frame $\{f_k\}_{k=1}^{\infty}$ for $\mathcal{H}$ has the subframe property. Then a subfamily $\{f_k\}_{k \in J}$ is a Schauder basis for $\mathcal{H}$ if and only if $\{f_k\}_{k \in J}$ is a Riesz basis for $\mathcal{H}$.*

Proof. Assume that $\{f_k\}_{k \in J}$ is a Schauder basis for some $J \subseteq \mathbb{N}$. By the subframe property, $\{f_k\}_{k \in J}$ is also a frame, so we conclude by Theorem 7.1.1 that $\{f_k\}_{k \in J}$ is a Riesz basis. $\square$

Lemma 7.4.4 does not hold if the assumption of $\{f_k\}_{k=1}^{\infty}$ having the subframe property is removed (Exercise 7.2).

7.5 A Frame Which Does Not Contain a Basis

Even if $\{f_k\}_{k=1}^{\infty}$ is not a Riesz frame, the proof of Theorem 7.4.3 shows that the set $\mathcal{M}$ in (7.6) will contain a maximal element for every lower frame bound A. However, this element need not constitute a Riesz basis. For example, let $\{e_k\}_{k=1}^{\infty}$ be an orthonormal basis and consider the frame from Example 5.1.4 (ii),

$$\{f_k\}_{k=1}^{\infty} := \left\{ e_1, \frac{1}{\sqrt{2}}e_2, \frac{1}{\sqrt{2}}e_2, \frac{1}{\sqrt{3}}e_3, \frac{1}{\sqrt{3}}e_3, \frac{1}{\sqrt{3}}e_3, \ldots \right\}. \qquad (7.8)$$

No matter how small A is chosen, the set $\mathcal{M}$ does not contain a Riesz basis for this frame. In fact, for any $\epsilon > 0$ and any $\{f_k\}_{k\in J} \in \mathcal{M}$, the condition

$$\epsilon \, \|f\|^2 \leq \sum_{k\in J} |\langle f, f_k \rangle|^2, \ \forall f \in \mathcal{H}$$

implies that $\frac{1}{\sqrt{n}}e_n$ appears more than once in $\{f_k\}_{k\in J}$ for large values of n. Thus $\mathcal{M}$ does not contain an ω-independent subset. Actually $\{f_k\}_{k=1}^{\infty}$ does not contain a Riesz basis at all. The only candidate would be $\{\frac{1}{\sqrt{k}}e_k\}_{k=1}^{\infty}$, which is a Schauder basis but not a Riesz basis.

Now we want to show that it even is possible to construct a frame which consists of vectors that are norm bounded below but which does not contain a Schauder basis; in particular, it does not contain a Riesz basis. The example was constructed by Casazza and Christensen and appeared in [120]. It is considerably more complicated than (7.8), and we need some preparation before the proof.

Lemma 7.5.1 *Let $\{e_k\}_{k=1}^n$ be an orthonormal basis for a finite-dimensional Hilbert space $\mathcal{H}_n$. Define the vectors*

$$\begin{cases} f_k = e_k - \dfrac{1}{n}\displaystyle\sum_{j=1}^{n} e_j, \ k = 1, \ldots, n; \\[4mm] f_{n+1} = \dfrac{1}{\sqrt{n}}\displaystyle\sum_{j=1}^{n} e_j. \end{cases}$$

Then the following hold:

(i) *$\{f_k\}_{k=1}^{n+1}$ is a tight frame for $\mathcal{H}_n$ with frame bound $A = 1$.*

(ii) *Assume that $n > 2$, and let $\{f_{k_i}\}_{i\in I}$ be any subset of $\{f_k\}_{k=1}^{n+1}$ for which $\mathrm{span}\{f_{k_i}\}_{i\in I} = \mathcal{H}_n$. Then, for an arbitrary ordering of the elements, $\{f_{k_i}\}_{i\in I}$ has basis constant greater than or equal to $\frac{1}{4}\sqrt{n-2}$.*

Proof. To prove (i), let $f \in \mathcal{H}_n$ and write

$$f = \sum_{k=1}^{n} a_k e_k, \text{ where } a_k = \langle f, e_k \rangle.$$

Letting P denote the orthogonal projection onto the unit vector $\frac{1}{\sqrt{n}} \sum_{k=1}^{n} e_k$,

$$Pf = \left\langle f, \frac{1}{\sqrt{n}} \sum_{k=1}^{n} e_k \right\rangle \frac{1}{\sqrt{n}} \sum_{k=1}^{n} e_k = \frac{\sum_{k=1}^{n} a_k}{\sqrt{n}} \frac{1}{\sqrt{n}} \sum_{k=1}^{n} e_k.$$

Therefore

$$\|Pf\|^2 = \frac{1}{n} \left| \sum_{k=1}^{n} a_k \right|^2 = |\langle f, f_{n+1} \rangle|^2.$$

Also,

$$
\begin{aligned}
\|(I - P)f\|^2 &= \left\| \sum_{k=1}^{n} a_k e_k - \frac{1}{n} \sum_{j=1}^{n} a_j \sum_{k=1}^{n} e_k \right\|^2 \\
&= \left\| \sum_{k=1}^{n} \left(a_k - \frac{1}{n} \sum_{j=1}^{n} a_j \right) e_k \right\|^2 \\
&= \sum_{k=1}^{n} \left| a_k - \frac{1}{n} \sum_{j=1}^{n} a_j \right|^2 \\
&= \sum_{k=1}^{n} |\langle f, f_k \rangle|^2.
\end{aligned}
$$

Putting the two last results together, we obtain that

$$\|f\|^2 = \|Pf\|^2 + \|(I - P)f\|^2 = \sum_{k=1}^{n+1} |\langle f, f_k \rangle|^2.$$

This proves (i). To prove (ii), we note that $\sum_{k=1}^{n} f_k = 0$, i.e., the vectors $\{f_k\}_{k=1}^{n}$ are linearly dependent. Therefore any subset of the frame $\{f_k\}_{k=1}^{n+1}$ which spans $\mathcal{H}_n$ must contain f_{n+1} and at least $n-1$ of the terms $\{f_k\}_{k=1}^{n}$. The basis constant for an arbitrary sequence is by its definition in (3.10) larger than or equal to the basis constant for any subsequence; thus, it is enough to prove the following:

Claim: For any family of the type

$$\{f_k\}_{k \in \Delta \cup \{n+1\}}, \text{ where } \Delta \subset \{1, 2, \ldots, n\}, \ |\Delta| = n - 1, \tag{7.9}$$

there exist an index set $\Lambda \subseteq \Delta \cup \{n+1\}$ and scalars $\{c_k\}_{k \in \Delta \cup \{n+1\}}$ such that

$$0 \neq \left\| \sum_{k \in \Lambda} c_k f_k \right\| \geq \frac{1}{4} \sqrt{n-2} \left\| \sum_{k \in \Delta \cup \{n+1\}} c_k f_k \right\|. \tag{7.10}$$

Note that this takes care of the fact that $\{f_k\}_{k \in \Delta \cup \{n+1\}}$ can be ordered in an arbitrary way.

Given a set Δ as in (7.9), let $\Delta^c = \{\ell\}$ denote its complement in $\{1, \ldots, n\}$. Then, since $\sum_{k=1}^n f_k = 0$, we have

$$\left\| \sum_{k \in \Delta} f_k \right\| = \|f_\ell\| = \sqrt{\frac{n-1}{n}}.$$

Note that $f_{n+1} \perp f_k$ for all $k = 1, \ldots, n$. Therefore

$$\left\| \sum_{k \in \Delta \cup \{n+1\}} f_k \right\| = \left(\|f_{n+1}\|^2 + \left\| \sum_{k \in \Delta} f_k \right\|^2 \right)^{1/2}$$

$$= \left(1 + \frac{n-1}{n} \right)^{1/2}$$

$$= \; \leq \sqrt{2}. \tag{7.11}$$

Let $\lfloor \frac{n-1}{2} \rfloor$ denote the integer part of $\frac{n-1}{2}$ (see page 645). For any subset $\Gamma \subset \{1, 2, \ldots, n\}$ with $|\Gamma| = \lfloor \frac{n-1}{2} \rfloor$, we have

$$|\Gamma| \leq \frac{n}{2} \quad \text{and} \quad |\Gamma| \geq \frac{n-1}{2} - \frac{1}{2} = \frac{n}{2} - 1;$$

therefore

$$\left\| \sum_{k \in \Gamma} f_k \right\| = \left\| \sum_{k \in \Gamma} \left(e_k - \frac{1}{n} \sum_{j=1}^n e_j \right) \right\|$$

$$= \left\| \sum_{k \in \Gamma} \left(1 - \frac{|\Gamma|}{n} \right) e_k + \sum_{k \notin \Gamma} \frac{-|\Gamma|}{n} e_k \right\|$$

$$\geq \left(\sum_{k \in \Gamma} \left(1 - \frac{|\Gamma|}{n} \right)^2 \right)^{1/2}$$

$$= |\Gamma|^{1/2} \left(1 - \frac{|\Gamma|}{n} \right)$$

$$\geq \frac{|\Gamma|^{1/2}}{2}$$

$$\geq \frac{1}{2} \sqrt{\frac{n}{2} - 1}. \tag{7.12}$$

We now consider a subset $\{f_k\}_{k\in\Lambda}$ of $\{f_{n+1}\}\cup\{f_k\}_{k\in\Delta}$. We choose Λ such that $\{f_k\}_{k\in\Lambda}$ contains exactly $\lfloor\frac{n-1}{2}\rfloor$ of the elements of the set $\{f_k\}_{k=1}^n$. Now, there are two possibilities.

Case 1 $f_{n+1}\notin\{f_k\}_{k\in\Lambda}$:

Then, as we saw in the estimate (7.12),

$$\left\|\sum_{k\in\Lambda}f_k\right\|\geq\frac{1}{2}\sqrt{\frac{n}{2}-1},$$

while by (7.11)

$$\left\|\sum_{k\in\Delta\cup\{n+1\}}f_k\right\|\leq\sqrt{2}.$$

Therefore

$$\left\|\sum_{k\in\Lambda}f_k\right\| \geq \frac{1}{2}\sqrt{\frac{n}{2}-1}\,\frac{1}{\sqrt{2}}\left\|\sum_{k\in\Delta\cup\{n+1\}}f_k\right\|$$

$$= \frac{1}{4}\sqrt{n-2}\left\|\sum_{k\in\Delta\cup\{n+1\}}f_k\right\|.$$

Hence, (7.10) is satisfied.

Case 2 $f_{n+1}\in\{f_k\}_{k\in\Lambda}$:

Now, since $f_{n+1}\perp f_k, k=1\ldots,n$, we have

$$\left\|\sum_{k\in\Lambda}f_k\right\| = \left(\|f_{n+1}\|^2+\left\|\sum_{k\in\Lambda\setminus\{n+1\}}f_k\right\|^2\right)^{1/2}$$

$$\geq \left(\left\|\frac{1}{\sqrt{n}}\sum_{j=1}^n e_j\right\|^2+\left[\frac{1}{2}\sqrt{\frac{n}{2}-1}\right]^2\right)^{1/2}$$

$$\geq \frac{1}{2}\sqrt{\frac{n}{2}-1},$$

while we still have $\|\sum_{k\in\Delta\cup\{n+1\}}f_k\|\leq\sqrt{2}$. Thus (7.10) is again satisfied. This completes the proof. $\qquad\square$

We are now ready to prove the existence of a tight frame which is norm bounded below and for which no subfamily is a Schauder basis. The exact meaning of this is that no matter how a subset of the frame is ordered, it is not a basis.

Theorem 7.5.2 *In every separable infinite-dimensional Hilbert space $\mathcal{H}$, there exists a tight frame which is norm bounded below and which does not contain a Schauder basis for $\mathcal{H}$.*

Proof. Since all infinite-dimensional separable Hilbert spaces are isometrically isomorphic, it is enough to construct an example in one particular Hilbert space. Let $\{e_k\}_{k=1}^{\infty}$ be an orthonormal basis for a Hilbert space $\mathcal{K}$ and define an infinite collection of finite-dimensional vector spaces by $\mathcal{H}_1 = \text{span}\{e_1\}$, $\mathcal{H}_2 = \text{span}\{e_2, e_3\}$, and in general

$$\mathcal{H}_n := \text{span}\left\{e_{\frac{(n-1)n}{2}+1}, e_{\frac{(n-1)n}{2}+2}, \dots, e_{\frac{(n-1)n}{2}+n}\right\}.$$

Consider the direct sum

$$\mathcal{H} = \left(\sum_{n=1}^{\infty} \oplus \mathcal{H}_n\right)_{\ell^2}$$

as defined in (A.4). In each space $\mathcal{H}_n$ we construct the sequence $\{f_k^n\}_{k=1}^{n+1}$ as in Lemma 7.5.1, starting with the orthonormal basis

$$\left\{e_{\frac{(n-1)n}{2}+1}, e_{\frac{(n-1)n}{2}+2}, \dots, e_{\frac{(n-1)n}{2}+n}\right\} = \left\{e_{\frac{(n-1)n}{2}+k}\right\}_{k=1}^{n}.$$

Specifically, given $n \in \mathbb{N}$,

$$\begin{cases} f_k^n = e_{\frac{(n-1)n}{2}+k} - \dfrac{1}{n}\sum_{j=1}^{n} e_{\frac{(n-1)n}{2}+j}, & 1 \le k \le n; \\[4mm] f_{n+1}^n = \dfrac{1}{\sqrt{n}}\sum_{j=1}^{n} e_{\frac{(n-1)n}{2}+j}. \end{cases}$$

We now show that $\{f_k^n\}_{k=1,n=1}^{n+1,\infty}$ is a tight frame for $\mathcal{H}$ with frame bound $A = 1$. Write $g \in \mathcal{H}$ as

$$g = (g_1, g_2, \dots), \quad g_n \in \mathcal{H}_n.$$

We identify elements in a space $\mathcal{H}_n$ with their counterpart in $\mathcal{H}$, i.e., we do not distinguish between $f \in \mathcal{H}_n$ and the sequence in $\mathcal{H}$ having f in the n-th entry and otherwise zero. Given $n \in \mathbb{N}$, it is clear that

$$\langle g, f_k^n \rangle_{\mathcal{H}} = \langle g_n, f_k^n \rangle_{\mathcal{H}_n} \text{ for } k = 1, \dots, n+1.$$

It now follows from Lemma 7.5.1 that

$$\sum_{n=1}^{\infty}\sum_{k=1}^{n+1} |\langle g, f_k^n \rangle_{\mathcal{H}}|^2 = \sum_{n=1}^{\infty}\sum_{k=1}^{n+1} |\langle g_n, f_k^n \rangle_{\mathcal{H}_n}|^2 = \sum_{n=1}^{\infty} \|g_n\|_{\mathcal{H}_n}^2$$

$$= \|g\|_{\mathcal{H}}^2.$$

That is, $\{f_k^n\}_{k=1,n=1}^{n+1,\infty}$ is a tight frame for $\mathcal{H}$ as claimed. Note that for $n \geq 2$ and $1 \leq k \leq n$,

$$\|f_k^n\| = \left\|\left(1 - \frac{1}{n}\right) e_{\frac{(n-1)n}{2}+k} - \frac{1}{n}\sum_{j\neq k} e_{\frac{(n-1)n}{2}+j}\right\|$$

$$= \sqrt{\left(1 - \frac{1}{n}\right)^2 + \frac{n-1}{n^2}} \geq \frac{1}{4},$$

while

$$f_1^1 = 0 \text{ and } \|f_n^{n+1}\| = 1, \ \forall n \in \mathbb{N}.$$

Removing f_1^1, we thus obtain a tight frame which is norm-bounded below.

Now let $\{h_k\}_{k=1}^\infty$ be any spanning subset of the frame $\{f_k^n\}_{k=1,n=1}^{n+1,\infty}$, ordered in an arbitrary way. The basis constant for $\{h_k\}_{k=1}^\infty$ is by definition greater than or equal to the basis constant for any subsequence of $\{h_k\}_{k=1}^\infty$. Now, for any $n > 2$, there exists $N \in \mathbb{N}$ such that $\{h_k\}_{k=1}^N$ contains a subsequence which spans $\mathcal{H}_n$; this follows from the special construction of $\mathcal{H}$ as an orthogonal sum of the spaces $\mathcal{H}_n$ and the choice of the frame $\{f_k^n\}_{k=1,n=1}^{n+1,\infty}$. By Lemma 7.5.1(ii), this implies that the basis constant for $\{h_k\}_{k=1}^N$ is at least $\frac{1}{4}\sqrt{n-2}$; since $n > 2$ was arbitrary, this implies that the basis constant for $\{h_k\}_{k=1}^\infty$ is infinite. Thus, by Theorem 3.1.4, $\{h_k\}_{k=1}^\infty$ is not a Schauder basis for $\mathcal{H}$. □

We note that a frame which is norm-bounded below satisfies inequalities like

$$0 < \inf_k \|f_k\| \leq \sup_k \|f_k\| < \infty;$$

thus we have a slight variation of Theorem 7.5.2 (Exercise 7.3):

Corollary 7.5.3 *In every separable infinite-dimensional Hilbert space, there exists a normalized frame which does not contain a Schauder basis.*

Vershynin has obtained a generalization of Theorem 7.5.2 [625]: there exists a frame $\{f_k\}_{k=1}^\infty$ which does not contain a *basis with brackets*, i.e., a subfamily $\{x_n\}_{n=1}^\infty$ for which there exist numbers $1 < n_1 < n_2 < \cdots$ such that every $f \in \mathcal{H}$ has a unique representation $f = \lim_j \sum_{n=1}^{n_j} a_n x_n$. Bases with brackets only require the convergence of some special partial sums, and it is thus a more general concept than a basis.

7.6   A Moment Problem

Let $\{f_k\}_{k=1}^{\infty}$ be a sequence in a Hilbert space $\mathcal{H}$, and let $\{a_k\}_{k=1}^{\infty} \in \ell^2(\mathbb{N})$. It is natural to ask whether we can find $f \in \mathcal{H}$ such that

$$\langle f, f_k \rangle = a_k, \ \forall k \in \mathbb{N}; \tag{7.13}$$

a problem of this type is called a *moment problem*. It is clear that there are cases where no solution exists: if, for example, $f_k = f_j$ for some $k \neq j$, a solution can only exist if $a_k = a_j$. More generally, if $\{f_k\}_{k=1}^{\infty}$ is ω-dependent, we can find coefficients $\{a_k\}_{k=1}^{\infty}$ (not all zero) such that $\sum_{k=1}^{\infty} a_k f_k = 0$. If, for example, $a_j \neq 0$, then $f_j = -\sum_{k\neq j} \frac{a_k}{a_j} f_k$; thus, (7.13) can only have a solution if a_j is equal to the corresponding linear combination of $\{a_k\}_{k\neq j}$.

If the moment problem (7.13) has a solution, it is unique if and only if $\{f_k\}_{k=1}^{\infty}$ is complete (Exercise 7.4). We now present one more equivalent condition for a frame to be a Riesz basis; it is formulated in terms of the adjoint of the synthesis operator T:

Theorem 7.6.1 *Let $\{f_k\}_{k=1}^{\infty}$ be a frame for $\mathcal{H}$. Then $\{f_k\}_{k=1}^{\infty}$ is a Riesz basis if and only if the analysis operator $T^* : \mathcal{H} \to \ell^2(\mathbb{N})$ is surjective.*

Proof. First assume that $\{f_k\}_{k=1}^{\infty}$ is a Riesz basis. Let $\{g_k\}_{k=1}^{\infty}$ be the biorthogonal sequence. For $\{a_k\}_{k=1}^{\infty} \in \ell^2(\mathbb{N})$, Theorem 3.2.3 shows that $f := \sum_{k=1}^{\infty} a_k g_k$ is well defined; furthermore,

$$T^* f = \{\langle f, f_k \rangle\}_{k=1}^{\infty} = \{a_k\}_{k=1}^{\infty}.$$

Thus T^* is surjective. On the other hand, if we assume that T^* is surjective, $\mathcal{R}_{T^*} = \ell^2(\mathbb{N})$, then $\mathcal{N}_T = \mathcal{R}_{T^*}^{\perp} = \{0\}$. Now the conclusion follows from Theorem 7.1.1. $\qquad\square$

An alternative formulation of Theorem 7.6.1 is that a frame $\{f_k\}_{k=1}^{\infty}$ is a Riesz basis if and only if the moment problem (7.13) has a solution for all $\{a_k\}_{k=1}^{\infty} \in \ell^2(\mathbb{N})$. Young proves a stronger result in [622]: for an arbitrary sequence $\{f_k\}_{k=1}^{\infty}$ in a Hilbert space, the range of the operator $f \mapsto \{\langle f, f_k \rangle\}_{k=1}^{\infty}$ contains $\ell^2(\mathbb{N})$ if and only if there exists a constant $A > 0$ such that

$$A \sum_{k=1}^{n} |c_k|^2 \leq \left\| \sum_{k=1}^{n} c_k f_k \right\|^2$$

for all finite sequences $\{c_k\}_{k=1}^{\infty}$. A sequence $\{f_k\}_{k=1}^{\infty}$ satisfying these conditions is called a *Riesz–Fischer sequence*.

For a frame which is not a Riesz basis, it follows from Theorem 7.6.1 that there exist sequences $\{a_k\}_{k=1}^{\infty} \in \ell^2(\mathbb{N})$ such that (7.13) has no solution. But we can still ask for a *best approximative solution*; here we have to give a precise definition of how we want to approximate the given sequence

$\{a_k\}_{k=1}^{\infty}$. Since we are working with $\ell^2(\mathbb{N})$, it is a natural question whether we can find an element in $\mathcal{H}$ which minimizes the functional

$$f \mapsto \sum_{k=1}^{\infty} |a_k - \langle f, f_k \rangle|^2 = || \{a_k - \langle f, f_k \rangle\}_{k=1}^{\infty} ||^2_{\ell^2(\mathbb{N})}.$$

The answer turns out to be yes:

Theorem 7.6.2 *Let* $\{f_k\}_{k=1}^{\infty}$ *be a frame for* $\mathcal{H}$ *with frame operator* S, *and let* $\{a_k\}_{k=1}^{\infty} \in \ell^2(\mathbb{N})$. *Then there exists a unique vector in* $\mathcal{H}$ *which minimizes the functional*

$$f \mapsto \sum_{k=1}^{\infty} |a_k - \langle f, f_k \rangle|^2;$$

this vector is $f = \sum_{k=1}^{\infty} a_k S^{-1} f_k$.

Proof. By the result in Exercise 5.18, the orthogonal projection of a sequence $\{c_k\}_{k=1}^{\infty} \in \ell^2(\mathbb{N})$ onto the range of T^* is given by

$$P\{c_k\}_{k=1}^{\infty} = \left\{ \langle \sum_{k=1}^{\infty} c_k S^{-1} f_k, f_j \rangle \right\}_{j=1}^{\infty}. \tag{7.14}$$

The functional $f \mapsto \sum_{k=1}^{\infty} |a_k - \langle f, f_k \rangle|^2$ is minimized when

$$\{\langle f, f_k \rangle\}_{k=1}^{\infty} = P\{a_k\}_{k=1}^{\infty},$$

which is the case for $f = \sum_{k=1}^{\infty} a_k S^{-1} f_k$; by completeness of $\{f_k\}_{k=1}^{\infty}$, the minimizer is unique. □

Corollary 7.6.3 *Assume that* $\{f_k\}_{k=1}^{\infty}$ *is a Riesz basis for* $\mathcal{H}$, *and let* $\{a_k\}_{k=1}^{\infty} \in \ell^2(\mathbb{N})$. *Then the moment problem (7.13) has a unique solution, which is given by*

$$f = \sum_{k=1}^{\infty} a_k S^{-1} f_k = \sum_{j=1}^{\infty} \left(\sum_{k=1}^{\infty} (T^*T)_{j,k}^{-1} a_k \right) f_j;$$

here $(T^*T)_{j,k}^{-1}$ *is the jk-th entry in the matrix representation for* T^*T *with respect to the canonical basis.*

Proof. The moment problem has a solution f when $\{f_k\}_{k=1}^{\infty}$ is a Riesz basis, and the representation of f via S^{-1} follows from Theorem 7.6.2. By Theorem 2.5.3, the solution can also be expressed as

$$f = (T^*)^{\dagger} \{a_k\}_{k=1}^{\infty} = T(T^*T)^{-1} \{a_k\}_{k=1}^{\infty} = \sum_{j=1}^{\infty} \left(\sum_{k=1}^{\infty} (T^*T)_{j,k}^{-1} a_k \right) f_j.$$

□

7.7 The Feichtinger Conjecture

Even though Section 7.5 shows that there exist frames with no relationship to a basis, it is still fair to think about a frame as an "overcomplete basis." A natural question is to analyze how this redundancy can be interpreted. For example, can any frame be split into a finite number of Riesz sequences? In 2002, Feichtinger formulated the following conjecture (see Exercise 7.5 for a comment on the assumptions):

The Feichtinger conjecture: *Let* $\{f_k\}_{k=1}^\infty$ *be a frame with the property that* $\inf_{k\in\mathbb{N}}\|f_k\| > 0$. *Then* $\{f_k\}_{k=1}^\infty$ *can be partitioned into a finite union of Riesz sequences.*

Relatively soon, the first positive partial results were published in [342] and [130]. However, the general question turned out to be very difficult. Around 2005 it was shown by Casazza and Tremain that the Feichtinger conjecture is equivalent to the Kadison–Singer conjecture from 1959, in the sense that either both conjectures are true or both are false. Later, Casazza related the conjecture to several other open problems in the literature. We refer to [132] and [117] for detailed descriptions of these conjectures. The Feichtinger conjecture was finally confirmed in 2013, by Marcus, Spielman, and Srivastava [517]. The paper verifies one of the equivalent formulations of the conjecture rather than the frame formulation.

7.8 Exercises

7.1 Give an example of a Riesz frame which is not a near-Riesz basis.

7.2 Find a frame which contains a Schauder basis which is not a Riesz basis (hint: use Example 5.1.4).

7.3 Prove Corollary 7.5.3.

7.4 Let $\{f_k\}_{k=1}^\infty$ be a sequence in $\mathcal{H}$, and let $\{a_k\}_{k=1}^\infty \in \ell^2(\mathbb{N})$. Prove that $\{f_k\}_{k=1}^\infty$ is complete in $\mathcal{H}$ if and only if (7.13) has at most one solution.

7.5 Show that Feichtinger's conjecture would be false if the assumption of $\{f_k\}_{k=1}^\infty$ being norm-bounded below is removed.

8

Selected Topics in Frame Theory

The content in Chapters 5–7 forms the fundament for frame theory, but much more is known. We will not be able to describe all the interesting directions of frame theory, but the purpose of this chapter is to give short presentations of certain topics that appear repeatedly in the literature. Section 8.1 deals with the theory for g-frames as developed by Sun; it "lifts" frame theory from a condition dealing with vectors in a Hilbert space to a condition dealing with operators on the Hilbert space and hereby provides more general ways of obtaining "frame-like" decompositions. Section 8.2 discusses localized frames; they were introduced by Gröchenig with the purpose to identify frames that not only lead to series expansions in the underlying Hilbert space but also in a class of associated Banach spaces. By now, localized frames appear in many papers where frames with "particularly good properties" are needed. Section 8.3 deals with the so-called R-dual $\{\omega_j\}_{j=1}^\infty$ of a sequence $\{f_k\}_{k=1}^\infty$ in a Hilbert space, originally introduced by Casazza, Kutyniok, and Lammers. The R-dual provides a way of checking that $\{f_k\}_{k=1}^\infty$ is a frame, by checking the (conceptually more accessible) condition that $\{\omega_j\}_{j=1}^\infty$ forms a Riesz sequence. The construction mimics the result in Proposition 1.4.3 and is strongly related to the duality principle in Gabor analysis, to which we return in Section 13.1. In Section 8.4 we consider a generalization of frame theory to the case where the upper frame condition is violated. In this case, the synthesis operator is unbounded, but under certain conditions it is still possible to develop a frame-like theory. Finally, Section 8.5 relates frame theory to signal processing and signal transmission. Much more can be said about this, but this should be done by the people who are deeply involved in these applications.

© Springer International Publishing Switzerland 2016   183
O. Christensen, *An Introduction to Frames and Riesz Bases*,
Applied and Numerical Harmonic Analysis,
DOI 10.1007/978-3-319-25613-9_8

8.1   G-Frames

Over the years, various extensions of the frame theory have been investigated. Several of these are contained as special cases of the elegant theory for g-frames that was introduced by W. Sun in [593]. Note that originally the name *g-frames* was used as a short form of *generalized frames,* a terminology that was also used by Kaiser as he defined what is now called continuous frames. For this reason, we will solely stick to the short name "g-frames."

In order to reduce repetitions, we will now list the standing assumptions for the entire section.

General setup: Consider two Hilbert spaces $\mathcal{H}$ and $\widetilde{\mathcal{H}}$. Let I denote a countable index set, and let $\{\mathcal{V}_k\}_{k\in I}$ be a sequence of closed subspaces of $\widetilde{\mathcal{H}}$. For each $k \in I$, let $\Lambda_k : \mathcal{H} \to \mathcal{V}_k$ be a bounded linear operator.

Definition 8.1.1 *Under the assumptions in the general setup, the sequence of operators $\{\Lambda_k\}_{k\in I}$ is called a g-frame for $\mathcal{H}$ with respect to the spaces $\{\mathcal{V}_k\}_{k\in I}$ if there exist constants $A, B > 0$ such that*

$$A\,||f||^2 \le \sum_{k\in I}||\Lambda_k f||^2 \le B\,||f||^2, \forall f \in \mathcal{H}. \tag{8.1}$$

The numbers A, B are called g-frame bounds or simply "bounds."

The relation to frames is evident: if $\{f_k\}_{k\in I}$ is a frame for $\mathcal{H}$ with bounds A, B, then the linear functionals

$$\Lambda_k : \mathcal{H} \to \mathbb{C}, \ \Lambda_k f := \langle f, f_k \rangle \tag{8.2}$$

satisfy (8.1), i.e., $\{\Lambda_k\}_{k\in I}$ is a g-frame for $\mathcal{H}$ with respect to $\mathbb{C}$. On the other hand, any bounded linear functional Λ_k on $\mathcal{H}$ has the form (8.2) for some $f_k \in \mathcal{H}$: this shows that the "classical" frames in Definition 5.1.1 can be identified with the g-frames with respect to the particular space $\mathbb{C}$.

The interest in g-frames arises from the fact that there is a large freedom in the choices of the spaces $\{\mathcal{V}_k\}_{k\in I}$ and corresponding operators $\{\Lambda_k\}_{k\in I}$. For example, any invertible operator $\Lambda : \mathcal{H} \to \mathcal{H}$ is a g-frame for $\mathcal{H}$ with respect to $\mathcal{H}$ (Exercise 8.1). The following example shows that any fusion frame is a g-frame.

Example 8.1.2 Let $(\{\mathcal{V}_k\}_{k=1}^m, \{w_k\}_{k=1}^m)$ be a fusion frame for $\mathbb{C}^n$; see (1.39). Denote the orthogonal projection of $\mathbb{C}^n$ onto $\mathcal{V}_k$ by P_k, and let $\Lambda_k := \sqrt{w_k}\, P_k$. Then the operators $\{\Lambda_k\}_{k=1}^m$ form a g-frame for $\mathbb{C}^n$ with respect to the spaces $\{\mathcal{V}_k\}_{k=1}^m$. Note that fusion frames for $\mathbb{C}^n$ have an immediate generalization to infinite-dimensional spaces, which is covered by the theory for g-frames in the same way. Prior to the general introduction of fusion frames, the special case corresponding to $w_k = 1, \forall k,$ was known under the name *frame of subspaces;* see [140, 19].   □

We will now show that g-frames lead to series expansions that are very similar to the frame decomposition (5.1.6); the reader who checks the proofs will observe that the techniques and ideas are also similar.

Theorem 8.1.3 *Let $\{\Lambda_k\}_{k\in I}$ be a g-frame for $\mathcal{H}$ with respect to the spaces $\{\mathcal{V}_k\}_{k\in I}$. Then the following hold:*

(i) The sequence $\sum_{k\in I}\Lambda_k^*\Lambda_k f$ converges unconditionally for all $f \in \mathcal{H}$, and the linear map

$$S : \mathcal{H} \to \mathcal{H}, \ Sf := \sum_{k\in I}\Lambda_k^*\Lambda_k f \qquad (8.3)$$

defines a bounded, invertible, self-adjoint, and positive operator.

(ii) The operators $\{\widetilde{\Lambda}_k\}_{k\in I} := \{\Lambda_k S^{-1}\}_{k\in I}$ form a g-frame for $\mathcal{H}$ with respect to $\{\mathcal{V}_k\}_{k\in I}$; denoting the bounds for $\{\Lambda_k\}_{k\in I}$ by A, B, the g-frame $\{\widetilde{\Lambda}_k\}_{k\in I}$ has the bounds $1/B, 1/A$.

(iii) Each $f \in \mathcal{H}$ has the decompositions

$$f = \sum_{k\in I}\Lambda_k^*\widetilde{\Lambda}_k f = \sum_{k\in I}(\widetilde{\Lambda}_k)^*\Lambda_k f. \qquad (8.4)$$

Proof. (i) For notational convenience, assume that $I = \mathbb{N}$. Consider n, $m \in \mathbb{N}, n > m$. Then, for any $f \in \mathcal{H}$, a similar approach as in the proof of Theorem 3.2.3 yields that

$$\left\|\sum_{k=1}^{n}\Lambda_k^*\Lambda_k f - \sum_{k=1}^{m}\Lambda_k^*\Lambda_k f\right\| = \left\|\sum_{k=m+1}^{n}\Lambda_k^*\Lambda_k f\right\|$$

$$= \sup_{\|g\|=1}\left|\langle\sum_{k=m+1}^{n}\Lambda_k^*\Lambda_k f, g\rangle\right| = \sup_{\|g\|=1}\left|\langle\sum_{k=m+1}^{n}\Lambda_k f, \Lambda_k g\rangle\right|$$

$$\leq \sup_{\|g\|=1}\sum_{k=m+1}^{n}\|\Lambda_k f\|\,\|\Lambda_k g\|$$

$$\leq \left(\sum_{k=m+1}^{n}\|\Lambda_k f\|^2\right)^{1/2}\sup_{\|g\|=1}\left(\sum_{k=m+1}^{n}\|\Lambda_k g\|^2\right)^{1/2}$$

$$\leq \sqrt{B}\left(\sum_{k=m+1}^{n}\|\Lambda_k f\|^2\right)^{1/2}.$$

This implies that the sequence $\{\sum_{k=1}^{n}\Lambda_k^*\Lambda_k f\}_{n\in\mathbb{N}}$ is a Cauchy sequence in $\mathcal{H}$ and hence convergent; once the convergence is established, a similar argument proves that the operator S in (8.3) is bounded and that $\|S\| \leq B$. We leave it to the reader to check that S is self-adjoint and positive. Now, note that

$$\langle Sf, f \rangle = \sum_{k=1}^{\infty} ||\Lambda_k f||^2, \ \forall f \in \mathcal{H};$$

using the g-frame condition this implies that

$$A\,||f|| \le ||Sf||, \ \forall f \in \mathcal{H}. \tag{8.5}$$

Hence, S is injective and the range $\mathcal{R}_S$ is closed. Denoting the kernel of S by $\mathcal{N}_S$, we have $\mathcal{R}_S = \overline{\mathcal{R}_S} = \mathcal{N}_{S^*}^{\perp} = \mathcal{N}_S^{\perp} = \mathcal{H}$, i.e., S is in fact surjective. We conclude that S is invertible, which concludes the proof of (i).

Let us now show (iii). Using the notation $\widetilde{\Lambda}_k := \Lambda_k S^{-1}$, the result in (i) implies that for $f \in \mathcal{H}$,

$$f = SS^{-1}f = \sum_{k=1}^{\infty} \Lambda_k^* \Lambda_k S^{-1} f = \sum_{k=1}^{\infty} \Lambda_k^* \widetilde{\Lambda}_k f;$$

similarly,

$$f = S^{-1}Sf = S^{-1} \sum_{k=1}^{\infty} \Lambda_k^* \Lambda_k f = \sum_{k=1}^{\infty} S^{-1} \Lambda_k^* \Lambda_k f = \sum_{k=1}^{\infty} (\widetilde{\Lambda}_k)^* \Lambda_k f,$$

as desired.

In order to show (ii), let $f \in \mathcal{H}$. Then

$$
\begin{aligned}
\sum_{k=1}^{\infty} ||\widetilde{\Lambda}_k f||^2 &= \sum_{k=1}^{\infty} \langle \Lambda_k S^{-1}f, \Lambda_k S^{-1}f \rangle = \sum_{k=1}^{\infty} \langle \Lambda_k^* \Lambda_k S^{-1}f, S^{-1}f \rangle \\
&= \langle \sum_{k=1}^{\infty} \Lambda_k^* \Lambda_k S^{-1}f, S^{-1}f \rangle = \langle f, S^{-1}f \rangle.
\end{aligned}
$$

Using (8.5) we see that $||S^{-1}|| \le 1/A$, so we can now conclude that

$$\sum_{k=1}^{\infty} ||\widetilde{\Lambda}_k f||^2 \le \frac{1}{A} ||f||^2, \ \forall f \in \mathcal{H},$$

i.e., that $\{\widetilde{\Lambda}_k\}_{k \in I}$ satisfies the upper frame condition with bound $1/A$. In order to prove the lower bound, we use the second decomposition in (iii) and get

$$||f||^2 = \langle f, f \rangle = \langle \sum_{k=1}^{\infty} (\widetilde{\Lambda}_k)^* \Lambda_k f, f \rangle = \sum_{k=1}^{\infty} \langle \Lambda_k f, \widetilde{\Lambda}_k f \rangle;$$

using Cauchy–Schwarz' inequality twice, this implies that

$$||f||^4 \le \sum_{k=1}^{\infty} ||\Lambda_k f||^2 \sum_{k=1}^{\infty} ||\widetilde{\Lambda}_k f||^2 \le B||f||^2 \sum_{k=1}^{\infty} ||\widetilde{\Lambda}_k f||^2.$$

Thus, the lower g-frame condition is satisfied with the claimed bound. $\square$

Following the terminology in frame theory, the operator S in (8.3) is called the *g-frame operator* , and (8.4) are the *g-frame decompositions*. The sequence $\{\widetilde{\Lambda}_k\}_{k \in I}$ introduced in Theorem 8.1.3 (ii) is called the *canonical dual g-frame* of $\{\Lambda_k\}_{k \in I}$; and any sequence $\{\Gamma\}_{k \in I}$ of bounded linear operators $\Gamma_k : \mathcal{H} \to \mathcal{V}_k$ such that

$$f = \sum_{k \in I} \Lambda_k^* \Gamma_k f, \ \forall f \in \mathcal{H}$$

is called a *dual g-frame* of $\{\Lambda_k\}_{k \in I}$.

We have already seen that the "classical frames" considered so far are special cases of the g-frames. We will now show that all g-frames actually are related to the classical frames.

Proposition 8.1.4 *For each $k \in I$, let $\{e_{j,k}\}_{j \in J_k}$ denote an orthonormal basis for $\mathcal{V}_k$, and let $\Lambda_k : \mathcal{H} \to \mathcal{V}_k$ be a bounded linear operator. Then the following hold:*

(i) For each $k \in I, j \in J_k$ there exists a unique element $u_{j,k} \in \mathcal{H}$ such that

$$\langle f, u_{j,k} \rangle = \langle \Lambda_k f, e_{j,k} \rangle, \ \forall f \in \mathcal{H}; \tag{8.6}$$

(ii) With $u_{j,k}$ defined as in (i), $\{\Lambda_k\}_{k \in I}$ is a g-frame for $\mathcal{H}$ with respect to $\{\mathcal{V}_k\}_{k \in I}$ if and only if $\{u_{j,k}\}_{k \in I, j \in J_k}$ is a frame for $\mathcal{H}$.

Proof.   For the proof of (i), we just note that for each $k \in I, j \in J_k$, the mapping $f \mapsto \langle \Lambda_k f, e_{j,k} \rangle$ is a bounded linear functional on $\mathcal{H}$; thus, the result is a consequence of Riesz' representation theorem. In order to prove (ii), let $f \in \mathcal{H}$; then, using the representation of $\Lambda_k f \in \mathcal{V}_k$ in terms of the orthonormal basis $\{e_{j,k}\}_{j \in J_k}$ for $\mathcal{V}_k$,

$$\Lambda_k f = \sum_{j \in J_k} \langle \Lambda_k f, e_{j,k} \rangle e_{j,k} = \sum_{j \in J_k} \langle f, u_{j,k} \rangle e_{j,k}.$$

Thus,

$$\|\Lambda_k f\|^2 = \sum_{j \in J_k} |\langle f, u_{j,k} \rangle|^2,$$

which immediately yields the desired conclusion.   □

It is evident from the above statements and proofs that large parts of frame theory carry over to g-frames with only minor modifications. In order not to be too repetitive, we will therefore not go more into the theory and just refer to the literature for more information about g-frames.

8.2   Localization of Frames

A new direction in frame theory was launched by Gröchenig in [341]. One of the motivations was to find a way to describe what could be called "good frames." Whether a frame is good or not clearly depends on the context, but one of the typical requirements is that the canonical dual frame inherits attractive properties from the frame itself. We will be more precise about this once we have described the theory.

We begin with the original definition of a localized frame that appears in [341].

Definition 8.2.1 *Let $\{f_k\}_{k=1}^\infty$ be a frame for a Hilbert space $\mathcal{H}$, and let $\{\psi_k\}_{k=1}^\infty$ be a Riesz basis with dual Riesz basis $\{\widetilde{\psi_k}\}_{k=1}^\infty$. Then $\{f_k\}_{k=1}^\infty$ is polynomially localized with respect to $\{\psi_k\}_{k=1}^\infty$, with decay rate $s > 1$, if there exists a constant $C > 0$ such that*

$$\begin{cases} |\langle f_k, \psi_j\rangle| \le C(1 + |k - j|)^{-s}, \ \forall k, j \in \mathbb{N}; \\ |\langle f_k, \widetilde{\psi_j}\rangle| \le C(1 + |k - j|)^{-s}, \ \forall k, j \in \mathbb{N}. \end{cases}$$

Likewise, a frame $\{f_k\}_{k=1}^\infty$ is *exponentially localized* with respect to a Riesz basis $\{\psi_k\}_{k=1}^\infty$, with decay rate $\alpha > 0$, if there exists a constant $C > 0$ such that

$$\begin{cases} |\langle f_k, \psi_j\rangle| \le Ce^{-\alpha|k-j|}, \ \forall k, j \in \mathbb{N}; \\ |\langle f_k, \widetilde{\psi_j}\rangle| \le Ce^{-\alpha|k-j|}, \ \forall k, j \in \mathbb{N}. \end{cases}$$

One of the main results in [341] says that the canonical dual of a frame $\{f_k\}_{k=1}^\infty$ inherits the localization property:

Theorem 8.2.2 *Assume that $\{f_k\}_{k=1}^\infty$ is a frame with frame operator S. Then the following hold:*

(i) If $\{f_k\}_{k=1}^\infty$ is polynomially localized with respect to a Riesz basis $\{\psi_k\}_{k=1}^\infty$, then the canonical dual frame $\{S^{-1}f_k\}_{k=1}^\infty$ is also polynomially localized with respect to $\{\psi_k\}_{k=1}^\infty$, with the same decay rate.

(ii) If $\{\psi_k\}_{k=1}^\infty$ is exponentially localized, then $\{S^{-1}f_k\}_{k=1}^\infty$ is also exponentially localized, possibly with a different decay rate.

The concept of localization as in Definition 8.2.1 clearly depends on the choice of the Riesz basis $\{\psi_k\}_{k=1}^\infty$. An alternative concept was defined by Gröchenig in [342] and studied in detail in his paper [304] with Fornasier. Two names for the concept appear in the literature: it was called *intrinsically localized frames* in [342] and *self-localized frames* in [304].

Definition 8.2.3 *A frame $\{f_k\}_{k=1}^{\infty}$ is self-localized with decay rate $s > 1$ if there exists a constant $C > 0$ such that*

$$|\langle f_k, f_j \rangle| \leq C(1 + |k - j|)^{-s}, \ \forall k, j \in \mathbb{N}. \tag{8.7}$$

Note that the definition of self-localization is a decay condition on the entries of the Gram matrix associated with $\{f_k\}_{k=1}^{\infty}$, outside the main diagonal. As we have seen in Corollary 3.5.5, the condition (8.7) by itself actually implies that $\{f_k\}_{k=1}^{\infty}$ is a Bessel sequence.

Let us collect some of the key relationships between the two types of localization; (i) was first proved in [342], and (ii)+(iii) in [304].

Lemma 8.2.4 *Let $\{f_k\}_{k=1}^{\infty}$ be a frame for $\mathcal{H}$, with frame operator S. Then the following hold:*

(i) *If $\{f_k\}_{k=1}^{\infty}$ is polynomially localized with respect to a Riesz basis, then $\{f_k\}_{k=1}^{\infty}$ is self-localized, with the same decay rate.*

(ii) *If $\{f_k\}_{k=1}^{\infty}$ is self-localized, then the canonical dual $\{S^{-1}f_k\}_{k=1}^{\infty}$ is also self-localized, with the same decay rate.*

(iii) *If $\{f_k\}_{k=1}^{\infty}$ is self-localized with decay rate $s > 1$, there exists a constant $C > 0$ such that*

$$|\langle f_k, S^{-1}f_j \rangle| \leq C(1 + |k - j|)^{-s}, \ \forall k, j \in \mathbb{N}.$$

Proof. We will not give a full proof, but just explain the main steps of the proof of (i). Assume that $\{f_k\}_{k=1}^{\infty}$ is polynomially localized with respect to the Riesz basis $\{\psi_k\}_{k=1}^{\infty}$, with decay rate $s > 1$. By Theorem 3.6.2, we can expand any of the frame elements f_k in terms of the Riesz basis $\{\psi_k\}_{k=1}^{\infty}$ and its dual Riesz basis $\{\widetilde{\psi_k}\}_{k=1}^{\infty}$,

$$f_k = \sum_{\ell=1}^{\infty} \langle f_k, \widetilde{\psi_\ell} \rangle \psi_\ell;$$

thus, for $k, j \in \mathbb{N}$,

$$|\langle f_k, f_j \rangle| = \left| \sum_{\ell=1}^{\infty} \langle f_k, \widetilde{\psi_\ell} \rangle \langle \psi_\ell, f_j \rangle \right| \leq \sum_{\ell=1}^{\infty} |\langle f_k, \widetilde{\psi_\ell} \rangle| \, |\langle \psi_\ell, f_j \rangle|$$

$$\leq C^2 \sum_{\ell=1}^{\infty} (1 + |k - \ell|)^{-s}(1 + |\ell - j|)^{-s}. \tag{8.8}$$

Now, elementary (but slightly tedious) estimates on the sum in (8.8) show that it is bounded by a (new) constant times $(1 + |k - j|)^{-s}$, as desired. $\square$

We will meet two applications of localized frames in the following chapters. In Section 23.3 we will use self-localized frames to obtain estimates for the speed of convergence of a method for approximation of the inverse frame operator, and in Section 24.3 we will see that localized frames for a Hilbert space automatically lead to series expansions in a class of Banach spaces associated with the given Hilbert space.

Note that Balan, Casazza, Heil, and Landau have introduced a different notation for localization of frames in the paper [33]. The applications of this concept go in a different direction compared with [341]; in fact, the main content of [33] is concerned with localization in relation to density and overcompleteness issues for the given frame. We also note that the concept of localization of frames has been generalized to continuous frames by Fornasier and Rauhut [303].

8.3   The R-Duals of a Frame

To check that a given sequence $\{f_k\}_{k=1}^{\infty}$ in a Hilbert space $\mathcal{H}$ is a frame is usually a nontrivial matter: we have to check the validity of the frame condition

$$A\,||f||^2 \le \sum_{k=1}^{\infty} |\langle f, f_k \rangle|^2 \le B\,||f||^2$$

for all $f \in \mathcal{H}$ (or at least on a dense subspace), with uniform bounds $A, B > 0$. In particular, the lower frame condition is notoriously complicated. Therefore it is natural to look for simpler ways to check that $\{f_k\}_{k=1}^{\infty}$ is a frame. A reduction appears in Proposition 5.1.10, which shows that the frame property can be checked via a calculation of a countable collection of numbers.

We note that at least conceptually, it is easier to check that a sequence in $\mathcal{H}$ is a Riesz sequence than to check the frame property. In order to explain this, consider the finite scalar sequences in the unit sphere of $\ell^2(\mathbb{N})$, i.e., let

$$\mathcal{S} := \left\{ \{c_j\}_{j=1}^{\infty} \in \ell^2(\mathbb{N}) \,\middle|\, \{c_j\}_{j=1}^{\infty} \text{ is finite and } \sum_{j=1}^{\infty} |c_j|^2 = 1 \right\}. \qquad (8.9)$$

Then, as shown in Corollary 3.7.4 (see also Exercise 3.10), there exists a countable collection of sequences $\{c_j^{\ell}\}_{j=1}^{\infty} \in \mathcal{S}$, $\ell \in \mathbb{N}$, with the following property: a given sequence $\{\omega_j\}_{j=1}^{\infty}$ in $\mathcal{H}$ is a Riesz sequence with bounds A, B if

$$0 < A \le \left|\left| \sum_{j=1}^{\infty} c_j^{\ell} \omega_j \right|\right|^2 \le B < \infty, \ \forall \ell \in \mathbb{N}. \qquad (8.10)$$

For each $\ell \in \mathbb{N}$ the series in (8.10) is a finite sum, so the condition (8.10) is quite explicit. Of course, the verification is still complicated because we need to obtain uniform bounds A, B that are valid for all $\ell \in \mathbb{N}$, but nevertheless the condition is significantly less involved than the frame condition.

In fact, even the seemingly parallel result in Proposition 5.1.10 (ii) requires that we have an appropriate orthonormal basis for the relevant Hilbert space $\mathcal{H}$ at hand – and it also involves an infinite series.

The frame literature contains several results relating frames and Riesz sequences; we saw the first such result in the finite-dimensional setting in Proposition 1.4.3. One of the most prominent connections, the *duality principle* in Gabor analysis, will be considered in Theorem 13.1.1. Partly motivated by the duality principle, Casazza, Kutyniok, and Lammers introduced the so-called R-duals in general Hilbert spaces in the paper [142]. Before we state the general definition and the key properties we will motivate the definition by considering sequences in finite-dimensional spaces.

Let $\{f_k\}_{k=1}^m$ denote a sequence of m vectors in $\mathbb{C}^n$; it will be implicit from the following discussion that we are interested in the case where $m \geq n$. Writing the vectors on the form

$$f_k = \begin{pmatrix} f_{1k} \\ f_{2k} \\ \cdot \\ \cdot \\ f_{nk} \end{pmatrix}, \ k = 1, \ldots, m,$$

we have seen in (1.18) that the synthesis operator for $\{f_k\}_{k=1}^m$ can be identified with the matrix

$$T = \begin{pmatrix} | & | & \cdot & \cdot & | \\ f_1 & f_2 & \cdot & \cdot & f_m \\ | & | & \cdot & \cdot & | \end{pmatrix} = \begin{pmatrix} f_{11} & f_{12} & \cdot & \cdot & f_{1m} \\ f_{21} & f_{22} & \cdot & \cdot & f_{2m} \\ \cdot & \cdot & \cdot & \cdot & \cdot \\ \cdot & \cdot & \cdot & \cdot & \cdot \\ f_{n1} & f_{n2} & \cdot & \cdot & f_{nm} \end{pmatrix}. \tag{8.11}$$

Now let $\{e_k\}_{k=1}^n$ denote the canonical orthonormal basis for $\mathbb{C}^n$ and $\{h_k\}_{k=1}^m$ the canonical orthonormal basis for $\mathbb{C}^m$. Define the vectors $\{w_j\}_{j=1}^n$ in $\mathbb{C}^m$ by

$$w_j = \sum_{k=1}^m \langle f_k, e_j \rangle h_k, \ j = 1, \ldots, n. \tag{8.12}$$

Then direct calculation shows that

$$w_j = \sum_{k=1}^m f_{jk} h_k = \begin{pmatrix} f_{j1} \\ f_{j2} \\ \cdot \\ \cdot \\ f_{jm} \end{pmatrix}, \ j = 1, \ldots, n.$$

The geometric interpretation is that the vectors f_k constitute the columns in the matrix T and that the vectors w_j form the rows in T.

Using Proposition 1.4.3 on the transposed of the matrix in (8.11), we conclude that $\{f_k\}_{k=1}^m$ is a frame for $\mathbb{C}^n$ if and only if the vectors $\{\omega_j\}_{j=1}^n$ form a basis for their span in $\mathbb{C}^m$ (i.e., if and only if the vectors $\{\omega_j\}_{j=1}^n$ are linearly independent).

Motivated by this relation between the sequences $\{f_k\}_{k=1}^m$ and $\{\omega_j\}_{j=1}^n$, we will now turn our attention to the infinite-dimensional setting and state the definition of the R-duals of a given sequence of vectors $\{f_k\}_{k=1}^\infty$ in a general Hilbert space $\mathcal{H}$, as proposed in [142].

Definition 8.3.1 *Let $\{f_k\}_{k=1}^\infty$ be any sequence in a separable Hilbert space $\mathcal{H}$. Let $\{e_k\}_{k=1}^\infty$ and $\{h_k\}_{k=1}^\infty$ denote orthonormal bases for $\mathcal{H}$, and assume that*

$$\sum_{k=1}^\infty |\langle f_k, e_j\rangle|^2 < \infty, \ \forall j \in \mathbb{N}. \tag{8.13}$$

The R-dual of $\{f_k\}_{k=1}^\infty$ with respect to the orthonormal bases $\{e_k\}_{k=1}^\infty$ and $\{h_k\}_{k=1}^\infty$ is the sequence $\{\omega_j\}_{j=1}^\infty$ in $\mathcal{H}$ given by

$$\omega_j = \sum_{k=1}^\infty \langle f_k, e_j\rangle h_k, \ j \in \mathbb{N}. \tag{8.14}$$

Note that (8.13) is a technical condition that is needed to ensure convergence of the series in (8.14). We will now state and prove the main results from [142].

Theorem 8.3.2 *Given a sequence $\{f_k\}_{k=1}^\infty$ in $\mathcal{H}$, choose $\{e_k\}_{k=1}^\infty$ and $\{h_k\}_{k=1}^\infty$ as in Definition 8.3.1 and consider the R-dual $\{\omega_j\}_{j=1}^\infty$ in (8.14). Then the following hold:*

 (i) *$\{f_k\}_{k=1}^\infty$ is a Bessel sequence with bound B if and only if $\{\omega_j\}_{j=1}^\infty$ is a Bessel sequence with bound B.*

 (ii) *Assume that $\{f_k\}_{k=1}^\infty$ is a Bessel sequence. Then $\{f_k\}_{k=1}^\infty$ satisfies the lower frame condition with bound A if and only if $\{\omega_j\}_{j=1}^\infty$ satisfies the lower Riesz sequence condition with bound A.*

 (iii) *$\{f_k\}_{k=1}^\infty$ is a frame for $\mathcal{H}$ with bounds A, B if and only if $\{\omega_j\}_{j=1}^\infty$ is a Riesz sequence with bounds A, B.*

Proof. In order to prove (i), consider any finite scalar sequence $\{c_j\}_{j=1}^\infty$. Then

$$\sum_{j=1}^\infty c_j \omega_j = \sum_{j=1}^\infty c_j \sum_{k=1}^\infty \langle f_k, e_j\rangle h_k = \sum_{k=1}^\infty \langle f_k, \sum_{j=1}^\infty \overline{c_j} e_j\rangle h_k;$$

it follows that

$$\left\| \sum_{j=1}^\infty c_j \omega_j \right\|^2 = \sum_{k=1}^\infty \left| \langle f_k, \sum_{j=1}^\infty \overline{c_j} e_j\rangle \right|^2. \tag{8.15}$$

Now assume that $\{f_k\}_{k=1}^{\infty}$ is a Bessel sequence with bound B. Then it follows from (8.15) that for any finite sequence $\{c_j\}_{j=1}^{\infty}$

$$\left\|\sum_{j=1}^{\infty} c_j \omega_j\right\|^2 \leq B \left\|\sum_{j=1}^{\infty} \overline{c_j} e_j\right\|^2 = B \sum_{j=1}^{\infty} |c_j|^2;$$

now Exercise 3.13 implies that $\{\omega_j\}_{j=1}^{\infty}$ is a Bessel sequence with bound B. Similarly, if $\{\omega_j\}_{j=1}^{\infty}$ is a Bessel sequence with bound B, then (8.15) shows that for any finite scalar sequence $\{c_j\}_{j=1}^{\infty}$,

$$\sum_{k=1}^{\infty} \left|\langle f_k, \sum_{j=1}^{\infty} \overline{c_j} e_j\rangle\right|^2 \leq B \sum_{j=1}^{\infty} |c_j|^2 = B \left\|\sum_{j=1}^{\infty} c_j e_j\right\|^2;$$

since the set of finite linear combinations of the vectors e_j is dense in $\mathcal{H}$, the conclusion now follows from Lemma 3.2.6.

In order to show (ii), we now assume that $\{f_k\}_{k=1}^{\infty}$ is a Bessel sequence. Then the continuity of the synthesis operators associated with $\{f_k\}_{k=1}^{\infty}$ and $\{\omega_j\}_{j=1}^{\infty}$ implies that (8.15) holds for all $\{c_j\}_{j=1}^{\infty} \in \ell^2(\mathbb{N})$. Now (ii) follows immediately from the fact that each $f \in \mathcal{H}$ has a representation $f = \sum_{j=1}^{\infty} \langle f, e_j \rangle e_j$ and that

$$\left\|\sum_{j=1}^{\infty} c_j e_j\right\|^2 = \sum_{j=1}^{\infty} |c_j|^2, \forall \{c_j\}_{j=1}^{\infty} \in \ell^2(\mathbb{N}).$$

The result in (iii) follows by combining (i) and (ii). □

Note that, at least theoretically, Theorem 8.3.2 (iii) does what we want: we can check the frame property for the sequence $\{f_k\}_{k=1}^{\infty}$ by verifying the Riesz sequence property for the associated sequence $\{\omega_j\}_{j=1}^{\infty}$. In practice, it might of course be problematic to calculate the vectors $\{\omega_j\}_{j=1}^{\infty}$ explicitly and check this condition.

Let us state one more result from [142]; it gives a characterization of two frames being dual frames, expressed in terms of the associated R-duals.

Theorem 8.3.3 Let $\{f_k\}_{k=1}^{\infty}$ and $\{g_k\}_{k=1}^{\infty}$ denote frames for $\mathcal{H}$, and let $\{\omega_j\}_{j=1}^{\infty}$ and $\{\gamma_j\}_{j=1}^{\infty}$ denote the associated R-duals with respect to the orthonormal bases $\{e_k\}_{k=1}^{\infty}, \{h_k\}_{k=1}^{\infty}$. Then the following are equivalent:

(i) $\{f_k\}_{k=1}^{\infty}$ and $\{g_k\}_{k=1}^{\infty}$ are dual frames.

(ii) $\langle \omega_j, \gamma_k \rangle = \delta_{j,k}, \forall j, k \in \mathbb{N}$.

It is an interesting open problem whether the theory for R-duals generalizes the duality principle in Gabor analysis. See the papers [142, 166, 580] and the discussion in Section 13.1.

8.4 Frame Theory via Unbounded Operators

Frame theory as presented in Chapter 5 can be generalized in various ways. In this section, we will describe an extension involving unbounded operators. We will not state the necessary definitions and results about unbounded operators in a formal way, but just give appropriate references to the literature.

In order to motivate the extension, consider a frame $\{f_k\}_{k=1}^{\infty}$ in a Hilbert space $\mathcal{H}$. The role of the Bessel condition is to guarantee that the synthesis operator is a bounded operator. Intuitively, the Bessel condition ensures that the vectors $\{f_k\}_{k=1}^{\infty}$ "do not contain too much information in any particular direction." For example, we know that the condition excludes the possibility that a certain nonzero vector is repeated infinitely often in the frame. However, from this point of view it is clear that "too much information in a particular direction" should not make it impossible to derive a reconstruction formula: it will just make the process more complicated because we might have to deal with an unbounded synthesis operator.

In this section we will give a short introduction to results from [153], showing that this intuition is correct. Consider a sequence $\{f_k\}_{k=1}^{\infty}$ in $\mathcal{H}$, and define the possibly unbounded operator $T : \mathcal{D}(T) \subseteq \ell^2(\mathbb{N}) \to \mathcal{H}$ by

$$T\{c_k\}_{k=1}^{\infty} := \sum_{k=1}^{\infty} c_k f_k, \quad \mathcal{D}(T) = \left\{ \{c_k\}_{k=1}^{\infty} \in \ell^2(\mathbb{N}) \,\Big|\, \sum_{k=1}^{\infty} c_k f_k \text{ is convergent} \right\}.$$

Note that the finite sequences are dense in $\ell^2(\mathbb{N})$ and contained in $\mathcal{D}(T)$; thus, the operator T is automatically densely defined. As in our study of Bessel sequences, we will call T the *synthesis operator*.

The key to the results in [153] is that to any densely defined, closed, and surjective operator from $\ell^2(\mathbb{N})$ into $\mathcal{H}$, we can associate a unique pseudo-inverse operator $T^{\dagger} : \mathcal{H} \to \ell^2(\mathbb{N})$ such that

$$\mathcal{N}_{T^{\dagger}} = \mathcal{R}_T^{\perp}; \quad \overline{\mathcal{R}_{T^{\dagger}}} = \mathcal{N}_T^{\perp}; \quad \text{and} \quad TT^{\dagger}f = f, \forall f \in \mathcal{R}_T.$$

We refer to Lemma 1.1 in [59] for a proof. Corollary 1.2 in the same paper shows that the pseudo-inverse is a bounded operator.

Theorem 8.4.1 *Let $\{f_k\}_{k=1}^{\infty}$ be a sequence in $\mathcal{H}$, and assume that the synthesis operator T is closed and surjective. Then the following hold:*

(i) There exists a Bessel sequence $\{g_k\}_{k=1}^{\infty}$ such that

$$f = \sum_{k=1}^{\infty} \langle f, g_k \rangle f_k, \forall f \in \mathcal{H};$$

(ii) The sequence $\{f_k\}_{k=1}^{\infty}$ satisfies the lower frame condition; in fact,

$$\frac{1}{||T^{\dagger}||^2} \, ||f||^2 \le \sum_{k=1}^{\infty} |\langle f, f_k \rangle|, \forall f \in \mathcal{H}.$$

Proof. Let $T^\dagger$ denote the pseudo-inverse of the synthesis operator T. Given $f \in \mathcal{H}$, write the sequence $T^\dagger f$ as $T^\dagger f = \{(T^\dagger f)_k\}_{k=1}^\infty$. According to the properties of the pseudo-inverse $T^\dagger$, we can decompose f as

$$f = TT^\dagger f = \sum_{k=1}^\infty (T^\dagger f)_k f_k; \qquad (8.16)$$

furthermore,

$$||T^\dagger f||^2 = \sum_{k=1}^\infty |(T^\dagger f)_k|^2 \leq ||T^\dagger||^2 ||f||^2. \qquad (8.17)$$

It follows that for each $k \in \mathbb{N}$, the mapping $f \mapsto (T^\dagger f)_k$ is a bounded linear functional on $\mathcal{H}$; thus, by Riesz' representation theorem, there exists an element $g_k \in \mathcal{H}$ such that $(T^\dagger f)_k = \langle f, g_k \rangle$. This proves (i). The result in (ii) also follows from (8.16) and (8.17); in fact,

$$||f||^2 = \langle f, f \rangle = \sum_{k=1}^\infty (T^\dagger f)_k \langle f_k, f \rangle,$$

which leads to the desired result by an application of Cauchy-Schwarz' inequality. □

The technical complication in relation to Theorem 8.4.1 is to show that the operator T is closed. Let us elaborate a little on this difficulty. In analogue with the definition of the synthesis operator, one can define the *analysis operator* associated with any sequence $\{f_k\}_{k=1}^\infty$ by

$$U : \mathcal{D}(U) \subseteq \mathcal{H} \to \ell^2(\mathbb{N}), \ Uf := \{\langle f, f_k \rangle\}_{k=1}^\infty,$$

where

$$\mathcal{D}(U) = \left\{ f \in \mathcal{H} \mid \{\langle f, f_k \rangle\}_{k=1}^\infty \in \ell^2(\mathbb{N}) \right\}.$$

For Bessel sequences $\{f_k\}_{k=1}^\infty$, we know that the operators T and U are related by $T = U^*$. In the general case discussed here, the situation is more complicated:

Lemma 8.4.2 *Consider any sequence $\{f_k\}_{k=1}^\infty$ in $\mathcal{H}$. Then the following hold:*

(i) *The analysis operator U is densely defined if*

$$\sum_{k=1}^\infty |\langle f_j, f_k \rangle| < \infty, \ \forall j \in \mathbb{N}.$$

(ii) *If the analysis operator U is densely defined, the (unbounded) adjoint operator U^* is an extension of the synthesis operator, i.e., $T \subseteq U^*$.*

(iii) *If the analysis operator U is densely defined, the (unbounded) adjoint operator U^* is closed.*

The results in (i) and (ii) are taken from the paper [153], and (iii) is a general result from functional analysis; see Theorem 13.9 in [566]. In words, the conclusion of the lemma is that even if we assume that the operator U is densely defined, we can only conclude that the synthesis operator T is *contained* in a closed operator, not that T itself is closed.

8.5   Frames and Signal Processing

The frame theory described so far takes place in an ideal world, which can hardly be realized in, e.g., signal processing. In this section, we describe some of the steps that have to be taken in order to apply the abstract results in practice. Much more can of course be said about this important subject, and we refer to the books [509] by Mallat and [626] by Vetterli and Kovačević for more detailed information.

Some of the problems appear before one even thinks about frames. In fact, even the most basic ingredient in mathematics, the real numbers, is disturbed when we move away from the abstract level: every number has to be replaced by a number with finitely many digits before any processing can take place. In practice, this means that we represent all numbers in an interval (e.g., $[1, 1 + 10^{-18}]$) by the same number (in this case probably the number 1). The consequence is an inaccuracy, which is called the *quantization error;* using a slightly different wording, we discussed this already on page 27.

The basic limitation in applications of the frame results is that any type of signal processing has to be performed on finite sequences of numbers. For example, this implies that the frame representation (5.1.6) has to be *truncated:* we can only aim at calculating a finite number of frame coefficients, say, $\{\langle f, S^{-1} f_k \rangle\}_{k=1}^{N}$, and the exact representation in (5.1.6) has to be replaced by

$$f \sim \sum_{k=1}^{N} \langle f, S^{-1} f_k \rangle f_k.$$

Even calculation of the frame coefficients $\langle f, S^{-1} f_k \rangle$ can in general only be done with finite precision. That is, the outcome of a calculation will be

$$\langle f, S^{-1} f_k \rangle + w_k \tag{8.18}$$

for some (hopefully small) error term w_k. All types of transmission or further processing will introduce extra inaccuracies. One says that the frame coefficients $\langle f, S^{-1} f_k \rangle$ have been *contaminated by the noise* w_k.

Already on page 40, we gave a rather intuitive argument that overcompleteness of frames might reduce the influence of noise, compared with the use of an orthonormal basis. To support this further, we now discuss a result that is borrowed from [509].

Let us again use the example of signal transmission, as on page 40. That is, we assume that one wants to transmit the signal f from $\mathcal{A}$ to $\mathcal{R}$ by sending the frame coefficients $\{\langle f, S^{-1}f_k\rangle\}_{k=1}^{\infty}$. Because of quantization, the coefficients will be contaminated by some noise $\{w_k\}_{k=1}^{\infty}$, and $\mathcal{R}$ will receive the coefficients $\{\langle f, S^{-1}f_k\rangle + w_k\}_{k=1}^{\infty}$; we assume that $\{w_k\}_{k=1}^{\infty} \in \ell^2(\mathbb{N})$. The receiver $\mathcal{R}$ will believe that the transmitted function was

$$\sum_{k=1}^{\infty} \left(\langle f, S^{-1}f_k\rangle + w_k\right) f_k = f + \sum_{k=1}^{\infty} w_k f_k$$

rather than f.

Note that $\mathcal{R}$ actually knows that the transmitted sequence was supposed to be a sequence of frame coefficients, i.e., a sequence of the form $\{\langle g, f_k\rangle\}_{k=1}^{\infty}$ for some $g \in \mathcal{H}$ (viz., $g = S^{-1}f$); that is, the sequence belongs to the range of the operator T^*. This might not be the case for the perturbed coefficients $\{\langle f, S^{-1}f_k\rangle + w_k\}_{k=1}^{\infty}$, so it is natural to compensate for this by projecting that sequence onto the range of the operator T^*. Denoting the projection operator by Q, Exercise 5.18 shows that the outcome is

$$Q\{\langle f, S^{-1}f_k\rangle + w_k\}_{k=1}^{\infty} = \{\langle f, S^{-1}f_k\rangle\}_{k=1}^{\infty} + Q\{w_k\}_{k=1}^{\infty}. \qquad (8.19)$$

Let $w = \{w_k\}_{k=1}^{\infty}$. Based on (8.19), $\mathcal{R}$ will reconstruct the transmitted signal as

$$\sum_{k=1}^{\infty} \left(\langle f, S^{-1}f_k\rangle + (Qw)_k\right) f_k = f + \sum_{k=1}^{\infty} (Qw)_k f_k.$$

We will assume that the quantization error is *white noise*. This means that the components w_k are random variables with zero mean, variance σ^2 independent of k, and that

$$E\left[w_j w_\ell\right] = \sigma^2 \delta_{j,\ell}. \qquad (8.20)$$

We now prove that increased redundancy of the frame, measured by a larger lower frame bound, will decrease the energy of the coefficients in the "projected noise" Qw, i.e., the mean of the random variables $|(Qw)_k|^2$. We return to this result in a concrete setting in Section 9.7.

Proposition 8.5.1 *Suppose that the frame $\{f_k\}_{k=1}^{\infty}$ has lower frame bound A and consists of normalized vectors. If w is white noise, then for each $k \in \mathbb{N}$,*

$$E|(Qw)_k|^2 \leq \frac{\sigma^2}{A},$$

with equality if $\{f_k\}_{k=1}^{\infty}$ is a tight frame.

Proof. According to Exercise 5.18, the kth component of Qw is given by

$$(Qw)_k = \sum_{j=1}^{\infty} w_j \langle S^{-1} f_j, f_k \rangle.$$

Via (8.20), this implies that

$$
\begin{aligned}
E|(Qw)_k|^2 &= E\left[\sum_{j=1}^{\infty} w_j \langle S^{-1} f_j, f_k \rangle \overline{\sum_{\ell=1}^{\infty} w_\ell \langle S^{-1} f_\ell, f_k \rangle} \right] \\
&= \sum_{j=1}^{\infty} \sum_{\ell=1}^{\infty} E\left[w_j \overline{w_\ell} \right] \langle S^{-1} f_j, f_k \rangle \overline{\langle S^{-1} f_\ell, f_k \rangle} \\
&= \sum_{\ell=1}^{\infty} E|w_\ell|^2 |\langle S^{-1} f_\ell, f_k \rangle|^2 = \sigma^2 \sum_{\ell=1}^{\infty} |\langle S^{-1} f_\ell, f_k \rangle|^2.
\end{aligned}
$$

Using that $\{S^{-1} f_\ell\}_{\ell=1}^{\infty}$ is a frame with upper frame bound A^{-1}, we finally arrive at

$$E|(Qw)_k|^2 \le \frac{\sigma^2}{A} \|f_k\|^2 = \frac{\sigma^2}{A};$$

the inequality is an equality if $\{S^{-1} f_k\}_{k=1}^{\infty}$ is a tight frame. By Lemma 5.1.5, the frame operator for $\{S^{-1} f_k\}_{k=1}^{\infty}$ is S^{-1}; thus, $\{S^{-1} f_k\}_{k=1}^{\infty}$ being a tight frame is equivalent with S^{-1} being a multiple of the identity. But this is equivalent with S being a multiple of the identity, i.e., with $\{f_k\}_{k=1}^{\infty}$ being a tight frame. $\square$

Quantization errors and noise during transmission are just some of the obstacles for frames in real life. Depending on the underlying Hilbert space $\mathcal{H}$, there might be additional complications. In many cases, $\mathcal{H}$ will be a function space like $L^2(\mathbb{R})$, and even finite-dimensional subspaces hereof cannot be processed directly: a discretization step is needed in order to transfer the setting to a vector space where the elements are numbers of sequences of numbers, e.g., $\mathbb{C}^n$. This is exactly the point where the importance of the Gram matrix becomes clear: whereas the frame operator $S = TT^*$ maps $\mathcal{H}$ onto itself, the Gram matrix T^*T is an operator on the sequence space $\ell^2(\mathbb{N})$, i.e., the only step that is needed is a truncation. For this reason, it is an advantage to formulate algorithms involving frames in terms of the Gram matrix rather than the frame operator, if possible.

8.6 Exercises

8.1 Let $\mathcal{H}$ denote a Hilbert space. Show that any invertible operator $\Lambda : \mathcal{H} \to \mathcal{H}$ is a g-frame for $\mathcal{H}$ with respect to $\mathcal{H}$.

9

Frames of Translates

The previous chapters have concentrated on general frame theory. We have only seen a few concrete frames, and most of them were constructed via manipulations on an orthonormal basis for an arbitrary separable Hilbert space. An advantage of this approach is that we obtain universal constructions, valid in all Hilbert spaces.

In order to apply frames in signal processing or any other branch of engineering, it is necessary to be more specific and construct frames in concrete Hilbert spaces consisting of functions or sequences. This will be the central theme in the following chapters. The central Hilbert space will be $L^2(\mathbb{R})$, but we will also consider periodic functions in $L^2(0, L)$ as well as the sequence spaces $\ell^2(\mathbb{Z})$ and $\mathbb{C}^n$. All frames will be *coherent,* i.e., all elements in a given frame will have a *common structure.* The exact meaning of this will become clear as soon as we define the frames, but the idea is that each element in the frame $\{f_k\}_{k=1}^\infty$ appears by the action of an operator (belonging to a special class) on a single element f in the Hilbert space. This feature is essential for applications: it simplifies manipulations on the frame, and makes it easier to store information about the frame.

In this chapter we consider the case where the operators act by translation. That is, we consider families of the form $\{\phi(\cdot - \lambda_k)\}_{k\in\mathbb{Z}}$, where $\{\lambda_k\}_{k\in\mathbb{Z}}$ is a sequence in $\mathbb{R}$ and $\phi \in L^2(\mathbb{R})$. Using the translation operators defined in Section 2.9, we can write $\{\phi(\cdot - \lambda_k)\}_{k\in\mathbb{Z}} = \{T_{\lambda_k}\phi\}_{k\in\mathbb{Z}}$. Recall also the modulation operator E_{λ_k} from Section 2.9: since translation of a function corresponds to modulation of its Fourier transform, i.e., $\mathcal{F}T_{\lambda_k}\phi = E_{-\lambda_k}\mathcal{F}\phi$, frames of translates are closely related to frames

© Springer International Publishing Switzerland 2016   199
O. Christensen, *An Introduction to Frames and Riesz Bases*,
Applied and Numerical Harmonic Analysis,
DOI 10.1007/978-3-319-25613-9_9

consisting of complex exponentials $\{e^{i\lambda_k x}\}_{k\in\mathbb{Z}}$; a section is devoted to this type of frames, too.

Frames of translates are natural examples of frame sequences. In fact, we will prove that $\{T_{\lambda_k}\phi\}_{k\in\mathbb{Z}}$ at most can be a frame for a proper subspace of $L^2(\mathbb{R})$. In the next chapters, we will see that sequences of translates (respectively, complex exponentials) are of fundamental importance also for construction of frames for $L^2(\mathbb{R})$.

This short introduction already indicates that the translation operators and modulation operators will play a central role, together with the Fourier transform.

This chapter brings us close to the origin of frames: historically, Duffin and Schaeffer introduced frames in the context of sequences of complex exponential functions. Their paper [262] from 1952 contains the general definition of frames, but the core subject is nonharmonic Fourier series. Young has given an outstanding presentation of complex exponentials and nonharmonic Fourier series in his book [622]. We will concentrate on results that are directly related to frames and mainly discuss those that appeared after the first edition of [622] from 1980. Several results will be presented without proofs, and the reader who knows [622] will understand why: a deeper analysis requires advanced complex analysis and would bring us too far away from the main theme.

9.1   Sequences in $\mathbb{R}^d$

The current chapter deals with sequences in $L^2(\mathbb{R})$ of the form $\{T_{\lambda_k}\phi\}_{k\in\mathbb{Z}}$, where $\{\lambda_k\}_{k\in\mathbb{Z}}$ is a sequence in $\mathbb{R}$. It turns out that certain conditions on the distribution of the points λ_k are necessary in order for $\{T_{\lambda_k}\phi\}_{k\in\mathbb{Z}}$ to be a frame sequence. We will consider some of these conditions here; since we later need higher-dimensional versions of the concepts, we already now consider sequences in $\mathbb{R}^d$.

Definition 9.1.1 *Let I be a countable index set and $\{\lambda_k\}_{k\in I}$ a sequence in $\mathbb{R}^d$. We say that*

(i) *A point $\lambda \in \mathbb{R}^d$ is an accumulation point for $\{\lambda_k\}_{k\in I}$ if every open ball in $\mathbb{R}^d$ centered at λ contains infinitely many λ_k.*

(ii) *$\{\lambda_k\}_{k\in I}$ is separated if $\inf_{j\neq k}|\lambda_j - \lambda_k| > 0$; a constant $\delta > 0$ such that $|\lambda_j - \lambda_k| \geq \delta$ for all $j \neq k$ is called a separation constant.*

(iii) *$\{\lambda_k\}_{k\in I}$ is relatively separated if it is a finite union of separated sequences.*

A relatively separated sequence can repeat the same point N times for some $N \in \mathbb{N}$, but it cannot have an accumulation point.

Example 9.1.2 We will consider two sequences in the space $\mathbb{R}$:

(i) The sequence $\{\frac{1}{k}\}_{k \in \mathbb{Z} \setminus \{0\}}$ has zero as accumulation point.

(ii) The sequence $\{k, k + \frac{1}{|k|+1}\}_{k \in \mathbb{Z}}$ has no accumulation point and is not separated. However, it is relatively separated.   □

Several characterizations of relatively separated sequences are known. Probably the most important one is in terms of the upper Beurling density, which we now introduce. For the sake of short notation, denote the given sequence by $\Lambda = \{\lambda_k\}_{k \in \mathbb{Z}}$. For $x \in \mathbb{R}^d$ and $h > 0$, we let $Q_h(x)$ denote the half-open cube in $\mathbb{R}^d$ centered at x and with side lengths h, i.e.,

$$Q_h(x) = \prod_{j=1}^{d} [x_j - h/2, x_j + h/2[, \quad \text{where } x = (x_1, \dots, x_d).$$

Note that $\{Q_h(hn)\}_{n \in \mathbb{Z}^d}$ is a disjoint cover of $\mathbb{R}^d$ for any $h > 0$. Let $\nu^+(h)$ and $\nu^-(h)$ denote the largest and smallest numbers of points from Λ that lie in any cube $Q_h(x)$, i.e.,

$$\nu^+(h) = \sup_{x \in \mathbb{R}^d} \natural\,(\Lambda \cap Q_h(x)), \quad \nu^-(h) = \inf_{x \in \mathbb{R}^d} \natural\,(\Lambda \cap Q_h(x)).$$

The *upper and lower Beurling densities* of Λ are now defined as

$$D^+(\Lambda) = \limsup_{h \to \infty} \frac{\nu^+(h)}{h^d}, \quad \text{respectively,} \quad D^-(\Lambda) = \liminf_{h \to \infty} \frac{\nu^-(h)}{h^d}. \quad (9.1)$$

In case $D^+(\Lambda) = D^-(\Lambda)$, we say that Λ has *uniform Beurling density*

$$D(\Lambda) = D^+(\Lambda) = D^-(\Lambda).$$

Lemma 9.1.3 *Let* $\Lambda = \{\lambda_k\}_{k \in \mathbb{Z}}$ *be a sequence in* $\mathbb{R}^d$. *Then the following are equivalent:*

(i) $D^+(\Lambda) < \infty.$

(ii) Λ *is relatively separated.*

(iii) *For some (and therefore every)* $h > 0$, *there is a natural number* N_h *such that each cube* $Q_h(hn)$, $n \in \mathbb{Z}^d$, *contains at most* N_h *points from* Λ, *i.e.,*

$$\sup_{n \in \mathbb{Z}^d} \natural\,(\Lambda \cap Q_h(hn)) < \infty.$$

Proof. (i)$\Rightarrow$ (iii). If (i) is satisfied, there exists a constant N such that for all sufficiently large $h > 0$,

$$\frac{\nu^+(h)}{h^d} \le N;$$

this gives (iii) for large values of h (as a consequence, it holds for all $h > 0$).

(iii) $\Rightarrow$ (ii). Let $h > 0$ be chosen such that (iii) is satisfied. We will show explicitly how Λ can be split into a number of h-separated sequences. Let $e_1, \ldots, e_{2^d}$ denote the vertices of the unit cube $[0, 1]^d$, and consider the sets

$$ Z_j = (2\mathbb{Z})^d + e_j, \ j = 1, \ldots, 2^d. $$

Note that $\mathbb{Z}^d$ is the disjoint union of the sets $Z_1, \ldots, Z_{2^d}$. Since $\{Q_h(hn)\}_{n \in \mathbb{Z}^d}$ is a disjoint cover of $\mathbb{R}^d$, this implies that $\mathbb{R}^d$ is the disjoint union of the sets

$$ B_j = \bigcup_{n \in Z_j} Q_h(hn), \ j = 1, \ldots, 2^d. $$

Let us analyze the cubes $Q_h(hn)$ which form a given set B_j. If we consider some $m, n \in Z_j$ with $m \neq n$, the distance between the cubes $Q_h(hn)$ and $Q_h(hm)$ is at least h, i.e., the distance between arbitrary elements in $Q_h(hn)$ and $Q_h(hm)$ is at least h. By the assumption in (iii), each cube $Q_h(hn)$ contains at most N_h elements from Λ, i.e., $\natural (\Lambda \cap Q_h(hn)) \leq N_h$; since

$$ \Lambda \cap B_j = \bigcup_{n \in Z_j} (\Lambda \cap Q_h(hn)), $$

it follows that $\Lambda \cap B_j$ can be split into N_h sets which are h-separated. Therefore Λ can be split into $2^d N_h$ sequences which are h-separated.

(ii)$\Rightarrow$(i). Assume that (ii) is satisfied; by definition, this means that we can choose a partition of Λ into a finite number of sequences, say, $\Lambda_1, \ldots, \Lambda_r$, such that each sequence Λ_k is separated with separation constant, say, $\delta_k < 1$. Let

$$ \delta := \min \left\{ \frac{\delta_1}{2\sqrt{d}}, \ldots, \frac{\delta_r}{2\sqrt{d}} \right\}. $$

The maximal distance between points in any cube $Q_\delta(x)$ is $\min\{\delta_1, \ldots, \delta_r\}$, so the cube contains at most one point from each sequence Λ_k and therefore at most r points from Λ. Thus, if h is any positive number, then $Q_{h\delta}(x)$ contains at most $r(h+1)^d$ elements from Λ. Via the definition of $D^+(\Lambda)$, this implies that

$$ D^+(\Lambda) = \limsup_{h \to \infty} \frac{\nu^+(h)}{h^d} = \limsup_{h \to \infty} \frac{\nu^+(h\delta)}{(h\delta)^d} $$

$$ \leq \limsup_{h \to \infty} \frac{r(h+1)^d}{(h\delta)^d} = \frac{r}{\delta^d} $$

$$ < \infty. $$

$\square$

9.2   Frames of Translates

We are now ready to discuss the frame properties of sequences consisting of translates of a function $\phi \in L^2(\mathbb{R})$. The main question is: which conditions on a real sequence $\{\lambda_k\}_{k\in\mathbb{Z}}$ and a function $\phi \in L^2(\mathbb{R})$ will imply that $\{T_{\lambda_k}\phi\}_{k\in\mathbb{Z}}$ is a frame?

If we by "frame" mean "frame for $L^2(\mathbb{R})$," the answer is simple – and disappointing:

Theorem 9.2.1 *A system of the form $\{T_{\lambda_k}\phi\}_{k\in\mathbb{Z}}$ is never a frame for $L^2(\mathbb{R})$, regardless of the choice of the function $\phi \in L^2(\mathbb{R})$ and the sequence $\{\lambda_k\}_{k\in\mathbb{Z}}$.*

Theorem 9.2.1 will be a direct consequence of a result proved later, in Theorem 9.6.1. However, frame sequences of the form $\{T_{\lambda_k}\phi\}_{k\in\mathbb{Z}}$ exist. With a slight abuse of the language, we will usually skip the word "sequence" and refer to $\{T_{\lambda_k}\phi\}_{k\in\mathbb{Z}}$ as a *frame of translates*.

Note that the "no-go" result in Theorem 9.2.1 is particular for the Hilbert space $L^2(\mathbb{R})$: in $\ell^2(\mathbb{Z})$ it is easy to construct frames consisting of translates of a single vector – we can even construct orthonormal bases consisting of integer-translates of a single vector (Exercise 14.2).

The theory for frames of translates is far from being fully developed, and for a given sequence $\{\lambda_k\}_{k\in\mathbb{Z}}$ and a function ϕ, it is often difficult to find out whether $\{T_{\lambda_k}\phi\}_{k\in\mathbb{Z}}$ is a frame or not. The interplay between $\{\lambda_k\}_{k\in\mathbb{Z}}$ and ϕ is complicated: certain conditions on the density of $\{\lambda_k\}_{k\in\mathbb{Z}}$ are necessary for $\{T_{\lambda_k}\phi\}_{k\in\mathbb{Z}}$ to be a frame (Exercise 9.1 or Theorem 9.6.1), but if they are satisfied the final answer still depends heavily on the choice of ϕ.

We begin with the special case where $\lambda_k = kb$ for some $b > 0$. Because the points $\{kb\}_{k\in\mathbb{Z}}$ are equidistantly distributed, a frame $\{T_{kb}\phi\}_{k\in\mathbb{Z}}$ is sometimes called a *regular frame of translates* in contrast to the *irregular* frames $\{T_{\lambda_k}\phi\}_{k\in\mathbb{Z}}$.

Let $\phi \in L^2(\mathbb{R})$. Our first goal is to prove a result by Benedetto and Li [51], which shows that the frame properties in the regular case $\{T_{kb}\phi\}_{k\in\mathbb{Z}}$ can be completely described in terms of the function

$$\Phi : \mathbb{R} \to \mathbb{R}, \quad \Phi(\gamma) = \sum_{k\in\mathbb{Z}} \left| \hat{\phi}\left(\frac{\gamma + k}{b}\right) \right|^2. \tag{9.2}$$

Note that the definition (9.2) is slightly imprecise, in the sense that the defined series might be divergent for some $\gamma \in \mathbb{R}$. However,

$$\int_0^1 \sum_{k\in\mathbb{Z}} \left| \hat{\phi}\left(\frac{\gamma + k}{b}\right) \right|^2 d\gamma = \sum_{k\in\mathbb{Z}} \int_0^1 \left| \hat{\phi}\left(\frac{\gamma + k}{b}\right) \right|^2 d\gamma$$

$$= \int_{-\infty}^{\infty} \left| \hat{\phi}\left(\frac{\gamma}{b}\right) \right|^2 d\gamma < \infty;$$

this implies that $\sum_{k\in\mathbb{Z}}\left|\hat{\phi}\left(\frac{\gamma+k}{b}\right)\right|^2$ is convergent for almost all $\gamma \in \mathbb{R}$ and that $\Phi \in L^1(0,1)$.

We begin with some lemmas, which will be used repeatedly. The first one will be needed as a tool to analyze series expansions consisting of translates of a function ϕ. It is enough for our purpose to consider the case $b = 1$.

Lemma 9.2.2 *Let $\phi \in L^2(\mathbb{R})$ and assume that $\{T_k\phi\}_{k\in\mathbb{Z}}$ is a Bessel sequence. Let $\{c_k\}_{k\in\mathbb{Z}} \in \ell^2(\mathbb{Z})$. Then $\sum_{k\in\mathbb{Z}} c_k T_k \phi$ converges in $L^2(\mathbb{R})$ and $\sum_{k\in\mathbb{Z}} c_k E_{-k}$ converges in $L^2(0,1)$, and*

$$\mathcal{F}\sum_{k\in\mathbb{Z}}c_kT_k\phi = \left(\sum_{k\in\mathbb{Z}}c_kE_{-k}\right)\hat{\phi}. \qquad (9.3)$$

Proof. That $\sum_{k\in\mathbb{Z}} c_k T_k \phi$ and $\sum_{k\in\mathbb{Z}} c_k E_{-k}$ converge as described follows from Theorem 3.2.3, so we only have to prove (9.3). We first observe that by the result in Exercise 9.2

$$\left(\sum_{k\in\mathbb{Z}}c_kE_{-k}\right)\hat{\phi} \in L^2(\mathbb{R}). \qquad (9.4)$$

Also, note that

$$\mathcal{F}\sum_{k\in\mathbb{Z}}c_kT_k\phi = \sum_{k\in\mathbb{Z}}c_k\mathcal{F}T_k\phi = \sum_{k\in\mathbb{Z}}(c_kE_{-k}\hat{\phi}),$$

where the series on the right-hand side converges in $L^2(\mathbb{R})$. We have to prove that

$$\sum_{k\in\mathbb{Z}}(c_kE_{-k}\hat{\phi}) = \left(\sum_{k\in\mathbb{Z}}c_kE_{-k}\right)\hat{\phi},$$

i.e., that

$$\left\|\sum_{|k|\le N}c_kE_{-k}\hat{\phi} - \left(\sum_{k\in\mathbb{Z}}c_kE_{-k}\right)\hat{\phi}\right\|_{L^2(\mathbb{R})} \to 0 \text{ as } N \to \infty.$$

Now,

$$\left\|\sum_{|k|\le N}c_kE_{-k}\hat{\phi} - \left(\sum_{k\in\mathbb{Z}}c_kE_{-k}\right)\hat{\phi}\right\|_{L^2(\mathbb{R})}$$

$$= \left(\int_{-\infty}^{\infty}\left|\sum_{|k|\le N}c_kE_{-k}(\gamma)\hat{\phi}(\gamma) - \left(\sum_{k\in\mathbb{Z}}c_kE_{-k}(\gamma)\right)\hat{\phi}(\gamma)\right|^2 d\gamma\right)^{1/2} = (*).$$

Because $\sum_{k\in\mathbb{Z}} c_k E_{-k}$ is 1-periodic,

$$(*) = \left(\int_0^1 \left| \sum_{|k|\leq N} c_k E_{-k}(\gamma) - \sum_{k\in\mathbb{Z}} c_k E_{-k}(\gamma) \right|^2 \sum_{k\in\mathbb{Z}} |\widehat{\phi}(\gamma+k)|^2 d\gamma \right)^{1/2}$$

$$\leq \sqrt{B} \left\| \sum_{|k|\leq N} c_k E_{-k} - \sum_{k\in\mathbb{Z}} c_k E_{-k} \right\|_{L^2(0,1)};$$

here, B denotes a Bessel bound for $\{T_k\phi\}_{k\in\mathbb{Z}}$. The last term converges to 0 as $N\to\infty$, and the proof is completed. $\square$

The following lemma will be used throughout the book. For this reason, we state it slightly more general than needed in the current section.

Lemma 9.2.3 *Let $a > 0$ be given and $f : \mathbb{R} \to \mathbb{C}$ be a bounded, a-periodic, and measurable function. Then, for $g \in L^1(\mathbb{R})$,*

$$\int_{-\infty}^{\infty} f(x)g(x)dx = \int_0^a f(x) \sum_{k\in\mathbb{Z}} g(x-ka)dx.$$

Proof. We first show that

$$\int_0^a |f(x)| \sum_{k\in\mathbb{Z}} |g(x-ka)|dx < \infty. \qquad (9.5)$$

For positive functions, sums and integrals can be interchanged, so

$$\int_0^a |f(x)| \sum_{k\in\mathbb{Z}} |g(x-ka)|dx = \sum_{k\in\mathbb{Z}} \int_0^a |f(x)|\,|g(x-ka)|dx = (*).$$

Using that f is assumed to be a-periodic,

$$(*) = \sum_{k\in\mathbb{Z}} \int_0^a |f(x-ka)|\,|g(x-ka)|dx = \int_{-\infty}^{\infty} |f(x)|\,|g(x)|dx,$$

which is finite because f is bounded and $g \in L^1(\mathbb{R})$. This proves (9.5); the result now follows from Lebesgue' dominated convergence theorem. $\square$

We will also need the following variant of the result (Exercise 9.3).

Lemma 9.2.4 *Let $a > 0$ be given and $f,g \in L^2(\mathbb{R})$. Then the series $\sum_{k\in\mathbb{Z}} f(x-ka)g(x-ka)$ is absolutely convergent for a.e. $x \in \mathbb{R}$, and*

$$\int_{-\infty}^{\infty} f(x)g(x)\,dx = \int_0^a \sum_{k\in\mathbb{Z}} f(x-ka)g(x-ka)\,dx.$$

We are now ready for the announced characterization of frame properties for $\{T_{kb}\phi\}_{k\in\mathbb{Z}}$. The result is stated in terms of properties of the function Φ in (9.2).

Theorem 9.2.5 *Let $\phi \in L^2(\mathbb{R})$ and $b > 0$ be given. For any $A, B > 0$, the following characterizations hold:*

(i) *$\{T_{kb}\phi\}_{k\in\mathbb{Z}}$ is a Bessel sequence with bound B if and only if*

$$\Phi(\gamma) \leq bB, \ a.e. \ \gamma \in [0,1].$$

(ii) *$\{T_{kb}\phi\}_{k\in\mathbb{Z}}$ is an orthonormal sequence if and only if*

$$\Phi(\gamma) = b, \ a.e. \ \gamma \in [0,1].$$

(iii) *$\{T_{kb}\phi\}_{k\in\mathbb{Z}}$ is a Riesz sequence with bounds A, B if and only if*

$$bA \leq \Phi(\gamma) \leq bB, \ a.e. \ \gamma \in [0,1].$$

(iv) *$\{T_{kb}\phi\}_{k\in\mathbb{Z}}$ is a frame sequence with bounds A, B if and only if*

$$bA \leq \Phi(\gamma) \leq bB, \ a.e. \ \gamma \ \in [0,1] \setminus N,$$

where $N = \{\gamma \in [0,1] \mid \Phi(\gamma) = 0\}$.

Proof. To prove (i) we note that without any Bessel assumption, the synthesis operator

$$T : \{c_k\}_{k\in\mathbb{Z}} \to \sum_{k\in\mathbb{Z}} c_k T_{kb}\phi$$

is well defined as a map from all *finite sequences* in $\ell^2(\mathbb{Z})$ to $L^2(\mathbb{R})$. Given a finite sequence $\{c_k\}_{k\in\mathbb{Z}}$, we consider the trigonometric polynomial in $L^2(0,1)$ given by $f(\gamma) = \sum_{k\in\mathbb{Z}} c_k e^{-2\pi i k\gamma}$. Then, using that the Fourier transform is unitary and the commutator relation $\mathcal{F}T_{kb} = E_{-kb}\mathcal{F}$,

$$
\begin{aligned}
||T\{c_k\}_{k\in\mathbb{Z}}||^2 &= \left|\left|\sum_{k\in\mathbb{Z}} c_k T_{kb}\phi\right|\right|^2 = \left|\left|\mathcal{F}\sum_{k\in\mathbb{Z}} c_k T_{kb}\phi\right|\right|^2 \\
&= \int_{-\infty}^{\infty} \left|\sum_{k\in\mathbb{Z}} c_k e^{-2\pi i k b\gamma} \widehat{\phi}(\gamma)\right|^2 d\gamma = \int_{-\infty}^{\infty} |f(b\gamma)|^2 \left|\widehat{\phi}(\gamma)\right|^2 d\gamma \\
&= \frac{1}{b}\int_{-\infty}^{\infty} |f(\gamma)|^2 \left|\widehat{\phi}\left(\frac{\gamma}{b}\right)\right|^2 d\gamma.
\end{aligned}
$$

Via Lemma 9.2.3 we can continue with

$$
\begin{aligned}
||T\{c_k\}_{k\in\mathbb{Z}}||^2 &= \frac{1}{b}\int_0^1 |f(\gamma)|^2 \sum_{k\in\mathbb{Z}} \left|\widehat{\phi}\left(\frac{k+\gamma}{b}\right)\right|^2 d\gamma \\
&= \frac{1}{b}\int_0^1 |f(\gamma)|^2 \Phi(\gamma) d\gamma. \qquad (9.6)
\end{aligned}
$$

If $\Phi(\gamma) \leq bB$ for a.e. $\gamma \in \mathbb{R}$, it follows that

$$||T\{c_k\}_{k\in\mathbb{Z}}||^2 \leq B\int_0^1 |f(\gamma)|^2 \, d\gamma = B\sum_{k\in\mathbb{Z}} |c_k|^2;$$

via Exercise 3.13, this implies that $\{T_{kb}\phi\}_{k\in\mathbb{Z}}$ is a Bessel sequence with bound B. To prove the opposite implication in (i), we note that if $\{T_{kb}\phi\}_{k\in\mathbb{Z}}$ is a Bessel sequence, then by Lemma 9.2.2 the calculations leading to (9.6) hold for all $\{c_k\}_{k\in\mathbb{Z}} \in \ell^2(\mathbb{Z})$. Denoting the Bessel bound by B, we conclude from (9.6) and Theorem 3.2.3 that

$$\frac{1}{b}\int_0^1 |f(\gamma)|^2\,\Phi(\gamma)d\gamma \;\leq\; B\sum_{k\in\mathbb{Z}}|c_k|^2$$

$$= \; B\int_0^1 |f(\gamma)|^2\,d\gamma, \;\; \forall\{c_k\}_{k\in\mathbb{Z}} \in \ell^2(\mathbb{Z}).$$

This implies that $\Phi(\gamma) \leq bB$ for a.e. $\gamma \in \mathbb{R}$ and concludes the proof of (i).

We now prove (iv) via Lemma 5.5.5. Since we have proved (i), we will assume that $\{T_{kb}\phi\}_{k\in\mathbb{Z}}$ is a Bessel sequence and concentrate our analysis on the lower bounds. Consequently, the equality

$$||T\{c_k\}_{k\in\mathbb{Z}}||^2 = \frac{1}{b}\int_0^1 |f(\gamma)|^2\,\Phi(\gamma)d\gamma \qquad (9.7)$$

now holds for all sequences $\{c_k\}_{k\in\mathbb{Z}} \in \ell^2(\mathbb{Z})$; it follows that the kernel $\mathcal{N}_T$ of the operator T is

$$\mathcal{N}_T = \left\{\{c_k\}_{k\in\mathbb{Z}} \in \ell^2(\mathbb{Z}) \;\Big|\; \sum_{k\in\mathbb{Z}} c_k e^{-2\pi i k\gamma} = 0 \text{ on } [0,1]\setminus N\right\}. \qquad (9.8)$$

For arbitrary sequences $\{c_k\}_{k\in\mathbb{Z}}, \{d_k\}_{k\in\mathbb{Z}} \in \ell^2(\mathbb{Z})$, the fact that $\{e^{2\pi i k x}\}_{k\in\mathbb{Z}}$ is an orthonormal basis for $L^2(0,1)$ implies that

$$\langle\{c_k\}_{k\in\mathbb{Z}},\{d_k\}_{k\in\mathbb{Z}}\rangle_{\ell^2(\mathbb{Z})} = 0 \Leftrightarrow \left\langle\sum_{k\in\mathbb{Z}} c_k e^{-2\pi i k x}, \sum_{k\in\mathbb{Z}} d_k e^{-2\pi i k x}\right\rangle_{L^2(0,1)} = 0;$$

it follows that (Exercise 9.4)

$$\mathcal{N}_T^{\perp} = \left\{\{c_k\}_{k\in\mathbb{Z}} \in \ell^2(\mathbb{Z}) \;\Big|\; \sum_{k\in\mathbb{Z}} c_k e^{-2\pi i k\gamma} = 0 \text{ on } N\right\}. \qquad (9.9)$$

So for $\{c_k\}_{k\in\mathbb{Z}} \in \mathcal{N}_T^{\perp}$,

$$\sum_{k\in\mathbb{Z}}|c_k|^2 = \int_{[0,1]} \left|\sum_{k\in\mathbb{Z}} c_k e^{-2\pi i k\gamma}\right|^2 d\gamma = \int_{[0,1]\setminus N} \left|\sum_{k\in\mathbb{Z}} c_k e^{-2\pi i k\gamma}\right|^2 d\gamma;$$

using (9.7), the left-hand condition in (5.19) in Lemma 5.5.5 is therefore equivalent with

$$A \int_{[0,1] \setminus N} \left| \sum_{k \in \mathbb{Z}} c_k e^{-2\pi i k \gamma} \right|^2 d\gamma$$

$$\leq \frac{1}{b} \int_{[0,1] \setminus N} \left| \sum_{k \in \mathbb{Z}} c_k e^{-2\pi i k \gamma} \right|^2 \Phi(\gamma) d\gamma, \ \forall \{c_k\}_{k \in \mathbb{Z}} \in \mathcal{N}_T^{\perp}.$$

This, in turn, is equivalent with (Exercise 9.4)

$$bA \leq \Phi(\gamma) \ a.e. \ \gamma \in [0,1] \setminus N. \tag{9.10}$$

This proves (iv). For the rest of the proof, recall that the Riesz bounds and the frame bounds coincide for Riesz sequences. By Theorem 3.6.6, $\{T_{kb}\phi\}_{k \in \mathbb{Z}}$ is a Riesz sequence if and only if the inequalities (5.19) hold for all $\{c_k\}_{k \in \mathbb{Z}} \in \ell^2(\mathbb{Z})$; this is the case if and only if $\mathcal{N}_T = \{0\}$, i.e., by (9.8), if and only if N is a null set; this gives (iii). The result in (ii) now follows from Proposition 3.4.8. □

As a very important application of Theorem 9.2.5, we now prove that the integer-translates of any B-spline form a Riesz sequence. We formulate the result for the symmetric B-splines B_n defined in (A.15), but the same result holds for the B-splines $\widetilde{B}_n$ in (A.18); this remark in fact applies to all results concerning B-splines in the current chapter.

Theorem 9.2.6 *For each $n \in \mathbb{N}$, the sequence $\{T_k B_n\}_{k \in \mathbb{Z}}$ is a Riesz sequence.*

Proof. For $n = 1$, $\{T_k B_1\}_{k \in \mathbb{Z}}$ is an orthonormal system and therefore a Riesz sequence. In order to prove the result for $n > 1$, we apply Theorem 9.2.5 (ii) to B_1; this shows that

$$\sum_{k \in \mathbb{Z}} |\widehat{B_1}(\gamma + k)|^2 = 1, \quad a.e. \ \gamma \in \mathbb{R}.$$

Since $|\widehat{B_1}(\gamma)| \leq 1$ for all $\gamma \in \mathbb{R}$ and $\widehat{B_n}(\gamma) = (\widehat{B_1}(\gamma))^n$ by Corollary A.8.2, it immediately follows that

$$\sum_{k \in \mathbb{Z}} |\widehat{B_n}(\gamma + k)|^2 \leq \sum_{k \in \mathbb{Z}} |\widehat{B_1}(\gamma + k)|^2 = 1, \quad a.e. \ \gamma \in \mathbb{R}.$$

Thus, $\{T_k B_n\}_{k \in \mathbb{Z}}$ is a Bessel sequence. In order to prove that $\{T_k B_n\}_{k \in \mathbb{Z}}$ satisfies the lower Riesz basis condition, we again use Corollary A.8.2: it shows that, for a.e. $\gamma \in \mathbb{R}$,

$$\sum_{k \in \mathbb{Z}} |\widehat{B_n}(\gamma + k)|^2 \geq \inf_{\gamma \in [-\frac{1}{2}, \frac{1}{2}]} |\widehat{B_n}(\gamma)|^2 = \left(\frac{\sin(\pi/2)}{\pi/2} \right)^{2n} = \left(\frac{2}{\pi} \right)^{2n}. \tag{9.11}$$

The result now follows from Theorem 9.2.5. □

Example 9.2.7 Let $\alpha \in]0, \frac{1}{2}[$ and define $\phi \in L^2(\mathbb{R})$ via its Fourier transform as $\widehat{\phi}(\gamma) = \chi_{[-\alpha,\alpha[}(\gamma)$. Take $b = 1$. Then, for $\gamma \in [-\frac{1}{2}, \frac{1}{2}[$,

$$\Phi(\gamma) = \chi_{[-\alpha,\alpha[}(\gamma).$$

Theorem 9.2.5 shows that $\{T_k\phi\}_{k\in\mathbb{Z}}$ is a frame sequence with frame bounds $A = B = 1$. Note that $\{T_k\phi\}_{k\in\mathbb{Z}}$ does not form a Riesz sequence, i.e., it is an example of an overcomplete frame of translates. $\square$

The structure provided by a system of translates $\{T_k\phi\}_{k\in\mathbb{Z}}$ (or any other of the systems in the subsequent chapters) allows us to return to the chapters on general frame theory and look at the results from a new angle. Remember, e.g., the important analysis in Example 5.4.6, which shows that if $\{e_k\}_{k=1}^{\infty}$ is an orthonormal basis for a Hilbert space $\mathcal{H}$, then the sequence $\{e_k + e_{k+1}\}_{k=1}^{\infty}$ cannot be a frame; with a different index set, Exercise 5.16 shows that if $\{e_k\}_{k\in\mathbb{Z}}$ is an orthonormal basis or a Riesz basis, then $\{e_k + e_{k+1}\}_{k\in\mathbb{Z}}$ cannot be a frame. Via Theorem 9.2.5, we can now prove that the situation changes if we allow $\{e_k\}_{k\in\mathbb{Z}}$ to be a frame.

Proposition 9.2.8 *Let $\phi \in L^2(\mathbb{R})$ and assume that $\{T_k\phi\}_{k\in\mathbb{Z}}$ is a frame for $V := \overline{span}\{T_k\phi\}_{k\in\mathbb{Z}}$. Then the following are equivalent:*

(i) $\{T_k\phi + T_{k+1}\phi\}_{k\in\mathbb{Z}}$ is a frame for V.

(ii) $\Phi = 0$ on a neighborhood of $\gamma = \frac{1}{2}$.

Proof. A slight modification of the argument in Example 5.4.6 gives that $V = \overline{span}\{T_k\phi + T_{k+1}\phi\}_{k\in\mathbb{Z}}$. Letting $\widetilde{\phi} := \phi + T_1\phi$, we can write $T_k\phi + T_{k+1}\phi = T_k\widetilde{\phi}$. Via the Fourier transform, $\mathcal{F}\widetilde{\phi} = (1 + E_{-1})\widehat{\phi}$, so

$$\widetilde{\Phi}(\gamma) := \sum_{k\in\mathbb{Z}} |\mathcal{F}\widetilde{\phi}(\gamma + k)|^2$$

$$= \sum_{k\in\mathbb{Z}} \left|1 + e^{-2\pi i(\gamma+k)}\right|^2 |\widehat{\phi}(\gamma + k)|^2$$

$$= \left|1 + e^{-2\pi i\gamma}\right|^2 \sum_{k\in\mathbb{Z}} |\widehat{\phi}(\gamma + k)|^2.$$

Now the result follows from Theorem 9.2.5. $\square$

Concrete examples of frame sequences $\{T_k\phi\}_{k\in\mathbb{Z}}$ satisfying condition (ii) can be found via a slight modification of Example 9.2.7, e.g., by defining $\phi \in L^2(\mathbb{R})$ via $\widehat{\phi} = \chi_{[-\alpha,-\epsilon]\cup[\epsilon,\alpha[}$ for some $0 < \epsilon < \alpha < 1/2$.

9.3   Frames of Integer-Translates

In this section, we will derive some consequences of Theorem 9.2.5. Let us first state a technical result, which allows us to transfer results between two systems of translates with different translation parameters via a scaling (recall the definition of the dilation operator D_a in Section 2.9).

Lemma 9.3.1 *Let $\phi \in L^2(\mathbb{R})$ and $b > 0$ be given. Assume that $\{T_{kb}\phi\}_{k\in\mathbb{Z}}$ is a frame sequence. Given $a > 0$, let $\phi_a := D_a\phi$. Then $\{T_{kba}\phi_a\}_{k\in\mathbb{Z}}$ is a frame sequence with the same frame bounds as $\{T_{kb}\phi\}_{k\in\mathbb{Z}}$.*

Proof. We just notice that by the commutator relations in (2.26),

$$D_a T_{kb} = T_{kba} D_a;$$

the rest follows from Lemma 5.3.3. $\square$

Mainly for notational convenience we will now consider frames of translations with $b = 1$; such frames also play an important role in the theory for frame multiresolution analysis, as we will see in Chapter 17. Given $\phi \in L^2(\mathbb{R})$ the function in (9.2) now becomes

$$\Phi(\gamma) = \sum_{k\in\mathbb{Z}} |\widehat{\phi}(\gamma + k)|^2. \tag{9.12}$$

We will now prove that when $\{T_k\phi\}_{k\in\mathbb{Z}}$ is a frame sequence, membership of $\overline{\text{span}}\{T_k\phi\}_{k\in\mathbb{Z}}$ can be characterized in terms of the Fourier transform of the function ϕ. A more general version of this result will appear in Lemma 17.2.1.

Lemma 9.3.2 *Assume that $\phi \in L^2(\mathbb{R})$ and that $\{T_k\phi\}_{k\in\mathbb{Z}}$ is a frame sequence. Then a function $f \in L^2(\mathbb{R})$ belongs to $\overline{\text{span}}\{T_k\phi\}_{k\in\mathbb{Z}}$ if and only if there exists a 1-periodic function F whose restriction to $[0,1[$ belongs to $L^2(0,1)$, such that*

$$\widehat{f} = F\widehat{\phi}.$$

Let us now return to the characterization of the frame properties of $\{T_k\phi\}_{k\in\mathbb{Z}}$ in Theorem 9.2.5. In order to apply this result, it is essential to be able to control the function Φ in (9.12). For this purpose it is useful to express Φ in terms of its Fourier series:

Lemma 9.3.3 *Let $\phi \in L^2(\mathbb{R})$. Then the Fourier coefficients for the 1-periodic function $\Phi \in L^1(0,1)$ in (9.12) are*

$$c_k = \int_{-\infty}^{\infty} \phi(x)\overline{\phi(x+k)}dx, \ k \in \mathbb{Z}. \tag{9.13}$$

Proof. Using the 1-periodicity of the modulation operator E_k, the Fourier coefficients for Φ can be expressed by

$$c_k = \int_0^1 \Phi(\gamma)e^{-2\pi ik\gamma}d\gamma = \int_0^1 \sum_{n\in\mathbb{Z}}\left(|\widehat{\phi}(\gamma+n)|^2 e^{-2\pi ik(\gamma+n)}\right)d\gamma.$$

Via Lebesgue's dominated convergence theorem, we can interchange the sum and the integral; thus,

$$c_k = \int_{-\infty}^{\infty}|\widehat{\phi}(\gamma)|^2 e^{-2\pi ik\gamma}d\gamma = \langle\widehat{\phi}, E_k\widehat{\phi}\rangle = \langle\phi, T_{-k}\phi\rangle.$$

This completes the proof. $\qquad\qquad\qquad\qquad\qquad\qquad\qquad\qquad\square$

In applications of systems of the form $\{T_k\phi\}_{k\in\mathbb{Z}}$, it is often important that the generator ϕ has compact support. This excludes $\{T_k\phi\}_{k\in\mathbb{Z}}$ from being an overcomplete frame sequence.

Proposition 9.3.4 *Assume that $\phi \in L^2(\mathbb{R})$ has compact support. Then the following hold:*

(i) $\{T_k\phi\}_{k\in\mathbb{Z}}$ is a Bessel sequence.

(ii) $\{T_k\phi\}_{k\in\mathbb{Z}}$ cannot be an overcomplete frame sequence.

Proof. Let $\{c_k\}_{k\in\mathbb{Z}}$ be the Fourier coefficients for the function Φ in (9.12). Because of the compact support of ϕ, (9.13) in Lemma 9.3.3 shows that there is an $N \in \mathbb{N}$ such that $c_k = 0$ if $|k| > N$. Thus, the associated function Φ in (9.2) is a trigonometric polynomial and therefore continuous. Now Theorem 9.2.5 shows that $\{T_k\phi\}_{k\in\mathbb{Z}}$ is a Bessel sequence and that overcompleteness of $\{T_k\phi\}_{k\in\mathbb{Z}}$ is impossible. $\qquad\qquad\square$

Thus, if $\phi \in L^2(\mathbb{R})$ has compact support, then $\{T_k\phi\}_{k\in\mathbb{Z}}$ can at most be a Riesz sequence. In cases where the concrete frame bounds are irrelevant, we can check whether this is the case or not via the following consequence of Theorem 9.2.5. We ask the reader to give the proof in Exercise 9.5.

Corollary 9.3.5 *Assume that $\phi \in L^2(\mathbb{R})$ is compactly supported. Then the following are equivalent:*

(i) $\{T_k\phi\}_{k\in\mathbb{Z}}$ is a Riesz sequence.

(ii) For every $\gamma \in \mathbb{R}$, there exists an $k \in \mathbb{Z}$ such that $\widehat{\phi}(\gamma+k) \neq 0$.

The assumption of $\phi \in L^2(\mathbb{R})$ having compact support has the additional benefit that we can find an explicit expression for the associated function Φ in (9.12):

Lemma 9.3.6 *Assume that $\phi \in L^2(\mathbb{R})$ has compact support in an interval of length N for some $N \in \mathbb{N}$ and is real-valued. Let $\{c_k\}_{k\in\mathbb{Z}}$ denote the Fourier coefficients for the function Φ in (9.12). Then Φ is a trigonometric polynomial of the form*

$$\Phi(\gamma) = c_0 + 2\sum_{k=1}^{N} c_k \cos(2\pi k\gamma).$$

Proof.   Via Lemma 9.3.3, the assumption that ϕ is real-valued implies that $c_k = c_{-k}$ for all $k \in \mathbb{Z}$, and the compact support implies that $c_k = 0$ if $|k| > N$. Expressing Φ via its Fourier series, we see that

$$
\begin{aligned}
\Phi(\gamma) = \sum_{|k|\leq N} c_k e^{2\pi i k\gamma} &= c_0 + \sum_{k=1}^{N} c_k(e^{2\pi i k\gamma} + e^{-2\pi i k\gamma}) \\
&= c_0 + 2\sum_{k=1}^{N} c_k \cos(2\pi k\gamma),
\end{aligned}
$$

as desired. $\square$

Example 9.3.7 Let $\phi = \chi_{[-1,2[}$. By Lemma 9.3.3, the Fourier coefficients for the 1-periodic function Φ are

$$
c_k = \begin{cases}
3 & \text{if } k = 0, \\
2 & \text{if } k = \pm 1, \\
1 & \text{if } k = \pm 2, \\
0 & \text{otherwise.}
\end{cases}
$$

Thus, by Lemma 9.3.6,

$$\Phi(\gamma) = 3 + 4\cos(2\pi\gamma) + 2\cos(4\pi\gamma).$$

Note that Φ is continuous and that Φ has two isolated zeros on $[0, 1[$: $\Phi(\gamma) = 0$ for $\gamma = \frac{1}{3}$ and for $\gamma = \frac{2}{3}$. By Theorem 9.2.5, it follows that $\{T_k\phi\}_{k\in\mathbb{Z}}$ is not a frame sequence. $\square$

Without referring to the Fourier transform, it is difficult to find a function ϕ such that $\{T_k\phi\}_{k\in\mathbb{Z}}$ is an overcomplete frame sequence. More restrictions are given in the following proposition:

Proposition 9.3.8 *Let $\phi \in L^2(\mathbb{R})$ and assume that $\{T_k\phi\}_{k\in\mathbb{Z}}$ is an overcomplete frame sequence. Then the following hold:*

(i) *The function Φ is discontinuous.*

(ii) *Either $\phi \notin L^1(\mathbb{R})$ or there is no constant $C > 0$ for which*

$$|\widehat{\phi}(\gamma)| \leq C\left(\frac{1}{1+|\gamma|^2}\right)^{1/2}. \tag{9.14}$$

Proof. (i) follows directly from Theorem 9.2.5. For the proof of (ii), assume that an estimate of the type (9.14) is available, and let $\gamma \in [0, 1[$. Then, for all $N \in \mathbb{N}$,

$$\left| \Phi(\gamma) - \sum_{|k| \leq N} |\widehat{\phi}(\gamma + k)|^2 \right| = \sum_{|k| > N} |\widehat{\phi}(\gamma + k)|^2$$

$$\leq C^2 \sum_{|k| > N} \frac{1}{1 + |\gamma + k|^2}$$

$$\leq 2C^2 \sum_{k=N}^{\infty} \frac{1}{1 + k^2}.$$

Since $\sum_{k=N}^{\infty} \frac{1}{1+k^2} \to 0$ as $N \to \infty$, this shows that the series

$$\sum_{k \in \mathbb{Z}} |\widehat{\phi}(\gamma + k)|^2$$

is uniformly convergent. Thus, if $\widehat{\phi}$ was continuous, then Φ would be continuous as a uniform limit of continuous functions. This would contradict (i), so $\widehat{\phi}$ cannot be continuous, and therefore $\phi \notin L^1(\mathbb{R})$. $\square$

The interpretation of Proposition 9.3.8 (ii) is that a function ϕ generating an overcomplete frame sequence $\{T_k\phi\}_{k \in \mathbb{Z}}$ has bad time–frequency localization: either ϕ does not decay fast, or its Fourier transform has slow decay.

9.4 The Canonical Dual Frame

We now turn the focus to the canonical dual frame associated with a frame of translates. Assuming that $\{T_k\phi\}_{k \in \mathbb{Z}}$ is a frame sequence, we will in the entire section consider the space

$$V := \overline{\text{span}}\{T_k\phi\}_{k \in \mathbb{Z}}. \tag{9.15}$$

The assumption that $\{T_k\phi\}_{k \in \mathbb{Z}}$ is a frame implies by general frame theory (Lemma 5.1.5) that the frame operator

$$S : V \to V, \; Sf = \sum_{k \in \mathbb{Z}} \langle f, T_k\phi \rangle T_k\phi$$

is invertible. Again, the fact that we now consider a frame with a certain structure adds new aspects to the general frame theory considered in Chapter 5. In the setting discussed here, we can now prove that the frame operator S and its inverse S^{-1} commute with integer translation, whenever the operators are restricted to the subspace V.

Lemma 9.4.1 *Let $\phi \in L^2(\mathbb{R})$ and assume that $\{T_k\phi\}_{k\in\mathbb{Z}}$ is a frame sequence, i.e., a frame for the space V. Then*

$$ST_k = T_k S \quad and \quad S^{-1}T_k = T_k S^{-1} \quad on \ V, \ \forall k \in \mathbb{Z}.$$

Proof. Given $f \in V$ and $k \in \mathbb{Z}$, we have

$$ST_k f = \sum_{k'\in\mathbb{Z}} \langle T_k f, T_{k'}\phi\rangle T_{k'}\phi = \sum_{k'\in\mathbb{Z}} \langle f, T_{k'-k}\phi\rangle T_{k'}\phi.$$

Replacing the summation index k' by $k' + k$ gives

$$
\begin{aligned}
ST_k f &= \sum_{k'\in\mathbb{Z}} \langle f, T_{k'}\phi\rangle T_{k'+k}\phi \\
&= T_k S f.
\end{aligned}
$$

Thus, $ST_k = T_k S$ on the subspace V. Since the operator S is invertible on V, the second part of the result follows from this.   □

By the general definition, the canonical dual frame associated with $\{T_k\phi\}_{k\in\mathbb{Z}}$ is given by $\{S^{-1}T_k\phi\}_{k\in\mathbb{Z}}$. Using Lemma 9.4.1, we see that

$$\{S^{-1}T_k\phi\}_{k\in\mathbb{Z}} = \{T_k S^{-1}\phi\}_{k\in\mathbb{Z}}.$$

This result is very useful for calculation of the canonical dual frame. In order to find $\{S^{-1}T_k\phi\}_{k\in\mathbb{Z}}$, we would have to compute the action of S^{-1} on the infinite family of functions $\{T_k\phi\}_{k\in\mathbb{Z}}$. On the other hand, calculation of $\{T_k S^{-1}\phi\}_{k\in\mathbb{Z}}$ only requires that we find $S^{-1}\phi$; the rest of the functions in the family are obtained by translation. This is certainly a simplification, but we are still left with the question of finding $S^{-1}\phi$. The problem is that S is an operator on V, which is an infinite-dimensional Hilbert space; theoretically, we know that S is invertible, but this is different from being able to find the inverse explicitly! For general frames, we return to this problem in Chapter 23. In the present context, we are able to express $S^{-1}\phi$ in terms of its Fourier transform:

Proposition 9.4.2 *Let $\phi \in L^2(\mathbb{R})$ and assume that $\{T_k\phi\}_{k\in\mathbb{Z}}$ is a frame for its closed linear span V, with frame operator S. Let*

$$D := \{\gamma \in \mathbb{R} \,|\, \Phi(\gamma) \neq 0\},$$

and define the function θ via its Fourier transform by

$$\widehat{\theta}(\gamma) := \begin{cases} \frac{\widehat{\phi}(\gamma)}{\Phi(\gamma)} & if \ \gamma \in D, \\ 0 & if \ \gamma \notin D. \end{cases} \tag{9.16}$$

Then $\theta = S^{-1}\phi$, and the canonical dual frame of $\{T_k\phi\}_{k\in\mathbb{Z}}$ is given by $\{T_k\theta\}_{k\in\mathbb{Z}}$.

Proof. The function

$$\gamma \mapsto \begin{cases} \frac{1}{\Phi(\gamma)} & \text{if } \gamma \in D, \\ 0 & \text{if } \gamma \notin D \end{cases}$$

is 1-periodic, and its restriction to $]0,1[$ belongs to $L^2(0,1)$. Thus, Lemma 9.3.2 shows that the function θ defined by (9.16) belongs to V. Using the definition of the frame operator, properties of the Fourier transform, and Lemma 9.2.2, we have

$$\begin{aligned}
\mathcal{F}S\theta &= \sum_{k \in \mathbb{Z}} \langle \theta, T_k \phi \rangle \mathcal{F}T_k \phi \\
&= \sum_{k \in \mathbb{Z}} \langle \hat{\theta}, \mathcal{F}T_k \phi \rangle \mathcal{F}T_k \phi \\
&= \left(\sum_{k \in \mathbb{Z}} \langle \hat{\theta}, E_{-k}\hat{\phi} \rangle E_{-k} \right) \hat{\phi}.
\end{aligned} \qquad (9.17)$$

Now, using that the exponential functions $\gamma \mapsto e^{2\pi i k \gamma}$ are 1-periodic and the definition of θ,

$$\begin{aligned}
\langle \hat{\theta}, E_{-k}\hat{\phi} \rangle &= \int_{-\infty}^{\infty} \hat{\theta}(\gamma) \overline{\hat{\phi}(\gamma)} e^{2\pi i k \gamma} d\gamma \\
&= \int_0^1 \sum_{n \in \mathbb{Z}} \left(\hat{\theta}(\gamma + n) \overline{\hat{\phi}(\gamma + n)} e^{2\pi i (k+n)\gamma} \right) d\gamma \\
&= \int_0^1 \sum_{n \in \mathbb{Z}} \frac{|\hat{\phi}(\gamma + n)|^2}{\Phi(\gamma + n)} \chi_D(\gamma + n) e^{2\pi i k \gamma} d\gamma \\
&= \int_0^1 \chi_{D \cap [0,1[}(\gamma) \overline{E_{-k}(\gamma)} d\gamma,
\end{aligned}$$

which is the $(-k)$-th Fourier coefficient for the function $\chi_{D \cap [0,1[}$ in $L^2(0,1)$. Therefore,

$$\sum_{k \in \mathbb{Z}} \langle \hat{\theta}, E_{-k}\hat{\phi} \rangle E_{-k} = \chi_{D \cap [0,1[} \quad \text{on } [0,1].$$

Since χ_D is 1-periodic, it follows that

$$\sum_{k \in \mathbb{Z}} \langle \hat{\theta}, E_{-k}\hat{\phi} \rangle E_{-k} = \chi_D \quad \text{on } \mathbb{R}.$$

Noting that $\chi_D(\gamma) \neq 0$ if $\hat{\phi}(\gamma) \neq 0$, (9.17) now implies that

$$\mathcal{F}S\theta = \chi_D \hat{\phi} = \hat{\phi}.$$

Therefore $S\theta = \phi$, and since S is an invertible operator on V, the proof is over. $\qquad \square$

Example 9.4.3 Consider the B-spline B_n for $n \geq 2$. As we saw in Theorem 9.2.6, $\{T_k B_n\}_{k \in \mathbb{Z}}$ is a Riesz sequence. By Proposition 9.4.2, the canonical dual frame is given by $\{T_k \theta\}_{k \in \mathbb{Z}}$, where

$$\widehat{\theta}(\gamma) = \frac{\widehat{B_n}(\gamma)}{\sum_{k \in \mathbb{Z}} |\widehat{B_n}(\gamma + k)|^2}.$$

Denoting the Fourier coefficients for the function $\left(\sum_{k \in \mathbb{Z}} |\widehat{B_n}(\gamma + k)|^2 \right)^{-1}$ by $\{c_k\}_{k \in \mathbb{Z}}$, this implies that

$$\widehat{\theta}(\gamma) = \sum_{k \in \mathbb{Z}} c_k e^{2\pi i k \gamma} \widehat{B_n}(\gamma) = \mathcal{F} \sum_{k \in \mathbb{Z}} c_k T_{-k} B_n(\gamma),$$

i.e., that

$$\theta = \sum_{k \in \mathbb{Z}} c_k T_{-k} B_n.$$

It follows from Lemma 9.3.3 that $\Phi(\gamma) = \sum_{k \in \mathbb{Z}} |\widehat{B_n}(\gamma + k)|^2$ is a trigonometric polynomial. Because Φ is positive and not constant, the inverse $\Phi(\gamma)^{-1}$ cannot be a trigonometric polynomial (Exercise 9.7). Therefore, the sequence $\{c_k\}_{k \in \mathbb{Z}}$ of Fourier coefficients is infinite. In particular, the generator θ does not have compact support. □

For the case where an orthonormal basis is preferred, Daubechies proved in [242] that any Riesz sequence $\{T_k \phi\}_{k \in \mathbb{Z}}$ can be transferred to an orthonormal sequence which spans the same space. The *orthonormalization trick* applied to a frame sequence will lead to a tight frame for the same space:

Proposition 9.4.4 *Let $\phi \in L^2(\mathbb{R})$, and assume that $\{T_k \phi\}_{k \in \mathbb{Z}}$ is a frame sequence. Define the function $\phi^\sharp$ via its Fourier transform by*

$$\mathcal{F}\phi^\sharp(\gamma) := \begin{cases} \widehat{\phi}(\gamma) \Phi^{-1/2}(\gamma) & \text{if } \widehat{\phi}(\gamma) \neq 0, \\ 0 & \text{if } \widehat{\phi}(\gamma) = 0. \end{cases} \tag{9.18}$$

Then $\{T_k \phi^\sharp\}_{k \in \mathbb{Z}}$ is a tight frame sequence, and

$$\overline{span}\{T_k \phi^\sharp\}_{k \in \mathbb{Z}} = \overline{span}\{T_k \phi\}_{k \in \mathbb{Z}}.$$

If $\{T_k \phi\}_{k \in \mathbb{Z}}$ is a Riesz sequence, then $\{T_k \phi^\sharp\}_{k \in \mathbb{Z}}$ is an orthonormal sequence.

Proof. The reader can check that $\phi^\sharp$ is well defined, i.e., that $\Phi(\gamma) \neq 0$ if $\widehat{\phi}(\gamma) \neq 0$. Define

$$\Phi^\sharp(\gamma) := \sum_{k \in \mathbb{Z}} |\mathcal{F}\phi^\sharp(\gamma + k)|^2.$$

By Theorem 9.2.5, we want to prove that $\Phi^\sharp$ is bounded above and below away from its zero set. Let $\gamma \in [0,1[$. If $\widehat{\phi}(\gamma + k) = 0$ for all $k \in \mathbb{Z}$, then $\Phi^\sharp(\gamma) = 0$, so we now assume that $\widehat{\phi}(\gamma + k) \neq 0$ for some $k \in \mathbb{Z}$. Then for *all* $k \in \mathbb{Z}$, we have that

$$0 \neq \Phi(\gamma + k) = \Phi(\gamma).$$

Therefore the definition of $\Phi^\sharp$ gives that

$$
\begin{aligned}
\Phi^\sharp(\gamma) &= \sum_{k \in \mathbb{Z}} \frac{|\widehat{\phi}(\gamma + k)|^2}{\Phi(\gamma + k)} \\
&= \frac{1}{\Phi(\gamma)} \sum_{k \in \mathbb{Z}} |\widehat{\phi}(\gamma + k)|^2 \\
&= 1.
\end{aligned}
$$

Since $\Phi^\sharp$ only assumes the values zero and 1, Theorem 9.2.5 implies that $\{T_k \phi^\sharp\}_{k \in \mathbb{Z}}$ is a tight frame. In the special case where $\{T_k \phi\}_{k \in \mathbb{Z}}$ is a Riesz sequence, we can for a.e. $\gamma \in [0,1]$ find $k \in \mathbb{Z}$ such that $\widehat{\phi}(\gamma + k) \neq 0$; thus, $\Phi^\sharp = 1$ a.e., i.e., $\{T_k \phi^\sharp\}_{k \in \mathbb{Z}}$ is an orthonormal sequence.

In order to prove that $\{T_k \phi^\sharp\}_{k \in \mathbb{Z}}$ spans the same space as $\{T_k \phi\}_{k \in \mathbb{Z}}$, we note that

$$\overline{\text{span}}\{T_k \phi^\sharp\}_{k \in \mathbb{Z}} = \left\{ \sum_{k \in \mathbb{Z}} c_k T_k \phi^\sharp \ \bigg| \ \{c_k\}_{k \in \mathbb{Z}} \in \ell^2(\mathbb{Z}) \right\}.$$

Taking the Fourier transform of the functions in this space and letting

$$F(\gamma) := \begin{cases} \Phi^{-1/2}(\gamma) & \text{if } \widehat{\phi}(\gamma) \neq 0, \\ 1 & \text{if } \widehat{\phi}(\gamma) = 0 \end{cases}$$

yield

$$\left\{ \sum_{k \in \mathbb{Z}} c_k E_{-k} \mathcal{F} \phi^\sharp \ \bigg| \ \{c_k\}_{k \in \mathbb{Z}} \in \ell^2(\mathbb{Z}) \right\} = \left\{ F \sum_{k \in \mathbb{Z}} c_k E_{-k} \widehat{\phi} \ \bigg| \ \{c_k\}_{k \in \mathbb{Z}} \in \ell^2(\mathbb{Z}) \right\}.$$

The function F is bounded above and below, so

$$\left\{ F \sum_{k \in \mathbb{Z}} c_k E_{-k} \widehat{\phi} \ \bigg| \ \{c_k\}_{k \in \mathbb{Z}} \in \ell^2(\mathbb{Z}) \right\} = \left\{ \sum_{k \in \mathbb{Z}} c_k E_{-k} \widehat{\phi} \ \bigg| \ \{c_k\}_{k \in \mathbb{Z}} \in \ell^2(\mathbb{Z}) \right\};$$

this final space equals the space of Fourier transforms of the functions in $\overline{\text{span}}\{T_k \phi\}_{k \in \mathbb{Z}}$. Thus $\mathcal{F}(\overline{\text{span}}\{T_k \phi^\sharp\}_{k \in \mathbb{Z}}) = \mathcal{F}(\overline{\text{span}}\{T_k \phi\}_{k \in \mathbb{Z}})$, and the result follows. $\qquad \square$

9.5 Frames of Translates and Oblique Duals

In the discussion of general frame theory in Section 5.1, we have given arguments that it often is an advantage to search for other dual frames than the canonical dual frame. Assume that $\{T_k\phi\}_{k\in\mathbb{Z}}$ is an overcomplete frame sequence, i.e., a frame for

$$V = \overline{\text{span}}\{T_k\phi\}_{k\in\mathbb{Z}}; \tag{9.19}$$

then Theorem 6.3.1 tells us that there exist various choices of sequences of functions $\{g_k\}_{k\in\mathbb{Z}} \subset V$ such that

$$f = \sum_{k\in\mathbb{Z}}\langle f, g_k\rangle T_k\phi, \ \forall f \in V. \tag{9.20}$$

It is natural to insist on the dual frame $\{g_k\}_{k\in\mathbb{Z}}$ having the same structure as $\{T_k\phi\}_{k\in\mathbb{Z}}$, i.e., being translates of a single function. Unfortunately, Corollary 9.5.2 will show us that this removes all the freedom: the canonical dual frame $\{T_k S^{-1}\phi\}_{k\in\mathbb{Z}}$ is the only dual frame, which consists of translates of a single function and belongs to $\overline{\text{span}}\{T_k\phi\}_{k\in\mathbb{Z}}$.

In order to gain flexibility, we will remove the constraint that the elements of the dual frame should belong to $\overline{\text{span}}\{T_k\phi\}_{k\in\mathbb{Z}}$. In fact, we will just search for *some* function $\widetilde{\phi} \in L^2(\mathbb{R})$ such that

$$f = \sum_{k\in\mathbb{Z}}\langle f, T_k\widetilde{\phi}\rangle T_k\phi, \ \forall f \in V. \tag{9.21}$$

A family $\{T_k\widetilde{\phi}\}_{k\in\mathbb{Z}}$ for which (9.21) holds will be called an *oblique dual* of $\{T_k\phi\}_{k\in\mathbb{Z}}$. Note that we do not require $\{T_k\widetilde{\phi}\}_{k\in\mathbb{Z}}$ to be a frame sequence; this is the reason that we use the name "oblique dual" rather than "oblique dual frame." See the general discussion of oblique duals on page 128.

Given two Bessel sequences $\{T_k\phi\}_{k\in\mathbb{Z}}$ and $\{T_k\widetilde{\phi}\}_{k\in\mathbb{Z}}$, the following theorem provides a necessary and sufficient condition on the generators ϕ and $\widetilde{\phi}$ such that $\{T_k\widetilde{\phi}\}_{k\in\mathbb{Z}}$ is an oblique dual of $\{T_k\phi\}_{k\in\mathbb{Z}}$. Again, the function Φ defined in (9.12) will play a role; the result appeared in [132].

Theorem 9.5.1 *Let $\phi, \widetilde{\phi} \in L^2(\mathbb{R})$, and assume that $\{T_k\phi\}_{k\in\mathbb{Z}}$ and $\{T_k\widetilde{\phi}\}_{k\in\mathbb{Z}}$ are Bessel sequences. Then the following are equivalent:*

(i) $f = \sum_{k\in\mathbb{Z}}\langle f, T_k\widetilde{\phi}\rangle T_k\phi, \ \forall f \in V.$

(ii) $\sum_{k\in\mathbb{Z}}\widehat{\phi}(\gamma + k)\overline{\widehat{\widetilde{\phi}}(\gamma + k)} = 1$ a.e. on $\{\gamma \in [0,1] \mid \Phi(\gamma) \neq 0\}$.

In case the equivalent conditions are satisfied, $\{T_k\phi\}_{k\in\mathbb{Z}}$ is a frame sequence.

Proof. First, consider an arbitrary function $f \in L^2(\mathbb{R})$ for which the map

$$\gamma \mapsto \sum_{k\in\mathbb{Z}}|\widehat{f}(\gamma + k)|^2$$

is bounded. Since we have assumed that $\{T_k\widetilde{\phi}\}_{k\in\mathbb{Z}}$ is a Bessel sequence, Theorem 9.2.5 and Cauchy–Schwarz' inequality imply that

$$[\gamma \mapsto \sum_{k\in\mathbb{Z}} \widehat{f}(\gamma+k)\overline{\widehat{\widetilde{\phi}}(\gamma+k)}] \in L^2(0,1).$$

Now observe that, via Lemma 9.2.4 and Lemma 9.2.2,

$$\mathcal{F}\sum_{k\in\mathbb{Z}}\langle f, T_k\widetilde{\phi}\rangle T_k\phi(\gamma)$$

$$= \sum_{k\in\mathbb{Z}}\left(\int_{-\infty}^{\infty}\widehat{f}(\mu)\overline{\widehat{\widetilde{\phi}}(\mu)}e^{2\pi ik\mu}\,d\mu\right)\,e^{-2\pi ik\gamma}\widehat{\phi}(\gamma)$$

$$= \widehat{\phi}(\gamma)\sum_{k\in\mathbb{Z}}\left(\int_0^1 \sum_{n\in\mathbb{Z}}\widehat{f}(\mu+n)\overline{\widehat{\widetilde{\phi}}(\mu+n)}e^{2\pi ik\mu}d\mu\right)\,e^{-2\pi ik\gamma}$$

$$= \widehat{\phi}(\gamma)\sum_{n\in\mathbb{Z}}\widehat{f}(\gamma+n)\overline{\widehat{\widetilde{\phi}}(\gamma+n)}. \tag{9.22}$$

Assuming that (i) holds and letting $f = \phi$, it follows that

$$\sum_{k\in\mathbb{Z}}\widehat{\phi}(\gamma+k)\overline{\widehat{\widetilde{\phi}}(\gamma+k)} = 1 \text{ a.e. on } \{\gamma\in[0,1] \mid \widehat{\phi}(\gamma)\neq 0\}.$$

Using the above calculation with γ replaced by $\gamma+m$ for some $m\in\mathbb{Z}$ (and using the periodicity of $\gamma\mapsto\sum_{k\in\mathbb{Z}}\widehat{\phi}(\gamma+k)\overline{\widehat{\widetilde{\phi}}(\gamma+k)}$), we even arrive at

$$\sum_{k\in\mathbb{Z}}\widehat{\phi}(\gamma+k)\overline{\widehat{\widetilde{\phi}}(\gamma+k)} = 1 \text{ a.e. on } \{\gamma\in[0,1] \mid \widehat{\phi}(\gamma+m)\neq 0\}, \forall m\in\mathbb{Z}.$$

This proves (ii). On the other hand, assuming (ii), our calculation (9.22) shows that for $m\in\mathbb{Z}$,

$$\mathcal{F}\sum_{k\in\mathbb{Z}}\langle T_m\phi, T_k\widetilde{\phi}\rangle T_k\phi(\gamma) = \widehat{\phi}(\gamma)\sum_{n\in\mathbb{Z}}\mathcal{F}T_m\phi(\gamma+n)\overline{\widehat{\widetilde{\phi}}(\gamma+n)}$$

$$= \widehat{\phi}(\gamma)\sum_{n\in\mathbb{Z}}\widehat{\phi}(\gamma+n)e^{-2\pi im(\gamma+n)}\overline{\widehat{\widetilde{\phi}}(\gamma+n)}$$

$$= \widehat{\phi}(\gamma)e^{-2\pi im\gamma} = \mathcal{F}T_m\phi(\gamma).$$

This shows that (i) holds for all functions $T_m\phi, m\in\mathbb{Z}$ and hence for all functions $f\in\text{span}\{T_k\phi\}_{k\in\mathbb{Z}}$. Now, because $\{T_k\phi\}_{k\in\mathbb{Z}}$ and $\{T_k\widetilde{\phi}\}_{k\in\mathbb{Z}}$ are Bessel sequences, the operator

$$f\mapsto\sum_{k\in\mathbb{Z}}\langle f, T_k\widetilde{\phi}\rangle T_k\phi$$

is continuous; in fact, it is a composition of the synthesis operator associated with $\{T_k\phi\}_{k\in\mathbb{Z}}$ and the analysis operator associated with $\{T_k\widetilde{\phi}\}_{k\in\mathbb{Z}}$; see page 122. Therefore, (i) holds for all $f \in \overline{\operatorname{span}}\{T_k\phi\}_{k\in\mathbb{Z}}$.

Now, assume that the equivalent conditions hold. In order to show that $\{T_k\phi\}_{k\in\mathbb{Z}}$ is a frame sequence, we need to show that the lower frame bound is satisfied. Via (i), for all $f \in V$, we have

$$\|f\|^2 = \sum_{k\in\mathbb{Z}}\langle f, T_k\widetilde{\phi}\rangle\langle T_k\phi, f\rangle;$$

that $\{T_k\phi\}_{k\in\mathbb{Z}}$ is a frame sequence now follows from Cauchy–Schwarz inequality and the assumption that $\{T_k\widetilde{\phi}\}_{k\in\mathbb{Z}}$ is a Bessel sequence.   □

One important consequence of Theorem 9.5.1 is that for a frame sequence $\{T_k\phi\}_{k\in\mathbb{Z}}$, the canonical dual frame is the only dual frame that consists of integer-translates of a single function; that is, there only exists one dual frame that has same structure as $\{T_k\phi\}_{k\in\mathbb{Z}}$ itself. This was first proved in [132]:

Corollary 9.5.2 *Let $\phi \in L^2(\mathbb{R})$ and assume that $\{T_k\phi\}_{k\in\mathbb{Z}}$ is a frame sequence. Then there is a unique function $\widetilde{\phi} \in \overline{\operatorname{span}}\{T_k\phi\}_{k\in\mathbb{Z}}$ such that*

$$f = \sum_{k\in\mathbb{Z}}\langle f, T_k\widetilde{\phi}\rangle T_k\phi, \ \forall f \in \overline{\operatorname{span}}\{T_k\phi\}_{k\in\mathbb{Z}}, \tag{9.23}$$

namely, $\widetilde{\phi} = S^{-1}\phi$.

Proof. The condition $\widetilde{\phi} \in \overline{\operatorname{span}}\{T_k\phi\}_{k\in\mathbb{Z}}$ implies by Lemma 9.3.2 that $\widehat{\widetilde{\phi}} = F\widehat{\phi}$ for some 1-periodic function $F \in L^2(0,1)$. Now, if (9.23) holds, condition (ii) in Theorem 9.5.1 implies that

$$\overline{F(\gamma)}\sum_{k\in\mathbb{Z}}\widehat{\phi}(\gamma+k)\overline{\widehat{\phi}(\gamma+k)} = 1, \text{ a.e. on } \{\gamma \in [0,1] \mid \Phi(\gamma) \neq 0\},$$

i.e., that

$$\overline{F(\gamma)}\sum_{k\in\mathbb{Z}}\left|\widehat{\phi}(\gamma+k)\right|^2 = 1, \text{ a.e. on } \{\gamma \in [0,1] \mid \Phi(\gamma) \neq 0\}.$$

This defines the function F uniquely, except on the zero set for Φ. For γ such that $\Phi(\gamma) = 0$, we can define $F(\gamma)$ arbitrarily, but regardless of the choice, we arrive at $\widehat{\widetilde{\phi}}(\gamma) = F(\gamma)\widehat{\phi}(\gamma) = 0$. Thus $\widehat{\widetilde{\phi}}$ is uniquely defined, and so is $\widetilde{\phi}$.   □

The role of Theorem 9.5.1 is that it might be used to construct oblique duals with better or more convenient properties than the canonical dual frame. Later in this section, we will show that one might be able to find oblique duals generated by a compactly supported function, even in cases where the canonical dual frame is generated by a function supported on $\mathbb{R}$. For now, we will show how to construct oblique duals $\{T_k\widetilde{\phi}\}_{k\in\mathbb{Z}}$ with generators $\widetilde{\phi}$ belonging to prescribed subspaces. In fact, the following consequence of Theorem 9.5.1 shows that if $\{T_k\phi\}_{k\in\mathbb{Z}}$ is a frame sequence, then certain conditions imply that we can find an oblique dual $\{T_k\widetilde{\phi}\}_{k\in\mathbb{Z}}$ with a generator $\widetilde{\phi}$ belonging to a space of the form $\overline{span}\{T_k\phi_1\}_{k\in\mathbb{Z}}$ for some $\phi_1 \in L^2(\mathbb{R})$. We ask the reader to provide the proof in Exercise 9.6.

Corollary 9.5.3 *Let* $\phi, \phi_1 \in L^2(\mathbb{R})$, *and assume that* $\{T_k\phi\}_{k\in\mathbb{Z}}$ *and* $\{T_k\phi_1\}_{k\in\mathbb{Z}}$ *are frame sequences. If there exists a constant* $A > 0$ *such that*

$$\left|\sum_{k\in\mathbb{Z}} \widehat{\phi}(\gamma+k)\overline{\widehat{\phi_1}(\gamma+k)}\right| \geq A \ a.e. \ on \ \{\gamma \in [0,1] \mid \Phi(\gamma) \neq 0\}, \qquad (9.24)$$

then there exists a function $\widetilde{\phi} \in \overline{span}\{T_k\phi_1\}_{k\in\mathbb{Z}}$ *such that*

$$f = \sum_{k\in\mathbb{Z}}\langle f, T_k\widetilde{\phi}\rangle T_k\phi, \ \forall f \in \overline{span}\{T_k\phi\}_{k\in\mathbb{Z}}; \qquad (9.25)$$

one choice of $\widetilde{\phi} \in \overline{span}\{T_k\phi_1\}_{k\in\mathbb{Z}}$ *satisfying* (9.25) *is given in the Fourier domain by*

$$\widehat{\widetilde{\phi}}(\gamma) = \begin{cases} \left(\sum_{k\in\mathbb{Z}} \widehat{\phi}(\gamma+k)\overline{\widehat{\phi_1}(\gamma+k)}\right)^{-1}\widehat{\phi_1}(\gamma) & on \ \{\gamma \in \mathbb{R} \mid \Phi(\gamma) \neq 0\}, \\ 0 & on \ \{\gamma \in \mathbb{R} \mid \Phi(\gamma) = 0\}. \end{cases}$$

Corollary 9.5.3 can be used to "tailor" an oblique dual: that is, if the canonical dual frame does not satisfy the requirements for a specific application, we might search for an oblique dual that does. As an example, we show that one might be able to construct oblique duals of arbitrary smoothness, even if the canonical dual frame consists of noncontinuous functions:

Example 9.5.4 Consider the B-spline B_n for some $n \in \mathbb{N}$. By Theorem 9.2.6, we know that $\{T_kB_n\}_{k\in\mathbb{Z}}$ is a Riesz sequence; in particular, $\{T_kB_n\}_{k\in\mathbb{Z}}$ has a unique dual frame consisting of elements in $\overline{span}\{T_kB_n\}_{k\in\mathbb{Z}}$. Now, fix any $m \in \mathbb{N}$; we will show that there exists an oblique dual $\{T_k\widetilde{\phi}\}_{k\in\mathbb{Z}}$ of $\{T_kB_n\}_{k\in\mathbb{Z}}$ belonging to $\overline{span}\{T_kB_{n+2m}\}_{k\in\mathbb{Z}}$, i.e., such that

$$\widetilde{\phi} \in \overline{span}\{T_kB_{n+2m}\}_{k\in\mathbb{Z}}.$$

For any $\gamma \in \mathbb{R}$, the argument in (9.11) shows that

$$\sum_{k\in\mathbb{Z}} \widehat{B_n}(\gamma+k)\overline{\widehat{B_{n+2m}}(\gamma+k)} = \sum_{k\in\mathbb{Z}} \left(\frac{\sin\pi(\gamma+k)}{\pi(\gamma+k)}\right)^{2(m+n)} \geq \left(\frac{2}{\pi}\right)^{2(m+n)}.$$

By Corollary 9.5.3, there exists a function $\widetilde{\phi} \in \overline{span}\{T_kB_{n+2m}\}_{k\in\mathbb{Z}}$ that generates an oblique dual of $\{T_kB_n\}_{k\in\mathbb{Z}}$. That is, for an arbitrary spline B_n, we can find an oblique dual $\{T_k\widetilde{\phi}\}_{k\in\mathbb{Z}}$ for which the generator $\widetilde{\phi}$ has pre-scribed smoothness. In contrast, the canonical dual of $\{T_kB_n\}_{k\in\mathbb{Z}}$ has the same smoothness as B_n itself; for example, the canonical dual frame of $\{T_kB_1\}_{k\in\mathbb{Z}}$ is generated by B_1, which is not even continuous. $\qquad\square$

In the rest of this section we will restrict our attention to frame sequences $\{T_k\phi\}_{k\in\mathbb{Z}}$ generated by compactly supported functions. We have already seen in Example 9.4.3 that this does not imply that the canonical dual frame necessarily is generated by a compactly supported function. Also, Corollary 9.5.2 shows that no other dual frame is generated by translates of a single function. Nevertheless, we will now show that it often is possi-ble to find *oblique duals* $\{T_k\widetilde{\phi}\}_{k\in\mathbb{Z}}$ for which the function $\widetilde{\phi}$ has compact (and small) support. For convenience, we will search for an oblique dual supported on the interval $[0,1]$, but the same considerations work on any other interval.

In Theorem 9.5.1, we characterized the oblique duals $\{T_k\widetilde{\phi}\}_{k\in\mathbb{Z}}$ associ-ated with a given frame of translates $\{T_k\phi\}_{k\in\mathbb{Z}}$. We first show that this result has a much simpler version if we assume that the functions ϕ and $\widetilde{\phi}$ are compactly supported, a result that first appeared in [189]:

Lemma 9.5.5 *Assume that the functions* $\phi, \widetilde{\phi} \in L^2(\mathbb{R})$ *have compact support, and define the space* V *as in* (9.19). *Then the following are equivalent:*

(i) $f = \sum_{k\in\mathbb{Z}}\langle f, T_k\widetilde{\phi}\rangle T_k\phi, \ \forall f \in V.$

(ii) $\langle\phi, T_k\widetilde{\phi}\rangle = \delta_{k,0}.$

Proof. If (i) holds, then Theorem 9.5.1 shows that $\{T_k\phi\}_{k\in\mathbb{Z}}$ is a frame for V. Because of the compact support of ϕ, this implies by Proposi-tion 9.3.4 that $\{T_k\phi\}_{k\in\mathbb{Z}}$ is a Riesz sequence. Using (i) on $f = \phi$, the statement in (ii) follows because the expansion coefficients in terms of a Riesz basis are unique. On the other hand, if (ii) holds, then (i) holds for $f = \phi$. A change of the summation index proves that then (i) holds for any translate $T_k\phi$ and therefore on span$\{T_k\phi\}_{k\in\mathbb{Z}}$; finally, by continuity of the operator $f \mapsto \sum_{k\in\mathbb{Z}}\langle f, T_k\widetilde{\phi}\rangle T_k\phi$, we obtain that (i) holds for all $f \in V$. $\square$

Lemma 9.5.5 shows that, with the given assumptions, the question of finding an oblique dual of a frame $\{T_k\phi\}_{k\in\mathbb{Z}}$ can be formulated as a moment problem; see the general discussion of such problems in Section 7.6.

The following result essentially provides conditions such that a Riesz sequence $\{T_k\phi\}_{k\in\mathbb{Z}}$ has an oblique dual $\{T_k\widetilde{\phi}\}_{k\in\mathbb{Z}}$, where $\widetilde{\phi}$ has the form

$$\widetilde{\phi}(x) = \left(\sum_{\ell=0}^{N-1} d_\ell\phi(x+\ell)\right)\chi_{[0,1]}(x). \qquad (9.26)$$

For practical reasons, we will formulate a more general version of the result. The reason is that even if ϕ is smooth, the multiplication with the characteristic function $\chi_{[0,1]}$ in the expression for $\widetilde{\phi}$ in (9.26) might lead to a function that is discontinuous at $x = 0$ or $x = 1$. On the other hand, multiplying the function in (9.26) with a function of the type $x^p(1-x)^q$ for some $p, q \in \mathbb{N}$ will lead to a continuous function if ϕ is continuous; and if ϕ is smooth, any desired (finite) smoothness can be obtained by choosing the parameters p, q sufficiently large. We will show that functions of that type can be used as generators as well; the result is taken from [189].

Theorem 9.5.6 *Assume that $\phi \in L^2(\mathbb{R})$ is a real-valued function with support on an interval $[0, N]$ for some $N \in \mathbb{N}$ and that $\{T_k\phi\}_{k\in\mathbb{Z}}$ is a Riesz sequence. Assume that*

$$\sum_{k=0}^{N-1} c_k \phi(x + k) = 0, \forall x \in [0, 1] \Rightarrow c_0 = 0. \tag{9.27}$$

Then, for any $p, q \in \{0\} \cup \mathbb{N}$, $\{T_k\phi\}_{k\in\mathbb{Z}}$ has an oblique dual $\{T_k\widetilde{\phi}\}_{k\in\mathbb{Z}}$, where $\widetilde{\phi}$ has the form

$$\widetilde{\phi}(x) = x^p(1-x)^q \left(\sum_{\ell=0}^{N-1} d_\ell \phi(x + \ell)\right) \chi_{[0,1]}(x) \tag{9.28}$$

for some coefficients $d_0, \ldots, d_{N-1} \in \mathbb{R}$.

Proof.   We use Lemma 9.5.5 and search for a function $\widetilde{\phi}$ such that $\langle \phi, T_k\widetilde{\phi}\rangle = \delta_{k,0}$. First, for any function $\widetilde{\phi} \in L^2(\mathbb{R})$ with support on $[0, 1]$, we have that

$$\langle \phi, T_k\widetilde{\phi}\rangle = \int_{-\infty}^{\infty} \phi(x)\overline{\widetilde{\phi}(x - k)}\, dx \quad = \quad \int_{-\infty}^{\infty} \phi(x + k)\overline{\widetilde{\phi}(x)}\, dx$$

$$= \quad \int_0^1 \phi(x + k)\overline{\widetilde{\phi}(x)}\, dx.$$

Assuming that ϕ has support on $[0, N]$, this shows that

$$\langle \phi, T_k\widetilde{\phi}\rangle = 0 \text{ if } k \notin \{0, 1, \ldots, N - 1\}.$$

Thus, the equations $\langle \phi, T_k\widetilde{\phi}\rangle = \delta_{k,0}$ can be reduced to the moment problem

$$\langle \widetilde{\phi}, T_{-k}\phi\rangle = \delta_{k,0}, \ k = 0, 1, \ldots, N - 1. \tag{9.29}$$

Because of the assumption (9.27), we know that if

$$\sum_{k=0}^{N-1} c_k \phi(x + k)x^{p/2}(1 - x)^{q/2} = 0 \text{ for all } x \in [0, 1],$$

then $c_0 = 0$. Thus, according to Lemma 2.6.1 with $\mathcal{H} = L^2(0, 1)$ and f_k corresponding to the functions $x \mapsto \phi(x+k)x^{p/2}(1-x)^{q/2}, k = 0, \ldots, N-1$, the moment problem

$$
\begin{cases}
1 = \displaystyle\int_0^1 \phi(x)x^{p/2}(1-x)^{q/2}h(x)\,dx \\[2mm]
0 = \displaystyle\int_0^1 \phi(x+1)x^{p/2}(1-x)^{q/2}h(x)\,dx \\
\quad . \\
\quad . \\
\quad . \\
0 = \displaystyle\int_0^1 \phi(x+N-1)x^{p/2}(1-x)^{q/2}h(x)\,dx
\end{cases}
$$

has a solution h of the form

$$
h(x) = \left(\sum_{\ell=0}^{N-1} d_\ell \phi(x+\ell)x^{p/2}(1-x)^{q/2} \right) \chi_{[0,1]}(x).
$$

This means that the function

$$
\widetilde{\phi}(x) := x^{p/2}(1-x)^{q/2}h(x) = \left(\sum_{\ell=0}^{N-1} d_\ell \phi(x+\ell) \right) x^p(1-x)^q \chi_{[0,1]}(x)
$$

solves the moment problem (9.29). □

Note that the coefficients $d_0, \ldots, d_{N-1}$ in (9.28) are determined by the conditions in (9.29), i.e., by the equations

$$
\sum_{\ell=0}^{N-1} \int_0^1 d_\ell \phi(x+k)\phi(x+\ell)x^p(1-x)^q\,dx = \delta_{k,0}, \ \ k = 0, \ldots, N-1. \ (9.30)
$$

On matrix form, this takes the form

$$
M\mathbf{d} = \mathbf{e},
$$

where M is the $N \times N$ symmetric matrix with entries

$$
M_{k,\ell} = \int_0^1 x^p(1-x)^q \phi(x+k)\phi(x+\ell)\,dx, \ \ k,\ell = 0, \ldots, N-1
$$

and

$$
\mathbf{d} = \begin{pmatrix} d_0 \\ d_1 \\ . \\ . \\ . \\ d_{N-1} \end{pmatrix}, \ \mathbf{e} = \begin{pmatrix} 1 \\ 0 \\ . \\ . \\ . \\ 0 \end{pmatrix}.
$$

Recall that the parameters p and q in (9.28) were introduced in order to ensure higher-order derivatives of $\widetilde{\phi}$ to exist. We see that this only affects the integrals in the entries of matrix M but not the size of the matrix; thus, the computational complexity does not increase in a drastic way whenever generators with higher regularity are constructed.

Let us apply Theorem 9.5.6 to the (shifted) B-splines $\widetilde{B}_n$, $n \in \mathbb{N}$ introduced in (A.18). We remember that $\mathrm{supp}\,\widetilde{B}_n = [0, n]$. Using Theorem 9.5.6, we will be able to find an oblique dual of any $\{T_k \widetilde{B}_n\}_{k \in \mathbb{Z}}$, which is generated by a compactly supported function. In contrast, we saw in Example 9.4.3 that for $n \geq 2$, the generator for the canonical dual frame of $\{T_k \widetilde{B}_n\}_{k \in \mathbb{Z}}$ never has compact support.

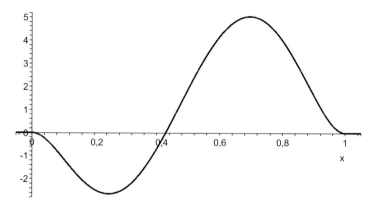

Figure 9.1. The generator $\widetilde{\phi}$ in (9.28) corresponding to $\phi = \widetilde{B}_2, p = q = 2$.

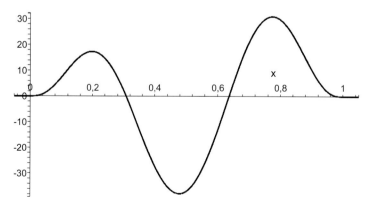

Figure 9.2. The generator $\widetilde{\phi}$ in (9.28) corresponding to $\phi = \widetilde{B}_3, p = q = 3$.

Example 9.5.7 For any $n \in \mathbb{N}$, Lemma A.8.6 shows that the functions $\widetilde{B_n}(\cdot + k), k = 0, \ldots, n - 1$, are linearly independent on the interval $[0, 1]$. Thus, the condition (9.27) is satisfied. Therefore, for any $p, q \in \mathbb{N} \cup \{0\}$, the frame sequence $\{T_k \widetilde{B_n}\}_{k \in \mathbb{Z}}$ has an oblique dual $\{T_k \widetilde{\phi}\}_{k \in \mathbb{Z}}$ with a function $\widetilde{\phi}$ of the form

$$\widetilde{\phi}(x) = \left(\sum_{\ell=0}^{n-1} d_\ell \widetilde{B_n}(x + \ell) \right) x^p (1 - x)^q \chi_{[0,1]}(x).$$

Note that on the interval $[0, 1]$, this function is a polynomial of degree $p + q + n - 1$.

Figures 9.1 and 9.2 show some oblique dual generators for Riesz sequences generated by the B-splines $\widetilde{B_2}$ and $\widetilde{B_3}$, for various values of p and q. We ask the reader to do the calculations (Exercise 9.16). □

9.6   Irregular Frames of Translates

We now return to the general case of a system of translates $\{T_{\lambda_k} \phi\}_{k \in \mathbb{Z}}$, where $\{\lambda_k\}_{k \in \mathbb{Z}}$ is an arbitrary sequence in $\mathbb{R}$. We have already noted that it is difficult to prove whether $\{T_{\lambda_k} \phi\}_{k \in \mathbb{Z}}$ is a frame or not, except under very special conditions on ϕ (Exercise 9.9).

The first part of the following theorem shows that if we want $\{T_{\lambda_k} \phi\}_{k \in \mathbb{Z}}$ to be a frame sequence, then we have to assume that $\{\lambda_k\}_{k \in \mathbb{Z}}$ is relatively separated; otherwise, $\{T_{\lambda_k} \phi\}_{k \in \mathbb{Z}}$ cannot be a Bessel sequence in $L^2(\mathbb{R})$ for *any* function $\phi \in L^2(\mathbb{R}) \setminus \{0\}$. The second part shows that $\{\lambda_k\}_{k \in \mathbb{Z}}$ being relatively separated *excludes* $\{T_{\lambda_k} \phi\}_{k \in \mathbb{Z}}$ from having a lower frame bound in $L^2(\mathbb{R})$. Put together, the conclusion is that $\{T_{\lambda_k} \phi\}_{k \in \mathbb{Z}}$ never can be a frame for all of $L^2(\mathbb{R})$ (but it can very well be a frame sequence). The result is due to Christensen, Deng, and Heil [168].

Theorem 9.6.1 *Let $\Lambda = \{\lambda_k\}_{k \in \mathbb{Z}}$ be a sequence in $\mathbb{R}$ and $\phi \in L^2(\mathbb{R}) \setminus \{0\}$. Then the following hold:*

(i) *If $\{T_{\lambda_k} \phi\}_{k \in \mathbb{Z}}$ is a Bessel sequence, then $D^+(\Lambda) < \infty$.*

(ii) *If $\{T_{\lambda_k} \phi\}_{k \in \mathbb{Z}}$ satisfies the lower frame condition in $L^2(\mathbb{R})$, then $D^+(\Lambda) = \infty$.*

In particular, $\{T_{\lambda_k} \phi\}_{k \in \mathbb{Z}}$ can at most be a frame for a proper subspace of $L^2(\mathbb{R})$.

Proof.   Recall from Lemma 9.1.3 that $D^+(\Lambda) < \infty$ is equivalent to Λ being relatively separated. For the proof of (i), we assume that Λ is not relatively separated; we have to prove that then $\{T_{\lambda_k} \phi\}_{k \in \mathbb{Z}}$ is not a Bessel sequence. Consider the function $x \mapsto \langle \phi, T_x \phi \rangle$, $x \in \mathbb{R}$. Since the function is continuous by Lemma 2.9.2 and nonzero for $x = 0$, there exists an interval $]-h, h[$, $h > 0$, such that

$$\mu := \inf_{x \in]-h,h[} |\langle \phi, T_x \phi \rangle| > 0.$$

Consider an arbitrary $N \in \mathbb{N}$. By Lemma 9.1.3, there exists an interval $]a-h, a+h[$, $a \in \mathbb{R}$, which contains at least N elements from the sequence Λ. Now, letting

$$\Lambda_N := \{k \in \mathbb{Z} : \lambda_k \in]a - h, a + h[\} = \{k \in \mathbb{Z} : \lambda_k - a \in]-h, h[\},$$

we have

$$\begin{aligned}
\sum_{k \in \mathbb{Z}} |\langle T_a \phi, T_{\lambda_k} \phi \rangle|^2 &\geq \sum_{k \in \Lambda_N} |\langle T_a \phi, T_{\lambda_k} \phi \rangle|^2 \\
&= \sum_{k \in \Lambda_N} |\langle \phi, T_{\lambda_k - a} \phi \rangle|^2 \\
&\geq N\mu^2 \\
&= \frac{N\mu^2}{||\phi||^2} \, ||T_a \phi||^2.
\end{aligned}$$

Since $N \in \mathbb{N}$ was arbitrary, it follows that $\{T_{\lambda_k} \phi\}_{k \in \mathbb{Z}}$ is not a Bessel sequence in $L^2(\mathbb{R})$. This proves (i).

For the proof of (ii), we assume that $D^+(\Lambda) < \infty$; thus, we have to prove that $\{T_{\lambda_k} \phi\}_{k \in \mathbb{Z}}$ does not satisfy the lower frame condition for any $A > 0$. By Lemma 9.1.3, $\{\lambda_k\}_{k \in \mathbb{Z}}$ is a finite union of separated sets, i.e., we can write

$$\{\lambda_k\}_{k \in \mathbb{Z}} = \bigcup_{j=1}^{s} \{\lambda_k\}_{k \in I_j},$$

where each set $\{\lambda_k\}_{k \in I_j}$ is separated. Choose a constant $\delta > 0$, which is a separation constant for each sequence $\{\lambda_k\}_{k \in I_j}$, $j = 1, \ldots, s$, and consider $h \in]0, \delta/2[$. With $I := [-h, h]$,

$$\begin{aligned}
\sum_{k \in \mathbb{Z}} |\langle \chi_I, T_{\lambda_k} \phi \rangle|^2 &= \sum_{j=1}^{s} \sum_{k \in I_j} |\langle \chi_I, \chi_I T_{\lambda_k} \phi \rangle|^2 \\
&\leq \sum_{j=1}^{s} \sum_{k \in I_j} ||\chi_I||^2 \, ||\chi_I T_{\lambda_k} \phi||^2. \quad (9.31)
\end{aligned}$$

By the choice of I, the intervals $\{I - \lambda_k\}_{k \in I_j}$ are disjoint; by defining

$$\Delta_j := \bigcup_{k \in I_j} (I - \lambda_k),$$

we have

$$\sum_{k \in I_j} ||\chi_I T_{\lambda_k} \phi||^2 = \sum_{k \in I_j} \int_I |\phi(x - \lambda_k)|^2 dx = \int_{\Delta_j} |\phi(x)|^2 dx.$$

Thus, via (9.31),

$$\sum_{k\in\mathbb{Z}} |\langle \chi_I, T_{\lambda_k}\phi\rangle|^2 \;\le\; ||\chi_I||^2 \sum_{j=1}^{s} \int_{\Delta_j} |\phi(x)|^2 dx.$$

An application of Lebesgue's dominated convergence theorem shows that for each fixed $j = 1, \ldots, s$,

$$\int_{\Delta_j} |\phi(x)|^2 dx \to 0 \text{ as } h \to 0;$$

thus, $\{T_{\lambda_k}\phi\}_{k\in\mathbb{Z}}$ does not have a lower frame bound in $L^2(\mathbb{R})$.   □

A more general result can be found in [168]: no union of arbitrary translates of a finite collection of functions $g_1, \ldots, g_M$ can be a frame for $L^2(\mathbb{R})$. The proof is almost identical to the above proof, just with a more involved notation. As a consequence of this result, a collection of functions of the form $\{T_{na}E_{mb}g\}_{n\in\mathbb{Z},m=1,\ldots,M}$ cannot be a frame for $L^2(\mathbb{R})$ for any choice of $g \in L^2(\mathbb{R})$ and any constants $a, b > 0$. However, frames of the type $\{T_{na}E_{mb}g\}_{m,n\in\mathbb{Z}}$ exist; they will be the topic of Chapters 11–13.

Note that provided that the sequence $\{\lambda_k\}_{k\in\mathbb{Z}}$ consists of distinct numbers, the set of translates $\{T_{\lambda_k}\phi\}_{k\in\mathbb{Z}}$ are linearly independent for any function $\phi \in L^2(\mathbb{R}) \setminus \{0\}$:

Proposition 9.6.2 *Assume that $\{\lambda_k\}_{k\in\mathbb{Z}}$ is a sequence for which $\lambda_k \ne \lambda_j$ for $k \ne j$. If $\phi \in L^2(\mathbb{R}) \setminus \{0\}$, then the functions $\{T_{\lambda_k}\phi\}_{k\in\mathbb{Z}}$ are linearly independent.*

Proof. Let $\mathcal{F} \subset \mathbb{Z}$ be a finite set, and assume that for some coefficients $\{c_k\}_{k\in\mathcal{F}}$,

$$\sum_{k\in\mathcal{F}} c_k T_{\lambda_k}\phi = 0.$$

Via the Fourier transform,

$$\sum_{k\in\mathcal{F}} c_k E_{-\lambda_k}\widehat{\phi} = 0. \tag{9.32}$$

Choose a bounded nonempty interval I on which $\widehat{\phi}$ is not identically zero. Assume that not all coefficients $\{c_k\}_{k\in\mathcal{F}}$ are zero; then, the function $\gamma \mapsto \sum_{k\in\mathcal{F}} c_k E_{-\lambda_k}(\gamma)$ is only zero for finitely many $\gamma \in I$, and then $\sum_{k\in\mathcal{F}} c_k E_{-\lambda_k}\widehat{\phi}$ is not the zero function. This contradicts (9.32); we conclude that $c_k = 0$ for all $k \in \mathcal{F}$ and that the functions $\{T_{\lambda_k}\phi\}_{k\in\mathbb{Z}}$ are linearly independent.   □

9.7 Sampling Theory and Applications

In Section 3.10, we gave a short introduction to sampling theory in the Paley–Wiener space PW. In particular, Theorem 3.10.1 shows that each continuous function $f \in PW$ can be recovered from its samples $\{f(k)\}_{k \in \mathbb{Z}}$. More information can be extracted from the proof of Theorem 3.10.1: the samples $\{f(-k)\}_{k \in \mathbb{Z}}$ are in fact the Fourier coefficients for the Fourier transform $\widehat{f}$, so

$$\sum_{k \in \mathbb{Z}} |f(k)|^2 = ||\widehat{f}||^2 = ||f||^2, \, \forall f \in PW. \tag{9.33}$$

Based on our knowledge of Riesz bases of translates, we will now look at sampling from a more general perspective, using the general setup of a reproducing kernel Hilbert space (RKHS). The key condition can be considered as a generalization of (9.33): we will replace the equality in (9.33) by a lower and upper bound, and we will place the equidistant sampling points k, $k \in \mathbb{Z}$, by a general sequence $\{\lambda_k\}_{k \in \mathbb{Z}}$.

Definition 9.7.1 *Consider a reproducing kernel Hilbert space $\mathcal{H} \subset L^2(\mathbb{R})$ consisting of continuous functions. A sequence $\{\lambda_k\}_{k \in \mathbb{Z}} \subset \mathbb{R}$ is a set of sampling for $\mathcal{H}$ if there exist constants $A, B > 0$ such that*

$$A \, ||f||^2 \leq \sum_{k \in \mathbb{Z}} |f(\lambda_k)|^2 \leq B \, ||f||^2, \, \forall f \in \mathcal{H}. \tag{9.34}$$

Note that in terms of the reproducing kernel K (see (5.29)), the condition (9.34) implies that $\{K_{\lambda_k}\}_{k \in \mathbb{Z}}$ is a frame for $\mathcal{H}$. Let us now introduce the relevant RKHS.

Lemma 9.7.2 *Let $\phi \in L^2(\mathbb{R})$ be a continuous function such that*

(i) $\sum_{k \in \mathbb{Z}} ||\phi \chi_{[k,k+1]}||_\infty < \infty$.

(ii) $\{T_k \phi\}_{k \in \mathbb{Z}}$ *is a Riesz sequence, with canonical dual* $\{T_k \widetilde{\phi}\}_{k \in \mathbb{Z}}$.

Then the vector space

$$\mathcal{H} := \left\{ \sum_{k \in \mathbb{Z}} c_k T_k \phi \mid \{c_k\}_{k \in \mathbb{Z}} \in \ell^2(\mathbb{Z}) \right\} \tag{9.35}$$

is an RKHS consisting of continuous functions, with reproducing kernel

$$K(x,y) = \sum_{k \in \mathbb{Z}} \overline{\phi(y-k)} \widetilde{\phi}(x-k).$$

Proof. Let $a_j := \sup_{x \in [0,1]} |\phi(x+j)|$, $j \in \mathbb{Z}$. By assumption $\{a_j\}_{j \in \mathbb{Z}} \in \ell^1(\mathbb{Z})$; thus, also $\{a_j\}_{j \in \mathbb{Z}} \in \ell^2(\mathbb{Z})$.

The assumption that $\{T_k \phi\}_{k \in \mathbb{Z}}$ is a Riesz sequence implies that the functions in $\mathcal{H}$ are well-defined in $L^2(\mathbb{R})$-sense and that $\{T_k \phi\}_{k \in \mathbb{Z}}$ is a Riesz basis for $\mathcal{H}$. In order to show that the functions in $\mathcal{H}$ are continuous, let $\{c_k\}_{k \in \mathbb{Z}} \in \ell^2(\mathbb{Z})$ be given. Then, for any given $\ell \in \mathbb{Z}$,

$$\sum_{k \in \mathbb{Z}} \sup_{x \in [\ell, \ell+1]} |c_k T_k \phi(x)| \leq \left(\sum_{k \in \mathbb{Z}} |c_k|^2 \right)^{1/2} \left(\sum_{k \in \mathbb{Z}} \sup_{x \in [\ell, \ell+1]} |\phi(x - k)|^2 \right)^{1/2}$$

$$\leq \left(\sum_{k \in \mathbb{Z}} |c_k|^2 \right)^{1/2} \left(\sum_{k \in \mathbb{Z}} a_k^2 \right)^{1/2} < \infty.$$

This implies that the function

$$f(x) := \sum_{k \in \mathbb{Z}} c_k T_k \phi(x) \tag{9.36}$$

is continuous on each interval $[\ell, \ell + 1]$, $\ell \in \mathbb{Z}$, and hence continuous on $\mathbb{R}$.

In order to show that $\mathcal{H}$ is an RKHS, consider a function $f \in \mathcal{H}$, written on the form (9.36) for some $\{c_k\}_{k \in \mathbb{Z}} \in \ell^2(\mathbb{Z})$. Then for any $\ell \in \mathbb{Z}$,

$$\sup_{x \in [0,1]} |f(x + \ell)| \leq \sum_{k \in \mathbb{Z}} |c_k| \sup_{x \in [0,1]} |\phi(x + \ell - k)| = (|c| * a)_\ell,$$

where $(|c| * a)_\ell$ denotes the ℓth coordinate of the convolution between the sequences $|c| := \{|c_k|\}_{k \in \mathbb{Z}}$ and $a := \{a_k\}_{k \in \mathbb{Z}}$. Given any $x \in \mathbb{R}$,

$$|f(x)|^2 \leq \sum_{\ell \in \mathbb{Z}} \sup_{x \in [0,1]} |f(x + \ell)|^2;$$

using Young's inequality and letting A denote a lower bound for $\{T_k \phi\}_{k \in \mathbb{Z}}$,

$$|f(x)|^2 \leq ||c * a||^2_{\ell^2(\mathbb{Z})} \leq ||c||^2_{\ell^2(\mathbb{Z})} ||a||^2_{\ell^1(\mathbb{Z})} \leq \frac{1}{A} ||a||^2_{\ell^1(\mathbb{Z})} ||f||^2.$$

This proves that the point evaluation $f \mapsto f(x)$ is continuous from $\mathcal{H}$ to $\mathbb{C}$, i.e., $\mathcal{H}$ is indeed an RKHS. Using the frame decomposition (5.8) on the Riesz sequence $\{T_k \phi\}_{k \in \mathbb{Z}}$ and the canonical dual $\{T_k \widetilde{\phi}\}_{k \in \mathbb{Z}}$, the reproducing kernel is

$$K(x, y) = K_y(x) = \sum_{k \in \mathbb{Z}} \langle K_y, T_k \phi \rangle T_k \widetilde{\phi}(x)$$

$$= \sum_{k \in \mathbb{Z}} \overline{T_k \phi(y)} T_k \widetilde{\phi}(x)$$

$$= \sum_{k \in \mathbb{Z}} \overline{\phi(y - k)} \widetilde{\phi}(x - k).$$

This completes the proof.   □

Note that the condition (i) in Lemma 9.7.2 will play a role in the treatment of Gabor systems; see Section 11.5.

We will now prove that under the conditions in Lemma 9.7.2, the functions $f \in \mathcal{H}$ can indeed be reconstructed based on a set of sampling $\{f(\lambda_k)\}_{k \in \mathbb{Z}}$.

Theorem 9.7.3 *Under the setup in Lemma 9.7.2, assume further that* $\{\lambda_k\}_{k\in\mathbb{Z}}$ *is a set of sampling for the Hilbert space* $\mathcal{H}$ *in* (9.35). *Then:*

(i) *The frame* $\{K_{\lambda_k}\}_{k\in\mathbb{Z}}$ *is given explicitly by*

$$K_{\lambda_k}(x) = \sum_{n\in\mathbb{Z}} \overline{\phi(\lambda_k - n)}\widetilde{\phi}(x - n). \tag{9.37}$$

(ii) *Let* S *denote the frame operator for* $\{K_{\lambda_k}\}_{k\in\mathbb{Z}}$. *Then any* $f \in \mathcal{H}$ *can be reconstructed from the samples* $\{f(\lambda_k)\}_{k\in\mathbb{Z}}$ *via*

$$f = \sum_{k\in\mathbb{Z}} f(\lambda_k) S^{-1} K_{\lambda_k}.$$

Proof.   The result in (i) is a direct consequence of Lemma 9.7.2. Furthermore, the frame decomposition (5.8) shows that for any $f \in \mathcal{H}$,

$$f = \sum_{k\in\mathbb{Z}} \langle f, K_{\lambda_k}\rangle S^{-1} K_{\lambda_k} = \sum_{k\in\mathbb{Z}} f(\lambda_k) S^{-1} K_{\lambda_k};$$

this proves (ii).   □

We refer to the papers [634, 494] by Walter et. al. for results about sampling in wavelet subspaces and to the survey paper [7] by Aldroubi and Gröchenig for more information about sampling in shift-invariant spaces.

In the rest of this section we will describe an application of frame theory which takes place in the Paley–Wiener space PW. Already in Theorem 3.10.1 we saw that any continuous function $f \in PW$ can be recovered from its samples $\{f(k)\}_{k\in\mathbb{Z}}$ via

$$f(x) = \sum_{k\in\mathbb{Z}} f(k)\operatorname{sinc}(x - k). \tag{9.38}$$

As discussed in Section 8.5, one will always encounter quantization errors when this result is applied in practice. We will now show that the effect of the quantization error can be reduced via *oversampling*, i.e., by invoking samples $\{f(k/M)\}_{k\in\mathbb{Z}}$ for some $M \in \mathbb{N}, M > 1$. Note that for the sequence $\{f(k/M)\}_{k\in\mathbb{Z}}$, the distance between two consecutive samples is $1/M$.

First, as noticed in Theorem 3.10.1, the expansion (9.38) is really an expansion in terms of the orthonormal basis $\{\operatorname{sinc}(\cdot - k)\}_{k\in\mathbb{Z}}$ for the Paley–Wiener space PW. Given any $M \in \mathbb{N}$, it follows that for each $m = 0, \ldots, M - 1$, the family $\{\operatorname{sinc}(\cdot - k - m/M)\}_{k\in\mathbb{Z}}$ also forms an orthonormal basis for PW. The union of these bases, i.e.,

$$\bigcup_{m=0}^{M-1} \{\operatorname{sinc}(\cdot - k - m/M)\}_{k\in\mathbb{Z}} = \bigcup_{m=0}^{M-1} \{\operatorname{sinc}(\cdot - k/M)\}_{k\in\mathbb{Z}} = \bigcup_{m=0}^{M-1} \{T_{\frac{k}{M}}\operatorname{sinc}\}_{k\in\mathbb{Z}},$$

therefore forms a tight frame for PW with frame bound $A = M$. Thus, for each $f \in PW$,

$$f = \frac{1}{M} \sum_{k \subset \mathbb{Z}} \langle f, \mathrm{sinc}(\cdot - k/M) \rangle \, \mathrm{sinc}(\cdot - k/M). \tag{9.39}$$

Note that because $\{\mathit{sinc}(\cdot - k)\}_{k \in \mathbb{Z}}$ forms an orthonormal basis for PW, we know from (9.38) that $f(k) = \langle f, T_k \mathrm{sinc} \rangle$; this implies that

$$\langle f, \mathrm{sinc}(\cdot - k/M) \rangle \;=\; \int_{-\infty}^{\infty} f(x) \, \mathrm{sinc}(x - k/M) \; dx$$

$$=\; \int_{-\infty}^{\infty} T_{-k/M} f(x) \, \mathrm{sinc}(x) \; dx = T_{-k/M} f(0) = f(k/M).$$

Thus, the expansion (9.39) takes the form

$$f = \frac{1}{M} \sum_{k \in \mathbb{Z}} f(k/M) \, \mathrm{sinc}(\cdot - k/M). \tag{9.40}$$

We now adopt the model for quantization errors discussed in Section 8.5. In particular, Proposition 8.5.1 shows that oversampling with the factor M, i.e., the use of (9.40) instead of (9.38), reduces the energy of the quantization noise by a factor M:

$$E|(Qw)_k|^2 = \frac{\sigma^2}{A} = \frac{\sigma^2}{M}.$$

The practical relevance of this result is that it usually is easier to increase the redundancy of the frame than to increase the quantization precision.

9.8   Frames of Exponentials

Recall that the complex exponential functions $\left\{ \frac{1}{\sqrt{2\pi}} e^{ikx} \right\}_{k \in \mathbb{Z}}$ constitute an orthonormal basis for $L^2(-\pi, \pi)$. Thus, $\{e^{ikx}\}_{k \in \mathbb{Z}}$ is a frame for $L^2(-\pi, \pi)$ with bounds $A = B = 2\pi$. More generally, given an interval $I \subset \mathbb{R}$ and a real sequence $\{\lambda_k\}_{k \in \mathbb{Z}}$, a frame for $L^2(I)$ of the form $\{e^{i\lambda_k x}\}_{k \in \mathbb{Z}}$ is called a *frame of exponentials* or a *Fourier frame*. Note that the exponentials are not square integrable on an unbounded interval, so we necessarily have $|I| < \infty$. An expansion

$$f(x) = \sum c_k e^{i\lambda_k x}$$

in $L^2(I)$ is called a *nonharmonic Fourier series*. As noticed at the beginning of the chapter, this is the context in which frames were originally defined.
    For a given sequence $\Lambda = \{\lambda_k\}_{k \in \mathbb{Z}}$, the *frame radius* is defined by

$$R(\Lambda) = \sup \left\{ R > 0 \mid \{e^{i\lambda_k x}\}_{k \in \mathbb{Z}} \text{ is a frame for } L^2(-R, R) \right\}.$$

If $\{e^{i\lambda_k x}\}_{k\in\mathbb{Z}}$ is a frame for $L^2(-R,R)$ for some $R>0$, it is automatically a frame for $L^2(-R',R')$ for all $R'\in]0,R]$ (Exercise 9.15). Writing

$$\mathbb{R}^+=]0,R(\Lambda)[\ \cup\ \{R(\Lambda)\}\ \cup\ ]R(\Lambda),\infty[,$$

we therefore have that:

- $\{e^{i\lambda_k x}\}_{k\in\mathbb{Z}}$ is a frame for $L^2(-R,R)$ whenever $R\in]0,R(\Lambda)[$.
- $\{e^{i\lambda_k x}\}_{k\in\mathbb{Z}}$ is not a frame for $L^2(-R,R)$ if $R\in]R(\Lambda),\infty[$.

The case $R=R(\Lambda)$ itself is critical: there are cases where $\{e^{i\lambda_k x}\}_{k\in\mathbb{Z}}$ is a frame for $L^2(-R(\Lambda),R(\Lambda))$ and cases where it is not.

A separated sequence $\{\lambda_k\}_{k\in\mathbb{Z}}$ is said to have *uniform density* $d>0$ if there exists a number $L>0$ such that

$$\left|\lambda_k-\frac{k}{d}\right|\le L,\ \forall k\in\mathbb{Z}. \tag{9.41}$$

Duffin and Schaeffer proved the following impressive theorem. We encourage the reader to consult the original paper [262] for the proof:

Theorem 9.8.1 *Assume that $\{\lambda_k\}_{k\in\mathbb{Z}}$ is a separated sequence with uniform density $d>0$. Then $\{\lambda_k\}_{k\in\mathbb{Z}}$ has frame radius at least πd.*

Theorem 9.8.1 can naturally be considered as a perturbation result. In fact, if we consider a fixed $d>0$, then $\{e^{ikx/d}\}_{k\in\mathbb{Z}}$ is a frame for $L^2(-R,R)$ for any $R\in]0,\pi d]$ (Exercise 9.10); Theorem 9.8.1 now tells us that if $\{\lambda_k\}_{k\in\mathbb{Z}}$ is separated and (9.41) is satisfied, then $\{e^{i\lambda_k x}\}_{k\in\mathbb{Z}}$ is a frame for $L^2(-R,R)$ for any $R\in]0,\pi d[$. It is immediately clear that we cannot expect $\{e^{i\lambda_k x}\}_{k\in\mathbb{Z}}$ to be a frame for $L^2(-R,R)$ for $R>\pi d$ and that it also might fail for $R=\pi d$ is more subtle.

Our purpose is to find conditions on a sequence $\{\lambda_k\}_{k\in\mathbb{Z}}$ and an interval I such that $\{e^{i\lambda_k x}\}_{k\in\mathbb{Z}}$ is a frame for $L^2(I)$. We begin with the Bessel condition.

Lemma 9.8.2 *Let $\{\lambda_k\}_{k\in\mathbb{Z}}$ be a real sequence. Then the following are equivalent:*

(i) *$\{\lambda_k\}_{k\in\mathbb{Z}}$ is relatively separated.*

(ii) *$\{e^{i\lambda_k x}\}_{k\in\mathbb{Z}}$ is a Bessel sequence in $L^2(-\pi,\pi)$.*

(iii) *$\{e^{i\lambda_k x}\}_{k\in\mathbb{Z}}$ is a Bessel sequence in $L^2(I)$ for any bounded interval $I\subset\mathbb{R}$.*

Proof. A proof of (ii) $\Leftrightarrow$(iii) is outlined in Exercise 9.11. That (ii)$\Rightarrow$(i) can be proved by an argument similar to the one used in Theorem 9.6.1; alternatively, it *follows* from Proposition 9.6.1 (Exercise 9.12). A proof of (i) $\Rightarrow$ (iii) can be found in [622]. We also note that this implication actually follows from Theorem 9.8.1. In fact, it is enough to prove that $\{e^{i\lambda_k x}\}_{k\in\mathbb{Z}}$

is a Bessel sequence if $\{\lambda_k\}_{k\in\mathbb{Z}}$ is separated. Assuming that $\{\lambda_k\}_{k\in\mathbb{Z}}$ is separated, we order $\{\lambda_k\}_{k\in\mathbb{Z}}$ increasingly; we denote the reordered sequence by $\{\lambda_k\}_{k\in K}$. Depending on the given sequence $\{\lambda_k\}_{k\in\mathbb{Z}}$, the index set K can be either $\mathbb{Z}, \mathbb{N}$, or $\mathbb{N}^- = \{-1, -2, \dots\}$. Now,

$$|\lambda_{k+1} - \lambda_k| \geq \delta > 0,$$

for some δ; therefore,

$$\left|\frac{\lambda_{k+1}}{\delta} - \frac{\lambda_k}{\delta}\right| \geq 1.$$

By enlarging $\{\lambda_k\}_{k\in K}$ if necessary, we can obtain a separated sequence $\{\mu_k\}_{k\in\mathbb{Z}}$, which, by choosing the ordering and indexing appropriately, satisfies that

$$\left|k - \frac{\mu_k}{\delta}\right| \leq 1, \ \forall k \in \mathbb{Z}.$$

By Theorem 9.8.1, the sequence $\{e^{i\mu_k x/\delta}\}_{k\in\mathbb{Z}}$ is a frame for $L^2(I)$ when I is sufficiently small. Therefore, $\{e^{i\mu_k x}\}_{k\in\mathbb{Z}}$ is a frame for $L^2(I)$ when I is sufficiently small (Exercise 9.13); since $\{\lambda_k\}_{k\in\mathbb{Z}}$ is a subsequence of $\{\mu_k\}_{k\in\mathbb{Z}}$, this implies that $\{e^{i\lambda_k x}\}_{k\in\mathbb{Z}}$ is a Bessel sequence in $L^2(I)$.   □

We are now ready to prove a characterization of exponential frames due to Jaffard [414].

Theorem 9.8.3 *Let $\{\lambda_k\}_{k\in\mathbb{Z}}$ be a real sequence. Then the following are equivalent:*

(i) *There exists an interval I such that $\{e^{i\lambda_k x}\}_{k\in\mathbb{Z}}$ is a frame for $L^2(I)$.*

(ii) *$\{\lambda_k\}_{k\in\mathbb{Z}}$ is the disjoint union of a sequence $\{\lambda_k\}_{k\in I_1}$ with a uniform density $d_1 > 0$ and a relatively separated sequence $\{\lambda_k\}_{k\in\mathbb{Z}\setminus I_1}$.*

If (ii) holds, then $\{e^{i\lambda_k x}\}_{k\in\mathbb{Z}}$ is a frame for $L^2(I)$ for any interval I with $|I| < 2\pi d_1$.

Proof. Assume that $\{e^{i\lambda_k x}\}_{k\in\mathbb{Z}}$ is a frame for $L^2(I)$ for some interval I. Then $\{\lambda_k\}_{k\in\mathbb{Z}}$ is relatively separated by Lemma 9.8.2; by Lemma 9.1.3, this implies that for each integer $N \in \mathbb{N}$ we can find a finite number C_N such that each interval of the type $[kN, (k+1)N[, \ k \in \mathbb{Z}$ contains at most C_N elements from $\{\lambda_k\}_{k\in\mathbb{Z}}$. By choosing N sufficiently large, we can assure that each interval $[kN, (k+1)N[, \ k \in \mathbb{Z}$ contains *at least* one element from $\{\lambda_k\}_{k\in\mathbb{Z}}$; this is not trivial, but here is an argument. Assume the opposite, i.e., that for each $N \in \mathbb{N}$ we could find an interval $[\ell N, (\ell + 1)N[, \ \ell \in \mathbb{Z}$, which does not contain any element from $\{\lambda_k\}_{k\in\mathbb{Z}}$. Letting

$$f_N(x) := e^{i(\ell+1/2)Nx},$$

it follows from Exercise 9.14 that

$$
\begin{aligned}
|\langle f_N, e^{i\lambda_k x}\rangle|^2 &= \left| \int_I e^{i((\ell+1/2)N - \lambda_k)x} dx \right|^2 \\
&= \left| \frac{2}{\lambda_k - (\ell+1/2)N} \sin\left(\frac{\lambda_k - (\ell+1/2)N}{2} |I| \right) \right|^2 \\
&\le \frac{4}{|\lambda_k - (\ell+1/2)N|^2}.
\end{aligned}
$$

Now consider an interval $[n, n+1[$, $n \in \mathbb{Z}$. If $\lambda_k \in [n, n+1[$ for some $k \in \mathbb{Z}$, then the opposite triangle inequality shows that

$$
\begin{aligned}
|\lambda_k - (\ell+1/2)N| &= |(n - (\ell+1/2)N) - (n - \lambda_k)| \\
&\ge |n - (\ell+1/2)N| - 1.
\end{aligned}
$$

Using the above notation, at most C_1 elements from $\{\lambda_k\}_{k\in\mathbb{Z}}$ belong to an interval $[n, n+1[$. Also, if $N > 4$ and $|n - (\ell+1/2)N| < \frac{N}{4}$, the interval $[n, n+1[$ is contained in $[\ell N, (\ell+1)N[$, and therefore $[n, n+1[$ contains by assumption no element from $\{\lambda_k\}_{k\in\mathbb{Z}}$ in this case. Putting all information together, we obtain that for $N > 4$,

$$
\begin{aligned}
\sum_{k\in\mathbb{Z}} |\langle f_N, e^{i\lambda_k x}\rangle|^2 &= \sum_{n\in\mathbb{Z}} \sum_{\{k:\ \lambda_k\in[n,n+1[\}} |\langle f_N, e^{i\lambda_k x}\rangle|^2 \\
&\le \sum_{\{n:\ |n-(\ell+1/2)N|\ge\frac{N}{4}\}} \sum_{\{k:\ \lambda_k\in[n,n+1[\}} \frac{4}{|\lambda_k - (\ell+1/2)N|^2} \\
&\le \sum_{\{n:\ |n-(\ell+1/2)N|\ge\frac{N}{4}\}} \frac{4C_1}{(|n - (\ell+1/2)N| - 1)^2}.
\end{aligned}
$$

So for $N > 8$,

$$
\sum_{k\in\mathbb{Z}} |\langle f_N, e^{i\lambda_k x}\rangle|^2 \le 4C_1 \sum_{\{n:\ |n|\ge N/4\}} \frac{1}{(|n| - 2)^2}
$$
$$
\to 0 \text{ as } N \to \infty.
$$

Since $\|f_N\| = \sqrt{|I|}$ for all $N \in \mathbb{N}$, it follows that the lower frame condition is violated. However, this contradicts our starting hypothesis that $\{e^{i\lambda_k x}\}_{k\in\mathbb{Z}}$ is a frame for $L^2(I)$. This proves the claim that for N chosen sufficiently large, each interval $[kN, (k+1)N[$, $k \in \mathbb{Z}$ contains at least one element from $\{\lambda_k\}_{k\in\mathbb{Z}}$.

Based on this, we can now pick a subsequence of $\{\lambda_k\}_{k\in\mathbb{Z}}$ having a uniform density. In fact, by choosing N large enough we can for each interval of the form $[2kN, (2k+1)N[$, $k \in \mathbb{Z}$ pick one element from $\{\lambda_k\}_{k\in\mathbb{Z}}$ belonging to the interval; this way we obtain a sequence $\{\mu_k\}_{k\in\mathbb{Z}} = \{\lambda_k\}_{k\in I_1}$ where the elements are separated by N and for which

$$
|\mu_k - 2kN| \le N, \ \forall k \in \mathbb{Z}.
$$

Finally, we have to prove that the remaining sequence $\{\lambda_k\}_{k\in\mathbb{Z}} \setminus \{\mu_k\}_{k\in\mathbb{Z}}$ is relatively separated. One way to obtain a separated subsequence from $\{\lambda_k\}_{k\in\mathbb{Z}} \setminus \{\mu_k\}_{k\in\mathbb{Z}}$ is, for each $k \in \mathbb{Z}$, to pick one element from the sequence belonging to each interval $[2kN, (2k+1)N[$ (if there is any); another separated subsequence is obtained by picking one element from each interval $[(2k+1)N, (2k+2)N[, k \in \mathbb{Z}$. After repeating these two procedures at most C_N times, no more elements from $\{\lambda_k\}_{k\in\mathbb{Z}} \setminus \{\mu_k\}_{k\in\mathbb{Z}}$ are left. This proves that $\{\lambda_k\}_{k\in\mathbb{Z}} \setminus \{\mu_k\}_{k\in\mathbb{Z}}$ is relatively separated.

For the proof of (ii) $\Rightarrow$ (i), we assume that there is a partition

$$\mathbb{Z} = I_1 \cup I_2,$$

such that $\{\lambda_k\}_{k\in I_1}$ has a uniform density $d_1 > 0$ and $\{\lambda_k\}_{k\in I_2}$ is relatively separated. By Theorem 9.8.1, the sequence $\{e^{i\lambda_k x}\}_{k\in I_1}$ is a frame for $L^2(-R, R)$ if $R \in]0, \pi d_1[$, and by Lemma 9.8.2 $\{e^{i\lambda_k x}\}_{k\in I_2}$ is a Bessel sequence. Therefore, $\{e^{i\lambda_k x}\}_{k\in\mathbb{Z}}$ is a frame for $L^2(-R, R)$ when $R \in]0, \pi d_1[$. By Exercise 9.15, this implies that $\{e^{i\lambda_k x}\}_{k\in\mathbb{Z}}$ is a frame for $L^2(I)$ for any interval I with $|I| < 2\pi d_1$.   □

The formulation in Theorem 9.8.3 is very convenient in order to prove that $\{e^{i\lambda_k x}\}_{k\in\mathbb{Z}}$ is a frame for a given sequence $\{\lambda_k\}_{k\in\mathbb{Z}}$. In the original paper by Jaffard, Theorem 9.8.3 is just one step toward the main result, where the frame radius is determined for any sequence $\{\lambda_k\}_{k\in\mathbb{Z}}$. Seip [570] gave a very elegant version of this final result in terms of the lower density.

Theorem 9.8.4 *For $\{e^{i\lambda_k x}\}_{k\in\mathbb{Z}}$ to be a frame for $L^2(-\pi, \pi)$, it is necessary that $\{\lambda_k\}_{k\in\mathbb{Z}}$ is relatively separated and $D^-(\{\lambda_k\}_{k\in\mathbb{Z}}) \geq 1$, and it is sufficient that $\{\lambda_k\}_{k\in\mathbb{Z}}$ is relatively separated and $D^-(\{\lambda_k\}_{k\in\mathbb{Z}}) > 1$.*

Seip also proves that if $\{\lambda_k\}_{k\in\mathbb{Z}}$ is separated and $D^-(\{\lambda_k\}_{k\in\mathbb{Z}}) > 1$, then $\{e^{i\lambda_k x}\}_{k\in\mathbb{Z}}$ contains a Riesz basis.

Theorem 9.8.4 is optimal in the sense that no conclusion is possible if $D^-(\{\lambda_k\}_{k\in\mathbb{Z}}) = 1$. For example, Seip proves that the sequence

$$\{\lambda_k\} = \left\{ k(1 - |k|^{-1/2}) \right\}_{|k|>1}$$

has density 1 and that $\{e^{i\lambda_k x}\}$ is a frame for $L^2(-\pi, \pi)$. On the other hand, a famous example by Kadec, namely, the sequence $\{\lambda_k\}_{k\in\mathbb{Z}}$ given by

$$\lambda_k := \begin{cases} k - 1/4 & \text{if } k > 0 \\ k + 1/4 & \text{if } k < 0, \\ 0 & \text{if } k = 0 \end{cases} \tag{9.42}$$

also has density 1, however, without generating a frame for $L^2(-\pi, \pi)$. For a discussion of this example, we refer to [622].

For a sequence $\{f_k\}_{k=1}^{\infty}$ in a general Hilbert space $\mathcal{H}$, the upper and lower Riesz conditions in (3.29) are unrelated (Exercise 3.12). In the present context, the situation is different: for a sequence $\{\lambda_k\}_{k\in\mathbb{Z}}$ consisting of distinct points, the existence of a lower Riesz bound for $\{e^{i\lambda_k x}\}_{k\in\mathbb{Z}}$ in $L^2(-\pi,\pi)$ implies that $\{e^{i\lambda_k x}\}_{k\in\mathbb{Z}}$ is a Bessel sequence. That is, the lower condition is enough to guarantee that $\{e^{i\lambda_k x}\}_{k\in\mathbb{Z}}$ is a Riesz basis for its closed span. This result was originally discovered by Young [623]. An elegant direct argument was later given by Lindner [497], and we repeat it here.

Theorem 9.8.5 *Suppose that the sequence* $\{\lambda_k\}_{k\in\mathbb{Z}}$ *consists of distinct points and that there exists a constant* $A > 0$ *such that*

$$A\sum_{k\in\mathbb{Z}}|c_k|^2 \leq \left\|\sum_{k\in\mathbb{Z}}c_k e^{i\lambda_k x}\right\|^2_{L^2(-\pi,\pi)} \tag{9.43}$$

for all finite scalar sequences $\{c_k\}_{k\in\mathbb{Z}}$. *Then* $\{e^{i\lambda_k x}\}_{k\in\mathbb{Z}}$ *is a Riesz basis for its closed span in* $L^2(-\pi,\pi)$.

Proof. Consider λ_k, λ_j, where $k \neq j$. By assumption,

$$2A = A(|1|^2 + |-1|^2) \leq \left\|e^{i\lambda_k x} - e^{i\lambda_j x}\right\|^2$$
$$= \int_{-\pi}^{\pi}|1 - e^{i(\lambda_k-\lambda_j)x}|^2 dx. \tag{9.44}$$

Using the expansion

$$e^{iy} = \sum_{k=0}^{\infty}\frac{(iy)^k}{k!} = 1 + \sum_{k=1}^{\infty}\frac{(iy)^k}{k!}, \ y \in \mathbb{R},$$

it follows that for $x \in]-\pi,\pi[$,

$$|1 - e^{i(\lambda_k-\lambda_j)x}| = \left|-\sum_{k=1}^{\infty}\frac{[i(\lambda_k-\lambda_j)x]^k}{k!}\right|$$
$$\leq \sum_{k=1}^{\infty}\frac{|(\lambda_k-\lambda_j)\pi|^k}{k!} = e^{|\lambda_k-\lambda_j|\ \pi} - 1.$$

Therefore, (9.44) shows that $2A \leq 2\pi(e^{|\lambda_k-\lambda_j|\ \pi} - 1)^2$; this implies that

$$|\lambda_k - \lambda_j| \geq \frac{1}{\pi}\ln(\sqrt{\frac{A}{\pi}} + 1).$$

Thus, $\{\lambda_k\}_{k\in\mathbb{Z}}$ is separated, and therefore $\{e^{i\lambda_k x}\}_{k\in\mathbb{Z}}$ is a Bessel sequence in $L^2(-\pi,\pi)$ by Lemma 9.8.2. $\qquad\square$

The classical *Kadec's 1/4-theorem* states that if $\{\lambda_k\}_{k\in\mathbb{Z}}$ is a real sequence for which $\sup_{k\in\mathbb{Z}}|\lambda_k - k| < \frac{1}{4}$, then $\{e^{i\lambda_k x}\}_{k\in\mathbb{Z}}$ is a Riesz basis for $L^2(-\pi,\pi)$. The example (9.42) of Kadec shows that the conclusion fails if $\sup|\lambda_k - k| = \frac{1}{4}$. Combining the proof of Kadec's theorem in [622] with perturbation results for frames in Section 22.1, it is an easy matter to extend Kadec's theorem to frames; we refer to the original papers by Balan [26] and Christensen [159] for details.

Theorem 9.8.6 *Let* $\{\lambda_k\}_{k\in\mathbb{Z}}, \{\mu_k\}_{k\in\mathbb{Z}}$ *be real sequences. Suppose that* $\{e^{i\mu_k x}\}_{k\in\mathbb{Z}}$ *is a frame for* $L^2(-\pi,\pi)$ *with bounds* A, B*. If there exists a constant* $L < 1/4$ *such that*

$$|\mu_k - \lambda_k| \leq L \ \forall k \in \mathbb{Z}, \quad and \quad 1 - cos(\pi L) + sin(\pi L) < \sqrt{\frac{A}{B}},$$

then $\{e^{i\lambda_k x}\}_{k\in\mathbb{Z}}$ *is a frame for* $L^2(-\pi,\pi)$ *with bounds*

$$A\left(1 - \sqrt{\frac{B}{A}}(1 - cos(\pi L) + sin(\pi L))\right)^2, \quad B(2 - cos(\pi L) + sin(\pi L))^2.$$

Compared with the Kadec 1/4-theorem, the advantages of Theorem 9.8.6 are twofold: it applies to frames, and we obtain estimates for the frame bounds. Good values for the frame bounds are essential for estimates of the speed of convergence in algorithms involving frames, as we have seen already in Section 1.3. In the paper [384], He, Key, and Volkmer used Theorem 9.8.6 to construct Riesz bases for weighted L^2-spaces consisting of solutions to certain Sturm–Liouville problems.

Let us end this section with a few words about the lower frame bounds for finite sets of exponentials. The following result was proved in [193]; it is based on an impressive analysis by Lindner in [496], where lower bounds were obtained for certain infinite sets of exponentials.

Proposition 9.8.7 *Let* $\lambda_1,\ldots,\lambda_M$ *be a finite sequence of distinct real numbers. Choose a separation constant* $\delta \leq 1$*. Then* $\{e^{i\lambda_k(\cdot)}\}_{k=1}^M$ *is a Riesz basis for its linear span in* $L^2(-\pi,\pi)$*, with lower frame bound*

$$A_M = 1.6 \cdot 10^{-14} \left(\frac{\delta}{2}\right)^{2M+1} \frac{1}{((M+1)!)^8}. \tag{9.45}$$

Note that the bounds in (9.45) are extremely small and tend very fast to zero when the number M increases. We will return to this issue in the context of finite Gabor systems, where lower bounds are derived based on Proposition 9.8.7; see page 345.

9.9   Exercises

9.1 Let $\phi \in L^2(\mathbb{R}) \setminus \{0\}$ and let $\{\lambda_k\}_{k\in\mathbb{Z}}$ be a sequence in $\mathbb{R}$. Show by a direct argument that $\{T_{\lambda_k}\phi\}_{k\in\mathbb{Z}}$ cannot be a frame sequence if $\{\lambda_k\}_{k\in\mathbb{Z}}$ has an accumulation point.

9.2 Prove (9.4) in the proof of Lemma 9.2.2.

9.3 Prove Lemma 9.2.4.

9.4 In this exercise, we ask the reader to provide some details in the proof of Theorem 9.2.5.

(i) Prove (9.9).

(ii) Prove the equivalence between (9.10) and the statement preceding it.

9.5 Prove Corollary 9.3.5.

9.6 Prove Corollary 9.5.3. (Hint: use the argument from the proof of Corollary 9.5.2.)

9.7 Let Φ be a positive trigonometric polynomial. Show that if $\Phi(\cdot)^{-1}$ is a trigonometric polynomial, then Φ is a constant.

9.8 This exercise connects to Exercises 3.12, 5.2. Let $\mathcal{H}$ be a separable Hilbert space.

(i) Find a sequence $\{f_k\}_{k=1}^{\infty}$ in $\mathcal{H}$ which satisfies the lower frame condition but not the upper frame condition.

(ii) Find a sequence $\{f_k\}_{k=1}^{\infty}$ of vectors with norm 1, which satisfies the lower frame condition, but not the upper frame condition.

(iii) Suppose that $\{e^{i\lambda_k x}\}$ satisfies the lower frame condition in $L^2(-\pi, \pi)$. Does it follow that $\{e^{i\lambda_k x}\}$ is a frame? Compare with Theorem 9.8.5.

9.9 Assume that ϕ has compact support and that there exist constants $a, b > 0$ such that $a \leq |\phi(x)| \leq b$ for a.e. $x \in \text{supp } \phi$. Prove that $\{T_{\lambda_k}\phi\}_{k\in\mathbb{Z}}$ is an orthogonal sequence for all sequences $\{\lambda_k\}_{k\in\mathbb{Z}}$ for which the sets $\{\lambda_k + \text{supp } \phi\}_{k\in\mathbb{Z}}$ are disjoint. How can the assumptions be modified to obtain a Riesz sequence which is not orthogonal?

9.10 Let $d > 0$, and prove that $\{e^{ikx/d}\}_{k\in\mathbb{Z}}$ is a frame for $L^2(-R, R)$ if and only if $R \in]0, \pi d]$.

9.11 Let I and J be arbitrary bounded intervals in $\mathbb{R}$ and $\{\lambda_k\}_{k\in\mathbb{Z}}$ a real sequence. Our purpose is to prove that $\{e^{i\lambda_k x}\}_{k\in\mathbb{Z}}$ is a Bessel sequence in $L^2(I)$ if and only if it is a Bessel sequence in $L^2(J)$; that is, the Bessel condition is independent of the considered finite interval. One way to proceed is to assume that $\{e^{i\lambda_k x}\}_{k\in\mathbb{Z}}$ is a Bessel sequence in $L^2(I)$ and prove the following:

(i) $\{e^{i\lambda_k x}\}_{k\in\mathbb{Z}}$ is a Bessel sequence in $L^2(a+I)$ for any $a \in \mathbb{R}$.

(ii) $\{c^{i\lambda_k x}\}_{k\in\mathbb{Z}}$ is a Bessel sequence in $L^2(I_1)$ for any interval $I_1 \subset I$.

(iii) $\{e^{i\lambda_k x}\}_{k\in\mathbb{Z}}$ is a Bessel sequence in $L^2(I\cup(a+I))$ for any $a \in \mathbb{R}$. Covering J with a finite number of translates of I, we can now conclude that $\{e^{i\lambda_k x}\}_{k\in\mathbb{Z}}$ is a Bessel sequence in $L^2(J)$.

9.12 Let $\{\lambda_k\}_{k\in\mathbb{Z}}$ be a sequence in $\mathbb{R}$ and assume that $\{e^{i\lambda_k x}\}_{k\in\mathbb{Z}}$ is a Bessel sequence in $L^2(-\pi,\pi)$. Define the function ϕ through $\widehat{\phi} = \chi_{[-\pi,\pi]}$, and prove that:

(i) $\{E_{\frac{\lambda_k}{2\pi}}\widehat{\phi}\}_{k\in\mathbb{Z}}$ is a Bessel sequence in $L^2(\mathbb{R})$.

(ii) $\{T_{-\frac{\lambda_k}{2\pi}}\phi\}_{k\in\mathbb{Z}}$ is a Bessel sequence in $L^2(\mathbb{R})$.

(iii) $\{\lambda_k\}_{k\in\mathbb{Z}}$ is relatively separated.

9.13 Consider an interval $[b,c] \subset \mathbb{R}$, and identify $L^2(b,c)$ with a subspace of $L^2(\mathbb{R})$. For a given $a > 0$, let D_a be the dilation operator. Prove the following:

(i) $D_a L^2(b,c) = L^2(ab,ac)$.

(ii) If $\{e^{i\lambda_k x}\}_{k\in\mathbb{Z}}$ is a frame for $L^2(b,c)$, then $\{e^{i\lambda_k x/a}\}_{k\in\mathbb{Z}}$ is a frame for $L^2(ab,ac)$.

9.14 Consider a bounded interval $[b,c] \subset \mathbb{R}$. Let $a \neq 0$ and prove that

$$\left|\int_b^c e^{iax}\,dx\right| = \frac{2}{|a|}\left|\sin\left(\frac{a(c-b)}{2}\right)\right|.$$

9.15 Assume that $\{e^{i\lambda_k x}\}_{k\in\mathbb{Z}}$ is a frame for $L^2(I)$ for some interval I. Prove that $\{e^{i\lambda_k x}\}_{k\in\mathbb{Z}}$ is also a frame for $L^2(J)$ for any interval J with $|J| \leq |I|$.

9.16 Calculate the oblique duals of $\{T_k \widetilde{B_n}\}_{k\in\mathbb{Z}}$ in Example 9.5.7 for $n = p = q = 2$ and $n = p = q = 3$ (use Maple or a similar program).

10

Shift-Invariant Systems in $L^2(\mathbb{R})$

Chapter 9 dealt with systems of functions generated by integer-translates of a single function in $L^2(\mathbb{R})$. We will now generalize this setup and consider translates of a given countable family of functions rather than just one function. Such systems of functions are called shift-invariant systems. Our goal is to characterize various frame properties for shift-invariant systems, a subject that was treated first in the paper [559] by Ron and Shen. The presentation is inspired by the approach by Janssen in [430]. The derived results will play an important role in the analysis of Gabor systems in Chapter 11.

The theory for shift-invariant systems is based on two classes of operators on $L^2(\mathbb{R})$, namely,

Translation by $a \in \mathbb{R}$, $T_a : L^2(\mathbb{R}) \to L^2(\mathbb{R})$, $(T_a f)(x) = f(x - a)$;

Modulation by $b \in \mathbb{R}$, $E_b : L^2(\mathbb{R}) \to L^2(\mathbb{R})$, $(E_b f)(x) = e^{2\pi i b x} f(x)$.

Both classes of operators were introduced in Section 2.9; in particular, we will use their interaction with the Fourier transform, a subject that is also treated in Section 2.9. We will return to shift-invariant systems in the setting of $L^2(\mathbb{R}^d)$ in Chapter 20; see also Sections 21.6–21.7.

10.1 Frame Properties of Shift-Invariant Systems

Let $\{g_m\}_{m \in I}$ be a countable collection of functions in $L^2(\mathbb{R})$ and $a > 0$ be a given (shift) parameter. The *shift-invariant system* generated by $\{g_m\}_{m \in I}$

© Springer International Publishing Switzerland 2016
O. Christensen, *An Introduction to Frames and Riesz Bases*,
Applied and Numerical Harmonic Analysis,
DOI 10.1007/978-3-319-25613-9_10

and a is the collection of functions $\{g_m(\cdot - na)\}_{m\in I, n\in\mathbb{Z}}$. Formulated in terms of the translation operator, the system has the form $\{T_{na}g_m\}_{m\in I, n\in\mathbb{Z}}$. Usually, we will let $I = \mathbb{Z}$, in which case we simply write

$$\{g_{nm}\} := \{T_{na}g_m\}_{m,n\in\mathbb{Z}}. \tag{10.1}$$

As already mentioned our goal is to characterize various frame properties for systems of the form $\{g_{nm}\}$. The Fourier transform will be an important tool; in fact, the characterizations will be formulated in terms of certain conditions on the functions $\widehat{g_m}$.

In particular, we will present equivalent conditions for two systems $\{g_{nm}\}$ and $\{h_{nm}\}$ to form dual frames. Given shift-invariant Bessel systems $\{g_{nm}\}$, $\{h_{nm}\}$ and two functions $e, f \in L^2(\mathbb{R})$, the analysis of the function $\rho(e, f)$ defined by

$$\rho(e, f) : \mathbb{R} \to \mathbb{C}, \ \rho(e, f)(x) = \sum_{m,n\in\mathbb{Z}} \langle T_x e, g_{nm}\rangle\langle h_{nm}, T_x f\rangle \tag{10.2}$$

will play a central role. The reason for considering this function is apparent from our discussion of general dual frame pairs in Section 6.3: in fact, Lemma 6.3.2 shows that two Bessel sequences $\{g_{nm}\}$ and $\{h_{nm}\}$ form dual frames for $L^2(\mathbb{R})$ if and only if

$$\rho(e, f)(0) = \langle e, f\rangle, \ \forall e, f \in L^2(\mathbb{R}).$$

In order to get information about the point evaluation $\rho(e, f)(0)$ based on the function values $\rho(e, f)(x)$ for x close to 0, we need to show that the function $\rho(e, f)$ is continuous; a substantial part of the technicalities in the current chapter will deal with exactly this point.

We first derive a useful consequence of the Bessel condition.

Lemma 10.1.1 *Assume that $\{g_{nm}\}$ is a Bessel sequence with bound B. Then*

$$\sum_{m\in\mathbb{Z}} |\widehat{g_m}(\nu)|^2 \le aB, \ a.e. \ \nu \in \mathbb{R}. \tag{10.3}$$

Proof. Let $f \in L^2(\mathbb{R})$, and consider the function

$$\rho(f, f)(x) = \sum_{m,n\in\mathbb{Z}} |\langle T_x f, g_{nm}\rangle|^2; \tag{10.4}$$

it corresponds to the general expression in (10.2) in the case $h_m = g_m$. The assumption that $\{g_{nm}\}$ is a Bessel sequence with bound B implies that $\rho(f, f)$ is bounded: in fact,

$$\rho(f, f)(x) \le B\,||T_x f||^2 = B\,||f||^2, \ \forall x \in \mathbb{R}. \tag{10.5}$$

The shift-invariance of the system $\{g_{nm}\}$ implies that $\rho(f, f)$ is periodic with period a (Exercise 10.1), so we can consider its Fourier expansion in

$L^2(0, a)$. By definition, the Fourier coefficient c_0 is given by

$$c_0 = \frac{1}{a} \int_0^a \sum_{m,n \in \mathbb{Z}} |\langle T_x f, g_{nm} \rangle|^2 \, dx \qquad (10.6)$$

$$= \frac{1}{a} \sum_{m,n \in \mathbb{Z}} \int_0^a \left| \int_{-\infty}^{\infty} f(z - x)\overline{g_m(z - na)} \, dz \right|^2 dx$$

$$= \frac{1}{a} \sum_{m,n \in \mathbb{Z}} \int_0^a \left| \int_{-\infty}^{\infty} f(z - (x - na))\overline{g_m(z)} \, dz \right|^2 dx.$$

Introducing the functions $\Phi_m(x)$, $m \in \mathbb{Z}$, by

$$\Phi_m(x) := \left| \int_{-\infty}^{\infty} f(z - x)\overline{g_m(z)} \, dz \right|^2 = |\langle T_x f, g_m \rangle|^2, \quad x \in \mathbb{R},$$

this can be written as

$$c_0 = \frac{1}{a} \sum_{m \in \mathbb{Z}} \sum_{n \in \mathbb{Z}} \int_0^a \Phi_m(x - na) \, dx$$

$$= \frac{1}{a} \sum_{m \in \mathbb{Z}} \int_{-\infty}^{\infty} \Phi_m(x) \, dx = \frac{1}{a} \sum_{m \in \mathbb{Z}} \int_{-\infty}^{\infty} |\langle T_x f, g_m \rangle|^2 dx.$$

By direct calculation, we see that for arbitrary functions $e, \phi \in L^2(\mathbb{R})$,

$$\langle T_x e, \phi \rangle = \langle \mathcal{F} T_x e, \mathcal{F} \phi \rangle = \langle E_{-x}\widehat{e}, \widehat{\phi} \rangle$$

$$= \int_{-\infty}^{\infty} \widehat{e}(\nu)\overline{\widehat{\phi}(\nu)} e^{-2\pi i x \nu} \, d\nu$$

$$= \mathcal{F}\left(\widehat{e} \, \overline{\widehat{\phi}} \right)(x). \qquad (10.7)$$

Thus, via Parseval's equation,

$$c_0 = \frac{1}{a} \sum_{m \in \mathbb{Z}} \int_{-\infty}^{\infty} \left| \mathcal{F}(\widehat{f} \, \overline{\widehat{g_m}})(\nu) \right|^2 d\nu = \frac{1}{a} \sum_{m \in \mathbb{Z}} \int_{-\infty}^{\infty} |\widehat{f}(\nu)\overline{\widehat{g_m}(\nu)}|^2 d\nu$$

$$= \frac{1}{a} \int_{-\infty}^{\infty} |\widehat{f}(\nu)|^2 \sum_{m \in \mathbb{Z}} |\widehat{g_m}(\nu)|^2 d\nu. \qquad (10.8)$$

On the other hand, via the definition of c_0 used in (10.6), an estimate of c_0 can be obtained:

$$c_0 = \frac{1}{a} \int_0^a \sum_{m,n \in \mathbb{Z}} |\langle T_x f, g_{nm} \rangle|^2 \, dx \leq B \, \|f\|^2 = B \int_{-\infty}^{\infty} |\widehat{f}(\nu)|^2 d\nu.$$

Thus, via (10.8) we see that

$$\frac{1}{a} \int_{-\infty}^{\infty} |\widehat{f}(\nu)|^2 \sum_{m \in \mathbb{Z}} |\widehat{g_m}(\nu)|^2 d\nu \leq B \int_{-\infty}^{\infty} |\widehat{f}(\nu)|^2 d\nu.$$

Since this holds for all $f \in L^2(\mathbb{R})$, we conclude that

$$\frac{1}{a} \sum_{m \in \mathbb{Z}} |\widehat{g_m}(\nu)|^2 \leq B, \quad a.e. \ \nu \in \mathbb{R};$$

the desired result now follows. □

We will now analyze the function $\rho(e, f)$ in (10.2), in particular with regard to continuity.

Lemma 10.1.2 *Assume that two shift-invariant systems $\{g_{nm}\}$ and $\{h_{nm}\}$ are Bessel sequences and let $e, f \in L^2(\mathbb{R})$. Then the function*

$$\rho(e, f) : \mathbb{R} \to \mathbb{C}, \ \rho(e, f)(x) = \sum_{m,n \in \mathbb{Z}} \langle T_x e, g_{nm} \rangle \langle h_{nm}, T_x f \rangle$$

is continuous and has period a. Its Fourier series in $L^2(0, a)$ is

$$\rho(e, f)(x) = \sum_{k \in \mathbb{Z}} c_k e^{2\pi i k x / a}, \tag{10.9}$$

where

$$c_k = \frac{1}{a} \int_{-\infty}^{\infty} \widehat{e}(\nu) \overline{\widehat{f}(\nu + k/a)} \sum_{m \in \mathbb{Z}} \overline{\widehat{g_m}(\nu)} \widehat{h_m}(\nu + k/a) \, d\nu, \ k \in \mathbb{Z}. \tag{10.10}$$

Proof. Assume that $\{g_{nm}\}$ and $\{h_{nm}\}$ are Bessel sequences; then, an application of Cauchy–Schwarz' inequality shows that the series defining $\rho(e, f)(x)$ converges absolutely for all $x \in \mathbb{R}$. In particular, this demonstrates that the function $\rho(e, f)(x)$ is well defined. We will now prove that $\rho(e, f)$ is a continuous function. First, given $x, x_0 \in \mathbb{R}$,

$$|\rho(e, f)(x) - \rho(e, f)(x_0)|$$
$$= \left| \sum_{m,n \in \mathbb{Z}} (\langle T_x e, g_{nm} \rangle \langle h_{nm}, T_x f \rangle - \langle T_{x_0} e, g_{nm} \rangle \langle h_{nm}, T_{x_0} f \rangle) \right|$$
$$\leq \sum_{m,n \in \mathbb{Z}} |\langle T_x e, g_{nm} \rangle \langle h_{nm}, T_x f \rangle - \langle T_{x_0} e, g_{nm} \rangle \langle h_{nm}, T_{x_0} f \rangle|.$$

Writing $T_x e = T_x e - T_{x_0} e + T_{x_0} e$, we see that

$$\langle T_x e, g_{nm} \rangle \langle h_{nm}, T_x f \rangle - \langle T_{x_0} e, g_{nm} \rangle \langle h_{nm}, T_{x_0} f \rangle$$
$$= \langle T_x e - T_{x_0} e, g_{nm} \rangle \langle h_{nm}, T_x f \rangle + \langle T_{x_0} e, g_{nm} \rangle \langle h_{nm}, T_x f \rangle$$
$$\qquad\qquad - \langle T_{x_0} e, g_{nm} \rangle \langle h_{nm}, T_{x_0} f \rangle$$
$$= \langle T_x e - T_{x_0} e, g_{nm} \rangle \langle h_{nm}, T_x f \rangle + \langle T_{x_0} e, g_{nm} \rangle \langle h_{nm}, T_x f - T_{x_0} f \rangle;$$

thus, letting B denote a common Bessel bound for $\{g_{nm}\}$ and $\{h_{nm}\}$,

$$|\rho(e,f)(x) - \rho(e,f)(x_0)|$$

$$\leq \quad \sum_{m,n\in\mathbb{Z}} |\langle T_x e - T_{x_0} e, g_{nm}\rangle| \, |\langle h_{nm}, T_x f\rangle|$$

$$+ \sum_{m,n\in\mathbb{Z}} |\langle T_{x_0} e, g_{nm}\rangle| \, |\langle h_{nm}, T_x f - T_{x_0} f\rangle|$$

$$\leq \quad \left(\sum_{m,n\in\mathbb{Z}} |\langle T_x e - T_{x_0} e, g_{nm}\rangle|^2 \right)^{1/2} \left(\sum_{m,n\in\mathbb{Z}} |\langle h_{nm}, T_x f\rangle|^2 \right)^{1/2}$$

$$+ \left(\sum_{m,n\in\mathbb{Z}} |\langle T_{x_0} e, g_{nm}\rangle|^2 \right)^{1/2} \left(\sum_{m,n\in\mathbb{Z}} |\langle h_{nm}, T_x f - T_{x_0} f\rangle|^2 \right)^{1/2}$$

$$\leq \quad B \, \|T_x e - T_{x_0} e\| \, \|T_x f\| + B \, \|T_{x_0} e\| \, \|T_x f - T_{x_0} f\|$$

$$= \quad B \, \|T_{x - x_0} e - e\| \, \|f\| + B \, \|e\| \, \|T_{x - x_0} f - f\|.$$

The last expression converges to zero for $x \to x_0$ by Lemma 2.9.2; this proves the desired continuity.

The periodicity of $\rho(e,f)$ follows from the structure of the shift-invariant systems $\{g_{nm}\}$ and $\{h_{nm}\}$ (Exercise 10.2). For the computation of the Fourier coefficients, we first assume that $e, f \in C_c(\mathbb{R})$; this will justify the following interchanges of sums and integrals. The coefficients in the Fourier expansion with respect to $\{e^{2\pi i k x/a}\}_{k\in\mathbb{Z}}$ are given by (Exercise 10.2)

$$c_k = \frac{1}{a} \int_0^a \rho(e,f)(x) e^{-2\pi i k x/a} \, dx$$

$$= \frac{1}{a} \sum_{m\in\mathbb{Z}} \sum_{n\in\mathbb{Z}} \int_0^a \langle T_x e, g_m(\cdot - na)\rangle \langle h_m(\cdot - na), T_x f\rangle e^{-2\pi i k x/a} \, dx$$

$$= \frac{1}{a} \sum_{m\in\mathbb{Z}} \int_{-\infty}^{\infty} \langle T_x e, g_m\rangle \langle h_m, T_x f\rangle e^{-2\pi i k x/a} \, dx$$

$$= \frac{1}{a} \sum_{m\in\mathbb{Z}} \int_{-\infty}^{\infty} \langle T_x e, g_m\rangle \overline{\langle T_x f, h_m\rangle} e^{2\pi i k x/a} \, dx. \qquad (10.11)$$

Using (10.7), it follows that

$$\langle T_x f, h_m\rangle e^{2\pi i k x/a} = E_{k/a}\mathcal{F}\left(\widehat{f}\,\widetilde{\widehat{h}_m}\right)(x) = \mathcal{F}\left(T_{-k/a}(\widehat{f}\,\widetilde{\widehat{h}_m})\right)(x).$$

Inserting this and (10.7) in (10.11) leads to

$$c_k = \frac{1}{a} \sum_{m\in\mathbb{Z}} \int_{-\infty}^{\infty} \langle T_x e, g_m\rangle \overline{\langle T_x f, h_m\rangle e^{2\pi i k x/a}} \, dx$$

$$= \frac{1}{a} \sum_{m\in\mathbb{Z}} \int_{-\infty}^{\infty} \mathcal{F}\left(\widehat{e}\,\widetilde{\widehat{g}_m}\right)(x) \overline{\mathcal{F}\left(T_{-k/a}(\widehat{f}\,\widetilde{\widehat{h}_m})\right)(x)} \, dx;$$

thus, via Plancherel's equation,

$$c_k = \frac{1}{a} \sum_{m \in \mathbb{Z}} \int_{-\infty}^{\infty} \widehat{e}(\nu) \overline{\widehat{g_m}(\nu)} \, \overline{\widehat{f}(\nu + k/a)} \widehat{h_m}(\nu + k/a) \, d\nu.$$

The reader can check (Exercise 10.2) that the series

$$\int_{-\infty}^{\infty} |\widehat{e}(\nu)| \, |\widehat{f}(\nu + k/a)| \sum_{m \in \mathbb{Z}} |\widehat{g_m}(\nu)| \, |\widehat{h_m}(\nu + k/a)| \, d\nu \qquad (10.12)$$

is convergent. Thus, we can interchange the sum and the integral in the above expression for c_0. This proves the result for all functions $e, f \in C_c(\mathbb{R})$. The general case now follows by a density argument (Exercise 10.2). □

It is very complicated to check the frame conditions for a shift-invariant system directly via the definition. We will now derive equivalent conditions in terms of matrix-valued functions. However, some preparation is needed before we state the main results in Theorem 10.1.6 and Theorem 10.1.7.
    We begin with a definition.

Definition 10.1.3 *For $f \in L^2(\mathbb{R})$, the fiber of f at a point $\nu \in \mathbb{R}$ is defined as the sequence*

$$\widehat{\mathbf{f}}(\nu) := \{\widehat{f}(\nu - k/a)\}_{k \in \mathbb{Z}}.$$

The following lemma shows that the sequence $\widehat{\mathbf{f}}(\nu)$ belongs to $\ell^2(\mathbb{Z})$ for a.e. $\nu \in \mathbb{Z}$; for this reason, $\|\widehat{\mathbf{f}}(\nu)\|$ will always denote the ℓ^2-norm. We ask the reader to provide the proof (Exercise 10.3).

Lemma 10.1.4 *Let $f \in L^2(\mathbb{R})$. Then $\widehat{\mathbf{f}}(\nu) \in \ell^2(\mathbb{Z})$ for a.e. $\nu \in \mathbb{R}$. Furthermore, for any interval I of length $1/a$,*

$$\|f\|^2 = \int_I \sum_{k \in \mathbb{Z}} |\widehat{f}(\nu - k/a)|^2 \, d\nu = \int_I \|\widehat{\mathbf{f}}(\nu)\|^2 d\nu.$$

Given a shift-invariant system $\{g_{nm}\}$ as in (10.1), define the matrix-valued function

$$H(\nu) := (\widehat{g_m}(\nu - k/a))_{k,m \in \mathbb{Z}}, \quad \text{a.e. } \nu \in \mathbb{R}. \qquad (10.13)$$

Note that the columns in the matrix $H(\nu)$ consist of the fibers for the functions g_m. In case $H(\nu)$ defines a bounded operator on $\ell^2(\mathbb{Z})$ for some $\nu \in \mathbb{R}$, the adjoint operator will be denoted by $H(\nu)^*$; it is given by

$$H(\nu)^* = \left(\overline{\widehat{g_m}(\nu - k/a)}\right)_{m,k \in \mathbb{Z}}.$$

For technical reasons, several of the following results will first be proven for functions f belonging to a subspace of the *Schwartz space* $\mathcal{S}$ of rapidly decaying functions. Recall that $\mathcal{S}$ consists of all infinitely often differentiable functions f on $\mathbb{R}$, which decay faster than any inverse polynomial; that is, for any $\alpha, k \in \mathbb{N} \cup \{0\}$,

$$\sup_{x \in \mathbb{R}} \left| x^\alpha \frac{d^k f}{dx^k}(x) \right| < \infty.$$

We will need the following dense subspace $\mathcal{D}$ of $L^2(\mathbb{R})$:

$$\mathcal{D} := \left\{ f \in \mathcal{S} \mid \widehat{f} \text{ is compactly supported} \right\}. \tag{10.14}$$

Lemma 10.1.5 *Let* $\{g_{nm}\}$ *be a shift-invariant system in* $L^2(\mathbb{R})$, *and assume that for a.e.* $\nu \in \mathbb{R}$, *the matrix* $H(\nu)$ *defines a bounded operator on* $\ell^2(\mathbb{Z})$. *Then, for any interval* I *of length* $1/a$ *and any function* $f \in \mathcal{D}$,

$$\int_I \|H(\nu)^* \widehat{\mathbf{f}}(\nu)\|^2 d\nu = a \sum_{m,n \in \mathbb{Z}} |\langle f, g_{nm} \rangle|^2.$$

Proof. For the $\nu \in \mathbb{R}$ for which the matrix $H(\nu)$ is bounded, its adjoint is bounded, and $\|H(\nu)\| = \|H(\nu)^*\|$. Given any interval I of length $1/a$,

$$
\begin{aligned}
\int_I \|H(\nu)^* \widehat{\mathbf{f}}(\nu)\|^2 d\nu &= \int_I \sum_{m \in \mathbb{Z}} \left| \sum_{k \in \mathbb{Z}} \overline{\widehat{g_m}(\nu - k/a)} \widehat{f}(\nu - k/a) \right|^2 d\nu \\
&= \sum_{m \in \mathbb{Z}} \int_I \left| \sum_{k \in \mathbb{Z}} \overline{\widehat{g_m}(\nu - k/a)} \widehat{f}(\nu - k/a) \right|^2 d\nu.
\end{aligned}
$$

Now, for each $m \in \mathbb{Z}$, one can prove (Exercise 10.4) that the mapping

$$\nu \mapsto \sum_{k \in \mathbb{Z}} \overline{\widehat{g_m}(\nu - k/a)} \widehat{f}(\nu - k/a) \tag{10.15}$$

is well defined for a.e. $\nu \in \mathbb{R}$ and defines a function in $L^2(I)$ with Fourier series

$$\sum_{k \in \mathbb{Z}} \overline{\widehat{g_m}(\cdot - k/a)} \widehat{f}(\cdot - k/a) = \sum_{n \in \mathbb{Z}} a \langle f, g_{nm} \rangle e^{2\pi i n a(\cdot)}. \tag{10.16}$$

Parseval's equation (see (3.35)) shows that

$$a \int_I \left| \sum_{k \in \mathbb{Z}} \overline{\widehat{g_m}(\nu - k/a)} \widehat{f}(\nu - k/a) \right|^2 d\nu = a^2 \sum_{n \in \mathbb{Z}} |\langle f, g_{nm} \rangle|^2.$$

Thus,

$$a \sum_{m,n \in \mathbb{Z}} |\langle f, g_{nm} \rangle|^2 = \int_I \|H(\nu)^* \widehat{\mathbf{f}}(\nu)\|^2 d\nu,$$

as desired. □

We are now ready to state characterizations of several frame properties for shift-invariant systems. They are formulated in terms of the matrix-valued function $H(\nu)$ in (10.13), i.e., in terms of the fibers associated with the generators g_m. We begin with the upper frame condition.

Theorem 10.1.6 *A shift-invariant system $\{g_{nm}\}$ is a Bessel sequence with bound B if and only if the matrix $H(\nu)$ for a.e. $\nu \in \mathbb{R}$ defines a bounded operator on $\ell^2(\mathbb{Z})$ with norm at most $\sqrt{aB}$.*

Proof. Let us first assume that the matrix $H(\nu)$ for a.e. $\nu \in \mathbb{R}$ defines a bounded operator on $\ell^2(\mathbb{Z})$ with norm at most $\sqrt{aB}$. Fix any interval I of length $1/a$. Then, for any function f belonging to the space $\mathcal{D}$ defined in (10.14), Lemma 10.1.5 shows that

$$\sum_{m,n\in\mathbb{Z}} |\langle f, g_{nm}\rangle|^2 = \frac{1}{a}\int_I ||H(\nu)^*\widehat{\mathbf{f}}(\nu)||^2 d\nu \le \frac{aB}{a}\int_I ||\widehat{\mathbf{f}}(\nu)||^2 d\nu = B\,||f||^2.$$

Since $\mathcal{D}$ is dense in $L^2(\mathbb{R})$, it follows from Lemma 3.2.6 that $\{g_{nm}\}$ is a Bessel sequence with bound B.

Assume now that $\{g_{nm}\}$ is a Bessel sequence with bound B. We have to prove that for almost all $\nu \in \mathbb{R}$, the inequality

$$\sum_{k\in\mathbb{Z}}\left|\sum_{m\in\mathbb{Z}} \widehat{g_m}(\nu - k/a)c_m\right|^2 \le aB \sum_{m\in\mathbb{Z}} |c_m|^2 \qquad (10.17)$$

holds for all $\{c_m\}_{m\in\mathbb{Z}} \in \ell^2(\mathbb{Z})$. We first assume that $\{c_m\}_{m\in\mathbb{Z}}$ is a finite sequence. Given another finite sequence $\{d_n\}_{n\in\mathbb{Z}}$, we consider the trigonometric polynomial

$$\varphi(\nu) = \sum_{n\in\mathbb{Z}} d_n e^{-2\pi i n a\nu}.$$

Note that φ has period $1/a$. For any interval I of length $1/a$, Parseval's theorem (see (3.35)) shows that

$$\sum_{n\in\mathbb{Z}} |d_n|^2 = a\int_I |\varphi(\nu)|^2 d\nu.$$

The periodicity of φ implies that for any such interval I, we have that

$$(*) := \int_I |\varphi(\nu)|^2 \sum_{k\in\mathbb{Z}}\left|\sum_{m\in\mathbb{Z}} \widehat{g_m}(\nu - k/a)c_m\right|^2 d\nu$$

$$= \int_I \sum_{k\in\mathbb{Z}}\left|\sum_{m\in\mathbb{Z}} \varphi(\nu - k/a)\widehat{g_m}(\nu - k/a)c_m\right|^2 d\nu;$$

thus,

$$(*) \quad = \quad \int_{-\infty}^{\infty} \left| \sum_{m\in\mathbb{Z}} \varphi(\nu)\widehat{g_m}(\nu)c_m \right|^2 d\nu$$

$$= \quad \int_{-\infty}^{\infty} \left| \sum_{m\in\mathbb{Z}}\sum_{n\in\mathbb{Z}} d_n c_m e^{-2\pi i n a \nu} \widehat{g_m}(\nu) \right|^2 d\nu$$

$$= \quad \int_{-\infty}^{\infty} \left| \mathcal{F} \sum_{m\in\mathbb{Z}}\sum_{n\in\mathbb{Z}} d_n c_m g_m(\nu - na) \right|^2 d\nu$$

$$= \quad \left\| \sum_{m,n\in\mathbb{Z}} d_n c_m g_{nm} \right\|^2 .$$

Using that $\{g_{nm}\}$ is a Bessel sequence with bound B, we can estimate this term as follows:

$$\left\| \sum_{m,n\in\mathbb{Z}} d_n c_m g_{nm} \right\|^2 \leq B \sum_{m,n\in\mathbb{Z}} |d_n c_m|^2 = B \sum_{m\in\mathbb{Z}} |c_m|^2 \sum_{n\in\mathbb{Z}} |d_n|^2$$

$$= \quad aB \sum_{m\in\mathbb{Z}} |c_m|^2 \int_I |\varphi(\nu)|^2 d\nu.$$

Altogether, we arrive at the inequality

$$\int_I |\varphi(\nu)|^2 \sum_{k\in\mathbb{Z}} \left| \sum_{m\in\mathbb{Z}} \widehat{g_m}(\nu - k/a)c_m \right|^2 d\nu \leq aB \sum_{m\in\mathbb{Z}} |c_m|^2 \int_I |\varphi(\nu)|^2 d\nu.$$

Since this holds for all trigonometric polynomials φ with period $1/a$, we conclude that (10.17) holds for a.e. $\nu \in I$, for any given finite sequence $\{c_m\}_{m\in\mathbb{Z}}$; for reasons of periodicity, it therefore holds for a.e. $\nu \in \mathbb{R}$ for such sequences.

However, we have to prove that there is a null set $N \subset \mathbb{R}$ such that (10.17) holds for *all* $\{c_m\}_{m\in\mathbb{Z}} \in \ell^2(\mathbb{Z})$ if $\nu \in \mathbb{R} \setminus N$. In order to do so, let $V \subset \ell^2(\mathbb{Z})$ be a subset formed by a countable and dense collection of finite sequences $\{c_m\}_{m\in\mathbb{Z}}$. Note that a countable union of null sets again is a null set; this implies that there exists a null set $N \subset I$ such that (10.17) holds for all $\nu \in \mathbb{R} \setminus N$ and all $\{c_m\}_{m\in\mathbb{Z}} \in V$. Via Lemma 10.1.1, we might also assume that the inequality (10.3) holds for all $\nu \in \mathbb{R} \setminus N$. We now prove that for $\nu \in \mathbb{R} \setminus N$, the result in (10.17) actually holds for all $\{c_m\}_{m\in\mathbb{Z}} \in \ell^2(\mathbb{Z})$. In order to do so, let $\{c_m\}_{m\in\mathbb{Z}} \in \ell^2(\mathbb{Z})$ be given, and take a sequence of finite sequences $\{c_m^n\}_{m\in\mathbb{Z}} \in V$, $n \in \mathbb{N}$, such that

$$\{c_m^n\}_{m\in\mathbb{Z}} \to \{c_m\}_{m\in\mathbb{Z}} \text{ in } \ell^2(\mathbb{Z}) \text{ as } n \to \infty.$$

Now let $\nu \in \mathbb{R} \setminus N$. For all $k \in \mathbb{Z}$, we know that $\{\widehat{g_m}(\nu - k/a)\}_{m\in\mathbb{Z}} \in \ell^2(\mathbb{Z})$; thus,

$$\sum_{m\in\mathbb{Z}} \widehat{g_m}(\nu - k/a)c_m \;=\; \langle\{\widehat{g_m}(\nu - k/a)\}_{m\in\mathbb{Z}}, \{\overline{c_m}\}_{m\in\mathbb{Z}}\rangle_{\ell^2(\mathbb{Z})}$$

$$= \lim_{n\to\infty} \langle\{\widehat{g_m}(\nu - k/a)\}_{m\in\mathbb{Z}}, \{\overline{c_m^n}\}_{m\in\mathbb{Z}}\rangle_{\ell^2(\mathbb{Z})}$$

$$= \liminf_{n\to\infty} \langle\{\widehat{g_m}(\nu - k/a)\}_{m\in\mathbb{Z}}, \{\overline{c_m^n}\}_{m\in\mathbb{Z}}\rangle_{\ell^2(\mathbb{Z})}$$

$$= \liminf_{n\to\infty} \sum_{m\in\mathbb{Z}} \widehat{g_m}(\nu - k/a)c_m^n.$$

By Fatou's lemma (see Lemma 2.7.2), this implies that

$$\sum_{k\in\mathbb{Z}} \Big|\sum_{m\in\mathbb{Z}} \widehat{g_m}(\nu - k/a)c_m\Big|^2 \;=\; \sum_{k\in\mathbb{Z}} \liminf_{n\to\infty} \Big|\sum_{m\in\mathbb{Z}} \widehat{g_m}(\nu - k/a)c_m^n\Big|^2$$

$$\leq\; \liminf_{n\to\infty} \sum_{k\in\mathbb{Z}} \Big|\sum_{m\in\mathbb{Z}} \widehat{g_m}(\nu - k/a)c_m^n\Big|^2$$

$$\leq\; aB\liminf_{n\to\infty} \sum_{m\in\mathbb{Z}} |c_m^n|^2 = aB \sum_{m\in\mathbb{Z}} |c_m|^2.$$

This shows that for $\nu \in \mathbb{R}\setminus N$, the inequality (10.17) indeed holds for all $\{c_m\}_{m\in\mathbb{Z}} \in \ell^2(\mathbb{Z})$. This completes the proof.   □

We now state characterizations of various frame properties for the system $\{g_{nm}\}$ in terms of the matrices $H(\nu)$ and their adjoints. The first of these is formulated as a matrix inequality, involving certain positive bi-infinite matrices indexed by $\mathbb{Z}\times\mathbb{Z}$. Recall that for two such matrices M and N, the inequality $M \leq N$ means that for all sequences $\{c_k\}_{k\in\mathbb{Z}} \in \ell^2(\mathbb{Z})$,

$$\langle M\{c_k\}_{k\in\mathbb{Z}}, \{c_k\}_{k\in\mathbb{Z}}\rangle \leq \langle N\{c_k\}_{k\in\mathbb{Z}}, \{c_k\}_{k\in\mathbb{Z}}\rangle. \tag{10.18}$$

Theorem 10.1.7 *The following characterizations hold:*

(i) *A Bessel sequence $\{g_{nm}\}$ is a frame for $L^2(\mathbb{R})$ with lower frame bound A if and only if*

$$aAI \leq H(\nu)H(\nu)^*, \ a.e. \ \nu \in \mathbb{R}. \tag{10.19}$$

(ii) *$\{g_{nm}\}$ is a tight frame for $L^2(\mathbb{R})$ if and only if there is a constant $c > 0$ such that*

$$\sum_{m\in\mathbb{Z}} \overline{\widehat{g_m}(\nu)}\widehat{g_m}(\nu + k/a) = c\delta_{k,0}, \ k \in \mathbb{Z}, \ a.e. \ \nu \in \mathbb{R}. \tag{10.20}$$

In case (10.20) is satisfied, the frame bound is $A = c/a$.

(iii) *Two shift-invariant systems $\{g_{nm}\}$ and $\{h_{nm}\}$, which form Bessel sequences, are dual frames if and only if*

$$\sum_{m\in\mathbb{Z}} \overline{\widehat{g_m}(\nu)}\widehat{h_m}(\nu + k/a) = a\delta_{k,0}, \ k \in \mathbb{Z}, \ a.e. \ \nu \in \mathbb{R}. \tag{10.21}$$

Proof. To prove (i), first assume that (10.19) holds. According to (10.18), this means that

$$aA\,||\{c_k\}_{k\in\mathbb{Z}}||^2 \leq \langle H(\nu)H(\nu)^*\{c_k\}_{k\in\mathbb{Z}}, \{c_k\}_{k\in\mathbb{Z}}\rangle$$

for all $\{c_k\}_{k\in\mathbb{Z}} \in \ell^2(\mathbb{Z})$ or, equivalently, that

$$||H(\nu)^*\{c_k\}_{k\in\mathbb{Z}}||^2 \geq aA\,||\{c_k\}_{k\in\mathbb{Z}}||^2 \qquad (10.22)$$

holds for all $\{c_k\}_{k\in\mathbb{Z}} \in \ell^2(\mathbb{Z})$. Let $I \subset \mathbb{R}$ be an arbitrary interval of length $1/a$. Then, for any function $f \in \mathcal{D}$, Lemma 10.1.5 shows that

$$\sum_{m,n\in\mathbb{Z}} |\langle f, g_{nm}\rangle|^2 = \frac{1}{a}\int_I ||H(\nu)^*\widehat{\mathbf{f}}(\nu)||^2 d\nu \geq A\int_I ||\widehat{\mathbf{f}}(\nu)||^2 d\nu = A\,||f||^2;$$

by Lemma 5.1.9, this implies that the number A is a lower frame bound for $\{g_{nm}\}$.

The second part of the proof of (i) is more technical. Assume that A is a lower frame bound for $\{g_{nm}\}$; we want to prove that the inequality in (10.22) holds for all $\{c_k\}_{k\in\mathbb{Z}} \in \ell^2(\mathbb{Z})$. Like in the proof of Theorem 10.1.6, we first consider a finite sequence $\{c_k\}_{k\in\mathbb{Z}}$. For technical reasons that will become clear soon, we furthermore consider a function φ for which $\widehat{\varphi}$ is supported on an interval I of length $1/a$. We can associate the function φ to a fiber of a certain function f. In fact, define the function f in terms of its Fourier transform by the following: given $\nu \in \mathbb{R}$, choose the unique number $k \in \mathbb{Z}$ such that $\nu + k/a \in I$, and let

$$\widehat{f}(\nu) = c_k\widehat{\varphi}(\nu + k/a).$$

By definition, the ℓth coordinate in the fiber for a function f is $\widehat{f}(\nu - \ell/a)$. Thus, directly by the definition of the function f, we see that for $\nu \in I$,

$$\widehat{f}(\nu - \ell/a) = c_\ell\widehat{\varphi}(\nu).$$

Thus,

$$\widehat{\mathbf{f}}(\nu) = \widehat{\varphi}(\nu)\{c_k\}_{k\in\mathbb{Z}}, \ \nu \in I.$$

By construction and Lemma 10.1.4,

$$||f||^2 = \int_I ||\widehat{\mathbf{f}}(\nu)||^2 d\nu = \int_I |\widehat{\varphi}(\nu)|^2 d\nu \sum_{k\in\mathbb{Z}} |c_k|^2;$$

thus, Lemma 10.1.5 shows that

$$
\begin{aligned}
\int_I |\widehat{\varphi}(\eta)|^2 ||H^*(\nu)\{c_k\}_{k\in\mathbb{Z}}||^2 d\nu &= \int_I ||H^*(\nu)\widehat{\mathbf{f}}(\nu)||^2 d\nu \\
&\geq a \sum_{m,n\in\mathbb{Z}} |\langle f, g_{nm}\rangle|^2 \\
&\geq aA\,||f||^2 \\
&= aA \int_I |\widehat{\varphi}(\nu)|^2 d\nu \sum_{k\in\mathbb{Z}} |c_k|^2.
\end{aligned}
$$

Since this holds for all functions φ for which $\widehat{\varphi}$ is supported on an interval I of length $1/a$, we conclude that

$$
||H^*(\nu)\{c_k\}_{k\in\mathbb{Z}}||^2 \geq aA \sum_{k\in\mathbb{Z}} |c_k|^2, \ a.e. \ \nu \in I. \tag{10.23}
$$

The null set depends on the sequence $\{c_k\}_{k\in\mathbb{Z}}$. It remains to show that there is a null set N such that (10.23) holds for all $\nu \in \mathbb{R} \setminus N$ and all $\{c_k\}_{k\in\mathbb{Z}} \in \ell^2(\mathbb{Z})$; this part is left to the reader (Exercise 10.5).

We now prove (iii). The proof is based on the functions $\rho(e, f)$ from Lemma 10.1.2 and the derived expression for their Fourier coefficients. Recall from Lemma 6.3.2 that two Bessel sequences $\{g_{nm}\}$ and $\{h_{nm}\}$ are dual frames if and only if

$$
\langle e, f\rangle = \sum_{m,n\in\mathbb{Z}} \langle e, g_{nm}\rangle\langle h_{nm}, f\rangle, \ \forall e, f \in L^2(\mathbb{R}). \tag{10.24}
$$

If we assume that $\{g_{nm}\}, \{h_{nm}\}$ are dual frames, it follows from this identity that for all $e, f \in L^2(\mathbb{R})$,

$$
\rho(e, f)(x) = \sum_{m,n\in\mathbb{Z}} \langle T_x e, g_{nm}\rangle\langle h_{nm}, T_x f\rangle = \langle T_x e, T_x f\rangle = \langle e, f\rangle, \ x \in \mathbb{R}.
$$

Hence, the function $\rho(e, f)(x)$ and the constant $\langle e, f\rangle$ have the same Fourier coefficients in $L^2(0, a)$. Via Lemma 10.1.2, this implies that for all $k \in \mathbb{Z}$,

$$
\begin{aligned}
\frac{1}{a} \int_{-\infty}^{\infty} \widehat{e}(\nu)\overline{\widehat{f}(\nu+k/a)} \sum_{m\in\mathbb{Z}} \overline{\widehat{g_m}(\nu)}\widehat{h_m}(\nu+k/a)\, d\nu &= \delta_{k,0}\langle e, f\rangle \\
&= \delta_{k,0} \int_{-\infty}^{\infty} \widehat{e}(\nu)\overline{\widehat{f}(\nu)}\, d\nu.
\end{aligned}
$$

Since this holds for all $e \in L^2(\mathbb{R})$,

$$
\overline{\widehat{f}(\nu+k/a)} \sum_{m\in\mathbb{Z}} \overline{\widehat{g_m}(\nu)}\widehat{h_m}(\nu+k/a) = a\delta_{k,0}\overline{\widehat{f}(\nu)}, \ a.e. \ \nu \in \mathbb{R}, \ \forall f \in L^2(\mathbb{R}).
$$

For $k = 0$, this implies that

$$
\sum_{m\in\mathbb{Z}} \overline{\widehat{g_m}(\nu)}\widehat{h_m}(\nu) = a, \ a.e. \ \nu \in \mathbb{R};
$$

furthermore, the choices $\widehat{f} = \chi_{[\ell/a,(\ell+1)/a[}, \ell \in \mathbb{Z}$, lead to

$$\sum_{m\in\mathbb{Z}} \overline{\widehat{g_m}(\nu)}\widehat{h_m}(\nu + k/a) = 0, \quad a.e. \ \nu \in \mathbb{R} \text{ when } k \neq 0.$$

The opposite implication can be obtained by reversing the above steps: assuming that (10.21) is satisfied, it follows that the function $\rho(e, f)(x)$ and the constant $\langle e, f \rangle$ have the same Fourier coefficients, so by continuity

$$\rho(e, f)(x) = \langle e, f \rangle, \ \forall x \in \mathbb{R}.$$

Now take $x = 0$, and we obtain (10.24).

We now prove (ii). Via Lemma 6.3.8, the equivalence in (ii) follows directly from (iii); thus, we only need to prove the statement about the frame bound. Now, if (10.20) is satisfied, the entry in the kth row and ℓth column of the matrix $H(\nu)H(\nu)^*$ is

$$\sum_{m\in\mathbb{Z}} \widehat{g_m}(\nu - k/a)\overline{\widehat{g_m}(\nu - \ell/a)} = c\delta_{k,\ell};$$

by the result in (i), this implies that the frame bound is $A = c/a$. $\qquad\square$

Theorem 10.1.7 characterizes frames of the type $\{g_{nm}\}$ in terms of the Fourier transforms $\widehat{g_m}$. One speaks about characterizations in the *frequency domain*, as opposed to *time domain* characterizations directly in terms of the functions g_m.

Via the Fourier transform, frame properties for the shift-invariant system $\{g_{nm}\} = \{T_{na}g_m\}_{m,n\in\mathbb{Z}}$ can be transferred to the set of functions $\{E_{na}g_m\}_{m,n\in\mathbb{Z}}$. This leads to the following consequence of Theorem 10.1.7, stated in the paper [203]; we leave the proof to the reader (Exercise 10.6).

Corollary 10.1.8 *Two Bessel sequences* $\{E_{na}g_m\}_{m,n\in\mathbb{Z}}$ *and* $\{E_{na}h_m\}_{m,n\in\mathbb{Z}}$ *are dual frames for* $L^2(\mathbb{R})$ *if and only if for all* $k \in \mathbb{Z}$,

$$\sum_{m\in\mathbb{Z}} \overline{g_m\left(x + \frac{k}{a}\right)}h_m(x) = a\delta_{k,0}, \qquad a.e. \ x \in \mathbb{R}. \qquad (10.25)$$

Systems of functions of the form $\{E_{na}g_m\}_{m,n\in\mathbb{Z}}$ have been considered under various names in the literature: in [257] they are called *nonstationary Gabor systems*, and in [197] the name *Fourier-like system* is used. Explicit constructions of dual pairs of frames of this form, based on B-splines with irregularly distributed knots, are given in the paper [203].

10.2 Representations of the Frame Operator

The frame operator plays a key role in frame theory, and it is important to have various ways of representing it. In the context of shift-invariant systems, we will now provide a representation in the Fourier domain in terms of fibers. In the literature, the result is known as the *Walnut representation* of the frame operator.

Theorem 10.2.1 *Assume that the shift-invariant system $\{g_{nm}\}$ is a Bessel sequence. Define the functions d_k, $k \in \mathbb{Z}$, by*

$$d_k(\nu) := \sum_{m \in \mathbb{Z}} \widehat{g_m}(\nu)\overline{\widehat{g_m}(\nu - k/a)}, \ \nu \in \mathbb{R}.$$

Then the frame operator S associated with $\{g_{nm}\}$ has a representation in the Fourier domain, given by

$$\widehat{Sf}(\nu) = \frac{1}{a}\sum_{k \in \mathbb{Z}} d_k(\nu)\widehat{f}(\nu - k/a), \ f \in L^2(\mathbb{R}), \tag{10.26}$$

with absolute convergence for a.e. $\nu \in \mathbb{R}$.

Proof. We ask the reader (Exercise 10.7) to verify that the series defining $d_k(\nu)$ is convergent for a.e. $\nu \in \mathbb{R}$ and that

$$\sum_{k \in \mathbb{Z}} |d_k(\nu)|^2 \le (aB)^2, \ a.e. \ \nu \in \mathbb{R}. \tag{10.27}$$

For any $f \in L^2(\mathbb{R})$, Lemma 10.1.4 shows that

$$\sum_{k \in \mathbb{Z}} |\widehat{f}(\nu - k/a)|^2 < \infty, \ a.e. \ \nu \in \mathbb{R};$$

via Cauchy–Schwarz' inequality, this implies that the series on the right-hand side of (10.26) is absolutely convergent for a.e. $\nu \in \mathbb{R}$. In order to show that it represents the frame operator S in the Fourier domain, we first note that for $f, h \in L^2(\mathbb{R})$,

$$\langle Sf, h \rangle = \langle \sum_{m,n \in \mathbb{Z}} \langle f, g_{nm} \rangle g_{nm}, h \rangle = \sum_{m,n \in \mathbb{Z}} \langle f, g_{nm} \rangle \langle g_{nm}, h \rangle.$$

By Lemma 10.1.2, the function

$$\rho(f, h)(x) := \sum_{m,n \in \mathbb{Z}} \langle T_x f, g_{nm} \rangle \langle g_{nm}, T_x h \rangle$$

is continuous and has period a; its Fourier series is

$$\rho(f, h)(x) = \sum_{k \in \mathbb{Z}} c_k e^{2\pi i k x/a},$$

where

$$
\begin{aligned}
c_k &= \frac{1}{a}\int_{-\infty}^{\infty} \widehat{f}(\nu)\overline{\widehat{h}(\nu + k/a)} \sum_{m \in \mathbb{Z}} \overline{\widehat{g_m}(\nu)}\widehat{g_m}(\nu + k/a)\, d\nu \\
&= \frac{1}{a}\int_{-\infty}^{\infty} \widehat{f}(\nu)\overline{\widehat{h}(\nu + k/a)} d_k(\nu + k/a)\, d\nu, \ k \in \mathbb{Z}.
\end{aligned}
$$

Note that $\rho(f,h)(0) = \langle Sf, h \rangle$. We will use the Fourier expansion of $\rho(f,h)$ to calculate $\langle Sf, h \rangle$ for functions $h \in \mathcal{D}$, see (10.14). In order to do so, we first need to show that the Fourier expansion for $\rho(f,h)$ holds pointwise; according to Theorem 3.8.1, this is the case if we can show that the sequence of Fourier coefficients belongs to $\ell^1(\mathbb{Z})$. Now,

$$
\begin{aligned}
\sum_{k \in \mathbb{Z}} |c_k| &= \frac{1}{a} \sum_{k \in \mathbb{Z}} \left| \int_{-\infty}^{\infty} \widehat{f}(\nu) \overline{\widehat{h}(\nu + k/a)} d_k(\nu + k/a)\, d\nu \right| \\
&\leq \frac{1}{a} \sum_{k \in \mathbb{Z}} \int_{-\infty}^{\infty} |\widehat{f}(\nu)| \, |\widehat{h}(\nu + k/a)| \, |d_k(\nu + k/a)| \, d\nu \\
&= \frac{1}{a} \int_{-\infty}^{\infty} \sum_{k \in \mathbb{Z}} |\widehat{f}(\nu - k/a)| \, |\widehat{h}(\nu)| \, |d_k(\nu)| \, d\nu.
\end{aligned}
$$

Using Cauchy–Schwarz inequality on the sum leads to

$$
\begin{aligned}
\sum_{k \in \mathbb{Z}} |c_k| &\leq \frac{1}{a} \int_{-\infty}^{\infty} \left(\sum_{k \in \mathbb{Z}} |d_k(\nu)|^2 \right)^{1/2} \left(\sum_{k \in \mathbb{Z}} |\widehat{f}(\nu - k/a)|^2 \right)^{1/2} |\widehat{h}(\nu)| \, d\nu \\
&\leq B \int_{-\infty}^{\infty} \left(\sum_{k \in \mathbb{Z}} |\widehat{f}(\nu - k/a)|^2 \right)^{1/2} |\widehat{h}(\nu)| \, d\nu.
\end{aligned}
$$

Applying Cauchy–Schwarz' inequality on the integral now shows that

$$
\begin{aligned}
\sum_{k \in \mathbb{Z}} |c_k| &\leq B \left(\int_{\mathrm{supp}\,\widehat{h}} \sum_{k \in \mathbb{Z}} |\widehat{f}(\nu - k/a)|^2 d\nu \right)^{1/2} \left(\int_{\mathrm{supp}\,\widehat{h}} |\widehat{h}(\nu)|^2 d\nu \right)^{1/2} \\
&\leq B \, \|h\|^2 \left(\int_{\mathrm{supp}\,\widehat{h}} \sum_{k \in \mathbb{Z}} |\widehat{f}(\nu - k/a)|^2 d\nu \right)^{1/2}.
\end{aligned}
$$

The compact set $\mathrm{supp}\,\widehat{h}$ can be covered by a finite number of intervals of length $1/a$, so Lemma 10.1.4 implies that $\{c_k\}_{k \in \mathbb{Z}} \in \ell^1(\mathbb{Z})$. Thus, we conclude that the Fourier series for $\rho(f,h)$ is absolutely convergent for all $x \in \mathbb{R}$. In particular,

$$
\langle Sf, h \rangle = \rho(f,h)(0) = \sum_{k \in \mathbb{Z}} c_k = \frac{1}{a} \sum_{k \in \mathbb{Z}} \int_{-\infty}^{\infty} \widehat{f}(\nu) \overline{\widehat{h}(\nu + k/a)} d_k(\nu + k/a)\, d\nu.
$$

Thus,

$$
\begin{aligned}
\langle Sf, h \rangle &= \frac{1}{a} \sum_{k \in \mathbb{Z}} \int_{-\infty}^{\infty} \widehat{f}(\nu - k/a) d_k(\nu) \overline{\widehat{h}(\nu)} \, d\nu \\
&= \left\langle \frac{1}{a} \sum_{k \in \mathbb{Z}} d_k(\cdot) \widehat{f}(\cdot - k/a), \widehat{h}(\cdot) \right\rangle.
\end{aligned}
$$

Since $\langle Sf, h \rangle = \langle \widehat{Sf}, \widehat{h} \rangle$, we conclude that

$$\langle \frac{1}{a} \sum_{k \in \mathbb{Z}} d_k(\cdot) \widehat{f}(\cdot - k/a), \widehat{h}(\cdot) \rangle = \langle \widehat{Sf}, \widehat{h} \rangle.$$

This holds for all $h \in \mathcal{D}$; thus, $\widehat{Sf} = \frac{1}{a} \sum_{k \in \mathbb{Z}} d_k(\cdot) \widehat{f}(\cdot - k/a)$, as desired. $\square$

10.3   Exercises

10.1 Show that the function $\rho(f, f)$ in (10.4) has period a.

10.2 In this exercise, we ask the reader to provide some details in the proof of Lemma 10.1.2.

    (i) Verify that the function $\rho(e, f)$ has period a.

    (ii) Justify the calculations leading to (10.11).

    (iii) Prove that the series in (10.12) is convergent.

    (iv) Provide the density argument at the end of the proof.

10.3 Prove Lemma 10.1.4.

10.4 Complete the proof of Lemma 10.1.5 by showing that the infinite series in (10.15) converges absolutely for a.e. $\nu \in \mathbb{R}$ and that the resulting function has the Fourier expansion stated in (10.16). (Hint: use the periodicity of the function in (10.15) followed by an application of Fourier's inversion theorem.)

10.5 Complete the proof of Theorem 10.1.7(i) by showing that there is a null set N such that (10.23) holds for all $\nu \in \mathbb{R} \setminus N$ and all $\{c_k\}_{k \in \mathbb{Z}} \in \ell^2(\mathbb{Z})$.

10.6 Prove Corollary 10.1.8.

10.7 Prove that the series defining $d_k(\nu)$ in Theorem 10.2.1 converges for a.e. $\nu \in \mathbb{R}$, and that (10.27) holds. (Hint: use Lemma 10.1.1.)

11
Gabor Frames in $L^2(\mathbb{R})$

The mathematical theory for Gabor analysis in $L^2(\mathbb{R})$ is based on two classes of operators on $L^2(\mathbb{R})$, namely,

Translation by $a \in \mathbb{R}$, $T_a : \ L^2(\mathbb{R}) \to L^2(\mathbb{R}), (T_a f)(x) = f(x-a)$,

Modulation by $b \in \mathbb{R}$, $E_b : \ L^2(\mathbb{R}) \to L^2(\mathbb{R}), (E_b f)(x) = e^{2\pi i b x} f(x)$.

Gabor analysis aims at representing functions $f \in L^2(\mathbb{R})$ as superpositions of translated and modulated versions of a fixed function $g \in L^2(\mathbb{R})$. There are two ways one can think about this. The first is to ask for *integral representations* involving all possible translations and modulations, i.e., representations like

$$f(x) = \int_{-\infty}^{\infty} \int_{-\infty}^{\infty} c_f(a,b) e^{2\pi i b x} g(x-a) db\, da; \qquad (11.1)$$

here we have to search for a function c_f of two variables making this true. Note that we also have to specify in which sense we want (11.1) to be valid, i.e., how the integral shall be interpreted. The second approach is to restrict the translation and modulation parameters to a discrete subset $\Lambda \subset \mathbb{R}^2$ and ask for series representations of f in terms of the functions

$$\{e^{2\pi i b x} g(x-a)\}_{(a,b) \in \Lambda}. \qquad (11.2)$$

The key to the first approach is the short-time Fourier transform, which we define in Section 11.1. Concerning the second approach, the natural question is how we can choose $g \in L^2(\mathbb{R})$ and the set Λ such that the functions in (11.2) constitute a frame for $L^2(\mathbb{R})$. Formulated in this generality, the question is very difficult, and we will mainly discuss the case where Λ is

© Springer International Publishing Switzerland 2016   257
O. Christensen, *An Introduction to Frames and Riesz Bases*,
Applied and Numerical Harmonic Analysis,
DOI 10.1007/978-3-319-25613-9_11

a lattice in $\mathbb{R}^2$, i.e., $\Lambda = \{(na, mb)\}_{m,n\in\mathbb{Z}}$ for some fixed parameters $a, b > 0$; we actually saw an example of such a frame in Example 3.8.3, where we proved that $\{E_m T_n \chi_{[0,1]}\}_{m,n\in\mathbb{Z}}$ is an orthonormal basis for $L^2(\mathbb{R})$.

The basic idea goes back to Gabor [314], who considered a sequence of functions of the form $\{E_{mb}T_{na}g\}_{m,n\in\mathbb{Z}}$, where $a = b = 1$ and g is the Gaussian, $g(x) = e^{-x^2/2}$ (the same set of functions actually appeared already in the book [525] by von Neumann in 1932 in the context of quantum mechanics). It was only observed much later (see the papers [422, 423] by Janssen and [250] by Davis and Heller) that this particular Gabor system does not form a frame: it leads to unstable expansions and is inappropriate for most applications. We come back to the exact meaning later, and just note that Davis and Heller proposed to overcome the difficulty by choosing a, b such that $ab < 1$.

The papers [422, 423] by Janssen can be seen as the starting point for the mathematical analysis of Gabor systems. Around the same time, more engineering-oriented papers were published by Bastiaans; see, e.g., [42].

Gabor analysis took an entirely new direction with the fundamental paper [244] by Daubechies, Grossmann, and Meyer from 1986. Here one finds for the first time the idea of combining Gabor analysis with frame theory. The authors constructed tight frames for $L^2(\mathbb{R})$ having the form $\{E_{mb}T_{na}g\}_{m,n\in\mathbb{Z}}$, and this contribution was the beginning of an intense activity which is still ongoing.

Parallel to this development, Feichtinger and Gröchenig were studying expansions in Banach spaces in terms of coherent states (among which Gabor systems is a special case). In particular they obtained Gabor expansions in a large class of Banach spaces, which eventually lead Gröchenig to introduce the concept of *Banach frames*. We will postpone a discussion of this subject to Chapter 24 and confine ourselves to Gabor analysis in $L^2(\mathbb{R})$ at the moment.

This chapter and the following three chapters will all deal with Gabor systems and their properties. The current chapter contains the fundamentals, like equivalent conditions (and necessary, respectively sufficient conditions) for $\{E_{mb}T_{na}g\}_{m,n\in\mathbb{Z}}$ being a frame. Some of the results will be derived as consequences of more general results that are valid for the larger class of shift-invariant systems considered in Chapter 10. The material forms the platform for the further study of Gabor systems in the subsequent chapters, where we will consider duality issues as well as Gabor systems in other Hilbert spaces than $L^2(\mathbb{R})$.

We begin in Section 11.1 by considering continuous representations. Then, after introducing (discrete) Gabor systems in $L^2(\mathbb{R})$, we find necessary conditions for $\{E_{mb}T_{na}g\}_{m,n\in\mathbb{Z}}$ to be a frame in Section 11.3. Sufficient conditions are given in Section 11.4. Section 11.5 will deal with an important vector space in Gabor analysis, namely, the Wiener space.

Even when we restrict our attention to time–frequency shifts of the type $\{E_{mb}T_{na}g\}_{m,n\in\mathbb{Z}}$, it turns out to be very difficult to find the exact range of parameters a,b for which $\{E_{mb}T_{na}g\}_{m,n\in\mathbb{Z}}$ is a frame for a given function $g \in L^2(\mathbb{R})$. There are a few functions as well as a certain class of functions for which an exact answer is known; they are discussed in Section 11.6. Section 11.7 deals with Gabor frames generated by B-splines; for this class of functions, a complete characterization of the parameters a,b leading to a frame is not known yet.

It is clear from the setup that the operators E_b and T_a will play a crucial role in this chapter. Note that even though E_b is defined as an operator acting on $L^2(\mathbb{R})$, we will frequently use the same notation when the operator acts on another function space. For example, the symbol E_b alone will simply mean the function $x \mapsto e^{2\pi i b x}$.

The commutator relations (2.29) show that modulation in the time domain corresponds to translation in the Fourier domain. For this reason, functions $E_b T_a g$ are called *time–frequency shifts* of g; Gabor analysis is in fact a branch of what is called *time–frequency analysis*.

11.1   Continuous Representations

Let us begin by motivating the definition of the short-time Fourier transform. For a signal $f(x)$, the variable x is often interpreted as time, and the Fourier transform $\hat{f}$ evaluated at a point $\gamma > 0$ gives information about the content of oscillations with frequency γ. In practice, it is a problem that the time information is lost in the Fourier transform, i.e., there is no information about which frequencies appear at which time. A way to try to overcome this problem is to "look at the signal at a small time interval and take the Fourier transform here." Mathematically this loose formulation means that we multiply the signal f with a *window function* g, which is constant on a small interval I and decays fast and smooth to zero outside I. By taking the Fourier transform of this product, we gain insight about the frequency content of f on the interval I. In order to obtain information about f on the entire time axis, we repeat the process with translated versions of the window function.

This discussion leads to the definition of the *short-time Fourier transform*, also called the *continuous Gabor transform*:

Definition 11.1.1 *Fix a function $g \in L^2(\mathbb{R}) \setminus \{0\}$. Furthermore, let $f \in L^2(\mathbb{R})$. The short-time Fourier transform of f with respect to g is defined as the function $\Psi_g(f)$ of two variables, given by*

$$\Psi_g(f)(y,\gamma) = \int_{-\infty}^{\infty} f(x)\overline{g(x-y)}e^{-2\pi i x \gamma}dx, \ y,\gamma \in \mathbb{R}.$$

Note that in terms of the modulation operators and translation operators,

$$\Psi_g(f)(y, \gamma) = \langle f, E_\gamma T_y g \rangle, \ y, \gamma \in \mathbb{R}. \tag{11.3}$$

The short-time Fourier transform is the key to obtain a representation of the type (11.1):

Proposition 11.1.2 *Let $f_1, f_2, g_1, g_2 \in L^2(\mathbb{R})$. Then*

$$\int_{-\infty}^{\infty} \int_{-\infty}^{\infty} \Psi_{g_1}(f_1)(a, b) \overline{\Psi_{g_2}(f_2)(a, b)} db da = \langle f_1, f_2 \rangle \langle g_2, g_1 \rangle,$$

i.e.,

$$\int_{-\infty}^{\infty} \int_{-\infty}^{\infty} \langle f_1, E_b T_a g_1 \rangle \langle E_b T_a g_2, f_2 \rangle db da = \langle f_1, f_2 \rangle \langle g_2, g_1 \rangle. \tag{11.4}$$

Proof. Assume first that $g_1, g_2 \in C_c(\mathbb{R})$. By definition,

$$\Psi_{g_1}(f_1)(a, b) = \langle f_1, E_b T_a g_1 \rangle = \int_{-\infty}^{\infty} f_1(x) e^{-2\pi i b x} \overline{g_1(x - a)} dx.$$

Consider for a moment a fixed value for a. Then (11.3) shows that $\Psi_{g_1}(f_1)(a, b)$ is the Fourier transform of the function $F_1(x) = f_1(x) \overline{g_1(x - a)}$, evaluated at the point b. By introducing F_2 similarly and using Plancherel's and Fubini's theorems, we have

$$\int_{-\infty}^{\infty} \int_{-\infty}^{\infty} \Psi_{g_1}(f_1)(a, b) \overline{\Psi_{g_2}(f_2)(a, b)} db da$$

$$= \int_{-\infty}^{\infty} \int_{-\infty}^{\infty} \widehat{F_1}(b) \overline{\widehat{F_2}(b)} db da = \int_{-\infty}^{\infty} \int_{-\infty}^{\infty} F_1(b) \overline{F_2(b)} db da$$

$$= \int_{-\infty}^{\infty} \int_{-\infty}^{\infty} f_1(b) \overline{g_1(b - a)} \overline{f_2(b)} g_2(b - a) db da$$

$$= \int_{-\infty}^{\infty} f_1(b) \overline{f_2(b)} \left(\int_{-\infty}^{\infty} \overline{g_1(b - a)} g_2(b - a) da \right) db = \langle f_1, f_2 \rangle \langle g_2, g_1 \rangle.$$

The extension to general functions in $L^2(\mathbb{R})$ is standard. $\square$

We now show how one can obtain integral representations like (11.1). Fix $f \in L^2(\mathbb{R})$; then Proposition 11.1.2 shows that the map

$$f_2 \mapsto \int_{-\infty}^{\infty} \int_{-\infty}^{\infty} \langle f, E_b T_a g_1 \rangle \langle E_b T_a g_2, f_2 \rangle db da$$

is a conjugated linear functional on $L^2(\mathbb{R})$. By Riesz' representation theorem, there exists a unique element in $L^2(\mathbb{R})$ – we call it

$$\int_{-\infty}^{\infty} \int_{-\infty}^{\infty} \langle f, E_b T_a g_1 \rangle E_b T_a g_2 db da$$

– such that for all $f_2 \in L^2(\mathbb{R})$,

$$\left\langle \int_{-\infty}^{\infty} \int_{-\infty}^{\infty} \langle f, E_b T_a g_1 \rangle E_b T_a g_2 db da, f_2 \right\rangle$$
$$= \int_{-\infty}^{\infty} \int_{-\infty}^{\infty} \langle f, E_b T_a g_1 \rangle \langle E_b T_a g_2, f_2 \rangle db da = \langle f, f_2 \rangle \langle g_2, g_1 \rangle.$$

These considerations lead to

Corollary 11.1.3 *Choose* $g_1, g_2 \in L^2(\mathbb{R})$ *such that* $\langle g_2, g_1 \rangle \neq 0$. *Then every* $f \in L^2(\mathbb{R})$ *has the representation*

$$f = \frac{1}{\langle g_2, g_1 \rangle} \int_{-\infty}^{\infty} \int_{-\infty}^{\infty} \langle f, E_b T_a g_1 \rangle E_b T_a g_2 db da, \qquad (11.5)$$

where the integral is interpreted in the weak sense.

Thus, we have obtained representations like (11.1) and explained how they have to be interpreted. Note that the function $f \in L^2(\mathbb{R})$ is represented as a superposition of time–frequency shifts of *one* function $g_2 \in L^2(\mathbb{R})$, with coefficients given by the short-time Fourier transformation of possibly *another* function g_1.

Proposition 11.1.2 immediately reveals a connection between the short-time Fourier transform and continuous frames:

Corollary 11.1.4 *Let* $g \in L^2(\mathbb{R}) \backslash \{0\}$. *Then* $\{E_b T_a g\}_{a,b\in\mathbb{R}}$ *is a continuous tight frame for* $L^2(\mathbb{R})$ *with respect to* $M = \mathbb{R}^2$ *equipped with the Lebesgue measure; the frame bound is* $A = ||g||^2$.

More generally, (11.4) shows that whenever $\langle g_2, g_1 \rangle \neq 0$, the systems $\{E_b T_a g_1\}_{a,b\in\mathbb{R}}$ and $\{\langle g_2, g_1 \rangle^{-1} E_b T_a g_2\}_{a,b\in\mathbb{R}}$ form a pair of dual continuous frames for $L^2(\mathbb{R})$.

Note that the weakly defined integral over $\mathbb{R}^2$ in (11.5) can be approximated by integrals over growing compact subsets of $\mathbb{R}^2$ that are well-defined pointwise, see, e.g., [340]:

Lemma 11.1.5 *Choose* $g_1, g_2 \in L^2(\mathbb{R})$ *such that* $\langle g_2, g_1 \rangle \neq 0$. *Let* $\{K_n\}_{n=1}^{\infty}$ *be a family of compact subsets of* $\mathbb{R}^2$ *for which*

$$K_1 \subset K_2 \subset \cdots K_n \subset \cdots \text{ and } \bigcup_{n=1}^{\infty} K_n = \mathbb{R}^2;$$

let $f \in \mathcal{H}$ *and define*

$$f_n := \frac{1}{\langle g_2, g_1 \rangle} \int_{K_n} \langle f, E_b T_a g_1 \rangle E_b T_a g_2 db da.$$

Then $||f - f_n|| \to 0$ *as* $n \to \infty$.

11.2 Gabor Frames $\{E_{mb}T_{na}g\}_{m,n\in\mathbb{Z}}$ for $L^2(\mathbb{R})$

We are now ready to define the main subject for this chapter.

Definition 11.2.1 *A Gabor frame is a frame for $L^2(\mathbb{R})$ of the form $\{E_{mb}T_{na}g\}_{m,n\in\mathbb{Z}}$, where $a,b>0$ and $g\in L^2(\mathbb{R})$ is a fixed function.*

Frames of this type are also called *Weyl–Heisenberg frames*. The function g is called the *window function* or the *generator*. Explicitly,

$$E_{mb}T_{na}g(x)=e^{2\pi imbx}g(x-na),\ x\in\mathbb{R}.$$

Note the convention, which is implicit in our definition: when speaking about a Gabor frame, it is understood that it is a frame for all of $L^2(\mathbb{R})$, i.e., we will not deal with frames for subspaces at the moment.

The Gabor system $\{E_{mb}T_{na}g\}_{m,n\in\mathbb{Z}}$ only involves translates with parameters na, $n\in\mathbb{Z}$ and modulations with parameters mb, $m\in\mathbb{Z}$. The points $\{(na,mb)\}_{m,n\in\mathbb{Z}}$ form a so-called *lattice* in $\mathbb{R}^2$, and for this reason one frequently calls $\{E_{mb}T_{na}g\}_{m,n\in\mathbb{Z}}$ a *regular Gabor frame*. Later we will consider more general sets of time–frequency shifts; in fact, we will let $\{(\mu_n,\lambda_n)\}_{n\in I}$ be an arbitrary countable subset of $\mathbb{R}^2$ and investigate the frame properties for sets of functions of the type

$$\{e^{2\pi i\lambda_n x}g(x-\mu_n)\}_{n\in I}. \tag{11.6}$$

To distinguish between the cases, we will call a frame of the form (11.6) an *irregular Gabor frame*. We postpone the discussion of such frames until Section 13.4.

The following technical lemma will be needed repeatedly.

Lemma 11.2.2 *Let $f,g\in L^2(\mathbb{R})$ and $a,b>0$ be given. Then, for any $n\in\mathbb{N}$ the following hold:*

(i) The series

$$\sum_{k\in\mathbb{Z}}f(x-k/b)\overline{g(x-na-k/b)},\quad x\in\mathbb{R}, \tag{11.7}$$

converges absolutely for a.e. $x\in\mathbb{R}$.

(ii) The mapping $x\mapsto\sum_{k\in\mathbb{Z}}|f(x-k/b)\overline{g(x-na-k/b)}|$ belongs to $L^1(0,1/b)$.

(iii) The $1/b$-periodic function $F_n\in L^1(0,1/b)$ defined by

$$F_n(x)=\sum_{k\in\mathbb{Z}}f(x-k/b)\overline{g(x-na-k/b)} \tag{11.8}$$

has the Fourier coefficients

$$c_m=b\,\langle f,E_{mb}T_{na}g\rangle,\ m\in\mathbb{Z}.$$

Proof. Since $f, T_{na}g \in L^2(\mathbb{R})$, we have $f\overline{T_{na}g} \in L^1(\mathbb{R})$ for all $n \in \mathbb{Z}$. Thus,

$$\int_0^{1/b} \sum_{n\in\mathbb{Z}} |f(x-k/b)\overline{g(x-na-k/b)}|\, dx \;\;=\;\; \int_{-\infty}^{\infty} |f(x)\overline{g(x-na)}|\, dx$$

$$< \infty.$$

This proves (ii) and also implies that $\sum_{n\in\mathbb{Z}} |f(x-k/b)\overline{g(x-na-k/b)}|$ converges for a.e. $x \in [0, 1/b]$; for reasons of periodicity, it therefore converges for a.e. $x \in \mathbb{R}$. As a consequence, the series in (11.7) converges for a.e. $x \in \mathbb{R}$ and defines a function with period $1/b$. For part (iii) concerning the Fourier coefficients for F_n, note that

$$\begin{aligned}
\langle f, E_{mb}T_{na}g\rangle &= \int_{-\infty}^{\infty} f(x)\overline{g(x-na)}e^{-2\pi imbx}\, dx \\
&= \sum_{k\in\mathbb{Z}} \int_0^{1/b} f(x-k/b)\overline{g(x-na-k/b)}e^{-2\pi imbx}\, dx \\
&= \int_0^{1/b} \left(\sum_{k\in\mathbb{Z}} f(x-k/b)\overline{g(x-na-k/b)} \right) e^{-2\pi imbx}\, dx \\
&= \int_0^{1/b} F_n(x)e^{-2\pi imbx}\, dx.
\end{aligned}$$

We leave it to the reader to justify the manipulations; the result now follows from the definition of the Fourier coefficients in (3.34). □

There is a close connection between Gabor systems and the shift-invariant systems considered in Chapter 10. In fact,

$$T_{na}E_{mb}g(x) = e^{-2\pi imnab}e^{2\pi imbx}g(x-na) = e^{-2\pi imnab}E_{mb}T_{na}g(x); \quad (11.9)$$

thus, the functions in the shift-invariant system $\{T_{na}E_{mb}g\}_{m,n\in\mathbb{Z}}$ only differ from the functions in the Gabor system $\{E_{mb}T_{na}g\}_{m,n\in\mathbb{Z}}$ by some complex factors of absolute value one. This implies that the shift-invariant system $\{T_{na}E_{mb}g\}_{m,n\in\mathbb{Z}}$ is a frame if and only if $\{E_{mb}T_{na}g\}_{m,n\in\mathbb{Z}}$ is a frame.

Our first goal is to prove a characterization of Gabor frames by Ron and Shen, which in fact is a consequence of the stated relation between Gabor frames and shift-invariant systems. As for the shift-invariant case in Chapter 10, the characterization is formulated in terms of a matrix inequality, which should be interpreted as explained on page 250. Given a function $g \in L^2(\mathbb{R})$ and two numbers $a, b > 0$, consider the matrix-valued function

$$M(x) := (g(x-na-m/b))_{m,n\in\mathbb{Z}}, \quad x \in \mathbb{R}. \quad (11.10)$$

That is, $M(x)$ is the bi-infinite matrix, whose entry in the m-th row and n-th column is

$$M_{m,n}(x) = g(x-na-m/b).$$

Letting $M(x)^*$ denote the conjugated transpose of $M(x)$, we formally consider the matrix product

$$M(x)M(x)^*, \tag{11.11}$$

whose entry in the m-th row and k-th column is

$$G_{m,k}(x) = \sum_{n\in\mathbb{Z}} g(x - na - m/b)\overline{g(x - na - k/b)}. \tag{11.12}$$

Note that by Lemma 11.2.2 the series defining $G_{m,k}(x)$ is convergent for a.e. $x \in \mathbb{R}$. The functions $G_{m,k}$ will also play a role in later sections.

When $\{c_k\}_{k\in\mathbb{Z}}$ is a finite sequence, we can formally define the matrix product $M(x)M(x)^*\{c_k\}_{k\in\mathbb{Z}}$; it is the sequence $\{d_k\}_{k\in\mathbb{Z}}$, whose m-th entry is

$$\sum_{n\in\mathbb{Z}}\sum_{k\in\mathbb{Z}} g(x - na - m/b)\overline{g(x - na - k/b)}c_k.$$

It turns out to be a necessary condition for $\{E_{mb}T_{na}g\}_{m,n\in\mathbb{Z}}$ being a Gabor frame that $M(x)M(x)^*$ defines a bounded operator that maps $\ell^2(\mathbb{Z})$ into $\ell^2(\mathbb{Z})$. Assuming that this is the case, we consider again a finite sequence $\{c_k\}_{k\in\mathbb{Z}}$ and obtain that

$$\langle M(x)M(x)^*\{c_k\}, \{c_k\}\rangle$$
$$= \sum_{n\in\mathbb{Z}}\sum_{k\in\mathbb{Z}}\sum_{m\in\mathbb{Z}} g(x - na - m/b)\overline{g(x - na - k/b)}c_k\overline{c_m}$$
$$= \sum_{n\in\mathbb{Z}}\left|\sum_{k\in\mathbb{Z}} \overline{g(x - na - k/b)}c_k\right|^2 \geq 0.$$

Thus, $M(x)M(x)^*$ is a positive operator on $\ell^2(\mathbb{Z})$; in operator terms,

$$M(x)M(x)^* \geq 0 \text{ on } \ell^2(\mathbb{Z}).$$

The characterization of Gabor frames by Ron and Shen [560] reads as follows:

Theorem 11.2.3 *Let $A, B > 0$ and the Gabor system $\{E_{mb}T_{na}g\}_{m,n\in\mathbb{Z}}$ be given. Then the following hold:*

(i) *$\{E_{mb}T_{na}g\}_{m,n\in\mathbb{Z}}$ is a Bessel sequence with bound B if and only if the matrix $M(x)$ in (11.10) for a.e. $x \in \mathbb{R}$ defines a bounded operator on $\ell^2(\mathbb{Z})$ with norm at most $\sqrt{bB}$.*

(ii) *Assuming that $\{E_{mb}T_{na}g\}_{m,n\in\mathbb{Z}}$ is a Bessel sequence, it is a frame for $L^2(\mathbb{R})$ with lower frame bound A if and only if*

$$bAI \leq M(x)M(x)^*, \text{ a.e. } x \in \mathbb{R}, \tag{11.13}$$

where I is the identity operator on $\ell^2(\mathbb{Z})$.

Proof.   We derive the result from Theorem 10.1.7. First we note that the Fourier transform $\mathcal{F}$ is unitary; thus, Lemma 5.3.3 shows that

$\{E_{mb}T_{na}g\}_{m,n\in\mathbb{Z}}$ is a frame if and only if $\{\mathcal{F}^{-1}E_{mb}T_{na}g\}_{m,n\in\mathbb{Z}}$ is a frame. The commutator relations (2.29) imply that

$$\mathcal{F}^{-1}E_{mb}T_{na}g = T_{-mb}E_{na}\mathcal{F}^{-1}g,$$

which is a shift-invariant system based on the translation parameter b and the functions $E_{na}\mathcal{F}^{-1}g$, $n \in \mathbb{Z}$. Consider the matrix H in (10.13) corresponding to this system; denoting the variable by x rather than ν, the k, n-th entry is

$$\mathcal{F}E_{na}\mathcal{F}^{-1}g(x - k/b) = T_{na}g(x - k/b) = g(x - na - k/b).$$

That is, $H(x)$ equals the matrix $M(x)$ in (11.10), and the result follows from Theorem 10.1.6 and Theorem 10.1.7. □

Note that Theorem 11.2.3 can be formulated directly in terms of the function g and the parameters a, b : in fact, $\{E_{mb}T_{na}g\}_{m,n\in\mathbb{Z}}$ is a frame with bounds $A, B > 0$ if and only if for almost all $x \in \mathbb{R}$ the inequalities

$$bA\sum_{k\in\mathbb{Z}}|c_k|^2 \le \sum_{n\in\mathbb{Z}}\left|\sum_{k\in\mathbb{Z}}\overline{g(x - na - k/b)}c_k\right|^2 \le bB\sum_{k\in\mathbb{Z}}|c_k|^2 \qquad (11.14)$$

hold for all $\{c_k\}_{k\in\mathbb{Z}} \in \ell^2(\mathbb{Z})$. Once the upper inequality in (11.14) has been established, a continuity argument shows that it is enough to consider finite sequences in order to prove the lower inequality.

In many cases, it is convenient to assume that either the translation parameter or the modulation parameter in a Gabor frame is equal to 1. Given an arbitrary Gabor frame $\{E_{mb}T_{na}g\}_{m,n\in\mathbb{Z}}$, this can be obtained by a scaling of g, i.e., by replacing g with a function of the type

$$D_c g(x) = \frac{1}{c^{1/2}}g(x/c).$$

Proposition 11.2.4 *Let $g \in L^2(\mathbb{R})$ and $a, b, c > 0$ be given, and assume that $\{E_{mb}T_{na}g\}_{m,n\in\mathbb{Z}}$ is a frame. Then, with $g_c := D_c g$, the Gabor family $\{E_{mb/c}T_{nac}g_c\}_{m,n\in\mathbb{Z}}$ is a frame with the same frame bounds as $\{E_{mb}T_{na}g\}_{m,n\in\mathbb{Z}}$.*

Proof. Operators of the type D_c, $c > 0$, are studied in Section 2.9, and they are unitary. By Corollary 5.3.4 it follows that $\{D_c E_{mb}T_{na}g\}_{m,n\in\mathbb{Z}}$ is a frame with the same frame bounds as $\{E_{mb}T_{na}g\}_{m,n\in\mathbb{Z}}$. Using the commutator relations in Section 2.9, $D_c E_{mb}T_{na} = E_{mb/c}D_c T_{na} = E_{mb/c}T_{nac}D_c$, and the proposition follows. □

Depending on the given candidate g for a window of a Gabor frame, it might be more convenient to work with the Fourier transform of g than g itself. The next result shows that this can be done if we interchange the parameters a, b :

Proposition 11.2.5 *Let* $g \in L^2(\mathbb{R})$ *and* $a, b > 0$ *be given. Then* $\{E_{mb}T_{na}g\}_{m,n\in\mathbb{Z}}$ *is a frame with bounds* A, B *if and only if* $\{E_{na}T_{mb}\widehat{g}\}_{m,n\in\mathbb{Z}}$ *is a frame with bounds* A, B.

Proof. By the commutations in (2.29) and (2.25), we have

$$\mathcal{F}E_{mb}T_{na}g = T_{mb}E_{-na}\mathcal{F}g = e^{2\pi imbna}E_{-na}T_{mb}\widehat{g}. \qquad (11.15)$$

Since the Fourier transform is a unitary operator (and the factor $e^{2\pi imbna}$ in (11.2.5) is just a complex number of absolute value 1), the result now follows from Corollary 5.3.4. □

11.3   Necessary Conditions

We now move to the question of how to obtain Gabor frames $\{E_{mb}T_{na}g\}_{m,n\in\mathbb{Z}}$ for $L^2(\mathbb{R})$. One of the fundamental results says that the product ab decides whether it is possible for $\{E_{mb}T_{na}g\}_{m,n\in\mathbb{Z}}$ to be a frame for $L^2(\mathbb{R})$ for some choice of $g \in L^2(\mathbb{R})$:

Theorem 11.3.1 *Let* $g \in L^2(\mathbb{R})$ *and* $a, b > 0$ *be given. Then the following hold:*

(i) *If* $ab > 1$, *then* $\{E_{mb}T_{na}g\}_{m,n\in\mathbb{Z}}$ *is not a frame for* $L^2(\mathbb{R})$.

(ii) *If* $\{E_{mb}T_{na}g\}_{m,n\in\mathbb{Z}}$ *is a frame for* $L^2(\mathbb{R})$, *then*

$$ab = 1 \iff \{E_{mb}T_{na}g\}_{m,n\in\mathbb{Z}} \text{ is a Riesz basis.}$$

The proof of Theorem 11.3.1 will use some of the results developed in this chapter and the next, so we delay the proof till page 301. We note that Lemma 11.3.2 will state a stronger result than (i): for $ab > 1$ the family $\{E_{mb}T_{na}g\}_{m,n\in\mathbb{Z}}$ is not even complete in $L^2(\mathbb{R})$.

Theorem 11.3.1 can be formulated in an alternative way. The result shows that it is only possible for $\{E_{mb}T_{na}g\}_{m,n\in\mathbb{Z}}$ to be a frame if $ab \le 1$; and, assuming that $\{E_{mb}T_{na}g\}_{m,n\in\mathbb{Z}}$ is a frame, it is *overcomplete* if and only if $ab < 1$. A result by Balan, Casazza, Heil, and Landau [32] states that an overcomplete frame $\{E_{mb}T_{na}g\}_{m,n\in\mathbb{Z}}$ always has infinite excess:

Lemma 11.3.2 *Let* $g \in L^2(\mathbb{R})$ *and* $a, b > 0$ *be given. Then the following hold.*

(i) *If* $ab < 1$ *and* $\{E_{mb}T_{na}g\}_{m,n\in\mathbb{Z}}$ *is a frame, then* $\{E_{mb}T_{na}g\}_{m,n\in\mathbb{Z}}$ *has infinite excess: infinitely many elements can be deleted while the remaining sequence is still a frame for* $L^2(\mathbb{R})$.

(ii) *If* $ab > 1$, *then* $\{E_{mb}T_{na}g\}_{m,n\in\mathbb{Z}}$ *has infinite deficit, i.e.,*

$$\dim(\overline{span}\{E_{mb}T_{na}g\}_{m,n\in\mathbb{Z}}^{\perp}) = \infty.$$

Lemma 11.3.2 shows that the excess defined in (7.5) does not give much information about the overcompleteness of a Gabor frame $\{E_{mb}T_{na}g\}_{m,n\in\mathbb{Z}}$. As a quantitative measure of the overcompleteness, the following definition of the *redundancy* is used in the literature:

Definition 11.3.3 *Given a Gabor frame* $\{E_{mb}T_{na}g\}_{m,n\in\mathbb{Z}}$, *the number* $(ab)^{-1}$ *is called the redundancy.*

With this definition, a Gabor Riesz basis $\{E_{mb}T_{na}g\}_{m,n\in\mathbb{Z}}$ has redundancy one, and the Gabor frame $\{E_m T_{n/2}\chi_{[0,1]}\}_{m,n\in\mathbb{Z}}$, which can be considered as a union of the two orthonormal bases $\{E_m T_n\chi_{[0,1]}\}_{m,n\in\mathbb{Z}}$ and $\{E_m T_n T_{1/2}\chi_{[0,1]}\}_{m,n\in\mathbb{Z}}$, has redundancy two.

Note that the assumption $ab \leq 1$ is not enough for $\{E_{mb}T_{na}g\}_{m,n\in\mathbb{Z}}$ to be a frame, even if $g \neq 0$. For example, if $a \in]1/2, 1[$, the functions $\{E_m T_{na}\chi_{[0,\frac{1}{2}]}\}_{m,n\in\mathbb{Z}}$ are not complete in $L^2(\mathbb{R})$ and cannot form a frame.

The following proposition gives a necessary condition for $\{E_{mb}T_{na}g\}_{m,n\in\mathbb{Z}}$ to be a frame for $L^2(\mathbb{R})$. It depends on the interplay between the function g and the translation parameter a and is expressed in terms of the function $G_{0,0}$ defined in (11.12); since this function will be used often, we simply write

$$G(x) = \sum_{n\in\mathbb{Z}} |g(x - na)|^2. \tag{11.16}$$

Proposition 11.3.4 *Let* $g \in L^2(\mathbb{R})$ *and* $a, b > 0$ *be given, and assume that* $\{E_{mb}T_{na}g\}_{m,n\in\mathbb{Z}}$ *is a frame with bounds* A, B. *Then*

$$bA \leq \sum_{n\in\mathbb{Z}} |g(x - na)|^2 \leq bB, \ a.e.\, x \in \mathbb{R}, \tag{11.17}$$

and

$$aA \leq \sum_{n\in\mathbb{Z}} |\hat{g}(\nu - nb)|^2 \leq aB, \ a.e.\, \nu \in \mathbb{R}. \tag{11.18}$$

More precisely, if the upper bound in (11.17) *or* (11.18) *is violated, then* $\{E_{mb}T_{na}g\}_{m,n\in\mathbb{Z}}$ *is not a Bessel sequence with bound* B; *if the lower bound in* (11.17) *or* (11.18) *is violated, then* $\{E_{mb}T_{na}g\}_{m,n\in\mathbb{Z}}$ *does not satisfy the lower frame condition with bound* A.

Proof. The proof of (i) is by contradiction. Assume that the upper condition in (11.17) is violated. Then there exists a measurable set $\Delta \subseteq \mathbb{R}$ with positive measure such that $G(x) = \sum_{n\in\mathbb{Z}} |g(x - na)|^2 > bB$ on Δ. We can assume that Δ is contained in an interval of length $\frac{1}{b}$. By letting

$$\Delta_0 = \{x \in \Delta \mid G(x) \geq 1 + bB\},$$
$$\Delta_k = \{x \in \Delta \mid (k+1)^{-1} + bB \leq G(x) < k^{-1} + bB\}, \ k \in \mathbb{N},$$

we obtain a partition of Δ into disjoint measurable sets. At least one of them, say, $\Delta_{k'}$, has positive measure. Now consider the function $f = \chi_{\Delta_{k'}}$, and note that $||f||^2 = |\Delta_{k'}|$. For $n \in \mathbb{Z}$, the function $f\,\overline{T_{na}g}$ has support in $\Delta_{k'}$; since $\Delta_{k'}$ is contained in an interval of length $1/b$ and the functions $\{\sqrt{b}E_{mb}\}_{m \in \mathbb{Z}}$ constitute an orthonormal basis for $L^2(I)$ for every interval I of length $1/b$, we have

$$\sum_{m \in \mathbb{Z}} |\langle f, E_{mb}T_{na}g\rangle|^2 = \sum_{m \in \mathbb{Z}} |\langle f\overline{T_{na}g}, E_{mb}\rangle|^2 = \frac{1}{b}\int_{-\infty}^{\infty} |f(x)|^2\,|g(x-na)|^2 dx.$$

Thus,

$$
\begin{aligned}
\sum_{m,n \in \mathbb{Z}} |\langle f, E_{mb}T_{na}g\rangle|^2 &= \frac{1}{b}\sum_{n \in \mathbb{Z}}\int_{-\infty}^{\infty} |f(x)|^2\,|g(x-na)|^2 dx \\
&= \frac{1}{b}\int_{\Delta_{k'}} G(x)dx \geq \frac{1}{b}\left(\frac{1}{k'+1} + bB\right)||f||^2 \\
&= \left(B + \frac{1}{b(k'+1)}\right)||f||^2.
\end{aligned}
$$

But then B cannot be an upper frame bound for $\{E_{mb}T_{na}g\}_{m,n \in \mathbb{Z}}$. A similar proof shows that if the lower condition in (11.17) is violated, then A cannot be a lower frame bound for $\{E_{mb}T_{na}g\}_{m,n \in \mathbb{Z}}$. The result in (11.18) follows from what we just proved and Proposition 11.2.5.   □

Proposition 11.3.4 implies that a function g generating a Gabor frame $\{E_{mb}T_{na}g\}_{m,n \in \mathbb{Z}}$ necessarily is bounded. Note also that Proposition 11.3.4 gives a relationship between the frame bounds and the lower and upper bounds for the function G in (11.16).

11.4   Sufficient Conditions

Sufficient conditions for $\{E_{mb}T_{na}g\}_{m,n \in \mathbb{Z}}$ to be a frame for $L^2(\mathbb{R})$ have been known since 1988. The basic insight was provided by Daubechies [241]; an improvement was obtained by Heil and Walnut in [395]. We present a more general result in Theorem 11.4.2, which is based on the following identity.

Lemma 11.4.1 *Let $a, b > 0$ be given. Suppose that f is a bounded measurable function with compact support and that g is a measurable function for which the associated function G defined by (11.16) is bounded. Then*

$$\sum_{m,n \in \mathbb{Z}} |\langle f, E_{mb}T_{na}g\rangle|^2$$

$$= \frac{1}{b}\int_{-\infty}^{\infty} |f(x)|^2 \sum_{n \in \mathbb{Z}} |g(x-na)|^2\,dx$$

$$+ \frac{1}{b}\sum_{k \neq 0}\int_{-\infty}^{\infty} \overline{f(x)}f(x-k/b)\sum_{n \in \mathbb{Z}} g(x-na)\overline{g(x-na-k/b)}\,dx.$$

Proof. We first notice that the assumptions imply that $g \in L^2(\mathbb{R})$ (Exercise 11.1). Now, let $n \in \mathbb{Z}$, and consider the $\frac{1}{b}$-periodic function F_n defined in (11.8). We have already given a general argument for F_n being well-defined pointwise almost everywhere, but our present assumptions give more; in fact, for a given $x \in \mathbb{R}$, the compact support of f implies that $f(x - k/b)$ only can be nonzero for finitely many k-values. The number of k-values for which $f(x - k/b) \neq 0$ is uniformly bounded, i.e., there is a constant C such that at most C k-values appear, independently of the chosen x. It follows that F_n is bounded, so $F_n \in L^1(0, 1/b) \cap L^2(0, 1/b)$; in fact, even

$$\left[x \mapsto \sum_{k \in \mathbb{Z}} |f(x - k/b)\overline{g(x - na - k/b)}| \right] \in L^1(0, 1/b) \cap L^2(0, 1/b).$$

By Lemma 11.2.2, for all $m, n \in \mathbb{Z}$,

$$\langle f, E_{mb} T_{na} g \rangle = \int_0^{1/b} F_n(x) e^{-2\pi i m b x} \, dx. \tag{11.19}$$

Parseval's theorem (see (3.35)) gives that

$$\sum_{m \in \mathbb{Z}} \left| \int_0^{1/b} F_n(x) e^{-2\pi i m b x} \, dx \right|^2 = \frac{1}{b} \int_0^{1/b} |F_n(x)|^2 \, dx. \tag{11.20}$$

The assumption on f being a bounded measurable function with compact support will justify all interchanges of integration and summation in the final calculation. This follows from the observation that

$$\sum_{k \in \mathbb{Z}} \int_{-\infty}^{\infty} |\overline{f(x)} f(x - k/b)| \sum_{n \in \mathbb{Z}} |g(x - na)\overline{g(x - na - k/b)}| \, dx < \infty. \tag{11.21}$$

The verification of (11.21) and the proof that this is exactly what we need is left to the reader (Exercise 11.9). Now, via (11.19) and (11.20),

$$\sum_{n \in \mathbb{Z}} \sum_{m \in \mathbb{Z}} |\langle f, E_{mb} T_{na} g \rangle|^2 = \sum_{n \in \mathbb{Z}} \sum_{m \in \mathbb{Z}} \left| \int_0^{1/b} F_n(x) e^{-2\pi i m b x} \, dx \right|^2$$

$$= \frac{1}{b} \sum_{n \in \mathbb{Z}} \int_0^{1/b} |F_n(x)|^2 \, dx.$$

Writing

$$|F_n(x)|^2 = \overline{F_n(x)} F_n(x) = \sum_{\ell \in \mathbb{Z}} \overline{f(x - \ell/b)} g(x - na - \ell/b) F_n(x),$$

and using that F_n is $1/b$-periodic, Lemma 9.2.4 finally implies that

$$\sum_{n\in\mathbb{Z}}\sum_{m\in\mathbb{Z}}|\langle f, E_{mb}T_{na}g\rangle|^2$$

$$= \frac{1}{b}\sum_{n\in\mathbb{Z}}\int_0^{1/b}\sum_{\ell\in\mathbb{Z}}\overline{f(x-\ell/b)}g(x-na-\ell/b)F_n(x)\,dx$$

$$= \frac{1}{b}\sum_{n\in\mathbb{Z}}\int_{-\infty}^{\infty}\overline{f(x)}g(x-na)F_n(x)\,dx$$

$$= \frac{1}{b}\sum_{n\in\mathbb{Z}}\int_{-\infty}^{\infty}\overline{f(x)}g(x-na)\sum_{k\in\mathbb{Z}}f(x-k/b)\overline{g(x-na-k/b)}\,dx \quad (11.22)$$

$$= \frac{1}{b}\int_{-\infty}^{\infty}|f(x)|^2\sum_{n\in\mathbb{Z}}|g(x-na)|^2\,dx$$

$$+ \frac{1}{b}\sum_{k\neq 0}\int_{-\infty}^{\infty}\overline{f(x)}f(x-k/b)\sum_{n\in\mathbb{Z}}g(x-na)\overline{g(x-na-k/b)}\,dx. \qquad \square$$

The proof of Lemma 11.4.1 relies strongly on summing over *all* $m \in \mathbb{Z}$: we need that $\{\sqrt{b}E_{mb}\}_{m\in\mathbb{Z}}$ forms an orthonormal basis for $L^2(0,1/b)$. On the other hand, the proof did not use that we were summing over all $n \in \mathbb{Z}$, so the assumptions actually imply (see (11.22)) that for *all* index sets $I \subseteq \mathbb{Z}$,

$$\sum_{n\in I}\sum_{m\in\mathbb{Z}}|\langle f, E_{mb}T_{na}g\rangle|^2$$

$$= \frac{1}{b}\sum_{n\in I}\int_{-\infty}^{\infty}\overline{f(x)}g(x-na)\sum_{k\in\mathbb{Z}}\overline{f(x-k/b)}g(x-na-k/b)\,dx. \quad (11.23)$$

We will now state the announced sufficient condition for $\{E_{mb}T_{na}g\}_{m,n\in\mathbb{Z}}$ to be a Gabor frame for $L^2(\mathbb{R})$. Note that various generalizations to Gabor systems in $L^2(\mathbb{R}^d)$ are given in Section 20.5.

Theorem 11.4.2 *Let $g \in L^2(\mathbb{R})$, $a,b > 0$ and suppose that*

$$B := \frac{1}{b}\sup_{x\in[0,a]}\sum_{k\in\mathbb{Z}}\left|\sum_{n\in\mathbb{Z}}g(x-na)\overline{g(x-na-k/b)}\right| < \infty. \quad (11.24)$$

Then $\{E_{mb}T_{na}g\}_{m,n\in\mathbb{Z}}$ is a Bessel sequence with bound B. If also

$$A := \quad (11.25)$$

$$\frac{1}{b}\inf_{x\in[0,a]}\left[\sum_{n\in\mathbb{Z}}|g(x-na)|^2 - \sum_{k\neq 0}\left|\sum_{n\in\mathbb{Z}}g(x-na)\overline{g(x-na-k/b)}\right|\right] > 0,$$

then $\{E_{mb}T_{na}g\}_{m,n\in\mathbb{Z}}$ is a frame for $L^2(\mathbb{R})$ with bounds A, B.

Proof.   Consider a function $f \in L^2(\mathbb{R})$ which is continuous and has compact support. By Lemma 11.4.1,

$$\sum_{m,n\in\mathbb{Z}} |\langle f, E_{mb}T_{na}g\rangle|^2$$

$$= \frac{1}{b}\int_{-\infty}^{\infty} |f(x)|^2 \sum_{n\in\mathbb{Z}} |g(x-na)|^2 dx$$

$$+ \frac{1}{b}\sum_{k\neq 0}\int_{-\infty}^{\infty} \overline{f(x)}f(x-k/b)\sum_{n\in\mathbb{Z}} g(x-na)\overline{g(x-na-k/b)}dx. \quad (11.26)$$

We want to estimate (11.26). For $k \in \mathbb{Z}$, let

$$H_k(x) := \sum_{n\in\mathbb{Z}} T_{na}g(x)\overline{T_{na+k/b}g(x)}; \quad (11.27)$$

we observe that H_k is well defined a.e. by Lemma 11.2.2. Now,

$$\sum_{k\neq 0}|T_{-k/b}H_k(x)| = \sum_{k\neq 0}\left|T_{-k/b}\sum_{n\in\mathbb{Z}} T_{na}g(x)\overline{T_{na+k/b}g(x)}\right|$$

$$= \sum_{k\neq 0}\left|\sum_{n\in\mathbb{Z}} T_{na-k/b}g(x)\overline{T_{na}g(x)}\right|.$$

Replacing k with $-k$ (which is allowed since we sum over all $k \neq 0$) and complex conjugating all terms, we arrive at

$$\sum_{k\neq 0}|T_{-k/b}H_k(x)| = \sum_{k\neq 0}\left|\sum_{n\in\mathbb{Z}} T_{na+k/b}g(x)\overline{T_{na}g(x)}\right|$$

$$= \sum_{k\neq 0}\left|\sum_{n\in\mathbb{Z}} \overline{T_{na+k/b}g(x)}T_{na}g(x)\right|$$

$$= \sum_{k\neq 0}|H_k(x)|.$$

So

$$\left|\sum_{k\neq 0}\int_{-\infty}^{\infty} \overline{f(x)}f(x-k/b)\sum_{n\in\mathbb{Z}} g(x-na)\overline{g(x-na-k/b)}dx\right|$$

$$\leq \sum_{k\neq 0}\int_{-\infty}^{\infty} |f(x)|\,|T_{k/b}f(x)|\,|H_k(x)|dx$$

$$= \sum_{k\neq 0}\int_{-\infty}^{\infty} |f(x)|\sqrt{|H_k(x)|}\,|T_{k/b}f(x)|\sqrt{|H_k(x)|}dx$$

$$= (*).$$

Using Cauchy–Schwarz' inequality twice, first on the integral and then on the sum over $k \neq 0$,

$$(*) \leq \sum_{k \neq 0} \left(\int_{-\infty}^{\infty} |f(x)|^2 |H_k(x)| dx \right)^{1/2} \left(\int_{-\infty}^{\infty} |T_{k/b} f(x)|^2 |H_k(x)| dx \right)^{1/2}$$

$$\leq \left(\sum_{k \neq 0} \int_{-\infty}^{\infty} |f(x)|^2 |H_k(x)| dx \right)^{1/2}$$

$$\times \left(\sum_{k \neq 0} \int_{-\infty}^{\infty} |T_{k/b} f(x)|^2 |H_k(x)| dx \right)^{1/2}$$

$$= \left(\int_{-\infty}^{\infty} |f(x)|^2 \sum_{k \neq 0} |H_k(x)| dx \right)^{1/2}$$

$$\times \left(\int_{-\infty}^{\infty} |f(x)|^2 \sum_{k \neq 0} |T_{-k/b} H_k(x)| dx \right)^{1/2}$$

$$= \int_{-\infty}^{\infty} |f(x)|^2 \sum_{k \neq 0} |H_k(x)| dx.$$

Note that the expression

$$\sum_{k \neq 0} |H_k(x)| = \sum_{k \neq 0} \left| \sum_{n \in \mathbb{Z}} T_{na} g(x) \overline{T_{na+k/b} g(x)} \right|$$

defines a periodic function with period a. By (11.26) and the condition (11.24), we now have

$$\sum_{m,n \in \mathbb{Z}} |\langle f, E_{mb} T_{na} g \rangle|^2$$

$$\leq \frac{1}{b} \int_{-\infty}^{\infty} \left(|f(x)|^2 \right.$$

$$\times \left[\sum_{n \in \mathbb{Z}} |g(x-na)|^2 + \sum_{k \neq 0} \left| \sum_{n \in \mathbb{Z}} g(x-na) \overline{g(x-na-k/b)} \right| \right] \right) dx$$

$$= \frac{1}{b} \int_{-\infty}^{\infty} |f(x)|^2 \sum_{k \in \mathbb{Z}} \left| \sum_{n \in \mathbb{Z}} g(x-na) \overline{g(x-na-k/b)} \right| dx$$

$$\leq B \, \|f\|^2.$$

Since this estimate holds on a dense subset of $L^2(\mathbb{R})$, it holds on $L^2(\mathbb{R})$ by Lemma 3.2.6. This proves the first part. If also (11.25) is satisfied, we again consider a continuous function f with compact support and obtain that

$$
\sum_{m,n\in\mathbb{Z}} |\langle f, E_{mb}T_{na}g\rangle|^2
$$

$$
\geq \frac{1}{b}\int_{-\infty}^{\infty} |f(x)|^2
$$

$$
\times \left[\sum_{n\in\mathbb{Z}} |g(x-na)|^2 - \sum_{k\neq 0}\left|\sum_{n\in\mathbb{Z}} g(x-na)\overline{g(x-na-k/b)}\right|\right] dx
$$

$$
\geq A\,\|f\|^2.
$$

By Lemma 5.1.9 the lower frame condition actually holds for all $f \in L^2(\mathbb{R})$. This completes the proof. □

The condition (11.24) is called *condition (CC)* in the literature. It leads to an easy sufficient condition for $\{E_{mb}T_{na}g\}_{m,n\in\mathbb{Z}}$ to be a Bessel sequence (Exercise 11.2):

Corollary 11.4.3 *Let $g \in L^2(\mathbb{R})$ be bounded and compactly supported. Then $\{E_{mb}T_{na}g\}_{m,n\in\mathbb{Z}}$ is a Bessel sequence for any choice of $a, b > 0$.*

Proposition 11.5.2 and Exercise 11.7 will show that windows in the Wiener space (in particular, functions in the Feichtinger algebra $\mathcal{S}_0$) satisfy condition (CC) and thus generate Bessel sequences for all choices of $a, b > 0$. One can show that the condition (CC) is not necessary for $\{E_{mb}T_{na}g\}_{m,n\in\mathbb{Z}}$ to be a Bessel sequence (cf. [123]). In Section 12.1, we relate condition (CC) to other conditions used in Gabor analysis.

We note that Theorem 11.4.2 can be extended to a result concerning frame sequences. We have seen that if the function

$$
G(x) = \sum_{n\in\mathbb{Z}} |g(x-na)|^2 \tag{11.28}
$$

is not bounded below, then $\{E_{mb}T_{na}g\}_{m,n\in\mathbb{Z}}$ cannot be a frame for $L^2(\mathbb{R})$. However, it can still be a frame for its closed linear span: in [125] it is proved that if the conditions in Theorem 11.4.2 hold with the infimum over $x \in [0, a]$ in (11.25) replaced with the infimum over $N_g := \{x \in \mathbb{R}\,|\,G(x) \neq 0\}$, then $\{E_{mb}T_{na}g\}_{m,n\in\mathbb{Z}}$ is a frame for $L^2(N_g)$. This gives a way to construct *multi-window Gabor frames*: if $g_1, g_2, \ldots, g_k$ is a collection of functions which satisfy the conditions in this extended version of Theorem 11.4.2, then $\{E_{mb}T_{na}g_k\}_{m,n\in\mathbb{Z},k=1,\ldots,K}$ is a frame for $L^2(\cup_{k=1}^{K} N_{g_k})$. In particular, if $\cup_{k=1}^{K} N_{g_k} = \mathbb{R}$, then we obtain a frame for $L^2(\mathbb{R})$.

Example 11.4.4 Let $a = b = 1$ and define

$$g(x) = \begin{cases} 1+x & \text{if } x \in]0,1], \\ \frac{1}{2}x & \text{if } x \in]1,2], \\ 0 & \text{otherwise.} \end{cases}$$

Consider for $n, k \in \mathbb{Z}$ the function $x \mapsto g(x-n)g(x-n-k)$ for $x \in]0,1]$. Due to the compact support of g, it can only be nonzero if $n \in \{-1,0\}$; for $n = -1$, it can only be nonzero for $k \in \{0,1\}$, and for $n = 0$, it can only be nonzero for $k \in \{-1,0\}$. Therefore,

$$\sum_{n\in\mathbb{Z}} g(x-n)g(x-n-k) = \begin{cases} g(x)g(x+1) & \text{if } k = -1, \\ g(x)^2 + g(x+1)^2 & \text{if } k = 0, \\ g(x+1)g(x) & \text{if } k = 1, \\ 0 & \text{otherwise,} \end{cases}$$

$$= \begin{cases} \frac{1}{2}(1+x)^2 & \text{if } k = -1, \\ \frac{5}{4}(1+x)^2 & \text{if } k = 0, \\ \frac{1}{2}(1+x)^2 & \text{if } k = 1, \\ 0 & \text{otherwise.} \end{cases}$$

So $G(x) = \sum_{n\in\mathbb{Z}} |g(x-n)|^2 = \frac{5}{4}(x+1)^2$ for $x \in]0,1]$, and

$$\sum_{k\neq 0} |H_k(x)| = \sum_{k\neq 0} \left| \sum_{n\in\mathbb{Z}} g(x-n)\overline{g(x-n-k)} \right| = (1+x)^2, \ x \in]0,1].$$

By Theorem 11.4.2 $\{E_m T_n g\}_{m,n\in\mathbb{Z}}$ is a frame for $L^2(\mathbb{R})$ with frame bounds $A = \frac{1}{4}, B = 9$. $\square$

In Corollary 11.4.3 we have seen that $\{E_{mb}T_{na}g\}_{m,n\in\mathbb{Z}}$ is a Bessel sequence for all $a, b > 0$ if the function g is bounded and compactly supported. Stronger conditions are necessary for $\{E_{mb}T_{na}g\}_{m,n\in\mathbb{Z}}$ to be a frame: as shown in Proposition 11.3.4, the associated function G in (11.16) needs to be bounded below and above (see Exercise 11.3). On the other hand, for a function g with compact support, the condition that the function G is bounded below and above for some $a > 0$ is enough for $\{E_{mb}T_{na}g\}_{m,n\in\mathbb{Z}}$ to be a frame for sufficiently small values of b. We also obtain explicit expressions for the frame operator and its inverse in this case. In fact, they are multiplication operators:

Corollary 11.4.5 Let $a, b > 0$. Suppose that $g \in L^2(\mathbb{R})$ has support in an open interval of length $\frac{1}{b}$ and that the function G satisfies (11.17) for some $A, B > 0$. Then $\{E_{mb}T_{na}g\}_{m,n\in\mathbb{Z}}$ is a frame for $L^2(\mathbb{R})$ with bounds A, B. The frame operator and its inverse are given by

$$Sf = \frac{G}{b}f, \ S^{-1}f = \frac{b}{G}f, \ f \in L^2(\mathbb{R}).$$

Proof. That $\{E_{mb}T_{na}g\}_{m,n\in\mathbb{Z}}$ is a frame follows directly from Theorem 11.4.2 because

$$\sum_{n\in\mathbb{Z}} g(x-na)\overline{g(x-na-k/b)} = 0 \text{ for all } k \neq 0.$$

Given a continuous function f with compact support, Lemma 11.4.1 implies that

$$\langle Sf, f\rangle = \sum_{m,n\in\mathbb{Z}} |\langle f, E_{mb}T_{na}g\rangle|^2 = \frac{1}{b}\int_{-\infty}^{\infty} |f(x)|^2 G(x)dx;$$

by continuity of S, this expression even holds for all $f \in L^2(\mathbb{R})$. Via Lemma 2.4.4, it follows that S acts by multiplication with the function $\frac{G}{b}$. $\qquad\square$

In the following special case we can be even more explicit:

Corollary 11.4.6 *Suppose that $g \in L^2(\mathbb{R})$ is a continuous function with support on an interval I and that $|g(x)| > 0$ on the interior of I. Then $\{E_{mb}T_{na}g\}_{m,n\in\mathbb{Z}}$ is a frame for all $(a,b) \in]0, |I|[\times]0, \frac{1}{|I|}[$.*

Proof. By Corollary 11.4.5 it is enough to prove that the function G is bounded above and below for the given values of a, b. For the upper bound, we observe that since g has support in an interval of length $\frac{1}{b}$, the function $g(x-na)$ can at most be nonzero for $\lfloor\frac{1}{ab}\rfloor+1$ values of $n \in \mathbb{Z}$, independently of the choice of $x \in \mathbb{R}$. Thus,

$$G(x) \leq \left(\lfloor\frac{1}{ab}\rfloor + 1\right)||g||_\infty^2.$$

For the lower bound, let J be the subinterval of I which has the same center as I and length a. Then, for any given $x \in \mathbb{R}$, we can find $n \in \mathbb{Z}$ such that $x - na \in J$; thus, $G(x) \geq \inf_{y\in J}|g(y)|^2 > 0$, as desired. $\qquad\square$

11.5 The Wiener Space W

Gabor analysis makes use of a number of *window classes*. Among these classes we find Feichtinger's algebra S_0 discussed in Section A.6 and the Wiener space. Given $a > 0$, the *Wiener space* is defined by

$$W := \left\{g : \mathbb{R} \to \mathbb{C} \;\middle|\; g \text{ measurable and } \sum_{k\in\mathbb{Z}}||g\chi_{[ka,(k+1)a[}||_\infty < \infty\right\}.\tag{11.29}$$

In the literature the space W is also called a *Wiener amalgam space* and is often denoted by $W(L^\infty, \ell^1)$. One can prove that W is a Banach space with respect to the norm

$$\|g\|_{W,a} = \sum_{k\in\mathbb{Z}} \|g\chi_{[ka,(k+1)a[}\|_\infty. \tag{11.30}$$

The space W is independent of the choice of a, and different choices give equivalent norms; both statements follow from the fact that if $0 < a \le b$, then (Exercise 11.4)

$$\sum_{k\in\mathbb{Z}} \|g\chi_{[kb,(k+1)b[}\|_\infty \;\le\; 2\sum_{k\in\mathbb{Z}} \|g\chi_{[ka,(k+1)a[}\|_\infty \tag{11.31}$$

$$\le\; 2\left(\lfloor \tfrac{b}{a}\rfloor + 2\right)\sum_{k\in\mathbb{Z}} \|g\chi_{[kb,(k+1)b[}\|_\infty.$$

That $g \in W$ means that g is bounded and decays so fast that the "local maximum function" $k \mapsto \|g\chi_{[ka,(k+1)a]}\|_\infty$ belongs to $\ell^1(\mathbb{Z})$. In Exercise 11.4 we ask the reader to prove that $W \subset L^1(\mathbb{R})\cap L^2(\mathbb{R})$. Furthermore, as stated in Section A.6, the Wiener space is related to the Schwartz space $\mathcal{S}$ and Feichtinger's algebra $\mathcal{S}_0$ by the inclusions

$$\mathcal{S} \subset \mathcal{S}_0 \subset W.$$

The condition for being in W is strong enough to exclude many of the pathological functions, which play a role for the understanding of functions in $L^2(\mathbb{R})$ but are of little practical interest (like functions in $L^2(\mathbb{R})$ which do not decay to zero whenever $|x| \to \infty$). As a consequence of the following lemma, we will show that functions in $g \in W$ generate Bessel sequences $\{E_{mb}T_{na}g\}_{m,n\in\mathbb{Z}}$ for all choices of the parameters a,b.

Lemma 11.5.1 *Let $g \in W$ and $a > 0$ be given. Then*

$$\sum_{n\in\mathbb{Z}} |g(x-na)| \le \|g\|_{W,a}, \quad a.e.\ x \in \mathbb{R}.$$

If also $h \in W$ and $b \in]0,\tfrac{1}{a}]$, then

$$\sum_{k\in\mathbb{Z}}\left|\sum_{n\in\mathbb{Z}} g(x-na)\overline{h(x-na-k/b)}\right| \le 2\,\|g\|_{W,a}\|h\|_{W,a}, \quad a.e.\ x \in \mathbb{R}.$$

Proof. For the first part, fix $x \in \mathbb{R}$, and observe that for any given $n \in \mathbb{Z}$, there exists exactly one value of $k \in \mathbb{Z}$ such that

$$x - na \in [ka, (k+1)a[;$$

furthermore, different values of n lead to different values for k. Therefore,

$$\sum_{n\in\mathbb{Z}} |g(x-na)| \le \sum_{k\in\mathbb{Z}} \|g\chi_{[ka,(k+1)a[}\|_\infty = \|g\|_{W,a}, \quad a.e.\ x \in \mathbb{R}.$$

For the second part, we have

$$\sum_{k\in\mathbb{Z}}\left|\sum_{n\in\mathbb{Z}}g(x-na)\overline{h(x-na-k/b)}\right| \le \sum_{n\in\mathbb{Z}}|g(x-na)|\sum_{k\in\mathbb{Z}}|h(x-na-k/b)|.$$

The first part of the lemma (applied to the function h and the translation parameter $\frac{1}{b}$) combined with (11.31) gives that

$$\sum_{k\in\mathbb{Z}}|h(x-na-k/b)| \le \|h\|_{W,\frac{1}{b}} \le 2\,\|h\|_{W,a}$$

and the lemma follows. □

We now prove that the Gabor system $\{E_{mb}T_{na}g\}_{m,n\in\mathbb{Z}}$ automatically is a Bessel sequence for windows $g\in W$.

Proposition 11.5.2 *If $g\in W$ and $a,b>0$, then $\{E_{mb}T_{na}g\}_{m,n\in\mathbb{Z}}$ is a Bessel sequence. If $ab\le 1$, then $B:=\frac{2}{b}\,\|g\|_{W,a}^2$ is an upper frame bound.*

Proof. The case $ab\le 1$ follows immediately from Lemma 11.5.1 combined with Theorem 11.4.2. In case $ab>1$, we can choose $N\in\mathbb{N}$ such that $ab/N\le 1$; this implies that $\{E_{mb}T_{na/N}g\}_{m,n\in\mathbb{Z}}$ is a Bessel sequence, and therefore the subsequence $\{E_{mb}T_{na}g\}_{m,n\in\mathbb{Z}}$ is also a Bessel sequence. □

An intuitive explanation of Proposition 11.5.2 is that functions in W decay relatively fast: given $x\in[0,a]$, the values of the functions

$$n,k\mapsto g(x-na)\overline{g(x-na-k/b)}$$

are small enough to make $\sum_{k\in\mathbb{Z}}\left|\sum_{n\in\mathbb{Z}}g(x-na)\overline{g(x-na-k/b)}\right|$ convergent, with a bound independent of x. Inspired by the necessary condition in Proposition 11.3.4, we can use this intuition to obtain a sufficient condition for $\{E_{mb}T_{na}g\}_{m,n\in\mathbb{Z}}$ to be a frame whenever $g\in W$:

Proposition 11.5.3 *Let $g\in W$ and $a>0$ be given. Assume that there exists a constant $C>0$ such that*

$$C\le \sum_{n\in\mathbb{Z}}|g(x-na)|^2, \quad a.e.\,x\in\mathbb{R}.$$

Then $\{E_{mb}T_{na}g\}_{m,n\in\mathbb{Z}}$ is a frame for $L^2(\mathbb{R})$ for all sufficiently small $b>0$. As $b\to 0$, the ratio between the frame bounds in Theorem 11.4.2 converges to

$$\frac{\sup_{x\in[0,a]}\sum_{n\in\mathbb{Z}}|g(x-na)|^2}{\inf_{x\in[0,a]}\sum_{n\in\mathbb{Z}}|g(x-na)|^2}.$$

Proof. Proposition 11.5.2 shows that $\{E_{mb}T_{na}g\}_{m,n\in\mathbb{Z}}$ is a Bessel sequence for all $b>0$. Fix $\epsilon>0$ (the value will be decided later) and choose

$N \in \mathbb{N}$ such that $\sum_{|n| \geq N} \left\| g \chi_{[na,(n+1)a[} \right\|_\infty < \epsilon$. Letting $g_0 := g \chi_{[-aN,aN]}$ and $g_1 := g - g_0$, we have

$$
\begin{aligned}
\|g_1\|_{W,a} &= \sum_{n \in \mathbb{Z}} \left\| (g - g \chi_{[-aN,aN]}) \chi_{[na,(n+1)a[} \right\|_\infty \\
&\leq \sum_{|n| \geq N} \left\| g \chi_{[na,(n+1)a[} \right\|_\infty \\
&< \epsilon.
\end{aligned}
$$

Now,

$$
\sum_{k \neq 0} \left| \sum_{n \in \mathbb{Z}} g(x - na) \overline{g(x - na - k/b)} \right|
$$

$$
= \sum_{k \neq 0} \left| \sum_{n \in \mathbb{Z}} (g_0 + g_1)(x - na) \overline{(g_0 + g_1)(x - na - k/b)} \right|
$$

$$
\leq \sum_{k \neq 0} \left| \sum_{n \in \mathbb{Z}} g_0(x - na) \overline{g_0(x - na - k/b)} \right|
$$

$$
+ \sum_{k \neq 0} \left| \sum_{n \in \mathbb{Z}} g_0(x - na) \overline{g_1(x - na - k/b)} \right|
$$

$$
+ \sum_{k \neq 0} \left| \sum_{n \in \mathbb{Z}} g_1(x - na) \overline{g_0(x - na - k/b)} \right|
$$

$$
+ \sum_{k \neq 0} \left| \sum_{n \in \mathbb{Z}} g_1(x - na) \overline{g_1(x - na - k/b)} \right|.
$$

The function g_0 has support in an interval of length $2aN$, so if we choose b so small that $\frac{1}{b} > 2aN$, then the first of the above four terms is zero. Using Lemma 11.5.1 on the remaining terms, we get

$$
\sum_{k \neq 0} \left| \sum_{n \in \mathbb{Z}} g(x - na) \overline{g(x - na - k/b)} \right| \leq 4 \|g_0\|_{W,a} \|g_1\|_{W,a} + 2 \|g_1\|_{W,a}^2
$$

$$
\leq 4\epsilon \|g_0\|_{W,a} + 2\epsilon^2
$$

$$
\leq 4\epsilon \|g\|_{W,a} + 2\epsilon^2.
$$

If ϵ is chosen such that $4\epsilon \|g\|_{W,a} + 2\epsilon^2 < C$, then the condition in Theorem 11.4.2 is satisfied, and $\{E_{mb} T_{na} g\}_{m,n \in \mathbb{Z}}$ is a frame.

The statement about the ratio of the frame bounds follows directly from the expression for the frame bounds in Theorem 11.4.2.   □

Proposition 11.5.3 is formulated as an existence result, and the usable values of b are somewhat hidden in the proof: after choosing ϵ such that $4\epsilon \|g\|_{W,a} + 2\epsilon^2 < C$, we have to choose N "large enough," and then all $b < \frac{1}{2aN}$ will lead to a frame. However, for a concrete function g, we will often be able to follow the proof directly and find an interval $]0, b_0]$ explicitly such that $\{E_{mb}T_{na}g\}_{m,n\in\mathbb{Z}}$ is a frame for all $b \in]0, b_0]$ (Exercise 11.5). But the same exercise shows that in practice it might be preferable to work directly with Theorem 11.4.2 because the estimates used in the proof of Proposition 11.5.3 make b_0 unnecessarily small.

The second part of Proposition 11.5.3 is very relevant for applications: it shows that if $g \in W$ and $\sum_{n\in\mathbb{Z}} |g(x - na)|^2$ is "almost constant," then g generates "almost tight" frames for small values of b. If we consider the function $g(x) = \frac{1}{1+x^2}$ from Exercise 11.5, then

$$1.52 \le \sum_{n\in\mathbb{Z}} |g(x - n)|^2 \le 1.62, \quad \forall x \in \mathbb{R};$$

therefore, as $b \to 0$, the ratio between the frame bounds in Proposition 11.5.3 will be of the size $\frac{1.621}{1.52} \sim 1.08$.

We will mainly use the Wiener space as a technical tool. However, we note that for continuous windows in W an elegant characterization of the frame property of $\{E_{mb}T_{na}g\}_{m,n\in\mathbb{Z}}$ (for the case where ab is irrational) was given in [346]. It is formulated in terms of the bi-infinite matrix-valued function $(G_{m,k}(\cdot))_{m,k\in\mathbb{Z}}$, whose entries are given in (11.12)

Proposition 11.5.4 *Assume that g is a continuous function belonging to the Wiener space W and that $ab \notin \mathbb{Q}$. Let $(G_{m,k}(\cdot))_{m,k\in\mathbb{Z}}$ denote the bi-infinite matrix-valued function, whose entries are given in (11.12). Then $\{E_{mb}T_{na}g\}_{m,n\in\mathbb{Z}}$ is a frame if and only if there exists a single $\xi \in [0, a]$ such that $(G_{m,k}(\xi))_{m,k\in\mathbb{Z}}$ is invertible on $\ell^2(\mathbb{Z})$.*

11.6 The Frame Set and Special Functions

In Gabor analysis, the *frame set* for a function $g \in L^2(\mathbb{R})$ is defined as the set

$$\mathcal{F}_g := \{(a, b) \in \mathbb{R}_+^2 \mid \{E_{mb}T_{na}g\}_{m,n\in\mathbb{Z}} \text{ is a frame for } L^2(\mathbb{R})\}. \quad (11.32)$$

It is usually difficult to determine the exact set $\mathcal{F}_g$ for a given function $g \in L^2(\mathbb{R})$. From Theorem 11.3.1 we know that a necessary condition for $\{E_{mb}T_{na}g\}_{m,n\in\mathbb{Z}}$ to be a frame for $L^2(\mathbb{R})$ is that $ab \le 1$; and for a given value of $a > 0$, we can often use the proof of Proposition 11.5.3 to find an interval $]0, b_0[$ such that $\{E_{mb}T_{na}g\}_{m,n\in\mathbb{Z}}$ is a frame for all $b \in]0, b_0[$. However, Proposition 11.5.3 is based on Theorem 11.4.2, and the estimates used in the proofs make the results suboptimal. Also, the general characterization of Gabor frames in Theorem 11.2.3 is usually too difficult to apply directly.

In this section we discuss some functions and classes of functions for which the frame set is known.

1) The Gaussian $g(x) = e^{-x^2}$.

The Gaussian was the first function for which the frame set was characterized:

Theorem 11.6.1 *Let $a, b > 0$ and consider $g(x) = e^{-x^2}$. Then the Gabor system $\{E_{mb}T_{na}g\}_{m,n\in\mathbb{Z}}$ is a frame if and only if $ab < 1$.*

The case $ab = 1$ is the easiest part. In fact, if $\{E_{mb}T_{na}g\}_{m,n\in\mathbb{Z}}$ was a frame for $ab = 1$, it would be a Riesz basis by Theorem 11.3.1; by the Balian–Low theorem, this is clearly not the case. That the Gaussian generates a frame if $ab < 1$ was proved in 1991 by Lyubarskii [505] and independently by Seip and Wallsten [571, 575] (that a fixed value of $a > 0$ will lead to a frame $\{E_{mb}T_{na}g\}_{m,n\in\mathbb{Z}}$ for sufficiently small values of b is clear from Proposition 11.5.3, but this is a much weaker statement). A historical note: Daubechies and Grossmann [243] proved around 1987 that $\{E_{mb}T_{na}g\}_{m,n\in\mathbb{Z}}$ is a frame whenever $ab < 0.994$ and conjectured the general result. The original proofs of the full result are complicated and use advanced complex analysis. Janssen gave later a shorter proof in [425]. A further analysis of the limiting case $ab = 1$ is given by Lyubarskii and Seip in the paper [507].

An interesting analysis of the behavior of the frame bounds near the critical density $ab = 1$ was performed in [72] by Borichev, Gröchenig, and Lyubarskii. Note that the result deals with the case $a = b$:

Proposition 11.6.2 *Let $g(x) = e^{-x^2}$. There exist constants $c, C > 0$ such that for each $a \in]1/2, 1[$ the frame bounds $A(a)$ and $B(a)$ for the frame $\{E_{ma}T_{na}g\}_{m,n\in\mathbb{Z}}$ satisfy that*

$$c(1 - a^2) \le A(a) \le C(1 - a^2), \qquad c < B(a) < C.$$

Note that Proposition 11.6.2 implies that the condition number for the frame operator for $\{E_{ma}T_{na}g\}_{m,n\in\mathbb{Z}}$ belongs to the interval

$$\left[\frac{c}{C}(1 - a^2)^{-1}, \frac{C}{c}(1 - a^2)^{-1} \right]$$

for each $a \in]1/2, 1[$.

The results in [505, 571], and [575] are actually much more general than described here: they also cover irregular Gabor systems; see Section 13.4.

2) The hyperbolic secant $g(x) = \frac{1}{\cosh(\pi x)}$

The hyperbolic secant was studied by Janssen and Strohmer [435], who proved that $\{E_{mb}T_{na}g\}_{m,n\in\mathbb{Z}}$ is a frame whenever $ab < 1$. The proof is based on Theorem 11.2.3 and the Zak transform, which we introduce in Section 13.2. The estimates for the frame bounds that were obtained for the Gaussian in Theorem 11.6.2 also hold for the hyperbolic secant.

The hyperbolic secant does not generate a frame when $ab = 1$.

3) The characteristic function $g(x) = \chi_{[0,c[}, \ c > 0$

For obvious reason, the question of characterizing the parameters $a, b, c > 0$ for which $\{E_{mb}T_{na}\chi_{[0,c]}\}_{m,n\in\mathbb{Z}}$ is a frame is called the *abc problem* in the literature. A scaling of a characteristic function is again (a multiple of) a characteristic function, so by Proposition 11.2.4 we can assume that $b = 1$. A detailed analysis performed by Janssen [432] shows that

(i) $\{E_m T_{na}\chi_{[0,c]}\}_{m,n\in\mathbb{Z}}$ is not a frame if $c < a$ or $a > 1$.

(ii) $\{E_m T_{na}\chi_{[0,c]}\}_{m,n\in\mathbb{Z}}$ is a frame if $1 \geq c \geq a$.

(iii) $\{E_m T_{na}\chi_{[0,c]}\}_{m,n\in\mathbb{Z}}$ is not a frame if $a = 1$ and $c > 1$.

Assuming now that $a < 1, c > 1$, we further have

(iv) $\{E_m T_{na}\chi_{[0,c]}\}_{m,n\in\mathbb{Z}}$ is a frame if $a \notin \mathbb{Q}$ and $c \in]1,2[$.

(v) $\{E_m T_{na}\chi_{[0,c]}\}_{m,n\in\mathbb{Z}}$ is not a frame if $a = p/q \in \mathbb{Q}$, $gcd(p,q) = 1$, and $2 - \frac{1}{q} < c < 2$.

(vi) $\{E_m T_{na}\chi_{[0,c]}\}_{m,n\in\mathbb{Z}}$ is not a frame if $a > \frac{3}{4}$ and $c = L - 1 + L(1-a)$ with $L \in \mathbb{N}, L \geq 3$.

(vii) $\{E_m T_{na}\chi_{[0,c]}\}_{m,n\in\mathbb{Z}}$ is a frame if $|c - \lfloor c \rfloor - \frac{1}{2}| < \frac{1}{2} - a$.

The graphical illustration of this result is known as *Janssen's tie*. The surprisingly complicated structure of the frame set for the characteristic functions indicates how complicated it is to find the exact range of parameters a, b which generate a frame for a given function g.

The full abc problem was finally solved in 2012 by X. R. Dai and Q. Sun; see [239] for this very impressive work. Note that the proof is done via a decision tree with many branches, which in each case either lead to a positive conclusion or a negative conclusion. Thus, the conditions on the parameters a, b, c are not explicit, as in the cases treated by Janssen.

4) The exponential functions $g(x) = e^{-|x|}$ **and** $g(x) = e^{-x}\chi_{[0,\infty[}(x)$

Janssen treated the function $g(x) = e^{-|x|}$ in [433] and showed that $\{E_{mb}T_{na}g\}_{m,n\in\mathbb{Z}}$ is a frame for all $a, b > 0$ such that $ab < 1$. The one-sided exponential $g(x) = e^{-x}\chi_{[0,\infty[}(x)$ was considered by Janssen in [428]; in this case $\{E_{mb}T_{na}g\}_{m,n\in\mathbb{Z}}$ is a frame if and only if $ab \leq 1$.

5) The totally positive functions

In 2012 Gröchenig and Stöckler [357] were able to identify a class of functions for which the frame set can be characterized. The paper is based on the following concepts:

Definition 11.6.3 *Consider a function $g : \mathbb{R} \to \mathbb{R}$.*

(i) *Assume that for any $N \in \mathbb{N}$ and any sequences $\{x_k\}_{k=1}^N, \{y_k\}_{k=1}^N$ of real numbers for which*

$$x_1 < x_2 < \cdots < x_N, \qquad y_1 < y_2 < \cdots < y_N,$$

the determinant of the $N \times N$ matrix with entries $g(x_j - y_k), j, k = 1, \ldots, N$, is nonnegative. Then the function g is said to be totally positive.

(ii) *A totally positive function $g \in L^2(\mathbb{R})$ for which the Fourier transform has the form*

$$\widehat{g}(\gamma) = C \prod_{\nu=1}^M (1 + 2\pi i \gamma a_\nu)^{-1} \tag{11.33}$$

for some $M \in \mathbb{N}, a_\nu \in \mathbb{R} \setminus \{0\}, C > 0$, is said to be of finite type M.

Among the totally positive functions, we find several of the functions discussed above, namely,

$$g(x) = e^{-x^2}, \ g(x) = e^{-|x|}, \ g(x) = e^{-x}\chi_{[0,\infty[}(x).$$

We also note that the function $g(x) = e^{-x}\chi_{[0,\infty[}(x)$ is of finite type: in fact, for this function

$$\widehat{g}(\gamma) = \int_0^\infty e^{-x} e^{-2\pi i \gamma x} \, dx = \frac{1}{1 + 2\pi i \gamma},$$

corresponding to (11.33) with $C = 1, M = 1$, and $a_1 = 1$. A similar calculation shows that $g(x) = e^{-|x|}$ is of finite type $M = 2$. In contrast, the Gaussian $g(x) = e^{-x^2}$ is obviously not of finite type. In general, a partial fraction decomposition of the expression in (11.33) shows that the totally positive functions of finite type are finite linear combinations of one-sided exponential functions multiplied with monomials. It can be shown that the totally positive functions of finite type $M \geq 2$ belong to Feichtinger's algebra $\mathcal{S}_0$.

The main result in [357] reads as follows:

Theorem 11.6.4 *Assume that $g \in L^2(\mathbb{R})$ is a totally positive function of finite type $M \geq 2$. Then*

$$\mathcal{F}_g = \left\{ (a, b) \in \mathbb{R}_+^2 \,\middle|\, ab < 1 \right\}.$$

The proof of Theorem 11.6.4 in [357] shows that if $ab < 1$ and g is a totally positive function of finite type $M \geq 2$, the Gabor frame $\{E_{mb}T_{na}g\}_{m,n\in\mathbb{Z}}$ has a dual frame $\{E_{mb}T_{na}h\}_{m,n\in\mathbb{Z}}$ for a suitable function $h \in L^2(\mathbb{R})$; such a function h is called a *dual window*; see page 298. In the setting of Theorem 11.6.4, it can be chosen to have compact support.

In general we can only expect a given function $g \in L^2(\mathbb{R})$ to generate a frame $\{E_{mb}T_{na}g\}_{m,n\in\mathbb{Z}}$ for some of the parameters $a, b > 0$ satisfying the inequality $ab \leq 1$. In such a case, it is interesting to identify the *obstructions* to the frame property. An interesting result concerning odd functions g in the Feichtinger algebra $\mathcal{S}_0$ was proved by Lyubarskii and Nes in [506]: such a function does not generate a Gabor frame $\{E_{mb}T_{na}g\}_{m,n\in\mathbb{Z}}$ whenever $ab = \frac{n-1}{n}$ for some $n = 2, 3, \ldots$. Note that these obstructions lie on hyperbolic curves in the (a, b)-plane; a similar shape of obstruction curves was later reported for a class of sign-changing windows; see [188].

Note that the terminology *frame set* has been used in a different sense in the literature: in analogue with the name *wavelet set* (see Section 16.3), a measurable set $K \subset \mathbb{R}$ is called a *Gabor frame set* if $\{E_m T_n \chi_K\}_{m,n\in\mathbb{Z}}$ is a frame for $L^2(\mathbb{R})$. It is highly nontrivial to classify such sets: Casazza and Kalton [137] have proved that this problem is equivalent to the longstanding problem of classifying the integer sets $\{n_1 < n_2 < \cdots < n_k\}$ for which the function $f(z) = \sum_{j=1}^{k} z^{n_j}$ does not have any zero on the unit circle in the complex plane.

11.7 Gabor Frames Generated by B-Splines

The purpose of this section is to discuss the frame set for the B-splines B_N, $N \in \mathbb{N}$ (see Section A.8). Recall that with our definition of the B-splines, B_N is supported on $[-N/2, N/2]$. Exactly the same results hold for the translated B-spline $\widetilde{B_N}$ in (A.18), and in some of the proofs, we will shift freely between these two versions.

The case $N = 1$ is covered by the discussion of the characteristic functions on page 281, so we focus on the case $N \geq 2$. For the B-splines B_N, $N \geq 2$, the exact frame set is not known. We hope the presentation will stimulate the research on this interesting problem.

By Corollary 11.4.3, any B-spline B_N generates a Bessel sequence $\{E_{mb}T_{na}B_N\}_{m,n\in\mathbb{Z}}$ for all $a, b > 0$; thus, in order to obtain a frame, we only need to verify the lower frame condition. We know that the Gabor system $\{E_{mb}T_{na}B_N\}_{m,n\in\mathbb{Z}}$ only can be a frame if $ab \leq 1$; furthermore, $B_N \in C_c(\mathbb{R})$ for $N \geq 2$, so in this case we even need that $ab < 1$ by Proposition 4.2.2.

We first state a sufficient condition for the Gabor system $\{E_{mb}T_{na}B_N\}_{m,n\in\mathbb{Z}}$ to be a frame.

Corollary 11.7.1 *For $N \in \mathbb{N}$, the B-splines B_N generate a Gabor frame $\{E_{mb}T_{na}B_N\}_{m,n\in\mathbb{Z}}$ for all $(a, b) \in]0, N[\times]0, 1/N]$.*

Proof. For $a \in]0, N[$ and $b \in]0, 1/N[$, the result is an immediate consequence of Corollary 11.4.6. For $N = 1$, the case $b = 1/N = 1$ is contained in the analysis by Janssen; see page 281. For $N > 1$, we will analyze the case $b = 1/N$ via Theorem 11.4.2; since supp $B_N = [-N/2, N/2]$ and $B_N(-N/2) = B_N(N/2)$, we see that for $b = 1/N$, $a < N$ and $k \in \mathbb{Z} \setminus \{0\}$,

$$\sum_{n\in\mathbb{Z}} B_N(x - na)\overline{B_N(x - na - k/b)} = 0.$$

Since the function $\sum_{k\in\mathbb{Z}} |B_N(x - ka)|^2$ is bounded below, it follows that $\{E_{mb}T_{na}B_N\}_{m,n\in\mathbb{Z}}$ is indeed a frame for $b = 1/N, 0 < a < N$.   □

Let us collect some of the known obstructions to the frame property of $\{E_{mb}T_{na}B_N\}_{m,n\in\mathbb{Z}}$ within the region determined by the inequalities $0 < ab < 1$. One of the results is quite surprising: it shows that the values $b \in \{2, 3, \dots\}$ are excluded from the frame set, regardless of the considered value for the translation parameter a!

Proposition 11.7.2 *For the B-splines B_N, $N \geq 2$, the following hold:*

(i) $\{E_{mb}T_{na}B_N\}_{m,n\in\mathbb{Z}}$ *is not a frame if $N \leq a$;*

(ii) $\{E_{mb}T_{na}B_N\}_{m,n\in\mathbb{Z}}$ *is not a frame if $b \in \mathbb{N} \setminus \{1\}$.*

Proof. The result in (i) easily follows from Proposition 11.3.4: in fact, since B_N is a continuous function supported on $[-N/2, N/2]$, the lower bound in (11.17) is violated if $N \leq a$. In order to prove (ii), it follows from Corollary A.8.2 that for $b \in \mathbb{N} \setminus \{1\}$, $n \in \mathbb{Z}$, and $\nu = 1$,

$$\widehat{B_N}(1 - nb) = \left(\frac{\sin(\pi(1 - nb))}{\pi(1 - nb)}\right)^N = 0;$$

thus, the continuous map

$$\nu \mapsto \sum_{n\in\mathbb{Z}} |\widehat{B_N}(\nu - nb)|^2$$

vanishes at $\nu = 1$, and the lower condition in (11.18) is violated.   □

The constraint in Proposition 11.7.2 was originally discovered by Del Prete [251]. A more general result was later given by Gröchenig, Janssen, Kaiblinger, and Pfander [347]: it says that if $g \in C_c(\mathbb{R})$ satisfies the *partition of unity condition*

$$\sum_{n\in\mathbb{Z}} g(x - n) = 1, \; x \in \mathbb{R},$$

then $\{E_{mb}T_{na}g\}_{m,n\in\mathbb{Z}}$ is not a Gabor frame for $b \in \mathbb{N} \setminus \{1\}$.

Some further partial results were recently shown by Kloos and Stöckler [459] and Christensen, Kim, and Kim [188].

Proposition 11.7.3 *Let $N \in \mathbb{N} \setminus \{1\}$, and consider $a, b > 0$ such that $ab < 1$. Then the following hold:*

(i) $\{E_{mb}T_{na}B_N\}_{m,n\in\mathbb{Z}}$ *is a frame if there exists $k \in \mathbb{N}$ such that*

$$1/N < b < 2/N, \ N/2 \le ak < 1/b. \tag{11.34}$$

(ii) $\{E_{mb}T_{na}B_N\}_{m,n\in\mathbb{Z}}$ *is a frame if $b \in \{1, \frac{1}{2}, \dots, \frac{1}{N-1}\}$.*

(iii) $\{E_{mb}T_{na}B_N\}_{m,n\in\mathbb{Z}}$ *is a frame if $a = \frac{k}{p}$ for some $k = 1, \dots, N-1$, $p \in \mathbb{N}$, and $b < 1/k$.*

The result in (i) appeared in [188], and (ii) in [459]. For $p = 1$, the result (iii) also appeared in [459]; the case of $p \ge 2$ in (iii) yields an oversampling of the case $p = 1$ and therefore also a frame.

For some time it was conjectured that $\{E_{mb}T_{na}B_N\}_{m,n\in\mathbb{Z}}$ for $N > 1$ is a frame whenever $ab < 1, a < N$, and $b \ne 2, 3, \dots$. This was recently disproved by Lemvig and Nielsen [486]; for example, the parameter choice $a = 1/3, b = 5/2$ does not lead to a frame whenever $N = 2$ or $N = 3$. The proof is based on Zak transform methods; see the discussion after Corollary 13.2.7. The results in [486] close the hope that a "simple" characterization of the frame set is possible for the B-splines B_N for $N > 2$: such a characterization will have the same complexity as the one we know for the characteristic functions. In short: it remains a very interesting and challenging problem to characterize the frame set for the B-splines B_N for $N \ge 2$.

11.8 Exercises

11.1 Let $a > 0$ and let g denote a measurable function for which the associated function G defined by (11.16) is bounded. Show that then $g \in L^2(\mathbb{R})$.

11.2 Prove Corollary 11.4.3.

11.3 Show that for the B-spline B_2, the system $\{E_{mb}T_{2n}B_2\}_{m,n\in\mathbb{Z}}$ cannot be a frame for any $b > 0$.

11.4 Let W denote the Wiener space.

(i) Prove (11.31).

(ii) Prove that $W \subset L^1(\mathbb{R}) \cap L^2(\mathbb{R})$.

(iii) Prove that every bounded measurable function with compact support belongs to W and that $\|g\|_{W,1} \le (|\mathrm{supp}(g)| + 1)\|g\|_\infty$.

11.5 Consider the function $g(x) = \frac{1}{1+x^2}$.

  (i) Show that $g \in W$ and find an estimate for $\|g\|_{W,1}$.

  (ii) Find a constant C such that

$$\sum_{n \in \mathbb{Z}} |g(x-n)|^2 \geq C, \ \forall x \in \mathbb{R}.$$

  (iii) Show that for all $N \in \mathbb{N}$,

$$\sum_{|n| \geq N} \|g\chi_{[n,n+1]}\|_\infty \leq \pi - 2\arctan(N-1) + \frac{1}{1+(N-1)^2}.$$

  (iv) Find via the proof of Proposition 11.5.3 a value $b_0 > 0$ such that $\{E_{mb}T_n g\}_{m,n \in \mathbb{Z}}$ is a Gabor frame for all $b \in]0, b_0]$.

  (v) Estimate numerically via Theorem 11.4.2 the range of b for which $\{E_{mb}T_n g\}_{m,n \in \mathbb{Z}}$ is a Gabor frame.

11.6 Consider a measurable function $g : \mathbb{R} \to \mathbb{C}$ satisfying the decay condition

$$|g(x)| \leq \frac{C}{1+x^2}, \ \forall x \in \mathbb{R},$$

for some $C \geq 0$. Show that $g \in W$.

11.7 Show that condition (CC) is satisfied for all $a, b > 0$ if $g \in W$.

11.8 Show by an example (maybe with $a = b = 1$) that the necessary condition in Proposition 11.3.4 does not suffice for $\{E_{mb}T_{na}g\}_{m,n \in \mathbb{Z}}$ being a Gabor frame. Similar statements with g replaced with $\widehat{g}$ (and separate discussions of the lower respectively upper conditions) can be found in [48].

11.9 Prove (11.21) under the assumptions in Lemma 11.4.1 and justify all the following interchanges of summation and integration in the proof.

11.10 Prove that $\{E_m T_{na} \chi_{[0,1]}\}_{m,n \in \mathbb{Z}}$ is a frame for $L^2(\mathbb{R})$ for all $a \in]0, 1]$.

12

Gabor Frames and Duality

The main issue in Chapter 11 was to state necessary and/or sufficient conditions for a Gabor system $\{E_{mb}T_{na}g\}_{m,n\in\mathbb{Z}}$ in $L^2(\mathbb{R})$ to form a frame. We will now take the next step and consider Gabor frames that are convenient to apply in practice. From the general frame theory in Chapter 5 and Chapter 6 we know that frames are particularly useful when the frame decomposition takes a simple form, which is the case if either the frame is tight or we have access to a convenient dual frame.

Section 12.1 gives a presentation of some conditions on a Gabor system $\{E_{mb}T_{na}g\}_{m,n\in\mathbb{Z}}$ which appear repeatedly, and a discussion of the connections among them. Some of the conditions will be needed already in Section 12.2, which deals with various representations of the frame operator. In Section 12.3 we prove that the canonical dual frame of a Gabor frame is itself a Gabor frame, and provide various characterizations of all the dual frames that have Gabor structure. More results about the canonical dual frame are given in Section 12.4. In Section 12.5 explicit constructions of dual pairs of Gabor frames generated by functions with compact support are provided; they can, e.g., be applied to any B-spline. Starting with a B-spline of sufficiently high order, any (finite) regularity of the elements in the frame and the dual frame can be obtained, but at the price of increased support size. An alternative construction is presented in Section 12.6, where arbitrary regularity can be combined with a small support size. In Section 12.7 we return to the topic treated in Section 6.4, where we showed that any pair of Bessel sequences in a separable Hilbert space can be extended to a pair of dual frames. We will now show that if the given sequences have Gabor structure and $ab \leq 1$, the extension can also be

© Springer International Publishing Switzerland 2016   287
O. Christensen, *An Introduction to Frames and Riesz Bases*,
Applied and Numerical Harmonic Analysis,
DOI 10.1007/978-3-319-25613-9_12

performed with Gabor systems. Section 12.8 shows that the deviation from equality in the duality condition (Theorem 12.3.4) in a direct way yields an expression for the reconstruction error. Finally, Section 12.9 provides characterizations and explicit constructions of tight Gabor frames.

12.1 Popular Gabor Conditions

In this section we will state a few conditions on Gabor systems $\{E_{mb}T_{na}g\}_{m,n\in\mathbb{Z}}$ in $L^2(\mathbb{R})$ that appear repeatedly in the literature, and discuss their interrelations. Some of the results will be needed already in Section 12.2.

For a Gabor system $\{E_{mb}T_{na}g\}_{m,n\in\mathbb{Z}}$, the set $\{(na, mb)\}_{m,n\in\mathbb{Z}} \subset \mathbb{R}^2$ is called the *time-frequency lattice*. It will be clear from Section 13.1 that there are close relationships between frame properties for g with respect to the lattice $\{(na, mb)\}_{m,n\in\mathbb{Z}}$, and frame properties with respect to the *dual lattice*, which is defined as the set $\{(n/b, m/a)\}_{m,n\in\mathbb{Z}}$. The first condition on a Gabor system we want to mention is related to this, and was introduced by Tolimieri and Orr [615] in 1995. A Gabor system $\{E_{mb}T_{na}g\}_{m,n\in\mathbb{Z}}$ is said to satisfy *condition (A)* if

$$\sum_{m,n\in\mathbb{Z}} |\langle g, E_{m/a}T_{n/b}g\rangle| < \infty. \tag{12.1}$$

Condition (A) is often needed in order to guarantee certain convergence properties of infinite series appearing in Gabor analysis. However, as observed by Gröchenig [340] it is preferable to avoid the condition if possible. For example, condition (A) is very sensitive to the choice of the lattice parameters: even for a simple function like $g = \chi_{[0,1]}$ and an arbitrary translation parameter $a > 0$, it is only satisfied for $b = 1/q, q \in \mathbb{N}$! Note that $\chi_{[0,1]}$ belongs to the Wiener space W treated in Section 11.5, i.e., stronger conditions are needed in order to avoid this kind of obstacle. One such condition is membership in Feichtinger's algebra: in [340] it is proved that condition (A) is satisfied for all $a, b > 0$ if $g \in S_0$.

Janssen introduced another condition in [429], which is frequently used in Gabor analysis. In contrast to condition (A), it only involves the function g and not the actual parameters a, b. We say that a function $g \in L^2(\mathbb{R})$ satisfies *condition (R)* if

$$\lim_{\epsilon\downarrow 0} \sum_{k\in\mathbb{Z}} \frac{1}{\epsilon} \int_{-\frac{1}{2}\epsilon}^{\frac{1}{2}\epsilon} |g(k+x) - g(k)|^2\, dx = 0. \tag{12.2}$$

Condition (R) might look restrictive, but it is actually satisfied for a dense class of functions in $L^2(\mathbb{R})$ (see Exercise 12.2 and page 371).

As we already proved in Theorem 11.4.2, a Gabor system $\{E_{mb}T_{na}g\}_{m,n\in\mathbb{Z}}$ is a Bessel sequence if *condition (CC)* is satisfied, i.e., if

$$\sup_{x\in[0,a]}\sum_{k\in\mathbb{Z}}\left|\sum_{n\in\mathbb{Z}}g(x-na)\overline{g(x-na-k/b)}\right|<\infty. \qquad (CC)$$

A variant of condition (CC) was used in [127]. We say that $\{E_{mb}T_{na}g\}_{m,n\in\mathbb{Z}}$ satisfies *condition (UCC)* or the *uniform condition (CC)*, if $\{E_{mb}T_{na}g\}_{m,n\in\mathbb{Z}}$ is a Bessel sequence and for any given $\epsilon>0$ there exists $K\in\mathbb{N}$ such that

$$\sup_{x\in[0,a]}\sum_{|k|\geq K}\left|\sum_{n\in\mathbb{Z}}g(x-na)\overline{g(x-na-k/b)}\right|<\epsilon. \qquad (UCC)$$

We emphasize that the Bessel condition is part of the definition of condition (UCC). As stated, condition (UCC) is strictly stronger than condition (CC), see [127]; this is no longer true if the Bessel assumption is removed (Exercise 12.4).

A slight modification of the proof of Proposition 11.5.2 shows that condition (CC) is satisfied for all $a,b>0$ if $g\in W$ (Exercise 11.7). A detailed analysis of the relationship between the mentioned conditions is given in [127], where the following results are proved (the references are to the page numbers, etc. in [127]):

- If $g\in W$, then condition (UCC) is satisfied for all $a,b>0$ (p.110).

- Membership in the Wiener space is a stronger condition than condition (CC), i.e., there are functions satisfying condition (CC) which are not in the Wiener space (Ex. 14.2).

- If $g\in L^2(\mathbb{R})$ is positive and real-valued, then $\{E_mT_ng\}_{m,n\in\mathbb{Z}}$ is a Bessel sequence if and only if g satisfies condition (CC) (Cor. 3.7). The equivalence does not hold if the condition of g being positive is removed (Ex. 3.8).

- If $\{E_{mb}T_{na}g\}_{m,n\in\mathbb{Z}}$ is a Bessel sequence and g satisfies condition (A), then g satisfies condition (UCC) (Prop. 4.12).

- There is a Gabor system $\{E_{mb}T_{na}g\}_{m,n\in\mathbb{Z}}$ satisfying condition (UCC) but not condition (A) (Ex. 4.13).

- If $ab\in\mathbb{Q}$ and $\{E_{mb}T_{na}g\}_{m,n\in\mathbb{Z}}$ is a frame satisfying condition (UCC), then also $S^{-1}g$ satisfies condition (UCC) (Th. 4.14).

12.2 Representations of the Gabor Frame Operator and Duality

The structure of a Gabor frame turns out to have important implications for its frame operator, which can be rewritten in several ways. Many central frame results are based on the obtained representations of the frame operator.

Walnut was the first to rewrite the frame operator S associated with a Gabor frame $\{E_{mb}T_{na}g\}_{m,n\in\mathbb{Z}}$. In his thesis [630] from 1989 (see also [631]) he obtained what is now known as the *Walnut representation:* it expresses Sf in terms of the functions

$$G_k(x) = \sum_{n\in\mathbb{Z}} g(x - na)\overline{g(x - na - k/b)}, \ k \in \mathbb{Z}. \qquad (12.3)$$

Note that these functions appear at several occasions in Chapter 11. By Lemma 11.2.2 the series defining $G_k(x)$ converges unconditionally for a.e. $x \in \mathbb{R}$.

Several variants of the Walnut representation are available in the literature. We will need the following version for the mixed frame operator associated with two functions $g, h \in L^2(\mathbb{R})$ in the Wiener space W; we ask the reader to compare with the version for shift-invariant systems in Theorem 10.2.1. Given any parameters $a, b > 0$, we know by Proposition 11.5.2 that $\{E_{mb}T_{na}g\}_{m,n\in\mathbb{Z}}$ and $\{E_{mb}T_{na}h\}_{m,n\in\mathbb{Z}}$ are Bessel sequences; denote the corresponding synthesis operators by T, respectively, U, i.e.,

$$T : \ell^2(\mathbb{Z}^2) \to L^2(\mathbb{R}), \ T\{c_{m,n}\}_{m,n\in\mathbb{Z}} = \sum_{m,n\in\mathbb{Z}} c_{m,n}E_{mb}T_{na}g, \qquad (12.4)$$

and

$$U : \ell^2(\mathbb{Z}^2) \to L^2(\mathbb{R}), \ U\{c_{m,n}\}_{m,n\in\mathbb{Z}} = \sum_{m,n\in\mathbb{Z}} c_{m,n}E_{mb}T_{na}h. \qquad (12.5)$$

Recall that the *mixed frame operator* associated with two Bessel sequences $\{E_{mb}T_{na}h\}_{m,n\in\mathbb{Z}}$ and $\{E_{mb}T_{na}g\}_{m,n\in\mathbb{Z}}$ is given by

$$UT^*f = \sum_{m,n\in\mathbb{Z}} \langle f, E_{mb}T_{na}g\rangle E_{mb}T_{na}h, \ f \in L^2(\mathbb{R}). \qquad (12.6)$$

Theorem 12.2.1 *Assume that $g, h \in W$ and let $a, b > 0$ be given. Define the functions $H_k : \mathbb{R} \to \mathbb{C}, k \in \mathbb{Z}$, by*

$$H_k(x) = \sum_{n\in\mathbb{Z}} h(x - na)\overline{g(x - na - k/b)}. \qquad (12.7)$$

Then the mixed frame operator associated with $\{E_{mb}T_{na}h\}_{m,n\in\mathbb{Z}}$ and $\{E_{mb}T_{na}g\}_{m,n\in\mathbb{Z}}$ has the representation

$$UT^*f = \frac{1}{b}\sum_{k\in\mathbb{Z}}(T_{k/b}f)H_k.$$

The series converges unconditionally in $L^2(\mathbb{R})$ for all $f \in L^2(\mathbb{R})$.

Proof. Let us write the mixed frame operator as

$$UT^*f = \sum_{n\in\mathbb{Z}} \left(\sum_{m\in\mathbb{Z}} \langle f, E_{mb}T_{na}g \rangle E_{mb} \right) T_{na}h. \qquad (12.8)$$

We will first consider the inner summand for fixed $n \in \mathbb{Z}$, i.e., the Fourier series

$$\sum_{m\in\mathbb{Z}} \langle f, E_{mb}T_{na}g \rangle e^{2\pi imbx} = \sum_{m\in\mathbb{Z}} \mathcal{F}(f\overline{T_{na}g})(mb)e^{2\pi imbx}. \qquad (12.9)$$

Let us now assume that f is bounded and has compact support. Then the Poisson summation formula applied to (12.9) yields that

$$\sum_{m\in\mathbb{Z}} \langle f, E_{mb}T_{na}g \rangle e^{2\pi imbx} = \sum_{m\in\mathbb{Z}} (f\overline{T_{na}g})(x+m/b)$$

$$= \sum_{m\in\mathbb{Z}} T_{m/b}(f\overline{T_{na}g})(x) \qquad (12.10)$$

for a.e. $x \in \mathbb{R}$ (see [340] for some technical details concerning the assumptions for the application of the Poisson summation formula). Using now (12.8) and an interchange of the summation (which is allowed because the summation in (12.10) is finite for each x), we arrive at

$$UT^*f(x) = \sum_{n\in\mathbb{Z}} \left(\sum_{m\in\mathbb{Z}} T_{m/b}(f\overline{T_{na}g})(x) \right) T_{na}h(x)$$

$$= \sum_{m\in\mathbb{Z}} (T_{m/b}f(x)) \left(\sum_{n\in\mathbb{Z}} \overline{T_{na+m/b}g}(x)T_{na}h(x) \right)$$

$$= \sum_{m\in\mathbb{Z}} (T_{m/b}f(x))H_m(x).$$

A density argument now extends the result to $L^2(\mathbb{R})$. □

   We refer to [127] for a detailed analysis of the Walnut representation and its convergence properties.

   We have already mentioned that there are close relationships between properties of a Gabor system $\{E_{mb}T_{na}g\}_{m,n\in\mathbb{Z}}$ and the Gabor system $\{E_{m/a}T_{n/b}g\}_{m,n\in\mathbb{Z}}$ with respect to the dual lattice. Results of that type were obtained in a more general context by Rieffel [554]; for Gabor systems, they were investigated almost at the same time by three groups of researchers, namely Daubechies, Landau & Landau [249]; Janssen [427]; and Ron & Shen [559]. There is a large overlap between their results, but their methods are quite different. A basic result is the following lemma, which actually constitutes "half of" the important duality principle, to which we return in Section 13.1. The proof is a slight modification of an argument by Jakobsen & Lemvig.

Lemma 12.2.2 *Let $g \in L^2(\mathbb{R})$ and $a, b > 0$ be given. Then $\{E_{mb}T_{na}g\}_{m,n\in\mathbb{Z}}$ is a Bessel sequence with bound B if and only if $\{E_{m/a}T_{n/b}g\}_{m,n\in\mathbb{Z}}$ is a Bessel sequence with bound abB.*

Proof. Let us first assume that $\{E_{mb}T_{na}g\}_{m,n\in\mathbb{Z}}$ is a Bessel sequence with bound B. Given any finite scalar sequence $\{c_{m,n}\}_{m,n\in\mathbb{Z}}$, consider the functions

$$\varphi_n : \mathbb{R} \to \mathbb{C}, \; \varphi_n(x) = \sum_{m\in\mathbb{Z}} c_{m,n}e^{2\pi imx/a}, \; n \in \mathbb{Z}.$$

In terms of the functions φ_n we obtain that

$$\left\| \sum_{m,n\in\mathbb{Z}} c_{m,n}E_{m/a}T_{n/b}g \right\|^2 = \int_{-\infty}^{\infty} \left| \sum_{m,n\in\mathbb{Z}} c_{m,n}e^{2\pi imx/a}g(x-n/b) \right|^2 dx$$

$$= \int_{-\infty}^{\infty} \left| \sum_{n\in\mathbb{Z}} \varphi_n(x)g(x-n/b) \right|^2 dx.$$

Using that the functions φ_n are periodic with period a now yields that

$$\left\| \sum_{m,n\in\mathbb{Z}} c_{m,n}E_{m/a}T_{n/b}g \right\|^2 = \int_0^a \sum_{m\in\mathbb{Z}} \left| \sum_{n\in\mathbb{Z}} \varphi_n(x)g(x-n/b-ma) \right|^2 dx.$$

Note that for each $x \in \mathbb{R}$, the function values $\{\varphi_n(x)\}_{n\in\mathbb{Z}}$ form a finite sequence. Using Theorem 11.2.3 (i) (or the explicit version in (11.14)), it follows that for a.e. $x \in \mathbb{R}$,

$$\sum_{m\in\mathbb{Z}} \left| \sum_{n\in\mathbb{Z}} \varphi_n(x)g(x-n/b-ma) \right|^2 \leq bB \sum_{n\in\mathbb{Z}} |\varphi_n(x)|^2;$$

thus,

$$\left\| \sum_{m,n\in\mathbb{Z}} c_{m,n}E_{m/a}T_{n/b}g \right\|^2 \leq \int_0^a bB \sum_{n\in\mathbb{Z}} |\varphi_n(x)|^2.$$

By Parseval's equation, see (3.35), $\int_0^a |\varphi_n(x)|^2 = a\sum_{m\in\mathbb{Z}}|c_{m,n}|^2$. Thus, we conclude that $\left\| \sum_{m,n\in\mathbb{Z}} c_{m,n}E_{m/a}T_{n/b}g \right\|^2 \leq abB\sum_{m\in\mathbb{Z}}|c_{m,n}|^2$, which implies that $\{E_{m/a}T_{n/b}g\}_{m,n\in\mathbb{Z}}$ is a Bessel sequence with bound abB (Exercise 3.13). The other implication follows by applying what we just showed to the Gabor system $\{E_{m/a}T_{n/b}g\}_{m,n\in\mathbb{Z}}$. □

In the following results we will need the synthesis operators associated with Gabor systems with respect to different generators and different parameters. For this reason we need a more detailed notation than before. We will denote the synthesis operator for $\{E_{mb}T_{na}g\}_{m,n\in\mathbb{Z}}$ by $T_{g;a,b}$ instead of just T. We first state a result from [249].

Proposition 12.2.3 *Let* $f, y, h \in L^2(\mathbb{R})$ *and* $a, b > 0$ *be given. If* $\{E_{mb}T_{na}g\}_{m,n\in\mathbb{Z}}, \{E_{mb}T_{na}f\}_{m,n\in\mathbb{Z}}$ *and* $\{E_{mb}T_{na}h\}_{m,n\in\mathbb{Z}}$ *are Bessel sequences, then*

$$T_{f;a,b}T^*_{g;a,b}h = \frac{1}{ab}T_{h;1/b,1/a}T^*_{g;1/b,1/a}f. \tag{12.11}$$

Proof. The complete proof in [249] is technical, and we will not provide all details. The main purpose of the following argument is to clarify how the dual lattice comes into play. We will prove Proposition 12.2.3 under the additional assumptions that f and h are compactly supported and bounded; this makes all needed interchanges of summations and integrals legal. First, let $\phi \in L^2(\mathbb{R})$. Then

$$T^*_{f;a,b}\phi = \{\langle \phi, E_{mb}T_{na}f\rangle\}_{m,n\in\mathbb{Z}}.$$

By Lemma 11.2.2,

$$\langle \phi, E_{mb}T_{na}f\rangle = \int_0^{1/b} \left(\sum_{k\in\mathbb{Z}} \phi(x - k/b)\overline{f(x - na - k/b)}\right) e^{-2\pi imbx} dx.$$

The interpretation of this equation in Lemma 11.2.2 in terms of Fourier coefficients together with Lemma 3.8.2 now gives that

$$\langle T_{f;a,b}T^*_{g;a,b}h, \phi\rangle = \langle T^*_{g;a,b}h, T^*_{f;a,b}\phi\rangle$$

$$= \sum_{n\in\mathbb{Z}}\sum_{m\in\mathbb{Z}} \langle h, E_{mb}T_{na}g\rangle\overline{\langle \phi, E_{mb}T_{na}f\rangle}$$

$$= \frac{1}{b}\sum_{n\in\mathbb{Z}} \left\langle \sum_{l\in\mathbb{Z}} h(\cdot - l/b)\overline{g(\cdot - na - l/b)}, \sum_{k\in\mathbb{Z}} \phi(\cdot - k/b)\overline{f(\cdot - na - k/b)}\right\rangle,$$

where the inner product in the last line is in $L^2(0, 1/b)$. When we write it out, we arrive at

$$\langle T_{f;a,b}T^*_{g;a,b}h, \phi\rangle = \frac{1}{b}\sum_{n\in\mathbb{Z}} \int_0^{1/b} \left(\sum_{l\in\mathbb{Z}} h(x - l/b)\overline{g(x - na - l/b)}\right.$$

$$\left. \times \sum_{k\in\mathbb{Z}} \overline{\phi(x - k/b)}f(x - na - k/b)\right) dx$$

$$= \frac{1}{b}\sum_{n\in\mathbb{Z}}\sum_{l\in\mathbb{Z}} \int_{-\infty}^{\infty} h(x - l/b)\overline{g(x - na - l/b)\phi(x)}f(x - na)dx.$$

If we apply this calculation with other choices of the generators and the parameters $1/b, 1/a$ instead of a, b, we obtain that

$$\langle T_{h;1/b,1/a}T^*_{g;1/b,1/a}f, \phi\rangle$$

$$= a\sum_{k\in\mathbb{Z}}\sum_{m\in\mathbb{Z}} \int_{-\infty}^{\infty} h(x - m/b)\overline{g(x - ka - m/b)\phi(x)}f(x - ka)dx.$$

This shows that $\langle T_{f;a,b}T^*_{g;a,b}h,\phi\rangle = \frac{1}{ab}\langle T_{h;1/b,1/a}T^*_{g;1/b,1/a}f,\phi\rangle$; since this holds for all $\phi \in L^2(\mathbb{R})$, the conclusion follows.   □

Written in terms of the involved sequences, (12.11) says that

$$\sum_{m,n\in\mathbb{Z}} \langle h, E_{mb}T_{na}g\rangle E_{mb}T_{na}f = \frac{1}{ab}\sum_{m,n\in\mathbb{Z}} \langle f, E_{m/a}T_{n/b}g\rangle E_{m/a}T_{n/b}h. (12.12)$$

The right-hand side of (12.12) converges unconditionally in $L^2(\mathbb{R})$ because $\{E_{m/a}T_{n/b}h\}_{m,n\in\mathbb{Z}}$ and $\{E_{m/a}T_{n/b}g\}_{m,n\in\mathbb{Z}} \in \ell^2(\mathbb{Z}^2)$ are Bessel sequences, see Lemma 12.2.2. We state some consequences of Proposition 12.2.3.

Corollary 12.2.4 *Let $g \in L^2(\mathbb{R})$ and $a, b > 0$ be given, and assume that $\{E_{mb}T_{na}g\}_{m,n\in\mathbb{Z}}$ is a frame with frame operator S. Then the following hold:*

(i) If $h \in L^2(\mathbb{R})$ and $\{E_{mb}T_{na}h\}_{m,n\in\mathbb{Z}}$ is a Bessel sequence, then

$$Sh = \frac{1}{ab}\sum_{m,n\in\mathbb{Z}} \langle g, E_{m/a}T_{n/b}g\rangle E_{m/a}T_{n/b}h.$$

(ii) $S^{-1}g = \frac{1}{ab}\sum_{m,n\in\mathbb{Z}}\langle S^{-1}g, E_{m/a}T_{n/b}S^{-1}g\rangle E_{m/a}T_{n/b}g.$

Both follow from (12.12): for the proof of the first part, let $f = g$; for the second part, replace h by g and replace g and f by $S^{-1}g$.

Janssen obtained similar results with slightly different assumptions in [427] (Theorem 12.2.5 below). One result only assumes that $\{E_{mb}T_{na}g\}_{m,n\in\mathbb{Z}}$ is a Bessel sequence, and delivers weak convergence of the frame operator for certain $f \in L^2(\mathbb{R})$; the second result requires that $\{E_{mb}T_{na}g\}_{m,n\in\mathbb{Z}}$ satisfies condition (A), and we obtain an unconditionally convergent representation.

Theorem 12.2.5 *Assume that $\{E_{mb}T_{na}g\}_{m,n\in\mathbb{Z}}$ is a Bessel sequence with frame operator S. Then, for any $f, h \in L^2(\mathbb{R})$ for which*

$$\sum_{m,n\in\mathbb{Z}} |\langle E_{m/a}T_{n/b}f, h\rangle|^2 < \infty$$

we have

$$\langle Sf, h\rangle = \frac{1}{ab}\sum_{m,n\in\mathbb{Z}} \langle g, E_{m/a}T_{n/b}g\rangle\langle E_{m/a}T_{n/b}f, h\rangle;$$

the series converges unconditionally. If $\{E_{mb}T_{na}g\}_{m,n\in\mathbb{Z}}$ also satisfies condition (A), then for all $f \in L^2(\mathbb{R})$,

$$Sf = \frac{1}{ab}\sum_{m,n\in\mathbb{Z}} \langle g, E_{m/a}T_{n/b}g\rangle E_{m/a}T_{n/b}f, \qquad (12.13)$$

with unconditional convergence in $L^2(\mathbb{R})$.

Condition (A) even implies that we have the representation

$$S = \frac{1}{ab} \sum_{m,n\in\mathbb{Z}} \langle g, E_{m/a}T_{n/b}g\rangle E_{m/a}T_{n/b},$$

with absolute convergence of the series in operator norm. The expression (12.13) is called the *Janssen representation of the frame operator;* by the properties of Feichtinger's algebra it is available if $g \in \mathcal{S}_0$.

12.3 The Duals of a Gabor Frame

For any frame which is not a Riesz basis, we know from Lemma 6.3.1 that there exist other dual frames than the canonical dual frame. When we consider a structured frame like a Gabor frame, other questions arises naturally. For example - does the canonical dual frame have Gabor structure as well? And do there exist other dual frames with Gabor structure?

In this section we prove that the canonical dual frame of a Gabor frame indeed has Gabor structure, and we provide various characterizations of all dual frames of this form. In Section 12.4 we return to a discussion of specific properties of the canonical dual frame.

For a Gabor frame $\{E_{mb}T_{na}g\}_{m,n\in\mathbb{Z}}$ with associated frame operator S, the frame decomposition, see Theorem 5.1.6, shows that

$$f = \sum_{m,n\in\mathbb{Z}} \langle f, S^{-1}E_{mb}T_{na}g\rangle E_{mb}T_{na}g, \ \forall f \in L^2(\mathbb{R}). \tag{12.14}$$

In order to use the frame decomposition, we need to be able to calculate the canonical dual frame $\{S^{-1}E_{mb}T_{na}g\}_{m,n\in\mathbb{Z}}$. This is usually difficult. Via the following lemma, we will be able to obtain a simplification.

Lemma 12.3.1 *Let $g \in L^2(\mathbb{R})$ and $a,b > 0$ be given, and assume that $\{E_{mb}T_{na}g\}_{m,n\in\mathbb{Z}}$ is a Bessel sequence with frame operator S. Then the following hold:*

(i) $SE_{mb}T_{na} = E_{mb}T_{na}S$ for all $m, n \in \mathbb{Z}$.

(ii) If $\{E_{mb}T_{na}g\}_{m,n\in\mathbb{Z}}$ is a frame for $L^2(\mathbb{R})$, then

$$S^{-1}E_{mb}T_{na} = E_{mb}T_{na}S^{-1}, \ \forall m, n \in \mathbb{Z}.$$

Proof. Let $f \in L^2(\mathbb{R})$, and assume that $\{E_{mb}T_{na}g\}_{m,n\in\mathbb{Z}}$ is a Bessel sequence. Using the commutator relations (2.25),

$$
\begin{aligned}
SE_{mb}T_{na}f &= \sum_{m',n'\in\mathbb{Z}} \langle E_{mb}T_{na}f, E_{m'b}T_{n'a}g\rangle E_{m'b}T_{n'a}g \\
&= \sum_{m',n'\in\mathbb{Z}} \langle f, T_{-na}E_{(m'-m)b}T_{n'a}g\rangle E_{m'b}T_{n'a}g \\
&= \sum_{m',n'\in\mathbb{Z}} \langle f, e^{2\pi i na(m'-m)b}E_{(m'-m)b}T_{(n'-n)a}g\rangle E_{m'b}T_{n'a}g.
\end{aligned}
$$

Performing the change of variables $m' \to m' + m$, $n' \to n' + n$ and using the commutator relations again,

$$
\begin{aligned}
SE_{mb}T_{na}f \\
= \sum_{m',n'\in\mathbb{Z}} e^{-2\pi i nam'b}\langle f, E_{m'b}T_{n'a}g\rangle E_{(m'+m)b}T_{(n'+n)a}g \\
= \sum_{m',n'\in\mathbb{Z}} e^{-2\pi i nam'b}\langle f, E_{m'b}T_{n'a}g\rangle e^{2\pi i nam'b} E_{mb}T_{na}E_{m'b}T_{n'a}g \\
= E_{mb}T_{na}Sf.
\end{aligned}
$$

This proves (i). In order to prove (ii) we recall that the frame operator is invertible whenever $\{E_{mb}T_{na}g\}_{m,n\in\mathbb{Z}}$ is a frame; now the result follows by applying the operator S^{-1} to both sides of the equality in (i). $\qquad\square$

Lemma 12.3.1 has a natural extension where the frame operator is replaced by the mixed frame operator associated with two Gabor systems (Exercise 12.3). The result also has important consequences for the structure of the canonical dual frame of a Gabor frame:

Theorem 12.3.2 *Let $g \in L^2(\mathbb{R})$ and $a, b > 0$ be given, and assume that $\{E_{mb}T_{na}g\}_{m,n\in\mathbb{Z}}$ is a Gabor frame with frame operator S. Then the following hold:*

(i) *The canonical dual frame also has the Gabor structure and is given by $\{E_{mb}T_{na}S^{-1}g\}_{m,n\in\mathbb{Z}}$.*

(ii) *The canonical tight frame associated with $\{E_{mb}T_{na}g\}_{m,n\in\mathbb{Z}}$ is given by $\{E_{mb}T_{na}S^{-1/2}g\}_{m,n\in\mathbb{Z}}$.*

Proof. The result in (i) is an immediate consequence of Lemma 12.3.1. Furthermore, Lemma 2.4.5 shows that the operator $S^{-1/2}$ is a limit of polynomials in S^{-1} in the strong operator topology; therefore, $S^{-1/2}$ commutes with $E_{mb}T_{na}$. Thus, according to the definition, the canonical tight frame associated with $\{E_{mb}T_{na}g\}_{m,n\in\mathbb{Z}}$ is given by

$$
\{S^{-1/2}E_{mb}T_{na}g\}_{m,n\in\mathbb{Z}} = \{E_{mb}T_{na}S^{-1/2}g\}_{m,n\in\mathbb{Z}};
$$

this proves (ii). $\qquad\square$

The function $S^{-1}g$ is called the *canonical dual window* or the *canonical dual generator*. Via Theorem 12.3.2, the frame decomposition (12.14) associated with a Gabor frame $\{E_{mb}T_{na}g\}_{m,n\in\mathbb{Z}}$ takes the form

$$f = \sum_{m,n\in\mathbb{Z}} \langle f, E_{mb}T_{na}S^{-1}g\rangle E_{mb}T_{na}g, \ \forall f \in L^2(\mathbb{R}). \qquad (12.15)$$

The version (12.15) of the frame decomposition is much more convenient than (12.14): instead of calculating the *double infinite* family $\{S^{-1}E_{mb}T_{na}g\}_{m,n\in\mathbb{Z}}$, it is enough to determine the function $S^{-1}g$ and then apply the modulation and translation operators. The result also gives a reason that even if $\{E_{mb}T_{na}g\}_{m,n\in\mathbb{Z}}$ contains a Riesz basis as a subfamily, it might not be an advantage to remove elements from $\{E_{mb}T_{na}g\}_{m,n\in\mathbb{Z}}$: the computational benefits from the lattice structure of $\{(na, mb)\}_{m,n\in\mathbb{Z}}$ will be lost, the operators $E_{mb}T_{na}$ will in general no longer commute with the frame operator, and it will be much more complicated to compute the elements in the canonical dual frame.

For overcomplete frames we know from Lemma 6.3.1 that there always exist other dual frames than the canonical dual frame. If the given frame has Gabor structure, the dual frames with Gabor structure are of course of particular interest. Given a frame $\{E_{mb}T_{na}g\}_{m,n\in\mathbb{Z}}$, any function $h \in L^2(\mathbb{R})$ such that $\{E_{mb}T_{na}h\}_{m,n\in\mathbb{Z}}$ is a dual frame is called a *dual window* or a *dual generator*. We also refer to $\{E_{mb}T_{na}g\}_{m,n\in\mathbb{Z}}$ and $\{E_{mb}T_{na}h\}_{m,n\in\mathbb{Z}}$ as a *pair of dual Gabor frames*. According to general frame theory (Lemma 6.3.2) the associated frame decompositions take the form

$$f = \sum_{m,n\in\mathbb{Z}} \langle f, E_{mb}T_{na}h\rangle E_{mb}T_{na}g \qquad (12.16)$$

$$= \sum_{m,n\in\mathbb{Z}} \langle f, E_{mb}T_{na}g\rangle E_{mb}T_{na}h, \ \forall f \in L^2(\mathbb{R}).$$

In this section we will provide various characterizations of the dual windows associated with a given Gabor frame. Having settled the question of existence of dual pairs of Gabor frames, the next issue is whether it is possible to construct such pairs with desirable properties. Of course, "desirable properties" depend on the concrete context, but natural candidates are compact support of the windows, high regularity, or membership of certain attractive window classes. All these issues will be addressed in the current chapter.

We know from Theorem 11.3.1 that if $\{E_{mb}T_{na}g\}_{m,n\in\mathbb{Z}}$ is a frame and $ab = 1$, then $\{E_{mb}T_{na}g\}_{m,n\in\mathbb{Z}}$ is actually a Riesz basis; by Proposition 4.2.2 this excludes that $g \in C_c(\mathbb{R})$. For the case $ab < 1$ we now show that there always exist Gabor frames with compactly supported windows of arbitrary smoothness, with dual windows enjoying the same properties:

Example 12.3.3 Assume that $ab < 1$. Take $\epsilon \in [0, 2^{-1}a]$ such that $a + 2\epsilon < 1/b$, and choose a function $g \in L^2(\mathbb{R})$ such that

- $\operatorname{supp} g \subseteq [0, a + 2\epsilon]$;

- $g = 1$ on $[\epsilon, a + \epsilon]$;

- $g \in C^\infty(\mathbb{R})$;

- $\|g\|_\infty = 1$.

Then the function $G(x) := \sum_{n \in \mathbb{Z}} |g(x - na)|^2$ is bounded below by 1 and bounded above by 3. Using Corollary 11.4.5 it follows that $\{E_{mb}T_{na}g\}_{m,n \in \mathbb{Z}}$ is a frame with bounds $1/b, 3/b$. Letting S denote the frame operator, the canonical dual frame is given by $\{E_{mb}T_{na}S^{-1}g\}_{m,n \in \mathbb{Z}}$, where $S^{-1}g = \frac{b}{G}g$. By construction, $S^{-1}g$ is compactly supported and belongs to $C^\infty(\mathbb{R})$. □

The pairs of dual frames are characterized in the following consequence of Theorem 10.1.7. The result is due to Ron and Shen, and it will play the key role in the construction of explicitly given dual pairs of Gabor frames in Sections 12.5–12.6.

Theorem 12.3.4 *Let* $g, h \in L^2(\mathbb{R})$ *and* $a, b > 0$ *be given. Two Bessel sequences* $\{E_{mb}T_{na}g\}_{m,n \in \mathbb{Z}}$ *and* $\{E_{mb}T_{na}h\}_{m,n \in \mathbb{Z}}$ *form dual frames if and only if for all* $n \in \mathbb{Z}$,

$$\sum_{k \in \mathbb{Z}} \overline{g(x - ka - n/b)} h(x - ka) = b\delta_{n,0}, \ \ a.e. \ x \in [0, a]. \qquad (12.17)$$

We leave the proof to the reader (Exercise 12.5). Often it is convenient to split the conditions (12.17) in the case $n = 0$, yielding

$$\sum_{k \in \mathbb{Z}} \overline{g(x - ka)} h(x - ka) = b, \ \ a.e. \ x \in [0, a], \qquad (12.18)$$

and the case $n \in \mathbb{Z} \setminus \{0\}$, yielding

$$\sum_{k \in \mathbb{Z}} \overline{g(x - ka - n/b)} h(x - ka) = 0 \ a.e. \ x \in [0, a]. \qquad (12.19)$$

Theorem 12.3.4 plays an important role in frame theory, not only in the study of dual pairs of frames. For example, it is often verified that a Gabor system $\{E_{mb}T_{na}g\}_{m,n \in \mathbb{Z}}$ is a frame by constructing a suitable function $h \in L^2(\mathbb{R})$ satisfying the conditions in Theorem 12.3.4; this is, e.g., the case for the proof of Theorem 11.6.4.

For the rest of this section we will consider various characterizations of the dual Gabor frames associated with a given Gabor frame. The general characterization of all dual frames in Theorem 6.3.7 of course also applies to Gabor frames, but if $\{E_{mb}T_{na}g\}_{m,n \in \mathbb{Z}}$ is an overcomplete frame, not all of these duals have the Gabor structure (Exercise 12.6). The duals with Gabor

structure are characterized in the famous *Wexler–Raz Theorem* [636]; we
will derive the result as a consequence of Theorem 10.1.7.

Theorem 12.3.5 *Let* $g, h \in L^2(\mathbb{R})$ *and* $a, b > 0$ *be given. Then, if
the two Gabor systems* $\{E_{mb}T_{na}g\}_{m,n\in\mathbb{Z}}$ *and* $\{E_{mb}T_{na}h\}_{m,n\in\mathbb{Z}}$ *are Bessel
sequences, they are dual frames if and only if*

$$\langle h, E_{m/a}T_{n/b}g \rangle = 0 \text{ for all } (m,n) \neq (0,0) \text{ and } \langle h, g \rangle = ab. \quad (12.20)$$

Proof. The Bessel sequences $\{E_{mb}T_{na}g\}_{m,n\in\mathbb{Z}}$ and $\{E_{mb}T_{na}h\}_{m,n\in\mathbb{Z}}$ are
dual frames if and only if the shift-invariant systems $\{T_{na}E_{mb}g\}_{m,n\in\mathbb{Z}}$ and
$\{T_{na}E_{mb}h\}_{m,n\in\mathbb{Z}}$ are dual frames. The generators for the two latter systems
are $g_m = E_{mb}g$ and $h_m = E_{mb}h$; by Theorem 10.1.7, they generate dual
frames if and only if

$$\sum_{m\in\mathbb{Z}} \overline{\widehat{g_m}(\nu)}\widehat{h}_m(\nu + k/a) = a\delta_{k,0}, \ k \in \mathbb{Z}, \ a.e. \ \nu \in \mathbb{R}.$$

In terms of the functions g and h this is equivalent to

$$\sum_{m\in\mathbb{Z}} \overline{\widehat{g}(\nu - mb)}\widehat{h}(\nu + k/a - mb) = a\delta_{k,0}, \ k \in \mathbb{Z}, \ a.e. \ \nu \in \mathbb{R}. \quad (12.21)$$

We can express this condition in terms of the Fourier coefficients in the
Fourier expansion with respect to $\{e^{2\pi in\nu/b}\}_{n\in\mathbb{Z}}$ for the b-periodic functions

$$\phi_k(\nu) := \sum_{m\in\mathbb{Z}} \overline{\widehat{g}(\nu - mb)}\widehat{h}(\nu + k/a - mb), \ k \in \mathbb{Z} :$$

in fact, (12.21) is equivalent to all Fourier coefficients for $\phi_k, k \neq 0$, being
zero and the Fourier coefficients $c_n, n \in \mathbb{Z}$, for ϕ_0 being zero for $n \neq 0$ and
equal to a for $n = 0$. The Wexler–Raz theorem is now a consequence of
the following computation, which yields the n-th Fourier coefficient for the
function ϕ_k in the Fourier expansion with respect to $\{e^{2\pi in\nu/b}\}_{n\in\mathbb{Z}}$:

$$\frac{1}{b}\int_0^b \phi_k(\nu)e^{-2\pi in\nu/b}d\nu$$

$$= \frac{1}{b}\int_0^b \sum_{m\in\mathbb{Z}} \overline{\widehat{g}(\nu - mb)}\widehat{h}(\nu + k/a - mb)e^{-2\pi in\nu/b}d\nu$$

$$= \frac{1}{b}\int_{-\infty}^{\infty} \overline{\widehat{g}(\nu)}\widehat{h}(\nu + k/a)e^{-2\pi in\nu/b}d\nu$$

$$= \frac{1}{b}\langle T_{-k/a}\widehat{h}, E_{n/b}\widehat{g}\rangle = \frac{1}{b}\langle \widehat{h}, T_{k/a}E_{n/b}\widehat{g}\rangle$$

$$= \frac{1}{b}\langle \mathcal{F}h, \mathcal{F}E_{k/a}T_{-n/b}g\rangle = \frac{1}{b}\langle h, E_{k/a}T_{-n/b}g\rangle.$$

$\square$

Note that the assumption of $\{E_{mb}T_{na}g\}_{m,n\in\mathbb{Z}}$ and $\{E_{mb}T_{na}h\}_{m,n\in\mathbb{Z}}$ being Bessel sequences is necessary in Theorem 12.3.5: as shown by Daubechies [241] there exist functions $g, h \in L^2(\mathbb{R})$ satisfying (12.20) which do not generate Bessel sequences, and hence do not form dual frames.

In the terminology used in Section 7.6, the Wexler–Raz theorem characterizes the functions h generating a dual Gabor system of a frame $\{E_{mb}T_{na}g\}_{m,n\in\mathbb{Z}}$ as the solutions to a moment problem with respect to the sequence $\{E_{m/a}T_{n/b}g\}_{m,n\in\mathbb{Z}}$. A more constructive procedure to find dual windows was given by Li [489], and in [163, 381] it was shown that the construction actually yields all the dual windows:

Proposition 12.3.6 *Let* $g \in L^2(\mathbb{R})$ *and* $a, b > 0$ *be given, and assume that* $\{E_{mb}T_{na}g\}_{m,n\in\mathbb{Z}}$ *is a frame for* $L^2(\mathbb{R})$. *Then a Gabor system* $\{E_{mb}T_{na}h\}_{m,n\in\mathbb{Z}}$ *is a dual Gabor frame if and only if the function* h *has the form*

$$h = S^{-1}g + \varphi - \sum_{m,n\in\mathbb{Z}} \langle S^{-1}g, E_{mb}T_{na}g\rangle E_{mb}T_{na}\varphi \qquad (12.22)$$

for some function $\varphi \in L^2(\mathbb{R})$ *for which* $\{E_{mb}T_{na}\varphi\}_{m,n\in\mathbb{Z}}$ *is a Bessel sequence.*

Proof. Applying Lemma 6.3.7 we see that if $\{E_{mb}T_{na}\varphi\}_{m,n\in\mathbb{Z}}$ is a Bessel sequence, then $\{E_{mb}T_{na}g\}_{m,n\in\mathbb{Z}}$ has the dual frame $\{k_{m,n}\}_{m,n\in\mathbb{Z}}$ given by

$$
\begin{aligned}
k_{m,n} &= S^{-1}E_{mb}T_{na}g + E_{mb}T_{na}\varphi \\
&\quad - \sum_{m',n'\in\mathbb{Z}} \langle S^{-1}E_{mb}T_{na}g, E_{m'b}T_{n'a}g\rangle E_{m'b}T_{n'a}\varphi \\
&= E_{mb}T_{na}(S^{-1}g + \varphi) \\
&\quad - \sum_{m',n'\in\mathbb{Z}} \langle E_{mb}T_{na}S^{-1}g, E_{m'b}T_{n'a}g\rangle E_{m'b}T_{n'a}\varphi.
\end{aligned}
$$

Exactly as in the proof of Lemma 12.3.1 (Exercise 12.3) one shows that

$$
\sum_{m',n'\in\mathbb{Z}} \langle E_{mb}T_{na}S^{-1}g, E_{m'b}T_{n'a}g\rangle E_{m'b}T_{n'a}\varphi
$$
$$
= E_{mb}T_{na} \sum_{m',n'\in\mathbb{Z}} \langle S^{-1}g, E_{m'b}T_{n'a}g\rangle E_{m'b}T_{n'a}\varphi;
$$

thus

$$
k_{m,n} = E_{mb}T_{na}\left(S^{-1}g + \varphi - \sum_{m',n'\in\mathbb{Z}} \langle S^{-1}g, E_{m'b}T_{n'a}g\rangle E_{m'b}T_{n'a}\varphi \right).
$$

This shows that the dual frame $\{k_{m,n}\}_{m,n\in\mathbb{Z}}$ indeed has Gabor structure, with a window of the form (12.22). On the other hand, letting $\phi' \in L^2(\mathbb{R})$ denote any function such that $\{E_{mb}T_{na}\varphi'\}_{m,n\in\mathbb{Z}}$ is a dual frame of $\{E_{mb}T_{na}g\}_{m,n\in\mathbb{Z}}$, taking $\varphi := \varphi'$ in the formula (12.22) clearly yields $h = \varphi'$. □

One can say that the functions $\varphi \in L^2(\mathbb{R})$ generating Bessel sequences $\{E_{mb}T_{na}\varphi\}_{m,n\in\mathbb{Z}}$ give a parametrization of the class of dual frames of $\{E_{mb}T_{na}g\}_{m,n\in\mathbb{Z}}$ maintaining the Gabor structure.

Depending on the specific purpose, various properties of dual frames might be relevant. Given a Gabor frame $\{E_{mb}T_{na}g\}_{m,n\in\mathbb{Z}}$ it is often important to search for a dual Gabor frame generated by a function with short support. For Gabor frames $\{E_{mb}T_{n}g\}_{m,n\in\mathbb{Z}}$ with a window supported on $[-1,1]$ and translation parameter $a = 1$ the following result from [180] guarantees the existence of a dual window with a certain support size, depending on the parameter b.

Theorem 12.3.7 *Let* $b \in [1/2,1[$, *and choose* $N \in \mathbb{N}$ *such that* $\frac{N-1}{N} \leq b < \frac{N}{N+1}$. *Assume that* $g \in L^2(\mathbb{R})$ *is supported on* $[-1,1]$ *and that* $\{E_{mb}T_{n}g\}_{m,n\in\mathbb{Z}}$ *is a frame for* $L^2(\mathbb{R})$. *Then* $\{E_{mb}T_{n}g\}_{m,n\in\mathbb{Z}}$ *has a dual frame* $\{E_{mb}T_{n}h\}_{m,n\in\mathbb{Z}}$, *generated by a function* $h \in L^2(\mathbb{R})$ *with* supp $h \subseteq [-N, N]$.

Theorem 12.3.7 has an interesting interpretation in terms of the redundancy of a frame $\{E_{mb}T_{n}g\}_{m,n\in\mathbb{Z}}$ with a window supported on $[-1,1]$: if the redundancy is larger than $1 + 1/N$, then the frame has a dual window supported on $[-N, N]$. In other words, the guaranteed size of the support for the dual window is increasing whenever the redundancy decreases. It would be interesting to generalize the result to arbitrary translation parameters $a > 0$ and windows g with an arbitrary support size.

The flexibility in the choice of dual windows has been explored at several places in the literature, and will also be discussed throughout the current chapter. For now, remember that the canonical dual frame associated with any frame minimizes the ℓ^2-norm of the coefficients in the frame expansion, see Lemma 5.4.2. However, one might be interested in minimizing other norms than the ℓ^2-norm. A concrete case appears in the paper [249]: instead of searching for the dual minimizing the ℓ^2-norm of the expansion coefficients, the authors find, for a specific operator L on $L^2(\mathbb{R})$, a dual frame $\{E_{mb}T_{na}h\}_{m,n\in\mathbb{Z}}$, for which $||Lh|| \leq ||L\varphi||$ for all dual frames $\{E_{mb}T_{na}\varphi\}_{m,n\in\mathbb{Z}}$.

We end this section by the announced proof of Theorem 11.3.1. The proof uses Theorem 12.3.2:

Proof of Theorem 11.3.1: Recall from the general frame theory that we with any frame $\{f_k\}_{k=1}^{\infty}$ can associate a canonical tight frame, see Theorem 6.1.1. Now, assume that $\{E_{mb}T_{na}g\}_{m,n\in\mathbb{Z}}$ is a frame, and denote the frame operator by S. By Theorem 12.3.2 the canonical tight frame associated with $\{E_{mb}T_{na}g\}_{m,n\in\mathbb{Z}}$ can be rewritten as

$$\{S^{-1/2}E_{mb}T_{na}g\}_{m,n\in\mathbb{Z}} = \{E_{mb}T_{na}S^{-1/2}g\}_{m,n\in\mathbb{Z}}. \qquad (12.23)$$

We first derive an equation, which will play a crucial role in the proof. Proposition 11.3.4 applied to the function $S^{-1/2}g$ implies that

$$\sum_{n\in\mathbb{Z}} |S^{-1/2}g(x-na)|^2 = b \text{ for a.e. } x \in \mathbb{R}. \qquad (12.24)$$

Since

$$||S^{-1/2}g||^2 = \int_{-\infty}^{\infty} |S^{-1/2}g(x)|^2 dx = \int_0^a \sum_{n\in\mathbb{Z}} |S^{-1/2}g(x-na)|^2 dx,$$

we conclude that

$$||S^{-1/2}g||^2 = ab. \qquad (12.25)$$

Note that (12.25) is a general result: we only used that $\{E_{mb}T_{na}g\}_{m,n\in\mathbb{Z}}$ is a frame for $L^2(\mathbb{R})$ in the proof.

In order to prove (i), we will show that $ab \leq 1$ for the arbitrary given frame $\{E_{mb}T_{na}g\}_{m,n\in\mathbb{Z}}$. Now, since $\{E_{mb}T_{na}S^{-1/2}g\}_{m,n\in\mathbb{Z}}$ is a tight frame with frame bounds equal to 1, Exercise 3.6 implies that $||S^{-1/2}g|| \leq 1$. Combining with the equation (12.25), we obtain that $ab \leq 1$ as desired.

For the proof of (ii), assume first that $\{E_{mb}T_{na}g\}_{m,n\in\mathbb{Z}}$ is a Riesz basis. Then by definition $\{S^{-1/2}E_{mb}T_{na}g\}_{m,n\in\mathbb{Z}}$ is also a Riesz basis, i.e., $\{E_{mb}T_{na}S^{-1/2}g\}_{m,n\in\mathbb{Z}}$ is a Riesz basis. By construction, this family is also a tight frame with frame bound 1, so the Riesz bounds are $A = B = 1$ by Theorem 5.4.1; in particular, this implies by Theorem 3.6.6 that $||S^{-1/2}g|| = 1$. Again via the equation (12.25), we conclude that $ab = 1$ as desired.

For the other implication in (ii) we now assume that $ab = 1$. Then, via (12.25),

$$||S^{-1/2}g||^2 = ab = 1,$$

and therefore $||E_{mb}T_{na}S^{-1/2}g|| = 1$ for all $m, n \in \mathbb{Z}$. Using Exercise 3.6, we conclude that $\{E_{mb}T_{na}S^{-1/2}g\}_{m,n\in\mathbb{Z}}$ is an orthonormal basis for $\mathcal{H}$, and therefore the family

$$\{E_{mb}T_{na}g\}_{m,n\in\mathbb{Z}} = \{S^{1/2}E_{mb}T_{na}S^{-1/2}g\}_{m,n\in\mathbb{Z}}$$

is a Riesz basis by definition. ☐

12.4   The Canonical Dual Window

In this section we will consider a Gabor frame $\{E_{mb}T_{na}g\}_{m,n\in\mathbb{Z}}$ with frame operator S, and analyze the properties of the canonical dual frame,

$$\{S^{-1}E_{mb}T_{na}g\}_{m,n\in\mathbb{Z}} = \{E_{mb}T_{na}S^{-1}g\}_{m,n\in\mathbb{Z}};$$

We will first use the general results for moment problems to find an alternative description of the generator $S^{-1}g$ for the canonical dual frame. In the literature the result is known as "the Wexler–Raz dual equals the canonical frame dual."

Proposition 12.4.1 *Let* $\{E_{mb}T_{na}g\}_{m,n\in\mathbb{Z}}$ *be a frame with frame operator* S. *Then* $S^{-1}g$ *is the unique minimal-norm solution to the moment problem*

$$\langle h, E_{m/a}T_{n/b}g\rangle = \delta_{m,0}\delta_{n,0}ab. \tag{12.26}$$

Letting $\widetilde{S}$ *denote the frame operator for* $\{E_{m/a}T_{n/b}g\}_{m,n\in\mathbb{Z}}$, *we further have*

$$S^{-1}g = ab\widetilde{S}^{-1}g.$$

Proof. We will need a key result in Gabor analysis, namely, the duality principle which we will prove in Theorem 13.1.1. Since $\{E_{mb}T_{na}g\}_{m,n\in\mathbb{Z}}$ is a frame, it implies that $\{E_{m/a}T_{n/b}g\}_{m,n\in\mathbb{Z}}$ is a Riesz sequence, i.e., a Riesz basis for $\mathcal{H} := \overline{\mathrm{span}}\{E_{m/a}T_{n/b}g\}_{m,n\in\mathbb{Z}}$. By Theorem 7.6.1 and Exercise 7.4, the moment problem (12.26) has a unique solution belonging to $\mathcal{H}$. This solution is $S^{-1}g$: in fact, $S^{-1}g$ is a solution by Theorem 12.3.5, and $S^{-1}g \in \mathcal{H}$ by Corollary 12.2.4. On the other hand, letting $\widetilde{S}$ denote the frame operator for $\{E_{m/a}T_{n/b}g\}_{m,n\in\mathbb{Z}}$, Theorem 7.6.2 shows that

$$S^{-1}g = ab \sum_{m,n\in\mathbb{Z}} \delta_{m,0}\delta_{n,0}\widetilde{S}^{-1}E_{m/a}T_{n/b}g = ab\widetilde{S}^{-1}g. \tag{12.27}$$

All other solutions to (12.26) are obtained by adding an element $f \in \mathcal{H}^{\perp}$ to the solution in $\mathcal{H}$. Thus, the special choice (12.27) minimizes the norm among all solutions to (12.26). □

We can also express the equations in (12.20) via an operator equation. Let

$$H : L^2(\mathbb{R}) \to \ell^2(\mathbb{Z}^2), \ Hf = \{\langle f, E_{m/a}T_{n/b}g\rangle\}_{m,n\in\mathbb{Z}}. \tag{12.28}$$

Note that H is the analysis operator associated with the Gabor system $\{E_{m/a}T_{n/b}g\}_{m,n\in\mathbb{Z}}$. In terms of H, (12.20) is equivalent to

$$Hh = ab\{\delta_{m,0}\delta_{n,0}\}_{m,n\in\mathbb{Z}}. \tag{12.29}$$

Corollary 12.4.2 *Let* $g \in L^2(\mathbb{R})$ *and* $a,b > 0$ *be given, and assume that* $\{E_{mb}T_{na}g\}_{m,n\in\mathbb{Z}}$ *is a frame. Then*

$$S^{-1}g = abH^*(HH^*)^{-1}\{\delta_{m,0}\delta_{n,0}\}_{m,n\in\mathbb{Z}}. \tag{12.30}$$

Proof. We again use that $\{E_{m/a}T_{n/b}g\}_{m,n\in\mathbb{Z}}$ is a Riesz sequence; it implies by Theorem 7.6.1 that the operator H in (12.28) is surjective. Thus, we know from Theorem 2.5.3 that the minimal-norm solution to (12.29) can be expressed via the pseudo-inverse of H. Using (2.12), we obtain (12.30), as desired. □

Equation (12.30) is known as the *Janssen representation* of the function generating the canonical dual frame of $\{E_{mb}T_{na}g\}_{m,n\in\mathbb{Z}}$.

We can obtain a more concrete expression for $S^{-1}g$. First we note that for any sequence $\{c_{m,n}\}_{m,n\in\mathbb{Z}} \in \ell^2(\mathbb{Z}^2)$,

$$HH^*\{c_{m,n}\}_{m,n\in\mathbb{Z}} = \left\{\left\langle \sum_{m',n'\in\mathbb{Z}} c_{m',n'} E_{m'/a}T_{n'/b}g, E_{m/a}T_{n/b}g \right\rangle\right\}_{m,n\in\mathbb{Z}}.$$

Let $\{e_{m,n}\}_{m,n\in\mathbb{Z}}$ be the canonical basis for $\ell^2(\mathbb{Z}^2)$; that is, $e_{m,n}$ is the sequence in $\ell^2(\mathbb{Z}^2)$ given by

$$e_{m,n} = \{\delta_{m,m'}\delta_{n,n'}\}_{m',n'\in\mathbb{Z}}.$$

We now re-index $\{e_{m,n}\}_{m,n\in\mathbb{Z}}$ as $\{e_k\}_{k=1}^\infty$ in an arbitrary way such that e_1 corresponds to $e_{0,0}$ (Exercise 12.1); denote the corresponding re-indexing of $\{E_{m/a}T_{n/b}g\}_{m,n\in\mathbb{Z}}$ by $\{g_k\}_{k=1}^\infty$. We can then represent HH^* via its matrix with respect to $\{e_k\}_{k=1}^\infty$, i.e., the bi-infinite matrix whose jk-th entry is $\langle HH^*e_k, e_j\rangle$, and (12.30) takes the form (see Corollary 7.6.3)

$$S^{-1}g = ab \sum_{j=1}^\infty (HH^*)_{j,1}^{-1} g_j. \tag{12.31}$$

If $e_j = e_{m,n}$ and $e_k = e_{m',n'}$, then

$$\langle HH^*e_k, e_j\rangle = \langle E_{m'/a}T_{n'/b}g, E_{m/a}T_{n/b}g\rangle,$$

the Gram matrix for $\{E_{m/a}T_{n/b}g\}_{m,n\in\mathbb{Z}}$. We write for short

$$(HH^*)_{m,n,m',n'} = \langle E_{m'/a}T_{n'/b}g, E_{m/a}T_{n/b}g\rangle, \quad m,n,m',n' \in \mathbb{Z}; \tag{12.32}$$

with this notation,

$$S^{-1}g = ab \sum_{m,n\in\mathbb{Z}} [(HH^*)^{-1}]_{m,n,0,0} E_{m/a}T_{n/b}g. \tag{12.33}$$

Bölcskei and Janssen proved in [98] that the canonical dual window $S^{-1}g$ inherits attractive decay properties from the window g. This is based on a fundamental result by Jaffard [414], to which we will refer several times in the sequel:

Lemma 12.4.3 *Suppose that $\{A_{k,\ell}\}_{k,\ell\in\mathbb{N}}$ is an invertible matrix and that there exist constants $C, \lambda > 0$ such that*

$$|A_{k,\ell}| \le Ce^{-\lambda|k-\ell|}, \ \forall k, \ell \in \mathbb{N}.$$

Then there exist constants $C', \lambda' > 0$ such that

$$|(A^{-1})_{k,\ell}| \le C'e^{-\lambda'|k-\ell|}, \ \forall k, \ell \in \mathbb{N}.$$

The constants C', λ' only depend on $\inf_{||x||=1} ||Ax||$ and $\sup_{||x||=1} ||Ax||$.

We say that a function $g \in L^2(\mathbb{R})$ *decays exponentially* if there exist constants $C, \lambda > 0$ such that

$$|g(x)| \leq Ce^{-\lambda|x|}, \quad a.e. \ x \in \mathbb{R}.$$

In [98], Lemma 12.4.3 is used to prove that if g decays exponentially and generates an overcomplete Gabor frame $\{E_{mb}T_{na}g\}_{m,n\in\mathbb{Z}}$, then $S^{-1}g$ also decay exponentially (possibly with a different exponent λ'). Using Proposition 11.2.5 it follows that if g generates an overcomplete frame, then exponential decay of $\widehat{g}$ implies exponential decay of $\mathcal{F}(S^{-1}g)$.

The same results hold with $S^{-1}g$ replaced by $S^{-1/2}g$. In particular this leads to the following important statement about the canonical tight frame associated with $\{E_{mb}T_{na}g\}_{m,n\in\mathbb{Z}}$:

Proposition 12.4.4 *Let $g \in L^2(\mathbb{R})$, and assume that g and $\widehat{g}$ decay exponentially. Let $a, b > 0$ be given and assume that $\{E_{mb}T_{na}g\}_{m,n\in\mathbb{Z}}$ is an overcomplete frame with frame operator S. Then $\{E_{mb}T_{na}S^{-1/2}\}_{m,n\in\mathbb{Z}}$ is a Parseval frame, for which $S^{-1/2}g$ as well as $\widehat{S^{-1/2}g}$ decay exponentially.*

Example 12.4.5 Let $g(x) = e^{-x^2}$. By Theorem 11.6.1 $\{E_{mb}T_{na}g\}_{m,n\in\mathbb{Z}}$ is an overcomplete frame for arbitrary parameters $a, b > 0$ with $ab < 1$. Denoting the frame operator by S, the above discussion shows that all the functions

$$g, \ \ \widehat{g}, \ \ S^{-1}g, \ \ \widehat{S^{-1}g}$$

decay exponentially. That is, the window and its canonical dual window are well-localized in time and frequency. Also, by Proposition 12.4.4 the canonical tight frame $\{E_{mb}T_{na}S^{-1/2}g\}_{m,n\in\mathbb{Z}}$ has the property that the window $S^{-1/2}g$ as well as $\widehat{S^{-1/2}g}$ decay exponentially.

Thus, from the point of view of time-frequency localization the Gaussian and its associated "tight window" $S^{-1/2}g$ are attractive Gabor windows; from a practical point of view it is less attractive that they do not have compact support, and that $S^{-1/2}g$ and $S^{-1}g$ are not given by easy expressions in terms of elementary functions. □

Prior to [98], Bölcskei considered in [94] the case where $\{E_{mb}T_{na}g\}_{m,n\in\mathbb{Z}}$ is *rationally oversampled*, i.e., $ab = p/q$ for some $p, q \in \mathbb{N}, q \geq p$. He proved that if g is compactly supported, then $S^{-1}g$ is compactly supported if and only if the frame operator is a multiplication operator.

In [427] Janssen proved that if $\{E_{mb}T_{na}g\}_{m,n\in\mathbb{Z}}$ is a frame and the window g belongs to the Schwartz space $\mathcal{S}$, then also $S^{-1}g \in \mathcal{S}$. A similar result holds for the Feichtinger algebra $\mathcal{S}_0$: in [349] Gröchenig and Leinert proved that if $\{E_{mb}T_{na}g\}_{m,n\in\mathbb{Z}}$ is a frame and $g \in \mathcal{S}_0$, then also $S^{-1}g \in \mathcal{S}_0$.

12.5   Explicit Construction of Dual Frame Pairs

In this section and the next we will use Theorem 12.3.4 to construct pairs of dual Gabor frames with explicitly given windows. The key to the constructions is the observation that if g and h are compactly supported functions, the condition (12.17) is automatically satisfied for $n \in \mathbb{Z} \setminus \{0\}$ whenever the modulation parameter b is sufficiently small; the reason is that the term n/b in (12.17) for $n \neq 0$ makes the supports of the involved functions move apart from each other. Thus we can focus on the equation

$$\sum_{k \in \mathbb{Z}} \overline{g(x - ka)} h(x - ka) = b, \quad a.e. \ x \in [0, a]. \tag{12.34}$$

The key condition in our analysis is that the integer-translates of the window g forms a partition of unity , see (12.36) below. This condition is satisfied not only for any B-spline $B_N, N \in \mathbb{N}$, but also for many other functions.

We will formulate all the conditions exclusively in terms of conditions on the window g. We will restrict our attention to dual windows h that are finite linear combinations of shifts of the window g, i.e.,

$$h(x) = \sum_{n=-N+1}^{N-1} a_n g(x + n), \tag{12.35}$$

for appropriate real coefficients $a_{-N+1}, a_{-N+2}, \ldots, a_{N-1}$. This structure allows us to control several properties of the dual window h: for example, compact support of g implies compact support of h, and regularity properties of g immediately transfers to h. Also, if the window g belongs to the Schwartz space, the Wiener space, or Feichtinger's algebra $\mathcal{S}_0$, a dual window of the form (12.35) will belong to the same space. We now state the first explicit construction. It appeared in [191] as a generalization of the main result in [164]. Note that for convenience we restrict our attention to the case $a = 1$; for generalizations to arbitrary translation parameters, we refer to the original sources [164, 191].

Theorem 12.5.1 *Let $N \in \mathbb{N}$. Let $g \in L^2(\mathbb{R})$ be a real-valued bounded function with supp $g \subset [0, N]$, for which*

$$\sum_{n \in \mathbb{Z}} g(x - n) = 1. \tag{12.36}$$

Let $b \in]0, \frac{1}{2N-1}]$. Consider any real scalar sequence $\{a_n\}_{n=-N+1}^{N-1}$ for which

$$a_0 = b \quad and \quad a_n + a_{-n} = 2b, \ n = 1, 2, \ldots N - 1, \tag{12.37}$$

and define $h \in L^2(\mathbb{R})$ by (12.35). Then g and h generate dual frames $\{E_{mb}T_n g\}_{m,n \in \mathbb{Z}}$ and $\{E_{mb}T_n h\}_{m,n \in \mathbb{Z}}$ for $L^2(\mathbb{R})$.

Proof. Note that with the definition (12.35), we have

$$\text{supp } h \subset [-N + 1, 2N - 1].$$

Thus the condition (12.17) is satisfied for $n \subset \mathbb{Z} \setminus \{0\}$ if $b \subset]0, \frac{1}{2N-1}]$ and we only need to check that

$$b = \sum_{k \in \mathbb{Z}} g(x + k)h(x + k), \quad x \in [0, 1];$$

due to the compact support of g, this is equivalent to

$$b = \sum_{k=0}^{N-1} g(x + k)h(x + k), \quad x \in [0, 1]. \tag{12.38}$$

To check that (12.38) holds, let

$$g_n(x) := \sum_{k=0}^{N-1} g(x + k)g(x + k + n).$$

Note that for $x \in [0, 1]$ and $n = 1, 2, \ldots, N - 1$,

$$
\begin{aligned}
g_{-n}(x) &= \sum_{k=0}^{N-1} g(x + k)g(x + k - n) = \sum_{k=n}^{N-1} g(x + k)g(x + k - n) \\
&= \sum_{\ell=0}^{N-1-n} g(x + \ell + n)g(x + \ell) = \sum_{\ell=0}^{N-1} g(x + \ell)g(x + \ell + n) \\
&= g_n(x).
\end{aligned}
$$

Putting this and (12.35) into the right-hand side of (12.38), we have that for $x \in [0, 1]$,

$$
\begin{aligned}
\sum_{k=0}^{N-1} g(x + k)h(x + k) &= \sum_{k=0}^{N-1} g(x + k) \sum_{n=-N+1}^{N-1} a_n g(x + k + n) \\
&= \sum_{n=-N+1}^{N-1} a_n \sum_{k=0}^{N-1} g(x + k)g(x + k + n) \\
&= \sum_{n=-N+1}^{N-1} a_n g_n(x) \\
&= a_0 g_0(x) + \sum_{n=1}^{N-1} (a_n + a_{-n}) g_n(x) \\
&= b \left[g_0(x) + 2 \sum_{n=1}^{N-1} g_n(x) \right]. \tag{12.39}
\end{aligned}
$$

On the other hand, for $x \in [0, 1]$ the partition of unity property implies that

$$\sum_{n=-N+1}^{N-1} g_n(x) = \sum_{n=-N+1}^{N-1} \sum_{k=0}^{N-1} g(x+k)g(x+k+n)$$

$$= \sum_{k=0}^{N-1} g(x+k) \sum_{n=-N+1}^{N-1} g(x+k+n)$$

$$= \sum_{k=0}^{N-1} g(x+k) = 1.$$

Since $g_{-n}(x) = g_n(x)$ for $x \in [0, 1]$ and $n = 1, 2, \ldots, N-1$, it follows that $g_0(x) + 2\sum_{n=1}^{N-1} g_n(x) = 1$ for $x \in [0, 1]$. This together with (12.39) implies that

$$\sum_{k=0}^{N-1} g(x+k)h(x+k) = b \text{ for } x \in [0, 1].$$

Thus (12.38) holds and the proof is completed.   □

A special choice of the coefficients a_n leads to the dual window considered already in [164]. For a given window g, this dual window has the shortest support among the ones in Theorem 12.5.1:

Corollary 12.5.2 *Under the assumptions in Theorem 12.5.1, the function*

$$h(x) = 2bg(x) + b \sum_{n=1}^{N-1} g(x+n)$$

generates a dual frame of $\{E_{mb}T_n g\}_{m,n\in\mathbb{Z}}$.

Another choice of the coefficients a_n in (12.37) implies that the dual window h inherits symmetry properties from the window g:

Corollary 12.5.3 *Under the assumptions in Theorem 12.5.1, the function*

$$h(x) = b \sum_{n=-N+1}^{N-1} g(x+n) \tag{12.40}$$

generates a dual frame of $\{E_{mb}T_n g\}_{m,n\in\mathbb{Z}}$. *The function* h *satisfies that* $h = b$ *on the support of* g. *Furthermore, if* g *is symmetric, then* h *is symmetric.*

Proof. The function in (12.40) appears by the choice $a_n = b$ in (12.37). For $x \in [0, N]$, the partition of unity property together with the compact support of g implies that $h(x) = b$. It is clear that h is symmetric in case g is symmetric.   □

The conditions in Theorem 12.5.1 are tailored to the properties of the (shifted) B-splines $\widetilde{B_N}$, $N \in \mathbb{N}$, see Appendix A.8. The B-splines $\widehat{B_2}$ and $\widehat{B_3}$ are shown on Figure 12.1. In other words, Theorem 12.5.1 applies to $g := \widetilde{B_N}$ for any $N \in \mathbb{N}$, and we obtain dual pairs of Gabor frames $\{E_{mb}T_n g\}_{m,n\in\mathbb{Z}}$ and $\{E_{mb}T_n h\}_{m,n\in\mathbb{Z}}$ with the following properties:

- The window g and the dual window h are explicitly given splines with compact support;

- For $N \geq 2$, the window g and the dual window h belong to Feichtingers algebra $\mathcal{S}_0$.

- Arbitrary high regularity of the window g and the dual window h can be obtained by choosing N sufficiently large; however, this increases the support of the windows as well;

- The redundancy of the frame $\{E_{mb}T_n g\}_{m,n\in\mathbb{Z}}$ for the choice $g = \widetilde{B_N}$ is $b^{-1} \geq 2N - 1$; thus, for this particular construction high regularity of the window implies high redundancy of the frame $\{E_{mb}T_n g\}_{m,n\in\mathbb{Z}}$.

Let us look at a concrete example.

Example 12.5.4 For the B-spline

$$
\widetilde{B_2}(x) = \begin{cases} x, & x \in [0,1[, \\ 2 - x, & x \in [1,2[, \\ 0, & x \notin [0,2[, \end{cases}
$$

we can use Theorem 12.5.3 for $b \in]0, 1/3]$. For $b = 1/3$ we obtain the symmetric dual

$$
h_2(x) = \begin{cases} 1/3x + 1/3, & x \in [-1,0[, \\ 1/3, & x \in [0,2[, \\ 1 - 1/3x, & x \in [2,3[, \\ 0, & x \notin [-1,3[. \end{cases} \tag{12.41}
$$

See Figure 12.2(a). For the B-spline

$$
\widetilde{B_3}(x) = \begin{cases} 1/2\, x^2, & x \in [0,1[, \\ -3/2 + 3x - x^2, & x \in [1,2[, \\ 9/2 - 3x + 1/2\, x^2, & x \in [2,3[, \\ 0, & x \notin [0,3[, \end{cases}
$$

and $b = 1/5$, we obtain the symmetric dual

$$
h_3(x) = \begin{cases} 1/10\, x^2 + 2/5\, x + 2/5, & x \in [-2,-1[, \\ -1/10\, x^2 + 1/5, & x \in [-1,0[, \\ 1/5, & x \in [0,3[, \\ -1/10\, x^2 + 3/5\, x - 7/10, & x \in [3,4[, \\ 1/10\, x^2 - x + 5/2, & x \in [4,5[, \\ 0, & x \notin [0,5[. \end{cases} \tag{12.42}
$$

See Figure 12.2(b). □

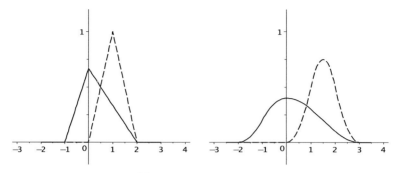

Figure 12.1. The B-spline $\widetilde{B}_2$ and the dual window h in Corollary 12.5.2 for $b = 1/3$ (figure to the left); and the B-spline $\widetilde{B}_3$ and the dual window h in Corollary 12.5.2 for $b = 1/5$.

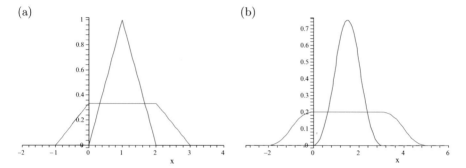

Figure 12.2. (a) The B-spline $\widetilde{B}_2$ and the dual window h_2 in (12.41). (b) The B-spline $\widetilde{B}_3$ and the dual window h_3 in (12.42).

12.6 Windows with Short Support and High Regularity

In Section 12.5 we saw a method for constructing dual pairs of frames based on a window that satisfies the partition of unity condition. We also saw that arbitrary regularity of the window and the dual window can be obtained by using B-splines of sufficiently high order; however, the price to pay for high regularity is that we obtain windows with large support, and also that the resulting frames are highly redundant.

In this section we will discuss two constructions that appeared in [185]. One of them yields classes of dual Gabor frames $\{E_{mb}T_n g\}_{m,n\in\mathbb{Z}}$, $\{E_{mb}T_n h\}_{m,n\in\mathbb{Z}}$ for which

- The window g and its dual window h are given explicitly;
- g and the dual window h are both supported on $[0,2]$;
- g and h can be constructed with arbitrary regularity;
- The redundancy of $\{E_{mb}T_n g\}_{m,n\in\mathbb{Z}}$ can be as low as 2.

The second construction has similar properties, except that the dual window h has slightly larger support and that the redundancy of the frames is at least 3. These constructions show that it actually is possible to obtain high regularity of the windows without having large support, and without introducing a high redundancy.

We will consider functions g, h of the form

$$g = G\chi_{[0,N]}, \quad h = H\chi_{[0,N]} \tag{12.43}$$

for some $N \in \mathbb{N}$ and some trigonometric polynomials G, H. By Corollary 11.4.3 the Gabor systems associated with such functions are Bessel sequences, regardless of the chosen translation and modulation parameters. The compact support of the functions g and h also implies that the duality condition (12.17) automatically is satisfied for $n \neq 0$ if we choose the modulation parameter b such that $b \leq 1/N$.

Thus, parallel to the case in Section 12.5 the main issue is that the function $P := \overline{G}H$ must satisfy the condition $\sum_{n\in\mathbb{Z}} P(x+n)\chi_{[0,N]}(x+n) = b$. Discarding the factor b then leads to the partition of unity constraint

$$\sum_{n\in\mathbb{Z}} P(x+n)\chi_{[0,N]}(x+n) = 1, \ x \in \mathbb{R}. \tag{12.44}$$

Motivated by this we will now analyze entire functions $P : \mathbb{C} \to \mathbb{C}$ satisfying the condition (12.44). We will first show that for such functions P the restriction to $\mathbb{R}$ is N-periodic. This implies that we have an extra tool at our disposal, namely Fourier expansions.

Lemma 12.6.1 *Let $N \in \mathbb{N}$. Then an entire function P satisfies (12.44) if and only if its restriction to $\mathbb{R}$ is N-periodic and the Fourier coefficients c_k in the expansion*

$$P(x) = \sum_{k\in\mathbb{Z}} c_k e^{2\pi i kx/N}, \ x \in \mathbb{R}, \tag{12.45}$$

satisfy that $c_k = \frac{1}{N}\delta_{k,0}$ for $k \in N\mathbb{Z}$.

Proof. Assume first that (12.44) holds. Then, for $x \in [0,1]$,

$$P(x) + P(x+1) + \cdots + P(x+N-1) = 1. \tag{12.46}$$

Since P is an entire function, (12.46) then holds for all $x \in \mathbb{R}$. Doing the similar calculation with x replaced by $x + 1$ and subtracting the two expressions shows that $P(x+N) = P(x)$, $\forall x \in [0, 1]$. The same calculation works with $[0, 1]$ replaced by any interval $[n, n+1]$, so we conclude that the restriction of P to $\mathbb{R}$ is N-periodic. Writing P as the Fourier series (12.45), the equation (12.46) takes the form

$$\sum_{k \in \mathbb{Z}} c_k \left[1 + e^{2\pi i k/N} + \cdots + \left(e^{2\pi i k/N} \right)^{N-1} \right] e^{2\pi i k x/N} = 1. \qquad (12.47)$$

We note that

$$1 + e^{2\pi i k/N} + \cdots + \left(e^{2\pi i k/N} \right)^{N-1} = \begin{cases} N, & k \in N\mathbb{Z} \\ 0, & k \notin N\mathbb{Z}. \end{cases} \qquad (12.48)$$

From (12.47) and (12.48), we see that $c_k = \frac{1}{N}\delta_{k,0}$ for $k \in N\mathbb{Z}$. Conversely, if P is N-periodic and satisfies that $c_k = \frac{1}{N}\delta_{k,0}$ for $k \in N\mathbb{Z}$, then for $x \in [0, 1]$,

$$\sum_{n \in \mathbb{Z}} P(x + n)\chi_{[0,N]}(x + n) = \sum_{n=0}^{N-1} P(x + n)$$

$$= \sum_{k \in \mathbb{Z}} c_k \left[1 + e^{2\pi i k/N} + \cdots + \left(e^{2\pi i k/N} \right)^{N-1} \right] e^{2\pi i k x/N} = 1$$

by (12.48). By periodicity (12.44) holds for all $x \in \mathbb{R}$.   □

Remember that our ultimate goal is to construct windows g, h of the form (12.43), with desired regularity. This means that we also need that the function $P\chi_{[0,N]} = \overline{G}H\chi_{[0,N]}$ satisfies certain regularity conditions. So far, Lemma 12.6.1 shows that $P\chi_{[0,N]}$ satisfies the partition of unity condition if we put restrictions on the Fourier coefficients c_k for the periodic function P for $k \in N\mathbb{Z}$. No restriction appears on the other Fourier coefficients – and this is exactly the freedom we will use in order to construct functions P such that $P\chi_{[0,N]}$ has desired regularity. The following result from [185] characterizes the regularity that can be obtained.

Theorem 12.6.2 *Let $N \in \mathbb{N}$. Assume that P is an N-periodic entire function satisfying that $c_k = \frac{1}{N}\delta_{0,k}, k \in N\mathbb{Z}$, and that the restriction of P to $\mathbb{R}$ is real-valued. Then the following hold.*

(a) *There does not exist P of this form such that $P\chi_{[0,N]} \in C^\infty(\mathbb{R})$;*

(b) *Fix $L \in \mathbb{N}$. Then $P\chi_{[0,N]} \in C^{L-1}(\mathbb{R})$ if and only if*

$$P(x) = \left(e^{\pi i x/N} \sin(\pi x/N) \right)^L A_L(x) \qquad (12.49)$$

for an N-periodic entire function $A_L(x) := \sum_{k \in \mathbb{Z}} a_k e^{2\pi i k x/N}$.

Proof. In order to prove (a), we note that if $P\chi_{[0,N]}$ belongs to $C^\infty(\mathbb{R})$, all
the derivatives at $x=0$ vanish. But P is an entire function and therefore
equal to its Taylor series, so this would imply that P is identically zero,
which is a contradiction. For the proof of (b), fix $L \in \mathbb{N}$. The "if" implication is clear, so suppose that $P\chi_{[0,N]} \in C^{L-1}(\mathbb{R})$. We use induction to
show (12.49). First, let D denote the differentiation operator and observe
that $P(0) = DP(0) = \cdots = D^{L-1}P(0) = 0$. Since $P(0) = \sum_{k\in\mathbb{Z}} c_k = 0$, we
have

$$P(x) = \sum_{k\neq 0} c_k(e^{2\pi kx/N} - 1).$$

Define P_+ and P_- by

$$P_+(x) := \sum_{k\in\mathbb{N}} c_k(e^{2\pi ikx/N} - 1), \quad P_-(x) := \sum_{k\in\mathbb{N}} c_{-k}(e^{-2\pi kix/N} - 1).$$

Then we see that

$$
\begin{aligned}
P_+(x) &= \sum_{k\in\mathbb{N}} c_k(e^{2\pi ix/N} - 1)\sum_{\ell=0}^{k-1} e^{2\pi i\ell x/N}\\
&= e^{\pi ix/N}\sin(\pi x/N)\left(2i\sum_{k\in\mathbb{N}} c_k \sum_{\ell=0}^{k-1} e^{2\pi i\ell x/N}\right)\\
&=: e^{\pi ix/N}\sin(\pi x/N)\Lambda_+(x).
\end{aligned}
$$

Similarly,

$$
\begin{aligned}
P_-(x) &= e^{\pi ix/N}\sin(\pi x/N)\left(-2i\sum_{k\in\mathbb{N}} c_{-k}\sum_{\ell=1}^{k} e^{-2\pi i\ell x/N}\right)\\
&=: e^{\pi ix/N}\sin(\pi x/N)\Lambda_-(x).
\end{aligned}
$$

Then we have

$$P(x) = P_+(x) + P_-(x) = e^{\pi ix/N}\sin(\pi x/N)A_1(x),$$

where $A_1(x) := \Lambda_+(x)+\Lambda_-(x)$ is an N-periodic function. In order to arrive
at (12.49) we will now inductively assume that, for some $1 \le \ell \le L-1$,

$$P(x) = \left(e^{\pi ix/N}\sin(\pi x/N)\right)^\ell A_\ell(x) \tag{12.50}$$

for an N-periodic entire function A_ℓ. By the Leibniz formula for the ℓ-th
derivative of a product, we have

$$D^\ell P(x) = \frac{1}{(2i)^\ell}\sum_{k=0}^{\ell}\binom{\ell}{k} D^k\left(e^{2\pi ix/N} - 1\right)^\ell D^{\ell-k}A_\ell(x). \tag{12.51}$$

Since $D^k\left(e^{2\pi ix/N} - 1\right)^\ell = \ell(\ell-1)\cdots(\ell-k+1)\left(e^{2\pi ix/N} - 1\right)^{\ell-k}\left(\frac{2\pi i}{N}\right)^k$,
we have $D^k\left(e^{2\pi ix/N} - 1\right)^\ell|_{x=0} = \ell!\left(\frac{2\pi i}{N}\right)^\ell \delta_{\ell,k}$. It follows from (12.51) that

$D^\ell P(0) = \frac{\ell!}{(2i)^\ell} \left(\frac{2\pi i}{N}\right)^\ell A_\ell(0)$. By assumption $D^\ell P(0) = 0$, so we conclude that $A_\ell(0) = 0$. By an argument similar to the case $P(0) = 0$, we see that

$$A_\ell(x) = e^{\pi i x/N} \sin(\pi x/N)\Lambda_{\ell+1}(x)$$

for an N-periodic entire function $\Lambda_{\ell+1}(x)$. This together with (12.50) leads to

$$P(x) = \left(e^{\pi i x/N} \sin(\pi x/N)\right)^{\ell+1} \Lambda_{\ell+1}(x),$$

which completes the induction. $\square$

Motivated by the desire to obtain windows with short support we will now restrict our attention to the case $N = 2$; the reader is referred to [185] for results covering the case $N \geq 3$. The following proposition shows that if we choose a trigonometric polynomial Q such that $Q\chi_{[0,2]}$ yields a partition of unity and belongs to $C^1(\mathbb{R})$, then we can generate polynomials that yield higher regularity: more precisely, for any $L \in \mathbb{N}$ we can find an explicitly given trigonometric polynomial P such that $P\chi_{[0,2]}$ has the partition of unity property and belongs to $C^{2L-1}(\mathbb{R})$.

Proposition 12.6.3 *Let* $N = 2$. *Consider a real-valued trigonometric polynomial* $Q(x) = \sum_k c_k e^{\pi i k x}$ *with* $c_k = \frac{1}{2}\delta_{k,0}$, $k \in 2\mathbb{Z}$. *Given* $L \in \mathbb{N}$, *define a trigonometric polynomial* P *by*

$$P(x) := Q^L(x) \sum_{k=0}^{L-1} \binom{2L-1}{k} Q^{L-1-k}(x)Q^k(x+1). \qquad (12.52)$$

Then $P\chi_{[0,2]}$ *satisfies the partition of unity property. If* $Q\chi_{[0,2]} \in C^1(\mathbb{R})$, *then* $P\chi_{[0,2]} \in C^{2L-1}(\mathbb{R})$.

Proof. Note that $Q\chi_{[0,2]}$ satisfies the partition of unity property by Lemma 12.6.1. Using the binomial formula, we have

$$1 = (Q(x)+Q(x+1))^{2L-1} = \sum_{k=0}^{2L-1} \binom{2L-1}{k} Q^{2L-1-k}(x)Q^k(x+1). \quad (12.53)$$

Take P as in (12.52). Then

$$P(x) = \sum_{k=0}^{L-1} \binom{2L-1}{k} Q^{2L-1-k}(x)Q^k(x+1).$$

Using the 2-periodicity of Q implies that

$$\begin{aligned} P(x+1) &= \sum_{k=0}^{L-1} \binom{2L-1}{k} Q^{2L-1-k}(x+1)Q^k(x) \\ &= \sum_{\ell=L}^{2L-1} \binom{2L-1}{\ell} Q^\ell(x+1)Q^{2L-1-\ell}(x). \end{aligned}$$

By (12.53), we have $P(x) + P(x + 1) - 1$, so $P\chi_{[0,2]}$ satisfies the partition of unity property, as desired. Furthermore, if $Q\chi_{[0,2]} \in C^1(\mathbb{R})$, then by Theorem 12.6.2 (b) we know that $Q(x) = \sin^2(\pi x/2)e^{\pi i x}A(x)$ for some 2-periodic entire function (actually a trigonometric polynomial) A. Using (12.52), it follows that

$$P(x) = \sin^{2L}(\pi x/2)e^{\pi i x L}\widetilde{A}(x)$$

for a 2-periodic entire function (trigonometric polynomial) $\widetilde{A}$. By Theorem 12.6.2 (b) we conclude that $P\chi_{[0,2]} \in C^{2L-1}(\mathbb{R})$.   □

In order to construct partition of unities based on functions with short support and high regularity, we just need to provide an example of a trigonometric polynomials Q satisfying the conditions in Proposition 12.6.3:

Example 12.6.4 Let

$$Q(x) := \sin^2(\pi x/2) = \left(\frac{e^{i\pi x/2} - e^{-i\pi x/2}}{2i}\right)^2 = -\frac{1}{4}e^{\pi i x} + \frac{1}{2} - \frac{1}{4}e^{-\pi i x}.$$

Then Q has the form described in Proposition 12.6.3, and $Q\chi_{[0,2]} \in C^1(\mathbb{R})$. Thus, for any $L \in \mathbb{N}$ we can use the procedure in Proposition 12.6.3 to construct real-valued trigonometric polynomials P such that $P\chi_{[0,2]}$ satisfies the partition of unity condition and belongs to $C^{2L-1}(\mathbb{R})$.   □

Let us now turn to the frame constructions. We first note that the construction of the trigonometric polynomial P in Proposition 12.6.3 implies that $P\chi_{[0,2]}$ has exactly the properties required in Theorem 12.5.1, combined with desired regularity. Thus, we immediately obtain a construction of a pair of dual Gabor frames:

Corollary 12.6.5 *Let* $L \in \mathbb{N}$ *take the trigonometric polynomial* Q *as in Proposition 12.6.3, and let*

$$P(x) := Q^L(x)\sum_{k=0}^{L-1}\binom{2L-1}{k}Q^{L-1-k}(x)Q^k(x+1).$$

Let $b \in]0, 1/3]$ *and assume that* $a_0 = b$, $a_1 + a_{-1} = 2b$. *Then the functions*

$$g(x) := (P\chi_{[0,2]})(x) \quad \text{and} \quad h(x) := \sum_{n=-1}^{1} a_n(P\chi_{[0,2]})(x+n) \quad (12.54)$$

belong to $C^{2L-1}(\mathbb{R})$ *and generate dual Gabor frames* $\{E_{mb}T_ng\}_{m,n\in\mathbb{Z}}$ *and* $\{E_{mb}T_nh\}_{m,n\in\mathbb{Z}}$ *for* $L^2(\mathbb{R})$.

The frames in Corollary 12.24 are generated by windows that are supported on $[0, 2]$, the dual windows have support within $[-1, 3]$ regardless of the desired regularity, and by taking $b = 1/3$ the redundancy is just 3. Let us consider a concrete example based on the function Q in Example 12.6.4:

Example 12.6.6 Let $L = 2, b \in]0, 1/3]$, and $Q(x) := \sin^2(\pi x/2)$. Define

$$P(x) = \sin^4(\pi x/2) \sum_{k=0}^{1} \binom{3}{k} \sin^{2(1-k)}(\pi x/2) \sin^{2k}(\pi(x+1)/2).$$

Then $g := P\chi_{[0,2]}$ and h defined as in (12.54) belong to $C^3(\mathbb{R})$ and generate dual Gabor frames $\{E_{mb}T_n g\}_{m,n \in \mathbb{Z}}$ and $\{E_{mb}T_n h\}_{m,n \in \mathbb{Z}}$ for $L^2(\mathbb{R})$. In Figure 12.3, we plot g and h for the choice $b = a_{-1} = a_0 = a_1 = 1/3$.   □

Our final construction shows that we can even obtain a pair of dual Gabor frames, generated by windows that are both supported on $[0, 2]$, and with redundancy as low as 2. We will formulate a concrete version of the result, where we take the function Q as in Example 12.6.4:

Corollary 12.6.7 *Let* $L_1, L_2 \in \mathbb{N}$, *and fix* $b \in]0, \frac{1}{2}]$. *Take* $Q(x) := \sin^2(\pi x/2)$. *Define*

$$g(x) = \sin^{2L_1}(\pi x/2)\chi_{[0,2]}(x)$$

and

$$h(x) = b\sin^{2L_2}(\pi x/2) \times$$
$$\left(\sum_{k=0}^{L_1+L_2-1} \binom{2L_1+2L_2-1}{k} Q^{L_1+L_2-1-k}(x)Q^k(x+1) \right) \chi_{[0,2]}(x).$$

Then $g \in C^{2L_1-1}(\mathbb{R})$, $h \in C^{2L_2-1}(\mathbb{R})$, *and the functions* $\{E_{mb}T_n g\}_{m,n \in \mathbb{Z}}$ *and* $\{E_{mb}T_n h\}_{m,n \in \mathbb{Z}}$ *form a pair of dual frames.*

Proof.   Given $L_1, L_2 \in \mathbb{N}$, let $L := L_1 + L_2$. Let $Q(x) := \sin^2(\pi x/2)$, and consider the trigonometric polynomial P in (12.52). Then $P\chi_{[0,2]} \in C^{2(L_1+L_2)-1}(\mathbb{R})$ and satisfies the partition of unity condition. Since $P = gh$, it follows that the duality condition (12.17) is satisfied for $n = 0$. The choice of b and the support sizes for g and h shows that (12.17) holds for $n \neq 0$ as well.   □

Figure 12.4 shows the windows g and h in Corollary 12.6.7 for $L_1 = L_2 = 2, b = 1/2$.

Note that construction of Gabor frames based on trigonometric polynomials appears at other places in the literature. In [244], Daubechies, Grossmann, and Meyer construct a tight Gabor frame based on the function $g(x) = \cos(x)\chi_{[-\pi/2,\pi/2]}(x)$, which is just a shifted and scaled version of the function $\sin(\pi x/2)\chi_{[0,2]}$. Also, in [107], the authors consider

(a)   (b)

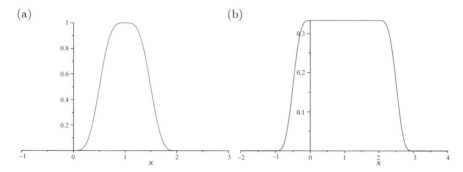

Figure 12.3. (a) The window g in Example 12.6.6. (b) The dual window h with $b = a_{-1} = a_0 = a_1 = 1/3$.

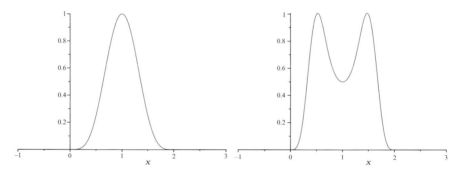

Figure 12.4. The windows g and h in Corollary 12.6.7 for $L_1 = L_2 = 2, b = 1/2$

frames generated by functions of the form $g_k(x) = \sin^k(\pi x/3)\,\chi_{[0,3]}(x)$ for parameters $k \in \mathbb{N}$. Interestingly, the results in [107] show that g_k generates a frame for all $b \in]0, 1/3]$ and all $k \in \mathbb{N}$; but only for $k < 6$ there is a dual Gabor frame $\{E_{mb}T_nh\}_{m,n\in\mathbb{Z}}$ for a function of the form $h(x) = a_0 g_k(x) + a_1 g_k(x+1) + a_2 g_k(x+2)$.

Also the work by I. Kim [453, 454] deals with construction of trigonometric pairs of dual frames. Let $m \in \mathbb{N}$ and consider the function $g(x) := \cos^{m+1}(\pi x/2)\chi_{[-1,1]}$. Then Kim shows that there exists a trigonometric polynomial h of degree $3m+1$ such that $h \in C^m(\mathbb{R})$ and g, h generate dual frames (with translation parameter $a = 1$ and sufficiently small modulation parameters b). Symmetry of g and h is built into the construction, and the redundancy can go down to two.

Thus, the construction in [453, 454] has similar features as the construction presented here, except that it is less explicit with regard to the dual window. Note that [453, 454] also contains results about the higher dimensional case.

The construction by I. Kim is similar to the work by Laugesen [479] with respect to the obtained properties. The main difference is that the constructions by Laugesen deal with polynomial splines rather than trigonometric splines.

12.7 Extension of Bessel Sequences to Dual Pairs

In Theorem 6.2.1 we saw that any Bessel sequence $\{f_k\}_{k=1}^\infty$ in a separable Hilbert space can be extended to a tight frame $\{f_k\}_{k=1}^\infty \cup \{p_j\}_{j\in J}$. If $\{f_k\}_{k=1}^\infty$ has a certain structure, it is natural to ask for the sequence $\{p_j\}_{j\in J}$ to have the same structure. In the paper [507], Li and W. Sun showed that if a Gabor system $\{E_{mb}T_{na}g_1\}_{m,n\in\mathbb{Z}}$ is a Bessel sequence in $L^2(\mathbb{R})$ and $ab \le 1$, then indeed there exists a Gabor system $\{E_{mb}T_{na}g_2\}_{m,n\in\mathbb{Z}}$ such that $\{E_{mb}T_{na}g_1\}_{m,n\in\mathbb{Z}}\cup\{E_{mb}T_{na}g_2\}_{m,n\in\mathbb{Z}}$ is a tight frame for $L^2(\mathbb{R})$. This naturally raises the next question: if the window g_1 has certain desirable properties, can we choose g_2 to have similar properties? For example, if g_1 has compact support, can we always choose g_2 to have compact support as well? In [507] a positive answer to this question is obtained, provided that the support size of the given window g_1 satisfies that $|\text{supp}\,g_1| \le b^{-1}$. Based on Theorem 6.4.1 and following [183] we will now consider the similar question of extension of a pair of Gabor Bessel sequences to a pair of dual Gabor frames. In particular, we will show that if the given windows have compact support and $ab \le 1$, we can *always* extend the Bessel systems to dual pairs of Gabor systems using windows with compact support; that is, the extension to a dual pair of frames removes the support condition that appears in the extension to a tight frame.

Theorem 12.7.1 *Let $\{E_{mb}T_{na}g_1\}_{m,n\in\mathbb{Z}}$ and $\{E_{mb}T_{na}h_1\}_{m,n\in\mathbb{Z}}$ be Bessel sequences in $L^2(\mathbb{R})$, and assume that $ab \le 1$. Then the following hold:*

(i) *There exists $g_2, h_2 \in L^2(\mathbb{R})$ such that $\{E_{mb}T_{na}g_1\}_{m,n\in\mathbb{Z}}\cup \{E_{mb}T_{na}g_2\}_{m,n\in\mathbb{Z}}$ and $\{E_{mb}T_{na}h_1\}_{m,n\in\mathbb{Z}} \cup \{E_{mb}T_{na}h_2\}_{m,n\in\mathbb{Z}}$ form a pair of dual frames for $L^2(\mathbb{R})$.*

(ii) *If g_1 and h_1 have compact support, the functions g_2 and h_2 can be chosen to have compact support.*

Proof. In order to prove (i), let T and U denote the synthesis operators for $\{E_{mb}T_{na}g_1\}_{m,n\in\mathbb{Z}}$ and $\{E_{mb}T_{na}h_1\}_{m,n\in\mathbb{Z}}$, respectively. Let $\{E_{mb}T_{na}r_1\}_{m,n\in\mathbb{Z}}$, $\{E_{mb}T_{na}r_2\}_{m,n\in\mathbb{Z}}$ denote any pair of dual frames for $L^2(\mathbb{R})$; for the case $ab < 1$, the existence of such a pair is guaranteed by Example 12.3.3, and for the case $ab - 1$ we leave the construction to the reader (Exercise 12.7). By the proof of Theorem 6.4.1, $\{E_{mb}T_{na}g_1\}_{m,n\in\mathbb{Z}} \cup \{(I-TU^*)E_{mb}T_{na}r_1\}_{m,n\in\mathbb{Z}}$ and $\{E_{mb}T_{na}h_1\}_{m,n\in\mathbb{Z}} \cup \{E_{mb}T_{na}r_2\}_{m,n\in\mathbb{Z}}$ are dual frames for $L^2(\mathbb{R})$. Thus we only need to show that $\{(I-TU^*)E_{mb}T_{na}r_1\}_{m,n\in\mathbb{Z}}$ has Gabor structure. Note that

$$(I-TU^*)f = f - \sum_{m,n\in\mathbb{Z}} \langle f, E_{mb}T_{na}h_1\rangle E_{mb}T_{na}g_1, \ f \in L^2(\mathbb{R}). \quad (12.55)$$

By a similar calculation as in the proof of Lemma 12.3.1(i) one can show (Exercise 12.3) that the operator $I - TU^*$ commutes with the time-frequency shift operators $E_{mb}T_{na}$. This proves that $\{(I-TU^*)E_{mb}T_{na}r_1\}_{m,n\in\mathbb{Z}}$ indeed has Gabor structure and concludes the proof of (i).

We now prove (ii). Take the functions r_1 and r_2 in the proof of (i) to be compactly supported; the possibility of this is again guaranteed by Example 12.3.3 and Exercise 12.7. Then we just need to show that the function $g_2 = (I - TU^*)r_1$ is compactly supported. Due to the compact support of the functions r_1 and h_1 there exists a number N such that

$$\langle r_1, E_{mb}T_{na}h_1\rangle = 0, \ \forall m \in \mathbb{Z} \text{ if } n \notin [-N, N].$$

Thus, by (12.55),

$$
\begin{aligned}
(I-TU^*)r_1 &= r_1 - \sum_{m,n\in\mathbb{Z}} \langle r_1, E_{mb}T_{na}h_1\rangle E_{mb}T_{na}g_1 \\
&= r_1 - \sum_{n=-N}^{N}\sum_{m\in\mathbb{Z}} \langle r_1, E_{mb}T_{na}h_1\rangle E_{mb}T_{na}g_1, \quad (12.56)
\end{aligned}
$$

which is clearly compactly supported.   □

In continuation of Theorem 12.7.1 one can show that if $g_1, h_1 \in C^\infty(\mathbb{R})$, then it is also possible to choose $g_2, h_2 \in C^\infty(\mathbb{R})$; we refer to [183] for a proof.

12.8 Approximately Dual Gabor Frames

Let us return to the general conditions for two Gabor systems being dual frames, see Theorem 12.3.4. It was observed by Christensen and Laugesen [192] that the *deviation* from equality in (12.18) and (12.19) gives a measure for "how far away" two Bessel sequences $\{E_{mb}T_{na}g\}_{m,n\in\mathbb{Z}}$ and $\{E_{mb}T_{na}h\}_{m,n\in\mathbb{Z}}$ are from being dual frames. Given two functions $g, h \in L^2(\mathbb{R})$ and two parameters $a, b > 0$ such that $\{E_{mb}T_{na}g\}_{m,n\in\mathbb{Z}}$ and $\{E_{mb}T_{na}h\}_{m,n\in\mathbb{Z}}$ are Bessel sequences, denote the corresponding synthesis operators by T, respectively, U, see (12.4) and (12.5). Then

$$\|I - UT^*\| = \sup_{f \in \mathcal{H}, \|f\|=1} \left\| f - \sum_{m,n \in \mathbb{Z}} \langle f, E_{mb}T_{na}g \rangle E_{mb}T_{na}h \right\|.$$

Proposition 12.8.1 *Let g, h be functions in the Wiener space W and consider two parameters $a, b > 0$. Then*

$$\|I - UT^*\| \le \frac{1}{b}\left[\left\| b - \sum_{k \in \mathbb{Z}} \overline{T_{ak}g}T_{ak}h \right\|_\infty + \sum_{n \ne 0} \left\| \sum_{k \in \mathbb{Z}} \overline{T_{n/b+ak}g}T_{ak}h \right\|_\infty \right].$$

Proof. The assumption of g, h belonging to the Wiener space implies by Proposition 11.5.2 that $\{E_{mb}T_{na}g\}_{m,n \in \mathbb{Z}}$ and $\{E_{mb}T_{na}h\}_{m,n \in \mathbb{Z}}$ are Bessel sequences. We will now consider the Walnut representation of the mixed frame operator associated with $\{E_{mb}T_{na}g\}_{m,n \in \mathbb{Z}}$ and $\{E_{mb}T_{na}h\}_{m,n \in \mathbb{Z}}$, see (12.6). According to Theorem 12.2.1, for any $f \in L^2(\mathbb{R})$,

$$UT^*f(\cdot) = \frac{1}{b}\sum_{n \in \mathbb{Z}}\left(\sum_{k \in \mathbb{Z}} \overline{T_{n/b+ak}g(\cdot)}T_{ak}h(\cdot) \right)T_{n/b}f(\cdot).$$

Thus, by pulling out the term corresponding to $n = 0$ and using the triangle inequality,

$$\|f - UT^*f\|$$
$$\le \left\| \left(1 - \frac{1}{b}\sum_{k \in \mathbb{Z}} \overline{T_{ak}g(\cdot)}T_{ak}h(\cdot)\right)f \right\|$$
$$+ \frac{1}{b}\left\| \sum_{n \ne 0}\left(\sum_{k \in \mathbb{Z}} \overline{T_{n/b+ak}g(\cdot)}T_{ak}h(\cdot) \right)T_{n/b}f \right\|$$
$$\le \frac{1}{b}\left\| b - \sum_{k \in \mathbb{Z}} \overline{T_{ak}g}T_{ak}h \right\|_\infty \|f\| + \frac{1}{b}\sum_{n \ne 0}\left\| \sum_{k \in \mathbb{Z}} \overline{T_{n/b+ak}g}T_{ak}h \right\|_\infty \|f\|,$$

which concludes the proof. $\square$

Note that Proposition 12.8.1 connects to the discussion of approximately dual frames in Section 6.5. For further results about Gabor systems and approximately dual frames, we refer to [192] and the paper [279] by Feichtinger, Grybos & Onchis; the more general case of a nonstationary Gabor system is treated in the paper [258] by Dörfler & Matusiak.

12.9   Tight Gabor frames

Already in Corollary 5.1.7 we saw that tight frames provide a way of obtaining convenient frame decompositions. Fortunately, tight Gabor frames exist and will be characterized in Theorem 12.9.2.

Lemma 12.9.1 *Let* $g, h \in L^2(\mathbb{R})$ *and* $a, b > 0$ *be given. Fix* $n \subset \mathbb{Z}$. *Then*

(i) h *is orthogonal to* $E_{mb}T_{na}g$ *for all* $m \neq 0$ *if and only if there is a constant* C *so that*

$$\sum_{k\in\mathbb{Z}} h(x - k/b)\overline{g(x - k/b - na)} = C, \quad a.e.\, x \in \mathbb{R}.$$

(ii) h *is orthogonal to* $E_{mb}T_{na}g$ *for all* $m \in \mathbb{Z}$ *if and only if*

$$\sum_{k\in\mathbb{Z}} h(x - k/b)\overline{g(x - k/b - na)} = 0, \quad a.e.\, x \in \mathbb{R}.$$

Proof. Lemma 11.2.2 shows that for any $m, n \in \mathbb{Z}$,

$$\langle h, E_{mb}T_{na}g \rangle = \int_0^{1/b} \sum_{k\in\mathbb{Z}} h(x - k/b)\overline{g(x - k/b - na)}e^{-2\pi imbx}\, dx.$$

Since $\{b^{1/2}e^{2\pi imb}\}_{m\in\mathbb{Z}}$ is an orthonormal basis for $L^2(0, 1/b)$, it follows that for a given $n \in \mathbb{Z}$, the equation $\langle h, E_{mb}T_{na}g \rangle = 0$ holds for all $m \in \mathbb{Z}$ if and only if $\sum_{k\in\mathbb{Z}} h(x - k/b)\overline{g(x - k/b - na)} = 0$, a.e. $x \in [0, 1/b]$. Since $x \mapsto \sum_{k\in\mathbb{Z}} h(x-k/b)\overline{g(x - k/b - na)}$ is periodic with period $1/b$ this proves (ii). The statement (i) also follows from the expression for $\langle h, E_{mb}T_{na}g \rangle$. $\square$

We now state equivalent conditions for $\{E_{mb}T_{na}g\}_{m,n\in\mathbb{Z}}$ being a tight frame. The equivalence (i)$\Leftrightarrow$(ii) below actually follows from the characterization in Theorem 10.1.7 of shift-invariant systems generating tight frames (Exercise 12.8), but we include a direct proof.

Theorem 12.9.2 *Let* $g \in L^2(\mathbb{R})$ *and* $a, b > 0$ *be given. The following are equivalent:*

(i) $\{E_{mb}T_{na}g\}_{m,n\in\mathbb{Z}}$ *is a tight frame for* $L^2(\mathbb{R})$ *with frame bound* $A = 1$.

(ii) *We have*

    (a) $G(x) := \sum_{n\in\mathbb{Z}} |g(x - na)|^2 = b$, a.e. $x \in \mathbb{R}$,

    (b) $G_k(x) := \sum_{n\in\mathbb{Z}} g(x - na)\overline{g(x - na - k/b)} = 0$, a.e. $x \in \mathbb{R}$ for all $k \neq 0$.

(iii) $g \perp E_{m/a}T_{n/b}g$ *for all* $(m, n) \neq (0, 0)$, *and* $\|g\|^2 = ab$.

(iv) $\{E_{m/a}T_{n/b}g\}_{m,n\in\mathbb{Z}}$ *is an orthogonal sequence and* $\|g\|^2 = ab$.

Moreover, when at least one of (i)–(iv) holds, $\{E_{mb}T_{na}g\}_{m,n\in\mathbb{Z}}$ *is an orthonormal basis for* $L^2(\mathbb{R})$ *if and only if* $\|g\| = 1$.

Proof. (i)$\Rightarrow$(ii): Assume $\{E_{mb}T_{na}g\}_{m,n\in\mathbb{Z}}$ is a tight frame for $L^2(\mathbb{R})$ with frame bound $A = 1$. For any function $f \in L^2(\mathbb{R})$ which is supported on an interval of length at most $1/b$ we see that $\overline{f(x)}f(x - k/b) = 0$ for all $x \in \mathbb{R}$ and all $k \in \mathbb{Z} \setminus \{0\}$. Via Lemma 11.4.1,

$$
\begin{aligned}
\int_{-\infty}^{\infty} |f(x)|^2 \, dx &= \sum_{m,n \in \mathbb{Z}} |\langle f, E_{mb}T_{na}g \rangle|^2 \\
&= \frac{1}{b} \int_{-\infty}^{\infty} |f(x)|^2 G(x) \, dx.
\end{aligned}
$$

Since this equality holds for all $f \in L^2(I)$, for any interval I of length at most $1/b$, it follows that $G(x) = b$ for a.e. $x \in \mathbb{R}$. Therefore

$$
\sum_{m,n \in \mathbb{Z}} |\langle f, E_{mb}T_{na}g \rangle|^2 = \frac{1}{b} \int_{-\infty}^{\infty} |f(x)|^2 G(x) \, dx
$$

for *all* functions $f \in L^2(\mathbb{R})$. Using Lemma 11.4.1 again, we have for all bounded, compactly supported $f \in L^2(\mathbb{R})$,

$$
\frac{1}{b} \sum_{k \neq 0} \int_{-\infty}^{\infty} \overline{f(x)}f(x - k/b) \sum_{n \in \mathbb{Z}} g(x - na)\overline{g(x - na - k/b)}dx = 0.
$$

A change of variable shows that the contribution in the above sum arising from any value of k is the complex conjugate of the contribution from the value $-k$. Therefore

$$
\sum_{k=1}^{\infty} \mathrm{Re} \left(\int_{-\infty}^{\infty} \overline{f(x)}f(x-k/b) \sum_{n \in \mathbb{Z}} g(x-na)\overline{g(x-na-k/b)}dx \right) = 0. \quad (12.57)
$$

Now fix $k_0 \geq 1$ and let I be any interval in $\mathbb{R}$ of length at most $1/b$. Let $\arg(G_{k_0}(x))$ denote an argument for the complex number $G_{k_0}(x)$ and define a function $f \in L^2(\mathbb{R})$ by $f(x) = e^{-i\arg(G_{k_0}(x))}$, $x \in I$, $f(x - k_0/b) = 1$ for $x \in I$, and $f(x) = 0$ otherwise. Then, by (12.57),

$$
\begin{aligned}
0 &= \sum_{k=1}^{\infty} \mathrm{Re} \left(\int_{-\infty}^{\infty} \overline{f(x)}f(x - k/b) \sum_{n \in \mathbb{Z}} g(x - na)\overline{g(x - na - k/b)} \, dx \right) \\
&= \mathrm{Re} \left(\int_{-\infty}^{\infty} \overline{f(x)}f(x - k_0/b)G_{k_0}(x) \, dx \right) = \int_{I} |G_{k_0}(x)| \, dx.
\end{aligned}
$$

It follows that $G_{k_0}(x) = 0$, a.e. on I. Since I was an arbitrary interval of length at most $1/b$, we conclude that $G_{k_0} = 0$. A direct computation shows that $G_{-k_0}(x) = \overline{G_{k_0}(x + k_0/b)} = 0$, and we have proved statement (b) in (ii) for all $k \neq 0$.

(ii)$\Rightarrow$ (i): The assumptions in (ii) imply, again by Lemma 11.4.1, that for all bounded, compactly supported $f \in L^2(\mathbb{R})$,

$$
\sum_{m,n \in \mathbb{Z}} |\langle f, E_{mb}T_{na}g \rangle|^2 = \frac{1}{b} \int_{-\infty}^{\infty} |f(x)|^2 \sum_{n \in \mathbb{Z}} |g(x - na)|^2 \, dx = \|f\|^2.
$$

Since the bounded compactly supported functions are dense in $L^2(\mathbb{R})$, the conclusion follows by Lemma 5.1.9.

(ii) $\Leftrightarrow$ (iii): By Lemma 12.9.1(ii), the statement (b) in (ii) is equivalent to $g \perp E_{m/a}T_{n/b}g$ for all $m \in \mathbb{Z}, n \neq 0$. Using Lemma 12.9.1(i) with $n = 0$, the function G is constant if and only if $g \perp E_{m/a}g$ for all $m \neq 0$; and if this is the case, the relationship between $\|g\|^2$ and $G(x)$ follows from

$$\|g\|^2 = \int_{-\infty}^{\infty} |g(x)|^2 \, dx = \int_0^a \sum_{n \in \mathbb{Z}} |g(x - na)|^2 \, dx$$

$$= \int_0^a G(x) \, dx = a \, G(x).$$

(iii) $\Leftrightarrow$ (iv): This follows from the observation that for all $m, n, \ell, k \in \mathbb{Z}$,

$$\langle E_{m/a}T_{n/b}g, E_{k/a}T_{\ell/b}g \rangle = e^{2\pi i \frac{m-k}{a}\frac{n}{b}} \langle g, E_{\frac{k-m}{a}}T_{\frac{\ell-n}{b}}g \rangle.$$

For the final part of the theorem, we just observe that if $\{E_{mb}T_{na}g\}_{m,n \in \mathbb{Z}}$ is a tight frame with frame bound 1, then for any $(m', n') \in \mathbb{Z}^2$,

$$\|E_{m'b}T_{n'a}g\|^2 = \sum_{m,n \in \mathbb{Z}} |\langle E_{m'b}T_{n'a}g, E_{mb}T_{na}g \rangle|^2$$

$$= \|E_{m'b}T_{n'a}g\|^4 + \sum_{(m,n) \neq (m',n')} |\langle E_{m'b}T_{n'a}g, E_{mb}T_{na}g \rangle|^2.$$

If $\|g\| = 1$ it follows from here that $\{E_{mb}T_{na}g\}_{m,n \in \mathbb{Z}}$ is an orthonormal system. $\qquad\square$

In general, it is not easy to construct functions g such that the conditions in Theorem 12.9.2 (ii) are satisfied for some given $a, b > 0$. A simplification occurs if we assume that g has compact support: in that case, the condition (b) in Theorem 12.9.2 (ii) is automatically satisfied for sufficiently small values of the parameter b. In particular, we obtain the following very useful sufficient condition for $\{E_{mb}T_{na}g\}_{m,n \in \mathbb{Z}}$ being a tight Gabor frame. We ask the reader to provide the proof in Exercise 12.9.

Corollary 12.9.3 *Let $a, b > 0$ be given. Assume that $\varphi \in L^2(\mathbb{R})$ is a real-valued non-negative function with support in an interval of length $1/b$, and that*

$$\sum_{n \in \mathbb{Z}} \varphi(x + na) = 1, \ a.e. \ x \in \mathbb{R}. \qquad (12.58)$$

Then the function

$$g(x) := \sqrt{b\varphi(x)}$$

generates a tight Gabor frame $\{E_{mb}T_{na}g\}_{m,n \in \mathbb{Z}}$ with frame bound $A = 1$.

Note that (12.58) is a partition of unity condition on the functions $\{T_{na}\varphi\}_{n\in\mathbb{Z}}$. Readers with knowledge of multiresolution analysis will notice that the associated scaling function satisfies this condition for $a = 1$; see, e.g., [633]. In particular, we can apply the result to the B-splines:

Example 12.9.4 For any $N \in \mathbb{N}$, the B-spline $\varphi = B_N$ defined in (A.15) satisfies the requirements in Corollary 12.9.3 with $a = 1$ and any $b \in {]}0, 1/N]$. Thus, for any $b \in {]}0, 1/N]$, the function

$$g(x) = \sqrt{bB_N(x)}$$

generates a tight Gabor frame $\{E_{mb}T_ng\}_{m,n\in\mathbb{Z}}$ with frame bound $A=1$. $\square$

We note that the windows in Example 12.9.4 have a number of attractive properties in the context of time-frequency analysis: they are given by an explicit formula, have compact support, and can be chosen with polynomial decay of any desired order in the frequency domain, simply by taking the parameter N sufficiently large. However, in contrast with the case for the construction in Section 12.6 the desire of high regularity forces the window to have a large support.

12.10   Exercises

12.1 Describe how a sequence $\{e_{m,n}\}_{m,n\in\mathbb{Z}}$ can be re-indexed as $\{e_k\}_{k=1}^{\infty}$.

12.2 This exercise concerns condition (R) and its relationship to Lebesgue points.

  (i) Assume that $g \in L^2(\mathbb{R})$ satisfies condition (R). Show that all integers are Lebesgue points for g.

  (ii) Assume that g is a bounded compactly supported function for which every integer is a Lebesgue point. Show that g satisfies condition (R).

  (iii) Prove via (ii) that condition (R) is satisfied on a dense subset of $L^2(\mathbb{R})$.

  (iv) Prove that the Gaussian $g(x) = e^{-\frac{1}{2}x^2}$ satisfies condition (R).

12.3 Consider two Bessel sequences $\{E_{mb}T_{na}g\}_{m,n\in\mathbb{Z}}, \{E_{mb}T_{na}h\}_{m,n\in\mathbb{Z}}$, and denote the corresponding synthesis operators by T and U, respectively. Show that

$$TU^*E_{mb}T_{na} = E_{mb}T_{na}TU^*, \ \forall m, n \in \mathbb{Z}. \tag{12.59}$$

12.4 Find a Gabor system $\{E_{mb}T_{na}g\}_{m,n\in\mathbb{Z}}$ which satisfies condition (UCC) *without the Bessel condition*, and which does not satisfy condition (CC).

12.5 Prove Theorem 12.3.4 via Theorem 10.1.7.

12.6 Assume that $\{E_{mb}T_{na}g\}_{m,n\in\mathbb{Z}}$ is an overcomplete frame. Show that there exist dual frames not having the Gabor structure. (Hint: check the proof of Lemma 6.3.1.)

12.7 Assume that $ab = 1$, and construct a pair of dual Gabor frame $\{E_{mb}T_{na}r_1\}_{m,n\in\mathbb{Z}}$, $\{E_{mb}T_{na}r_2\}_{m,n\in\mathbb{Z}}$ for which the functions r_1, r_2 have compact support (note that by Proposition 4.2.2 the functions r_1, r_2 can not be continuous).

12.8 Derive a characterization for tight Gabor system $\{E_{mb}T_{na}g\}_{m,n\in\mathbb{Z}}$ corresponding to Theorem 10.1.7 (for inspiration, see Theorem 12.9.2).

12.9 Prove Corollary 12.9.3.

12.10 Show that the B-splines B_2 and B_3 defined in (A.15) are given by

$$B_2(x) = \begin{cases} 1+x & \text{if } x \in [-1,0], \\ 1-x & \text{if } x \in [0,1], \\ 0 & \text{otherwise,} \end{cases}$$

$$B_3(x) = \begin{cases} \frac{1}{2}x^2 + \frac{3}{2}x + \frac{9}{8} & \text{if } x \in [-\frac{3}{2}, -\frac{1}{2}], \\ -x^2 + \frac{3}{4} & \text{if } x \in [-\frac{1}{2}, \frac{1}{2}], \\ \frac{1}{2}x^2 - \frac{3}{2}x + \frac{9}{8} & \text{if } x \in [\frac{1}{2}, \frac{3}{2}], \\ 0 & \text{otherwise.} \end{cases}$$

13

Selected Topics on Gabor Frames

Gabor analysis has now been a very active research field for about 30 years, and even a description of its connection to frame theory would cover an entire book. Based on the core material in Chapters 11–12, we will now present selected topics and tools. All of them are of general importance, in the sense that they find applications within several areas of time–frequency analysis. The sections are to a large extent independent of each other.

Section 13.1 deals with the duality principle, which is considered as one of the cornerstones in Gabor analysis; it characterizes Gabor frames $\{E_{mb}T_{na}g\}_{m,n\in\mathbb{Z}}$ in terms of Riesz sequences generated by the same window, but with the parameters a, b replaced by b^{-1}, a^{-1}.

Section 13.2 presents the Zak transform, which is one of the key tools in Gabor analysis. The Zak transform is particularly interesting for Gabor systems $\{E_{mb}T_{na}g\}_{m,n\in\mathbb{Z}}$ for which $ab \in \mathbb{Q}$; in this case, the central frame properties of $\{E_{mb}T_{na}g\}_{m,n\in\mathbb{Z}}$ can be expressed in terms of a matrix-valued function, the so-called Zibulski–Zeevi matrix.

Section 13.3 presents work by Feichtinger and Janssen and deals with the role of the parameters a, b in a Gabor system $\{E_{mb}T_{na}g\}_{m,n\in\mathbb{Z}}$. It is shown that for general functions $g \in L^2(\mathbb{R})$, the frame property of $\{E_{mb}T_{na}g\}_{m,n\in\mathbb{Z}}$ is very sensitive to the choice of these parameters, i.e., an arbitrarily small displacement can destroy the frame property. This is again an argument for using windows g belonging to Feichtinger's algebra $\mathcal{S}_0$ where such phenomena do not occur.

Section 13.4 is devoted to Gabor systems where the lattice $\{(na, mb)\}_{m,n\in\mathbb{Z}}$ is replaced by an irregular set $\{(\mu_n, \lambda_n)\}_{n\in I}$ in $\mathbb{R}^2$. The analysis of such systems is considerably more involved than its lattice

© Springer International Publishing Switzerland 2016
O. Christensen, *An Introduction to Frames and Riesz Bases*,
Applied and Numerical Harmonic Analysis,
DOI 10.1007/978-3-319-25613-9_13

counterpart. Some of the techniques used in Chapters 11–12 may be applied for special choices of the set $\{(\mu_n, \lambda_n)\}_{n \in I}$, but the general case is usually treated in terms of density conditions, similar to what we saw for irregular frames of translates in Section 9.6.

Section 13.5 returns to the concept of localization of frames, treated in Section 8.2. The main result shows that the key to obtain a localized Gabor frame is to select the window from a particular vector space.

In Section 13.6 we give a short presentation of Wilson bases. The functions in a Wilson bases are formed as linear combinations of the functions in a well-chosen Gabor system. The role of these systems is that they provide a way of obtaining orthonormal bases for $L^2(\mathbb{R})$ with good time–frequency localization; in that sense, Wilson bases yield an alternative way of overcoming the Balian–Low theorem.

Section 13.7 presents work by Daubechies, showing how the time–frequency localization of a function $f \in L^2(\mathbb{R})$ affects its representation in terms of a Gabor frame and its dual frame. The main result gives an estimate for the norm difference between the function f and the finite partial sums of the frame decomposition, expressed in terms of the norm of certain operators that directly reflects the behavior of f in the time domain and the frequency domain.

Finally, we mention a few applications of Gabor frames in Section 13.8. For more information about the role of Gabor frames in time–frequency analysis, we refer to the book [340] by Gröchenig. For a broader perspective on Gabor frames and their applications, the reader can consult the two books [291] and [292] edited by Feichtinger and Strohmer, which contain surveys and research articles covering several theoretical and applied aspects.

13.1   The Duality Principle

We now state one of the most fundamental and important results in Gabor analysis. It is known as the *duality principle* and was discovered almost simultaneously by three groups of researchers: Janssen [427], Daubechies, Landau, and Landau [249], and Ron and Shen [560]. The duality principle concerns the relationship between frame properties for a function g with respect to the lattice $\{(na, mb)\}_{m,n \in \mathbb{Z}}$ and with respect to the dual lattice $\{(n/b, m/a)\}_{m,n \in \mathbb{Z}}$. The proof below is a slight modification of a proof due to Jakobsen and Lemvig.

Theorem 13.1.1 *Let $g \in L^2(\mathbb{R})$ and $a, b > 0$ be given. Then the following are equivalent:*

(i) *$\{E_{mb}T_{na}g\}_{m,n \in \mathbb{Z}}$ is a frame for $L^2(\mathbb{R})$ with bounds A, B;*

(ii) *$\{E_{m/a}T_{n/b}g\}_{m,n \in \mathbb{Z}}$ is a Riesz sequence with bounds abA, abB.*

Proof. (i) $\Rightarrow$ (ii). Let us first assume that $\{E_{mb}T_{na}g\}_{m,n\in\mathbb{Z}}$ is a frame with bounds A, B. We know from Lemma 12.2.2 that then $\{E_{m/a}T_{n/b}g\}_{m,n\in\mathbb{Z}}$ is a Bessel sequence with bound abB, so we only need to show that $\{E_{m/a}T_{n/b}g\}_{m,n\in\mathbb{Z}}$ satisfies the lower Riesz sequence condition with the stated bound. Let S denote the frame operator for $\{E_{mb}T_{na}g\}_{m,n\in\mathbb{Z}}$. By Lemma 12.3.1, we know that the canonical dual frame of $\{E_{mb}T_{na}g\}_{m,n\in\mathbb{Z}}$ has the form

$$\{S^{-1}E_{mb}T_{na}g\}_{m,n\in\mathbb{Z}} = \{E_{mb}T_{na}S^{-1}g\}_{m,n\in\mathbb{Z}}; \tag{13.1}$$

using the Wexler–Raz theorem, it follows that the systems $\{E_{m/a}T_{n/b}g\}_{m,n\in\mathbb{Z}}$ and $\{E_{m/a}T_{n/b}(ab)^{-1}S^{-1}g\}_{m,n\in\mathbb{Z}}$ are biorthogonal (Exercise 13.1). We will now show that $\{E_{m/a}T_{n/b}(ab)^{-1}S^{-1}g\}_{m,n\in\mathbb{Z}}$ is a Bessel sequence with bound $(abA)^{-1}$; when this is done, Proposition 3.7.3 implies that $\{E_{m/a}T_{n/b}g\}_{m,n\in\mathbb{Z}}$ is a Riesz sequence with lower bound abA, as desired.

In order to proceed, we first note that by Lemma 5.1.5, the canonical dual frame $\{S^{-1}E_{mb}T_{na}g\}_{m,n\in\mathbb{Z}}$ of $\{E_{mb}T_{na}g\}_{m,n\in\mathbb{Z}}$ has the upper frame bound A^{-1}; thus, by (13.1), the upper frame bound for the Gabor system $\{E_{mb}T_{na}(ab)^{-1}S^{-1}g\}_{m,n\in\mathbb{Z}}$ is $(ab)^{-2}A^{-1}$. Using again Lemma 12.2.2, it follows that $\{E_{m/a}T_{n/b}(ab)^{-1}S^{-1}g\}_{m,n\in\mathbb{Z}}$ is a Bessel sequence with bound $(abA)^{-1}$, as desired.

(ii) $\Rightarrow$ (i). We now assume that the Gabor system $\{E_{m/a}T_{n/b}g\}_{m,n\in\mathbb{Z}}$ is a Riesz sequence with lower bound abA. As above, we only need to show that $\{E_{mb}T_{na}g\}_{m,n\in\mathbb{Z}}$ satisfies the lower frame condition with bound A. Now, it follows from Lemma 12.3.1 (i) that the dual Riesz sequence of $\{E_{m/a}T_{n/b}g\}_{m,n\in\mathbb{Z}}$ has the form $\{(ab)^{-1}E_{m/a}T_{n/b}h\}_{m,n\in\mathbb{Z}}$ for some $h \in L^2(\mathbb{R})$; note that for convenience, we have included the number $(ab)^{-1}$ explicitly in this expression. By general frame theory (Lemma 5.1.5), the sequence $\{(ab)^{-1}E_{m/a}T_{n/b}h\}_{m,n\in\mathbb{Z}}$ has the upper bound $(abA)^{-1}$; by scaling, this implies that $\{E_{m/a}T_{n/b}h\}_{m,n\in\mathbb{Z}}$ has the upper bound $(ab)^2(abA)^{-1} = abA^{-1}$. By Lemma 12.2.2, it follows that $\{E_{mb}T_{na}h\}_{m,n\in\mathbb{Z}}$ has the upper bound $(ab)^{-1}abA^{-1} = A^{-1}$. Finally, by the Wexler–Raz theorem, $\{E_{mb}T_{na}g\}_{m,n\in\mathbb{Z}}$ and $\{E_{mb}T_{na}h\}_{m,n\in\mathbb{Z}}$ are dual frames for $L^2(\mathbb{R})$; it follows that for $f \in L^2(\mathbb{R})$,

$$\|f\|^4 = \left| \sum_{m,n\in\mathbb{Z}} \langle f, E_{mb}T_{na}g\rangle\langle E_{mb}T_{na}h, f\rangle \right|^2$$

$$\leq \sum_{m,n\in\mathbb{Z}} |\langle f, E_{mb}T_{na}g\rangle|^2 \sum_{m,n\in\mathbb{Z}} |\langle f, E_{mb}T_{na}h\rangle|^2$$

$$\leq A^{-1}\|f\|^2 \sum_{m,n\in\mathbb{Z}} |\langle f, E_{mb}T_{na}g\rangle|^2.$$

Thus $\{E_{mb}T_{na}g\}_{m,n\in\mathbb{Z}}$ has the lower frame bound A, as claimed. $\qquad\square$

The importance of Theorem 13.1.1 lies in the fact that it often is easier to prove that $\{E_{m/a}T_{n/b}g\}_{m,n\in\mathbb{Z}}$ is a Riesz sequence than to prove directly that $\{E_{mb}T_{na}g\}_{m,n\in\mathbb{Z}}$ is a frame. We come back to an application of this idea in connection with various methods for approximation of the inverse frame operator; see Section 23.5 and Section 23.7. We also note that the duality principle was used in the proof of Proposition 12.4.1, which characterizes the canonical dual window associated with a frame $\{E_{mb}T_{na}g\}_{m,n\in\mathbb{Z}}$.

The duality principle has inspired a lot of activity in the literature. Feichtinger and Zimmermann [295] considered the duality principle for general lattices in $\mathbb{R}^2$. Dutkay, Han, and Larson considered a general duality principle for groups in [263], and Jakobsen and Lemvig extended the duality principle to co-compact Gabor systems on LCA groups in [418]; we state this generalization in Theorem 21.8.3.

Also, the theory for R-duals considered in Section 8.3 is directly inspired by the duality principle. Theorem 8.3.2 gives a relationship between a sequence $\{f_k\}_{k=1}^\infty$ and its R-duals $\{\omega_j\}_{j=1}^\infty$ which precisely match the relation between the Gabor systems $\{E_{mb}T_{na}g\}_{m,n\in\mathbb{Z}}$ and $\{\frac{1}{\sqrt{ab}}E_{m/a}T_{n/b}g\}_{m,n\in\mathbb{Z}}$ in Theorem 13.1.1. This makes it natural to expect that the theory for R-duals actually generalizes the duality principle, in the sense that the Gabor system $\{\frac{1}{\sqrt{ab}}E_{m/a}T_{n/b}g\}_{m,n\in\mathbb{Z}}$ can be realized as an R-dual of $\{E_{mb}T_{na}g\}_{m,n\in\mathbb{Z}}$. Considerable effort has been spent on this problem, but it remains open. Let us formulate it as such:

Question: Given any Gabor frame $\{E_{mb}T_{na}g\}_{m,n\in\mathbb{Z}}$ for $L^2(\mathbb{R})$, do there exist orthonormal bases $\{e_{m,n}\}_{m,n\in\mathbb{Z}}$ and $\{h_{m,n}\}_{m,n\in\mathbb{Z}}$ for $L^2(\mathbb{R})$ such that

$$\frac{1}{\sqrt{ab}}E_{m/a}T_{n/b}g = \sum_{m',n'\in\mathbb{Z}} \langle E_{m'b}T_{n'a}, e_{m,n}\rangle h_{m',n'}, \; \forall m,n \in \mathbb{Z}?$$

It is known that the answer is affirmative in case $\{E_{mb}T_{na}g\}_{m,n\in\mathbb{Z}}$ is tight or $ab = 1$; see [142]. We refer to the paper [166] for a more detailed discussion.

13.2 The Zak Transform

The *Zak transform* is a very useful tool to analyze Gabor systems $\{E_{mb}T_{na}g\}_{m,n\in\mathbb{Z}}$ in the case where $ab \in \mathbb{Q}$. For a classical survey on the Zak transform, we refer to the article [424] by Janssen; applications to Gabor analysis appear in, e.g., [340], [98], [48], [459].

For a fixed parameter $\lambda > 0$, the *Zak transform* $Z_\lambda f$ of a function $f \in L^2(\mathbb{R})$ is formally defined as a function of two real variables:

$$(Z_\lambda f)(t,\nu) := \lambda^{1/2} \sum_{k\in\mathbb{Z}} f(\lambda(t-k))e^{2\pi ik\nu}, \qquad t,\nu \in \mathbb{R}. \qquad (13.2)$$

In the case $\lambda = 1$, we simply write

$$(Zf)(t, \nu) = \sum_{k \in \mathbb{Z}} f(t - k)e^{2\pi ik\nu}, \qquad t, \nu \in \mathbb{R}. \tag{13.3}$$

For functions $f \in C_c(\mathbb{R})$, the Zak transform is defined pointwise and is continuous (the same holds whenever f belongs to the Feichtinger algebra $\mathcal{S}_0$), but for general functions in $L^2(\mathbb{R})$, we have to be more precise about how to interpret the definition. Letting $Q := [0, 1[\times[0, 1[$, we now prove that the series defining $Z_\lambda f$ in fact converges in $L^2(Q)$ for all $f \in L^2(\mathbb{R})$:

Lemma 13.2.1 *Given $\lambda > 0$, the Zak transform Z_λ is a unitary map of $L^2(\mathbb{R})$ onto $L^2(Q)$.*

Proof. We first consider the case $\lambda = 1$. Let $f \in L^2(\mathbb{R})$ be given. In order to show that Zf is well-defined as a function in $L^2(Q)$, we consider the functions

$$F_k(t, \nu) := f(t - k)e^{2\pi ik\nu}, \ k \in \mathbb{Z}.$$

These functions belong to $L^2(Q)$. Denoting their norm by $||F_k||_{L^2(Q)}$, we observe that

$$\sum_{k \in \mathbb{Z}} ||F_k||_{L^2(Q)}^2 = \sum_{k \in \mathbb{Z}} \int_0^1 \int_0^1 |F_k(t, \nu)|^2 d\nu dt = \sum_{k \in \mathbb{Z}} \int_0^1 |f(t - k)|^2 dt = ||f||^2.$$

Furthermore, for $j \neq k$,

$$\langle F_k, F_j \rangle_{L^2(Q)} = \int_0^1 f(t - k)\overline{f(t - j)} \left(\int_0^1 e^{2\pi i(k-j)\nu} d\nu \right) dt = 0. \tag{13.4}$$

Combining the obtained results shows that $\sum_{k \in \mathbb{Z}} F_k$ in fact converges in $L^2(Q)$ and that

$$\left|\left| \sum_{k \in \mathbb{Z}} F_k \right|\right|_{L^2(Q)}^2 = \sum_{k \in \mathbb{Z}} ||F_k||_{L^2(Q)}^2 = ||f||^2;$$

thus Z is an isometry from $L^2(\mathbb{R})$ into $L^2(Q)$.

For the rest of the proof, we use the Gabor basis $\{E_m T_n \chi_{[0,1]}\}_{m,n \in \mathbb{Z}}$ for $L^2(\mathbb{R})$ (cf. Example 3.8.3). By direct computation for $(t, \nu) \in Q$,

$$\begin{aligned} (ZE_m T_n \chi_{[0,1]})(t, \nu) &= \sum_{k \in \mathbb{Z}} e^{2\pi im(t-k)} \chi_{[0,1]}(t - n - k)e^{2\pi ik\nu} \\ &= e^{2\pi imt} e^{-2\pi in\nu} \sum_{k \in \mathbb{Z}} \chi_{[0,1]}(t - k)e^{2\pi ik\nu} \\ &= e^{2\pi imt} e^{-2\pi in\nu}. \end{aligned} \tag{13.5}$$

That is, the Zak transform maps the orthonormal basis $\{E_m T_n \chi_{[0,1]}\}_{m,n\in\mathbb{Z}}$ for $L^2(\mathbb{R})$ onto the orthonormal basis $\{e^{-2\pi i n\nu}e^{2\pi i m t}\}_{m,n\in\mathbb{Z}}$ for $L^2(Q)$. This implies that Z is unitary.

For the general case, we note that in terms of the unitary dilation operator $D_{\lambda^{-1}}$ defined in Section 2.9,

$$Z_\lambda f = Z(D_{\lambda^{-1}} f).$$

As a composition of unitary operators, Z_λ is itself unitary. $\qquad\square$

By Lemma 13.2.1, the Zak transform of a function $f \in L^2(\mathbb{R})$ converges almost everywhere on Q. An inspection of the expression (13.2) now reveals that $Z_\lambda f(t,\nu)$ converges almost everywhere on $\mathbb{R}^2$ and that the *quasiperiodicity* in Lemma 13.2.2(i) below holds. We collect some more properties of the Zak transform:

Lemma 13.2.2 *Consider the Zak transform Z_λ, $\lambda > 0$, and $f \in L^2(\mathbb{R})$. Then the following hold:*

(i) $Z_\lambda f(t+1,\nu) = e^{2\pi i\nu} Z_\lambda f(t,\nu), \quad Z_\lambda f(1,\nu+1) = Z_\lambda f(t,\nu).$

(ii) *If f is continuous and belongs to the Wiener space W, then $Z_\lambda f$ is continuous on $\mathbb{R}^2$.*

(iii) *If $Z_\lambda f$ is continuous on $\mathbb{R}^2$, then there exists $(t,\nu) \in \mathbb{R}^2$ such that $Z_\lambda f(t,\nu) = 0$.*

The proof of (ii) is elementary and follows from the uniform continuity on compact sets of the partial sums of $Z_\lambda f$ whenever f is continuous (Exercise 13.2); in particular, the result implies that $Z_\lambda f$ is continuous if f belongs to Feichtinger's algebra S_0. (iii) is proved by Janssen in [423] and in [395]. Note that the quasiperiodicity in (i) often leads to jump discontinuities on the lines $t = k, k \in \mathbb{Z}$: even if $Z_\lambda f$ is continuous on Q, it might not be continuous on $\mathbb{R}^2$. For a concrete example, take the function f whose Zak transform is equal to 1 on Q: in this case $Z_\lambda f$ is continuous on Q but not on $\mathbb{R}^2$.

If $g \in L^2(\mathbb{R})$ and $ab = 1$, a computation as in (13.5) shows that

$$Z_a E_{mb} T_{na} g = e^{2\pi i m t} e^{-2\pi i n\nu} Z_a g. \qquad (13.6)$$

The family $\{e^{2\pi i m t}e^{-2\pi i n\nu}\}_{m,n\in\mathbb{Z}}$ is an orthonormal basis for $L^2(Q)$, which we denote by $\{E_{(m,n)}\}_{m,n\in\mathbb{Z}}$. The equation (13.6) shows that $\{E_{mb} T_{na} g\}_{m,n\in\mathbb{Z}}$ is complete in $L^2(\mathbb{R})$ (respectively, an orthonormal basis for $L^2(\mathbb{R})$ or a Riesz basis) if and only if $\{E_{(m,n)} Z_a g\}_{m,n\in\mathbb{Z}}$ has the same property in $L^2(Q)$. This observation will be used in the following theorem, which expresses properties for a Gabor system $\{E_{mb} T_{na} g\}_{m,n\in\mathbb{Z}}$ with $ab = 1$ in terms of the Zak transform $Z_a g$. Remember from Theorem 11.3.1 that a Gabor system with $ab = 1$ is a frame if and only if it is a Riesz basis.

Proposition 13.2.3 *Let $g \in L^2(\mathbb{R})$ and $a, b > 0$ with $ab = 1$ be given Then the following hold:*

(i) $\{E_{mb}T_{na}g\}_{m,n\in\mathbb{Z}}$ *is complete in $L^2(\mathbb{R})$ if and only if $Z_a g \neq 0$, a.e.*

(ii) $\{E_{mb}T_{na}g\}_{m,n\in\mathbb{Z}}$ *is a Bessel sequence with bound B if and only if $|Z_a g|^2 \leq B$, a.e.*

(iii) $\{E_{mb}T_{na}g\}_{m,n\in\mathbb{Z}}$ *is a Riesz basis for $L^2(\mathbb{R})$ with bounds A, B if and only if $A \leq |Z_a g|^2 \leq B$, a.e.*

(iv) $\{E_{mb}T_{na}g\}_{m,n\in\mathbb{Z}}$ *is an orthonormal basis for $L^2(\mathbb{R})$ if and only if $|Z_a g|^2 = 1$, a.e.*

Proof. To prove (i), consider the subspace $V \subset L^2(\mathbb{R})$ given by

$$V = \{f \in L^2(\mathbb{R}) : Z_a f \text{ is bounded}\}.$$

The bounded functions are dense in $L^2(Q)$, so V is dense in $L^2(\mathbb{R})$ by Lemma 13.2.1. Now let $f \in V$. Then

$$
\begin{aligned}
\langle f, E_{mb}T_{na}g\rangle_{L^2(\mathbb{R})} &= \langle Z_a f, E_{(m,n)} Z_a g\rangle_{L^2(Q)} \\
&= \langle Z_a f \overline{Z_a g}, E_{(m,n)}\rangle_{L^2(Q)}. \quad (13.7)
\end{aligned}
$$

First assume that $Z_a g \neq 0$ a.e. If $f \neq 0$, then $Z_a f \overline{Z_a g}$ is *not* the zero function, and there exists $(m, n) \in \mathbb{Z}^2$ such that

$$\langle Z_a f \overline{Z_a g}, E_{(m,n)}\rangle_{L^2(Q)} \neq 0.$$

Therefore, (13.7) shows that $\{E_{mb}T_{na}g\}_{m,n\in\mathbb{Z}}$ is complete. For the other implication, assume that $Z_a g = 0$ on a measurable set $\Delta \subseteq Q$ with positive measure. We leave the slight modifications in the case $\Delta = Q$ to the reader and assume that $Q \setminus \Delta \neq \emptyset$. By choosing $f \in L^2(\mathbb{R})$ such that $Z_a f = \chi_{Q\setminus\Delta}$, it follows that $\langle f, E_{mb}T_{na}g\rangle = 0$ for all $m, n \in \mathbb{Z}$, so $\{E_{mb}T_{na}g\}_{m,n\in\mathbb{Z}}$ is incomplete in $L^2(\mathbb{R})$.

For the rest of the proof, we note that for any $F \in L^2(Q)$, we have $F Z_a g \in L^1(Q)$. Since $\{E_{(m,n)}\}_{m,n\in\mathbb{Z}}$ is an orthonormal basis for $L^2(Q)$,

$$
\begin{aligned}
\sum_{m,n\in\mathbb{Z}} |\langle F, E_{(m,n)} Z_a g\rangle_{L^2(Q)}|^2 &= \sum_{m,n\in\mathbb{Z}} \left| \int_Q (F \overline{Z_a g})\, \overline{E_{(m,n)}} \right|^2 \\
&= \int_Q |F \overline{Z_a g}|^2. \quad (13.8)
\end{aligned}
$$

(ii)–(iv) now follow by a standard argument (Exercise 13.3), yielding, e.g., that

$$\int_Q |F \overline{Z_a g}|^2 \leq B \, \|F\|_{L^2(Q)}^2, \quad \forall F \in L^2(Q) \Leftrightarrow |Z_a g|^2 \leq B, \text{ a.e.} \quad (13.9)$$

$\square$

Lemma 13.2.2 and Proposition 13.2.3 put restrictions on the functions g for which $\{E_{mb}T_{na}g\}_{m,n\in\mathbb{Z}}$ can be a Riesz basis for $ab = 1$. For example, the Gaussian $g(x) = e^{-x^2/2}$ has a continuous Zak transform with a zero, which by Proposition 13.2.3 implies that $\{E_{mb}T_{na}g\}_{m,n\in\mathbb{Z}}$ cannot be a Riesz basis. We will now use these results to prove that $\{E_{mb}T_{na}g\}_{m,n\in\mathbb{Z}}$ cannot be a Riesz basis for $L^2(\mathbb{R})$ whenever $g \in C_c(\mathbb{R})$; this result was stated already in Proposition 4.2.2 as a motivation for the need of overcomplete Gabor systems.

Proposition 13.2.4 *Let g be a continuous function with compact support. Then the following hold:*

(i) $\{E_{mb}T_{na}g\}_{m,n\in\mathbb{Z}}$ *cannot be an orthonormal basis for $L^2(\mathbb{R})$.*

(ii) $\{E_{mb}T_{na}g\}_{m,n\in\mathbb{Z}}$ *cannot be a Riesz basis for $L^2(\mathbb{R})$.*

Proof.   It is obviously enough to prove (ii). Let $g \in C_c(\mathbb{R})$. If $\{E_{mb}T_{na}g\}_{m,n\in\mathbb{Z}}$ is a Riesz basis for $L^2(\mathbb{R})$, then Theorem 11.3.1 shows that $ab = 1$. Now, Lemma 13.2.2(ii) implies that the Zak transform Z_ag is continuous and therefore has a zero by (iii) in the same lemma. But this contradicts Proposition 13.2.3 (iii). We conclude that $\{E_{mb}T_{na}g\}_{m,n\in\mathbb{Z}}$ cannot be a Riesz basis for $L^2(\mathbb{R})$. $\square$

For the rest of this section, we consider a *rationally oversampled* Gabor system $\{E_{mb}T_{na}g\}_{m,n\in\mathbb{Z}}$, i.e., we assume that

$$ab \in \mathbb{Q}, \ ab = \frac{p}{q} \text{ with } 1 \leq p \leq q. \tag{13.10}$$

We always choose p, q as small as possible, i.e., such that $\gcd(p,q) = 1$. We state results by Zibulski and Zeevi, resp. Janssen. The references for further information and proofs are [643], [430], [431], and [426].

In the special case considered here, the *Zibulski–Zeevi matrix* associated with a Gabor system $\{E_{mb}T_{na}g\}_{m,n\in\mathbb{Z}}$ is a useful tool. It is a $p \times q$ matrix, with entries depending on the variables $t, \nu \in \mathbb{R}$, and defined by

$$\Phi^g(t,\nu) = p^{-\frac{1}{2}}\left(\left(Z_{\frac{1}{b}}g\right)\left(t - \ell\frac{p}{q}, \nu + \frac{k}{p}\right)\right)_{k=0,\ldots,p-1;\ell=0,\ldots,q-1}, \ a.e. \ t, \nu \in \mathbb{R}.$$

In terms of this matrix, one can prove that $\{E_{mb}T_{na}g\}_{m,n\in\mathbb{Z}}$ is a Bessel sequence with bound B if and only if the matrices $\Phi^g(t,\nu)$, considered as bounded linear mappings of $\mathbb{C}^q$ into $\mathbb{C}^p$, for a.e. $t, \nu \in [0,1[$ have norms at most $B^{\frac{1}{2}}$. If we do not need the information about a specific Bessel bound, this result has a nice formulation:

Theorem 13.2.5 *A   rationally   oversampled   Gabor   system $\{E_{mb}T_{na}g\}_{m,n\in\mathbb{Z}}$ is a Bessel sequence if and only if there exists a constant $C > 0$ such that*

$$\left|Z_{\frac{1}{b}}g(t,\nu)\right| \leq C, \ a.e. \ t, \nu \in [0,1[.$$

Note that Theorem 13.2.5 generalizes Proposition 13.2.3(ii) to the case of rational oversampling.

We will now collect some of the key results proved in [426] and [643].

Theorem 13.2.6 *Assume that $\{E_{mb}T_{na}g\}_{m,n\in\mathbb{Z}}$ is a rationally oversampled Gabor system, see (13.10). Then the following hold:*

(i) *If $\{E_{mb}T_{na}g\}_{m,n\in\mathbb{Z}}$ is a Bessel sequence, then the frame operator is represented by*

$$\Phi^{Sf}(t,\nu) = \Phi^g(t,\nu)\left(\Phi^g(t,\nu)\right)^* \Phi^f(t,\nu), \quad a.e. \ t,\nu \in \mathbb{R}.$$

(ii) *The Gabor system $\{E_{mb}T_{na}g\}_{m,n\in\mathbb{Z}}$ is a frame with frame bounds $A, B > 0$ if and only if*

$$AI \leq \Phi^g(t,\nu)\left(\Phi^g(t,\nu)\right)^* \leq BI, \quad a.e. \ t,\nu \in \mathbb{R}.$$

Here I denotes the identity operator on $\mathbb{C}^p$.

(iii) *Two Bessel systems $\{E_{mb}T_{na}g\}_{m,n\in\mathbb{Z}}$ and $\{E_{mb}T_{na}h\}_{m,n\in\mathbb{Z}}$ are dual frames for $L^2(\mathbb{R})$ if and only if for a.e. $t,\nu \in [0,1[$ and all $k = 0,\ldots,p-1$,*

$$\frac{1}{p}\sum_{\ell=0}^{q-1}(Z_{\frac{1}{b}}g)\,(t-\ell p/q,\nu+k/p)\,\overline{(Z_{\frac{1}{b}}h)\,(t-\ell p/q,\nu)} = \delta_{k,0}.$$

Even though the Zibulski–Zeevi matrix is finite-dimensional, it is clearly a nontrivial matter to verify the conditions in Theorem 13.2.6. In case we are only interested in the frame property of $\{E_{mb}T_{na}g\}_{m,n\in\mathbb{Z}}$ and not the frame bounds, interesting simplifications were obtained by Lyubarskii and Nes in [506]. They involve windows belonging to Feichtinger's algebra S_0:

Corollary 13.2.7 *Let $g \in S_0$ and assume that $\{E_{mb}T_{na}g\}_{m,n\in\mathbb{Z}}$ is a rationally oversampled Gabor system, see (13.10). Then the following are equivalent:*

(i) *$\{E_{mb}T_{na}g\}_{m,n\in\mathbb{Z}}$ is a frame for $L^2(\mathbb{R})$;*

(ii) *For each $(t,\nu) \in [0,a/p[\times[0,1/a[$, the $p \times q$ matrix*

$$\left\{\sum_{n\in\mathbb{Z}} g(t+a\ell - aqn + k/b)e^{2\pi i n a q \nu}\right\}_{k=0,\ldots,p-1,\ell=0,\ldots,q-1} \tag{13.11}$$

has rank p.

Observe that the continuity requirement in Corollary 13.2.7 implies that the condition in (ii) must hold for all $(t,\nu) \in [0,a/p[\times[0,1/a[$ in order for $\{E_{mb}T_{na}g\}_{m,n\in\mathbb{Z}}$ to be a frame. Thus, in order to falsify the frame property it is enough to identify a single point (t,ν) where the condition breaks down. This is exactly the idea that was used by Lemvig and Nielsen to falsify the B-spline conjecture in [486] – see the discussion on page 285.

One might wonder if the assumption of $\{E_{mb}T_{na}g\}_{m,n\in\mathbb{Z}}$ being rationally oversampled is just for technical reasons or in the nature of Gabor analysis. It turns out to be the second option. Already in Section 12.1 in the discussion of condition (A), we saw a case where a rational parameter is essential, and there are several results for rationally oversampled systems which do not hold in the general case. Without going into details, the rationality opens for the use of Banach algebra techniques which are not available in the general case; see [349].

13.3   The Lattice Parameters

Whether a Gabor system $\{E_{mb}T_{na}g\}_{m,n\in\mathbb{Z}}$ forms a frame or not depends on a complicated interplay between the parameters a, b and the function g. Even by fixing the function $g \in L^2(\mathbb{R})$, the frame condition is in general very sensitive toward the choice of a, b. Recall, e.g., Janssen's results for the characteristic function in Section 11.6: taking $g = \chi_{[0,7/4]}$, they show, for example, that

- $\{E_m T_{na}g\}_{m,n\in\mathbb{Z}}$ is a frame if $a < 1$ and $a \neq \mathbb{Q}$;
- $\{E_m T_{na}g\}_{m,n\in\mathbb{Z}}$ is not a frame if $a \in \left\{\frac{1}{2}, \frac{1}{3}, \frac{2}{3}\right\}$.

The possibility of such a strange behavior for functions in $L^2(\mathbb{R})$ is one of the reasons for considering functions g belonging to selected window classes. For the sake of motivation, remember that Proposition 11.5.2 shows that $\{E_{mb}T_{na}g\}_{m,n\in\mathbb{Z}}$ is a Bessel sequence for any choice of $a, b > 0$ if g belongs to the Wiener space. Furthermore, a very reasonable condition in Proposition 11.5.3 shows that by fixing $a > 0$, we obtain a frame for all sufficiently small values of $b > 0$.

For a function $g \in L^2(\mathbb{R})$ generating a Bessel sequence $\{E_{mb}T_{na}g\}_{m,n\in\mathbb{Z}}$ for some $a, b > 0$, the following result by Feichtinger and Janssen [285] shows that the Bessel property is at least preserved if a, b are replaced by *rationally related parameters*. In the formulation of the result and the proof, we will need the Bessel bounds for various Gabor systems; as convention, we will denote the optimal bound for a Gabor system $\{E_{mb}T_{na}g\}_{m,n\in\mathbb{Z}}$ with parameters $a, b > 0$ by $B(a, b)$.

Proposition 13.3.1 *Let* $g \in L^2(\mathbb{R})$ *and* $a, b > 0$ *be given. If* $\{E_{mb}T_{na}g\}_{m,n\in\mathbb{Z}}$ *is a Bessel sequence, then* $\{E_{mbr/s}T_{nap/q}g\}_{m,n\in\mathbb{Z}}$ *is a Bessel sequence for any* $r, s, p, q \in \mathbb{N}$. *Furthermore, the optimal bounds are related by*

$$B(ap/q, br/s) \leq qsB(a, b) \leq pqrsB(ap/q, br/s). \tag{13.12}$$

Proof. Let $s \subset \mathbb{N}$. Note that if j runs through $0, \ldots, s-1$ and m runs through $\mathbb{Z}$, then $ms + j$ runs through $\mathbb{Z}$. Therefore, for $f \in L^2(\mathbb{R})$,

$$\sum_{m,n\in\mathbb{Z}} |\langle f, E_{mbr/s}T_{nap/q}g\rangle|^2$$

$$= \sum_{j=0}^{s-1}\sum_{l=0}^{q-1}\sum_{m,n\in\mathbb{Z}} |\langle f, E_{b(ms+j)r/s}T_{a(nq+l)p/q}g\rangle|^2$$

$$= \sum_{j=0}^{s-1}\sum_{l=0}^{q-1}\sum_{m,n\in\mathbb{Z}} |\langle E_{-bjr/s}T_{-alp/q}f, E_{mbr}T_{nap}g\rangle|^2$$

$$\leq \sum_{j=0}^{s-1}\sum_{l=0}^{q-1}\sum_{m,n\in\mathbb{Z}} |\langle E_{-bjr/s}T_{-alp/q}f, E_{mb}T_{na}g\rangle|^2$$

$$\leq \sum_{j=0}^{s-1}\sum_{l=0}^{q-1} B(a,b) \left\|E_{-bjr/s}T_{-alp/q}f\right\|^2 = qsB(a,b)\|f\|^2.$$

Thus, $\{E_{mbr/s}T_{nap/q}g\}_{m,n\in\mathbb{Z}}$ is a Bessel sequence with bound

$$B(ap/q, br/s) \leq qsB(a,b).$$

Using this result with a, b replaced by $ap/q, br/s$ and $p/q, r/s$ replaced by $q/p, s/r$ leads to

$$B(a,b) = B\left(a\frac{p}{q}\frac{q}{p}, b\frac{r}{s}\frac{s}{r}\right) \leq prB(ap/q, br/s). \qquad \square$$

Given a Gabor system $\{E_{mb}T_{na}g\}_{m,n\in\mathbb{Z}}$ and positive integers $M, N \in \mathbb{N}$, the Gabor system $\{E_{mb/M}T_{na/N}g\}_{m,n\in\mathbb{Z}}$ is called an *oversampling* of $\{E_{mb}T_{na}g\}_{m,n\in\mathbb{Z}}$. The terminology (coming from engineering) is natural from the point of view of group representations; consider, e.g., the way the two Gabor systems are related in terms of the Schrödinger representation in Example 24.1.2; see (24.2).

Proposition 13.3.1 implies that an oversampling of a Gabor frame again yields a frame (Exercise 13.5):

Corollary 13.3.2 *Assume that $\{E_{mb}T_{na}g\}_{m,n\in\mathbb{Z}}$ is a frame. Then, for any $M, N \in \mathbb{N}$, the Gabor oversampled system $\{E_{mb/M}T_{na/N}g\}_{m,n\in\mathbb{Z}}$ is also a frame.*

In order for the estimates for the Bessel bounds (13.12) in Proposition 13.3.1 to hold, it is crucial that we are speaking about the *optimal* bounds. We also note that the same Gabor system $\{E_{mbr/s}T_{nap/q}g\}_{m,n\in\mathbb{Z}}$ can appear by different choices of r, s, p, q; in fact, having one choice, we obtain another choice if we multiply all four numbers with the same $k \in \mathbb{N}$.

For the estimate (13.12) to be interesting, it is important to choose the smallest possible parameters, i.e., we take r, s, p, q such that

$$gcd(r,s) = gcd(p,q) = 1.$$

As a special case of Proposition 13.3.1 we note that if $\{E_m T_n g\}_{m,n \in \mathbb{Z}}$ is a Bessel sequence, then $\{E_{mr/s} T_{np/q} g\}_{m,n \in \mathbb{Z}}$ is a Bessel sequence for all $r, s, p, q \in \mathbb{N}$. The points $\{(p/q, r/s)\}_{r,s,p,q \in \mathbb{N}}$ are dense in $]0, \infty[\times]0, \infty[$, so it is natural to ask if $\{E_{mb} T_{na} g\}_{m,n \in \mathbb{Z}}$ automatically is a Bessel sequence for all $a, b > 0$ in this case. Feichtinger and Janssen proved that the answer is no:

Proposition 13.3.3 *Let $\alpha > 0$ be any irrational number. Then there exists a function $g \in C^\infty(\mathbb{R}) \cap L^2(\mathbb{R})$ with $|supp(g)| \leq 1$, for which*

(i) $\{E_{mb} T_{na} g\}_{m,n \in \mathbb{Z}}$ *is a Bessel sequence for all rational $a, b > 0$.*

(ii) $\{E_{mb} T_{nac} g\}_{m,n \in \mathbb{Z}}$ *is not a Bessel sequence if $c > 0$ is rational, regardless of the choice of $b > 0$.*

Proof. By Proposition 13.3.1, we obtain (i) if we construct g such that $\{E_m T_n g\}_{m,n \in \mathbb{Z}}$ is a Bessel sequence; by Proposition 13.2.3, this is the case if the Zak transform Zg is bounded, and a sufficient condition for this is that

$$\sup_{x \in [0,1]} \sum_{n \in \mathbb{Z}} |g(x+n)| < \infty. \tag{13.13}$$

Also, the negative conclusion in (ii) is by Proposition 11.3.4 and Proposition 13.3.1 obtained if

$$\sup_{x \in [0,1]} \sum_{n \in \mathbb{Z}} |g(x+n\alpha)|^2 = \infty. \tag{13.14}$$

In fact, in this case $\{E_{mb} T_{na} g\}_{m,n \in \mathbb{Z}}$ is not a Bessel sequence, and therefore $\{E_{mb} T_{nac} g\}_{m,n \in \mathbb{Z}}$ is not a Bessel sequence either when $c \in \mathbb{Q}$. We will now construct a smooth function, which satisfies (13.13) and (13.14).

We start with the given irrational number α, and observe that the set $\{n\alpha - \lfloor n\alpha \rfloor\}_{n=1}^\infty$ is dense in $]0, 1[$ (Exercise 13.4). We now construct a sequence of mutually disjoint intervals $\{I_k\}_{k=1}^\infty$ contained in $]0, 1[$ as follows. The interval I_1 is defined by

$$I_1 =]\alpha - \lfloor \alpha \rfloor - \epsilon_1, \alpha - \lfloor \alpha \rfloor + \epsilon_1[,$$

where $\epsilon_1 < \alpha/2$ is chosen so small that I_1 is in fact contained in $]0, 1[$. Since the set $\{n\alpha - \lfloor n\alpha \rfloor\}_{n=1}^\infty$ is dense in $]0, 1[$, we can now find an interval $I_2 \subset]0, 1[\setminus I_1$ of the form

$$I_2 =]n_2\alpha - \lfloor n_2\alpha \rfloor - \epsilon_2, n_2\alpha - \lfloor n_2\alpha \rfloor + \epsilon_2[.$$

In fact, we can take $n_2 > 1$ such that $n_2\alpha - \lfloor n_2\alpha \rfloor \in]0, 1[\setminus I_1$ and then choose $\epsilon_2 < \alpha/2$ so small that

$$]n_2\alpha - \lfloor n_2\alpha \rfloor - \epsilon_2, n_2\alpha - \lfloor n_2\alpha \rfloor + \epsilon_2[\subset]0, 1[\setminus I_1.$$

Continuing this process inductively, we obtain the desired intervals $\{I_k\}_{k=1}^\infty$, where each interval I_k has the form

$$I_k =]n_k\alpha - \lfloor n_k\alpha \rfloor - \epsilon_k, n_k\alpha - \lfloor n_k\alpha \rfloor + \epsilon_k[;$$

we choose the sequence $\{n_k\}_{k=1}^\infty$ to be increasing, and we take $\epsilon_k < \alpha/2$.

For each $k \in \mathbb{N}$, we now let J_k denote the middle third of I_k and choose a smooth "plateau" function g_k supported on I_k and satisfying

$$0 \le g_k \le 1 \text{ on } I_k, \ g_k = 1 \text{ on } J_k.$$

Finally, let

$$g(x) := \sum_{k=1}^\infty g_k(x - \lfloor n_k\alpha \rfloor).$$

The disjoint support of the functions g_k implies that g is well defined and smooth and has a support with measure at most 1. This also yields (13.13), so we only have to verify (13.14). By definition,

$$\sum_{n\in\mathbb{Z}} |g(x + n\alpha)|^2 = \sum_{n\in\mathbb{Z}} \left| \sum_{k=1}^\infty g_k(x + n\alpha - \lfloor n_k\alpha \rfloor) \right|^2.$$

By throwing positive terms away in the sum on the right-hand side, we see that for any $K \in \mathbb{N}$,

$$\sum_{n\in\mathbb{Z}} |g(x + n\alpha)|^2 \ge \sum_{j=1}^K \left| \sum_{k=1}^\infty g_k(x + n_j\alpha - \lfloor n_k\alpha \rfloor) \right|^2.$$

For all $x \in \bigcap_{k=1}^K] - \epsilon_k/3, \epsilon_k/3[$ and $j = 1, \ldots, K$,

$$\left| \sum_{k=1}^\infty g_k(x + n_j\alpha - \lfloor n_k\alpha \rfloor) \right| \ge g_j(x + n_j\alpha - \lfloor n_j\alpha \rfloor) = 1;$$

therefore, $\sum_{n\in\mathbb{Z}} |g(x + n\alpha)|^2 \ge K$. Since $K \in \mathbb{N}$ is arbitrary, this leads to (13.14). □

Another amazing example in [285] is a function g for which

(i) $\{E_{mb}T_{na}g\}_{m,n\in\mathbb{Z}}$ is a frame for all $a = \frac{1}{2k}$, $k \in \mathbb{N}$ and $b \in]0, 1[$.

(ii) $\{E_{mb}T_{na}g\}_{m,n\in\mathbb{Z}}$ is never a frame when $a = \frac{\ell}{3k}$, $k, \ell \in \mathbb{N}$ and $b \in]0, 1[$.

The results by Feichtinger and Janssen show that one has to be very careful when dealing with Gabor systems for general functions. For example, numerical calculations will always lead to round-off errors, which can change the properties of a Gabor system drastically. The way to avoid the problem is to use a class of "well-behaving" generators $g \in L^2(\mathbb{R})$ for which the undesired phenomena do not appear. It has been known for some years that when $g \in L^2(\mathbb{R}) \setminus \{0\}$ is in the Feichtinger algebra $\mathcal{S}_0$, then there exist constants $a_0, b_0 > 0$ such that $\{E_{mb}T_{na}g\}_{m,n\in\mathbb{Z}}$ is a frame for $L^2(\mathbb{R})$ for all $a \in]0, a_0], b \in]0, b_0]$; see Corollary 24.2.6. More recently, Feichtinger and Kaiblinger proved in [286] that the triples (g, a, b) in $\mathcal{S}_0 \times \mathbb{R}^+ \times \mathbb{R}^+$ which generate Gabor frames are open with respect to the product topology. In particular, for a fixed $g \in \mathcal{S}_0$, the set of points $(a, b) \in]0, \infty[\times]0, \infty[$ for which $\{E_{mb}T_{na}g\}_{m,n\in\mathbb{Z}}$ is a frame is open. Thus, the use of generators in $\mathcal{S}_0$ will lead to Gabor systems that are less sensitive toward round-off errors.

13.4   Irregular Gabor Systems

Until now we have only considered Gabor systems of the special form $\{E_{mb}T_{na}g\}_{m,n\in\mathbb{Z}}$, i.e., time–frequency shifts of the function g along a lattice $\{(na, mb)\}_{m,n\in\mathbb{Z}}$. By replacing the lattice with a countable sequence of points $\{(\mu_n, \lambda_n)\}_{n\in I} \subset \mathbb{R}^2$, we obtain a more general Gabor system of the form

$$\{E_{\lambda_n}T_{\mu_n}g(x)\}_{n\in I} = \{e^{2\pi i\lambda_n x}g(x - \mu_n)\}_{n\in I}. \qquad (13.15)$$

We call (13.15) an *irregular Gabor system*.

The analysis of irregular Gabor systems is complicated, and the theory is not fully developed. Especially the general case described here, where no structure on the sequence $\{(\mu_n, \lambda_n)\}_{n\in I}$ is assumed, causes difficulties. There are, however, some types of irregular Gabor systems which somehow are between the systems $\{E_{\lambda_n}T_{\mu_n}g\}_{n\in I}$ and the regular systems $\{E_{mb}T_{na}g\}_{m,n\in\mathbb{Z}}$. In fact, if we consider two countable sequences $\{\mu_n\}_{n\in\mathbb{Z}}, \{\lambda_m\}_{m\in\mathbb{Z}} \subset \mathbb{R}$, then the Gabor system

$$\{E_{\lambda_m}T_{\mu_n}g\}_{m,n\in\mathbb{Z}} \qquad (13.16)$$

still has some kind of lattice structure: if $\{\mu_n\}_{n\in\mathbb{Z}}, \{\lambda_m\}_{m\in\mathbb{Z}}$ are increasing and $\mu_n, \lambda_m \to \pm\infty$ for $m, n \to \pm\infty$, this Gabor system also splits the time–frequency plane $\mathbb{R}^2$ into boxes, but with varying size. We encourage the reader to make a sketch, based on, for example, $\mu_n = \lambda_n = n2^{|n|}$. If we further assume that $\lambda_m = mb$ for some $b > 0$ (or $\mu_n = an$), we are "close" to the regular case. This even holds in a more precise sense: several results for Gabor systems $\{E_{mb}T_{na}g\}_{m,n\in\mathbb{Z}}$ can be extended to the case where only one of the sequences $\{mb\}_{m\in\mathbb{Z}}, \{na\}_{n\in\mathbb{Z}}$ is replaced by an irregular sequence. In fact, if the translation is still along the set $\{na\}_{n\in\mathbb{Z}}$, the commutator relations between the modulation/translation operators show that the Gabor system can be considered as a shift-invariant system, and the results from Chapter 10 can be applied, and if modulation is along the set $\{mb\}_{m\in\mathbb{Z}}$, the Gabor system can be turned into a shift-invariant system via the Fourier transform. We return to this point in the analysis of generalized shift-invariant systems in Chapter 20; see in particular Corollary 20.5.5.

An important result was proved in the paper [604] by Sun and Zhou. It is based on the assumption that g as well as the function $x \mapsto xg(x)$ belongs to the Sobolev space $H_1(\mathbb{R})$; see (A.5). In order to avoid cumbersome notation, we allow us to write the latter assumption as $xg(x) \in H_1(\mathbb{R})$:

Theorem 13.4.1 *Assume that* $g, xg(x) \in H_1(\mathbb{R}) \setminus \{0\}$ *and that* $a, b > 0$ *are chosen such that*

$$\Delta := \frac{2a}{\pi}||g'|| + 4b\,||xg(x)|| + \frac{8ab}{\pi}||xg'(x) + g(x)|| < ||g||. \qquad (13.17)$$

Let $\{(\mu_{m,n}, \lambda_{m,n})\}_{m,n \in \mathbb{Z}} \subset \mathbb{R}^2$ *be chosen such that*

$$(\mu_{m,n}, \lambda_{m,n}) \in [na, (n+1)a[\times[mb, (m+1)b[, \ \forall m, n \in \mathbb{Z}. \qquad (13.18)$$

Then $\{E_{\lambda_{m,n}} T_{\mu_{m,n}} g\}_{m,n \in \mathbb{Z}}$ *is a frame for* $L^2(\mathbb{R})$ *with frame bounds*

$$\frac{1}{ab}(||g|| - \Delta)^2, \ \frac{1}{ab}(||g|| + \Delta)^2.$$

To avoid confusion, we note that all norms in (13.17) are $L^2(\mathbb{R})$-norms. We will not prove Theorem 13.4.1 (it can be considered as an explicit version of Corollary 24.2.6), but it is worth discussing the assumption (13.18). When $a, b > 0$ are given, the boxes

$$[na, (n+1)a[\times[mb, (m+1)b[, \ m, n \in \mathbb{Z},$$

form a partition of $\mathbb{R}^2$ into disjoint sets. Theorem 13.4.1 shows that by taking a, b small enough and picking exactly one point from each box, the associated time–frequency shifts of g will form a frame. Thus, Theorem 13.4.1 can be considered as a density result, saying that if the points $\{(\mu_{m,n}, \lambda_{m,n})\}_{m,n \in \mathbb{Z}}$ are "dense enough in $\mathbb{R}^2$ but not too close," then they generate a frame under the stated assumptions on the window g.

On the other hand, one can also consider Theorem 13.4.1 as a perturbation result. In fact, the conditions imply that $\{E_{mb}T_{na}g\}_{m,n \in \mathbb{Z}}$ itself is a frame; now the natural interpretation of condition (13.18) is that $\{E_{\lambda_{m,n}} T_{\mu_{m,n}} g\}_{m,n \in \mathbb{Z}}$ is a frame if the points $\{(\mu_{m,n}, \lambda_{m,n})\}_{m,n \in \mathbb{Z}}$ are sufficiently close to $\{(na, mb)\}_{m,n \in \mathbb{Z}}$. We return to perturbation of Gabor frames in Section 22.4.

Irregular Gabor systems were already considered around 1990 in a series of papers by Feichtinger and Gröchenig [280], [336]; this was in a more general context, to which we return in Section 24.2. The first direct approach to irregular Gabor systems in $L^2(\mathbb{R})$ was by Gröchenig [337] in 1993. Around the same time, Ramanathan and Steger [550] studied completeness properties of irregular Gabor systems in terms of the Beurling densities of $\{(\mu_n, \lambda_n)\}_{n \in I}$, defined in (9.1). In particular, they proved that the density must be exactly 1 in order for $\{E_{\lambda_n} T_{\mu_n} g\}_{n \in I}$ to be a Riesz basis; also, for

$\{E_{\lambda_n}T_{\mu_n}g\}_{n\in I}$ to be a frame, it is necessary that $D^-(\{(\mu_n,\lambda_n)\}) \geq 1$. For a regular Gabor system $\{E_{mb}T_{na}g\}_{m,n\in\mathbb{Z}}$, the latter assumption corresponds exactly to the condition $ab \leq 1$. We also note that Christensen, Deng, and Heil [168] proved that $\{E_{\lambda_n}T_{\mu_n}g\}_{n\in I}$ only can be a frame if $\{(\mu_n,\lambda_n)\}_{n\in I}$ is relatively separated.

As noted before, $\{E_{mb}T_{na}g\}_{m,n\in\mathbb{Z}}$ is not complete in $L^2(\mathbb{R})$ if $ab > 1$. One could therefore expect that $\{E_{\lambda_n}T_{\mu_n}g\}_{n\in I}$ must be incomplete whenever $D^-(\{(\mu_n,\lambda_n)\}) < 1$, and this appears as a conjecture in [550]. However, Benedetto, Heil and Walnut [48] have shown that this is false. We will discuss this in some detail, but leave most of the calculations to the reader (Exercise 13.6). The construction is based on a result by Landau [475]:

Lemma 13.4.2 *Let $\delta \in]0,\frac{1}{2}[$ and $K \in \mathbb{N}$ be given and consider the interval*

$$J := \bigcup_{n=0}^{K-1}]n - (\frac{1}{2} - \delta), n + \frac{1}{2} - \delta[. \tag{13.19}$$

Then, for any $\epsilon > 0$, there exists a symmetric real sequence $\{\lambda_m\}_{m\in\mathbb{Z}}$ for which $|\lambda_m - m| \leq \epsilon$ for all $m \in \mathbb{Z}$ and such that $\{e^{2\pi i\lambda_m x}\}_{m\in\mathbb{Z}}$ is complete in $C(J)$ with respect to the $||\cdot||_\infty$-norm.

Recall from Section 9.1 that a sequence $\{\lambda_k\}_{k\in\mathbb{Z}}$ for which the upper and lower Beurling densities coincide is said to have a Beurling density, which equals the upper and lower densities and is denoted by $D(\{\lambda_k\}_{k\in\mathbb{Z}})$.

The exact statement of the result by Benedetto et al. is as follows:

Proposition 13.4.3 *Given an arbitrary $\epsilon > 0$, there exist a sequence $\{(\mu_n,\lambda_n)\}_{n\in I} \subset \mathbb{R}^2$ and a function $g \in L^2(\mathbb{R})$ such that*

(i) $D(\{(\mu_n,\lambda_n)\}) \leq \epsilon$;

(ii) $\{E_{\lambda_n}T_{\mu_n}g\}$ is complete in $L^2(\mathbb{R})$.

Proof. We will leave some details to the reader – see Exercise 13.6. In order to apply Lemma 13.4.2, we fix $\epsilon \in]0, 1/2[$ and $\delta \in]0, 1/4[$. For a given $K \in \mathbb{N}$, we use the notation in Lemma 13.4.2 and choose a real sequence $\{\lambda_m\}_{m\in\mathbb{Z}}$ such that $|\lambda_m - m| \leq \epsilon$ and $\{e^{2\pi i\lambda_m x}\}_{m\in\mathbb{Z}}$ is complete in $C(J)$. Let

$$\Gamma := \{(Kn,\lambda_m)\}_{m,n\in\mathbb{Z}} \cup \{(Kn + \frac{1}{2},\lambda_m)\}_{m,n\in\mathbb{Z}}.$$

The Beurling densities of $\{\lambda_m\}_{m\in\mathbb{Z}}$ and $\{Kn\}_{n\in\mathbb{Z}}$ in $\mathbb{R}$ are

$$D(\{\lambda_m\}_{m\in\mathbb{Z}}) = 1, \quad D(\{Kn\}_{n\in\mathbb{Z}}) = \frac{1}{K}; \tag{13.20}$$

it follows from here that

$$D(\Gamma) = \frac{2}{K}. \tag{13.21}$$

We now prove that the irregular Gabor system associated to Λ and the function $g = \chi_J$, i.e.,

$$\{E_{\lambda_m} T_{Kn} \chi_J\}_{m,n\in\mathbb{Z}} \cup \{E_{\lambda_m} T_{Kn+\frac{1}{2}} \chi_J\}_{m,n\in\mathbb{Z}}, \tag{13.22}$$

is complete in $L^2(\mathbb{R})$. Suppose that $f \in L^2(\mathbb{R})$ is orthogonal to all the functions in (13.22). Considering an arbitrary fixed value of $n \in \mathbb{Z}$, we have

$$\begin{aligned}
0 = \langle f, E_{\lambda_m} T_{Kn} \chi_J \rangle &= \int_{-\infty}^{\infty} f(x) e^{-2\pi i \lambda_m x} \chi_J(x - Kn) dx \\
&= e^{-2\pi i \lambda_m Kn} \int_J f(x + Kn) e^{-2\pi i \lambda_m x} dx.
\end{aligned}$$

Since $\{e^{-2\pi i \lambda_m x}\}_{m\in\mathbb{Z}}$ is complete in $L^2(J)$, it follows that $f(x + Kn) = 0$ for a.e. $x \in J$. This holds for all $n \in \mathbb{Z}$. A similar argument gives that $f(x + Kn + 1/2) = 0$ for a.e. $x \in J$ and all $n \in \mathbb{Z}$. By the choice of J in (13.19),

$$\mathbb{R} = \left(\bigcup_{n\in\mathbb{Z}} (J + Kn) \right) \bigcup \left(\bigcup_{n\in\mathbb{Z}} (J + Kn + \frac{1}{2}) \right), \tag{13.23}$$

so we conclude that $f = 0$. Thus, the functions in (13.22) are complete. Since this construction is possible for all $K \in \mathbb{N}$, we are done.   □

It is of course particularly interesting to analyze irregular Gabor systems from the point of view of the shortcomings of the regular Gabor systems. For example, knowing from the Balian–Low theorem that a regular Gabor system $\{E_{mb} T_{na} g\}_{m,n\in\mathbb{Z}}$ cannot form a Riesz basis for $L^2(\mathbb{R})$ if the window g is well localized in time and frequency (see (4.8)), it is natural to ask the similar question for an irregular Gabor system. The answer is disappointing: it is shown in [352] that even an irregular Gabor system cannot form a Riesz basis if the window is well localized in time and frequency. A related result appears in [18]: a Gabor system with window belonging to Feichtinger's algebra $\mathcal{S}_0$ (see Section A.6) cannot form a Riesz basis for $L^2(\mathbb{R})$.

Let us now return to the conjecture by Heil, Ramanathan, and Topiwala [393], which we stated already on page 36:

The HRT Conjecture: Given any finite collection of distinct points $\{(\mu_n, \lambda_n)\}_{n\in\mathcal{F}}$ in $\mathbb{R}^2$ and a function $g \neq 0$, the Gabor system $\{E_{\lambda_n} T_{\mu_n} g\}_{n\in\mathcal{F}}$ is linearly independent.

Considerable effort has been invested in the conjecture (see, e.g., the article [387] for a description of the history). The conjecture is proved under some extra assumptions in [393]. Later, Linnell [499] was able to prove it in the case where $\{(\mu_n, \lambda_n)\}$ is a lattice (or a subset hereof), i.e., for

$$\{(\mu_n, \lambda_n)\} = \{(na, mb)\}_{n=1, m=1}^{N, M}.$$

Thus, finite subsets of a regular Gabor frame $\{E_{mb}T_{na}g\}_{m,n\in\mathbb{Z}}$ for $L^2(\mathbb{R})$ are linearly independent. More results can be found in the paper [88] by Bownik and Speegle. The general conjecture is still open.

The linear independence of the elements in a frame $\{E_{mb}T_{na}g\}_{m,n\in\mathbb{Z}}$ has the surprising consequence that the lower frame bounds for finite subfamilies $\{E_{mb}T_{na}g\}_{|m|,|n|\leq N}$ (consequently considered as frames for the span of the elements, i.e., as frame sequences) are forced to go to zero for $N \to \infty$ if $\{E_{mb}T_{na}g\}_{m,n\in\mathbb{Z}}$ is overcomplete:

Theorem 13.4.4 *Suppose that $ab < 1$ and that $\{E_{mb}T_{na}g\}_{m,n\in\mathbb{Z}}$ is a frame for $L^2(\mathbb{R})$. Let $\mathcal{E}_N$ denote a lower frame bound for the frame sequence $\{E_{mb}T_{na}g\}_{|m|,|n|\leq N}$. Then*

$$\mathcal{E}_N \to 0 \ as \ N \to \infty.$$

Proof. For convenience we will prove the result for the case where $\mathcal{E}_N$ denotes the optimal lower bound for $\{E_{mb}T_{na}g\}_{|m|,|n|\leq N}$ (this clearly implies that the result is also correct as stated in the theorem).

By Theorem 11.3.1, the assumption $ab < 1$ implies that $\{E_{mb}T_{na}g\}_{m,n\in\mathbb{Z}}$ is not a Riesz basis. By Linnell's result, $\{E_{mb}T_{na}g\}_{m,n\in\mathbb{Z}}$ is linearly independent, so for each $N \in \mathbb{N}$, $\{E_{mb}T_{na}g\}_{|m|,|n|\leq N}$ is a (Riesz) basis for its span. The optimal lower frame bound $\mathcal{E}_N$ for $\{E_{mb}T_{na}g\}_{|m|,|n|\leq N}$ coincides with the optimal lower Riesz bound by Theorem 5.4.1, and the sequence $\{\mathcal{E}_N\}_{N=1}^{\infty}$ is decreasing. Now the conclusion follows by Proposition 7.1.2. $\square$

Explicit estimates for the lower frame bound for certain finite Gabor systems have been carried out in [194]:

Proposition 13.4.5 *Assume that*

(i) *$\{\lambda_m\}_{m=1}^M$ and $\{\mu_n\}_{n=1}^N$ are two finite separated sequences of real numbers, the latter separated by $\varepsilon > 0$;*

(ii) *$g \in L^2(\mathbb{R})$ and $\operatorname{supp} g \subseteq]-\infty, c]$ for some $c \in \mathbb{R}$;*

(iii) *There is a nondegenerate interval $I \subseteq [c-\varepsilon, \varepsilon]$ and a positive number d such that*

$$|g(x)| \geq d \ \ \forall \, x \in I.$$

Denote a lower bound for $\{e^{2\pi i\lambda_m x}\}_{m=1}^M$ in $L^2(I)$ by A_M and an upper bound for $\{e^{2\pi i\lambda_m x}g(x)\}_{m=1}^M$ in $L^2(\mathbb{R})$ by B_M. Then the finite Gabor system $\{E_{\lambda_m}T_{\mu_n}g\}_{m=1,n=1}^{M,N}$ is linearly independent; considering the sequence as a frame sequence, the number

$$\mathcal{E}_{M,N} = d^2 A_M \left(\frac{d^2 A_M}{16 B_M}\right)^{N-1} \tag{13.24}$$

is a lower frame bound.

As we have seen in Proposition 9.8.7, the available lower bounds A_M for the exponentials $\{e^{2\pi i\lambda_m x}\}_{m=1}^M$ are already very small; this makes the bounds $\mathcal{E}_{M,N}$ in (13.24) extremely small. The small bounds are partly due to coarse estimates, but numerical calculations confirm that the bounds are indeed very small. Let us mention a recent result by Gröchenig [344], which estimates the lower bounds for finite Gabor systems with windows in certain modulation spaces:

Proposition 13.4.6 *Consider an irregular Gabor system* $\{E_{\lambda_n}T_{\mu_n}g\}_{n\in I}$ *in* $L^2(\mathbb{R})$ *and assume that*

(i) $v : \mathbb{R}^2 \to [0,\infty[$ *is a submultiplicative weight function such that* $\lim_{n\to\infty} v(nz)^{1/n} = 1$ *for all* $z \in \mathbb{R}^2$;

(ii) *The window* g *belongs to the modulation space* M_v^1;

(iii) $\{E_{\lambda_n}T_{\mu_n}g\}_{n\in I}$ *is an overcomplete frame for* $L^2(\mathbb{R})$.

For $N \in \mathbb{N}$, *let* $I_N := \{n \in I \,\big|\, |\mu_n|^2 + |\lambda_n|^2 \leq N^2\}$, *and let* $\mathcal{E}_N$ *denote a lower frame bound for* $\{E_{\lambda_n}T_{\mu_n}g\}_{n\in I_N}$. *Then there exists a constant* $C > 0$ *such that*

$$\mathcal{E}_N \leq C \, \frac{1}{\sup\left\{v(\mu_n,\lambda_n)^2 \,\big|\, |\mu_n|^2 + |\lambda_n|^2 > N\right\}}. \tag{13.25}$$

Already for a polynomial weight v (see (A.6)), the estimate in (13.25) forces $\mathcal{E}_N$ to tend very fast to 0 as $N \to \infty$, and for a sub-exponential weight the convergence is almost exponential. As pointed out by Gröchenig in [344], this leads to a certain discrepancy between the mathematical aspects and the numerical aspects whenever we apply Proposition 13.4.6 to a regular Gabor system $\{E_{mb}T_{na}g\}_{m,n\in\mathbb{Z}}$: *mathematically,* the finite subfamilies $\{E_{mb}T_{na}g\}_{\{m,n\in\mathbb{Z}|\,(mb)^2+(na)^2\leq N^2\}}$ are linearly independent for any $N \in \mathbb{N}$, but *numerically* they will be considered as linearly dependent, already for relatively small values of N. Clearly, this means that results obtained numerically have to be treated with care.

13.5 Localized Gabor Frames

We will now return to the concept of localization of frames, which we discussed in general Hilbert spaces in Section 8.2. We will also need results about modulation spaces (Appendix A.5). Knowledge about the harmonic analysis approach to Gabor analysis in Section 24.1 will be useful, but not strictly necessary.

In Section 8.2 we followed the convention of the general frame sections and considered frames $\{f_k\}_{k=1}^\infty$, indexed by the natural numbers. For our current purpose it is important to notice that the original papers by

Gröchenig and his coauthors formulated the definitions and results concerning localization of frames for sequences indexed by any countable subset Γ of $\mathbb{R}^d$. For example, a frame $\{f_k\}_{k \in \Gamma}$ is self-localized with decay rate $s > 1$ if there exists a constant C such that

$$|\langle f_k, f_j \rangle| \leq C(1 + |k - j|)^{-s}, \, \forall k, j \in \Gamma; \tag{13.26}$$

this is exactly the same as our definition in (8.7), except that $|k - j|$ now denotes the Euclidean norm on $\mathbb{R}^d$. In the sequel we will apply the results from Section 8.2 to such instances without further comments.

We will consider irregular Gabor frames, but for reasons that will be clear soon, we change the notation slightly. Given $\gamma = (\mu, \lambda) \in \mathbb{R}^2$, we define the operator

$$\pi(\gamma) : L^2(\mathbb{R}) \to L^2(\mathbb{R}), \, \pi(\gamma) := E_\lambda T_\mu.$$

The operators $\pi(\gamma)$, $\gamma \in \mathbb{R}^2$ are closely related with the Schrödinger representation; see Example 24.1.2. Let us now choose a countable sequence of points $\Gamma \subset \mathbb{R}^2$ and a window $g \in L^2(\mathbb{R})$. The following result by Fornasier and Gröchenig [304] yields conditions for the irregular Gabor system $\{\pi(\gamma)g\}_{\gamma \in \Gamma}$ being self-localized.

Proposition 13.5.1 *Let $g \neq 0$ belong to the modulation space M_v^∞, where v is a polynomial weight for some $s > 1$. Assume that $\{\pi(\gamma)g\}_{\gamma \in \Gamma}$ is a frame for $L^2(\mathbb{R})$. Then $\{\pi(\gamma)g\}_{\gamma \in \Gamma}$ and its canonical dual frame are both self-localized, with decay rate s.*

Proof. By (A.10) and Lemma A.5.3 (iii), the assumption $g \in M_v^\infty$ implies that for some $C > 0$,

$$|\langle g, \pi(\gamma)g \rangle| \, v(\gamma) \leq C, \, \forall \gamma \in \Gamma;$$

therefore,

$$|\langle g, \pi(\gamma)g \rangle| \leq Cv(\gamma)^{-1} = C \, (1 + |\gamma|)^{-s}, \, \forall \gamma \in \Gamma.$$

It follows that for any $\gamma_1, \gamma_2 \in \Gamma$,

$$|\langle \pi(\gamma_1)g, \pi(\gamma_2)g \rangle| = |\langle g, \pi(\gamma_2 - \gamma_1)g \rangle| \leq C \, (1 + |\gamma_1 - \gamma_2|)^{-s},$$

i.e., that $\{\pi(\gamma)g\}_{\gamma \in \Gamma}$ is self-localized with decay rate s. Using Lemma 8.2.4, we now conclude that the canonical dual frame is self-localized as well. $\square$

Localized Gabor frames play a central role, not only in $L^2(\mathbb{R})$ but also in the context of Gabor expansions in Banach spaces. We will return to this in Section 24.3.

13.6 Wilson Bases

In Chapter 4, we discussed some of the limitations on the function g if we want a Gabor system $\{E_{mb}T_{na}g\}_{m,n\in\mathbb{Z}}$ to be a Riesz basis for $L^2(\mathbb{R})$. In particular, the Balian–Low theorem shows that g cannot be well localized in both time and frequency. However, these properties can very well be combined with $\{E_{mb}T_{na}g\}_{m,n\in\mathbb{Z}}$ being a frame (the Gaussian is a concrete example), and this is just one motivation for the study of Gabor frames.

Daubechies, Jaffard, and Journé [245] proposed in 1991 another way to circumvent the Balian–Low theorem. They proved that if one is ready to give up the Gabor structure, it is possible to obtain a well-localized orthonormal basis: more precisely, if $g \in L^2(\mathbb{R})$ is chosen such that the Fourier transform $\widehat{g}$ is real-valued and $\{E_mT_{n/2}g\}_{m,n\in\mathbb{Z}}$ is a tight frame with bound $A = 2$, then the collection of functions $\{\psi_{\ell,k}\}_{\ell\geq 0,k\in\mathbb{Z}}$ defined by

$$\psi_{\ell,k}(x) = \begin{cases} g(x-k) & \text{for} \quad \ell = 0, \\ \sqrt{2}g(x-k/2)\cos(2\pi\ell x) & \text{for} \quad \ell > 0, k+\ell \text{ even}, \\ \sqrt{2}g(x-k/2)\sin(2\pi\ell x) & \text{for} \quad \ell > 0, k+\ell \text{ odd} \end{cases}$$

constitute an orthonormal basis for $L^2(\mathbb{R})$. A basis of the form $\{\psi_{\ell,k}\}_{\ell\geq 0,k\in\mathbb{Z}}$ is called a *Wilson basis*. In terms of the modulation operators and translation operators,

$$\psi_{\ell,k} = \begin{cases} E_0T_kg & \text{for} \quad \ell = 0, \\ \frac{1}{\sqrt{2}}(E_\ell T_{k/2}g + E_{-\ell}T_{k/2}g) & \text{for} \quad \ell > 0, k+\ell \text{ even}, \\ \frac{-i}{\sqrt{2}}(E_\ell T_{k/2}g - E_{-\ell}T_{k/2}g) & \text{for} \quad \ell > 0, k+\ell \text{ odd}, \end{cases}$$

i.e., the functions in the Wilson basis consist of linear combinations of the functions in the Gabor system $\{E_mT_{n/2}g\}_{m,n\in\mathbb{Z}}$. By choosing g such that the abovementioned conditions are satisfied and

$$\left(\int_{-\infty}^{\infty}|xg(x)|^2dx\right)\left(\int_{-\infty}^{\infty}|\gamma\widehat{g}(\gamma)|^2d\gamma\right) < \infty,$$

we have obtained an orthonormal basis circumventing the Balian–Low theorem. We refer to [21], [48], and [245] for more information, especially to [245] for a construction of a suitable function g.

Observe that the important feature of the system $\{\psi_{\ell,k}\}_{\ell\geq 0,k\in\mathbb{Z}}$ is that it is an orthonormal basis. It is not complicated to construct frames with a similar structure:

Proposition 13.6.1 *Let $g \in L^2(\mathbb{R})$ and $a,b > 0$ be given, and suppose that $\{E_{mb}T_{na}g\}_{m,n\in\mathbb{Z}}$ is a frame with upper bound B. Then the functions*

$$\{g(x-na)\}_{n\in\mathbb{Z}} \cup \{\cos(2\pi mbx)g(x-na), \sin(2\pi mbx)g(x-na)\}_{m\in\mathbb{N},n\in\mathbb{Z}}$$

constitute a frame for $L^2(\mathbb{R})$ with upper bound B.

For the proof, one can check (Exercise 13.8) that the synthesis operator corresponding to the functions in Proposition 13.6.1 is bounded and surjective, so the result follows by Theorem 5.5.1.

Riesz bases of Wilson type generated by B-splines are investigated by Trebels and Steidl in [616]; overcomplete Wilson expansions are studied in [95]. For surveys on Wilson bases, we refer to [63] and [62].

A more recent analysis of Wilson bases for general time–frequency lattices is given by Kutyniok and Strohmer in [470]. An interesting application of Wilson bases appears in [340], where they are used as a technical tool in the proof of a kernel theorem on modulation spaces.

13.7 Time–Frequency Localization of Gabor Expansions

It is well known that no function $g \neq 0$ can have compact support simultaneously in the time domain and the frequency domain. However, most signals appearing in practice are *essentially localized* in the time–frequency plane, meaning that the interesting part of the signal takes place on a finite time interval, with frequencies belonging to a certain finite interval. We will now analyze how this affects the Gabor frame expansion of such signals.

Given a number $T > 0$, define the operator

$$Q_T : L^2(\mathbb{R}) \to L^2(\mathbb{R}), \ (Q_T f)(x) = \chi_{[-T,T]}(x) f(x).$$

We will use $||(I - Q_T)f||$ as a measure for the content of the function f outside the interval $[-T, T]$. So, intuitively, to say that a function $f \in L^2(\mathbb{R})$ essentially is localized on the interval $[-T, T]$ means that $||(I - Q_T)f||$ is small compared with $||f||$. Similarly, for $\Omega > 0$ we introduce an operator P_Ω (the expression below defines the operator in the Fourier domain) by

$$P_\Omega : L^2(\mathbb{R}) \to L^2(\mathbb{R}), \ \widehat{P_\Omega f}(\nu) = \chi_{[-\Omega,\Omega]}(\nu)\widehat{f}(\nu);$$

the function $\widehat{f}$ being essentially localized on $[-\Omega, \Omega]$ means that $||(I-P_\Omega)f||$ is small compared with $||f||$.

Now assume that the function f is essentially localized in both domains, i.e., on $[-T, T] \times [-\Omega, \Omega]$ for some $T, \Omega > 0$. Let

$$B(T, \Omega) := \left\{(m, n) \in \mathbb{Z}^2 \mid mb \in [-\Omega, \Omega], \ na \in [-T, T]\right\}.$$

A natural question is how well frame decompositions capture the localization of the signal f. That is, considering the frame expansion of f in terms of dual Gabor frames $\{E_{mb}T_{na}g\}_{m,n\in\mathbb{Z}}$ and $\{E_{mb}T_{na}h\}_{m,n\in\mathbb{Z}}$,

$$f = \sum_{m,n\in\mathbb{Z}} \langle f, E_{mb}T_{na}h\rangle E_{mb}T_{na}g, \tag{13.27}$$

do we obtain a reasonable approximation of f if we replace the sum over $(m,n) \in \mathbb{Z}^2$ with a sum over $(m,n) \in B(T,\Omega)$?

Since the expansion (13.27) involves inner products between the function f and the functions in the Gabor system $\{E_{mb}T_{na}h\}_{m,n\in\mathbb{Z}}$, it is natural to consider windows h with good localization properties. We will prove a result by Daubechies [241]. It shows that under this assumption, the question has an affirmative answer, at least if we replace $B(T,\Omega)$ by a certain enlargement $B(T+\Lambda, \Omega+\Gamma)$. Daubechies formulated the result for the classical frame decomposition in terms of a Gabor frame and its canonical dual frame, but the same argument holds for dual Gabor frame pairs.

Theorem 13.7.1 *Assume that the Gabor systems $\{E_{mb}T_{na}g\}_{m,n\in\mathbb{Z}}$ and $\{E_{mb}T_{na}h\}_{m,n\in\mathbb{Z}}$ form a pair of dual frames for $L^2(\mathbb{R})$ with upper frame bounds B and D, respectively, and that for some constants $C > 0, \alpha > 1/2$, the decay conditions*

$$|h(x)| \leq C(1+x^2)^{-\alpha}, \ x \in \mathbb{R}, \quad |\hat{h}(\nu)| \leq C(1+\nu^2)^{-\alpha}, \ \nu \in \mathbb{R}, \quad (13.28)$$

hold. Then, for any $\epsilon > 0$, there exist numbers $\Lambda, \Gamma > 0$ such that for all $T, \Omega > 0$,

$$\left\| f - \sum_{(m,n)\in B(T+\Lambda, \Omega+\Gamma)} \langle f, E_{mb}T_{na}h \rangle E_{mb}T_{na}g \right\|$$
$$\leq \sqrt{BD}\left(\|(I - Q_T)f\| + \|(I - P_\Omega)f\| + \epsilon \|f\| \right)$$

for all $f \in L^2(\mathbb{R})$.

Proof. Let $f \in L^2(\mathbb{R})$, and consider some fixed numbers $T, \Omega > 0$. Then, for any given $\Lambda, \Gamma > 0$, the assumption of $\{E_{mb}T_{na}g\}_{m,n\in\mathbb{Z}}$ and $\{E_{mb}T_{na}h\}_{m,n\in\mathbb{Z}}$ being dual frames implies that

$$\left\| f - \sum_{(m,n)\in B(T+\Lambda, \Omega+\Gamma)} \langle f, E_{mb}T_{na}h \rangle E_{mb}T_{na}g \right\|$$
$$= \left\| \sum_{(m,n)\notin B(T+\Lambda, \Omega+\Gamma)} \langle f, E_{mb}T_{na}h \rangle E_{mb}T_{na}g \right\|.$$

Via Lemma 2.3.1(ii), it follows that

$$\left\| f - \sum_{(m,n)\in B(T+\Lambda, \Omega+\Gamma)} \langle f, E_{mb}T_{na}h \rangle E_{mb}T_{na}g \right\|$$
$$= \sup_{\|\varphi\|=1} \left| \left\langle \sum_{(m,n)\notin B(T+\Lambda, \Omega+\Gamma)} \langle f, E_{mb}T_{na}h \rangle E_{mb}T_{na}g, \varphi \right\rangle \right|$$
$$\leq \sup_{\|\varphi\|=1} \sum_{(m,n)\notin B(T+\Lambda, \Omega+\Gamma)} |\langle f, E_{mb}T_{na}h \rangle| \, |\langle E_{mb}T_{na}g, \varphi \rangle|.$$

Observe that

$$B(T + \Lambda, \Omega + \Gamma) \subseteq \{(m, n) \mid |na| > T + \Lambda\} \cup \{(m, n) \mid |mb| > \Omega + \Gamma\};$$

thus, we arrive at

$$\left\| f - \sum_{(m,n) \in B(T+\Lambda, \Omega+\Gamma)} \langle f, E_{mb} T_{na} h \rangle E_{mb} T_{na} g \right\| \tag{13.29}$$

$$\leq \sup_{\|\varphi\|=1} \sum_{\{(m,n) \mid |na|>T+\Lambda\}} |\langle f, E_{mb} T_{na} h \rangle| \, |\langle E_{mb} T_{na} g, \varphi \rangle| \tag{13.30}$$

$$+ \sup_{\|\varphi\|=1} \sum_{\{(m,n) \mid |mb|>\Omega+\Gamma\}} |\langle f, E_{mb} T_{na} h \rangle| \, |\langle E_{mb} T_{na} g, \varphi \rangle|. \tag{13.31}$$

We will now estimate the terms in (13.30) and (13.31) separately. For the term in (13.30), we use that $f = Q_T + (I - Q_T) f$; via the triangle inequality, this implies that

$$\sum_{\{(m,n) \mid |na|>T+\Lambda\}} |\langle f, E_{mb} T_{na} h \rangle| \, |\langle E_{mb} T_{na} g, \varphi \rangle|$$

$$= \sum_{\{(m,n) \mid |na|>T+\Lambda\}} |\langle Q_T + (I - Q_T) f, E_{mb} T_{na} h \rangle| \, |\langle E_{mb} T_{na} g, \varphi \rangle|$$

$$\leq \sum_{\{(m,n) \mid |na|>T+\Lambda\}} |\langle Q_T f, E_{mb} T_{na} h \rangle| \, |\langle E_{mb} T_{na} g, \varphi \rangle|$$

$$+ \sum_{\{(m,n) \mid |na|>T+\Lambda\}} |\langle (I - Q_T) f, E_{mb} T_{na} h \rangle| \, |\langle E_{mb} T_{na} g, \varphi \rangle|$$

$$\leq \left(\sum_{\{(m,n) \mid |na|>T+\Lambda\}}^{*} |\langle Q_T f, E_{mb} T_{na} h \rangle|^2 \right)^{1/2} \times$$

$$\left(\sum_{\{(m,n) \mid |na|>T+\Lambda\}} |\langle E_{mb} T_{na} g, \varphi \rangle|^2 \right)^{1/2}$$

$$+ \left(\sum_{\{(m,n) \mid |na|>T+\Lambda\}} |\langle (I - Q_T) f, E_{mb} T_{na} h \rangle|^2 \right)^{1/2} \times$$

$$\left(\sum_{\{(m,n) \mid |na|>T+\Lambda\}} |\langle E_{mb} T_{na} g, \varphi \rangle|^2 \right)^{1/2}.$$

Using that $\{E_{mb} T_{na} g\}_{m,n \in \mathbb{Z}}$ has the upper frame bound B and that the dual frame $\{E_{mb} T_{na} h\}_{m,n \in \mathbb{Z}}$ has the upper frame bound D, this implies that

$$\sup_{\|\varphi\|=1} \sum_{\{(m,n) \mid |na|>T+\Lambda\}} |\langle f, E_{mb} T_{na} h \rangle| \, |\langle E_{mb} T_{na} g, \varphi \rangle| \tag{13.32}$$

$$\leq \sqrt{B} \left(\sum_{\{(m,n)\mid\ |na|>T+\Lambda\}} |\langle Q_T f, E_{mb}T_{na}h\rangle|^2 \right)^{1/2} + \sqrt{BD}\, \|(I - Q_T)f\|.$$

In order to estimate the expression further, we will use the calculation in (11.23). For this reason, we now assume that f is bounded and has compact support. Then (11.23) implies that

$$\sum_{\{(m,n)\mid\ |na|>T+\Lambda\}} |\langle Q_T f, E_{mb}T_{na}h\rangle|^2$$

$$= \frac{1}{b} \left| \sum_{\{n\mid\ |na|>T+\Lambda\}} \int_{-\infty}^{\infty} \overline{Q_T f(x)} h(x - na) \times \right.$$

$$\left. \sum_{k\in\mathbb{Z}} Q_T f(x - k/b)\overline{h(x - na - k/b)}\, dx \right|$$

$$\leq \frac{1}{b} \sum_{\{n\mid\ |na|>T+\Lambda\}} \sum_{k\in\mathbb{Z}} \int_{-\infty}^{\infty} |Q_T f(x)|\, |Q_T f(x - k/b)| \times$$

$$|h(x - na)|\, |h(x - na - k/b)|\, dx.$$

Using that $Q_T f$ has support on $[-T, T]$ and the decay condition on h leads to

$$\sum_{\{(m,n)\mid\ |na|>T+\Lambda\}} |\langle Q_T f, E_{mb}T_{na}h\rangle|^2 \qquad (13.33)$$

$$\leq \frac{1}{b} \sum_{\{n\mid\ |na|>T+\Lambda\}} \sum_{k\in\mathbb{Z}} \sup_{|x|\leq T, |x-k/b|\leq T} |h(x - na)|\, |h(x - na - k/b)| \times$$

$$\int_{-\infty}^{\infty} |Q_T f(x)|\, |Q_T f(x - k/b)|\, dx$$

$$\leq \frac{\|Q_T f\|^2}{b} \sum_{\{n\mid\ |na|>T+\Lambda\}} \sum_{k\in\mathbb{Z}} \sup_{|x|\leq T, |x-k/b|\leq T} |h(x - na)|\, |h(x - na - k/b)|$$

$$\leq \frac{1}{b} \|f\|^2 \times$$

$$\sum_{\{n\mid\ |na|>T+\Lambda\}} \sum_{k\in\mathbb{Z}} \sup_{|x|\leq T, |x-k/b|\leq T} \frac{1}{(1 + (x - na)^2)^\alpha} \frac{1}{(1 + (x - na - k/b)^2)^\alpha}.$$

Now, a careful examination performed in [241] (we will skip it) shows that for some constant κ that is independent of T and Λ,

$$\sum_{\{n \mid |na|>T+\Lambda\}} \sum_{k \in \mathbb{Z}} \sup_{|x| \leq T, |x-k/b| \leq T} \frac{1}{(1+(x-na)^2)^\alpha} \frac{1}{(1+(x-na-k/b)^2)^\alpha}$$
$$\leq \kappa (1+\Lambda^2)^{-2\alpha+1}.$$

Together with the calculation in (13.33) and (13.32), this leads to the following estimate of the term (13.30):

$$\sup_{||\varphi||=1} \sum_{\{(m,n) \mid |na|>T+\Lambda\}} |\langle f, E_{mb}T_{na}h\rangle| \, |\langle E_{mb}T_{na}g, \varphi\rangle|$$
$$\leq \sqrt{\frac{B}{b}} \kappa (1+\Lambda^2)^{-2\alpha+1} ||f|| + \sqrt{BD} \, ||(I-Q_T)f||. \qquad (13.34)$$

We will now estimate the term (13.31). First,

$$\langle f, E_{mb}T_{na}h\rangle = \langle \widehat{f}, \mathcal{F}E_{mb}T_{na}h\rangle$$
$$= e^{-2\pi imbna} \langle \widehat{f}, E_{-na}T_{mb}\widehat{h}\rangle.$$

Using that $\widehat{f} = \widehat{P_\Omega f} + \mathcal{F}(I-P_\Omega)f$, calculations like before lead to

$$\sup_{||\varphi||=1} \sum_{\{(m,n) \mid |mb|>\Omega+\Gamma\}} |\langle f, E_{mb}T_{na}h\rangle| \, |\langle E_{mb}T_{na}g, \varphi\rangle|$$

$$= \sup_{||\varphi||=1} \sum_{\{(m,n) \mid |mb|>\Omega+\Gamma\}} |\langle \widehat{f}, E_{-na}T_{mb}\widehat{h}\rangle| \, |\langle E_{mb}T_{na}g, \varphi\rangle|$$

$$\leq \sup_{||\varphi||=1} \left(\sum_{\{(m,n) \mid |mb|>\Omega+\Gamma\}} |\langle \widehat{P_\Omega f}, E_{na}T_{mb}\widehat{h}\rangle|^2 \right)^{1/2} \times$$

$$\left(\sum_{\{(m,n) \mid |mb|>\Omega+\Gamma\}} |\langle E_{mb}T_{na}g, \varphi\rangle|^2 \right)^{1/2}$$

$$+ \sup_{||\varphi||=1} \left(\sum_{\{(m,n) \mid |mb|>\Omega+\Gamma\}} |\langle \mathcal{F}(I-P_\Omega)f, E_{na}T_{mb}\widehat{h}\rangle|^2 \right)^{1/2} \times$$

$$\left(\sum_{\{(m,n) \mid |mb|>\Omega+\Gamma\}} |\langle E_{mb}T_{na}g, \varphi\rangle|^2 \right)^{1/2}$$

$$\leq \sqrt{B} \left(\sum_{\{(m,n) \mid |mb|>\Omega+\Gamma\}} |\langle \widehat{P_\Omega f}, E_{na}T_{mb}\widehat{h}\rangle|^2 \right)^{1/2} + \sqrt{BD} \, ||(I-P_\Omega)f||.$$

The assumptions that f is bounded and has compact support imply that $\widehat{P_\Omega f}$ is bounded as well; also, by definition $\widehat{P_\Omega f}$ has compact support. Thus, exactly as before, we can use (11.23) to prove that

$$
\sum_{\{(m,n)\mid\,|mb|>\Omega+\Gamma\}} |\langle \widehat{P_\Omega f}, E_{na}T_{mb}\widehat{h}\rangle|^2 \le
$$

$$
\le \frac{1}{a}||\widehat{P_\Omega f}||^2 \sum_{\{m\mid\,|mb|>\Omega+\Gamma\}}\sum_{k\in\mathbb{Z}} \sup_{|\nu|\le\Omega,|\nu-k/a|\le\Omega} |\widehat{h}(\nu-nb)|\,|\widehat{h}(\nu-nb-k/a)|.
$$

The decay condition on $\widehat{h}$ together with the above calculations now leads to an estimate on (13.31), with some constant $\eta > 0$ that is independent of Ω and Γ:

$$
\sup_{||\varphi||=1}\sum_{\{(m,n)\mid\,|mb|>\Omega+\Gamma\}} |\langle f, E_{mb}T_{na}h\rangle|\,|\langle E_{mb}T_{na}g,\varphi\rangle|
$$

$$
\le \sqrt{\frac{B}{a}}\eta(1+\Gamma^2)^{-2\alpha+1}||f|| + \sqrt{BD}\,||(I-P_\Omega)f||. \qquad (13.35)
$$

Finally, inserting (13.35) and (13.34) in the calculation in (13.29) leads to

$$
\left\| f - \sum_{(m,n)\in B(T+\Lambda,\Omega+\Gamma)} \langle f, E_{mb}T_{na}h\rangle E_{mb}T_{na}g \right\|
$$

$$
\le \sqrt{\frac{B}{b}}\kappa(1+\Lambda^2)^{-2\alpha+1}||f|| + \sqrt{BD}\,||(I-Q_T)f||
$$

$$
+\sqrt{\frac{B}{a}}\eta(1+\Gamma^2)^{-2\alpha+1}||f|| + \sqrt{BD}\,||(I-P_\Omega)f||,
$$

valid for all bounded functions $f \in L^2(\mathbb{R})$ with compact support. For a given $\epsilon > 0$, one can now find $\Lambda, \Gamma > 0$ such that the conclusion in the theorem holds for all such functions f; because the set of bounded functions with compact support are dense in $L^2(\mathbb{R})$, the result actually holds for all $f \in L^2(\mathbb{R})$ by Lemma 3.2.6. $\qquad\square$

It is interesting that Theorem 13.7.1 only requires decay conditions on one of the windows for the dual frame pair. In practice, this means that we can apply the result to any Gabor frame $\{E_{mb}T_{na}h\}_{m,n\in\mathbb{Z}}$ for which the window h satisfies the conditions (13.28), simply by taking the function g as an arbitrary dual window. A good candidate for the function h would be to take a B-spline B_N of sufficiently high order $N \in \mathbb{N}$. In fact, these functions have compact support, and fast decay in the Fourier domain can be obtained by taking N sufficiently large. Furthermore, explicit expressions for a dual window are known for $a = 1$ and small values of b; see Section 12.5.

The conclusion is that under the assumptions in Theorem 13.7.1, the frame expansion indeed captures the time–frequency localization of a given signal $f \in L^2(\mathbb{R})$. That is, if we for a given "allowed deviation" $\epsilon > 0$ choose the constants Λ, Γ as in Theorem 13.7.1, and the function f essentially is localized on $[-T, T] \times [-\Omega, \Omega]$, then the function

$$\sum_{(m,n) \in B(T+\Lambda, \Omega+\Gamma)} \langle f, E_{mb} T_{na} h \rangle F_{mb} T_{na} g$$

yields a good approximation of f.

13.8   Applications of Gabor Frames

There is a large diversity of research fields where Gabor systems and frames play a role. We will mention a few of them as inspiration for further reading and provide a slightly more detailed discussion of an application to pseudodifferential operators.

We have already noticed that Gabor systems appeared already around 1930 in the context of quantum mechanics and that they also find use in the study of molecules [250]. A large diversity of applications appear in the two books [291] and [292], which contain articles by researchers in different fields. Among the applied papers in [291] are

- Gabor representation and signal detection (Zeira and Friedlander);

- Multi-window Gabor schemes in signal and image representation (Zeevi, Zibulski, Porat);

- Gabor kernels for affine-invariant object recognition (Ben-Arie and Wang);

- Gabor's signal expansion in optics (Bastiaans).

From [292] we mention

- Optimal stochastic approximations and encoding schemes using Weyl–Heisenberg sets (Balan and Daubechies);

- Orthogonal frequency division multiplexing based on offset-QAM (Bölcskei).

We also note that an application of Gabor frames to noise reduction appeared already in [522].

We now go a little more into detail with an application of Gabor frames to estimation of singular values of compact operators defined via the Weyl correspondence. In general, the *singular values* of a compact operator L on a Hilbert space $\mathcal{H}$ are defined as the eigenvalues of the compact self-adjoint operator $(L^*L)^{1/2}$. Alternatively, when the singular values are arranged decreasingly, the nth singular value is

$$s_n(L) = \inf\{\|L - F\| \; : \; F \text{ has finite rank and } \dim \mathcal{R}_F < n\}. \quad (13.36)$$

One says that L belongs to the *Schatten–von Neumann class* $S_p, 0 < p < \infty$, if

$$\sum_{n=1}^{\infty} s_n(L)^p < \infty.$$

To introduce the operators we will consider, we need the *Wigner distribution* of $f, g \in L^2(\mathbb{R})$, which is defined by

$$W(f,g)(\xi,x) = \int_{-\infty}^{\infty} e^{-2\pi i p\xi} f(x + \frac{p}{2})\overline{g(x - \frac{p}{2})}dp.$$

Let $\mathcal{S}(\mathbb{R}^d)$ denote the Schwarz space of rapidly decreasing functions on $\mathbb{R}^d$. One can show that if $f, g \in \mathcal{S}(\mathbb{R})$, then $W(f,g) \in \mathcal{S}(\mathbb{R}^2)$. The *Weyl correspondence* associates to each tempered distribution $\sigma \in \mathcal{S}'(\mathbb{R}^2)$ a *pseudodifferential operator* $L_\sigma : \mathcal{S}(\mathbb{R}) \to \mathcal{S}'(\mathbb{R})$, defined via

$$\langle L_\sigma f, g \rangle \;=\; \langle \sigma, W(g,f) \rangle, \; f, g \in \mathcal{S}(\mathbb{R}).$$

In case σ corresponds to a function,

$$\langle L_\sigma f, g \rangle = \int_{-\infty}^{\infty} \int_{-\infty}^{\infty} \sigma(\xi, x) W(f,g)(\xi, x) d\xi dx, \; f, g \in \mathcal{S}(\mathbb{R}).$$

One also writes

$$L_\sigma f(x) = \int_{-\infty}^{\infty} \int_{-\infty}^{\infty} \sigma(\xi, \frac{x+y}{2}) e^{2\pi i(x-y)\xi} f(y) dy d\xi.$$

It is known that L_σ defines a compact operator on $L^2(\mathbb{R})$ if $\sigma \in L^1(\mathbb{R})$. There is a rich literature concerned with estimates of the corresponding singular values. The first appearance of Gabor frames in this context was in the paper [556] by Rochberg and Tachizawa, where they were used to find conditions on σ implying that L_σ belongs to a given Schatten–von Neumann class (see also [609], where Wilson bases are used instead of Gabor frames). The same theme was taken up by Heil, Ramanathan, and Topiwala in [394] (and more recently by Heil in [385]), and we sketch their main idea here.

The characterization of the singular values in (13.36) indicates how frames can be used to estimate $s_n(L)$, even in the general case of an operator L on a general Hilbert space $\mathcal{H}$. In fact, letting $\{f_k\}_{k=1}^{\infty}$ be a frame for $\mathcal{H}$ with frame operator S, the finite partial sums of the frame decomposition (5.7) define operators of finite rank. More precisely, for any $n \in \mathbb{N}$, the bounded operator

$$F_n : \mathcal{H} \to \mathcal{H}, \; F_n f = \sum_{k=1}^{n} \langle f, S^{-1} f_k \rangle f_k$$

has rank at most n, so

$$s_n(L) \leq ||L - LF_{n-1}||.$$

The actual technique in [394] is a variation of this idea. In fact, the authors approximate L_σ by operators of the form L_{σ_n}, where $\sigma_n = F_n \sigma$ and $\{f_k\}_{k=1}^\infty$ is a Gabor frame for $L^2(\mathbb{R})$. With this alternative approach, it is not immediately clear that L_{σ_n} has finite rank (and how large it is), but the authors provide the necessary estimates. From the above short description, it is not clear why we need to use overcomplete frames in this application. Why not just use an orthonormal basis? The answer is that for technical reasons, one needs the generator g of the Gabor frame as well as its Fourier transform to be well localized, i.e., to decay fast. The exact condition collides with the Balian–Low theorem and thus cannot be combined with g generating a basis. For this reason, we have to use an overcomplete frame, where well-localized generators are possible; in fact, the authors use a Gaussian. The overcompleteness of the frame by itself is not directly used.

More recent references to Gabor analysis and pseudodifferential operators are [614] by Toft and [355] by Gröchenig and Rzeszotnik. A large number of papers deal with representation of operators using Gabor multipliers; see [319, 227, 260] and the references therein.

Among more recent papers dealing with applications of Gabor frames, we refer to [272, 589] by Strohmer et al., where linear time-varying operators (e.g., in mobile communication) are analyzed. Gabor methods for identification of sparse operators based on the response to a probing signal are given in [382] by Heckel and Bölcskei. Sampling of operators and applications was considered in the papers [544, 540] by Pfander et al. All of these papers are directed toward the engineering community and contain several additional references for applications; the above list is just a small sample of the literature.

13.9   Exercises

13.1 Let $\{E_{mb}T_{na}g\}_{m,n\in\mathbb{Z}}$ be a frame with frame operator S Complete the proof of Theorem 13.1.1 by showing that $\{E_{m/a}T_{n/b}g\}_{m,n\in\mathbb{Z}}$ and $\{E_{m/a}T_{n/b}(ab)^{-1}S^{-1}g\}_{m,n\in\mathbb{Z}}$ are biorthogonal.

13.2 Prove (i) and (ii) in Lemma 13.2.2.

13.3 Complete the proof of Proposition 13.2.3 by proving (13.9) and the similar statement for the lower bound.

13.4 Let $\alpha \notin \mathbb{Q}$ and prove that the set $\{n\alpha - \lfloor n\alpha \rfloor\}_{n=1}^{\infty}$ is dense in $[0,1]$; use, e.g., the following result attributed to Weyl: if f is a continuous 1-periodic function, then

$$\int_0^1 f(x)dx = \lim_{N\to\infty} \frac{1}{N} \sum_{n=1}^{N} f(n\alpha).$$

13.5 Prove Corollary 13.3.2.

13.6 In this exercise we ask the reader to provide some of the details in the proof of Proposition 13.4.3.

(i) Verify (13.20) and (13.21).

(ii) Choose the sequence $\{\lambda_m\}_{m\in\mathbb{Z}}$ as in Lemma 13.4.2. Prove that $\{e^{2\pi i\lambda_m}\}_{m\in\mathbb{Z}}$ is complete in $L^2(I)$.

(iii) Verify (13.23).

13.7 Let $Q = [0,1[\times[0,1[$. Prove that $L^2(Q) \subset L^1(Q)$, and find a function $f \in L^1(Q)$ which does not belong to $L^2(Q)$.

13.8 Prove Proposition 13.6.1.

14

Gabor Frames in $\ell^2(\mathbb{Z}), L^2(0, L), \mathbb{C}^L$

In concrete applications, a model involving a Gabor frame $\{E_{mb}T_{na}g\}_{m,n\in\mathbb{Z}}$ for $L^2(\mathbb{R})$ will ultimately have to be transferred into a model involving a finite number of vectors in a finite-dimensional space. In this chapter, we show that it indeed is possible to construct Gabor-type frames in $\mathbb{C}^L$ for $L \in \mathbb{N}$, based on certain Gabor frames for $L^2(\mathbb{R})$. As intermediate steps, we will construct frames for $\ell^2(\mathbb{Z})$ based on sampling of the frame $\{E_{mb}T_{na}g\}_{m,n\in\mathbb{Z}}$, as well as frames for the space $L^2(0, L)$ based on periodization. Each of the mentioned steps keeps the frame bounds; furthermore, dual pairs of Gabor frames in one space are turned into dual pairs in the other spaces as well.

The basic insight was provided by Janssen [429], who showed how to obtain Gabor-type frames for $\ell^2(\mathbb{Z})$ by sampling of a Gabor frame for $L^2(\mathbb{R})$. A large part of the chapter will deal with his results, but we will also consider more recent results by Kaiblinger [443], Søndergaard [607], and Lopez and Han [502].

The following elegant diagram, due to Søndergaard [607], illustrates the procedure and the involved spaces. The arrows to the left indicate sampling, and the arrows down indicate periodization:

$$
\begin{array}{ccc}
L^2(\mathbb{R}) & \xrightarrow{\text{sampling}} & \ell^2(\mathbb{Z}) \\
\downarrow{\scriptstyle\text{periodization}} & & \downarrow \\
L^2(0, L) & \longrightarrow & \mathbb{C}^L
\end{array}
$$

© Springer International Publishing Switzerland 2016   359
O. Christensen, *An Introduction to Frames and Riesz Bases*,
Applied and Numerical Harmonic Analysis,
DOI 10.1007/978-3-319-25613-9_14

We will provide a more detailed diagram whenever we have derived the frame constructions in the involved spaces; see page 377.

We begin in Section 14.1 by considering the definitions of the translation operator and the modulation operator on $\ell^2(\mathbb{Z})$; in particular, this section highlights an important difference between the modulation operators on $L^2(\mathbb{R})$ and $\ell^2(\mathbb{Z})$. The frame theory for Gabor systems in $\ell^2(\mathbb{Z})$ is very similar to the Gabor theory in $L^2(\mathbb{R})$, and we will only state a few central results in Section 14.2; more results, stated in the general framework of shift-invariant systems, can be extracted from Section 14.7. Section 14.3 shows how the well-developed duality theory for Gabor frames in $L^2(\mathbb{R})$ and $\ell^2(\mathbb{Z})$ yields an easy way to construct dual pairs of Gabor frames for $\ell^2(\mathbb{Z})$ through sampling of a pair of dual frames for $L^2(\mathbb{R})$. Section 14.4 presents work by Janssen and describes how to obtain Gabor frames for $\ell^2(\mathbb{Z})$ by sampling of Gabor frames for $L^2(\mathbb{R})$; the results in this section are based on weaker assumptions than in Section 14.3. Section 14.5 explains how to construct periodic Gabor frames for $L^2(0,L)$ based on periodization of Gabor frames for $L^2(\mathbb{R})$. Gabor systems in $\mathbb{C}^L$ (and how to construct them based on Gabor frames for $L^2(\mathbb{R})$) are discussed in Section 14.6. Finally, Section 14.8 connects frames for $\ell^2(\mathbb{Z})$ and filter banks, and Section 14.9 states a few key results about Gabor frames for $\ell^2(\mathbb{Z}^d)$.

Throughout the chapter, the reader will notice that the Gabor theory has a very similar structure in the four spaces $L^2(\mathbb{R}), \ell^2(\mathbb{Z}), L^2(0,L)$, and $\mathbb{C}^L$. The mathematical reason for this is that all of the sets $\mathbb{R}, \mathbb{Z}$, the torus $\mathbb{T}$, and the set $\mathbb{Z}_L$ of integers modulo L are examples of locally compact abelian (LCA) groups. The general theory for frame decompositions on LCA groups is presented in Chapter 21; see in particular Section 21.3, where it is shown that the Gabor systems on the mentioned groups are identical to the systems considered in the current chapter.

Note that a large part of the theory for discrete Gabor frames has been implemented by Søndergaard in his time–frequency toolbox (LTFAT); the toolbox is documented in the article [608].

14.1   Translation and Modulation on $\ell^2(\mathbb{Z})$

In this chapter, we will see that there is a strong similarity between frame results for Gabor systems in $L^2(\mathbb{R})$ and Gabor systems in $\ell^2(\mathbb{Z})$. This will be most apparent if we use a similar notation in the two case. Thus, for a sequence $g \in \ell^2(\mathbb{Z})$, we will denote the jth coordinate by $g(j)$ and write

$$g = (\dots, g(-1), g(0), g(1), \dots).$$

The notation $g(j), j \in \mathbb{Z}$, is of course similar to the standard notation $g(x), x \in \mathbb{R}$, used for point evaluation of a function $g : \mathbb{R} \to \mathbb{C}$.

We now want to define the *modulation operator* $E_b, b \in \mathbb{R}$, on $\ell^2(\mathbb{Z})$. That is, given $g \in \ell^2(\mathbb{Z})$, we want $E_b g$ to be a sequence in $\ell^2(\mathbb{Z})$; we define it to be the sequence whose jth coordinate is

$$E_b g(j) := e^{2\pi i b j} g(j),\, j \in \mathbb{Z}. \tag{14.1}$$

Even though the definition of E_b makes sense for all $b \in \mathbb{R}$, we will only use modulations of the form $E_{m/M}$, where $M \in \mathbb{N}$ is fixed and $m \in \mathbb{Z}$. In the terminology used for Gabor systems in $L^2(\mathbb{R})$, this corresponds to having the modulation parameter equal to $1/M$. There is, however, one important difference between the two settings: in the $L^2(\mathbb{R})$-setting, modulation operators with different parameters are necessarily different, but this is not the case in the discrete setting discussed here. In fact, with the definition (14.1), the operators E_b are 1-periodic in b, so,

$$E_{\frac{m}{M}} = E_{\frac{m}{M}+k} \text{ for all } k \in \mathbb{Z}.$$

Therefore, $\{E_{m/M}g\}_{m\in\mathbb{Z}}$ cannot be a Bessel sequence in $\ell^2(\mathbb{Z})$, except in the case $g = 0$. For this reason, we will only consider modulations $E_{m/M}$ with $m = 0, \ldots, M-1$.

We now introduce the *translation operator* on $\ell^2(\mathbb{Z})$. Given $n \in \mathbb{Z}$ and $g \in \ell^2(\mathbb{Z})$, we let $T_n g$ be the sequence in $\ell^2(\mathbb{Z})$ whose jth coordinate is

$$T_n g(j) = g(j - n),\, j \in \mathbb{Z}. \tag{14.2}$$

The *discrete Gabor system* generated by a sequence $g \in \ell^2(\mathbb{Z})$ and with modulation parameter $1/M$ and translation parameter N, $(M, N \in \mathbb{N})$ is now defined as the family of sequences $\{E_{m/M}T_{nN}g\}_{n\in\mathbb{Z}, m=0,\ldots,M-1}$; specifically, $E_{m/M}T_{nN}g$ is the sequence in $\ell^2(\mathbb{Z})$ whose jth coordinate is

$$E_{m/M}T_{nN}g(j) = e^{2\pi i j m/M}g(j - nN).$$

If $\{E_{m/M}T_{nN}g\}_{n\in\mathbb{Z}, m=0,\ldots,M-1}$ is a Bessel sequence in $\ell^2(\mathbb{Z})$ with frame operator $S : \ell^2(\mathbb{Z}) \to \ell^2(\mathbb{Z})$, one can repeat the proof of Lemma 12.3.1 and show that (Exercise 14.1)

$$SE_{m/M}T_{nN} = E_{m/M}T_{nN}S. \tag{14.3}$$

If $\{E_{m/M}T_{nN}g\}_{n\in\mathbb{Z}, m=0,\ldots,M-1}$ is a frame for $\ell^2(\mathbb{Z})$, this implies that the canonical dual frame is given by $\{E_{m/M}T_{nN}S^{-1}g\}_{n\in\mathbb{Z}, m=0,\ldots,M-1}$; that is, as for Gabor frames in $L^2(\mathbb{R})$, it consists of time–frequency shifts of a single function.

14.2 Dual Gabor Frames in $\ell^2(\mathbb{Z})$

In (14.3) we already saw the first similarity between Gabor analysis in $L^2(\mathbb{R})$ and in $\ell^2(\mathbb{Z})$. Another one is the duality conditions: compare the following characterization of dual Gabor frames for $\ell^2(\mathbb{Z})$ with the characterization of dual Gabor frames for $L^2(\mathbb{R})$ given in Theorem 12.3.4!

Theorem 14.2.1 *Let* $M, N \in \mathbb{N}$, *and consider two sequences* $g, h \in \ell^2(\mathbb{Z})$. *If* $\{E_{m/M}T_{nN}g\}_{n\in\mathbb{Z},m=0,\dots,M-1}$ *and* $\{E_{m/M}T_{nN}h\}_{n\in\mathbb{Z},m=0,\dots,M-1}$ *are Bessel sequences, they are dual frames for* $\ell^2(\mathbb{Z})$ *if and only if*

$$\sum_{k\in\mathbb{Z}} \overline{g(j-kN-nM)}h(j-kN) = \frac{1}{M}\,\delta_{n,0}, \forall j, n \in \mathbb{Z}. \tag{14.4}$$

A proof of Theorem 14.2.1 can be found in the paper [502] by Lopez and Han, which actually deals with the d-dimensional case (see Section 14.9). Alternatively Theorem 14.2.1 can be derived from a result stated much later in Theorem 21.7.10 (Exercise 21.5).

Also the sufficient conditions in Theorem 11.4.2 for $\{E_{mb}T_{na}g\}_{m,n\in\mathbb{Z}}$ being a Bessel sequence or a frame for $L^2(\mathbb{R})$ have similar versions for Gabor systems in $\ell^2(\mathbb{Z})$. The finite sequences are dense in $\ell^2(\mathbb{Z})$, and they will often play a similar role as the set of continuous functions with compact support does in $L^2(\mathbb{R})$. In particular, the Gabor system $\{E_{m/M}T_{nN}g\}_{n\in\mathbb{Z},m=0,\dots,M-1}$ associated with a finite sequence $g \in \ell^2(\mathbb{Z})$ is a Bessel sequence for all choices of $M, N \in \mathbb{N}$. On an abstract level, the reason for the similarity between Gabor analysis in $L^2(\mathbb{R})$ and $\ell^2(\mathbb{Z})$ is that the set $\mathbb{R}$ and the set $\mathbb{Z}$ are examples of locally compact abelian groups, which implies that the general theory developed in Chapter 21 applies to both cases. Alternatively, one can obtain the results for Gabor systems in $\ell^2(\mathbb{Z})$ simply by repeating the steps in the proofs for the Gabor systems in $L^2(\mathbb{R})$, without referring to the theory for locally compact groups (Exercise 14.3).

Other results for Gabor systems in $\ell^2(\mathbb{Z})$ are parallel with the case of Gabor systems in $L^2(\mathbb{R})$ as well. For example, a necessary condition for $\{E_{m/M}T_{nN}g\}_{n\in\mathbb{Z},m=0,\dots,M-1}$ to be a frame for $\ell^2(\mathbb{Z})$ is that $\frac{N}{M} \le 1$, and if $\{E_{m/M}T_{nN}g\}_{n\in\mathbb{Z},m=0,\dots,M-1}$ is a frame for $\ell^2(\mathbb{Z})$, it is a Riesz basis if and only if $M = N$. For a selection of papers on Gabor systems in $\ell^2(\mathbb{Z})$, we refer to [229, 230, 626, 502, 41], and the references therein.

14.3 Dual Gabor Frames in $\ell^2(\mathbb{Z})$ Through Sampling

The purpose of this section and the next is to relate Gabor frames for $L^2(\mathbb{R})$ and Gabor frames for $\ell^2(\mathbb{Z})$. To be more precise, we will show that certain conditions on a Gabor frame for $L^2(\mathbb{R})$ imply that the sequences formed by sampling at the integers yield a frame for $\ell^2(\mathbb{Z})$, and that sampling of a dual pair of frames for $L^2(\mathbb{R})$ also yields a pair of dual frames for $\ell^2(\mathbb{Z})$. We begin with the case of sampling of a dual pair of frames.

For a continuous function $f : \mathbb{R} \to \mathbb{C}$, *sampling* at a point $x \in \mathbb{R}$ simply means that we consider the function value $f(x)$. For functions $f \in L^2(\mathbb{R})$ that are not assumed to be continuous, we have to be careful with the meaning of sampling. By definition, $L^2(\mathbb{R})$ consists of equivalence classes of functions that are identical almost everywhere, so point evaluations do not immediately make sense. So when we speak about sampling a function $f \in L^2(\mathbb{R})$, we really mean that we evaluate a specific representative for the considered equivalence class. For functions that are not continuous we will in general assume that the chosen representative has all the integers among its Lebesgue points.

In general, given a function $f \in L^2(\mathbb{R})$, define the discrete sequence f^D by

$$f^D := \{f(j)\}_{j \in \mathbb{Z}}. \tag{14.5}$$

The starting point for our analysis is a Gabor system for $L^2(\mathbb{R})$; we assume it to have the form $\{E_{m/M} T_{nN} g\}_{m,n \in \mathbb{Z}}$, where $g \in L^2(\mathbb{R})$ and $M, N \in \mathbb{N}$. The first question is how one can construct sequences, indexed by $\mathbb{Z}$, based on the Gabor system in $L^2(\mathbb{R})$. For each $m, n \in \mathbb{Z}$, there are two natural ways of doing so. In fact, we can consider the sequence $E_{m/M} T_{nN}(f^D)$, obtained by letting the discrete Gabor system $E_{m/M} T_{nN}$ act on the sequence f^D (we have not yet discussed whether $f^D \in \ell^2(\mathbb{Z})$, but the expression for the involved operators make sense anyway); or we can consider the discrete sequence $\left(E_{m/M} T_{nN} f\right)^D$, obtained by sampling of the function $E_{m/M} T_{nN} f \in L^2(\mathbb{R})$. Fortunately, the two procedures lead to the same outcome, and we simply write

$$E_{m/M} T_{nN} f^D = \{e^{2\pi i j m/M} f(j - nN)\}_{j \in \mathbb{Z}}. \tag{14.6}$$

Let us now assume that $\{E_{m/M} T_{nN} g\}_{m,n \in \mathbb{Z}}$ is a frame for $L^2(\mathbb{R})$. The basic idea is to ask for conditions such that the family of all the sequences $E_{m/M} T_{nN} g^D$, where $m = 0, \dots, M-1, n \in \mathbb{Z}$, constitute a frame for $\ell^2(\mathbb{Z})$. Before we consider the frame condition by itself, we will state a simple and very satisfying result concerning dual pairs of Gabor frames. The reason for doing this is that the well-developed theory for duality in $L^2(\mathbb{R})$ and $\ell^2(\mathbb{Z})$ yields an easy way to construct dual pairs of Gabor frames for $\ell^2(\mathbb{Z})$ through sampling.

Theorem 14.3.1 *Let $M, N \in \mathbb{N}$ be given, and assume that*

(i) *The functions g and h belong to either $C_c(\mathbb{R})$ or the Feichtinger algebra $\mathcal{S}_0$;*

(ii) *The Gabor systems $\{E_{m/M} T_{nN} g\}_{m,n \in \mathbb{Z}}$ and $\{E_{m/M} T_{nN} h\}_{m,n \in \mathbb{Z}}$ are dual frames for $L^2(\mathbb{R})$.*

Then the discrete Gabor systems $\{E_{m/M} T_{nN} g^D\}_{n \in \mathbb{Z}, m=0, \dots, M-1}$ and $\{E_{m/M} T_{nN} h^D\}_{n \in \mathbb{Z}, m=0, \dots, M-1}$ are dual frames for $\ell^2(\mathbb{Z})$; in the case where $g, h \in C_c(\mathbb{R})$, these sequences are finite.

Proof. In the case where $g \in \mathcal{S}_0$ and h denotes a dual window in $\mathcal{S}_0$, the result was first reported by Søndergaard [607]. We will give the proof under the assumption that $g, h \in C_c(\mathbb{R})$ and leave the minor modifications in the case of $\mathcal{S}_0$ to the reader.

If $g, h \in C_c(\mathbb{R})$, the associated sequences g^D and h^D are finite, and therefore the discrete Gabor systems $\{E_{m/M} T_{nN} g^D\}_{n \in \mathbb{Z}, m=0,\ldots,M-1}$ and $\{E_{m/M} T_{nN} h^D\}_{n \in \mathbb{Z}, m=0,\ldots,M-1}$ are Bessel sequences. Since $\{E_{m/M} T_{nN} g\}_{m,n \in \mathbb{Z}}$ and $\{E_{m/M} T_{nN} h\}_{m,n \in \mathbb{Z}}$ are dual frames, we know from Theorem 12.3.4 that

$$\sum_{k \in \mathbb{Z}} \overline{g(x - kN - nM)} h(x - kN) = \frac{1}{M} \delta_{n,0} \tag{14.7}$$

for all $n \in \mathbb{Z}$ and a.e. $x \in \mathbb{R}$. Since g and h are continuous functions with compact support, it follows that (14.7) actually holds for all $x \in \mathbb{R}$; taking $x := j \in \mathbb{Z}$ now shows that (14.4) holds, and the desired result follows from Theorem 14.2.1. $\square$

Natural candidates for applications of Theorem 14.3.1 are obtained via Theorem 12.5.1 by restricting to continuous windows g; this applies, e.g., to the B-splines B_N with $N \geq 2$. This particular construction gives an easy way to obtain dual pairs of Gabor frames for $\ell^2(\mathbb{Z})$, with windows that are explicitly given and have finite support.

For a later application we need the following version of Theorem 14.3.1, which involves more general translation/modulation parameters as well as an extra "free parameter" $\ell \in \mathbb{N}$. The result is based on an application of the scaling operator D_c on $L^2(\mathbb{R})$, see (2.23).

Corollary 14.3.2 *Let $\ell \in \mathbb{N}$ and $a, b > 0$ be given and assume that $ab = N/M$ for some $M, N \in \mathbb{N}$. Consider dual pairs of frames $\{E_{mb} T_{na} g\}_{m,n \in \mathbb{Z}}$ and $\{E_{mb} T_{na} h\}_{m,n \in \mathbb{Z}}$ for $L^2(\mathbb{R})$, and assume that either $g, h \in C_c(\mathbb{R})$ or $g, h \in \mathcal{S}_0$. Then the discrete Gabor systems*

$$\{E_{m/(M\ell)} T_{nN\ell} (D_{bM\ell} g)^D\}_{n \in \mathbb{Z}, m=0,\ldots,M\ell-1}$$

and

$$\{E_{m/(M\ell)} T_{nN\ell} (D_{bM\ell} h)^D\}_{n \in \mathbb{Z}, m=0,\ldots,M\ell-1}$$

are dual Gabor frames for $\ell^2(\mathbb{Z})$.

Proof. Let us first consider a frame $\{E_{mb} T_{na} g\}_{m,n \in \mathbb{Z}}$. Applying the unitary scaling operator D_c with $c := bM\ell$, the "standard scaling trick" in Proposition 11.2.4 yields that the Gabor system $\{E_{m/(M\ell)} T_{nabM\ell} (D_{bM\ell} g)\}_{m,n \in \mathbb{Z}}$ is a frame for $L^2(\mathbb{R})$. Similarly, $\{E_{m/(M\ell)} T_{nabM\ell} (D_{bM\ell} g)\}_{m,n \in \mathbb{Z}}$ is a frame for $L^2(\mathbb{R})$, and by Lemma 6.3.3, these frames are in fact dual frames. Using that the spaces $C_c(\mathbb{R})$ and $\mathcal{S}_0$ are invariant under scaling, the result now follows from Theorem 14.3.1. $\square$

Note that on explicit form, the window $(D_{bM\ell}g)^D$ in Corollary 14.3.2 is the sequence

$$(D_{bM\ell}g)^D = \left\{ \frac{1}{\sqrt{bM\ell}} g\left(\frac{j}{bM\ell}\right) \right\}_{j\in\mathbb{Z}}.$$

14.4 Discrete Gabor Frames Through Sampling

In this section, we will discuss the sampling results for a Gabor system $\{E_{m/M}T_{nN}g\}_{m,n\in\mathbb{Z}}$ in $L^2(\mathbb{R})$, due to Janssen [429]. In contrast to the setup in Section 14.3, it is not assumed that the window g is continuous or has compact support. As already mentioned, we have to be careful with the meaning of "sampling" for noncontinuous functions; furthermore, without the assumption that the window g is compactly supported, the discrete sequence g^D might not be finite, and we cannot be sure that it belongs to $\ell^2(\mathbb{Z})$. Parallel to the discussion in Section 14.3, sampling results can be obtained in an easier way under stronger assumptions, e.g., for windows $g \in C_c(\mathbb{R})$ or $g \in \mathcal{S}_0$. In fact, the condition in (11.14) for a Gabor system $\{E_{mb}T_{na}g\}_{m,n\in\mathbb{Z}}$ to be a frame for $L^2(\mathbb{R})$ has a similar version in $\ell^2(\mathbb{Z})$ for the parameter values $a = 1/M, b = N$; thus, continuity and fast decay of the window g will imply that the sampled Gabor system forms a frame as well.

Following Janssen we will put minimal restrictions on the function $g \in L^2(\mathbb{R})$ and find conditions such that the sampled Gabor system $\{E_{m/M}T_{nN}g^D\}_{n\in\mathbb{Z},m=0,1,\dots,M-1}$ forms a frame for $\ell^2(\mathbb{Z})$. The first result gives conditions on the Gabor system $\{E_{m/M}T_{nN}g\}_{m,n\in\mathbb{Z}}$ in $L^2(\mathbb{R})$ which imply that the discrete time–frequency shifts of g^D belong to $\ell^2(\mathbb{Z})$.

Lemma 14.4.1 *Let $g \in L^2(\mathbb{R})$ and let $M, N \in \mathbb{N}$ be given. Assume that g contains all the integers among its Lebesgue points and that $\{E_{m/M}T_{nN}g\}_{m,n\in\mathbb{Z}}$ is a Bessel sequence in $L^2(\mathbb{R})$. Then*

$$\sum_{j\in\mathbb{Z}} |g(j)|^2 \leq \frac{BN}{M}.$$

In particular, $E_{m/M}T_{nN}g^D \in \ell^2(\mathbb{Z})$ for all $m, n \in \mathbb{Z}$.

Proof. Let $j \in \mathbb{Z}$ and $\epsilon > 0$ be given. Letting B denote an upper frame bound for the Gabor system $\{E_{m/M}T_{nN}g\}_{m,n\in\mathbb{Z}}$ in $L^2(\mathbb{R})$, we know from Proposition 11.3.4 that

$$\sum_{n\in\mathbb{Z}} |g(x+nN)|^2 \leq \frac{B}{M}, \quad a.e.\ x \in \mathbb{R}. \tag{14.8}$$

In particular, g is essentially bounded. Now let $j \in \mathbb{Z}$. The assumption that j is a Lebesgue point combined with (14.8) shows that $|g(j)| \leq \sqrt{\frac{B}{M}}$. Assuming that g is real-valued, it follows that

$$\left| \, |g(j)|^2 - |g(j+x)|^2 \, \right| = |(g(j) + g(j+x))(g(j) - g(j+x))|$$

$$\leq 2\sqrt{\frac{B}{M}} \, |g(j) - g(j+x)|, \quad a.e. \ x \in \mathbb{R}.$$

Using that j is assumed to be a Lebesgue point for g,

$$\left| |g(j)|^2 - \frac{1}{\epsilon} \int_{-\epsilon/2}^{\epsilon/2} |g(x+j)|^2 dx \right| = \left| \frac{1}{\epsilon} \int_{-\epsilon/2}^{\epsilon/2} \left(|g(j)|^2 - |g(x+j)|^2 \right) dx \right|$$

$$\leq 2\sqrt{\frac{B}{M}} \frac{1}{\epsilon} \int_{-\epsilon/2}^{\epsilon/2} |g(j) - g(j+x)| dx$$

$$\to 0 \text{ as } \epsilon \to 0.$$

We conclude that

$$|g(j)|^2 = \lim_{\epsilon \to 0} \frac{1}{\epsilon} \int_{-\epsilon/2}^{\epsilon/2} |g(x+j)|^2 dx, \quad \forall j \in \mathbb{Z}. \tag{14.9}$$

We have derived this under the assumption that g is real-valued, but it now follows that (14.9) also holds for complex-valued functions. In the rest of the proof, we can proceed without assuming g to be real-valued. Using Fatou's lemma on $X = \mathbb{Z}$ and the counting measure, followed by an application of (14.8),

$$\sum_{j \in \mathbb{Z}} |g(j)|^2 = \sum_{j \in \mathbb{Z}} \liminf_{\epsilon \to 0} \frac{1}{\epsilon} \int_{-\epsilon/2}^{\epsilon/2} |g(x+j)|^2 dx$$

$$\leq \liminf_{\epsilon \to 0} \sum_{j \in \mathbb{Z}} \frac{1}{\epsilon} \int_{-\epsilon/2}^{\epsilon/2} |g(x+j)|^2 dx$$

$$= \liminf_{\epsilon \to 0} \sum_{j=0}^{N-1} \frac{1}{\epsilon} \int_{-\epsilon/2}^{\epsilon/2} \sum_{n \in \mathbb{Z}} |g(x+j+nN)|^2 dx$$

$$\leq \frac{BN}{M}.$$

This proves that $g^D \in \ell^2(\mathbb{Z})$; now the lemma follows because the operators $E_{m/M}$ and T_{nN} on $\ell^2(\mathbb{Z})$ are norm-preserving. $\qquad \square$

The next lemma is an important step from Gabor systems in $L^2(\mathbb{R})$ to Gabor systems in $\ell^2(\mathbb{Z})$. It contains an identity involving functions in $L^2(\mathbb{R})$, which "approaches discrete sequences" for small values of ϵ:

Lemma 14.4.2 *Let $g \in L^2(\mathbb{R})$ and $M, N \in \mathbb{N}$ be given, and assume that $\{E_{m/M}T_{nN}g\}_{m,n\in\mathbb{Z}}$ is a Bessel sequence in $L^2(\mathbb{R})$. Given $\epsilon \in]0, \frac{1}{2}[$, let*

$$\delta^\epsilon = \frac{1}{\epsilon}\chi_{]-\frac{1}{2}\epsilon, \frac{1}{2}\epsilon[}.$$

Consider a finite linear combination of translates of δ^ϵ,

$$f^\epsilon = \sum_{j\in\mathbb{Z}} c_j T_j \delta^\epsilon. \tag{14.10}$$

Then

$$\sum_{m,n\in\mathbb{Z}} |\langle f^\epsilon, E_{m/M}T_{nN}g\rangle|^2$$

$$= \sum_{n\in\mathbb{Z}}\sum_{m=0}^{M-1}\sum_{j,k\in\mathbb{Z}} c_j\overline{c_k}\frac{1}{\epsilon^2}\int_{-\frac{1}{2}\epsilon}^{\frac{1}{2}\epsilon} \overline{E_{m/M}T_{nN}g(x+j)}E_{m/M}T_{nN}g(x+k)dx.$$

Proof. First, we use the definition of f^ϵ to write

$$\sum_{m,n\in\mathbb{Z}} |\langle f^\epsilon, E_{m/M}T_{nN}g\rangle|^2$$

$$= \sum_{m,n\in\mathbb{Z}}\sum_{j,k\in\mathbb{Z}} c_j\overline{c_k}\langle T_j\delta^\epsilon, E_{m/M}T_{nN}g\rangle\langle E_{m/M}T_{nN}g, T_k\delta^\epsilon\rangle$$

$$= \sum_{n\in\mathbb{Z}}\sum_{m=0}^{M-1}\sum_{\ell\in\mathbb{Z}}\sum_{j,k} c_j\overline{c_k}\langle T_j\delta^\epsilon, E_{\ell+m/M}T_{nN}g\rangle\langle E_{\ell+m/M}T_{nN}g, T_k\delta^\epsilon\rangle.$$

Now, via Lemma 11.2.2,

$$\langle T_j\delta^\epsilon, E_{\ell+m/M}T_{nN}g\rangle$$
$$= \langle E_{-m/M}T_j\delta^\epsilon, E_\ell T_{nN}g\rangle$$
$$= \int_0^1 \left(\sum_{r\in\mathbb{Z}} T_j\delta^\epsilon(x-r)\overline{E_{m/M}T_{nN}g(x-r)}\right)e^{-2\pi i\ell x}dx,$$

which is the ℓth Fourier coefficient of the 1-periodic function

$$\alpha_j(x) = \sum_{r\in\mathbb{Z}} T_j\delta^\epsilon(x-r)\overline{E_{m/M}T_{nN}g(x-r)}$$

in $L^2(0,1)$. Note that for $x \in [-1/2, 1/2]$,

$$\alpha_j(x) = \delta^\epsilon(x)\overline{E_{m/M}T_{nN}g(x+j)}$$
$$= \frac{1}{\epsilon}\chi_{]-\frac{1}{2}\epsilon, \frac{1}{2}\epsilon[}(x)\overline{E_{m/M}T_{nN}g(x+j)}.$$

Via Lemma 3.8.2,

$$\sum_{\ell\in\mathbb{Z}}\langle T_j\delta^\epsilon, E_{\ell+m/M}T_{nN}g\rangle\langle E_{\ell+m/M}T_{nN}g, T_k\delta^\epsilon\rangle$$

$$= \langle\alpha_j,\alpha_k\rangle = \int_{-\frac{1}{2}}^{\frac{1}{2}}\alpha_j(x)\overline{\alpha_k(x)}dx$$

$$= \frac{1}{\epsilon^2}\int_{-\frac{1}{2}\epsilon}^{\frac{1}{2}\epsilon}\overline{E_{m/M}T_{nN}g(x+j)}E_{m/M}T_{nN}g(x+k)dx,$$

and the result follows. □

If we impose stronger conditions on the window g, we can obtain a Gabor frame for $\ell^2(\mathbb{Z})$ by sampling of a Gabor frame $\{E_{m/M}T_{nN}g\}_{m,n\in\mathbb{Z}}$ for $L^2(\mathbb{R})$. We will use "condition (R)", see (12.2); this condition is satisfied, e.g., if the window g belongs to Feichtinger's algebra $\mathcal{S}_0$.

Theorem 14.4.3 *Let $M, N \in \mathbb{N}$. Assume that $g \in L^2(\mathbb{R})$ satisfies condition (R) and that $\{E_{m/M}T_{nN}g\}_{m,n\in\mathbb{Z}}$ is a frame for $L^2(\mathbb{R})$ with frame bounds A, B. Then the discrete Gabor system $\{E_{m/M}T_{nN}g^D\}_{n\in\mathbb{Z},m=0,...,M-1}$ is a frame for $\ell^2(\mathbb{Z})$ with frame bounds A, B.*

Proof. In order to prove that $\{E_{m/M}T_{nN}g^D\}_{n\in\mathbb{Z},m=0,...,M-1}$ is a frame for $\ell^2(\mathbb{Z})$, we consider a finite sequence $\{c_k\}_{k\in\mathbb{Z}}$. For $\epsilon\in]0,\frac{1}{2}[$, consider the function f^ϵ in (14.10); we have $||f^\epsilon||^2 = \frac{1}{\epsilon}\sum_{j\in\mathbb{Z}}|c_j|^2$. Applying the frame condition for $\{E_{m/M}T_{nN}g\}_{m,n\in\mathbb{Z}}$ on f^ϵ gives that for all $\epsilon\in]0,\frac{1}{2}[$,

$$A\sum_{j\in\mathbb{Z}}|c_j|^2 \le \epsilon\sum_{m,n\in\mathbb{Z}}|\langle f^\epsilon, E_{m/M}T_{nN}g\rangle|^2 \le B\sum_{j\in\mathbb{Z}}|c_j|^2.$$

For the proof of Theorem 14.4.3, it is therefore enough to show that

$$\liminf_{\epsilon\to 0}\epsilon\sum_{m,n\in\mathbb{Z}}|\langle f^\epsilon, E_{m/M}T_{nN}g\rangle|^2 = \sum_{n\in\mathbb{Z}}\sum_{m=0}^{M-1}\left|\sum_j c_j\overline{E_{m/M}T_{nN}g(j)}\right|^2. \quad (14.11)$$

In fact, if (14.11) is satisfied for all finite sequences $\{c_k\}_{k\in\mathbb{Z}}$, then $\{E_{m/M}T_{nN}g^D\}_{n\in\mathbb{Z},m=0,...,M-1}$ satisfies the frame condition in $\ell^2(\mathbb{Z})$ on a dense set and therefore on $\ell^2(\mathbb{Z})$ by Lemma 5.1.9.

First we note that

$$\sum_{n\in\mathbb{Z}}\sum_{m=0}^{M-1}\left|\sum_j c_j\overline{E_{m/M}T_{nN}g(j)}\right|^2$$

$$= \sum_{n\in\mathbb{Z}}\sum_{m=0}^{M-1}\sum_{j,k}c_j\overline{c_k}\overline{E_{m/M}T_{nN}g(j)}E_{m/M}T_{nN}g(k),$$

while by Lemma 14.4.2

$$\epsilon \sum_{m,n\in\mathbb{Z}} |\langle f^\epsilon, E_{m/M}T_{nN}g\rangle|^2$$

$$= \sum_{n\in\mathbb{Z}}\sum_{m=0}^{M-1}\sum_{j,k} c_j\overline{c_k}\frac{1}{\epsilon}\int_{-\frac{1}{2}\epsilon}^{\frac{1}{2}\epsilon} \overline{E_{m/M}T_{nN}g(x+j)}E_{m/M}T_{nN}g(x+k)dx.$$

Comparing the two expressions, we see that (14.11) follows if we can prove that

$$\sum_{n\in\mathbb{Z}}\frac{1}{\epsilon}\int_{-\frac{1}{2}\epsilon}^{\frac{1}{2}\epsilon} \overline{E_{m/M}T_{nN}g(x+j)}E_{m/M}T_{nN}g(x+k)dx$$

$$\to \sum_{n\in\mathbb{Z}} \overline{E_{m/M}T_{nN}g(j)}E_{m/M}T_{nN}g(k) \text{ as } \epsilon \to 0$$

for all $m = 0,\ldots,M-1$ and $j,k \in \mathbb{Z}$ (recall that the sums over j,k are finite). Now,

$$\left|\frac{1}{\epsilon}\int_{-\frac{1}{2}\epsilon}^{\frac{1}{2}\epsilon} \overline{E_{m/M}T_{nN}g(x+j)}E_{m/M}T_{nN}g(x+k)dx\right.$$

$$\left. - \overline{E_{m/M}T_{nN}g(j)}E_{m/M}T_{nN}g(k)\right|$$

$$\leq \frac{1}{\epsilon}\int_{-\frac{1}{2}\epsilon}^{\frac{1}{2}\epsilon} \left|\overline{E_{m/M}T_{nN}g(x+j)}E_{m/M}T_{nN}g(x+k)\right.$$

$$\left. -\overline{E_{m/M}T_{nN}g(j)}E_{m/M}T_{nN}g(k)\right|dx$$

$$= \frac{1}{\epsilon}\int_{-\frac{1}{2}\epsilon}^{\frac{1}{2}\epsilon} \left|\overline{g(x+j-nN)}g(x+k-nN) - \overline{g(j-nN)}g(k-nN)\right|dx$$

$$\leq \frac{1}{\epsilon}\int_{-\frac{1}{2}\epsilon}^{\frac{1}{2}\epsilon} \left|\overline{g(x+j-nN)} - \overline{g(j-nN)}\right| \left|g(x+k-nN)\right|dx$$

$$+\frac{1}{\epsilon}\int_{-\frac{1}{2}\epsilon}^{\frac{1}{2}\epsilon} \left|\overline{g(j-nN)}\right| \left|g(x+k-nN) - g(k-nN)\right|dx.$$

It follows that

$$\left|\sum_{n\in\mathbb{Z}}\frac{1}{\epsilon}\int_{-\frac{1}{2}\epsilon}^{\frac{1}{2}\epsilon} \overline{E_{m/M}T_{nN}g(x+j)}E_{m/M}T_{nN}g(x+k)dx\right.$$

$$\left. - \sum_{n\in\mathbb{Z}} \overline{E_{m/M}T_{nN}g(j)}E_{m/M}T_{nN}g(k)\right|$$

$$\leq \frac{1}{\epsilon} \sum_{n\in\mathbb{Z}} \int_{-\frac{\epsilon}{2}}^{\frac{\epsilon}{2}} \left|\overline{g(x+j-nN)} - \overline{g(j-nN)}\right|\ |g(x+k-nN)|\,dx \qquad (14.12)$$

$$+\frac{1}{\epsilon} \sum_{n\in\mathbb{Z}} \int_{-\frac{1}{2}\epsilon}^{\frac{1}{2}\epsilon} \left|\overline{g(j-nN)}\right|\ |g(x+k-nN) - g(k-nN)|\,dx. \qquad (14.13)$$

Both (14.12) and (14.13) converge to zero as $\epsilon \to 0$; we give the argument for (14.12). Applying Cauchy–Schwarz inequality twice,

$$\frac{1}{\epsilon} \sum_{n\in\mathbb{Z}} \int_{-\frac{1}{2}\epsilon}^{\frac{1}{2}\epsilon} \left|\overline{g(x+j-nN)} - \overline{g(j-nN)}\right|\ |g(x+k-nN)|\,dx$$

$$\leq \frac{1}{\epsilon} \sum_{n\in\mathbb{Z}} \left(\int_{-\frac{1}{2}\epsilon}^{\frac{1}{2}\epsilon} |g(x+j-nN) - g(j-nN)|^2 dx\right)^{1/2}$$

$$\times \left(\int_{-\frac{1}{2}\epsilon}^{\frac{1}{2}\epsilon} |g(x+k-nN)|^2\,dx\right)^{1/2}$$

$$\leq \frac{1}{\epsilon} \left(\sum_{n\in\mathbb{Z}} \int_{-\frac{1}{2}\epsilon}^{\frac{1}{2}\epsilon} |g(x+j-nN) - g(j-n\hat{N})|^2 dx\right)^{1/2}$$

$$\times \left(\sum_{n\in\mathbb{Z}} \int_{-\frac{1}{2}\epsilon}^{\frac{1}{2}\epsilon} |g(x+k-nN)|^2\,dx\right)^{1/2} = (*).$$

Via Lemma 14.4.1, the second term in (*) can be estimated by

$$\left(\sum_{n\in\mathbb{Z}} \int_{-\frac{1}{2}\epsilon}^{\frac{1}{2}\epsilon} |g(x+k-nN)|^2\,dx\right)^{1/2} \leq \sqrt{\frac{BN}{M}}\,\epsilon;$$

thus,

$$(*) \leq \sqrt{\frac{BN}{M}} \left(\frac{1}{\epsilon} \sum_{n\in\mathbb{Z}} \int_{-\frac{1}{2}\epsilon}^{\frac{1}{2}\epsilon} |g(x+j-nN) - g(j-nN)|^2 dx\right)^{1/2},$$

which converges to zero for $\epsilon \to 0$ because of condition (R); the proof is completed. $\square$

Following the proof of Corollary 14.3.2, it is easy to derive a version of Theorem 14.4.3 with additional freedom.

Corollary 14.4.4 *Let $\ell \in \mathbb{N}$ and $a, b > 0$ be given and assume that $ab = N/M$ for some $M, N \in \mathbb{N}$. Consider a frame $\{E_{mb}T_{na}g\}_{m,n\in\mathbb{Z}}$ for $L^2(\mathbb{R})$ for $L^2(\mathbb{R})$ with bounds A, B, and assume that $g \in S_0$. Then the discrete Gabor system $\{E_{m/(M\ell)}T_{nN\ell}(D_{bM\ell}g)^D\}_{n\in\mathbb{Z}, m=0,\dots,M\ell-1}$ is a frame for $\ell^2(\mathbb{Z})$, with bounds A, B.*

In continuation of the above results and with similar techniques, Janssen proves the following result in [427]:

Lemma 14.4.5 *Suppose that $g \in L^2(\mathbb{R})$ satisfies condition (R) and that $\{E_{m/M}T_{nN}g\}_{m,n\in\mathbb{Z}}$ is a Bessel sequence in $L^2(\mathbb{R})$ for some $M, N \in \mathbb{N}$. Then any function of the form*

$$\phi = \sum_{m,n\in\mathbb{Z}} c_{mn} E_{m/N} T_{nM} g, \ where \ \{c_{mn}\}_{m,n\in\mathbb{Z}} \in \ell^1(\mathbb{Z}^2) \qquad (14.14)$$

also satisfies condition (R).

Note that the lattice associated with the given Bessel sequence $\{E_{m/M}T_{nN}g\}_{m,n\in\mathbb{Z}}$ is $\{(nN, m/M)\}_{m,n\in\mathbb{Z}}$ and that the functions ϕ in (14.14) are linear combinations of the Gabor system with respect to the *dual lattice* $\{(nM, m/N)\}_{m,n\in\mathbb{Z}}$. Since the Gaussian $g(x) = e^{-\frac{1}{2}x^2}$ satisfies condition (R) (see Exercise 12.2) and $\{E_m T_n g\}_{m,n\in\mathbb{Z}}$ is complete in $L^2(\mathbb{R})$, Lemma 14.4.5 gives an alternative argument for condition (R) being satisfied on a dense set of functions in $L^2(\mathbb{R})$.

Lemma 14.4.5 has an interesting application to sampling of the frame operator S associated with a frame $\{E_{m/M}T_{nN}g\}_{m,n\in\mathbb{Z}}$ for $L^2(\mathbb{R})$. For the proof, we will use a result from [426], which uses "condition (A)"; see (12.1); this condition is satisfied, e.g., if the window g belongs to Feichtinger's algebra $\mathcal{S}_0$.

Lemma 14.4.6 *Let $g \in L^2(\mathbb{R})$, $M, N \in \mathbb{N}$, and assume that the Gabor system $\{E_{m/M}T_{nN}g\}_{m,n\in\mathbb{Z}}$ is a frame for $L^2(\mathbb{R})$, with frame operator S. If $\{E_{m/M}T_{nN}g\}_{m,n\in\mathbb{Z}}$ satisfies condition (A), then also $\{E_{m/M}T_{nN}S^{-1}g\}_{m,n\in\mathbb{Z}}$ satisfies condition (A).*

Proposition 14.4.7 *Let $g \in L^2(\mathbb{R})$, $M, N \in \mathbb{N}$, and assume that $\{E_{m/M}T_{nN}g\}_{m,n\in\mathbb{Z}}$ is a Bessel sequence and satisfies condition (A). Then, for any $f \in L^2(\mathbb{R})$ which satisfies condition (R) and for which $\{E_{m/M}T_{nN}f\}_{m,n\in\mathbb{Z}}$ is a Bessel sequence,*

$$Sf(j) = \frac{M}{N} \sum_{m,n\in\mathbb{Z}} \langle g, E_{m/N}T_{nM}g\rangle E_{m/N}T_{nM}f(j), \ j \in \mathbb{Z}. \qquad (14.15)$$

If furthermore g satisfies condition (R) and we denote the frame operator for $\{E_{m/M}T_{nN}g^D\}_{n\in\mathbb{Z},m=0,\ldots,M-1}$ by $S^D : \ell^2(\mathbb{Z}) \to \ell^2(\mathbb{Z})$, then

$$(Sf)^D = S^D f^D; \qquad (14.16)$$

if we also add the assumption that $\{E_{m/M}T_{nN}g\}_{m,n\in\mathbb{Z}}$ is a frame, then

$$(S^{-1}g)^D = (S^D)^{-1}g^D. \qquad (14.17)$$

Proof. By Theorem 12.2.5, the frame operator has the representation

$$Sf = \frac{M}{N} \sum_{m,n \in \mathbb{Z}} \langle g, E_{m/N} T_{nM} g \rangle E_{m/N} T_{nM} f, \ f \in L^2(\mathbb{R}).$$

If f satisfies condition (R) and $\{E_{m/M} T_{nN} f\}_{m,n \in \mathbb{Z}}$ is a Bessel sequence, then Sf satisfies condition (R) by Lemma 14.4.5. In particular we can sample Sf at the integers; this yields (14.15), with absolute convergence of the series because $\{\langle g, E_{m/N} T_{nM} g \rangle\}_{m,n \in \mathbb{Z}} \in \ell^1(\mathbb{Z}^2)$. For the proof that the extra assumption implies (14.16), we refer to [429]. Now, assume that $\{E_{m/M} T_{nN} g\}_{m,n \in \mathbb{Z}}$ is a frame. By Corollary 12.2.4(ii),

$$S^{-1}g = \frac{M}{N} \sum_{m,n \in \mathbb{Z}} \langle S^{-1}g, E_{m/N} T_{nM} S^{-1}g \rangle E_{m/N} T_{nM} g.$$

Since $\{E_{m/M} T_{nN} S^{-1}g\}_{m,n \in \mathbb{Z}}$ satisfies condition (A) by Lemma 14.4.6 and $\{E_{m/N} T_{nM} S^{-1}g\}_{m,n \in \mathbb{Z}}$ is a Bessel sequence, we can apply (14.16) to the function $f := S^{-1}g$, and (14.17) follows. $\square$

In words, (14.16) means that we can obtain knowledge about the frame operator for a Gabor system in $L^2(\mathbb{R})$ based on the frame operator for a Gabor system in $\ell^2(\mathbb{Z})$. Using functional calculus, Janssen has extended (14.16) considerably. In fact, there are conditions such that the sampling procedure can be generalized to operators $\varphi(S)$, where φ is an analytic function. Hereby the obtained sampling results are also applicable to the function $S^{-1/2}g$ which generates the canonical tight frame associated with $\{E_{m/M} T_{nN} g^D\}_{n \in \mathbb{Z}, m=0,\ldots,M-1}$. Janssen's proof is published in [201].

14.5   Gabor Frames for $L^2(0, L)$ via Periodization

This section will deal with construction of Gabor frames for $L^2(0, L)$, where $L \in \mathbb{N}$. We will consider the functions in $L^2(0, L)$ as L-periodic functions on $\mathbb{R}$. The construction will be based on a periodization of a Gabor frame for $L^2(\mathbb{R})$.

Most of the section is based on work by Søndergaard [607]. As stated in [607], Gabor frames in $L^2(0, L)$ are not widely used; but they play an important role as an intermediate step toward the results in Section 14.6, where a Gabor frame for $L^2(\mathbb{R})$ ultimately is turned into a discrete model in $\mathbb{C}^L$.

Let us first define the central operators on $L^2(0, L)$ for a fixed choice of $L \in \mathbb{N}$. For any $a \in \mathbb{R}$, we can define the *translation operator*

$$T_a : L^2(0, L) \to L^2(0, L), T_a f(x) = f(x - a), x \in \mathbb{R}. \qquad (14.18)$$

Note that the expression for T_a is well-defined, because we consider $L^2(0, L)$ as a space of L-periodic functions.

The *modulation operator* E_b on $L^2(0,L)$ is for $b \in L^{-1}\mathbb{Z}$ defined by

$$E_b : L^2(0,L) \to L^2(0,L), \ E_b f(x) = e^{2\pi i bx} f(x). \qquad (14.19)$$

The choice of the parameter b implies that $E_b f$ is indeed L-periodic whenever $f \in L^2(0,L)$. Exactly as the corresponding operators on $L^2(\mathbb{R})$, the translation operators and modulation operators are unitary operators on $L^2(0,L)$.

Still fixing $L \in \mathbb{N}$, let $b \in L^{-1}\mathbb{N}$ and choose $a \in \mathbb{N}$ such that $N' := L/a \in \mathbb{N}$. The corresponding *Gabor system* in $L^2(0,L)$, generated by a function $g \in L^2(0,L)$, is defined by

$$\{E_{mb}T_{na}g\}_{m\in\mathbb{Z}, n=0,\ldots,N'-1} := \{e^{2\pi i mbx}g(x-na)\}_{m\in\mathbb{Z}, n=0,\ldots,N'-1}. \ (14.20)$$

The *periodization operator* $\mathcal{P}_L$ on $L^2(\mathbb{R})$ is formally defined by

$$(\mathcal{P}_L f)(x) := \sum_{k\in\mathbb{Z}} f(x+kL), \ x \in \mathbb{R}. \qquad (14.21)$$

Note that we need stronger conditions than $f \in L^2(\mathbb{R})$ in order for $\mathcal{P}_L f$ to be well-defined; one such condition is that $f \in S_0$.

Finally, we will need the Fourier transform on $L^2(0,L)$, i.e., the operator that associates the sequence of Fourier coefficients to a given function $f \in L^2(0,L)$. Note that the functions $\left\{\frac{1}{\sqrt{L}}e^{2\pi i kx/L}\right\}_{k\in\mathbb{Z}}$ form an orthonormal basis for $L^2(0,L)$; see (3.31). We define the *Fourier transform* $\mathcal{F}_{[0,L]}$ by

$$\mathcal{F}_{[0,L]} : L^2(0,L) \to \ell^2(\mathbb{Z}), \ \mathcal{F}_{[0,L]}f := \left\{(\mathcal{F}_{[0,L]}f)(k)\right\}_{k\in\mathbb{Z}}, \qquad (14.22)$$

where

$$(\mathcal{F}_{[0,L]}f)(k) := \frac{1}{\sqrt{L}}\int_0^L f(x)e^{-2\pi i kx/L}\,dx, \ k \in \mathbb{Z}. \qquad (14.23)$$

Using (3.32) with the functions e_k defined by (3.31) [the parameter "*b*" in these expressions is different from the b in the current context and is chosen as the number L^{-1}], we see that any function $f \in L^2(0,L)$ has the expansion

$$f = \sum_{k\in\mathbb{Z}}\langle f, e_k\rangle e_k = \frac{1}{\sqrt{L}}\sum_{k\in\mathbb{Z}}(\mathcal{F}_{[0,L]}f)(k)e^{2\pi i kx/L};$$

thus, the inverse Fourier transform $\mathcal{F}_{[0,L]}^{-1} : \ell^2(\mathbb{Z}) \to L^2(0,L)$ is given by

$$\left(\mathcal{F}_{[0,L]}^{-1}\{c(k)\}_{k\in\mathbb{Z}}\right)(x) = \frac{1}{\sqrt{L}}\sum_{k\in\mathbb{Z}}c(k)e^{2\pi i kx/L}. \qquad (14.24)$$

The Fourier transform $\mathcal{F}_{[0,L]}$ and its inverse $\mathcal{F}_{[0,L]}^{-1}$ are unitary operators (we allow ourselves to use the name "unitary" even though the domain and the range spaces are different). Exactly as in the case of the Fourier transformation on $L^2(\mathbb{R})$, we can derive commutator relationships for compositions of $\mathcal{F}_{[0,L]}$ with the operators E_b and T_a. Note however that because the domain space and the range space of $\mathcal{F}_{[0,L]}$ are different, we need to be careful with the meaning of these operators; i.e., we must specify the space they act on. We leave the proof of the result to the reader (Exercise 14.4):

Lemma 14.5.1 *Let $L \in \mathbb{N}$ be given.*

(i) *For $b \in L^{-1}\mathbb{Z}$, let E_b denote the modulation operator on $L^2(0,L)$ and let T_{bL} be the translation operator on $\ell^2(\mathbb{Z})$. Then*

$$\mathcal{F}_{[0,L]}E_b = T_{bL}\mathcal{F}_{[0,L]} \ \text{on} \ L^2(0,L) \tag{14.25}$$

and

$$E_b\mathcal{F}_{[0,L]}^{-1} = \mathcal{F}_{[0,L]}^{-1}T_{bL} \ \text{on} \ \ell^2(\mathbb{Z}). \tag{14.26}$$

(ii) *For $a \in \mathbb{Z}$, let T_a denote the translation operator on $L^2(0,L)$ and let $E_{a/L}$ be the modulation operator on $\ell^2(\mathbb{Z})$. Then*

$$\mathcal{F}_{[0,L]}T_a = E_{-a/L}\mathcal{F}_{[0,L]} \ \text{on} \ L^2(0,L), \tag{14.27}$$

and

$$T_a\mathcal{F}_{[0,L]}^{-1} = \mathcal{F}_{[0,L]}^{-1}E_{-a/L} \ \text{on} \ \ell^2(\mathbb{Z}). \tag{14.28}$$

We note that *Poisson's summation formula* (see Lemma A.6.3) has a formulation in terms of the periodization operator $\mathcal{P}_L$ and the Fourier transform $\mathcal{F}_{[0,L]}$. Recall the notation f^D used for the sampling sequence associated with a function $f \in L^2(\mathbb{R})$; see (14.5).

Lemma 14.5.2 *Given $L \in \mathbb{N}$, let D_L denote the scaling operator on $L^2(\mathbb{R})$, see (2.23), and let as usual $\mathcal{F}$ denote the Fourier transform on $L^2(\mathbb{R})$. Then, given $f \in \mathcal{S}_0$,*

$$(\mathcal{P}_L f)(x) = \left(\mathcal{F}_{[0,L]}^{-1}(D_L\mathcal{F}f)^D\right)(x), \ \forall x \in \mathbb{R}.$$

We are now ready to show how to obtain Gabor frames and pairs of dual Gabor frames in $L^2(I)$ for certain intervals $I \subset \mathbb{R}$, by periodization of frames for $L^2(\mathbb{R})$.

Theorem 14.5.3 *Let $\ell, M, N \in \mathbb{N}$. Then the following hold:*

(i) *If $g \in \mathcal{S}_0$ and $\{E_{m/M}T_{nN}g\}_{m,n\in\mathbb{Z}}$ is a frame for $L^2(\mathbb{R})$ with bounds A, B, then the periodized Gabor system $\{E_{m/M}T_{nN}$ $\mathcal{P}_{NM\ell}g\}_{n\in\mathbb{Z}, m=0,...,M\ell-1}$ is a frame for $L^2(0, NM\ell)$ with bounds A, B.*

(ii) *Let $g, h \in \mathcal{S}_0$. If $\{E_{m/M}T_{nN}g\}_{m,n\in\mathbb{Z}}$ and $\{E_{m/M}T_{nN}h\}_{m,n\in\mathbb{Z}}$ are dual frames for $L^2(\mathbb{R})$, then the periodized Gabor systems $\{E_{m/M}T_{nN}$ $\mathcal{P}_{NM\ell}g\}_{n\in\mathbb{Z}, m=0,...,M\ell-1}$ and $\{E_{m/M}T_{nN}$ $\mathcal{P}_{NM\ell}h\}_{n\in\mathbb{Z}, m=0,...,M\ell-1}$ are dual frames for $L^2(0, NM\ell)$.*

Proof. Under the assumptions in (i), Proposition 11.2.5 shows that the Gabor system $\{E_{mN}T_{n/M}\widehat{g}\}_{m,n\in\mathbb{Z}}$ also is a frame for $L^2(\mathbb{R})$ with bounds A, B. The Feichtinger algebra S_0 is invariant under the Fourier transform, so by Corollary 14.4.4 the sampled Gabor system

$$\{E_{m/(M\ell)}T_{nN\ell}(D_{NM\ell}\widehat{g})^D\}_{n\in\mathbb{Z},m=0,\ldots,M\ell-1}$$

is a frame for $\ell^2(\mathbb{Z})$. Now, using Lemma 14.5.1,

$$\mathcal{F}^{-1}_{[0,NM\ell]}E_{m/(M\ell)}T_{nN\ell}(D_{NM\ell}\widehat{g})^D = T_{-mN}\mathcal{F}^{-1}_{[0,NM\ell]}T_{nN\ell}(D_{NM\ell}\widehat{g})^D$$

$$= T_{-mN}E_{n/M}\mathcal{F}^{-1}_{[0,NM\ell]}(D_{NM\ell}\widehat{g})^D.$$

Since $\mathcal{F}^{-1}_{[0,NM\ell]} : \ell^2(\mathbb{Z}) \to L^2(0, NM\ell)$ is unitary, it follows that $\{T_{-mN}E_{n/M}\mathcal{F}^{-1}_{[0,NM\ell]}(D_{NM\ell}\widehat{g})^D\}_{n\in\mathbb{Z},m=0,\ldots,M\ell-1}$ is a frame for $L^2(0, NM\ell)$ with bounds A, B. A moment thought shows that the Gabor system equals $\{T_{mN}E_{n/M}\mathcal{F}^{-1}_{[0,NM\ell]}(D_{NM\ell}\widehat{g})^D\}_{n\in\mathbb{Z},m=0,\ldots,M\ell-1}$. Note also that exchanging the order of the translation operator and the modulation operator just introduces an irrelevant complex factor of absolute value 1 and does not change the frame property or the frame bounds. Thus, $\{E_{n/M}T_{mN}\mathcal{F}^{-1}_{[0,NM\ell]}(D_{NM\ell}\widehat{g})^D\}_{n\in\mathbb{Z},m=0,\ldots,M\ell-1}$ is a frame for $L^2(0, NM\ell)$, with bounds A, B. Finally, using now Lemma 14.5.2, we conclude that $\{E_{n/M}T_{mN}P_{NM\ell}g\}_{n\in\mathbb{Z},m=0,\ldots,M\ell-1}$ is indeed a frame for $L^2(0, NM\ell)$. This proves (i); the proof of (ii) is similar, using Corollary 14.3.2 instead of Corollary 14.4.4. □

It is interesting to note that we proved Theorem 14.5.3 without having derived any sufficient conditions for a Gabor system being a frame in $L^2(0, L)$! The technical reason for this is that we could construct the systems by applying the inverse Fourier transform to a Gabor frame for $\ell^2(\mathbb{Z})$.

If necessary, it is possible to follow the approach in Chapters 11–13 and derive characterizations of Gabor frames and dual frames in $L^2(0, L)$. The reason that the results and their proofs and similar to the $L^2(\mathbb{R})$-case is that both cases are concrete manifestations of the general theory for Gabor systems on LCA groups; see Chapter 21, in particular Example 21.3.3.

14.6 Gabor Frames in $\mathbb{C}^L$

Gabor systems also have a natural version in the finite-dimensional spaces $\mathbb{C}^L, L \in \mathbb{N}$. We will describe this now; in order to connect with the results for frames in $\ell^2(\mathbb{Z})$, we will write a sequence $g \in \mathbb{C}^L$ as

$$g = (g(0), g(1), \ldots, g(L-1)).$$

The definition of the modulation operator on $\ell^2(\mathbb{Z})$ given in (14.1) also defines E_b as an operator on $\mathbb{C}^L$. In contrast, the definition (14.2) of the translation operator T_n, $n \in \mathbb{Z}$, does not immediately make sense on $\mathbb{C}^L$ because $j - n$ does not always belong to $\{0, 1, \ldots, L-1\}$. The natural way to solve this problem is to extend $g \in \mathbb{C}^L$ to a periodic sequence indexed by $\mathbb{Z}$. That is, we define

$$g(j + kL) = g(j) \text{ for } j = 0, \ldots, L-1, k \in \mathbb{Z}.$$

With this convention, we can apply the translation operators T_n, $n \in \mathbb{Z}$, to sequences in $\mathbb{C}^L$.

Given any $L \in \mathbb{N}$, let $M, N \in \mathbb{N}$ and assume that $M' := L/M \in \mathbb{N}$ and $N' := L/N \in \mathbb{N}$. Given a sequence $g \in \mathbb{C}^L$, we now define the associated Gabor system on $\mathbb{C}^L$ by

$$\{E_{m/M} T_{nN} g\}_{m=0,\ldots,M-1; n=0,\ldots,N'-1} \tag{14.29}$$
$$= \{e^{2\pi i n(\cdot)/M} g(\cdot - nN)\}_{m=0,\ldots,M-1; n=0,\ldots,N'-1}.$$

Thus, the Gabor system consists of MN' vectors in $\mathbb{C}^L$. Note that the definition in (14.29) corresponds precisely to the one that arises by considering $\mathbb{C}^L$ as an LCA group; see Example 21.3.2. A characterization of dual Gabor frames in $\mathbb{C}^L$ can be obtained either by direct calculations or as a consequence of Theorem 21.7.10 (Exercise 14.7):

Theorem 14.6.1 *Two Gabor systems $\{E_{m/M} T_{nN} g\}_{m=0,\ldots,M-1; n=0,\ldots,N'-1}$ and $\{E_{m/M} T_{nN} g\}_{m=0,\ldots,M-1; n=0,\ldots,N'-1}$ as above form dual frames for $\mathbb{C}^L$ if and only if*

$$\sum_{k=0}^{N'-1} \overline{g(j - kN - nM)} h(j - kN) = \frac{1}{M} \delta_{n,0}$$

for all $j \in \{0, \ldots, N-1\}, n \in \{0, \ldots, M'-1\}$.

The sampling procedures described in Section 14.4 transfer frames for $L^2(\mathbb{R})$ into frames for $\ell^2(\mathbb{Z})$ and hereby take an important step toward concrete applications. However, in signal and image processing, we ultimately need a finite model, with a finite number of vectors spanning a finite-dimensional space. Based on the material in the current chapter, there are two ways of transferring a Gabor frame $\{E_{mb} T_{na} g\}_{m,n \in \mathbb{Z}}$ in $L^2(\mathbb{R})$ into a finite frame:

- Applying the sampling results in Sections 14.3–14.4 to the frame $\{E_{mb} T_{na} g\}_{m,n \in \mathbb{Z}}$ yields a Gabor frame in $\ell^2(\mathbb{Z})$; using a similar periodization method as in Section 14.5 on this frame yields a finite Gabor frame in $\mathbb{C}^L$.

- Applying the periodization results in Section 14.5 to the frame $\{E_{mb} T_{na} g\}_{m,n \in \mathbb{Z}}$ yields a Gabor frame in $L^2(0, L)$; using a similar sampling method as in Sections 14.3–14.4 on this frame yields a finite frame in $\mathbb{C}^L$.

Note that for both methods, we have done "half the work": the steps from $L^2(\mathbb{R})$ to $\ell^2(\mathbb{Z})$ and from $L^2(\mathbb{R})$ to $L^2(0,L)$ have been described carefully in the previous sections, while the steps from $\ell^2(\mathbb{Z})$ to $\mathbb{C}^L$ and from $L^2(0,L)$ to $\mathbb{C}^L$ have not been discussed yet. Since these periodization/sampling steps are completely analog to the material in Sections 14.3–14.5, we will just state the final result and refer to the papers [443] by Kaiblinger and [607] by Søndergaard for details.

We will state the version where a function $g \in L^2(\mathbb{R})$ is first sampled and then periodized. Let us fix $L \in \mathbb{N}$. For a sequence $g^D \in \ell^2(\mathbb{Z})$, the *periodization* is formally defined by

$$\left(\mathcal{P}_L g^D\right)(j) := \sum_{n\in\mathbb{Z}} g^D(j-nL), \ j = 1, \ldots, L-1. \tag{14.30}$$

Thus, $\mathcal{P}_L$ is a sequence in $\mathbb{C}^L$ whenever it is well-defined; clearly this is the case at least whenever $g^D \in \ell^1(\mathbb{Z})$.

Skipping the intermediate steps mentioned above, the final transition from a Gabor system in $L^2(\mathbb{R})$ to a finite-dimensional Gabor system reads as follows:

Theorem 14.6.2 *Let $N, M, \ell \in \mathbb{N}$ be given. Then the following hold:*

(i) *If $g \in S_0$ and the Gabor system $\{E_{m/M}T_{nN}g\}_{m,n\in\mathbb{Z}}$ is a frame for $L^2(\mathbb{R})$ with bounds A, B, then the discrete Gabor system $\{E_{m/M}T_{nN}\mathcal{P}_{NM\ell}g^D\}_{m=0,\ldots,M-1,n=0,\ldots,M\ell-1}$ is a frame for $\mathbb{C}^{NM\ell}$ with bounds A, B.*

(ii) *If $g, h \in S_0$ and the Gabor systems $\{E_{m/M}T_{nN}g\}_{m,n\in\mathbb{Z}}$ and $\{E_{m/M}T_{nN}g\}_{m,n\in\mathbb{Z}}$ are dual frames for $L^2(\mathbb{R})$, then the discrete Gabor systems $\{E_{m/M}T_{nN}\mathcal{P}_{NM\ell}g^D\}_{m=0,\ldots,M-1,n=0,\ldots,M\ell-1}$ and $\{E_{m/M}T_{nN}\mathcal{P}_{NM\ell}g^D\}_{m=0,\ldots,M-1,n=0,\ldots,M\ell-1}$ are dual frames for $\mathbb{C}^{NM\ell}$.*

In Theorem 14.6.2, the notation indicates that we first sample the window g in order to obtain the sequence g^D, and then apply the periodization operator $\mathcal{P}_{NM\ell}$ on $\ell^2(\mathbb{Z})$ in order to arrive at the window $\mathcal{P}_{NM\ell}g^D$ for the discrete Gabor frame in $\mathbb{C}^{NM\ell}$. However, for any $L \in \mathbb{N}$,

$$\left(\mathcal{P}_L g^D\right)(j) = \sum_{n\in\mathbb{Z}} g(j-nL) = \left(\mathcal{P}_L g\right)^D(j); \tag{14.31}$$

thus, Theorem 14.6.2 has exactly the same form for the case where we first periodize the Gabor frame and then perform the sampling.

We can now provide a detailed diagram, which demonstrates the entire procedure of constructing finite-dimensional Gabor frames based on a Gabor frame $\{E_{mb}T_{na}g\}_{m,n\in\mathbb{Z}}$ for $L^2(\mathbb{R})$. Recall from Søndergaard's original diagram on page 359 that the arrows to the left indicate sampling and that the arrows down indicate periodization:

$$L^2(\mathbb{R}), \{E_{m/M}T_{nN}g\} \xrightarrow{\text{sampling}} \ell^2(\mathbb{Z}), \{E_{m/M}T_{nN}g^D\}$$

$$\downarrow \text{periodization} \qquad\qquad\qquad\qquad \downarrow$$

$$L^2(0, NM\ell), \{E_{m/M}T_{nN}\mathcal{P}_{NM\ell}g\} \longrightarrow \mathbb{C}^{NM\ell}, \{E_{m/M}T_{nN}\mathcal{P}_{NM\ell}g^D\}$$

For the sake of clarity the index sets of the various Gabor systems are eliminated in the diagram; we refer to Theorem 14.3.1, Theorem 14.5.3, and Theorem 14.6.2 for the details.

Note that the diagram states the Gabor frames and the involved spaces explicitly, including the free parameter ℓ. This parameter allows to obtain finite-dimensional models in spaces of various dimensions. Looking at the diagram, it is natural to ask about the relation between the given Gabor frame $\{E_{mb}T_{na}g\}_{m,n\in\mathbb{Z}}$ and its "finite-dimensional version" $\{E_{m/M}T_{nN}\mathcal{P}_{NM\ell}g^D\}_{m=0,\dots,M-1,n=0,\dots,M\ell-1}$ in $\mathbb{C}^{NM\ell}$ for large values of $\ell \in \mathbb{N}$. Kaiblinger [443] and Søndergaard [607] have indeed used Theorem 14.6.2 to develop methods to approximate the inverse frame operator associated with a frame $\{E_{mb}T_{na}g\}_{m,n\in\mathbb{Z}}$ for $L^2(\mathbb{R})$; in a vague form, one can say that the finite-dimensional frame $\{E_{m/M}T_{nN}\mathcal{P}_{NM\ell}g^D\}_{m=0,\dots,M-1,n=0,\dots,M\ell-1}$ for $\mathbb{C}^{NM\ell}$ approximates the frame $\{E_{mb}T_{na}g\}_{m,n\in\mathbb{Z}}$ "well" whenever the parameter ℓ is large.

The constructions in Theorem 14.3.1, Theorem 14.5.3, and Theorem 14.6.2 have a number of important features that are worth mentioning explicitly. In fact, since they ultimately transfer a frame $\{E_{mb}T_{na}g\}_{m,n\in\mathbb{Z}}$ for $L^2(\mathbb{R})$ into a finite frame, it is natural to examine how this frame perform compared to the "wish list" for finite frames; see page 43. Recall that for any frame with frame operator S, the *condition number* denotes the ratio between the optimal upper frame bound and the optimal lower frame bound; see the discussion in the finite-dimensional context in Section 1.3.

- Any Gabor frame is an equal-norm frame, i.e., the elements in the frame have the same norm. In particular, this holds for the given Gabor frame $\{E_{m/M}T_{nN}g\}_{m,n\in\mathbb{Z}}$ in $L^2(\mathbb{R})$ and the constructed Gabor frames in $\ell^2(\mathbb{Z}), L^2(0,L)$, and $\mathbb{C}^L$.

- As stated explicitly in the theorems concerning the steps of sampling and periodization, all transitions keep the frame bounds; the optimal bounds of the frames obtained by sampling and periodization might even be "tighter" than for the given frame. In particular, if the procedure is applied to a tight Gabor frame for $L^2(\mathbb{R})$, also the constructed frames in $\ell^2(\mathbb{Z}), L^2(0,L)$, and $\mathbb{C}^L$ will be tight.

- More generally, and a direct consequence of the preceding comment, the condition number for the frame operator associated with the Gabor frame $\{E_{m/M}T_{nN}g\}_{m,n\in\mathbb{Z}}$ in $L^2(\mathbb{R})$ is dominating the condition number for the constructed frames in $\ell^2(\mathbb{Z}), L^2(0,L)$, and $\mathbb{C}^L$.

In particular, if $\{E_{mb}T_{na}g\}_{m,n\in\mathbb{Z}}$ is *well conditioned*, i.c., if the associated frame operator has a low condition number, the same is true for the frames obtained by sampling and periodization. As stated above, sampling and periodization might even lead to frames with a better condition number.

- A natural candidate for concrete applications would be the B-splines B_N, $N \geq 2$. They are continuous, have compact support, and belong to $\mathcal{S}_0$. Furthermore, explicitly given dual windows with the same properties can be constructed via Theorem 12.5.1.

Now consider a general Gabor system $\{E_{m/M}\ T_{nN}g\}_{m=0,\ldots,M-1;n=0,\ldots,N'-1}$ in $\mathbb{C}^L$. It was shown by Lawrence, Pfander, and Walnut [480] that whenever L is prime, the Gabor system has full spark for a.e. $g \in \mathbb{C}^L$; it was proved by Malikiosis [508] that the assumption of L being prime can be removed, i.e., the result holds in full generality. On the other hand, Kutyniok proved in [466] that nontrivial subsets of the Gabor system $\{E_{m/M}T_{nN}g\}_{m=0,\ldots,M-1;n=0,\ldots,N'-1}$ cannot be linearly independent for all $g \in \mathbb{C}^L \setminus \{0\}$; that is, $\{E_{m/M}T_{nN}g\}_{m=0,\ldots,M-1;n=0,\ldots,N'-1}$ cannot have full spark for all choices of $g \in \mathbb{C}^L \setminus \{0\}$. It is interesting to compare this result with the fact that a Gabor system $\{E_{mb}T_{na}g\}_{m,n\in\mathbb{Z}}$ in $L^2(\mathbb{R})$ is linearly independent for all $g \in L^2(\mathbb{R}) \setminus \{0\}$. Also, compare with the discussion about linear independence of general Gabor systems in $L^2(\mathbb{R})$; see page 343.

Note that finite-dimensional systems can be obtained from a Gabor frame $\{E_{mb}T_{na}g\}_{m,n\in\mathbb{Z}}$ for $L^2(\mathbb{R})$ in several ways. For example, one could also consider the frame sequences $\{E_{mb}T_{na}g\}_{|m|,|n|\leq N}$ for some $N \in \mathbb{N}$. However, as we have seen in Section 13.4 and in particular in Theorem 13.4.4, this always leads to badly conditioned systems whenever $\{E_{mb}T_{na}g\}_{m,n\in\mathbb{Z}}$ is overcomplete; thus, from the numerical point of view, an application based on this system would not be attractive.

Let us end this section with a result by Qiu und Feichtinger [290] about the Gabor frame operator associated with a finite Gabor system $\{E_{m/M}T_{nN}g\}_{m=0,\ldots,M-1;n=0,\ldots,N'-1}$:

Theorem 14.6.3 *For $L \in \mathbb{N}$, let $M, N \in \mathbb{N}$ and assume that $M' := L/M \in \mathbb{N}$ and $N' := L/N \in \mathbb{N}$. Let $g \in \mathbb{C}^L$, and consider the Gabor system $\{E_{m/M}T_{nN}g\}_{m=0,\ldots,M-1;n=0,\ldots,N'-1}$ in $\mathbb{C}^L$ and its frame operator $S : \mathbb{C}^L \to \mathbb{C}^L$. Then the jkth entry in the matrix representation of the operator S w.r.t. the canonical orthonormal basis $\{e_k\}_{k=0}^{L-1}$ for $\mathbb{C}^L$ is given by*

$$\langle Se_k, e_j\rangle = \begin{cases} M \sum_{n=0}^{N'-1} \overline{T_{nN}g(k)}T_{nN}g(j) & \text{if } j - k \in M\mathbb{Z}, \\ 0 & \text{if } j - k \notin M\mathbb{Z}. \end{cases}$$

Proof. The jkth entry in the matrix representation for $S : \mathbb{C}^L \to \mathbb{C}^L$ is

$$
\begin{aligned}
\langle Se_k, e_j \rangle &= \sum_{n=0}^{N'-1} \sum_{m=0}^{M-1} \langle e_k, E_{m/M} T_{nN} g \rangle \langle E_{m/M} T_{nN} g, e_j \rangle \\
&= \sum_{n=0}^{N'-1} \sum_{m=0}^{M-1} \overline{E_{m/M} T_{nN} g(k)} E_{m/M} T_{nN} g(j) \\
&= \left(\sum_{n=0}^{N'-1} \overline{T_{nN} g(k)} T_{nN} g(j) \right) \left(\sum_{m=0}^{M-1} e^{2\pi i m(j-k)/M} \right).
\end{aligned}
$$

Since

$$
\left(\sum_{m=0}^{M-1} e^{2\pi i m(j-k)/M} \right) = \begin{cases} M & \text{if } j - k \in M\mathbb{Z}, \\ 0 & \text{if } j - k \notin M\mathbb{Z}, \end{cases}
$$

this proves the result. $\square$

In words, Theorem 14.6.3 says that only every Mth subdiagonal in the matrix representation of the frame operator S is nonzero. In [290] the result is used to find fast algorithms to calculate the dual frame. For further reading, we note that Qiu has a series of papers [548, 549] about the structure of Gabor systems in $\mathbb{C}^L$.

14.7 Shift-Invariant Systems

As in the $L^2(\mathbb{R})$-case, the discrete Gabor systems in $\ell^2(\mathbb{Z})$ are special cases of general shift-invariant systems. Given a collection of sequences $\{g_m\}_{m=0,\dots,M-1}$ in $\ell^2(\mathbb{Z})$ and a shift-parameter $N \in \mathbb{N}$, consider the sequence $g_{nm} \in \ell^2(\mathbb{Z})$ with entries

$$ g_{nm}(j) = g_m(j - nN), \ j \in \mathbb{Z}. $$

Similar to our notation for shift-invariant systems in $L^2(\mathbb{R})$, we will skip the indices and simply denote the system $\{g_{nm}\}_{n\in\mathbb{Z},m=0,\dots,M-1}$ by $\{g_{nm}\}$.

The results for continuous shift-invariant systems in Section 10.1 have discrete counterparts, which are stated in [430]. In order to formulate the results, we consider the Fourier transform of a sequence $h \in \ell^2(\mathbb{Z})$, given by

$$ \widehat{h}(\nu) = \sum_{j\in\mathbb{Z}} h(j) e^{-2\pi i j \nu}, \ a.e. \ \nu \in \mathbb{R}. $$

Given a shift-invariant system $\{g_{nm}\}$, we define, analogous to (10.13), the matrix-valued function

$$ H(\nu) = (\widehat{g}_m(\nu - k/N))_{k=0,\dots,N-1,m=0,\dots M-1}, \ a.e. \ \nu \in \mathbb{R}. $$

Observe that this is an $N \times M$ matrix.

Theorem 14.7.1 *In the setting above, the following hold:*

(i) $\{g_{nm}\}$ *is a Bessel sequence in $\ell^2(\mathbb{Z})$ with upper bound B if and only if the matrix $H(\nu)$ for a.e. $\nu \in \mathbb{R}$ defines a bounded linear mapping from $\mathbb{C}^M$ into $\mathbb{C}^N$ of norm at most $\sqrt{NB}$.*

(ii) $\{g_{nm}\}$ *is a frame for $\ell^2(\mathbb{Z})$ with frame bounds A, B if and only if*

$$NAI \le H(\nu)H(\nu)^* \le NBI, \; a.e. \; \nu \in \mathbb{R}.$$

(ii) $\{g_{nm}\}$ *is a tight frame for $\ell^2(\mathbb{Z})$ if and only if there is a constant $c > 0$ such that*

$$\sum_{m=0}^{M-1} \widehat{g}_m(\nu - k/N)\overline{\widehat{g}_m(\nu)} = c\delta_{k,0}, \; k \in \mathbb{Z}, \; a.e. \; \nu \in \mathbb{R}.$$

(iv) *Two shift-invariant systems $\{g_{nm}\}$ and $\{h_{nm}\}$, which form Bessel sequences in $\ell^2(\mathbb{Z})$, are dual frames if and only if*

$$\sum_{m=0}^{M-1} \widehat{g}_m(\nu - k/N)\overline{\widehat{h}_m(\nu)} = N\delta_{k,0}, \; k \in \mathbb{Z}, \; a.e. \; \nu \in \mathbb{R}.$$

Most proofs follow by repeating the arguments from the continuous setting, and we will not go into detail. Again, the statements have direct consequences for discrete Gabor frames in $\ell^2(\mathbb{Z})$ (Exercise 14.5).

14.8 Frames in $\ell^2(\mathbb{Z})$ and Filter Banks

Shift-invariant systems appear in signal processing, especially in connection with *filter banks*. We refer to the book by Vetterli and Kovačević [626] for a detailed description of this subject and its relationship to signal expansions. We will think about a filter bank as some kind of "black box," which performs some operations (i.e., processing) on a given input signal and then delivers an output. An example could be that the filter bank performs an analysis of the signal f via a shift-invariant system $\{g_{nm}\}$ and then a synthesis via another system $\{h_{nm}\}$; the outcome will be a sequence

$$\widetilde{f} = \sum_{m,n}\langle f, g_{nm}\rangle h_{nm}. \tag{14.32}$$

Note that the words "synthesis" and "analysis" correspond to the names of the operator T and its adjoint T^* (cf. page 122).

The study of the relations between frames and filter banks was initiated by Cvetković and Vetterli [229, 230] and has been elaborated on by many other authors (see, e.g., [96, 97, 587] and the references given there). We will not go into detail with any of the obtained results but only guide the reader to the terminology in filter bank theory and provide links to further reading.

The case of a *perfect reconstruction filter bank* corresponds to $\{g_{nm}\}$ and $\{h_{nm}\}$ being dual frames. In this case the outcome $\widetilde{f}$ in (14.32) equals f. A *paraunitary filter bank* corresponds to the special case where $\{g_{nm}\}$ is a tight frame, implying that we can take h_{nm} to be a multiple of g_{nm}. A *modulated filter bank* is a Gabor system, and if it is *oversampled* , we have an overcomplete Gabor frame.

In the signal processing literature the results are often formulated via the *polyphase representation*, which is called the *discrete Zak transform* by mathematicians. For a given signal $h \in \ell^2(\mathbb{Z})$ and a parameter $N \in \mathbb{N}$, it is defined by

$$(Zh)(j,\nu) = \sum_{\ell \in \mathbb{Z}} h(j - \ell N)e^{2\pi i \ell \nu}, \ j \in \mathbb{Z}, \ a.e. \ \nu \in \mathbb{R}.$$

An interpretation of the discrete Zak transform is that for a.e. $\nu \in \mathbb{R}$ it defines a sequence $(Zh)(\cdot, \nu)$. In terms of the polyphase representation one can now define a *polyphase matrix*, which plays a similar role for discrete Gabor systems as the Zibulski–Zeevi matrix in the continuous case. Among the results in [229, 230] are

- Characterizations of frames and tight frames in terms of the polyphase matrix.

- Conditions for the dual frame to consist of vectors in $\ell^2(\mathbb{Z})$ with finite length (i.e., only finitely many nonzero entries).

- Characterizations of tight Gabor frames in $\ell^2(\mathbb{Z})$, generated by a vector with finite length.

Some further results and extensions were later given by Bölcskei, Hlawatsch, and Feichtinger in [97]. Among their results are

- A parameterization of all synthesis filter banks providing perfect reconstruction for a given analysis filter bank.

- Methods for estimation of the frame bounds.

- Conditions for the shift-invariant system forming the analysis filter bank to be a frame.

- Construction of paraunitary filter banks from perfect reconstruction filter banks.

Let us finally mention the paper [587] by Strohmer, where he provides methods for approximation of the canonical dual frame associated to a shift-invariant frame.

14.9 Gabor frames in $\ell^2(\mathbb{Z}^d)$

Theorem 14.2.1 is just the one-dimensional version of a result by Lopez and Han that yields dual pair of discrete Gabo frames in $\ell^2(\mathbb{Z}^d)$. As in the original paper, we will describe the results using group-theoretic terms that will appear in a more general setting in Section 21.1.

Let $\mathcal{A}$ and $\mathcal{B}$ denote invertible $d\times d$ matrices with real entries, and assume that $\mathcal{B}^{-1}$ has integer entries. Consider the subgroup $G := \mathcal{B}^{-1}\mathbb{Z}^d$ of $\mathbb{Z}^d$, and let Ω denote a collection of coset representatives of the coset $\mathbb{Z}^d/G$; that is, $\mathbb{Z}^d$ is a disjoint union of the sets $G + \mathbf{m}$, where $\mathbf{m} \in \Omega$. It is well known that the number of elements in Ω is

$$|\Omega| = |\det(\mathcal{B}^{-1})| = \frac{1}{|\det \mathcal{B}|}.$$

Let us now fix any sequence $\{c(\mathbf{j})\}_{\mathbf{j}\in\mathbb{Z}^d} \in \ell^2(\mathbb{Z}^d)$. Consider the Gabor system in $\ell^2(\mathbb{Z}^d)$ generated by the sequence $\{c(\mathbf{j})\}_{\mathbf{j}\in\mathbb{Z}^d}$ and the matrices $\mathcal{A}, \mathcal{B}$, i.e., the collection of sequences $\{c_{\mathbf{m},\mathbf{n}}\}_{\mathbf{m},\mathbf{n}\in\mathbb{Z}^d} \subset \ell^2(\mathbb{Z}^d)$ given by

$$c_{\mathbf{m},\mathbf{n}}(\mathbf{j}) = e^{2\pi i \mathcal{B}\mathbf{m}\cdot\mathbf{j}}c(\mathbf{j} - \mathcal{A}\mathbf{n}), \; \mathbf{j} \in \mathbb{Z}^d. \qquad (14.33)$$

Given two sequences $\{c(\mathbf{j})\}_{\mathbf{j}\in\mathbb{Z}}, \{d(\mathbf{j})\}_{\mathbf{j}\in\mathbb{Z}} \in \ell^2(\mathbb{Z}^d)$ such that $\{c_{\mathbf{m},\mathbf{n}}\}_{\mathbf{m},\mathbf{n}\in\mathbb{Z}^d}$ and $\{d_{\mathbf{m},\mathbf{n}}\}_{\mathbf{m},\mathbf{n}\in\mathbb{Z}^d}$ are Bessel sequences, it was shown in Theorem 1.4 in [502] that $\{c_{\mathbf{m},\mathbf{n}}\}_{\mathbf{m},\mathbf{n}\in\mathbb{Z}^d}$ and $\{d_{\mathbf{m},\mathbf{n}}\}_{\mathbf{m},\mathbf{n}\in\mathbb{Z}^d}$ are dual frames for $\ell^2(\mathbb{Z}^d)$ if and only if

$$\sum_{k\in\mathbb{Z}^d} \overline{c(\mathbf{j} - \mathcal{A}\mathbf{k} - \mathcal{B}^\sharp\mathbf{n})}d(\mathbf{j} - \mathcal{A}\mathbf{k}) = |\det \mathcal{B}| \, \delta_{\mathbf{n},0} \qquad (14.34)$$

for all $\mathbf{j}, \mathbf{n} \in \mathbb{Z}^d$. Parallel to our discussion in Section 14.4, dual pairs of discrete Gabor frames can be obtained via sampling of dual pairs of Gabor frames for $L^2(\mathbb{R}^d)$, using the results in Section 20.5; see the paper [186].

Note also that characterizations of duality of two Gabor systems in $\ell^2(\mathbb{Z}^d)$, in terms of the Fourier transform on $\mathbb{Z}^d$, can be obtained via the group-theoretic approach in Chapter 21; see Theorem 21.6.4.

14.10 Exercises

14.1 Prove (14.3) under the stated assumptions.

14.2 Consider the sequence

$$\delta = (\cdots, 0, \ldots, 0, 1, 0, \ldots 0, \ldots),$$

and show that the sequences $\{T_k\delta\}_{k\in\mathbb{Z}}$ form an orthonormal basis for $\ell^2(\mathbb{Z})$. Compare with the result in Theorem 9.2.1!

14.3 Derive conditions for $\{E_{m/M}T_{nN}g\}_{n\in\mathbb{Z},m=0,...,M-1}$ being a Bessel sequence or a frame for $\ell^2(\mathbb{Z})$ by appropriate modifications of the proof of Theorem 11.4.2.

14.4 Prove Lemma 14.5.1.

14.5 Derive the consequences of Theorem 14.7.1 for discrete Gabor systems.

14.6 Here we ask the reader to prove an extension of Proposition 1.4.3. In fact, show that for a bi-infinite matrix $\Lambda = \{\lambda_{m,n}\}_{m,n\in\mathbb{Z}}$, the following are equivalent:

(i) There exist constants $A, B > 0$ such that

$$A\sum_{k\in\mathbb{Z}}|c_k|^2 \le ||\Lambda\{c_k\}_{k\in\mathbb{Z}}||^2 \le B\sum_{k\in\mathbb{Z}}|c_k|^2 \text{ for all finite}$$

$$\text{sequences } \{c_k\}_{k\in\mathbb{Z}}.$$

(ii) The columns in Λ constitute a Riesz basis for their closed span in $\ell^2(\mathbb{Z})$.

(iii) The rows in Λ constitute a frame for $\ell^2(\mathbb{Z})$.

14.7 Prove Theorem 14.6.1 based on Theorem 21.7.10.

15

General Wavelet Frames in $L^2(\mathbb{R})$

A fundamental question in wavelet analysis is what conditions we have to impose on a function ψ such that a given signal $f \in L^2(\mathbb{R})$ can be expanded via translated and scaled versions of ψ, i.e., via functions

$$\psi^{a,b}(x) := (T_b D_a \psi)(x) = \frac{1}{|a|^{1/2}} \psi(\frac{x-b}{a}), \ a \neq 0, \ b \in \mathbb{R}. \qquad (15.1)$$

Thus, there is a basic similarity between wavelet analysis and Gabor analysis: both concern sequences of functions defined by letting a special class of operators act on a fixed function, i.e., in both cases, we are dealing with coherent systems. The connections are even closer, and both can be seen as manifestations of the theory for generalized shift-invariant systems (Chapter 20), as well as the theory for decompositions in terms of group representations (Chapter 24). As in Gabor analysis there are two ways in which one can think about expansions of a signal f in terms of the functions $\psi^{a,b}$. One way is to ask for representations of f as integrals involving the functions $\psi^{a,b}$ over the entire parameter range $(\mathbb{R} \setminus \{0\}) \times \mathbb{R}$. Alternatively, one can restrict the parameters a, b to a discrete subset Λ of $\mathbb{R}^2$ and ask for series expansions of f in terms of the corresponding functions $\psi^{a,b}$. For applications, the latter is usually the most convenient choice, and most of this chapter will deal with the question of *how* we can choose the discrete subset Λ and ψ such that $\{\psi^{a,b}\}_{(a,b) \in \Lambda}$ is a frame for $L^2(\mathbb{R})$.

Collections of functions of the type (15.1) have been used in different contexts a long time before the wavelet era began; see, for example, the construction by Haar discussed in Example 3.9.1. Morlet was the first to propose representing signals as integrals involving $\psi^{a,b}$, and together with

© Springer International Publishing Switzerland 2016   385
O. Christensen, *An Introduction to Frames and Riesz Bases*,
Applied and Numerical Harmonic Analysis,
DOI 10.1007/978-3-319-25613-9_15

Grossmann, he introduced in [333] what is now known as the wavelet transform. Again it was Grossmann who brought the theory a big step forward by proposing to construct frames consisting of a countable number of functions $\psi^{a,b}$. Together with Daubechies and Meyer, he published the first constructions in [244].

Another breakthrough came with the concept of multiresolution analysis, as developed by Mallat and Meyer. As discussed in Section 3.9, its original purpose is to construct orthonormal bases for $L^2(\mathbb{R})$ of the form $\{2^{j/2}\psi(2^j x - k)\}_{j,k\in\mathbb{Z}}$. The importance of this new subject was immediately recognized by the mathematical as well as the engineering community, and very soon most of the effort in wavelet analysis was concentrated on construction of orthonormal bases with prescribed properties. Nowadays "wavelet analysis" is for many people almost synonymous with "multiresolution analysis," but wavelet analysis is in fact a much broader subject. For historical accuracy the reader is encouraged to consult the paper [241] by Daubechies, which was written around the time when multiresolution analysis was introduced. The paper contains a large number of important wavelet results, but multiresolution analysis is barely mentioned. The same remark applies to the excellent survey paper [395] by Heil and Walnut, which was published in 1989. At that time one could certainly not predict that soon almost all effort would go into constructions via multiresolution analysis.

This chapter and the following four chapters will deal with different aspects related to overcomplete collections of functions of the form (15.1). As discussed in Section 4.3, overcompleteness is introduced in order to obtain more flexibility and be able to make constructions which cannot be done with, e.g., orthonormal bases. We follow the historical development and begin by constructing frames without any multiresolution structure. Frames based on various multiresolution schemes are discussed in Chapters 17–18. We focus on one-dimensional wavelet systems; for the higher-dimensional case some of the key results are proved in Section 20.6.

A few words on terminology are needed. The word *wavelet* is usually reserved for a function ψ for which

$$\{2^{j/2}\psi(2^j x - k)\}_{j,k\in\mathbb{Z}} = \{\psi^{2^{-j},2^{-j}k}\}_{j,k\in\mathbb{Z}} \qquad (15.2)$$

is an orthonormal basis for $L^2(\mathbb{R})$. We will follow this tradition, but the word "wavelet" will appear in several constellations. Since we are interested in more general ways of choosing the translates and dilates than in (15.2), we will call *any* discrete family of the type $\{\psi^{a,b}\}_{(a,b)\in\Lambda}$, $\Lambda \subset \mathbb{R}^2$, a *wavelet system*.

We begin with a section on the continuous wavelet transform, which delivers integral representations of each $f \in L^2(\mathbb{R})$ of the type

$$f = \int_{-\infty}^{\infty} \int_{-\infty}^{\infty} c_f(a,b)\psi^{a,b}\,da\,db, \qquad (15.3)$$

provided that ψ satisfies some admissibility conditions and that the integral is interpreted in the right sense.

Then we move to construction of frames for $L^2(\mathbb{R})$ consisting of functions of the type $\{\psi^{a,b}\}_{(a,b)\in\Lambda}$. The obtained representations can be considered as discrete versions of (15.3), but our presentation does not rely on any result about the continuous wavelet transform. In Section 15.2 we consider the (regular) case, where the dilation parameter "two" in (15.2) is replaced by a number $a > 1$ and integer translation is replaced by translation with a step size $b > 0$, i.e., we consider wavelet systems of the form $\{a^{j/2}\psi(a^j x - kb)\}_{j,k\in\mathbb{Z}}$. Later, in Section 19.1 and Section 20.6, we discuss certain irregular choices of the discretization.

Section 15.3 deals with dual pairs of wavelet frames, a topic that will follow us throughout all the wavelet chapters. In contrast to the analogue issue for Gabor systems, we will see that a wavelet frame might not have a dual frame with the same structure; thus, the desire of obtaining dual pairs of wavelets puts additional restrictions on the function that is used to generate the wavelet frame.

15.1 The Continuous Wavelet Transform

Let $\psi \in L^2(\mathbb{R})$, and denote the Fourier transform of ψ by $\widehat{\psi}$. We say that ψ satisfies the *admissibility condition* if

$$C_\psi := \int_{-\infty}^{\infty} \frac{|\widehat{\psi}(\gamma)|^2}{|\gamma|} d\gamma < \infty. \tag{15.4}$$

We also say that ψ is *admissible*. Note that if $\widehat{\psi}$ is continuous in 0 [e.g., if $\psi \in L^1(\mathbb{R})$], then (15.4) can only be satisfied if $\widehat{\psi}(0) = 0$, i.e., if $\int_{-\infty}^{\infty} \psi(x)dx = 0$. But if this condition is satisfied, relatively weak decay conditions on $\widehat{\psi}$ imply that (15.4) is satisfied.

Now consider an admissible function $\psi \in L^2(\mathbb{R})$. Given $f \in L^2(\mathbb{R})$, the *continuous wavelet transform of f with respect to ψ* is defined as a function $W_\psi(f)$ of two variables, given by

$$\begin{aligned} W_\psi(f)(a,b) &:= \langle f, \psi^{a,b} \rangle \\ &= \int_{-\infty}^{\infty} f(x) \frac{1}{|a|^{1/2}} \overline{\psi(\frac{x-b}{a})} dx, \ (a,b) \in (\mathbb{R} \setminus \{0\}) \times \mathbb{R}. \end{aligned}$$

We will now prove that the wavelet transform has a similar property as what we saw for the short-time Fourier transform in Proposition 11.1.2. In fact, whenever $(\mathbb{R} \setminus \{0\}) \times \mathbb{R}$ is equipped with a certain weighted Lebesgue measure, the wavelet transform is a multiple of an isometry from $L^2(\mathbb{R})$ to $L^2((\mathbb{R} \setminus \{0\}) \times \mathbb{R})$. The role of the particular measure is explained in Example 24.1.3.

Proposition 15.1.1 *Assume that ψ is admissible. Then, for all functions $f, g \in L^2(\mathbb{R})$,*

$$\int_{-\infty}^{\infty} \int_{-\infty}^{\infty} W_\psi(f)(a,b)\overline{W_\psi(g)(a,b)}\frac{dadb}{a^2} = C_\psi\langle f, g\rangle. \tag{15.5}$$

Proof.   Assume first that ψ is admissible and $\widehat{\psi} \in C_c(\mathbb{R})$. Using the commutator relations for the Fourier transform and the operators T_b, D_a,

$$
\begin{aligned}
W_\psi(f)(a,b) &= \langle f, \psi^{a,b}\rangle = \langle \mathcal{F}f, \mathcal{F}T_b D_a\psi\rangle \\
&= \langle \widehat{f}, E_{-b}D_{1/a}\widehat{\psi}\rangle = \int_{-\infty}^{\infty} \widehat{f}(\gamma)e^{2\pi i b\gamma}|a|^{1/2}\overline{\widehat{\psi}(a\gamma)}d\gamma.
\end{aligned}
$$

If we for a moment consider a fixed value for a, this expression is the Fourier transform of the function

$$F_a(\gamma) = \widehat{f}(\gamma)|a|^{1/2}\overline{\widehat{\psi}(a\gamma)},$$

calculated in the point $-b$. If we define $G_a(\gamma)$ similarly, it follows that

$$
\begin{aligned}
\int_{-\infty}^{\infty} W_\psi(f)(a,b)\overline{W_\psi(g)(a,b)}db &= \int_{-\infty}^{\infty} \widehat{F_a}(-b)\overline{\widehat{G_a}(-b)}db \\
&= \langle \widehat{F_a}, \widehat{G_a}\rangle = \langle F_a, G_a\rangle \\
&= \int_{-\infty}^{\infty} \widehat{f}(\gamma)\overline{\widehat{g}(\gamma)}\,|a|\,|\widehat{\psi}(a\gamma)|^2 d\gamma.
\end{aligned}
$$

Inserting this expression in the left-hand side of (15.5) and using Fubini's theorem gives

$$
\begin{aligned}
&\int_{-\infty}^{\infty} \int_{-\infty}^{\infty} W_\psi(f)(a,b)\overline{W_\psi(g)(a,b)}\frac{dadb}{a^2} \\
&= \int_{-\infty}^{\infty} \int_{-\infty}^{\infty} \widehat{f}(\gamma)\overline{\widehat{g}(\gamma)}\,|a|\,|\widehat{\psi}(a\gamma)|^2 d\gamma \frac{da}{a^2} \\
&= \int_{-\infty}^{\infty} \left(\int_{-\infty}^{\infty} \frac{1}{|a|}|\widehat{\psi}(a\gamma)|^2 da\right) \widehat{f}(\gamma)\overline{\widehat{g}(\gamma)}d\gamma.
\end{aligned}
$$

For $\gamma \neq 0$ a change of variable shows that

$$\int_{-\infty}^{\infty} \frac{1}{|a|}|\widehat{\psi}(a\gamma)|^2 da = \int_{-\infty}^{\infty} \frac{1}{|a|}|\widehat{\psi}(a)|^2 da = C_\psi;$$

thus,

$$\int_{-\infty}^{\infty} \int_{-\infty}^{\infty} W_\psi(f)(a,b)\overline{W_\psi(g)(a,b)}\frac{dadb}{a^2} = C_\psi\langle\widehat{f}, \widehat{g}\rangle = C_\psi\langle f, g\rangle,$$

as desired. The extension to general admissible ψ is standard.   □

Similar to the Gabor case, we can write (15.5) as

$$f = \frac{1}{C_\psi} \int_{-\infty}^{\infty} \int_{-\infty}^{\infty} W_\psi(f)(a,b)\psi^{a,b} \frac{dadb}{a^2}, \quad f \in L^2(\mathbb{R}), \qquad (15.6)$$

where the integral is understood in the weak sense. Slightly stronger conditions imply that the weakly defined integral can be approximated by integrals over growing compact sets in $\mathbb{R}^2$; see [412].

As an immediate consequence of Proposition 15.1.1, we obtain that the wavelet system forms a continuous frame whenever the admissibility condition is satisfied:

Corollary 15.1.2 *If $\psi \in L^2(\mathbb{R})$ is admissible, then $\{\psi^{a,b}\}_{a\neq0,b\in\mathbb{R}}$ is a continuous tight frame for $L^2(\mathbb{R})$ with respect to $(\mathbb{R} \setminus \{0\}) \times \mathbb{R}$ equipped with the measure $\frac{1}{a^2}dadb$. The frame bound is $A = C_\psi$.*

15.2 Sufficient and Necessary Conditions

We now turn to the construction of (discrete) frames having the wavelet structure. We will first consider the case where the points (a,b) in (15.1) are restricted to discrete sets of the type $\{(a^j, kba^j)\}_{j,k\in\mathbb{Z}}$, where $a > 1, b > 0$ are fixed; a is the dilation parameter or scaling parameter, and b is the translation parameter. We hereby obtain the functions

$$(T_{kba^j} D_{a^j}\psi)(x) = (D_{a^j} T_{kb}\psi)(x) = \frac{1}{a^{j/2}}\psi(\frac{x}{a^j} - kb), \quad j, k \in \mathbb{Z}.$$

Re-indexing (i.e., replacing j by $-j$), we see that

$$\{T_{kba^j} D_{a^j}\psi\}_{j,k\in\mathbb{Z}} = \{a^{j/2}\psi(a^j x - kb)\}_{j,k\in\mathbb{Z}}. \qquad (15.7)$$

The re-indexing is purely introduced in order to get the wavelet system on the convenient form in (15.7). In the introductory sections, we will in general denote the wavelet systems by $\{a^{j/2}\psi(a^j x - kb)\}_{j,k\in\mathbb{Z}}$, but later we will use the operator notation $\{D_{a^j} T_{kb}\psi\}_{j,k\in\mathbb{Z}}$.

Definition 15.2.1 *Let $a > 1, b > 0$ and $\psi \in L^2(\mathbb{R})$. A frame for $L^2(\mathbb{R})$ of the form $\{a^{j/2}\psi(a^j x - kb)\}_{j,k\in\mathbb{Z}}$ is called a wavelet frame.*

The main purpose of this chapter is to present sufficient conditions for $\{a^{j/2}\psi(a^j x - kb)\}_{j,k\in\mathbb{Z}}$ to be a frame. The results will be stated in terms of the functions

$$G_0(\gamma) = \sum_{j\in\mathbb{Z}} |\widehat{\psi}(a^j\gamma)|^2, G_1(\gamma) = \sum_{k\neq0}\sum_{j\in\mathbb{Z}} |\widehat{\psi}(a^j\gamma)\widehat{\psi}(a^j\gamma + k/b)|, \gamma \in \mathbb{R}. (15.8)$$

Because we usually consider fixed values of a, b, the dependence of these parameters is suppressed in the notation. Note that

$$G_0(a\gamma) = G_0(\gamma), \quad G_1(a\gamma) = G_1(\gamma), \gamma \in \mathbb{R}.$$

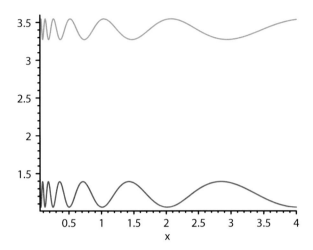

Figure 15.1. The functions G_0 (the upper graph) and G_1 based on the Mexican hat and the parameters $a = 2, b = 1.5$.

Geometrically this means that the graphs for G_0, G_1 for $|\gamma| \in [a^j, a^{j+1}]$ are stretched versions of their graphs for $|\gamma| \in [a^{j-1}, a^j]$. See Figure 15.1 for an illustration based on the Mexican hat wavelet described in Example A.7.2. It follows that

$$\sup_{\gamma \in \mathbb{R}} G_k(\gamma) = \sup_{|\gamma| \in [1,a]} G_k(\gamma), \ \inf_{\gamma \in \mathbb{R}} G_k(\gamma) = \inf_{|\gamma| \in [1,a]} G_k(\gamma), \ k = 0, 1.$$

The role played by the functions G_0 and G_1 in wavelet analysis corresponds to the role of the functions G and $\sum_{k \neq 0} |H_k(x)|$ defined in (11.16) and (11.27) in Gabor analysis. Similar to Proposition 11.3.4 (but technically more involved, especially for the lower bound), the following necessary condition for $\{a^{j/2}\psi(a^j x - kb)\}_{j,k \in \mathbb{Z}}$ to be a frame was proved by Chui and Shi [214].

Proposition 15.2.2 Let $a > 1, b > 0$ and $\psi \in L^2(\mathbb{R})$ be given. If $\{a^{j/2}\psi(a^j x - kb)\}_{j,k \in \mathbb{Z}}$ is a frame for $L^2(\mathbb{R})$ with frame bounds A, B, then

$$bA \leq \sum_{j \in \mathbb{Z}} |\widehat{\psi}(a^j \gamma)|^2 \leq bB, \ a.e. \, \gamma \in \mathbb{R}.$$

We will now state convenient sufficient conditions for a wavelet system $\{a^{j/2}\psi(a^j x - kb)\}_{j,k \in \mathbb{Z}}$ being a Bessel sequence or a frame. A generalization to wavelet systems in $L^2(\mathbb{R}^d)$ is proved in Theorem 20.6.1, using a different approach.

Theorem 15.2.3 *Let $a > 1, b > 0$ and $\psi \in L^2(\mathbb{R})$ be given. Suppose that*

$$B := \frac{1}{b} \sup_{|\gamma| \in [1,a]} \sum_{j,k \in \mathbb{Z}} \left| \widehat{\psi}(a^j \gamma) \widehat{\psi}(a^j \gamma + k/b) \right| < \infty. \tag{15.9}$$

Then $\{a^{j/2} \psi(a^j x - kb)\}_{j,k \in \mathbb{Z}}$ is a Bessel sequence with bound B, and for all functions $f \in L^2(\mathbb{R})$ for which $\widehat{f} \in C_c(\mathbb{R})$,

$$\sum_{j,k \in \mathbb{Z}} |\langle f, D_{a^j} T_{kb} \psi \rangle|^2 = \frac{1}{b} \int_{-\infty}^{\infty} |\widehat{f}(\gamma)|^2 \sum_{j \in \mathbb{Z}} |\widehat{\psi}(a^j \gamma)|^2 d\gamma \tag{15.10}$$

$$+ \frac{1}{b} \sum_{k \neq 0} \sum_{j \in \mathbb{Z}} \int_{-\infty}^{\infty} \widehat{f}(\gamma) \overline{\widehat{f}(\gamma - a^j k/b)} \widehat{\psi}(a^{-j}\gamma) \overline{\widehat{\psi}(a^{-j}\gamma - k/b)} d\gamma.$$

If furthermore

$$A := \frac{1}{b} \inf_{|\gamma| \in [1,a]} \left(\sum_{j \in \mathbb{Z}} |\widehat{\psi}(a^j \gamma)|^2 - \sum_{k \neq 0} \sum_{j \in \mathbb{Z}} |\widehat{\psi}(a^j \gamma) \widehat{\psi}(a^j \gamma + k/b)| \right) > 0,$$

$$\tag{15.11}$$

then $\{a^{j/2} \psi(a^j x - kb)\}_{j,k \in \mathbb{Z}}$ is a frame for $L^2(\mathbb{R})$ with bounds A, B.

Proof. Let $f \in L^2(\mathbb{R})$ and assume that $\widehat{f}$ is continuous and has compact support. We begin with some calculations leading to (15.14) below, which will be a key ingredient in the proof. Fix $j \in \mathbb{Z}$. Then

$$\int_0^{a^j/b} \sum_{k \in \mathbb{Z}} \left| \widehat{f}(\gamma - a^j k/b) \widehat{\psi}(a^{-j}\gamma - k/b) \right| d\gamma$$

$$= \sum_{k \in \mathbb{Z}} \int_0^{a^j/b} |\widehat{f}(\gamma - a^j k/b) \widehat{\psi}(a^{-j}\gamma - k/b)| d\gamma$$

$$= \sum_{k \in \mathbb{Z}} \int_{-a^j \frac{k}{b}}^{-a^j \frac{k}{b} + \frac{a^j}{b}} |\widehat{f}(\gamma) \widehat{\psi}(a^{-j}\gamma)| d\gamma$$

$$= \int_{-\infty}^{\infty} \left| \widehat{f}(\gamma) \widehat{\psi}(a^{-j}\gamma) \right| d\gamma$$

$$\leq \left(\int_{-\infty}^{\infty} |\widehat{f}(\gamma)|^2 d\gamma \right)^{1/2} \left(\int_{-\infty}^{\infty} |\widehat{\psi}(a^{-j}\gamma)|^2 d\gamma \right)^{1/2}$$

$$< \infty.$$

Thus we can define a function $F_j : \mathbb{R} \to \mathbb{C}$ by

$$F_j(\gamma) = \sum_{k \in \mathbb{Z}} \widehat{f}(\gamma - a^j k/b) \overline{\widehat{\psi}(a^{-j}\gamma - k/b)}, \ a.e. \ \gamma \in \mathbb{R}.$$

The function F_j is a^j/b-periodic, and the above argument gives that $F_j \in L^1(0, a^j/b)$. In fact, we even have $F_j \in L^2(0, a^j/b)$. To see this, we first note that

$$|F_j(\gamma)|^2 \leq \sum_{k\in\mathbb{Z}}' |\hat{f}(\gamma - a^j k/b)|^2 \sum_{k\in\mathbb{Z}} |\hat{\psi}(a^{-j}\gamma - k/b)|^2.$$

Since $\hat{f} \in C_c(\mathbb{R})$, the function $\gamma \to \sum_{k\in\mathbb{Z}} |\hat{f}(\gamma - a^j k/b)|^2$ is bounded; now an argument similar to above shows that $F_j \in L^2(0, a^j/b)$. We leave it to the reader to verify this and also that

$$\int_{-\infty}^{\infty} \hat{f}(\gamma)\overline{\hat{\psi}(a^{-j}\gamma)}e^{2\pi ima^{-j}b\gamma}d\gamma = \int_0^{a^j/b} F_j(\gamma)e^{2\pi ima^{-j}b\gamma}d\gamma. \quad (15.12)$$

Since $\{a^{-j/2}b^{1/2}e^{2\pi ima^{-j}b\gamma}\}_{m\in\mathbb{Z}}$ is an orthonormal basis for $L^2(0, a^j/b)$, Parseval's equality shows that

$$\sum_{m\in\mathbb{Z}} \left| \int_0^{a^j/b} F_j(\gamma)e^{2\pi ima^{-j}b\gamma}d\gamma \right|^2 = \frac{a^j}{b}\int_0^{a^j/b} |F_j(\gamma)|^2 d\gamma; \quad (15.13)$$

combining (15.12), (15.13), and the definition of F_j, we obtain that

$$\sum_{m\in\mathbb{Z}} \left| \int_{-\infty}^{\infty} \hat{f}(\gamma)\overline{\hat{\psi}(a^{-j}\gamma)}e^{2\pi ima^{-j}b\gamma}d\gamma \right|^2 \quad (15.14)$$

$$= \frac{a^j}{b}\int_0^{a^j/b} \left| \sum_{k\in\mathbb{Z}} \hat{f}(\gamma - a^j k/b)\overline{\hat{\psi}(a^{-j}\gamma - k/b)} \right|^2 d\gamma.$$

We now prove (15.10) for our special choice of the function f. The most delicate point in the proof is several interchanges of sums and integrals. In order to make the argument rigorous, we will first show that $\sum_{j,k\in\mathbb{Z}} |\langle f, D_{a^j}T_{kb}\psi\rangle|^2$ is finite by replacing all occurring functions by their absolute values. For positive functions, all interchanges are allowed. After showing that the sum is finite, all calculations can be repeated without absolute sign to get the exact expression (15.10). We will not perform the repetition, and for that reason, we keep all occurring complex conjugations in the first part of the calculation, even though they are superfluous for the first part.

The first step is to use the commutator relations for the Fourier transform and the operators D_a, T_b:

$$
\sum_{j,k\in\mathbb{Z}} |\langle f, D_{a^j} T_{kb}\psi\rangle|^2 = \sum_{j\in\mathbb{Z}}\sum_{m\in\mathbb{Z}} |\langle f, D_{a^j} T_{mb}\psi\rangle|^2
$$

$$
= \sum_{j\in\mathbb{Z}}\sum_{m\in\mathbb{Z}} |\langle \mathcal{F}f, \mathcal{F}D_{a^j} T_{mb}\psi\rangle|^2
$$

$$
= \sum_{j\in\mathbb{Z}}\sum_{m\in\mathbb{Z}} |\langle \widehat{f}, D_{a^{-j}} E_{-mb}\widehat{\psi}\rangle|^2
$$

$$
= \sum_{j\in\mathbb{Z}}\sum_{m\in\mathbb{Z}} |\langle \widehat{f}, E_{-ma^j b} D_{a^{-j}}\widehat{\psi}\rangle|^2
$$

$$
= \sum_{j\in\mathbb{Z}} a^j \sum_{m\in\mathbb{Z}} \left| \int_{-\infty}^{\infty} \widehat{f}(\gamma)\overline{\widehat{\psi}(a^j\gamma)} e^{2\pi i m a^j b\gamma} d\gamma \right|^2
$$

$$
= (*).
$$

Since we are summing over all $j \in \mathbb{Z}$, we can replace j by $-j$; doing so, and continuing using (15.14), we have

$$
\begin{aligned}
(*) \\
= \sum_{j\in\mathbb{Z}} a^{-j} \frac{a^j}{b} \int_0^{a^j/b} \left| \sum_{k\in\mathbb{Z}} \widehat{f}(\gamma - a^j k/b)\overline{\widehat{\psi}(a^{-j}\gamma - k/b)} \right|^2 d\gamma \qquad (15.15) \\
\leq \frac{1}{b} \sum_{j\in\mathbb{Z}} \int_0^{a^j/b} \left| \sum_{k\in\mathbb{Z}} |\widehat{f}(\gamma - a^j k/b)\overline{\widehat{\psi}(a^{-j}\gamma - k/b)}| \right|^2 d\gamma \\
= (**).
\end{aligned}
$$

Using that $|c|^2 = c\bar{c}$ for any complex number,

$$
\begin{aligned}
(**) \\
\leq \frac{1}{b} \sum_{j\in\mathbb{Z}} \int_0^{a^j/b} \Bigg(\sum_{\ell\in\mathbb{Z}} |\widehat{f}(\gamma - a^j \ell/b)\overline{\widehat{\psi}(a^{-j}\gamma - \ell/b)}| \\
\times \sum_{k\in\mathbb{Z}} |\overline{\widehat{f}(\gamma - a^j k/b)}\widehat{\psi}(a^{-j}\gamma - k/b)| \Bigg) d\gamma \\
= \frac{1}{b} \sum_{j\in\mathbb{Z}} \sum_{\ell\in\mathbb{Z}} \int_0^{a^j/b} \Bigg(|\widehat{f}(\gamma - a^j \ell/b)\overline{\widehat{\psi}(a^{-j}\gamma - \ell/b)}| \\
\times \sum_{k\in\mathbb{Z}} |\overline{\widehat{f}(\gamma - a^j k/b)}\widehat{\psi}(a^{-j}\gamma - k/b)| \Bigg) d\gamma \\
= (***).
\end{aligned}
$$

The function $\gamma \mapsto \sum_{k\in\mathbb{Z}} |\overline{\widehat{f}(\gamma - a^j k/b)\widehat{\psi}(a^{-j}\gamma - k/b)}|$ is a^j/b-periodic, so we can continue with

$$
\begin{aligned}
(***) &= \frac{1}{b}\sum_{j\in\mathbb{Z}}\int_{-\infty}^{\infty} |\widehat{f}(\gamma)\overline{\widehat{\psi}(a^{-j}\gamma)}| \cdot \sum_{k\in\mathbb{Z}}|\overline{\widehat{f}(\gamma - a^j k/b)}\widehat{\psi}(a^{-j}\gamma - k/b)|d\gamma \\
&= \frac{1}{b}\sum_{k\in\mathbb{Z}}\sum_{j\in\mathbb{Z}}\int_{-\infty}^{\infty} |\widehat{f}(\gamma)\overline{\widehat{f}(\gamma - a^j k/b)}\widehat{\psi}(a^{-j}\gamma)\overline{\widehat{\psi}(a^{-j}\gamma - k/b)}|d\gamma \\
&= \frac{1}{b}\int_{-\infty}^{\infty} |\widehat{f}(\gamma)|^2 \sum_{j\in\mathbb{Z}}|\widehat{\psi}(a^{-j}\gamma)|^2 d\gamma \\
&\quad + \frac{1}{b}\sum_{k\neq 0}\sum_{j\in\mathbb{Z}}\int_{-\infty}^{\infty}\left|\widehat{f}(\gamma)\overline{\widehat{f}(\gamma - a^j k/b)}\widehat{\psi}(a^{-j}\gamma)\overline{\widehat{\psi}(a^{-j}\gamma - k/b)}\right|d\gamma \\
&= \frac{1}{b}\int_{-\infty}^{\infty} |\widehat{f}(\gamma)|^2 \cdot \sum_{j\in\mathbb{Z}}|\widehat{\psi}(a^{-j}\gamma)|^2 d\gamma + \frac{1}{b}R,
\end{aligned}
$$

where

$$
R = \sum_{k\neq 0}\sum_{j\in\mathbb{Z}}\int_{-\infty}^{\infty} |\widehat{f}(\gamma)\overline{\widehat{f}(\gamma - a^j k/b)}\widehat{\psi}(a^{-j}\gamma)\overline{\widehat{\psi}(a^{-j}\gamma - k/b)}|d\gamma.
$$

We now estimate the term R. Using Cauchy–Schwarz' inequality twice, first on the integral and then on the sum over k, we obtain

$$
\begin{aligned}
R &\le \sum_{j\in\mathbb{Z}}\sum_{k\neq 0}\left(\int_{-\infty}^{\infty} |\widehat{f}(\gamma)|^2\, |\widehat{\psi}(a^{-j}\gamma)\,\widehat{\psi}(a^{-j}\gamma - k/b)|d\gamma\right)^{1/2} \\
&\quad \times \left(\int_{-\infty}^{\infty} |\widehat{f}(\gamma - a^j k/b)|^2\, |\widehat{\psi}(a^{-j}\gamma)\,\widehat{\psi}(a^{-j}\gamma - k/b)|d\gamma\right)^{1/2} \\
&\le \sum_{j\in\mathbb{Z}}\left(\sum_{k\neq 0}\int_{-\infty}^{\infty} |\widehat{f}(\gamma)|^2\, |\widehat{\psi}(a^{-j}\gamma)\,\widehat{\psi}(a^{-j}\gamma - k/b)|d\gamma\right)^{1/2} \\
&\quad \times \left(\sum_{k\neq 0}\int_{-\infty}^{\infty} |\widehat{f}(\gamma - a^j k/b)|^2\, |\widehat{\psi}(a^{-j}\gamma)\,\widehat{\psi}(a^{-j}\gamma - k/b)|d\gamma\right)^{1/2} \\
&= \sum_{j\in\mathbb{Z}}(*)(**),
\end{aligned}
$$

where

$$(*) \quad = \quad \left(\sum_{k \neq 0} \int_{-\infty}^{\infty} |\widehat{f}(\gamma)|^2 \; |\widehat{\psi}(a^{-j}\gamma) \; \widehat{\psi}(a^{-j}\gamma - k/b)| d\gamma \right)^{1/2},$$

$$(**) \quad = \quad \left(\sum_{k \neq 0} \int_{-\infty}^{\infty} |\widehat{f}(\gamma - a^j k/b)|^2 \; |\widehat{\psi}(a^{-j}\gamma) \; \widehat{\psi}(a^{-j}\gamma - k/b)| d\gamma \right)^{1/2}.$$

The terms (*) and (**) are actually identical; in fact, by the change of variable $\gamma \to \gamma + a^j k/b$ in (**),

$$(**) \quad = \quad \left(\sum_{k \neq 0} \int_{-\infty}^{\infty} |\widehat{f}(\gamma)|^2 \; |\widehat{\psi}(a^{-j}\gamma + k/b) \; \widehat{\psi}(a^{-j}\gamma)| d\gamma \right)^{1/2}$$

$$= \quad \left(\sum_{k \neq 0} \int_{-\infty}^{\infty} |\widehat{f}(\gamma)|^2 \; |\widehat{\psi}(a^{-j}\gamma) \; \widehat{\psi}(a^{-j}\gamma - k/b)| d\gamma \right)^{1/2}$$

$$= \quad (*).$$

Therefore,

$$R \quad \leq \quad \sum_{j \in \mathbb{Z}} \sum_{k \neq 0} \int_{-\infty}^{\infty} |\widehat{f}(\gamma)|^2 \; |\widehat{\psi}(a^{-j}\gamma) \; \widehat{\psi}(a^{-j}\gamma - k/b)| d\gamma.$$

It follows that

$$\sum_{j,k \in \mathbb{Z}} |\langle f, D_{a^j} T_{kb} \psi \rangle|^2$$

$$\leq \quad \frac{1}{b} \int_{-\infty}^{\infty} |\widehat{f}(\gamma)|^2 \sum_{j \in \mathbb{Z}} |\widehat{\psi}(a^{-j}\gamma)|^2 d\gamma$$

$$+ \frac{1}{b} \int_{-\infty}^{\infty} |\widehat{f}(\gamma)|^2 \sum_{k \neq 0} \sum_{j \in \mathbb{Z}} |\widehat{\psi}(a^{-j}\gamma)\widehat{\psi}(a^{-j}\gamma - k/b)| d\gamma$$

$$= \quad \frac{1}{b} \int_{-\infty}^{\infty} |\widehat{f}(\gamma)|^2 \sum_{k \in \mathbb{Z}} \sum_{j \in \mathbb{Z}} |\widehat{\psi}(a^{-j}\gamma)\widehat{\psi}(a^{-j}\gamma - k/b)| d\gamma.$$

Note that

$$\sum_{k \in \mathbb{Z}} \sum_{j \in \mathbb{Z}} |\widehat{\psi}(a^{-j}\gamma)\widehat{\psi}(a^{-j}\gamma - k/b)| = \sum_{k \in \mathbb{Z}} \sum_{j \in \mathbb{Z}} |\widehat{\psi}(a^j\gamma)\widehat{\psi}(a^j\gamma + k/b)|;$$

using the assumption (15.9), we therefore have

$$\sum_{j,k \in \mathbb{Z}} |\langle f, D_{a^j} T_{kb} \psi \rangle|^2 \quad \leq \quad B \, \|\widehat{f}\|^2 = B \, \|f\|^2.$$

Since this holds for all functions f for which $\widehat{f}$ is continuous and has compact support, it holds for all $f \in L^2(\mathbb{R})$ by Lemma 3.2.6; thus, the wavelet system $\{a^{j/2}\psi(a^j x - kb)\}_{j,k\in\mathbb{Z}}$ is a Bessel sequence with bound B. We can now go back and repeat all the above calculations without absolute sign, still with $\widehat{f} \in C_c(\mathbb{R})$, to get the announced expression (15.10) for $\sum_{j,k\in\mathbb{Z}} |\langle f, D_{a^j} T_{kb}\psi\rangle|^2$; since we can just skip the single inequality (15.15), we obtain an *exact* expression. If we also assume that (15.11) is satisfied, then

$$\sum_{j,k\in\mathbb{Z}} |\langle f, D_{a^j} T_{kb}\psi\rangle|^2$$

$$\geq \quad \frac{1}{b} \int_{-\infty}^{\infty} |\widehat{f}(\gamma)|^2 \sum_{j\in\mathbb{Z}} |\widehat{\psi}(a^j\gamma)|^2$$

$$- \left| \frac{1}{b} \sum_{k\neq 0} \sum_{j\in\mathbb{Z}} \int_{-\infty}^{\infty} \widehat{f}(\gamma)\overline{\widehat{f}(\gamma - a^j k/b)} \widehat{\psi}(a^{-j}\gamma)\overline{\widehat{\psi}(a^{-j}\gamma - k/b)} d\gamma \right|$$

$$\geq \quad \frac{1}{b} \int_{-\infty}^{\infty} |\widehat{f}(\gamma)|^2 \left(\sum_{j\in\mathbb{Z}} |\widehat{\psi}(a^j\gamma)|^2 - \sum_{k\neq 0}\sum_{j\in\mathbb{Z}} |\widehat{\psi}(a^j\gamma)\widehat{\psi}(a^j\gamma + k/b)| \right) d\gamma$$

$$\geq \quad A\, ||f||^2.$$

The proof is now completed via Lemma 5.1.9.   □

If $\widehat{\psi}$ is continuous in 0 [e.g., if $\psi \in L^1(\mathbb{R})$], the condition (15.9) can only be satisfied if $\widehat{\psi}(0) = 0$, because $\widehat{\psi}(a^j\gamma) \to \widehat{\psi}(0)$ as $j \to -\infty$. If this necessary condition is satisfied, then very reasonable conditions on $\widehat{\psi}$ will imply that $\{a^{j/2}\psi(a^j x - kb)\}_{j,k\in\mathbb{Z}}$ is a frame whenever b is sufficiently small. We need some lemmas before we prove a formal version of this statement in Proposition 15.2.6.

Lemma 15.2.4 *Let $x, y \in \mathbb{R}$. Then, for all $\delta \in [0,1]$,*

$$\frac{1}{1+(x+y)^2} \leq 2\left(\frac{1+x^2}{1+y^2}\right)^\delta.$$

Proof. Given $x, y \in \mathbb{R}$, the function $\delta \to 2\left(\frac{1+x^2}{1+y^2}\right)^\delta$ is monotone, so it is enough to prove the result for $\delta = 0$ and $\delta = 1$. The case $\delta = 0$ is clear; for $\delta = 1$, we use that $2ab \leq a^2 + b^2$ for all $a, b \in \mathbb{R}$ to obtain that

$$\begin{aligned} 1 + y^2 &= 1 + ((y+x) - x)^2 \\ &= 1 + (y+x)^2 + x^2 - 2x(y+x) \\ &\leq 1 + 2((y+x)^2 + x^2) \\ &\leq 2(1 + (y+x)^2)(1 + x^2). \end{aligned}$$

□

Lemma 15.2.5 *Let* $\psi \in L^2(\mathbb{R})$ *and assume that there exists a constant* $C > 0$ *such that*

$$|\widehat{\psi}(\gamma)| \leq C \frac{|\gamma|}{(1 + |\gamma|^2)^{3/2}} \quad a.e.\ \gamma \in \mathbb{R}.$$

Then, for all $a > 1$ *and* $b > 0$,

$$\sum_{k \neq 0} \sum_{j \in \mathbb{Z}} |\widehat{\psi}(a^j \gamma) \widehat{\psi}(a^j \gamma + k/b)|$$

$$\leq 16 C^2 b^{4/3} \left(\frac{a^2}{a - 1} + \frac{a}{a^{2/3} - 1} \right). \qquad (15.16)$$

Proof. The decay condition on ψ gives that

$$|\widehat{\psi}(a^j \gamma) \widehat{\psi}(a^j \gamma + k/b)| \leq C^2 \frac{|a^j \gamma|}{(1 + |a^j \gamma|^2)^{3/2}} \frac{|a^j \gamma + k/b|}{(1 + |a^j \gamma + k/b|^2)^{3/2}}$$

$$\leq C^2 \frac{|a^j \gamma|}{(1 + |a^j \gamma|^2)^{3/2}} \frac{(1 + |a^j \gamma + k/b|^2)^{1/2}}{(1 + |a^j \gamma + k/b|^2)^{3/2}}$$

$$= C^2 \frac{|a^j \gamma|}{(1 + |a^j \gamma|^2)^{3/2}} \frac{1}{1 + |a^j \gamma + k/b|^2}.$$

Applying Lemma 15.2.4 on $(1 + |a^j \gamma + k/b|^2)^{-1}$ with $\delta = \frac{2}{3}$ gives

$$|\widehat{\psi}(a^j \gamma) \widehat{\psi}(a^j \gamma + k/b)| \leq 2 C^2 \frac{|a^j \gamma|}{(1 + |a^j \gamma|^2)^{3/2}} \left(\frac{1 + |a^j \gamma|^2}{1 + |k/b|^2} \right)^{2/3}$$

$$\leq 2 C^2 \frac{|a^j \gamma|}{(1 + |a^j \gamma|^2)^{5/6}} \left(\frac{1}{1 + |k/b|^2} \right)^{2/3}.$$

In this last estimate, j and k appear in separate terms. Thus,

$$\sum_{k \neq 0} \sum_{j \in \mathbb{Z}} |\widehat{\psi}(a^j \gamma) \widehat{\psi}(a^j \gamma + k/b)|$$

$$\leq 2 C^2 \left(\sum_{j \in \mathbb{Z}} \frac{|a^j \gamma|}{(1 + |a^j \gamma|^2)^{5/6}} \right) \left(\sum_{k \neq 0} \left(\frac{1}{1 + |k/b|^2} \right)^{2/3} \right). \quad (15.17)$$

For the sum over $k \neq 0$,

$$\sum_{k \neq 0} \left(\frac{1}{1 + |k/b|^2} \right)^{2/3} = 2 \sum_{k=1}^{\infty} \frac{b^{4/3}}{(b^2 + k^2)^{2/3}}$$

$$\leq 2 b^{4/3} \sum_{k=1}^{\infty} \frac{1}{k^{4/3}} \leq 8 b^{4/3}.$$

In order to estimate the sum over $j \in \mathbb{Z}$ in (15.17), we define the function

$$f(\gamma) = \sum_{j \in \mathbb{Z}} \frac{|a^j \gamma|}{(1 + |a^j \gamma|^2)^{5/6}}, \quad \gamma \in \mathbb{R}.$$

We want to show that f is bounded. Note that $f(a\gamma) = f(\gamma)$ for all γ; it is therefore enough to consider $|\gamma| \in [1, a]$, so we can use that

$$|a^j\gamma| \le a^{j+1}, \ 1 + |a^j\gamma|^2 \ge 1 + a^{2j}.$$

Thus,

$$
\begin{aligned}
\sum_{j\in\mathbb{Z}} \frac{|a^j\gamma|}{(1+|a^j\gamma|^2)^{5/6}} &\le \sum_{j\in\mathbb{Z}} \frac{a^{j+1}}{(1+a^{2j})^{5/6}} \\
&= \sum_{j=-\infty}^{0} \frac{a^{j+1}}{(1+a^{2j})^{5/6}} + \sum_{j=1}^{\infty} \frac{a^{j+1}}{(1+a^{2j})^{5/6}} \\
&\le \sum_{j=-\infty}^{0} a^{j+1} + \sum_{j=1}^{\infty} \frac{a^{j+1}}{a^{5j/3}} \\
&= a \sum_{j=0}^{\infty} a^{-j} + a \sum_{j=1}^{\infty} (a^{-2/3})^j \\
&= \frac{a^2}{a-1} + \frac{a}{a^{2/3}-1}.
\end{aligned}
$$

That is, f is bounded as claimed. Putting all information together and using (15.17), we now arrive at

$$\sum_{k\ne 0}\sum_{j\in\mathbb{Z}} |\widehat{\psi}(a^j\gamma)\widehat{\psi}(a^j\gamma + k/b)|$$

$$\le 2C^2 \left(\sum_{j\in\mathbb{Z}} \frac{|a^j\gamma|}{(1+|a^j\gamma|^2)^{5/6}} \right) \left(\sum_{k\ne 0} \left(\frac{1}{1+|k/b|^2} \right)^{2/3} \right)$$

$$\le 16C^2 b^{4/3} \left(\frac{a^2}{a-1} + \frac{a}{a^{2/3}-1} \right),$$

as desired.   □

We are now ready to give sufficient conditions for $\{a^{j/2}\psi(a^jx-kb)\}_{j,k\in\mathbb{Z}}$ to be a frame for small values of b:

Proposition 15.2.6 *Let $\psi \in L^2(\mathbb{R})$ and $a > 1$ be given. Assume that*

(i) $\inf_{|\gamma|\in[1,a]} \sum_{j\in\mathbb{Z}} |\widehat{\psi}(a^j\gamma)|^2 > 0.$

(ii) There exists a constant $C > 0$ such that

$$|\widehat{\psi}(\gamma)| \le C \frac{|\gamma|}{(1+|\gamma|^2)^{3/2}}, \quad a.e. \ \gamma \in \mathbb{R}. \tag{15.18}$$

Then $\{a^{j/2}\psi(a^jx-kb)\}_{j,k\in\mathbb{Z}}$ is a frame for $L^2(\mathbb{R})$ for all sufficiently small translation parameters $b > 0$.

Proof. We first prove that $\{a^{j/2}\psi(a^j x - kb)\}_{j,k\in\mathbb{Z}}$ is a Bessel sequence for all $b > 0$. Arguments similar to the one used in the proof of Lemma 15.2.5 show that (Exercise 15.1)

$$\sum_{j\in\mathbb{Z}} |\widehat{\psi}(a^j\gamma)|^2 \le \left(\frac{1}{a^4 - 1} + \frac{a^4}{a^2 - 1}\right) C^2. \tag{15.19}$$

Via Lemma 15.2.5, it follows that

$$\sum_{k\in\mathbb{Z}}\sum_{j\in\mathbb{Z}} |\widehat{\psi}(a^j\gamma)\widehat{\psi}(a^j\gamma + k/b)|$$

$$\le 16C^2 b^{4/3}\left(\frac{a^2}{a - 1} + \frac{a}{a^{2/3} - 1}\right) + \left(\frac{1}{a^4 - 1} + \frac{a^4}{a^2 - 1}\right) C^2;$$

by Theorem 15.2.3, we conclude that $\{a^{j/2}\psi(a^j x - kb)\}_{j,k\in\mathbb{Z}}$ is a Bessel sequence. By choosing b sufficiently small, the assumption (i) implies that

$$\inf_{|\gamma|\in[1,a]}\left(\sum_{j\in\mathbb{Z}} |\widehat{\psi}(a^j\gamma)|^2 - 16C^2 b^{4/3}\left(\frac{a^2}{a - 1} + \frac{a}{a^{2/3} - 1}\right)\right) > 0, \tag{15.20}$$

and in this case, by Lemma 15.2.5,

$$\inf_{|\gamma|\in[1,a]}\left(\sum_{j\in\mathbb{Z}} |\widehat{\psi}(a^j\gamma)|^2 - \sum_{k\neq 0}\sum_{j\in\mathbb{Z}} |\widehat{\psi}(a^j\gamma)\widehat{\psi}(a^j\gamma + k/b)|\right) > 0.$$

Theorem 15.2.3 now gives the desired conclusion. □

The proof of Proposition 15.2.6 shows that $\{a^{j/2}\psi(a^j x - kb)\}_{j,k\in\mathbb{Z}}$ is a frame whenever $b > 0$ satisfies (15.20). In concrete cases, we can often use much larger values of b:

Example 15.2.7 Let $a = 2$ and consider the function

$$\psi(x) = \frac{2}{\sqrt{3}}\pi^{-1/4}(1 - x^2)e^{-\frac{1}{2}x^2}.$$

Due to its shape, ψ is called the *Mexican hat*. As proved in Example A.7.2,

$$\widehat{\psi}(\gamma) = 8\sqrt{\frac{2}{3}}\pi^{9/4}\gamma^2 e^{-2\pi^2\gamma^2}.$$

A numerical calculation shows that

$$\inf_{|\gamma|\in[1,2]}\sum_{j\in\mathbb{Z}} |\widehat{\psi}(2^j\gamma)|^2 > 3.27.$$

Also, (15.18) is satisfied for $C = 4$, so a direct calculation using (15.20) shows that $\{2^{j/2}\psi(2^j x - kb)\}_{j,k\in\mathbb{Z}}$ is a frame if $b < 0.0084$. This is far from being optimal: numerical calculations based on the expressions for A, B in

Theorem 15.2.3 gives that $\{2^{j/2}\psi(2^j x - kb)\}_{j,k\in\mathbb{Z}}$ is a frame if $b < 1.97$! The obtained frame bounds A, B for some selected values for b are as follows:

b	0.25	0.5	0.75	1	1.25	1.5	1.75	1.97
A	13.1	6.55	4.36	3.26	2.33	1.25	0.422	0.0069
B	14.2	7.1	4.73	3.57	3.09	3.13	3.5	3.54

For small values of b, the frame bounds are almost identical to the values obtained in [242] via a different criterion. For large values of b, the bounds above are sharper (for $b = 1.5$, the bounds given in [242] are $A = 0.325$ and $B = 4.221$). Furthermore, the criterion used in [242] suggests that the frame property breaks down already before $b = 1.75$.

In terms of the functions G_0 and G_1 in (15.8), Theorem 15.2.3 says that $\{2^{j/2}\psi(2^j x - kb)\}_{j,k\in\mathbb{Z}}$ is a frame with lower frame bound A if

$$A := \inf_{\gamma\in[1,2]} \frac{1}{b}\left(G_0(\gamma) - G_1(\gamma)\right) > 0. \tag{15.21}$$

As upper frame bound, we can use

$$B = \sup_{\gamma\in[1,2]} \frac{1}{b}\left(G_0(\gamma) + G_1(\gamma)\right).$$

□

The Fourier transform of the Mexican hat decays much faster than assumed in (15.18). Thus it is not a surprise that direct estimates via Theorem 15.2.3 give that $\{2^{j/2}\psi(2^j x - kb)\}_{j,k\in\mathbb{Z}}$ is a frame for larger values of b than suggested by Proposition 15.2.6. The same will happen for all functions ψ for which $\widehat{\psi}$ decay faster than assumed in (15.18). In fact, it is the decay of $\widehat{\psi}$ that will make $|\widehat{\psi}(a^j\gamma)\widehat{\psi}(a^j\gamma+k/b)|$ small, so when $\widehat{\psi}$ decays much faster than (15.18), it is clear that $\sum_{k\neq 0}\sum_{j\in\mathbb{Z}} |\widehat{\psi}(a^j\gamma)\widehat{\psi}(a^j\gamma+k/b)|$ will be significantly smaller than the bound in (15.16). Even for the function ψ given by

$$\widehat{\psi}(\gamma) = \frac{|\gamma|}{(1+|\gamma|^2)^{3/2}}, \tag{15.22}$$

the estimate in Proposition 15.2.6 is far from being sharp: a numerical estimate shows that

$$\inf_{|\gamma|\in[1,2]} \sum_{j\in\mathbb{Z}} |\widehat{\psi}(2^j\gamma)|^2 \sim 1.5,$$

which implies that $\{2^{j/2}\psi(2^j x - kb)\}_{j,k\in\mathbb{Z}}$ is a frame if $b < 0.037$, while numerical estimates based on Theorem 15.2.3 just give that $b < 0.24$ is sufficient (Exercise 15.2). Proposition 15.2.6 is mainly interesting because it gives the *existence* of an interval $]0, b_0[$ such that all translation parameters b belonging to the interval lead to a frame; it does not yield the maximal value of b_0.

There is one remarkable difference between Theorem 15.2.3 and Theorem 11.4.2 for Gabor frames: in the condition for the lower bound in the Gabor version, it is the sum over k of

$$\left| \sum_{n \in \mathbb{Z}} g(x - na)\overline{g(x - na - k/b)} \right|$$

that has to be subtracted from $\sum_{n \in \mathbb{Z}} |g(x - na)|^2$. That is, the absolute sign is *outside* the sum over n. This is in contrast to the condition in Theorem 15.2.3, where the absolute sign is *inside* the sums. The condition in the Gabor version is clearly the best, since the position of the absolute sign opens up for possible cancellations. For $a = 2$, it is known that the condition in Theorem 15.2.3 can be replaced with a condition where the absolute sign is outside (cf. [241], Theorem 2.9). A more detailed discussion of this phenomenon for other values of a was given by Laugesen in [478].

A sufficient condition for $\{a^{j/2}(\psi(a^j x - kb))\}_{j,k \in \mathbb{Z}}$ being a Bessel sequence was obtained by Chui and Shi in [212]:

Proposition 15.2.8 *Let* θ : $[0, \infty[\to [0, \infty[$ *be a function which is nondecreasing on* $[0, \frac{1}{2\pi}]$, *nonincreasing on* $[\frac{1}{2\pi}, \infty[$, *and for which*

$$\int_0^\infty \theta(\gamma)(1 + \frac{1}{\gamma})d\gamma < \infty.$$

Then every function $\psi \in L^2(\mathbb{R})$ *for which*

$$|\widehat{\psi}(\gamma)| \le \theta(|\gamma|), \ a.e. \ \gamma \in \mathbb{R},$$

generates a Bessel sequence $\{a^{j/2}(\psi(a^j x - kb)\}_{j,k \in \mathbb{Z}}$ *for any* $b > 0, a > 1$.

15.3 Dual Pairs of Wavelet Frames

From now on the operators used to generate a wavelet system will play a key role. For this reason we will primarily use the notation $\{D_{a^j} T_{kb} \psi\}_{j,k \in \mathbb{Z}}$ to denote a wavelet system.

Given a wavelet frame $\{D_{a^j} T_{kb} \psi\}_{j,k \in \mathbb{Z}}$ for $L^2(\mathbb{R})$ with frame operator S, the frame decomposition (5.7) takes the form

$$f = \sum_{j,k \in \mathbb{Z}} \langle f, S^{-1} D_{a^j} T_{kb} \psi \rangle D_{a^j} T_{kb} \psi, \tag{15.23}$$

In the analog case of a Gabor system, we were able to simplify the frame decomposition, due to the observation that the frame operator commutes with the involved modulation/translation operators. The situation is more complicated in the wavelet case, but at least the frame operator commutes with the scaling operator (Exercise 15.3):

Proposition 15.3.1 *Let $\{D_{a^j}T_{kb}\psi\}_{j,k\in\mathbb{Z}}$ be a wavelet frame for $L^2(\mathbb{R})$, with frame operator S. Then*

$$SD_a = D_aS \qquad (15.24)$$

and

$$S^{-1}D_a = D_aS^{-1}. \qquad (15.25)$$

The result in (15.25) implies that the canonical dual frame of a wavelet frame $\{D_{a^j}T_{kb}\psi\}_{j,k\in\mathbb{Z}}$ has the form

$$\{S^{-1}D_{a^j}T_{kb}\psi\}_{j,k\in\mathbb{Z}} = \{D_{a^j}S^{-1}T_{kb}\psi\}_{j,k\in\mathbb{Z}}.$$

Unfortunately Example 16.1.1 will show that in general the frame operator S and its inverse S^{-1} do not commute with the translation operators T_{kb}. This implies that in general the canonical dual frame will not have wavelet structure.

If $\{D_{a^j}T_{kb}\psi\}_{j,k\in\mathbb{Z}}$ is a Riesz basis, then the dual frame is unique by Theorem 3.6.2, and we cannot replace the sequence $\{S^{-1}D_{a^j}T_{kb}\psi\}_{j,k\in\mathbb{Z}}$ by another sequence in the frame decomposition (15.23). But in the case of a wavelet frame that is not a Riesz basis, we know from general theory (Lemma 6.3.1) that there exist other dual frames than the canonical dual frame. Thus, it is natural to exploit the freedom offered by the redundancy and examine whether there exists an alternative dual frame which has the wavelet structure. Daubechies and B. Han gave in [246] an example of a wavelet frame $\{D^jT_k\psi\}_{j,k\in\mathbb{Z}}$ for which the canonical dual does not have the wavelet structure; however, there exist infinitely many functions $\widetilde{\psi}$ for which $\{D^jT_k\widetilde{\psi}_{j,k}\}_{j,k\in\mathbb{Z}}$ is a dual frame (see the paper [82] for a technical correction). The generator of the frame $\{D^jT_k\psi\}_{j,k\in\mathbb{Z}}$ has the property that $\widehat{\psi} = \chi_k$ for a compact subset K of $\mathbb{R}$; frames of this type are the subject of Section 16.3.

A characterization of all pairs of dual wavelet frame pairs was obtained by Chui and Shi [217]. We will prove the result on page 515.

Theorem 15.3.2 *Given* $a > 1$, $b > 0$, *two Bessel sequences* $\{D_{a^j}\,T_{kb}\psi\}_{j,k\in\mathbb{Z}}$ *and* $\{D_{a^j}T_{kb}\widetilde{\psi}\}_{j,k\in\mathbb{Z}}$, *where* $\psi, \widetilde{\psi} \in L^2(\mathbb{R})$, *form dual wavelet frames for $L^2(\mathbb{R})$ if and only if the following two conditions are satisfied:*

(i) $\sum_{j\in\mathbb{Z}} \overline{\widehat{\psi}(a^j\gamma)}\widehat{\widetilde{\psi}}(a^j\gamma) = b$ *for a.e.* $\gamma \in \mathbb{R}$.

(ii) *For any number* $\alpha \neq 0$ *of the form* $\alpha = m/a^j$, $m,j \in \mathbb{Z}$,

$$\sum_{\{(j,m)\in\mathbb{Z}^2 \,|\, \alpha=m/a^j\}} \overline{\widehat{\psi}(a^j\gamma)}\widehat{\widetilde{\psi}}(a^j\gamma + m/b) = 0, \quad a.e.\ \gamma \in \mathbb{R}. \quad (15.26)$$

The condition (15.26) is clearly satisfied if

$$\overline{\widehat{\psi}(a^j\gamma)}\widehat{\widetilde{\psi}}(a^j\gamma + q/b) = 0, \text{ a.e. } \gamma \in \mathbb{R}, \ \forall q \in \mathbb{Z} \setminus \{0\}.$$

In particular, if $\widehat{\psi}$ and $\widehat{\widetilde{\psi}}$ have compact support, then for sufficiently small values of $b > 0$, the wavelet systems $\{D_{a^j}T_{kb}\psi\}_{j,k\in\mathbb{Z}}$ and $\{D_{a^j}T_{kb}\widetilde{\psi}\}_{j,k\in\mathbb{Z}}$ form dual frames for $L^2(\mathbb{R})$ when the condition (i) in Theorem 15.3.2 is satisfied. Lemvig used in [483] this observation to construct dual pairs of wavelet frames. The construction is parallel to the approach we saw for Gabor frames in Theorem 12.5.1: the basic condition (12.36) appearing in the Gabor case is replaced by a partition of unity condition of the form

$$\sum_{j\in\mathbb{Z}} \widehat{\psi}(a^j\gamma) = 1, \ \gamma \in \mathbb{R}. \tag{15.27}$$

The connections between the Gabor case and the wavelet case are even closer. As shown in [175] by Christensen and Goh, certain dual pairs of wavelet frames can be constructed directly based on the Gabor frames obtained via Theorem 12.5.1:

Example 15.3.3 Consider two Gabor Bessel sequences $\{E_{mb}T_n g\}_{m,n\in\mathbb{Z}}$ and $\{E_{mb}T_n h\}_{m,n\in\mathbb{Z}}$, generated by functions g, h that are supported in an interval $[M, N]$. Assume that

$$b \leq \min((N - M)^{-1}, 2^{-1}a^{-N}) \tag{15.28}$$

and that

$$\sum_{k\in\mathbb{Z}} \overline{g(x - k)}h(x - k) = b, \ x \in [0, 1].$$

Then $\{E_{mb}T_n g\}_{m,n\in\mathbb{Z}}$ and $\{E_{mb}T_n h\}_{m,n\in\mathbb{Z}}$ are dual Gabor frames for $L^2(\mathbb{R})$ by Theorem 12.3.4. Now, fix $a > 1$ and define the functions $\psi, \widetilde{\psi} \in L^2(\mathbb{R})$ by

$$\widehat{\psi}(\gamma) = g(\log_a(|\gamma|)), \ \widehat{\widetilde{\psi}}(\gamma) = h(\log_a(|\gamma|)), \gamma \neq 0. \tag{15.29}$$

Then, for a.e. $\gamma \in \mathbb{R}$,

$$\sum_{j\in\mathbb{Z}} \overline{\widehat{\psi}(a^j\gamma)}\widehat{\widetilde{\psi}}(a^j\gamma) = \sum_{j\in\mathbb{Z}} \overline{g(\log_a(|a^j\gamma|))}h(\log_a(|a^j\gamma|))$$

$$= \sum_{j\in\mathbb{Z}} \overline{g(j + \log_a(|\gamma|))}h(j + \log_a(|\gamma|)) = b,$$

i.e., the condition (i) in Theorem 15.3.2 is satisfied. The definition of the functions ψ and $\widetilde{\psi}$ shows that $\widehat{\psi}$ and $\widehat{\widetilde{\psi}}$ are supported on

$$[-a^N, -a^M] \cup [a^M, a^N] \subset [-a^N, a^N];$$

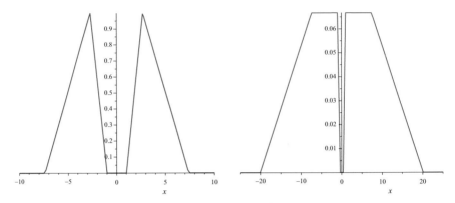

Figure 15.2. Plots of geometric splines $\widehat{\psi}$ and $\widetilde{\widehat{\psi}}$ obtained via the procedure in Example 15.3.3, based on the exponential spline of order 2 in Example A.9.3. The function $\widehat{\psi}$ is a geometric spline with knots at the points $\pm 1, \pm e, \pm e^2$; the function $\widetilde{\widehat{\psi}}$ is a geometric spline with knots at the points $\pm e^{-1}, \pm 1, \pm e^2, \pm e^3$.

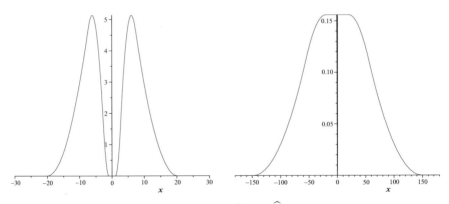

Figure 15.3. Plots of geometric splines $\widehat{\psi}$ and $\widetilde{\widehat{\psi}}$ obtained via the procedure in Example 15.3.3, based on the exponential spline of order 3 in Example A.9.3. The function $\widehat{\psi}$ is a geometric spline with knots at the points $\pm 1, \pm e, \pm e^2, \pm e^3$; the function $\widetilde{\widehat{\psi}}$ is a geometric spline with knots at the points $\pm e^{-2}, \pm e^{-1}, \pm 1, \pm e^3, \pm e^4, \pm e^5$. Note that $\widetilde{\widehat{\psi}}(\gamma) = 0$ for $|\gamma| \in [-e^{-2}, e^{-2}]$ and that $\widetilde{\widehat{\psi}}(\gamma) = 1$ for $|\gamma| \in [1, e^3]$. Arbitrary smoothness can be obtained by applying the procedure to higher-order exponential B-splines, at the price of increased support size.

thus, by the choice of b in (15.28) the condition (ii) in Theorem 15.3.2 is also satisfied. We conclude that the wavelet systems $\{D_{a^j}T_{kb}\psi\}_{j,k\in\mathbb{Z}}$ and $\{D_{a^j}T_{kb}\widetilde{\psi}\}_{j,k\in\mathbb{Z}}$ form dual frames for $L^2(\mathbb{R})$.

In [175] this approach is applied on dual Gabor frames $\{E_{mb}T_n g\}_{m,n\in\mathbb{Z}}$, $\{E_{mb}T_n h\}_{m,n\in\mathbb{Z}}$, obtained by using Theorem 12.3.4 on exponential B-splines. The outcome is a dual pair of wavelet frames, generated by functions $\psi, \widetilde{\psi}$, whose Fourier transforms are compactly supported splines with geometrically distributed knots and desired smoothness; see Figures 15.2 and 15.3.   □

The idea of using the log-transform to transfer results from the Gabor setting to the wavelet setting also appears in the paper [410] by Holighaus and Wiesmeyr.

Note also that the transformation in (15.29) yields a natural way to obtain functions ψ satisfying the partition of unity condition (15.27) for some $a > 1$. In fact, taking $g \in L^2(\mathbb{R})$ such that $\operatorname{supp} g \subseteq [M, N]$ and

$$\sum_{k\in\mathbb{Z}} g(x - k) = 1, \; x \in \mathbb{R},$$

the function ψ given by (15.29) will satisfy (15.27).

15.4 Exercises

15.1 Prove the inequality (15.19).

15.2 Consider the function ψ given by (15.22) and find, based on Proposition 15.2.6, respectively Theorem 15.2.3, values for b_0 such that $\{2^{j/2}\psi(2^j x - kb)\}_{j,k\in\mathbb{Z}}$ is a frame for all $b \in]0, b_0]$.

15.3 Prove Proposition 15.3.1.

16
Dyadic Wavelet Frames for $L^2(\mathbb{R})$

In this chapter we consider *dyadic wavelet systems*, i.e., wavelet systems for $L^2(\mathbb{R})$ with scaling parameter $a = 2$ and translation parameter $b = 1$. We will usually denote the resulting wavelet systems $\{2^{j/2}\psi(2^j x - k)\}_{j,k\in\mathbb{Z}}$ by $\{D^j T_k \psi\}_{j,k\in\mathbb{Z}}$ or $\{\psi_{j,k}\}_{j,k\in\mathbb{Z}}$. Recall that bases of this type were considered already in Section 3.9.

In Section 16.1 we state results concerning the structure of the canonical dual frame associated with a dyadic wavelet frame. In particular, we show that the canonical dual of a wavelet frame might not have the wavelet structure. In case the given frame is not a Riesz basis, we know from Lemma 6.3.1 that other duals exist, and this naturally leads to the question whether an alternative dual having the wavelet structure exists. We state the formal definition of dual (multi)wavelet frames [see below] and a characterization hereof; concrete constructions will appear in Chapters 17–18. In Section 16.3 we present results concerning wavelet frames generated by a function whose Fourier transform is a characteristic function.

In this chapter we will also consider wavelet frames generated by more than one function. That is, we consider a finite number of functions $\psi_1, \ldots, \psi_n \in L^2(\mathbb{R})$ and ask for

$$\{D^j T_k \psi_1\}_{j,k\in\mathbb{Z}} \cup \{D^j T_k \psi_2\}_{j,k\in\mathbb{Z}} \cup \cdots \cup \{D^j T_k \psi_n\}_{j,k\in\mathbb{Z}} \qquad (16.1)$$

to be a frame for $L^2(\mathbb{R})$. A frame of this type is called a *multi-wavelet frame* and will usually be denoted by $\{D^j T_k \psi_\ell\}_{j,k\in\mathbb{Z},\ell=1,\ldots,n}$ or $\{\psi_{\ell;j,k}\}_{j,k\in\mathbb{Z},\ell=1,\ldots,n}$. We will often skip the word "multi" and simply use the shorter name "wavelet frame."

© Springer International Publishing Switzerland 2016
O. Christensen, *An Introduction to Frames and Riesz Bases*,
Applied and Numerical Harmonic Analysis,
DOI 10.1007/978-3-319-25613-9_16

16.1 Wavelet Frames and Their Duals

The canonical dual frame associated to a wavelet frame $\{D^j T_k \psi\}_{j,k\in\mathbb{Z}}$ with frame operator S is given by $\{S^{-1} D^j T_k \psi\}_{j,k\in\mathbb{Z}}$. Due to the difficulty in inverting S explicitly it is usually hard to find the dual frame, but in the following example from [241] it can be done by direct computation. The example is important because it demonstrates that the dual of a wavelet frame might not have the wavelet structure. The example also shows that the frame operator for a wavelet frame $\{D^j T_k \psi\}_{j,k\in\mathbb{Z}}$ might not commute with the translation operators T_k, $k \in \mathbb{Z}$; see the discussion in Section 15.3.

Example 16.1.1 Let $\{D^j T_k \psi\}_{j,k\in\mathbb{Z}}$ be a wavelet orthonormal basis for $L^2(\mathbb{R})$. Given $\epsilon \in]0, 1[$, we define a function θ by

$$\theta = \psi + \epsilon D\psi.$$

We want to prove that $\{D^j T_k \theta\}_{j,k\in\mathbb{Z}}$ is a Riesz basis and compute the dual Riesz basis. It will be convenient to use the notation

$$\{\psi_{j,k}\}_{j,k\in\mathbb{Z}} = \{D^j T_k \psi\}_{j,k\in\mathbb{Z}}, \quad \{\theta_{j,k}\}_{j,k\in\mathbb{Z}} = \{D^j T_k \theta\}_{j,k\in\mathbb{Z}}.$$

The idea is to consider the function θ as a small perturbation of ψ and use a stability result for frames to conclude that $\{\theta_{j,k}\}_{j,k\in\mathbb{Z}}$ is a Riesz basis. First, the commutator relation (2.28) shows that

$$\psi_{j,k} - \theta_{j,k} = -\epsilon D^j T_k D\psi = -\epsilon D^{j+1} T_{2k}\psi; \tag{16.2}$$

thus, given any finite scalar sequence $\{c_{j,k}\}_{j,k\in\mathbb{Z}}$,

$$\left\| \sum_{j,k\in\mathbb{Z}} c_{j,k}(\psi_{j,k} - \theta_{j,k}) \right\|^2 = \epsilon^2 \left\| \sum_{j,k\in\mathbb{Z}} c_{j,k} D^{j+1} T_{2k}\psi \right\|^2$$

$$= \epsilon^2 \sum_{j,k\in\mathbb{Z}} |c_{j,k}|^2;$$

the last equality follows from $\{D^{j+1} T_{2k}\psi\}_{j,k\in\mathbb{Z}}$ being a subfamily of the orthonormal basis $\{\psi_{j,k}\}_{j,k\in\mathbb{Z}}$. Via the general perturbation result stated in Theorem 22.1.1, we see that $\{\theta_{j,k}\}_{j,k\in\mathbb{Z}}$ is a Riesz basis for $L^2(\mathbb{R})$. By the definition of a Riesz basis we can define a bounded invertible operator

$$U : L^2(\mathbb{R}) \to L^2(\mathbb{R}), \quad U\psi_{j,k} := \theta_{j,k}.$$

Via Exercise 5.20, the frame operator for $\{\theta_{j,k}\}_{j,k\in\mathbb{Z}}$ is $S = UU^*$, so the canonical dual associated to $\{\theta_{j,k}\}_{j,k\in\mathbb{Z}}$ is

$$\{S^{-1}\theta_{j,k}\}_{j,k\in\mathbb{Z}} = \{(U^*)^{-1} U^{-1}\theta_{j,k}\}_{j,k\in\mathbb{Z}} = \{(U^{-1})^* \psi_{j,k}\}_{j,k\in\mathbb{Z}}. \tag{16.3}$$

We want to obtain a more concrete expression for the canonical dual frame of $\{\theta_{j,k}\}_{j,k\in\mathbb{Z}}$. In terms of the operator U, (16.2) means that

$$(I - U)\psi_{j,k} = -\epsilon D^{j+1}T_{2k}\psi = -\epsilon\psi_{j+1,2k};$$

expanding an arbitrary $f \in L^2(\mathbb{R})$ in the orthonormal basis $\{\psi_{j,k}\}_{j,k\in\mathbb{Z}}$, it follows that

$$(I - U)f = -\epsilon \sum_{j,k\in\mathbb{Z}} \langle f, \psi_{j,k}\rangle \psi_{j+1,2k}.$$

Thus, for $f, g \in L^2(\mathbb{R})$,

$$\begin{aligned}
\langle f, (I - U)^* g\rangle &= \langle (I - U)f, g\rangle \\
&= -\epsilon \sum_{j,k\in\mathbb{Z}} \langle f, \psi_{j,k}\rangle \langle \psi_{j+1,2k}, g\rangle \\
&= \langle f, -\epsilon \sum_{j,k\in\mathbb{Z}} \langle g, \psi_{j+1,2k}\rangle \psi_{j,k}\rangle.
\end{aligned}$$

It follows that

$$\begin{aligned}
(I - U^*)g &= (I - U)^* g \\
&= -\epsilon \sum_{j,k\in\mathbb{Z}} \langle g, \psi_{j+1,2k}\rangle \psi_{j,k}. \qquad (16.4)
\end{aligned}$$

In particular, $||I-U^*|| = \epsilon < 1$, which implies that $(U^*)^{-1}$ can be expanded in a Neumann series,

$$(U^*)^{-1} = \sum_{n=0}^{\infty} (I - U^*)^n.$$

Now (16.3) implies that the dual Riesz basis of $\{\theta_{j,k}\}_{j,k\in\mathbb{Z}}$ is

$$\{S^{-1}\theta_{j,k}\}_{j,k\in\mathbb{Z}} = \left\{\sum_{n=0}^{\infty} (I - U^*)^n \psi_{j,k}\right\}_{j,k\in\mathbb{Z}}. \qquad (16.5)$$

We can go one step further. In fact, the action of $I - U^*$ on the functions $\psi_{j,k}, j, k \in \mathbb{Z}$ can be found via (16.4) using that $\{\psi_{j,k}\}_{j,k\in\mathbb{Z}}$ is an orthonormal basis; the outcome depends on k being even or odd:

$$(I - U^*)\psi_{j,2k} = -\epsilon\psi_{j-1,k}, \quad \text{while} \quad (I - U^*)\psi_{j,2k+1} = 0, \ \forall j, k \in \mathbb{Z}. \ (16.6)$$

In particular, via (16.5),

$$S^{-1}\theta_{j,2k+1} = \psi_{j,2k+1} \text{ for all } j, k \in \mathbb{Z}.$$

Also, for any $k \neq 0$, the equations in (16.6) show that there exists a value of $n \in \mathbb{N}$ for which

$$(I - U^*)^n \psi_{j,2k} = 0;$$

thus, $S^{-1}\theta_{j,2k}$ is a finite linear combination of functions $\{\psi_{j,k}\}_{j,k\in\mathbb{Z}}$,

$$
\begin{aligned}
S^{-1}\theta_{j,2k} &= \psi_{j,2k} + (I - U^*)\psi_{j,2k} + \cdots + (I - U^*)^n\psi_{j,2k} \\
&= \psi_{j,2k} - \epsilon\psi_{j-1,k} + \cdots + 0.
\end{aligned}
$$

For $k = 0$, we have

$$
S^{-1}\theta_{j,0} = \sum_{n=0}^{\infty}(-\epsilon)^n\psi_{j-n,0}. \tag{16.7}
$$

In particular, the canonical dual frame of $\{\theta_{j,k}\}_{j,k\in\mathbb{Z}}$ does *not* have the wavelet structure; the functions $\{S^{-1}\theta_{j,k}\}_{j,k\in\mathbb{Z}}$ do not even have the same norm. This is in contrast to the situation for Gabor frames and frames of translates, where we saw that the canonical dual has the same structure as the frame itself. We know from Proposition 15.3.1 that

$$
S^{-1}D^jT_k\theta = D^jS^{-1}T_k\theta; \tag{16.8}
$$

thus, the fact that the canonical dual frame of $\{\theta_{j,k}\}_{j,k\in\mathbb{Z}}$ does not have wavelet structure also implies that the operator S^{-1} cannot commute with all the operators T_k, $k \in \mathbb{Z}$; as a consequence, the frame operator S does not commute with all the operators T_k in this example.

The above calculations show that there are other properties which are not inherited by the canonical dual frame. For example, if we assume that the function ψ has compact support, then also θ has compact support, and all the functions $\{\theta_{j,k}\}_{j,k\in\mathbb{Z}}$ have compact support. If we look at the dual frame $\{S^{-1}\theta_{j,k}\}_{j,k\in\mathbb{Z}}$, then we obtain functions with compact support when $k \neq 0$ because the elements in the dual frame are finite linear combinations of the functions in $\{\psi_{j,k}\}_{j,k\in\mathbb{Z}}$ in this case. However, for $k = 0$ the expression (16.7) shows that the functions $S^{-1}\theta_{j,0}$ do not have compact support. $\square$

Example 16.1.1 is disappointing. Wavelet frames were introduced because their structure makes them convenient to work with, but as soon as we want to apply the frame decomposition we need to know the canonical dual frame. If the canonical dual frame has the wavelet structure, we can find it by calculating one single function (the generator) and then simply applying the operators $D^jT_k, j, k \in \mathbb{Z}$ to get the other elements. But if the canonical dual frame does not have the wavelet structure, we have to find it using (16.8). That is, we have to apply S^{-1} to the infinite set of functions $T_k\psi$; this is a much harder task than applying S^{-1} on a single function.

This is exactly the point where the flexibility of frame theory comes in again. We know that overcomplete frames have various dual frames, so if the canonical dual frame does not have wavelet structure, we might search for an alternative dual frame which has wavelet structure. The following formal definition expresses that we want to obtain constructions of dual pairs of frames, both having wavelet structure. For reasons that will become clear later (see, e.g., Theorem 18.5.1), we state it for multiwavelet frames.

Definition 16.1.2 *Consider two sequences of functions*

$$\psi_1, \ldots, \psi_n \in L^2(\mathbb{R}) \text{ and } \widetilde{\psi}_1, \ldots, \widetilde{\psi}_n \in L^2(\mathbb{R}).$$

We say that $\{D^j T_k \psi_\ell\}_{j,k\in\mathbb{Z},\ell=1,\ldots,n}$ *and* $\{D^j T_k \widetilde{\psi}_\ell\}_{j,k\in\mathbb{Z},\ell=1,\ldots,n}$ *are a pair of dual wavelet frames if both are Bessel sequences and*

$$f = \sum_{\ell=1}^{n} \sum_{j,k\in\mathbb{Z}} \langle f, D^j T_k \psi_\ell \rangle D^j T_k \widetilde{\psi}_\ell, \quad \forall f \in L^2(\mathbb{R}). \qquad (16.9)$$

That Bessel sequences $\{D^j T_k \psi_\ell\}_{j,k\in\mathbb{Z},\ell=1,\ldots,n}$ and $\{D^j T_k \widetilde{\psi}_\ell\}_{j,k\in\mathbb{Z},\ell=1,\ldots,n}$ are frames if they satisfy (16.9) follows from Lemma 6.3.2. A pair of dual wavelet frames is called *sibling frames* in [209] and *bi-frames* in [248].

A characterization of all dual dyadic wavelet frame pairs was obtained by Frazier et al. [309]. We will derive the result as a consequence of a characterization of dual GSI-systems, see page 515.

Theorem 16.1.3 *Let* $\psi_1, \ldots, \psi_n, \widetilde{\psi}_1, \ldots, \widetilde{\psi}_n \in L^2(\mathbb{R})$ *and assume that* $\{D^j T_k \psi_\ell\}_{j,k\in\mathbb{Z},\ell=1,\ldots,n}$ *and* $\{D^j T_k \widetilde{\psi}_\ell\}_{j,k\in\mathbb{Z},\ell=1,\ldots,n}$ *are Bessel sequences. Then* $\{D^j T_k \psi_\ell\}_{j,k\in\mathbb{Z},\ell=1,\ldots,n}$ *and* $\{D^j T_k \widetilde{\psi}_\ell\}_{j,k\in\mathbb{Z},\ell=1,\ldots,n}$ *are a pair of dual wavelet frames if and only if*

$$\begin{cases} \displaystyle\sum_{\ell=1}^{n} \sum_{j\in\mathbb{Z}} \widehat{\psi_\ell}(2^j\gamma)\overline{\widehat{\widetilde{\psi}_\ell}(2^j\gamma)} = 1, \ a.e. \ \gamma \in \mathbb{R}; \\[2em] \displaystyle\sum_{\ell=1}^{n} \sum_{j=0}^{\infty} \widehat{\psi_\ell}(2^j\gamma)\overline{\widehat{\widetilde{\psi}_\ell}(2^j(\gamma + q))} = 0 \ \text{for all odd integers } q, \ a.e. \ \gamma \in \mathbb{R}. \end{cases}$$

Constructions of dual wavelet frames will appear in Chapter 18; more constructions are given in the original papers [209, 248].

16.2 Tight Wavelet Frames

As we have discussed in Section 6.1, tight frames are very convenient because the frame decomposition can be applied without any cumbersome inversion of the frame operator. The dual of a tight frame $\{D^j T_k \psi\}_{j,k\in\mathbb{Z}}$ with frame bound A is simply $\{\frac{1}{A} D^j T_k \psi\}_{j,k\in\mathbb{Z}}$; that is, in contrast to the situation for general wavelet frames, the canonical dual of a tight wavelet frame automatically has wavelet structure. The frame decomposition in (5.7) takes the form

$$f = \frac{1}{A} \sum_{j,k\in\mathbb{Z}} \langle f, D^j T_k \psi \rangle D^j T_k \psi, \quad \forall f \in L^2(\mathbb{R}).$$

The functions ψ generating a tight wavelet frame can be characterized based on the characterization of dual wavelet pairs. We leave the details to the reader (Exercise 16.2) and also note that a direct proof can be found in, e.g., [400].

Theorem 16.2.1 *A function* $\psi \in L^2(\mathbb{R})$ *generates a tight wavelet frame* $\{D^j T_k \psi\}_{j,k\in\mathbb{Z}}$ *with frame bound A if and only if*

$$
\begin{cases}
\sum_{j\in\mathbb{Z}} |\widehat{\psi}(2^j\gamma)|^2 = A, \ a.e. \ \gamma \in \mathbb{R}; \\
\sum_{j=0}^{\infty} \widehat{\psi}(2^j\gamma)\overline{\widehat{\psi}(2^j(\gamma+q))} = 0 \ \ \text{for all odd integers } q, \ a.e. \ \gamma \in \mathbb{R}.
\end{cases}
$$

Note that the only difference between the conditions for $\{D^j T_k \psi\}_{j,k\in\mathbb{Z}}$ being a tight frame with frame bound $A = 1$ and the characterization of wavelets in Lemma 3.9.4 is that the condition $||\psi|| = 1$ does not appear in Theorem 16.2.1. We can, in fact, derive Lemma 3.9.4 as a consequence of Theorem 16.2.1. Indeed, if ψ is a wavelet, then $\{D^j T_k \psi\}_{j,k\in\mathbb{Z}}$ is a tight wavelet frame with bound $A = 1$ and $||\psi|| = 1$; thus, by Theorem 16.2.1 the conditions (3.53) and (3.54) in Lemma 3.9.4 are satisfied. On the other hand, the conditions (3.53) and (3.54) imply that $\{D^j T_k \psi\}_{j,k\in\mathbb{Z}}$ is a tight frame with frame bound $A = 1$; via Proposition 3.4.8 the assumption $||\psi|| = 1$ then implies that $\{D^j T_k \psi\}_{j,k\in\mathbb{Z}}$ is an orthonormal basis, i.e., ψ is a wavelet.

Chapter 18 will yield several constructions of tight wavelet frames based on an extension of the classical multiresolution analysis scheme.

16.3   Wavelet Frame Sets

Theorem 15.2.3 gives a sufficient condition for a function $\psi \in L^2(\mathbb{R})$ to generate a wavelet frame $\{D^j T_k \psi\}_{j,k\in\mathbb{Z}}$, expressed in terms of the Fourier transform $\widehat{\psi}$. For special classes of functions ψ we can give simpler conditions for ψ generating a wavelet frame; one natural choice is to consider functions ψ for which $\widehat{\psi}$ is a characteristic function for a Lebesgue measurable set K in $\mathbb{R}$. In order for χ_K to belong to $L^2(\mathbb{R})$, we assume that K has finite Lebesgue measure.

Definition 16.3.1 *A Lebesgue measurable set K in $\mathbb{R}$ is called a wavelet frame set if $|K| < \infty$ and the function ψ defined by $\widehat{\psi} = \chi_K$ generates a wavelet frame $\{D^j T_k \psi\}_{j,k\in\mathbb{Z}}$ for $L^2(\mathbb{R})$.*

We will give a short description of results obtained by D. Han [362] and Dai et al. [237], respectively. We begin with some definitions:

Definition 16.3.2 *Let K be a measurable set in $\mathbb{R}$ with finite measure. We say that*

(i) $x, y \in \mathbb{R}$ *are δ-equivalent if there is an $j \in \mathbb{Z}$ such that*

$$x = 2^j y.$$

For $x \in K$, the number of elements $y \in K$ which belong to its δ-equivalence class is denoted by $\delta_K(x)$. Finally, let

$$K(\delta, k) := \{x \in K : \delta_K(x) = k\}, \ \ k \in \mathbb{N}.$$

(ii) $x, y \in \mathbb{R}$ *are τ-equivalent if there is an $k \in \mathbb{Z}$ such that*

$$x = y + k.$$

For $x \in K$, the number of elements $y \in K$ which belong to its τ-equivalence class is denoted by $\tau_K(x)$. Finally, let

$$K(\tau, k) := \{x \in K : \tau_K(x) = k\}, \ \ k \in \mathbb{N}.$$

Using the above notation, Dai et al. were almost able to characterize frame wavelet sets:

Theorem 16.3.3 *Let K be a Lebesgue measurable set in $\mathbb{R}$ with finite measure. Then the following hold:*

(i) K *is a wavelet frame set if $\cup_{j \in \mathbb{Z}} 2^j K(\tau, 1) = \mathbb{R}$ (up to a null set) and there exists $M \in \mathbb{N}$ such that $K(\delta, m)$ and $K(\tau, m)$ are null sets for $m > M$; in this case one is a lower frame bound for $\{D^j T_k \psi\}_{j,k \in \mathbb{Z}}$ and $M^{5/2}$ is an upper frame bound.*

(ii) *If K is a wavelet frame set, then $\cup_{j \in \mathbb{Z}} 2^j K = \mathbb{R}$ (up to a null set), and there exists $M \in \mathbb{N}$ such that $K(\delta, m)$ and $K(\tau, m)$ are null sets for $m > M$.*

For wavelet frame sets generating a tight frame, a complete characterization is obtained:

Theorem 16.3.4 *A Lebesgue measurable set K in $\mathbb{R}$ with finite measure is a wavelet frame set generating a tight frame if and only if the following conditions hold:*

(i) $\cup_{j \in \mathbb{Z}} 2^j K = \mathbb{R}$ *(up to a null set);*

(ii) *for some $m \geq 1$ we have $K = K(\tau, 1) = K(\delta, m)$.*

In case (i) and (ii) are satisfied, the frame bound for $\{D^j T_k \psi\}_{j,k \in \mathbb{Z}}$ is $A = m$.

Let us show how the conditions in Theorem 16.3.4 can be reformulated. The condition $K = K(\tau, 1)$ means that for $\gamma \in \mathbb{R}$, the point $\gamma + k$ belongs to the set K for at most one value of $k \in \mathbb{Z}$ or, expressed differently, that

$$\sum_{k \in \mathbb{Z}} \chi_K(\gamma + k) \leq 1, \quad a.e. \ \gamma \in \mathbb{R}. \tag{16.10}$$

Now assume that

$$\bigcup_{j \in \mathbb{Z}} 2^j K = \mathbb{R}, \text{ and for some } m \in \mathbb{N}, \ K = K(\delta, m). \tag{16.11}$$

Then, given $\gamma \in \mathbb{R}$ there exists $j' \in \mathbb{Z}$ such that $2^{-j'}\gamma \in K$. The δ-equivalence class of $2^{-j'}\gamma$ contains exactly m elements, so

$$\sum_{j \in \mathbb{Z}} \chi_{2^j K}(\gamma) = m, \quad a.e. \ \gamma \in \mathbb{R}. \tag{16.12}$$

Similarly, one proves that if (16.12) holds for some $m \in \mathbb{N}$, then (16.11) holds. Thus, we have obtained an equivalent formulation of Theorem 16.3.4:

Theorem 16.3.5 *A Lebesgue measurable set K in $\mathbb{R}$ with finite measure is a wavelet frame set generating a tight frame if and only if (16.10) and (16.12) are satisfied for some $m \geq 1$.*

In the case where K is a finite union of closed intervals, the sufficiency of the conditions (16.10) and (16.12) was obtained by D. Han a couple of years before Dai et al. proved Theorem 16.3.4. We illustrate the use of Theorem 16.3.5 with some examples:

Example 16.3.6 (i) Let $K = [-\frac{1}{2}, -\frac{1}{4}[\cup [\frac{1}{4}, \frac{1}{2}]$. Then, except for $\gamma = 0$ there exists exactly one value of $j \in \mathbb{Z}$ such that $\gamma \in 2^j K$, so (16.12) is satisfied with $m = 1$. Equation (16.10) is also satisfied, so K is a frame wavelet set, which generates a tight frame with frame bound $A = 1$.

(ii) Similarly, for $n = 1, 2, \ldots$, the set $K = [-\frac{1}{2}, -\frac{1}{2^{n+1}}[\cup [\frac{1}{2^{n+1}}, \frac{1}{2}]$ is a wavelet frame set, which generates a tight frame with frame bound $A = n$.

(iii) Let $K = [-\frac{3}{4}, -\frac{1}{4}[\cup [\frac{1}{8}, \frac{1}{2}[$. Then

$$K(\tau, 1) = [-\frac{1}{2}, -\frac{1}{4}[\cup [\frac{1}{8}, \frac{1}{4}[,$$

and $\cup_{j \in \mathbb{Z}} 2^j K = \mathbb{R}$ up to a null set. Also, for $m \geq 2$ we have

$$K(\delta, m) = K(\tau, m) = 0.$$

Thus, by Theorem 16.3.3, K is a wavelet frame set. □

Note that wavelet frame sets in $\mathbb{R}^d$ have been considered in [238].

16.4   Frames and Multiresolution Analysis

The classical definition of multiresolution analysis was introduced with the purpose to construct orthonormal bases for $L^2(\mathbb{R})$. There are several natural ways to extend the scope to construction of Riesz bases or frames. The biorthogonal multiresolution by Cohen, Daubechies, and Feauveau [223] delivers dual Riesz bases $\{D^j T_k \psi\}_{j,k \in \mathbb{Z}}, \{D^j T_k \widetilde{\psi}\}_{j,k \in \mathbb{Z}}$ via the construction of two coupled multiresolution analyses. Another option is to consider Definition 3.9.2 without modifications and ask for the existence of a function $\psi \in V_0$ for which $\{D^j T_k \psi\}_{j,k \in \mathbb{Z}}$ is a Riesz basis or a frame for $L^2(\mathbb{R})$. This idea was elaborated by Zalik. In [638] he characterizes all functions ψ which appear via some multiresolution analysis and generate Riesz bases.

Other authors have constructed dyadic wavelet frames by modifying the original definition of a multiresolution analysis. In the next chapter we discuss frame multiresolution analysis as defined by Benedetto and Li. In Chapter 18 we give a treatise of another approach by Ron and Shen and its further development with contributions from many researchers. Common for these approaches is that they keep the conditions in Definition 3.9.2, except condition (v). Note also that the literature contains several other variations on the theme, e.g., the generalized multiresolution considered in the paper [24] and the references therein.

16.5   Exercises

16.1 Assume that K is a frame wavelet set, and let $\theta \in L^2(\mathbb{R})$ be a function with support on K. Assume that there exist constants $C, D > 0$ such that $C \leq |\theta| \leq D$. Prove that the function $\psi \in L^2(\mathbb{R})$ defined by $\widehat{\psi} = \theta \chi_K$ generates a wavelet frame.

16.2 Prove Theorem 16.2.1 via Theorem 16.1.3.

17

Frame Multiresolution Analysis

The introduction of multiresolution analysis by Mallat and Meyer was the beginning of a new era; the short descriptions in Section 3.9 and Section 4.3 only give a glimpse of the research activity based on this new tool, aiming at construction of orthonormal bases $\{\psi_{j,k}\}_{j,k\in\mathbb{Z}}$.

As described in Section 4.3 and Section 16.4, the "classical" theory has been extended in different ways with the purpose of removing some of its constraints; as an example, we mention the biorthogonal multiresolution analysis, which leads to constructions of Riesz bases and their duals. In this chapter and the next, we go one step further and extend the multiresolution scheme in such a way that we can construct overcomplete wavelet frames. It is important to notice that we insist on the key idea of multiscales; they make the constructions attractive from the computational aspect, as a reader familiar with multiresolution analysis will know.

This chapter will deal with frame multiresolution analysis as defined by Benedetto and Li; here the condition (v) in Definition 3.9.2 is simply replaced with the condition that $\{T_k\phi\}_{k\in\mathbb{Z}}$ is a frame for V_0. This seemingly innocent change causes many technical difficulties, but under certain conditions, we are actually able to construct wavelet frames from here.

Frame multiresolution analysis is not the most general way to obtain frames via multiscale techniques, but it provides us with a natural link from the classical constructions described in Section 3.9 to the more advanced theory presented in Chapter 18.

This chapter is independent of Chapters 15–16. However, the results for frames of translates (especially Theorem 9.2.5) will play an important role.

© Springer International Publishing Switzerland 2016

O. Christensen, *An Introduction to Frames and Riesz Bases*,
Applied and Numerical Harmonic Analysis,
DOI 10.1007/978-3-319-25613-9_17

17.1   Frame Multiresolution Analysis

Frame multiresolution analysis was introduced by Benedetto and Li [52], [51]. The purpose of the theory is to construct wavelet frames for $L^2(\mathbb{R})$ of the form $\{2^{j/2}\psi(2^j x - k)\}_{j,k \in \mathbb{Z}}$. We will use the notation introduced at the beginning of Chapter 16 and denote such a frame by $\{D^j T_k \psi\}_{j,k \in \mathbb{Z}}$ or $\{\psi_{j,k}\}_{j,k \in \mathbb{Z}}$.

We introduce some terminology before we define frame multiresolution analysis. The interval $]-\frac{1}{2}, \frac{1}{2}[$ is identified with the torus $\mathbb{T}$, and the class of 1-periodic functions on $\mathbb{R}$ whose restriction to $]-\frac{1}{2}, \frac{1}{2}[$ belongs to $L^2(-\frac{1}{2}, \frac{1}{2})$ is denoted by $L^2(\mathbb{T})$. Similarly, $L^\infty(\mathbb{T})$ consists of the bounded measurable 1-periodic functions on $\mathbb{R}$. With this notation, $L^\infty(\mathbb{T}) \subset L^2(\mathbb{T})$. We note that $L^2(\mathbb{T})$ and $L^\infty(\mathbb{T})$ actually consist of equivalence classes of functions which are identical almost everywhere, so when we speak about pointwise relationships between functions, it is understood that they can only be expected to hold almost everywhere. In the entire section, we will not mention this explicitly, i.e., we will not follow the equations by "a.e."

One of the main tools will be Fourier expansions of functions in $L^2(\mathbb{T})$. Using the complex exponentials $E_k(x) = e^{2\pi i k x}$, the Fourier series of a function $f \in L^2(\mathbb{T})$ will be written as

$$f = \sum_{k \in \mathbb{Z}} c_k E_k, \quad \text{where} \quad c_k = \int_{-\frac{1}{2}}^{\frac{1}{2}} f(x) E_{-k}(x) dx.$$

Formally, a frame multiresolution analysis is defined as a multiresolution analysis, with the condition "$\{T_k \phi\}_{k \in \mathbb{Z}}$ is a orthonormal basis for V_0" replaced by a frame condition:

Definition 17.1.1 *A frame multiresolution analysis for $L^2(\mathbb{R})$ consists of a sequence of closed subspaces $\{V_j\}_{j \in \mathbb{Z}}$ of $L^2(\mathbb{R})$ and a function $\phi \in V_0$ such that*

(i) $\cdots V_{-1} \subset V_0 \subset V_1 \cdots$.

(ii) $\overline{\cup_{j \in \mathbb{Z}} V_j} = L^2(\mathbb{R})$ *and* $\cap_{j \in \mathbb{Z}} V_j = \{0\}$.

(iii) $V_j = D^j V_0$.

(iv) $f \in V_0 \Rightarrow T_k f \in V_0, \ \forall k \in \mathbb{Z}$.

(v) $\{T_k \phi\}_{k \in \mathbb{Z}}$ *is a frame for V_0.*

If the conditions in Definition 17.1.1 are satisfied, it follows that

$$V_j := D^j(\overline{\text{span}}\{T_k \phi\}_{k \in \mathbb{Z}}) = \overline{\text{span}}\{D^j T_k \phi\}_{k \in \mathbb{Z}}, \ j \in \mathbb{Z}. \qquad (17.1)$$

The starting point for a construction of a frame multiresolution analysis is indeed to choose a function $\phi \in L^2(\mathbb{R})$ such that $\{T_k \phi\}_{k \in \mathbb{Z}}$ is a frame sequence. Defining the spaces V_j by (17.1), it follows that $\cap_{j \in \mathbb{Z}} V_j = \{0\}$

without any extra assumption; this is actually a classical result from multiresolution analysis! In [242], for example, it is proved that $\cap_{j\in\mathbb{Z}}V_j = \{0\}$
under the assumption that $\{T_k\phi\}_{k\in\mathbb{Z}}$ is a Riesz sequence, but the first step
in the proof is to observe that then $\{T_k\phi\}_{k\in\mathbb{Z}}$ is a frame sequence, and this
is all that is needed for the argument. A more general result obtained by
deBoor, DeVore, and Ron [71] shows that even the frame condition can be
removed:

Lemma 17.1.2 *Let $\phi \in L^2(\mathbb{R})$ and define the spaces V_j by (17.1). Then*
$\cap_{j\in\mathbb{Z}}V_j = \{0\}$.

With this in mind, it is convenient to formulate a shorter definition of a
frame multiresolution analysis, where the redundancy in Definition 17.1.1
is removed (see also Exercise 17.1):

Definition 17.1.3 *A function $\phi \in L^2(\mathbb{R})$ generates a frame multiresolution analysis if $\{T_k\phi\}_{k\in\mathbb{Z}}$ is a frame sequence and the spaces $\{V_j\}_{j\in\mathbb{Z}}$
defined by (17.1) satisfy the conditions*

(i) $\cdots V_{-1} \subset V_0 \subset V_1 \cdots$.

(ii) $\overline{\cup_{j\in\mathbb{Z}}V_j} = L^2(\mathbb{R})$.

We will consequently refer to this version of the definition.

Two major questions concerning frame multiresolution analysis are:

(i) Under which conditions does a function $\phi \in L^2(\mathbb{R})$ generate a frame
multiresolution analysis?

(ii) If $\phi \in L^2(\mathbb{R})$ generates a frame multiresolution analysis, can we construct a function ψ such that $\{2^{j/2}\psi(2^j x - k)\}_{k\in\mathbb{Z}}$ is a frame for
$L^2(\mathbb{R})$?

It turns out that sufficient conditions for ϕ to generate a frame multiresolution analysis can be found by small modifications of results concerning
"classical" multiresolution analysis, so we will only give a relatively
short description of this part in Section 17.2. The question (ii) is more
complicated, and we will treat it in detail in Section 17.4.

17.2 Sufficient Conditions

In this section we find sufficient conditions for a function $\phi \in L^2(\mathbb{R})$ to
generate a frame multiresolution analysis. We are mainly interested in
the case where $\{T_k\phi\}_{k\in\mathbb{Z}}$ is an overcomplete frame sequence; note that
Proposition 9.3.8 puts some restrictions on ϕ in order for this to happen.

As starting point, we will consider a function ϕ for which $\{T_k\phi\}_{k\in\mathbb{Z}}$ is a frame sequence; recall that Theorem 9.2.5 gives an equivalent condition for this in terms of the function

$$\Phi(\gamma) := \sum_{k\in\mathbb{Z}} |\widehat{\phi}(\gamma+k)|^2.$$

In order for the spaces V_j defined by (17.1) to satisfy (i) and (ii) in Definition 17.1.3, it is natural to follow the approach used in the classical multiresolution analysis. In [242] the density of $\cup V_j$ in $L^2(\mathbb{R})$ is obtained by assuming that

(i) $\{T_k\phi\}_{k\in\mathbb{Z}}$ is a Riesz sequence.

(ii) $\widehat{\phi}$ is bounded and continuous in 0 with $\widehat{\phi}(0) \neq 0$.

The first step in the proof in [242] is to notice that a Riesz sequence is a frame sequence; no special properties for Riesz sequences are needed. Thus, we can replace the word "Riesz sequence" in (i) by "frame sequence." Concerning (ii), we observe that by Theorem 9.2.5, the function $\widehat{\phi}$ is automatically bounded when $\{T_k\phi\}_{k\in\mathbb{Z}}$ is a frame sequence. Also, the condition of continuity of $\widehat{\phi}$ in 0 with a nonvanishing function value can be replaced by

(iii) $|\widehat{\phi}| > 0$ on a neighborhood of zero.

Now we only need a condition ensuring that $V_j \subset V_{j+1}$ for all $j \in \mathbb{Z}$. For this purpose, we first state some properties for V_j.

Lemma 17.2.1 *Let $\phi \in L^2(\mathbb{R})$ and assume that $\{T_k\phi\}_{k\in\mathbb{Z}}$ is a frame sequence with frame bounds A, B. With $V_j, j \in \mathbb{Z}$, defined as in (17.1), the following hold:*

(i) *$\{D^jT_k\phi\}_{k\in\mathbb{Z}}$ is a frame for V_j with frame bounds A, B.*

(ii) *A function $f \in L^2(\mathbb{R})$ belongs to V_j if and only if $f = \sum_{k\in\mathbb{Z}} c_k D^j T_k \phi$ for some $\{c_k\}_{k\in\mathbb{Z}} \in \ell^2(\mathbb{Z})$.*

(iii) *A function $f \in L^2(\mathbb{R})$ belongs to V_j if and only if there exists a 1-periodic function $F \in L^2(\mathbb{T})$ such that*

$$\widehat{f}(2^j\gamma) = F(\gamma)\widehat{\phi}(\gamma). \tag{17.2}$$

If $f \in V_j$, the function F is uniquely determined on all γ for which $\Phi(\gamma) \neq 0$; if $\Phi(\gamma) = 0$, one can choose $F(\gamma) = 0$.

Proof. Since D is unitary, (i) follows from Lemma 5.3.3. (ii) is a consequence of (i) combined with Theorem 5.5.1. For the proof of (iii), let $f \in V_j$; taking the Fourier transform of the expression in (ii) and using the commutator relations, we have (see Lemma 9.2.2)

$$\widehat{f} = \mathcal{F}D^j \sum_{k\in\mathbb{Z}} c_k T_k \phi = D^{-j} \sum_{k\in\mathbb{Z}} c_k E_{-k}\widehat{\phi}.$$

This implies that

$$\widehat{f}(2^j\gamma) = 2^{-j/2}(D^j\widehat{f})(\gamma) = 2^{-j/2}\sum_{k\in\mathbb{Z}}c_k E_{-k}(\gamma)\widehat{\phi}(\gamma).$$

Thus, we have the formula (17.2) with $F(\gamma) = 2^{-j/2}\sum_{k\in\mathbb{Z}}c_k E_{-k}(\gamma)$. On the other hand, if $f \in L^2(\mathbb{R})$ and $F \in L^2(\mathbb{T})$ satisfy (17.2) for some $j \in \mathbb{Z}$, let us denote its Fourier coefficients for F by $\{d_k\}_{k\in\mathbb{Z}}$ and define $c_k = 2^{j/2}d_k$; then $f = \sum_{k\in\mathbb{Z}}c_{-k}D^j T_k\phi \in V_j$.

For the last part of (iii), we note that if $\Phi(\gamma) \neq 0$ for some γ, then there exists $k \in \mathbb{Z}$ such that $\widehat{\phi}(\gamma+k) \neq 0$; since

$$\widehat{f}(2^j(\gamma+k)) = F(\gamma+k)\widehat{\phi}(\gamma+k),$$

and F is assumed to be 1-periodic, it follows that

$$F(\gamma) = \frac{\widehat{f}(2^j(\gamma+k))}{\widehat{\phi}(\gamma+k)}.$$

If $\Phi(\gamma) = 0$ for some γ, then $\widehat{\phi}(\gamma+k) = 0$ for all $k \in \mathbb{Z}$. The equation (17.2) is satisfied no matter how $F(\gamma)$ is defined, but if we want F to be 1-periodic, we must require $F(\gamma+k) = F(\gamma), k \in \mathbb{Z}$. One choice is to take $F(\gamma) = 0$ for all γ for which $\Phi(\gamma) = 0$. $\qquad\square$

Conditions for $V_j \subset V_{j+1}$ are given in the following lemma:

Lemma 17.2.2 *Assume that $\phi \in L^2(\mathbb{R})$ and that $\{T_k\phi\}_{k\in\mathbb{Z}}$ is a frame sequence. Define the spaces V_j by (17.1). Then the following conditions are equivalent:*

(i) $V_j \subset V_{j+1}$ *for all $j \in \mathbb{Z}$.*

(ii) $V_0 \subset V_1$.

(iii) *There exists a 1-periodic function $H_0 \in L^\infty(\mathbb{T})$ such that*

$$\widehat{\phi}(\gamma) = H_0(\gamma/2)\widehat{\phi}(\gamma/2). \tag{17.3}$$

If (17.3) is satisfied, the functions H_0 and Φ are related by

$$\Phi(\gamma) = |H_0(\gamma/2)|^2\Phi(\gamma/2) + |H_0(\gamma/2+1/2)|^2\Phi(\gamma/2+1/2). \tag{17.4}$$

Proof. (iii)$\Rightarrow$(i). Assume that (iii) is satisfied, and let $f \in V_j$. Using Lemma 17.2.1 (iii) and the assumption, there exists a function $F \in L^2(\mathbb{T})$ for which

$$\widehat{f}(2^{j+1}\gamma) = F(2\gamma)\widehat{\phi}(2\gamma) = F(2\gamma)H_0(\gamma)\widehat{\phi}(\gamma).$$

The function $\gamma \mapsto F(2\gamma)H_0(\gamma)$ is 1-periodic and belongs to $L^2(\mathbb{T})$ because $F \in L^2(\mathbb{T})$ and $H_0 \in L^\infty(\mathbb{T})$, so Lemma 17.2.1(iii) shows that $f \in V_{j+1}$.

For (i)$\Rightarrow$(ii) there is nothing to prove. To prove that (ii)$\Rightarrow$(iii), assume that (ii) is satisfied. Then $\phi \in V_0 \subset V_1$, and $D^{-1}\phi \in V_0$. By

Lemma 17.2.1(iii) there exists a 1-periodic function $H_0 \in L^2(\mathbb{T})$ such that $\mathcal{F}D^{-1}\phi(\gamma) = H_0(\gamma)\widehat{\phi}(\gamma)$; since $\mathcal{F}D^{-1}\phi(\gamma) = 2^{1/2}\widehat{\phi}(2\gamma)$, a slight redefining shows the existence of a 1-periodic function $H_0 \in L^2(\mathbb{T})$ such that

$$\widehat{\phi}(2\gamma) = H_0(\gamma)\widehat{\phi}(\gamma). \tag{17.5}$$

Let us choose H_0 such that $H_0(\gamma) = 0$ if $\Phi(\gamma) = 0$. We now prove (17.4) and that H_0 is bounded. First, (17.5) implies that

$$\Phi(\gamma) = \sum_{k \in \mathbb{Z}} |\widehat{\phi}(\gamma + k)|^2 = \sum_{k \in \mathbb{Z}} |H_0(\frac{\gamma + k}{2})\widehat{\phi}(\frac{\gamma + k}{2})|^2.$$

If we split the sum into sums over even integers $2k, k \in \mathbb{Z}$ and odd integers $2k + 1, k \in \mathbb{Z}$, and use the periodicity of H_0, we arrive at

$$\begin{aligned}
\Phi(\gamma) &= \sum_{k \in \mathbb{Z}} |H_0(\frac{\gamma + 2k}{2})\widehat{\phi}(\frac{\gamma + 2k}{2})|^2 \\
&+ \sum_{k \in \mathbb{Z}} |H_0(\frac{\gamma + 2k + 1}{2})\widehat{\phi}(\frac{\gamma + 2k + 1}{2})|^2 \\
&= |H_0(\gamma/2)|^2 \sum_{k \in \mathbb{Z}} |\widehat{\phi}(\gamma/2 + k)|^2 \\
&+ |H_0(\gamma/2 + 1/2)|^2 \sum_{k \in \mathbb{Z}} |\widehat{\phi}(\gamma/2 + 1/2 + k)|^2 \\
&= |H_0(\gamma/2)|^2 \, \Phi(\gamma/2) + |H_0(\gamma/2 + 1/2)|^2 \Phi(\gamma/2 + 1/2).
\end{aligned}$$

This proves (17.4). To show that H_0 is bounded, we consider $\gamma \in \mathbb{R}$ such that $\Phi(\gamma) \neq 0$. Let A, B denote frame bounds for $\{T_k\phi\}_{k \in \mathbb{Z}}$. By Theorem 9.2.5 we have $A \leq \Phi(\gamma) \leq B$, and via (17.4) this gives that

$$B \geq \Phi(2\gamma) \geq |H_0(\gamma)|^2 \, \Phi(\gamma) \geq A \, |H_0(\gamma)|^2.$$

Thus $H_0 \in L^\infty(\mathbb{T})$ as desired.   □

An equation of the type (17.3) is called a *refinement equation*; we say that ϕ is *refinable*. In the case of a classical multiresolution analysis, where $\{T_k\phi\}_{k \in \mathbb{Z}}$ is an orthonormal system or a Riesz sequence, the function H_0 satisfying (17.3) is unique; this follows from Lemma 17.2.1(iii) because Φ is bounded away from zero by Theorem 9.2.5. For a general frame multiresolution analysis, several choices for H_0 might be possible because $\Phi(\gamma)$ might be zero for γ belonging to a set with positive Lebesgue measure. A convenient choice is to define $H_0(\gamma) = 0$ if $\Phi(\gamma) = 0$, as we already did in the proof of Lemma 17.2.2; the function H_0 obtained this way is called the *two-scale symbol* or the *refinement mask* for the frame multiresolution analysis.

By collecting all the obtained information we obtain a sufficient condition for ϕ to generate a frame multiresolution analysis:

Theorem 17.2.3 *Suppose that $\phi \in L^2(\mathbb{R})$, that $\{T_k\phi\}_{k\in\mathbb{Z}}$ is a frame sequence, and that $|\widehat{\phi}| > 0$ on a neighborhood of zero. If there exists a function $H_0 \in L^\infty(\mathbb{T})$ such that*

$$\widehat{\phi}(\gamma) = H_0(\frac{\gamma}{2})\widehat{\phi}(\frac{\gamma}{2}), \tag{17.6}$$

then ϕ generates a frame multiresolution analysis.

The statement of Theorem 17.2.3 corresponds exactly to the formulation of the similar result for multiresolution analysis in [637], Theorem 2.13.

Example 17.2.4 Define as in Example 9.2.7 the function ϕ via its Fourier transform,

$$\widehat{\phi}(\gamma) = \chi_{[-\alpha,\alpha[}, \text{ for some } \alpha \in]0, \frac{1}{2}[.$$

We have already seen that $\{T_k\phi\}_{k\in\mathbb{Z}}$ is a frame sequence. Note that

$$\widehat{\phi}(2\gamma) = \chi_{[-\frac{\alpha}{2},\frac{\alpha}{2}[}(\gamma) = \chi_{[-\frac{\alpha}{2},\frac{\alpha}{2}[}(\gamma)\widehat{\phi}(\gamma).$$

For $|\gamma| < \frac{1}{2}$, let

$$H_0(\gamma) = \chi_{[-\frac{\alpha}{2},\frac{\alpha}{2}[};$$

extending H_0 to a 1-periodic function, we see that (17.6) is satisfied. By Theorem 17.2.3 we conclude that ϕ generates a frame multiresolution analysis.

Given a continuous nonvanishing function θ on $[-\alpha, \alpha]$, we can generalize the example by considering

$$\widehat{\phi}(\gamma) := \theta(\gamma)\chi_{[-\alpha,\alpha[}(\gamma).$$

Defining

$$H_0(\gamma) = \begin{cases} \frac{\theta(2\gamma)}{\theta(\gamma)} & \text{if } \gamma \in [-\frac{\alpha}{2}, \frac{\alpha}{2}[, \\ 0 & \text{if } \gamma \in [-\frac{1}{2}, -\frac{\alpha}{2}[\cup[\frac{\alpha}{2}, \frac{1}{2}[, \end{cases}$$

and extending H_0 periodically, it again follows that $\widehat{\phi}$ generates a frame multiresolution analysis. $\square$

17.3 Relaxing the Conditions

In Theorem 17.2.3 all the conditions in frame multiresolution analysis were derived on the basis of a function ϕ generating a frame sequence $\{T_k\phi\}_{k\in\mathbb{Z}}$. In Chapter 18 we will consider another multiresolution scheme, proposed by Ron and Shen [561], where it is not assumed that $\{T_k\phi\}_{k\in\mathbb{Z}}$ is a frame sequence. For this reason we now show that (i) and (ii) in Definition 17.1.3 can be satisfied without any frame assumption.

The basic idea of Ron and Shen is to replace the frame condition on $\{T_k\phi\}_{k\in\mathbb{Z}}$ with the condition that ϕ satisfies a refinement equation. Recall from Lemma 17.2.2 that if the spaces V_j in (17.1) are nested and $\{T_k\phi\}_{k\in\mathbb{Z}}$ is a frame sequence, then ϕ satisfies a refinement equation. We now prove that a refinement equation is enough to imply that V_j are nested, and Example 17.3.4 will show that nothing guarantees that $\{T_k\phi\}_{k\in\mathbb{Z}}$ is a frame sequence; thus, the idea of Ron and Shen is in fact more general.

Lemma 17.3.1 *Assume that $\phi \in L^2(\mathbb{R})$ and that $\{T_k\phi\}_{k\in\mathbb{Z}}$ is a Bessel sequence. Define the spaces V_j by (17.1). Then the following hold:*

(i) *If $\psi \in L^2(\mathbb{R})$ and there exists a function $F \in L^\infty(\mathbb{T})$ such that $\widehat{\psi}(2\gamma) = F(\gamma)\widehat{\phi}(\gamma)$, then $\psi \in V_1$.*

(ii) *If there exists a function $H_0 \in L^\infty(\mathbb{T})$ such that*

$$\widehat{\phi}(2\gamma) = H_0(\gamma)\widehat{\phi}(\gamma), \tag{17.7}$$

then $V_0 \subseteq V_1$.

Proof.   Let $\psi \in L^2(\mathbb{R})$ and assume that for some $F \in L^\infty(\mathbb{T})$, we have $\widehat{\psi}(2\gamma) = F(\gamma)\widehat{\phi}(\gamma)$. Writing the Fourier series of F as $F = \sum_{k\in\mathbb{Z}} c_k E_k$, we have (see Lemma 9.2.2)

$$\frac{1}{\sqrt{2}}D\widehat{\psi} = F\widehat{\phi} = \sum_{k\in\mathbb{Z}} c_k E_k \widehat{\phi} = \mathcal{F}\sum_{k\in\mathbb{Z}} c_k T_{-k}\phi.$$

Since $D\widehat{\psi} = \mathcal{F}D^{-1}\psi$, this shows that $D^{-1}\psi = \sqrt{2}\sum_{k\in\mathbb{Z}} c_k T_{-k}\phi \in V_0$, i.e., $\psi \in DV_0 = V_1$. This proves (i). (ii) follows from here because V_1 is closed and invariant under integer-translations.   □

The condition (ii) in Definition 17.1.3 can also be derived without assuming ϕ to be a frame sequence. This follows from a result by deBoor, DeVore, and Ron [71]:

Lemma 17.3.2 *Let $\phi \in L^2(\mathbb{R})$ and assume that the spaces V_j in (17.1) are nested. If $|\widehat{\phi}| > 0$ on a neighborhood of 0, then $\cup_j V_j$ is dense in $L^2(\mathbb{R})$.*

Via Lemma 17.3.1 and Lemma 17.3.2, we have:

Theorem 17.3.3 *Let $\phi \in L^2(\mathbb{R})$. Assume that (17.7) is satisfied for a function $H_0 \in L^\infty(\mathbb{T})$ and that $|\widehat{\phi}| > 0$ on a neighborhood of 0. Then the spaces V_j defined in (17.1) satisfy the conditions (i)–(iv) in Definition 17.1.1.*

In this chapter we will always assume that $\{T_k\phi\}_{k\in\mathbb{Z}}$ is a frame sequence, in which case Theorem 17.3.3 equals Theorem 17.2.3. The role of Theorem 17.3.3 will be clear in Chapter 18; for now, we just give an example, which shows that it can actually happen that all the conditions in Definition 17.1.1 except (v) are satisfied.

Example 17.3.4 Assume that $\phi \in L^2(\mathbb{R})$ generates a classical multiresolution analysis for $L^2(\mathbb{R})$, where $\{T_k\phi\}_{k\in\mathbb{Z}}$ is an orthonormal basis for V_0. Let

$$\widetilde{\phi} = \phi + T_1\phi.$$

Then

$$\{T_k\widetilde{\phi}\}_{k\in\mathbb{Z}} = \{T_{k+1}\phi + T_k\phi\}_{k\in\mathbb{Z}}.$$

From Example 5.4.6 we know that $V_0 = \overline{\text{span}}\{T_k\widetilde{\phi}\}_{k\in\mathbb{Z}}$ and that $\{T_k\widetilde{\phi}\}_{k\in\mathbb{Z}}$ is not a frame for V_0. However, all the other conditions for $\widetilde{\phi}$ generating a frame multiresolution analysis are satisfied.

The function $\widetilde{\phi}$ satisfies a refinement equation. In fact, since ϕ generates a multiresolution analysis, it is known from the classical theory or from Lemma 17.2.2 that there exists a function $H_0 \in L^\infty(\mathbb{T})$ such that

$$\widehat{\phi}(2\gamma) = H_0(\gamma)\widehat{\phi}(\gamma);$$

using that

$$\widehat{\widetilde{\phi}} = (1 + E_{-1})\widehat{\phi},$$

it follows that

$$
\begin{aligned}
\widehat{\widetilde{\phi}}(2\gamma) &= (1 + e^{-4\pi i\gamma})H_0(\gamma)\widehat{\phi}(\gamma) \\
&= \frac{1 + e^{-4\pi i\gamma}}{1 + e^{-2\pi i\gamma}} H_0(\gamma)\widehat{\widetilde{\phi}}(\gamma) \\
&=: \widetilde{H_0}(\gamma)\widehat{\widetilde{\phi}}(\gamma), \ \gamma \neq \frac{1}{2} + \mathbb{Z}.
\end{aligned}
$$

The reader can verify that $\widetilde{H_0} \in L^\infty(\mathbb{T})$. $\square$

17.4   Construction of Frames

The next question is whether a frame multiresolution analysis can be used to construct a frame for $L^2(\mathbb{R})$. We prove in Theorem 17.4.5 that an extra condition is needed in order to assure this.

Assume that $\phi \in L^2(\mathbb{R})$ generates a frame multiresolution analysis, and let W_j denote the orthogonal complement of V_j in V_{j+1}. Exactly as in the case of a multiresolution analysis, this gives the orthogonal decomposition

$$L^2(\mathbb{R}) = \bigoplus_{j\in\mathbb{Z}} W_j. \tag{17.8}$$

All we need in order to construct a frame for $L^2(\mathbb{R})$ is a function $\psi \in L^2(\mathbb{R})$ for which $\{T_k\psi\}_{k\in\mathbb{Z}}$ is a frame for W_0. This follows from the observation that the spaces W_j are related by the same dilation property as we have for V_j:

Lemma 17.4.1 *Assume that $\phi \in L^2(\mathbb{R})$ generates a frame multiresolution analysis. Then the following hold:*

(i) $W_j = D^j W_0,\ \forall j \in \mathbb{Z}$.

(ii) *If $\psi \in W_0$ generates a frame $\{T_k\psi\}_{k\in\mathbb{Z}}$ for W_0, then for all $j \in \mathbb{Z}$, the family $\{D^j T_k\psi\}_{k\in\mathbb{Z}}$ is a frame for W_j, and $\{D^j T_k\psi\}_{j,k\in\mathbb{Z}}$ is a frame for $L^2(\mathbb{R})$; these frames have the same frame bounds as $\{T_k\psi\}_{k\in\mathbb{Z}}$.*

Proof. For the proof of (i), let $f \in W_0$. Then $f \in V_1$, so $D^j f \in V_{j+1}$. Furthermore, $f \perp V_0$, so since D^j is unitary, $D^j f \perp D^j V_0 = V_j$. This proves that $D^j W_0 \subseteq W_j$; the proof of $W_j \subseteq D^j W_0$ is similar.

Now assume that $\{T_k\psi\}_{k\in\mathbb{Z}}$ is a frame for W_0 with frame bounds A, B. Then Lemma 5.3.3 shows that $\{D^j T_k\psi\}_{k\in\mathbb{Z}}$ is a frame for

$$\overline{\text{span}}\{D^j T_k\psi\}_{k\in\mathbb{Z}} = D^j W_0 = W_j,$$

also with frame bounds A, B. Let $f \in L^2(\mathbb{R})$. Denoting the orthogonal projection of $L^2(\mathbb{R})$ onto W_j by Q_j, we have by (17.8) that $f = \sum_{j\in\mathbb{Z}} Q_j f$ and

$$||f||^2 = \sum_{j\in\mathbb{Z}} ||Q_j f||^2.$$

The last part of the proof follows from this combined with the observation that

$$A\,||Q_j f||^2 \le \sum_{k\in\mathbb{Z}} |\langle Q_j f, D^j T_k\psi\rangle|^2 = \sum_{k\in\mathbb{Z}} |\langle f, D^j T_k\psi\rangle|^2 \le B\,||Q_j f||^2. \qquad \square$$

In the classical case where ϕ generates a multiresolution analysis, we know that there always exists a function $\psi \in W_0$ such that $\{T_k\psi\}_{k\in\mathbb{Z}}$ is an orthonormal basis for W_0 and $\{\psi_{j,k}\}_{j,k\in\mathbb{Z}}$ is an orthonormal basis for $L^2(\mathbb{R})$. The corresponding result for a frame multiresolution analysis is more complicated: there might not exist a function $\psi \in L^2(\mathbb{R})$ for which $\{T_k\psi\}_{k\in\mathbb{Z}}$ is a frame for W_0. Equivalent conditions for the existence of such a function were found by Benedetto and Treiber [57]; we need some preparation before we present their result in Theorem 17.4.5.

The first step is to characterize the space W_0. In Lemma 17.2.1, we have seen that if $\{T_k\phi\}_{k\in\mathbb{Z}}$ is a frame sequence and $F \in L^2(\mathbb{T})$, then the function f defined by $\hat{f}(2\gamma) = F(\gamma)\hat{\phi}(\gamma)$ belongs to V_1. An extra condition implies that $f \in W_0$.

Lemma 17.4.2 *Assume that $\phi \in L^2(\mathbb{R})$ generates a frame multiresolution analysis with two-scale symbol $H_0 \in L^\infty(\mathbb{T})$. Let $F \in L^2(\mathbb{T})$ and define $f \in V_1$ by $\hat{f}(2\gamma) = F(\gamma)\hat{\phi}(\gamma)$. Then the following hold:*

(i) $\langle f, T_k\phi \rangle = 2 \int_0^{\frac{1}{2}} \left[F\Phi^2 \overline{H_0} + T_{1/2}\left(F\Phi^2\overline{H_0}\right) \right] E_{2k}.$

(ii) $f \in W_0$ *if and only if*

$$\overline{H_0}F\Phi + T_{1/2}(\overline{H_0}F\Phi) = 0 \ \ on \ \left[0, \frac{1}{2}\right[. \tag{17.9}$$

Proof. For $k \in \mathbb{Z}$ we use the Fourier transform and (17.3) to obtain that

$$
\begin{aligned}
\langle f, T_k\phi \rangle &= \langle \hat{f}, E_{-k}\hat{\phi} \rangle \\
&= \langle F(\cdot/2)\hat{\phi}(\cdot/2), E_{-k}(\cdot)H_0(\cdot/2)\hat{\phi}(\cdot/2) \rangle \\
&= \int_{-\infty}^{\infty} F(\gamma/2)\hat{\phi}(\gamma/2)\overline{e^{-2\pi i k\gamma}H_0(\gamma/2)\hat{\phi}(\gamma/2)}d\gamma \\
&= 2\int_{-\infty}^{\infty} F(\gamma)|\hat{\phi}(\gamma)|^2 \overline{H_0(\gamma)}e^{4\pi i k\gamma}d\gamma.
\end{aligned}
$$

The function $\gamma \mapsto F(\gamma)\overline{H_0(\gamma)}e^{4\pi i k\gamma}$ is 1-periodic, so (we ask the reader to justify the rearrangements)

$$
\begin{aligned}
\langle f, T_k\phi \rangle &= 2\int_{-\frac{1}{2}}^{\frac{1}{2}} \sum_{n\in\mathbb{Z}} \left(F(\gamma+n)|\hat{\phi}(\gamma+n)|^2\overline{H_0(\gamma+n)}e^{4\pi i k(\gamma+n)} \right) d\gamma \\
&= 2\int_{-\frac{1}{2}}^{\frac{1}{2}} F(\gamma)|\Phi(\gamma)|^2\overline{H_0(\gamma)}e^{4\pi i k\gamma}d\gamma.
\end{aligned}
$$

Splitting the integral in two, we can continue with

$$
\begin{aligned}
\langle f, T_k\phi \rangle &= 2\int_0^{\frac{1}{2}} F(\gamma)|\Phi(\gamma)|^2\overline{H_0(\gamma)}e^{4\pi i k\gamma}d\gamma \\
&\quad + 2\int_0^{\frac{1}{2}} F(\gamma-1/2)|\Phi(\gamma-1/2)|^2\overline{H_0(\gamma-1/2)}e^{4\pi i k(\gamma-1/2)}d\gamma \\
&= 2\int_0^{\frac{1}{2}} \left[F(\gamma)|\Phi(\gamma)|^2\overline{H_0(\gamma)} + T_{1/2}\left(F(\gamma)|\Phi(\gamma)|^2\overline{H_0(\gamma)}\right) \right] E_{2k}(\gamma)d\gamma.
\end{aligned}
$$

This proves (i). By definition, $f \in V_1$, so in order to prove (ii), we have to prove that f is orthogonal to $V_0 = \overline{\text{span}}\{T_k\phi\}_{k\in\mathbb{Z}}$ if and only if (17.9) is satisfied, but this follows from $\{\sqrt{2}E_{2k}\}_{k\in\mathbb{Z}}$ being an orthonormal basis for $L^2(0, \frac{1}{2})$. $\qquad\square$

Note that the function $\overline{H_0}F\Phi + T_{1/2}(\overline{H_0}F\Phi)$ is $\frac{1}{2}$-periodic. Thus, if (17.9) holds on $[0, 1/2[$, then it holds on $\mathbb{R}$.

We can now give a condition for the existence of a frame $\{T_k\psi\}_{k\in\mathbb{Z}}$ for W_0. As standing notation in the rest of the chapter, we let $F \in L^2(\mathbb{T})$ and

$\psi \in V_1$ be related by

$$\widehat{\psi}(2\gamma) = F(\gamma)\widehat{\phi}(\gamma). \tag{17.10}$$

Also, let

$$\Psi(\gamma) = \sum_{k \in \mathbb{Z}} |\widehat{\psi}(\gamma + k)|^2.$$

Proposition 17.4.3 *Assume that $\phi \in L^2(\mathbb{R})$ generates a frame multiresolution analysis with two-scale symbol $H_0 \in L^\infty(\mathbb{T})$. Let $F \in L^\infty(\mathbb{T})$ and define $\psi \in V_1$ by $\widehat{\psi}(2\gamma) = F(\gamma)\widehat{\phi}(\gamma)$. If there exist $G_0, G_1 \in L^\infty(\mathbb{T})$ such that the three equations*

$$\overline{H_0}F\Phi + T_{1/2}(\overline{H_0}F\Phi) = 0, \tag{17.11}$$
$$H_0 G_0 \Phi + F G_1 \Phi = \Phi, \tag{17.12}$$
$$T_{1/2}(H_0\Phi)G_0 + T_{1/2}(F\Phi)G_1 = 0, \tag{17.13}$$

are satisfied on $\mathbb{T}$, then $\{T_k\psi\}_{k\in\mathbb{Z}}$ is a frame for W_0.

Proof. The first step is to show that $\{T_k\psi\}_{k\in\mathbb{Z}}$ is a frame sequence. Let A, B denote frame bounds for $\{T_k\phi\}_{k\in\mathbb{Z}}$. We note that an argument similar to the proof of (17.4) applies to Ψ; it gives that

$$\Psi(2\gamma) = |F(\gamma)|^2 \Phi(\gamma) + |F(\gamma + 1/2)|^2 \Phi(\gamma + 1/2). \tag{17.14}$$

We want to apply Theorem 9.2.5, so we have to show that outside its zero set, the function Ψ is bounded away from zero and above. Since we have assumed that F is bounded, it immediately follows from (17.14) and Theorem 9.2.5 that Ψ is bounded above. In order to prove that Ψ is bounded below, it is enough to estimate $\Psi(2\gamma)$ for $\gamma \in [0, \frac{1}{2}[$. We examine four cases separately, so let us define

$$\mathbb{T}_1 := \{\gamma \in [0, \frac{1}{2}[\mid \Phi(\gamma) = 0, \ T_{1/2}\Phi(\gamma) = 0\}.$$

$$\mathbb{T}_2 := \{\gamma \in [0, \frac{1}{2}[\mid \Phi(\gamma) > 0, \ T_{1/2}\Phi(\gamma) > 0\}.$$

$$\mathbb{T}_3 := \{\gamma \in [0, \frac{1}{2}[\mid \Phi(\gamma) > 0, \ T_{1/2}\Phi(\gamma) = 0\}.$$

$$\mathbb{T}_4 := \{\gamma \in [0, \frac{1}{2}[\mid \Phi(\gamma) = 0, \ T_{1/2}\Phi(\gamma) > 0\}.$$

If $\gamma \in \mathbb{T}_1$, then $\Psi(2\gamma) = 0$, so this case is okay. If $\gamma \in \mathbb{T}_2$, it follows by (17.14) that

$$\Psi(2\gamma) \geq A\left(|F(\gamma)|^2 + |F(\gamma + 1/2)|^2\right). \tag{17.15}$$

Furthermore, we see by (17.12) that the two equations

$$H_0(\gamma)G_0(\gamma) + F(\gamma)G_1(\gamma) = 1, \tag{17.16}$$
$$T_{1/2}(H_0 G_0)(\gamma) + T_{1/2}(F G_1)(\gamma) = 1, \tag{17.17}$$

hold.

Now assume that for some $\epsilon \in]0, (1 + ||H_0||_\infty)^{-1}[$ and some $\gamma \in \mathbb{T}_2$, we have

$$|T_{1/2}F(\gamma)| \leq \frac{\epsilon^2}{1 + ||G_1||_\infty}. \tag{17.18}$$

Then $|(T_{1/2}F)(\gamma)G_1(\gamma)| \leq \epsilon^2$, and (17.13) implies that

$$|(T_{1/2}H_0)(\gamma)G_0(\gamma)| \leq \epsilon^2. \tag{17.19}$$

Therefore, at least one of the following two options holds:

(i) $|T_{1/2}H_0(\gamma)| \leq \epsilon$;

(ii) $|G_0(\gamma)| \leq \epsilon$.

We will use (17.15) to obtain a lower bound for $\Psi(2\gamma)$ in each of these cases separately. In case (i), (17.17) gives

$$|T_{1/2}(FG_1)(\gamma)| = \left|1 - (T_{1/2}H_0(\gamma))(T_{1/2}G_0(\gamma))\right| \geq 1 - \epsilon||G_0||_\infty,$$

and therefore

$$|T_{1/2}F(\gamma)| \geq \frac{1 - \epsilon||G_0||_\infty}{1 + ||G_1||_\infty}. \tag{17.20}$$

The equations (17.18) and (17.20) give a contradiction if

$$\frac{1 - \epsilon||G_0||_\infty}{1 + ||G_1||_\infty} > \frac{\epsilon^2}{1 + ||G_1||_\infty},$$

i.e., if

$$\epsilon < \frac{-||G_0||_\infty + \sqrt{||G_0||_\infty^2 + 4}}{2}.$$

Thus, in the case (i) the inequality (17.18) shows that we have a lower bound on $|T_{1/2}F|$; by (17.15) this gives a lower bound on $\Psi(2\gamma)$. In the case (ii), we apply (17.16) to get

$$|F(\gamma)G_1(\gamma)| = |1 - H_0(\gamma)G_0(\gamma)| \geq 1 - |H_0(\gamma)|\,|G_0(\gamma)| \geq 1 - \epsilon||H_0||_\infty,$$

which by the choice of ϵ implies that

$$|F(\gamma)| \geq \frac{1 - \epsilon||H_0||_\infty}{1 + ||G_1||_\infty} > \frac{1}{1 + ||H_0||_\infty}\frac{1}{1 + ||G_1||_\infty}.$$

Thus, $|F(\gamma)|$ is bounded below in case (ii), and again we conclude via (17.15) that $\Psi(2\gamma)$ is bounded below.

If $\gamma \in \mathbb{T}_3$, then

$$\Psi(2\gamma) = |F(\gamma)|^2\Phi(\gamma).$$

By (17.12) and (17.11), we have

$$H_0(\gamma)G_0(\gamma) + F(\gamma)G_1(\gamma) = 1, \quad \overline{H_0(\gamma)}F(\gamma) = 0. \tag{17.21}$$

The last equation is satisfied if $F(\gamma) = 0$ (leading to $\Psi(2\gamma) = 0$) or if $H_0(\gamma) = 0$; in the latter case, the first equation in (17.21) gives $F(\gamma)G_1(\gamma) = 1$, and therefore $|F(\gamma)| \geq \frac{1}{||G_1||_\infty}$. Thus,

$$\Psi(2\gamma) \geq \frac{A}{||G_1||_\infty^2},$$

as desired.

Now let $\gamma \in \mathbb{T}_4$. Then $\Psi(2\gamma) = |F(\gamma + 1/2)|^2 \Phi(\gamma + 1/2)$. Translating (17.12), we obtain that

$$T_{1/2}(H_0 G_0 \Phi)(\gamma) + T_{1/2}(F G_1 \Phi)(\gamma) = T_{1/2}\Phi(\gamma). \qquad (17.22)$$

By (17.11) we have $T_{1/2}(\overline{H_0}F)(\gamma) = 0$. If $T_{1/2}F(\gamma) = 0$ we have $\Psi(2\gamma) = 0$; otherwise, $T_{1/2}H_0(\gamma) = 0$, and (17.22) gives that $T_{1/2}(FG_1)(\gamma) = 1$. As in the previous case, this leads to

$$\Psi(2\gamma) \geq \frac{A}{||G_1||_\infty^2}.$$

This completes the analysis of the four separate cases: now we know that Ψ is bounded away from zero outside the set where it is equal to zero, so $\{T_k\psi\}_{k\in\mathbb{Z}}$ is a frame sequence. The rest of the proof will show that $\overline{\text{span}}\{T_k\psi\}_{k\in\mathbb{Z}} = W_0$. For this purpose we first rewrite the equations (17.12) and (17.13). If $\widehat{\phi}(\gamma) \neq 0$, then $\Phi(\gamma) \neq 0$, and we can multiply (17.12) with $\frac{\widehat{\phi}(\gamma)}{\Phi(\gamma)}$; the obtained equation clearly also holds if $\widehat{\phi}(\gamma) = 0$, i.e., we have

$$H_0(\gamma)G_0(\gamma)\widehat{\phi}(\gamma) + F(\gamma)G_1(\gamma)\widehat{\phi}(\gamma) = \widehat{\phi}(\gamma), \ \gamma \in \mathbb{R}. \qquad (17.23)$$

We can rewrite (17.13) in the same way to obtain

$$T_{1/2}(H_0\phi)(\gamma)G_0(\gamma) + T_{1/2}(F\phi)(\gamma)G_1(\gamma) = 0, \ \gamma \in \mathbb{R};$$

applying the operator $T_{-1/2}$ this can also be written

$$H_0(\gamma)\phi(\gamma)G_0(\gamma + 1/2) + F(\gamma)\widehat{\phi}(\gamma)G_1(\gamma + 1/2) = 0, \ \gamma \in \mathbb{R}. \qquad (17.24)$$

Let $\{c_k\}_{k\in\mathbb{Z}}$ and $\{d_k\}_{k\in\mathbb{Z}}$ denote the Fourier coefficients for G_0 and G_1; then

$$G_0(\gamma) = \sum_{k\in\mathbb{Z}} c_k e^{2\pi ik\gamma}, \quad G_1(\gamma) = \sum_{k\in\mathbb{Z}} d_k e^{2\pi ik\gamma}.$$

Note that

$$G_0(\gamma + 1/2) = \sum_{k\in\mathbb{Z}} c_k e^{2\pi ik(\gamma+1/2)} = \sum_{k\in\mathbb{Z}} c_k(-1)^k e^{2\pi ik\gamma}$$

with a similar calculation valid for G_1. Inserting these expressions in (17.23) and (17.24) and recalling (17.10) and that $H_0(\gamma)\widehat{\phi}(\gamma) = \widehat{\phi}(2\gamma)$ by (17.3), we obtain the equations

$$\widehat{\phi}(\gamma) \;=\; \widehat{\phi}(2\gamma)\sum_{k\in\mathbb{Z}} c_k e^{2\pi i k\gamma} + \widehat{\psi}(2\gamma)\sum_{k\in\mathbb{Z}} d_k e^{2\pi i k\gamma}$$

and

$$0 \;=\; \widehat{\phi}(2\gamma)\sum_{k\in\mathbb{Z}} c_k(-1)^k e^{2\pi i k\gamma} + \widehat{\psi}(2\gamma)\sum_{k\in\mathbb{Z}} d_k(-1)^k e^{2\pi i k\gamma}.$$

By addition (respectively subtraction), it follows that

$$\widehat{\phi}(\gamma) \;=\; 2\widehat{\phi}(2\gamma)\sum_{k\in\mathbb{Z}} c_{2k} e^{2\pi i 2k\gamma} + 2\widehat{\psi}(2\gamma)\sum_{k\in\mathbb{Z}} d_{2k} e^{2\pi i 2k\gamma}$$

and

$$\widehat{\phi}(\gamma) \;=\; 2\widehat{\phi}(2\gamma)\sum_{k\in\mathbb{Z}} c_{2k+1} e^{2\pi i(2k+1)\gamma} + 2\widehat{\psi}(2\gamma)\sum_{k\in\mathbb{Z}} d_{2k+1} e^{2\pi i(2k+1)\gamma};$$

these two equations imply that

$$\widehat{\phi}(\gamma) \;=\; 2\widehat{\phi}(2\gamma)\sum_{k\in\mathbb{Z}} c_{2k+n} e^{2\pi i(2k+n)\gamma}$$
$$+ 2\widehat{\psi}(2\gamma)\sum_{k\in\mathbb{Z}} d_{2k+n} e^{2\pi i(2k+n)\gamma},\ \forall n\in\mathbb{Z}.$$

We can rewrite this as

$$\frac{1}{2}\widehat{\phi}(\gamma/2)e^{-2\pi i n\gamma/2} = \widehat{\phi}(\gamma)\sum_{k\in\mathbb{Z}} c_{2k+n} e^{2\pi i k\gamma} + \widehat{\psi}(\gamma)\sum_{k\in\mathbb{Z}} d_{2k+n} e^{2\pi i k\gamma},$$

or, in operator notation (see Lemma 9.2.2),

$$\frac{1}{\sqrt{2}}D^{-1}\left(E_{-n}\widehat{\phi}\right) = \sum_{k\in\mathbb{Z}} c_{2k+n} E_k\widehat{\phi} + \sum_{k\in\mathbb{Z}} d_{2k+n} E_k\widehat{\psi}. \tag{17.25}$$

Using the commutator relations for the Fourier transformation and the operators E_k, D,

$$\frac{1}{\sqrt{2}}D^{-1}(E_{-n}\widehat{\phi}) = \frac{1}{\sqrt{2}}D^{-1}\mathcal{F}T_n\phi = \frac{1}{\sqrt{2}}\mathcal{F}DT_n\phi;$$

thus, applying the inverse Fourier transform to (17.25) gives that

$$DT_n\phi = \sqrt{2}\sum_{k\in\mathbb{Z}} c_{2k+n} T_{-k}\phi + \sqrt{2}\sum_{k\in\mathbb{Z}} d_{2k+n} T_{-k}\psi. \tag{17.26}$$

Note that the two terms on the right-hand side are orthogonal: the first belongs to V_0, while the second belongs to W_0 by Lemma 17.4.2. Now let $f\in W_0$. Since $W_0\subset V_1$ and $\{DT_k\phi\}_{k\in\mathbb{Z}}$ is a frame for V_1, we can for each given $\epsilon>0$ find a finite sequence $\{b_n\}_{n\in\mathcal{F}}$ such that

$$\left\|\sum_{n\in\mathcal{F}} b_n DT_n\phi - f\right\|^2 \le \epsilon.$$

Via (17.26), this implies that

$$\left\|\sqrt{2}\sum_{n\in\mathcal{F}}b_n\sum_{k\in\mathbb{Z}}c_{2k+n}T_{-k}\phi\right\|^2 + \left\|\sqrt{2}\sum_{n\in\mathcal{F}}b_n\sum_{k\in\mathbb{Z}}d_{2k+n}T_{-k}\psi - f\right\|^2 \le \epsilon.$$

Thus, also

$$\left\|\sqrt{2}\sum_{n\in\mathcal{F}}b_n\sum_{k\in\mathbb{Z}}d_{2k+n}T_{-k}\psi - f\right\|^2 \le \epsilon,$$

and we conclude that $\overline{\text{span}}\{T_k\psi\}_{k\in\mathbb{Z}}$ is dense in W_0. Since we already know that $\{T_k\psi\}_{k\in\mathbb{Z}}$ is a frame sequence, it follows that $\{T_k\psi\}_{k\in\mathbb{Z}}$ is a frame for W_0.   □

We now need to find conditions such that the three equations in Proposition 17.4.3 can be solved. It turns out that the set

$$\Gamma := \{\gamma \in \mathbb{T} \mid \Phi(2\gamma) = 0, \Phi(\gamma) > 0, \Phi(\gamma + 1/2) > 0\} \qquad (17.27)$$

will play the key role. Recall that if A is a lower frame bound for $\{T_k\phi\}_{k\in\mathbb{Z}}$, Theorem 9.2.5 shows that for all $\gamma \in \Gamma$,

$$\Phi(\gamma) \ge A, \quad \Phi(\gamma + 1/2) \ge A.$$

In case Γ has positive Lebesgue measure, we can define some functions in W_0 via their Fourier transforms:

Lemma 17.4.4 *Assume that $\phi \in L^2(\mathbb{R})$ generates a frame multiresolution analysis with two-scale symbol H_0. Assume that the set Γ in (17.27) has positive Lebesgue measure and define $F_1, F_2 \in L^\infty(\mathbb{T})$ by*

$$F_1(\gamma) = \chi_\Gamma(\gamma), \quad F_2(\gamma) = \chi_{\Gamma\cap[0,\frac{1}{2}[}(\gamma) - \chi_{\Gamma\cap[-\frac{1}{2},0[}(\gamma), \quad \gamma \in \left[-\frac{1}{2}, \frac{1}{2}\right[.$$

Then the functions f_1, f_2 defined by

$$\widehat{f_i}(2\gamma) = F_i(\gamma)\widehat{\phi}(\gamma), \quad i = 1, 2, \qquad (17.28)$$

belong to W_0.

Proof. Lemma 17.2.1(iii) shows that $f_i \in V_1$, $i = 1, 2$. Furthermore, they are not identically zero; in fact, if $\gamma \in \Gamma$, then

$$\sum_{k\in\mathbb{Z}}|\widehat{f_i}(2\gamma + 2k)|^2 = \sum_{k\in\mathbb{Z}}|F_i(\gamma + k)\widehat{\phi}(\gamma + k)|^2$$

$$= |F_i(\gamma)|^2\Phi(\gamma)$$

$$\ge A,$$

where A is a lower frame bound for $\{T_k\phi\}_{k\in\mathbb{Z}}$. To prove that the functions are orthogonal to $V_0 = \overline{\text{span}}\{T_k\phi\}_{k\in\mathbb{Z}}$, let $k \in \mathbb{Z}$; then, by Lemma 17.4.2,

$$\langle f_i, T_k\phi\rangle$$
$$= 2\int_0^{\frac{1}{2}} \left[F_i(\gamma)|\Phi(\gamma)|^2\overline{H_0(\gamma)} + T_{1/2}\left(F_i(\gamma)|\Phi(\gamma)|^2\overline{H_0(\gamma)}\right)\right] E_{2k}(\gamma)d\gamma.$$

For $\gamma \in \Gamma$, (17.4) shows that

$$0 = \Phi(2\gamma) \ge |H_0(\gamma)|^2\Phi(\gamma),$$

which implies that $H_0(\gamma) = 0$. For $\gamma \in [-\frac{1}{2}, \frac{1}{2}[\backslash\Gamma$, we have

$$F_1(\gamma) = F_2(\gamma) = 0.$$

It follows that

$$\langle f_i, T_k\phi\rangle \quad = \quad 0, \ \forall k \in \mathbb{Z},$$

i.e., $f_1, f_2 \in W_0$.   □

The three equations in Proposition 17.4.3 will be the key to the next step: if they have solutions $F, G_1, G_2 \in L^\infty(\mathbb{T})$, then there exists a function $\psi \in W_0$ such that $\{T_k\psi\}_{k\in\mathbb{Z}}$ is a frame for W_0 . In contrast to the case of a classical multiresolution analysis, the equations cannot always be solved:

Theorem 17.4.5 *Assume that $\phi \in L^2(\mathbb{R})$ generates a frame multiresolution analysis, and let*

$$\Gamma = \{\gamma \in \mathbb{T}: \ \Phi(2\gamma) = 0, \Phi(\gamma) > 0, \Phi(\gamma + 1/2) > 0\}.$$

Then the following hold:

(i) *If Γ has positive Lebesgue measure, there does not exist a function $\psi \in W_0$ such that $\{T_k\psi\}_{k\in\mathbb{Z}}$ is a frame for W_0.*

(ii) *If Γ has vanishing Lebesgue measure, then there exists a function $\psi \in W_0$ such that $\{T_k\psi\}_{k\in\mathbb{Z}}$ is a frame for W_0, and $\{D^jT_k\psi\}_{j,k\in\mathbb{Z}}$ is a frame for $L^2(\mathbb{R})$.*

Proof. The proof of (i) is by contradiction, so we assume that $|\Gamma| > 0$ and that there exists a function $\psi \in W_0$ such that $\{T_k\psi\}_{k\in\mathbb{Z}}$ is a frame for W_0. We can now apply Lemma 17.2.1 (iii) (with $j = 0$) on the frame $\{T_k\psi\}_{k\in\mathbb{Z}}$ for W_0 and the functions $f_1, f_2 \in W_0$ defined in Lemma 17.4.4; thus, we obtain the existence of functions $C_1, C_2 \in L^2(\mathbb{T})$ such that

$$\widehat{f_i}(\gamma) = C_i(\gamma)\widehat{\psi}(\gamma), \ i = 1, 2. \tag{17.29}$$

Since $\psi \in W_0 \subset V_1$, we can apply Lemma 17.2.1 again, this time on the frame $\{T_k\phi\}_{k\in\mathbb{Z}}$ for V_0 and with $j = 1$; we obtain the existence of a function $F \in L^2(\mathbb{T})$ for which

$$\widehat{\psi}(2\gamma) = F(\gamma)\widehat{\phi}(\gamma). \tag{17.30}$$

Combining (17.29) and (17.30) gives

$$\widehat{f}_i(2\gamma) = C_i(2\gamma)\widehat{\psi}(2\gamma) = C_i(2\gamma)F(\gamma)\widehat{\phi}(\gamma), \ i = 1, 2. \tag{17.31}$$

For $\gamma \in \Gamma$ we know that $\Phi(\gamma) > 0$. It follows that there exists $k \in \mathbb{Z}$ such that $\widehat{\phi}(\gamma + k) \neq 0$. Via (17.31) and the definition of $\widehat{f}_i$ in terms of F_i in (17.28) we obtain that

$$C_i(2(\gamma + k))F(\gamma + k)\widehat{\phi}(\gamma + k) = F_i(\gamma + k)\widehat{\phi}(\gamma + k),$$

which implies that

$$F_i(\gamma) = C_i(2\gamma)F(\gamma), \ \gamma \in \Gamma, \ i = 1, 2. \tag{17.32}$$

We will show that (17.32) leads to a contradiction. For this purpose, we first note that

$$\gamma \in \Gamma \cap]0, \frac{1}{2}[\Leftrightarrow \gamma - \frac{1}{2} \in \Gamma \cap]-\frac{1}{2}, 0[.$$

Since we have assumed that Γ has positive measure, also $\Gamma \cap]0, \frac{1}{2}[$ has positive measure. Let $\gamma \in \Gamma \cap]0, \frac{1}{2}[$. Then $F_1(\gamma) = F_1(\gamma + 1/2) = 1$, so by (17.32) and the 1-periodicity of C_i we have the equations

$$C_1(2\gamma)F(\gamma) = 1, \quad C_1(2\gamma)F(\gamma + 1/2) = 1.$$

It follows that

$$0 \neq F(\gamma) = F(\gamma + 1/2), \ \gamma \in \Gamma \cap]0, \frac{1}{2}[. \tag{17.33}$$

We now show that a different result is obtained by looking at the function F_2. If $\gamma \in \Gamma \cap [0, 1/2[$, then $F_2(\gamma) = 1, F_2(\gamma - 1/2) = -1$. So again via (17.32),

$$C_2(2\gamma)F(\gamma) = 1, \quad C_2(2\gamma)F(\gamma - 1/2) = -1.$$

Adding those equations gives

$$F(\gamma) = -F(\gamma + 1/2), \ \gamma \in \Gamma \cap]0, \frac{1}{2}[. \tag{17.34}$$

The equations (17.33) and (17.34) give a contradiction. Thus, if $|\Gamma| > 0$, there does not exist a function $\psi \in W_0$ such that $\{T_k\psi\}_{k \in \mathbb{Z}}$ is a frame for W_0. This concludes the proof of (i).

For the proof of (ii), we now assume that Γ is a null set. We want to use Proposition 17.4.3 and find bounded 1-periodic functions F, G_0 and G_1 such that the equations (17.11), (17.12), and (17.13) are satisfied. For simplicity of the formulation, we only define the functions on $\mathbb{T}$, with the understanding that we extend them periodically. For convenience we restate the three key equations here:

$$\overline{H_0}F\Phi + T_{1/2}(\overline{H_0}F\Phi) = 0, \tag{13.11}$$

$$H_0G_0\Phi + FG_1\Phi = \Phi, \tag{13.12}$$

$$T_{1/2}(H_0\Phi)G_0 + T_{1/2}(F\Phi)G_1 = 0. \tag{13.13}$$

We split the set $\mathbb{T}$ into four sets which we examine separately, as we already did in the proof of Proposition 17.4.3:

$$
\begin{aligned}
\mathbb{T}_1 &:= \{\gamma \in \mathbb{T} \mid \Phi(\gamma) = 0,\ T_{1/2}\Phi(\gamma) = 0\}. \\
\mathbb{T}_2 &:= \{\gamma \in \mathbb{T} \mid \Phi(\gamma) > 0,\ T_{1/2}\Phi(\gamma) > 0\}. \\
\mathbb{T}_3 &:= \{\gamma \in \mathbb{T} \mid \Phi(\gamma) > 0,\ T_{1/2}\Phi(\gamma) = 0\}. \\
\mathbb{T}_4 &:= \{\gamma \in \mathbb{T} \mid \Phi(\gamma) = 0,\ T_{1/2}\Phi(\gamma) > 0\}.
\end{aligned}
$$

In the entire proof, we ignore null sets. When needed, we let as usual A, B denote frame bounds for $\{T_k\phi\}_{k\in\mathbb{Z}}$. First, we consider $\gamma \in \mathbb{T}_1$. Then the equations (17.11), (17.12), and (17.13) hold for all choices of F, G_0 and G_1, so we can define them to be arbitrary bounded functions on $\mathbb{T}_1$; in particular, we can let

$$F(\gamma) = G_0(\gamma) = G_1(\gamma) = 0,\ \gamma \in \mathbb{T}_1. \tag{17.35}$$

Now, let $\gamma \in \mathbb{T}_2$. Since Γ is a null set, we have $\Phi(2\gamma) \neq 0$, which by Theorem 9.2.5 implies that $\Phi(2\gamma) \geq A$. Therefore, via (17.4),

$$
\begin{aligned}
A &\leq \Phi(2\gamma) \\
&= |H_0(\gamma)|^2\Phi(\gamma) + |H_0(\gamma + 1/2)|^2\Phi(\gamma + 1/2) \\
&\leq \left(|H_0(\gamma)|^2 + |H_0(\gamma + 1/2)|^2\right)B,
\end{aligned}
$$

and

$$
\begin{aligned}
&|H_0(\gamma)|^2 + |H_0(\gamma + 1/2)|^2 \\
&\leq \frac{1}{A}\left(|H_0(\gamma)|^2\Phi(\gamma) + |H_0(\gamma + 1/2)|^2\Phi(\gamma + 1/2)\right) \\
&= \frac{\Phi(2\gamma)}{A} \\
&\leq \frac{B}{A}.
\end{aligned}
$$

Altogether this shows that

$$\frac{A}{B} \leq |H_0(\gamma)|^2 + |H_0(\gamma + 1/2)|^2 \leq \frac{B}{A}. \tag{17.36}$$

In order to reduce the number of unknown functions in our three equations, we now define the function F on $\mathbb{T}_2$ by

$$F(\gamma) := T_{1/2}(\overline{H_0\Phi})(\gamma)E_{-1}(\gamma),\ \gamma \in \mathbb{T}_2. \tag{17.37}$$

As a product of bounded functions, it is clear that F is bounded. With our choice of F, the equation (17.11) is automatically satisfied. In fact, observing that

$$T_{1/2}E_{-1}(\gamma) = e^{-2\pi i(\gamma - \frac{1}{2})} = -E_{-1}(\gamma),$$

we have

$$
\begin{aligned}
& \overline{H_0}F\Phi + T_{1/2}(\overline{H_0}F\Phi) \\
&= \overline{H_0}T_{1/2}(\overline{H_0}\Phi)E_{-1}\Phi + T_{1/2}\left(\overline{H_0}T_{1/2}(\overline{H_0}\Phi)E_{-1}\Phi\right) \\
&= \overline{H_0}T_{1/2}(\overline{H_0}\Phi)E_{-1}\Phi - \left(T_{1/2}\overline{H_0}\right)\overline{H_0}\Phi E_{-1}T_{1/2}\Phi \\
&= 0.
\end{aligned}
$$

The equations (17.12) and (17.13) are now two linear equations in G_0 and G_1; the determinant of the equation system is

$$
\begin{aligned}
\Delta &= H_0\Phi T_{1/2}(F\Phi) - T_{1/2}(H_0\Phi)F\Phi \\
&= H_0\Phi T_{1/2}\left(T_{1/2}(\overline{H_0}\Phi)E_{-1}\Phi\right) - T_{1/2}(H_0\Phi)T_{1/2}(\overline{H_0}\Phi)E_{-1}\Phi \\
&= -H_0\Phi\overline{H_0}\Phi E_{-1}T_{1/2}\Phi - T_{1/2}(H_0\Phi)T_{1/2}(\overline{H_0}\Phi)E_{-1}\Phi \\
&= -\Phi T_{1/2}\Phi\left(|H_0|^2\Phi + T_{1/2}(|H_0|^2\Phi)\right)E_{-1}.
\end{aligned}
$$

By (17.36) and the fact that $\Phi \geq A, T_{1/2}\Phi \geq A$ on $\mathbb{T}_2$,

$$
|\Delta| \geq A^3(|H_0|^2 + T_{1/2}|H_0|^2) \geq \frac{A^4}{B} > 0.
$$

Thus, the set of equations (17.12) and (17.13) has a unique solution. Via Cramer's rule,

$$
G_0 = \frac{\begin{vmatrix} \Phi & F\Phi \\ 0 & T_{1/2}(F\Phi) \end{vmatrix}}{\Delta} = \frac{\Phi T_{1/2}(F\Phi)}{\Delta},
$$

so

$$
|G_0(\gamma)| \leq \frac{B^2\|F\|_\infty}{A^4/B} = \frac{B^3}{A^4}\|F\|_\infty, \ \gamma \in \mathbb{T}_2.
$$

Thus G_0 is bounded. A similar calculation gives G_1 and that also this function is bounded.

Now let $\gamma \in \mathbb{T}_3$. Then (17.13) is automatically satisfied. In order to solve (17.11) and (17.12), we first prove that if $H_0(\gamma) \neq 0$, then

$$
\sqrt{\frac{A}{B}} \leq |H_0(\gamma)| \leq \sqrt{\frac{B}{A}}. \tag{17.38}
$$

For $\gamma \in \mathbb{T}_3$ we have $\Phi(\gamma + 1/2) = 0$, so (17.4) implies that

$$
\Phi(2\gamma) = |H_0(\gamma)|^2\Phi(\gamma). \tag{17.39}
$$

Also, $\Phi(\gamma) \neq 0$ for $\gamma \in \mathbb{T}_3$; by Theorem 9.2.5, this implies that $\Phi(\gamma) \geq A$ on $\mathbb{T}_3$, so via (17.39)

$$
|H_0(\gamma)|^2 = \frac{\Phi(2\gamma)}{\Phi(\gamma)} \leq \frac{B}{A}.
$$

This proves the right-hand inequality in (17.38). Also, (17.39) shows that if $\gamma \in \mathbb{T}_3$ and $H_0(\gamma) \neq 0$, then $\Phi(2\gamma) \neq 0$; thus,

$$A \leq \Phi(2\gamma) = |H_0(\gamma)|^2 \Phi(\gamma) \leq B \, |H_0(\gamma)|^2,$$

which gives the left-hand inequality in (17.38).

Now we return to the equations (17.11) and (17.12). For $\gamma \in \mathbb{T}_3$ they reduce to

$$\overline{H_0}F = 0, \quad H_0 G_0 + F G_1 = 1.$$

In case $H_0(\gamma) = 0$, the first equation is satisfied, and the second equation reduces to $F(\gamma)G_1(\gamma) = 1$; this can be obtained by defining

$$F(\gamma) = G_1(\gamma) = 1.$$

In case $H_0(\gamma) \neq 0$, the first equation shows that we are forced to define $F(\gamma) = 0$. Therefore, the second equation simplifies to $H_0(\gamma)G_0(\gamma) = 1$, which can be obtained for a bounded function G_0 because of (17.38); this concludes the proof for $\gamma \in \mathbb{T}_3$. We note that one choice for the function F is

$$F(\gamma) = \begin{cases} 1 & \text{if } \gamma \in \mathbb{T}_3 \text{ and } H_0(\gamma) = 0, \\ 0 & \text{if } \gamma \in \mathbb{T}_3 \text{ and } H_0(\gamma) \neq 0. \end{cases} \tag{17.40}$$

The proof for $\gamma \in \mathbb{T}_4$ is similar. In this case, (17.12) is automatically satisfied, and (17.11) and (17.13) reduce to

$$T_{1/2}(\overline{H_0}F) = 0, \quad (T_{1/2}H_0)G_0 + (T_{1/2}F)G_1 = 0. \tag{17.41}$$

Since $\gamma \in \mathbb{T}_4$, we know that $\gamma - \frac{1}{2} \in \mathbb{T}_3$ (or, at least, its "periodic extension"). If $H_0(\gamma - \frac{1}{2}) \neq 0$, the first equation in (17.41) forces us to define $F(\gamma - \frac{1}{2}) = 0$; this is consistent with (17.40). The second equation in (17.41) simplifies to $(T_{1/2}H_0)(\gamma)G_0(\gamma) = 0$, which is satisfied if we let $G_0(\gamma) = 0$. If $H_0(\gamma - \frac{1}{2}) = 0$, the first equation in (17.41) is satisfied, and the second equation gives $T_{1/2}F(\gamma)G_1(\gamma) = 0$; this can be obtained by letting $G_1(\gamma) = 0$. This completes the analysis of the case $\gamma \in \mathbb{T}_4$. Note in particular that no condition on $F(\gamma)$ is needed for $\gamma \in \mathbb{T}_4$, except that we want F to be bounded. In particular, we can define

$$F(\gamma) = 0, \quad \gamma \in \mathbb{T}_4. \tag{17.42}$$

$\square$

In order to conclude that there are frame multiresolution analyses which do not lead to construction of wavelet frames for $L^2(\mathbb{R})$, we have to know that it is actually possible for Γ to have positive measure. An example where this happens is given in [57].

In case $|\Gamma| = 0$, it is worth noticing that (17.35), (17.37), (17.40), and (17.42) show how one can define F such that $\{T_k\psi\}_{k \in \mathbb{Z}}$ is a frame for W_0 when ψ is defined by $\widehat{\psi}(2\gamma) = F(\gamma)\widehat{\phi}(\gamma)$; in fact, we can take

$$F(\gamma) = \begin{cases} T_{1/2}(\overline{H_0}\Phi)(\gamma)E_{-1}(\gamma) & \text{if } \gamma \in \mathbb{T}_2, \\ 1 & \text{if } \gamma \in \mathbb{T}_3 \text{ and } H_0(\gamma) = 0, \\ 0 & \text{otherwise.} \end{cases} \quad (17.43)$$

The proof of Theorem 17.4.5 also shows how other choices of F can be made. In particular, nothing forces us to define F on $\mathbb{T}_2$ as in (17.43); this choice was only made in order to simplify the calculations. Also for $\gamma \in \mathbb{T}_3$, we have a useful freedom: on the set of $\gamma \in \mathbb{T}_3$ for which $H_0(\gamma) = 0$, the only condition is that $F(\gamma)G_1(\gamma) = 1$ for some bounded function G_1, so we can choose F to be any function which is bounded above and below on this set. In contrast, the freedom in the definition on $\mathbb{T}_1 \cup \mathbb{T}_4$ is not helpful. For $\gamma \in \mathbb{T}_1 \cup \mathbb{T}_4$ (or its 1-periodic extension), we have $\Phi(\gamma) = 0$ and therefore $\widehat{\phi}(\gamma) = 0$; that is, different choices of F will not change the function ψ.

Before we exploit the freedom in the choice of F further, we give an example, where we solve the three equations in Proposition 17.4.3 by direct calculations:

Example 17.4.6 We continue Example 17.2.4, and consider the function ϕ given by $\widehat{\phi}(\gamma) = \chi_{[-\frac{1}{4},\frac{1}{4}[}$. We have already seen that ϕ generates a frame multiresolution analysis. Recall that for $|\gamma| < \frac{1}{2}$,

$$H_0(\gamma) = \chi_{[-\frac{1}{8},\frac{1}{8}[}, \text{ and } \Phi(\gamma) = \chi_{[-\frac{1}{4},\frac{1}{4}[}.$$

It is clear that Γ is an empty set, so we know that we can construct a wavelet frame via the frame multiresolution analysis generated by ϕ. We will use the equations (17.11), (17.12), and (17.13) directly to find $F \in L^\infty(\mathbb{T})$ such that the function ψ defined by

$$\widehat{\psi}(\gamma) := F(\gamma/2)\widehat{\phi}(\gamma/2) \quad (17.44)$$

generates a frame for W_0. In our search for F, we only consider $\gamma \in [-\frac{1}{2},\frac{1}{2}[$. In order to simplify the calculations, we will restrict our search to functions F for which

$$F = \chi_I \text{ for some set } I \text{ with } I \cap [-\frac{1}{8},\frac{1}{8}[= \emptyset.$$

This simplification immediately implies that (17.11) is satisfied. (17.12) is automatically satisfied outside the support of Φ, i.e., for $\gamma \notin [-\frac{1}{4},\frac{1}{4}[$. For $\gamma \in [-\frac{1}{4},\frac{1}{4}[$, it reduces to

$$\chi_{[-\frac{1}{8},\frac{1}{8}[}G_0 + FG_1 = 1. \quad (17.45)$$

We can satisfy (17.45) by requiring that

$$G_0 = 1 \text{ on } [-\frac{1}{8},\frac{1}{8}[\quad (17.46)$$

and

$$F = G_1 = 1 \text{ on } [-\frac{1}{4}, -\frac{1}{8}[\cup [\frac{1}{8}, \frac{1}{4}[. \qquad (17.47)$$

We now rewrite (17.13) as

$$H_0(\gamma)\Phi(\gamma)G_0(\gamma + \frac{1}{2}) + F(\gamma)\Phi(\gamma)G_1(\gamma + \frac{1}{2}) = 0;$$

using our information about F this is equivalent to

$$\chi_{[-\frac{1}{8},\frac{1}{8}[}(\gamma)G_0(\gamma + \frac{1}{2}) + F(\gamma)\chi_{[-\frac{1}{4}-\frac{1}{8}[\cup[\frac{1}{8},\frac{1}{4}[}G_1(\gamma + \frac{1}{2}) = 0. \qquad (17.48)$$

In order to satisfy this, we would like both terms to vanish. The first term will vanish if $G_0(\gamma + \frac{1}{2}) = 0$ on $[-\frac{1}{8}, \frac{1}{8}[$, i.e., if

$$G_0(\gamma) = 0, \ \gamma \in [-\frac{1}{2}, -\frac{1}{2} + \frac{1}{8}[\cup [\frac{1}{2} - \frac{1}{8}, \frac{1}{2}[;$$

this choice can be made without conflict with (17.46). For the second term in (17.48) to vanish, we want $F(\gamma)G_1(\gamma + \frac{1}{2})$ to vanish on $[-\frac{1}{4}, -\frac{1}{8}[\cup [\frac{1}{8}, \frac{1}{4}[$; this is obtained by defining

$$G_1 = 0 \text{ on } [-\frac{1}{2} + \frac{1}{8}, -\frac{1}{4}[\cup [\frac{1}{4}, \frac{1}{2} - \frac{1}{8}[.$$

Again, this choice is allowed. The construction gives no conditions on F on

$$[-\frac{1}{2}, -\frac{1}{4}[\cup [\frac{1}{4}, \frac{1}{2}[. \qquad (17.49)$$

However, different choices of F on this set will lead to the same function ψ in (17.44) because $\widehat{\phi} = \chi_{[-\frac{1}{4},\frac{1}{4}]}$ and $\widehat{\psi}(2\gamma) = F(\gamma)\widehat{\phi}(\gamma)$. Note that this result is in accordance with our proof of Theorem 17.4.5: in the considered example $\mathbb{T}_1 = \emptyset$, and the set in (17.49) equals $\mathbb{T}_4$. □

The choice of F in (17.43) implies that $\{T_k\psi\}_{k\in\mathbb{Z}}$ is tight in case $\{T_k\phi\}_{k\in\mathbb{Z}}$ itself is a tight frame with frame bound equal to 1:

Corollary 17.4.7 *Assume that $\phi \in L^2(\mathbb{R})$ generates a frame multiresolution analysis and that $\{T_k\phi\}_{k\in\mathbb{Z}}$ is a tight frame with frame bound equal to 1. If $|\Gamma| = 0$, there exists a function $\psi \in W_0$ such that $\{T_k\psi\}_{k\in\mathbb{Z}}$ is a tight frame for W_0 and $\{D^j T_k\psi\}_{j,k\in\mathbb{Z}}$ is a tight frame for $L^2(\mathbb{R})$.*

Proof. We first prove that $\{T_k\psi\}_{k\in\mathbb{Z}}$ is a tight frame when ψ is defined via the choice of F in (17.43). By Theorem 9.2.5 it is enough to prove that the function Ψ is constant outside its zero set. Recall that

$$\Psi(2\gamma) = |F(\gamma)|^2\Phi(\gamma) + |F(\gamma + 1/2)|^2\Phi(\gamma + 1/2).$$

As before, we split $\mathbb{T}$ into the sets $\mathbb{T}_i, i = 1, .., 4$. First, we compute $\Psi(2\gamma)$ for $\gamma \in \mathbb{T}_2$. For $\gamma \in \mathbb{T}_2$, we have $\Phi(\gamma) > 0$ and $\Phi(\gamma+1/2) > 0$; since $|\Gamma| = 0$, we can assume that $\Phi(2\gamma) > 0$. Now,

$$\Phi(2\gamma) = \Phi(\gamma) = T_{-1/2}\Phi(\gamma) = 1,$$

so equation (17.4) shows that

$$|H_0(\gamma)|^2 + |H_0(\gamma + 1/2)|^2 = 1.$$

Thus, by (17.43),

$$\Psi(2\gamma) = |F(\gamma)|^2 + |F(\gamma + 1/2)|^2 = T_{1/2}|H_0(\gamma)|^2 + |H_0(\gamma)|^2 = 1.$$

Now consider $\gamma \in \mathbb{T}_3$. In this case,

$$\Psi(2\gamma) = |F(\gamma)|^2\Phi(\gamma) = \begin{cases} 1 & \text{if } H_0(\gamma) = 0, \\ 0 & \text{if } H_0(\gamma) \neq 0. \end{cases} \tag{17.50}$$

If $\gamma \in \mathbb{T}_4$, then $\Psi(2\gamma) = |F(\gamma + 1/2)|^2\Phi(\gamma + 1/2)$ and $\gamma + \frac{1}{2} \in \mathbb{T}_3$ (or its "periodic extension"); thus (17.50) shows that $\Psi(2\gamma)$ only assumes the values 0 and 1. Finally, for $\gamma \in \mathbb{T}_1$ we have $\Psi(2\gamma) = 0$.

We have now proved that Ψ only assumes the values 0 and 1, so $\{T_k\psi\}_{k\in\mathbb{Z}}$ is a tight frame sequence; the choice of F guarantees by Proposition 17.4.3 that it is a frame for W_0. □

As a special case we obtain the classical result already mentioned in (3.47) and (3.48) for construction of an orthonormal basis based on a multiresolution analysis:

Corollary 17.4.8 *Assume that $\phi \in L^2(\mathbb{R})$ generates a multiresolution analysis with two-scale symbol H_0. Let $F := (T_{1/2}\overline{H_0})E_{-1}$ and define the function $\psi \in V_1$ by $\psi(2\gamma) := F(\gamma)\widehat{\phi}(\gamma)$. Then ψ generates an orthonormal basis $\{D^jT_k\psi\}_{j,k\in\mathbb{Z}}$ for $L^2(\mathbb{R})$.*

Proof. In the case of a multiresolution analysis, we have $\mathbb{T}_2 = \mathbb{T}$, and the proof of Corollary 17.4.7 shows that $\Psi = 1$. Thus, by Theorem 9.2.5, the functions $\{T_k\psi\}_{k\in\mathbb{Z}}$ constitute an orthonormal basis for W_0. □

Using the freedom in the choice of F, we now prove that if ϕ generates a frame multiresolution analysis and $|\Gamma| = 0$, then we can construct a tight frame $\{T_k\psi\}_{k\in\mathbb{Z}}$ for W_0 without assuming that $\{T_k\phi\}_{k\in\mathbb{Z}}$ itself is tight. We again refer to the splitting $\mathbb{T} = \cup_{i=1}^4 \mathbb{T}_i$ from the proof of Theorem 17.4.5.

Theorem 17.4.9 *Assume that $\phi \in L^2(\mathbb{R})$ generates a frame multiresolution analysis and that $|\Gamma| = 0$. Let $K \in L^\infty(\mathbb{T})$ be a $\frac{1}{2}$-periodic function which is bounded below, and define $F \in L^\infty(\mathbb{T})$ by*

$$F(\gamma) = \begin{cases} T_{1/2}(\overline{H_0}\Phi)(\gamma)E_{-1}(\gamma)K(\gamma) & \text{if } \gamma \in \mathbb{T}_2, \\ \dfrac{1}{\sqrt{\Phi(\gamma)}} & \text{if } \gamma \in \mathbb{T}_3 \text{ and } H_0(\gamma) = 0, \\ 0 & \text{otherwise.} \end{cases} \tag{17.51}$$

Then, with $\psi \in V_1$ defined by $\widehat{\psi}(2\gamma) = F(\gamma)\widehat{\phi}(\gamma)$, the following hold:

(i) $\{T_k\psi\}_{k\in\mathbb{Z}}$ is a frame for W_0 and $\{D^j T_k\psi\}_{j,k\in\mathbb{Z}}$ is a frame for $L^2(\mathbb{R})$.

(ii) Assume that K is chosen such that on $\mathbb{T}_2$ we have

$$|K|^2 \left(\Phi T_{1/2}\Phi(T_{1/2}(|H_0|^2\Phi) + |H_0|^2\Phi\right) = 1. \qquad (17.52)$$

Then $\{T_k\psi\}_{k\in\mathbb{Z}}$ is a tight frame for W_0 and $\{D^j T_k\psi\}_{k\in\mathbb{Z}}$ is a tight frame for $L^2(\mathbb{R})$; both have frame bounds equal to 1.

Proof. Compared to (17.43) we have only changed F on

$$\mathbb{T}_2 \cup \{\gamma \in \mathbb{T}_3 : H_0(\gamma) = 0\};$$

that is, to prove (i) it is enough to show that the three equations in Proposition 17.4.3 can be solved on this set, with the new choice of F. We have already on page 438 argued that the choice of F on $\mathbb{T}_3$ given in (17.51) is allowed, because Φ is bounded above and below on $\mathbb{T}_3$. For use in (ii), we note that with this choice,

$$\Psi(2\gamma) = |F(\gamma)|^2\Phi(\gamma) = 1 \text{ if } \gamma \in \mathbb{T}_3 \text{ and } H_0(\gamma) = 0. \qquad (17.53)$$

Now we check that the equations (17.11), (17.12), and (17.13) can be satisfied with the new choice of F on $\mathbb{T}_2$. Using that K is $\frac{1}{2}$-periodic, we see that

$$\overline{H_0}F\Phi + T_{1/2}(\overline{H_0}F\Phi)$$
$$= \overline{H_0}T_{1/2}(\overline{H_0}\Phi)E_{-1}K\Phi + T_{1/2}\left(\overline{H_0}T_{1/2}(\overline{H_0}\Phi)E_{-1}K\Phi\right)$$
$$= K\left(\overline{H_0}T_{1/2}(\overline{H_0}\Phi)E_{-1}\Phi - (T_{1/2}\overline{H_0})\,\overline{H_0}\Phi E_{-1}T_{1/2}\Phi\right)$$
$$= 0.$$

Similarly, we can repeat the rest of the proof of Theorem 17.4.5; the fact that $|K|$ is bounded above and below will again imply that the determinant of the set of equations determining G_0 and G_1 is nonzero and that the obtained solutions are bounded (Exercise 17.2). This concludes the proof of (i). To prove (ii), we first argue that one can actually choose K such that (17.52) is satisfied. Letting A, B denote frame bounds for $\{T_k\phi\}_{k\in\mathbb{Z}}$, it follows from (17.36) that

$$\frac{A^4}{B} \leq \Phi T_{1/2}\Phi \left(T_{1/2}(|H_0|^2\Phi) + |H_0|^2\Phi\right) \leq \frac{B^4}{A} \text{ on } \mathbb{T}_2.$$

Since the function $\Phi T_{1/2}\Phi(T_{1/2}(|H_0|^2\Phi) + |H_0|^2\Phi)$ is $\frac{1}{2}$-periodic, we can therefore choose a $\frac{1}{2}$-periodic function K such that (17.52) is satisfied, and K is bounded below and above. The next step is to show that with the choice (17.52), the function Ψ will only assume the values 0 and 1. The case $\gamma \in \mathbb{T}_1$ is trivial, and the case $\gamma \in \mathbb{T}_3, H(\gamma) = 0$ is considered in (17.53).

On $\mathbb{T}_2$, we get

$$
\begin{aligned}
\Psi(2\cdot) &= |F(\cdot)|^2\Phi(\cdot) + |F(\cdot + 1/2)|^2\Phi(\cdot + 1/2) \\
&= |K|^2 \left(T_{1/2}(|H_0\Phi|^2)\ \Phi + |H_0\Phi|^2\ T_{1/2}\Phi \right) \\
&= |K|^2 \left(\Phi T_{1/2}\Phi(T_{1/2}(|H_0|^2\Phi) + |H_0|^2\Phi \right) = 1.
\end{aligned}
$$

In the rest of the cases, the function F is unchanged, so the proof of Corollary 17.4.7 gives the rest.   □

17.5   Frames with Two Generators

In light of the fact that we cannot always associate a wavelet frame $\{D^j T_k\psi\}_{j,k\in\mathbb{Z}}$ to a frame multiresolution analysis, it is interesting to notice that we can always construct a multiwavelet frame. We need a lemma before we present the result in Theorem 17.5.2.

Lemma 17.5.1 *Assume that $\phi \in L^2(\mathbb{R})$ generates a frame multiresolution analysis. For $j \in \mathbb{Z}$, let $S_j : V_j \to V_j$ denote the frame operator for $\{D^j T_k\phi\}_{k\in\mathbb{Z}}$ and let $P_j : L^2(\mathbb{R}) \to V_j$ denote the orthogonal projection onto V_j. Then:*

(i) *For any $j, k \in \mathbb{Z}$, the following identities hold on V_j:*

$$
S_j = D^j T_k S_0 T_{-k} D^{-j} \ and \ S_j^{-1} = D^j T_k S_0^{-1} T_{-k} D^{-j}. \qquad (17.54)
$$

(ii) *For all $j, k \in \mathbb{Z}$,*

$$
P_j D^{j+1} T_{2k} = D^j T_k P_0 D \ on \ V_0.
$$

Proof. Let us fix $j \in \mathbb{Z}$. Then, for all $k \in \mathbb{Z}$ and $f \in V_0$,

$$
\begin{aligned}
S_j D^j T_k f &= \sum_{k'\in\mathbb{Z}} \langle D^j T_k f, D^j T_{k'}\phi\rangle D^j T_{k'}\phi = D^j \sum_{k'\in\mathbb{Z}} \langle T_k f, T_{k'}\phi\rangle T_{k'}\phi \\
&= D^j \sum_{k'\in\mathbb{Z}} \langle f, T_{k'-k}\phi\rangle T_{k'}\phi = D^j \sum_{k'\in\mathbb{Z}} \langle f, T_{k'}\phi\rangle T_{k'+k}\phi \\
&= D^j T_k S_0 f.
\end{aligned}
$$

Thus $S_j D^j T_k = D^j T_k S_0$ on V_0, and therefore $S_j = D^j T_k S_0 T_{-k} D^{-j}$ on V_j; the second equality in (17.54) follows from this. In order to prove (ii), we apply Proposition 5.2.3 on $f \in V_0$ and obtain that

$$
P_j D^{j+1} T_{2k} f = \sum_{k'\in\mathbb{Z}} \langle D^{j+1} T_{2k} f, S_j^{-1} D^j T_{k'}\phi\rangle D^j T_{k'}\phi;
$$

via the commutator relation $DT_{2k} = T_k D$ and a change of the summation index, we continue with

$$
\begin{aligned}
P_j D^{j+1} T_{2k} f &= D^j \sum_{k' \in \mathbb{Z}} \langle D^j T_k Df, D^j T_{k'} S_0^{-1} \phi \rangle T_{k'} \phi \\
&= D^j \sum_{k' \in \mathbb{Z}} \langle Df, T_{k'-k} S_0^{-1} \phi \rangle T_{k'} \phi \\
&= D^j T_k \sum_{k' \in \mathbb{Z}} \langle Df, T_{k'} S_0^{-1} \phi \rangle T_{k'} \phi \\
&= D^j T_k \sum_{k' \in \mathbb{Z}} \langle Df, S_0^{-1} T_{k'} \phi \rangle T_{k'} \phi \\
&= D^j T_k P_0 Df.
\end{aligned}
$$

$\square$

Theorem 17.5.2 *Assume that $\phi \in L^2(\mathbb{R})$ generates a frame multiresolution analysis, and let Q_j denote the orthogonal projection onto W_j. Then*

$$
\{D^j T_k Q_0 D\phi\}_{j,k\in\mathbb{Z}} \cup \{D^j T_k Q_0 DT_1\phi\}_{j,k\in\mathbb{Z}}
$$

is a multiwavelet frame for $L^2(\mathbb{R})$.

Proof. Let A, B denote frame bounds for $\{T_k\phi\}_{k\in\mathbb{Z}}$, and P_j be the orthogonal projection onto V_j. For each $j \in \mathbb{Z}$, we know that $\{D^{j+1} T_k\phi\}_{k\in\mathbb{Z}}$ is a frame for V_{j+1} with frame bounds A, B. Since W_j is a subspace of V_{j+1}, it follows by Proposition 5.2.3 that $\{Q_j D^{j+1} T_k\phi\}_{k\in\mathbb{Z}}$ is a frame for W_j, also with frame bounds A, B. Given $f \in L^2(\mathbb{R})$, we can write $f = \sum_{j\in\mathbb{Z}} Q_j f$, and $||f||^2 = \sum_{j\in\mathbb{Z}} ||Q_j f||^2$. Since $Q_j f \in W_j$, we have

$$
\begin{aligned}
A \, ||Q_j f||^2 &\le \sum_{k\in\mathbb{Z}} |\langle Q_j f, Q_j D^{j+1} T_k\phi \rangle|^2 \\
&= \sum_{k\in\mathbb{Z}} |\langle f, Q_j D^{j+1} T_k\phi \rangle|^2 \\
&\le B \, ||Q_j f||^2.
\end{aligned}
$$

Summing over $j \in \mathbb{Z}$, we obtain that

$$
A \, ||f||^2 \le \sum_{j,k\in\mathbb{Z}} |\langle f, Q_j D^{j+1} T_k\phi \rangle|^2 \le B \, ||f||^2,
$$

which shows that $\{Q_j D^{j+1} T_k\phi\}_{j,k\in\mathbb{Z}}$ is a frame for $L^2(\mathbb{R})$. We now split this family in two by considering translations with $2k, k \in \mathbb{Z}$ and $2k+1, k \in \mathbb{Z}$ separately. Observing that $Q_j = P_{j+1} - P_j$, Lemma 17.5.1 implies that for any $f \in V_0$,

$$
\begin{aligned}
Q_j D^{j+1} T_{2k} f &= (P_{j+1} - P_j) D^{j+1} T_{2k} f = D^{j+1} T_{2k} f - P_j D^{j+1} T_{2k} f \\
&= D^j T_k (Df - P_0 Df) = D^j T_k Q_0 Df;
\end{aligned}
$$

in the last equality, we used that $Df = P_1 Df$ because $Df \in V_1$. Thus, applying the result to $f = \phi$ and $f = T_1\phi$ yields

$$
\begin{aligned}
\{Q_j D^{j+1} T_k \phi\}_{j,k\in\mathbb{Z}} &= \{Q_j D^{j+1} T_{2k}\phi\}_{j,k\in\mathbb{Z}} \cup \{Q_j D^{j+1} T_{2k+1}\phi\}_{j,k\in\mathbb{Z}} \\
&= \{D^j T_k Q_0 D\phi\}_{j,k\in\mathbb{Z}} \cup \{D^j T_k Q_0 DT_1\phi\}_{j,k\in\mathbb{Z}}.
\end{aligned}
$$

$\square$

17.6 Some Limitations

There are some restrictions on which frames one can obtain via frame multiresolution analysis. If $\{\psi_{j,k}\}_{j,k\in\mathbb{Z}}$ is constructed via a frame multiresolution analysis, the orthogonal decomposition (17.8) together with the fact that $\{\psi_{j,k}\}_{k\in\mathbb{Z}}$ for a given value of $j \in \mathbb{Z}$ is a frame for W_j implies that

$$
\psi_{j,k} \perp \psi_{j',k'} \text{ whenever } j \neq j', \text{ for all } k, k' \in \mathbb{Z}.
$$

For this reason, the frame $\{\psi_{j,k}\}_{k\in\mathbb{Z}}$ is said to be *semiorthogonal*. Note that in [534], Weiss et al. also proposed a multiresolution analysis scheme which leads to a construction of wavelet frames; the obtained class contains the tight frames in Theorem 17.4.9, but the frames $\{\psi_{j,k}\}_{k\in\mathbb{Z}}$ are not necessarily semiorthogonal. On the other hand, the freedom in the choice of the function F in the proof of Theorem 17.4.5 also makes it possible to construct non-tight frames via frame multiresolution analysis.

We have already mentioned that Proposition 9.3.8 restricts the class of functions ϕ which can generate an overcomplete frame $\{T_k\psi\}_{k\in\mathbb{Z}}$. We also note that in the case of a classical multiresolution analysis, where $\{T_k\phi\}_{k\in\mathbb{Z}}$ is assumed to be an orthonormal basis for V_0, no frame at all can be constructed. In fact, in this case we have $\Phi = 1$ on $\mathbb{R}$, and $\mathbb{T} = \mathbb{T}_2$. The proof of Proposition 17.4.3 shows that if $F \in L^\infty(\mathbb{T})$ satisfies the three key equations and we define $\widehat{\psi}(2\gamma) = F(\gamma)\widehat{\phi}(\gamma)$ as usual, then Ψ is bounded above and below. That is, $\{T_k\psi\}_{k\in\mathbb{Z}}$ will be a Riesz sequence, and $\{D^j T_k\psi\}_{j,k\in\mathbb{Z}}$ is a Riesz basis for $L^2(\mathbb{R})$. The conclusion is that no overcomplete frame for W_0 can be constructed this way, and we cannot use the multiresolution analysis to obtain an overcomplete frame for $L^2(\mathbb{R})$. The same happens if ϕ generates a frame multiresolution analysis and $\{T_k\phi\}_{k\in\mathbb{Z}}$ is a Riesz sequence.

17.7 Exercises

17.1 Prove that the assumptions in Definition 17.1.3 are enough to make $\{V_j, \phi\}$ a frame multiresolution analysis.

17.2 Provide the details in the proof of Theorem 17.4.9.

18

Wavelet Frames via Extension Principles

Frame multiresolution analysis is just one way to construct wavelet frames via multiscale techniques. We already mentioned in Section 17.3 that the conditions can be weakened further, and the purpose of this chapter is to show how one can still construct frames.

We will follow a fundamental idea by Ron and Shen, which (in its first version) appeared in [561]. As discussed in Section 17.3, the idea is to modify the classical multiresolution analysis setup in Definition 3.9.2 by requiring ϕ to satisfy a scaling equation instead of $\{T_k\phi\}_{k\in\mathbb{Z}}$ being an orthonormal sequence. The other conditions will be stated in the general setup in the next section; they imply that the spaces V_j defined by

$$V_j = D^j \overline{\operatorname{span}}\{T_k\phi\}_{k\in\mathbb{Z}} \tag{18.1}$$

satisfy that

$$V_j \subset V_{j+1} \ \forall j \in \mathbb{Z}, \text{ and } \overline{\cup_j V_j} = L^2(\mathbb{R}). \tag{18.2}$$

Thus, the multiscale idea is integrated in the setup, although it will not appear explicitly in the constructions. The multiscale feature is very important because of all its computational advances.

In contrast to frame multiresolution analysis, the purpose is no longer to construct frames for the orthogonal complement of V_1 in V_0. In fact, we will construct functions $\psi_1, \ldots, \psi_n$ belonging to V_1, such that the multiwavelet system $\{D^j T_k \psi_\ell\}_{j,k\in\mathbb{Z},\ell=1,\ldots,n}$ forms a tight frame for $L^2(\mathbb{R})$. In practice, one usually wishes to have as few generators as possible, and we show how to construct frames with two or three generators (and explain why one generator does not suffice).

© Springer International Publishing Switzerland 2016
O. Christensen, *An Introduction to Frames and Riesz Bases*,
Applied and Numerical Harmonic Analysis,
DOI 10.1007/978-3-319-25613-9_18

After presenting the general setup in Section 18.1, we prove the original unitary extension principle of Ron and Shen in Section 18.2; this is the cornerstone for the constructions of tight wavelet frames. This is followed by a discussion of more recent results, which facilitate the search for frames with prescribed properties and also lead to frames with better approximation properties. Finally, a relatively small modification of the setup leads to constructions of pairs of dual wavelet frames. This construction is actually easier than its tight counterpart. Throughout the chapter the results will be applied to construct multiwavelet frames with B-spline generators.

18.1   The General Setup

We now present the setup for the general multiresolution analysis of Ron and Shen, which enables us to construct tight frames for $L^2(\mathbb{R})$ of the form

$$\{\psi_{\ell;j,k}\}_{j,k\in\mathbb{Z},\ell=1,\ldots,n} = \{D^j T_k \psi_1\}_{j,k\in\mathbb{Z}} \cup \cdots \cup \{D^j T_k \psi_n\}_{j,k\in\mathbb{Z}}. \quad (18.3)$$

As noted before, a frame of this type is called a multiwavelet frame. The functions $\psi_1,\ldots,\psi_n$ will be constructed on the basis of a function ψ_0 satisfying a scaling equation. Since we will work with all these functions simultaneously, it is convenient to change our previous notation slightly and denote the refinable function by ψ_0 instead of ϕ. Except this, we keep the notation from Chapters 16–17; note in particular that $L^2(\mathbb{T}), L^\infty(\mathbb{T})$ are introduced on page 418. We now list the standing assumptions and conventions for this chapter.

General setup: Let $\psi_0 \in L^2(\mathbb{R})$ and assume that

(i) There exists a function $H_0 \in L^\infty(\mathbb{T})$ (the *refinement mask*) such that

$$\widehat{\psi_0}(2\gamma) = H_0(\gamma)\widehat{\psi_0}(\gamma), \ \gamma \in \mathbb{R}. \quad (18.4)$$

(ii) $\lim_{\gamma\to 0} \widehat{\psi_0}(\gamma) = 1.$

Further, let $H_1,\ldots,H_n \in L^\infty(\mathbb{T})$, and define $\psi_1,\ldots,\psi_n \in L^2(\mathbb{R})$ by

$$\widehat{\psi_\ell}(2\gamma) = H_\ell(\gamma)\widehat{\psi_0}(\gamma), \ \ell = 1,\ldots,n. \quad (18.5)$$

The functions $H_1,\ldots,H_n$ are called *masks*. Let H denote the $(n+1)\times 2$ matrix-valued function defined by

$$H(\gamma) = \begin{pmatrix} H_0(\gamma) & T_{1/2}H_0(\gamma) \\ H_1(\gamma) & T_{1/2}H_1(\gamma) \\ \cdot & \cdot \\ \cdot & \cdot \\ H_n(\gamma) & T_{1/2}H_n(\gamma) \end{pmatrix}, \ \gamma \in \mathbb{R}. \quad (18.6)$$

We will frequently suppress the dependence on γ and simply speak about the matrix H.

With this setup, our purpose is to find conditions on the functions $H_1, \ldots, H_n$ such that $\psi_1, \ldots, \psi_n$ defined by (18.5) generate a multiwavelet frame for $L^2(\mathbb{R})$. By Theorem 17.3.3, the spaces

$$V_j := \overline{\operatorname{span}}\{D^j T_k \psi_0\}_{k \in \mathbb{Z}}, \ j \in \mathbb{Z}$$

automatically satisfy the conditions for a multiresolution analysis in Definition 3.9.2, except (v). In Lemma 18.2.5, we prove that the general setup implies that $\{T_k \psi_0\}_{k \in \mathbb{Z}}$ is a Bessel sequence, so by Lemma 17.3.1, $\psi_1, \ldots, \psi_n \in V_1$. In the literature, frames constructed on the basis of the general setup are frequently said to be MRA based.

Ron and Shen gave in [561] a complete characterization of the tight frames which can be obtained via the general setup. It uses the *periodization* of a function $f : \mathbb{R} \to \mathbb{C}$, which formally is defined as

$$\mathcal{P}f(\gamma) = \sum_{n \in \mathbb{Z}} f(\gamma + n).$$

If $f \in L^1(\mathbb{R})$, then

$$\int_{-\frac{1}{2}}^{\frac{1}{2}} \sum_{n \in \mathbb{Z}} |f(\gamma + n)| d\gamma = \int_{-\infty}^{\infty} |f(\gamma)| d\gamma < \infty,$$

so $\sum_{n \in \mathbb{Z}} f(\gamma + n)$ is absolutely convergent for almost all $\gamma \in \mathbb{R}$. That is, $\mathcal{P}f$ is a well-defined 1-periodic function, and the above argument shows that $\mathcal{P}f \in L^1(\mathbb{T})$.

Theorem 18.1.1 *Let* $\{\psi_\ell, H_\ell\}_{\ell=0}^n$ *be as in the general setup, and define the function*

$$\Theta(\gamma) := \sum_{j=0}^{\infty} \sum_{\ell=1}^{n} |H_\ell(2^j \gamma)|^2 \prod_{m=0}^{j-1} |H_0(2^m \gamma)|^2$$

with the convention $\prod_{m=0}^{-1} |H_0(2^m \gamma)|^2 = 1$. *Then the following are equivalent:*

(i) $\{D^j T_k \psi_\ell\}_{j,k \in \mathbb{Z}, \ell=1,\ldots,n}$ *is a tight frame.*

(ii) For almost all γ *for which* $\mathcal{P}(|\widehat{\psi_0}|^2)(\gamma) > 0$, *we have*

$$\lim_{j \to -\infty} \Theta(2^j \gamma) = 1$$

and

$$\overline{H_0(\gamma)} H_0(\gamma + \frac{1}{2}) \Theta(2\gamma) + \sum_{\ell=1}^{n} \overline{H_\ell(\gamma)} H_\ell(\gamma + \frac{1}{2}) = 0.$$

We will not prove Theorem 18.1.1 but instead give direct proofs of the unitary extension principle and its variants.

18.2   The Unitary Extension Principle

The purpose of this section is to prove the unitary extension principle of Ron and Shen. It is based on the general setup in Section 18.1. We state the main result in Theorem 18.2.6, but we need some preparation first. We follow the approach by Benedetto and Treiber [57]. Recall that E_k is used as notation for the modulation operator on $L^2(\mathbb{R})$, and also for the function $x \mapsto e^{2\pi i k x}$.

Lemma 18.2.1 *Let $g, \psi_0 \in L^2(\mathbb{R})$ and assume that $\mathcal{P}(g\overline{\widehat{\psi_0}}) \in L^2(\mathbb{T})$. Then*

$$\mathcal{P}(g\overline{\widehat{\psi_0}}) = \sum_{k \in \mathbb{Z}} \langle g, \widehat{\psi_0} E_k \rangle E_k \qquad (18.7)$$

and

$$\int_{-\frac{1}{2}}^{\frac{1}{2}} |\mathcal{P}(g\overline{\widehat{\psi_0}})(\gamma)|^2 d\gamma = \sum_{k \in \mathbb{Z}} |\langle g, \widehat{\psi_0} E_k \rangle|^2. \qquad (18.8)$$

Proof. Since $g, \psi_0 \in L^2(\mathbb{R})$, we know that $g\overline{\widehat{\psi_0}} \in L^1(\mathbb{R})$, so $\mathcal{P}(g\overline{\widehat{\psi_0}})$ is well defined. Now,

$$
\begin{aligned}
\langle g, \widehat{\psi_0} E_k \rangle &= \int_{-\infty}^{\infty} g(\gamma)\overline{\widehat{\psi_0}(\gamma)} e^{-2\pi i k \gamma} d\gamma \\
&= \int_{-\frac{1}{2}}^{\frac{1}{2}} \sum_{n \in \mathbb{Z}} \left(\overline{\widehat{\psi_0}(\gamma+n)} g(\gamma+n) e^{-2\pi i (k+n)\gamma} \right) d\gamma \\
&= \int_{-\frac{1}{2}}^{\frac{1}{2}} \left(\sum_{n \in \mathbb{Z}} \overline{\widehat{\psi_0}(\gamma+n)} g(\gamma+n) \right) e^{-2\pi i k \gamma} d\gamma,
\end{aligned}
$$

which is the kth Fourier coefficient for the function $\sum_{n \in \mathbb{Z}} \overline{\widehat{\psi_0}}(\cdot+n)g(\cdot+n)$. Since this function belongs to $L^2(\mathbb{T})$ by assumption, the lemma follows: (18.7) is just the expansion of $\mathcal{P}(g\overline{\widehat{\psi_0}})$ in a Fourier series, and (18.8) is Parseval's equation.   □

The first main result, proved in Theorem 18.2.6, will show that a condition on the matrix H in (18.6) implies that the multiwavelet system in (18.3) is a tight frame for $L^2(\mathbb{R})$. In the proof of this, it is enough to show that the frame condition is satisfied on a dense subset of $L^2(\mathbb{R})$ (cf. Lemma 5.1.9). Already in the following lemmas, we work with functions $f \in L^2(\mathbb{R})$ for which $\widehat{f}$ is a continuous function with compact support, $\widehat{f} \in C_c(\mathbb{R})$.

Lemma 18.2.2 *Let $\psi_0 \in L^2(\mathbb{R})$ and assume that $\lim_{\gamma \to 0} \widehat{\psi}_0(\gamma) = 1$. Let $f \in L^2(\mathbb{R})$ be any function for which $\widehat{f} \in C_c(\mathbb{R})$. Then, for any $\epsilon > 0$, there exists $J \in \mathbb{Z}$ such that*

$$(1 - \epsilon)||f||^2 \leq \sum_{k \in \mathbb{Z}} |\langle f, D^j T_k \psi_0 \rangle|^2 \leq (1 + \epsilon)||f||^2 \text{ for all } j \geq J.$$

Proof. Let $j \in \mathbb{Z}$, $f \in L^2(\mathbb{R})$, and assume that $\widehat{f} \in C_c(\mathbb{R})$. As a product of $L^2(\mathbb{R})$-functions, $(D^j \widehat{f})\overline{\widehat{\psi}_0} \in L^1(\mathbb{R})$; thus, $\mathcal{P}((D^j \widehat{f})\overline{\widehat{\psi}_0})$ is well defined. When we only consider $\gamma \in \mathbb{T}$, $\mathcal{P}((D^j \widehat{f})\overline{\widehat{\psi}_0})$ can be bounded by a finite linear combination of translates of $\overline{\widehat{\psi}_0}$, so $\mathcal{P}((D^j \widehat{f})\overline{\widehat{\psi}_0}) \in L^2(\mathbb{T})$. Via the Fourier transform,

$$\langle f, D^j T_k \psi_0 \rangle = \langle \mathcal{F}f, \mathcal{F}D^j T_k \psi_0 \rangle = \langle D^j \widehat{f}, E_{-k} \widehat{\psi}_0 \rangle; \tag{18.9}$$

therefore, Lemma 18.2.1 shows that

$$\sum_{k \in \mathbb{Z}} |\langle f, D^j T_k \psi_0 \rangle|^2 = \sum_{k \in \mathbb{Z}} |\langle D^j \widehat{f}, E_{-k} \widehat{\psi}_0 \rangle|^2$$

$$= \int_{-\frac{1}{2}}^{\frac{1}{2}} \left| \sum_{n \in \mathbb{Z}} (D^j \widehat{f})(\gamma + n) \overline{\widehat{\psi}_0(\gamma + n)} \right|^2 d\gamma.$$

Now let $\epsilon > 0$ be given. By assumption, we can choose $b \in]0, 1/2[$ such that $1 - \epsilon \leq |\widehat{\psi}_0(\gamma)|^2 \leq 1 + \epsilon$ whenever $|\gamma| \leq b$. By taking $J \in \mathbb{Z}$ such that $D^j \widehat{f}$ has support in $[-b, b]$ for $j > J$, we obtain that for all $j > J$,

$$\int_{-\frac{1}{2}}^{\frac{1}{2}} \left| \sum_{n \in \mathbb{Z}} (D^j \widehat{f})(\gamma + n) \overline{\widehat{\psi}_0(\gamma + n)} \right|^2 d\gamma = \int_{-b}^{b} |(D^j \widehat{f})(\gamma) \widehat{\psi}_0(\gamma)|^2 d\gamma.$$

Therefore,

$$(1 - \epsilon)||D^j \widehat{f}||^2 \leq \sum_{k \in \mathbb{Z}} |\langle f, D^j T_k \psi_0 \rangle|^2 \leq (1 + \epsilon)||D^j \widehat{f}||^2.$$

Since D^j and the Fourier transform are unitary operators, the lemma follows. $\qquad\square$

In the rest of this section we assume that $\{\psi_\ell, H_\ell\}_{\ell=0}^n$ is as in the general setup. For a function f with $\widehat{f} \in C_c(\mathbb{R})$, an argument as in the proof of Lemma 18.2.2 shows that

$$\{\langle f, D^j T_k \psi_\ell \rangle\}_{k \in \mathbb{Z}} \in \ell^2(\mathbb{Z}) \text{ for all } \ell = 1, \ldots, n \tag{18.10}$$

(Exercise 18.1). We can therefore define a family of functions $F_{j,\ell} \in L^2(\mathbb{T})$ by the Fourier series

$$F_{j,\ell} := \sum_{k \in \mathbb{Z}} \langle f, D^j T_k \psi_\ell \rangle E_{-k}, \quad j \in \mathbb{Z}, \ \ell = 0, 1, \ldots, n. \tag{18.11}$$

Since $F_{j,\ell}$ is defined in terms of ψ_ℓ, which is again defined via ψ_0 and H_ℓ, it is natural to search for an expression for $F_{j,\ell}$ in terms of $F_{j,0}$ and H_ℓ. For convenience, we work with $F_{j-1,\ell}$:

Lemma 18.2.3 *Let $\{\psi_\ell, H_\ell\}_{\ell=0}^n$ be as in the general setup. Then, for all $j \in \mathbb{Z}, \ell = 0, 1, \ldots, n$,*

$$F_{j-1,\ell}(\gamma) = 2^{-1/2}(\overline{H_\ell}F_{j,0} + T_{1/2}(\overline{H_\ell}F_{j,0}))(\gamma/2).$$

Proof. The commutator relations show that

$$
\begin{aligned}
\langle f, D^{j-1}T_k\psi_\ell \rangle &= \langle D^{-j}f, D^{-1}T_k\psi_\ell \rangle = \langle D^{-j}f, T_{2k}D^{-1}\psi_\ell \rangle \\
&= \langle \mathcal{F}D^{-j}f, \mathcal{F}T_{2k}D^{-1}\psi_\ell \rangle = \langle D^j\hat{f}, E_{-2k}D\widehat{\psi_\ell} \rangle.
\end{aligned}
$$

By (18.5), we can continue with

$$
\begin{aligned}
\langle f, D^{j-1}T_k\psi_\ell \rangle &= \langle D^j\hat{f}, E_{-2k}2^{1/2}H_\ell\widehat{\psi_0} \rangle \\
&= 2^{1/2}\int_{-\infty}^{\infty}(D^j\hat{f})\overline{H_\ell\widehat{\psi_0}}E_{2k} = 2^{1/2}\int_{-\frac{1}{2}}^{\frac{1}{2}}\mathcal{P}((D^j\hat{f})\overline{H_\ell\widehat{\psi_0}})E_{2k} \\
&= 2^{1/2}\int_0^{\frac{1}{2}}\left(\mathcal{P}((D^j\hat{f})\overline{H_\ell\widehat{\psi_0}})E_{2k} + T_{1/2}\mathcal{P}((D^j\hat{f})\overline{H_\ell\widehat{\psi_0}})\,T_{1/2}E_{2k}\right) \\
&= 2^{1/2}\int_0^{\frac{1}{2}}\left(\mathcal{P}((D^j\hat{f})\overline{H_\ell\widehat{\psi_0}}) + T_{1/2}\mathcal{P}((D^j\hat{f})\overline{H_\ell\widehat{\psi_0}})\right)E_{2k}.
\end{aligned}
$$

This calculation shows that $\langle f, D^{j-1}T_k\psi_\ell \rangle$ is the $-k$th coefficient in the Fourier expansion for the $\frac{1}{2}$-periodic function

$$\mathcal{P}((D^j\hat{f})\overline{H_\ell\widehat{\psi_0}}) + T_{1/2}\mathcal{P}((D^j\hat{f})\overline{H_\ell\widehat{\psi_0}})$$

with respect to the orthonormal basis $\{2^{1/2}E_{2k}\}_{k\in\mathbb{Z}} = \{2^{1/2}e^{4\pi ik(\cdot)}\}_{k\in\mathbb{Z}}$ for $L^2(0, \frac{1}{2})$. Using the definition of $F_{j-1,\ell}$, it follows that

$$
\begin{aligned}
F_{j-1,\ell}(\gamma) &= 2^{-1/2}\sum_{k\in\mathbb{Z}}\langle f, D^{j-1}T_k\psi_\ell \rangle 2^{1/2}E_{-2k}(\gamma/2) \\
&= 2^{-1/2}\left(\mathcal{P}((D^j\hat{f})\overline{H_\ell\widehat{\psi_0}}) + T_{1/2}\mathcal{P}((D^j\hat{f})\overline{H_\ell\widehat{\psi_0}})\right)(\gamma/2). \quad (18.12)
\end{aligned}
$$

The function H_ℓ is 1-periodic, so

$$\mathcal{P}((D^j\hat{f})\overline{H_\ell\widehat{\psi_0}}) = \overline{H_\ell}\mathcal{P}((D^j\hat{f})\overline{\widehat{\psi_0}}). \quad (18.13)$$

Also, by the calculation in (18.9) we have $\langle f, D^jT_k\psi_0 \rangle = \langle D^j\hat{f}, E_{-k}\widehat{\psi_0} \rangle$; via Lemma 18.2.1 (check the assumptions),

$$F_{j,0} = \sum_{k\in\mathbb{Z}}\langle D^j\hat{f}, E_{-k}\widehat{\psi_0} \rangle E_{-k} = \mathcal{P}((D^j\hat{f})\overline{\widehat{\psi_0}}). \quad (18.14)$$

Inserting (18.13) and (18.14) in the expression (18.12) for $F_{j-1,\ell}$ finally gives the result. $\qquad\square$

In terms of the matrix H in (18.6), Lemma 18.2.3 shows that

$$\begin{pmatrix} F_{j-1,0}(\gamma) \\ F_{j-1,1}(\gamma) \\ \cdot \\ \cdot \\ F_{j-1,n}(\gamma) \end{pmatrix} = 2^{-1/2}\overline{H(\tfrac{\gamma}{2})}\begin{pmatrix} F_{j,0}(\tfrac{\gamma}{2}) \\ T_{1/2}F_{j,0}(\tfrac{\gamma}{2}) \end{pmatrix}, \ \gamma \in \mathbb{R}. \qquad (18.15)$$

In the rest of the chapter the 2×2 matrix $H(\gamma)^*H(\gamma)$ will play the central role; the key condition turns out to be that this matrix equals the identity matrix I for a.e. $\gamma \in \mathbb{T}$.

Lemma 18.2.4 *Let $\{\psi_\ell, H_\ell\}_{\ell=0}^n$ be as in the general setup, and assume that $H(\gamma)^*H(\gamma) = I$ for a.e. $\gamma \in \mathbb{T}$. Then, for all $j \in \mathbb{Z}$ and all $f \in L^2(\mathbb{R})$ for which $\hat{f} \in C_c(\mathbb{R})$,*

$$\sum_{k \in \mathbb{Z}} |\langle f, D^j T_k \psi_0 \rangle|^2 = \sum_{\ell=0}^n \sum_{k \in \mathbb{Z}} |\langle f, D^{j-1} T_k \psi_\ell \rangle|^2.$$

Proof. The definition of $F_{j-1,\ell}$ and Parseval's equation show that

$$\sum_{\ell=0}^n \sum_{k \in \mathbb{Z}} |\langle f, D^{j-1} T_k \psi_\ell \rangle|^2 = \sum_{\ell=0}^n \int_{-\frac{1}{2}}^{\frac{1}{2}} |F_{j-1,\ell}|^2. \qquad (18.16)$$

The assumption on the matrix $H(\gamma)$ means that we can consider $H(\gamma)$ as an isometry from $\mathbb{C}^2$ into $\mathbb{C}^{n+1}$ for a.e. $\gamma \in \mathbb{T}$. Using this together with (18.15), it follows from (18.16) that

$$\begin{aligned} \sum_{\ell=0}^n \sum_{k \in \mathbb{Z}} |\langle f, D^{j-1} T_k \psi_\ell \rangle|^2 &= 2^{-1} \int_{-\frac{1}{2}}^{\frac{1}{2}} \left\| H(\tfrac{\gamma}{2})\begin{pmatrix} F_{j,0}(\gamma/2) \\ T_{1/2}F_{j,0}(\gamma/2) \end{pmatrix} \right\|_{\mathbb{C}^{n+1}}^2 d\gamma \\ &= 2^{-1} \int_{-\frac{1}{2}}^{\frac{1}{2}} \left\| \begin{pmatrix} F_{j,0}(\gamma/2) \\ T_{1/2}F_{j,0}(\gamma/2) \end{pmatrix} \right\|_{\mathbb{C}^2}^2 d\gamma \\ &= 2^{-1} \int_{-\frac{1}{2}}^{\frac{1}{2}} \left(|F_{j,0}(\gamma/2)|^2 + |T_{1/2}F_{j,0}(\gamma/2)|^2 \right) d\gamma \\ &= \int_{-\frac{1}{2}}^{\frac{1}{2}} |F_{j,0}(\gamma)|^2 d\gamma = \sum_{k \in \mathbb{Z}} |\langle f, D^j T_k \psi_0 \rangle|^2. \end{aligned}$$

This completes the proof. $\qquad\qquad\qquad\qquad\qquad\qquad\qquad\qquad\qquad\quad \square$

The next result shows that the scaling function in the general setup always generates a Bessel sequence.

Lemma 18.2.5 *Let $\{\psi_\ell, H_\ell\}_{\ell=0}^n$ be as in the general setup, and assume that $H(\gamma)^* H(\gamma) = I$ for a.e. $\gamma \in \mathbb{T}$. Then the following hold:*

(i) $\{T_k \psi_0\}_{k \in \mathbb{Z}}$ is a Bessel sequence with bound 1, i.e.,

$$\mathcal{P}(|\hat{\psi}_0|^2) \leq 1.$$

(ii) If $f \in L^2(\mathbb{R})$, then $\lim_{j \to -\infty} \sum_{k \in \mathbb{Z}} |\langle f, D^j T_k \psi_0 \rangle|^2 = 0.$

Proof. Consider a function f for which $\hat{f} \in C_c(\mathbb{R})$. Lemma 18.2.4 shows that for any $j \in \mathbb{Z}$,

$$\sum_{k \in \mathbb{Z}} |\langle f, D^{j-1} T_k \psi_0 \rangle|^2 \leq \sum_{k \in \mathbb{Z}} |\langle f, D^j T_k \psi_0 \rangle|^2. \tag{18.17}$$

Let $\epsilon > 0$ be given. Via Lemma 18.2.2, we can find $j > 0$ such that

$$\sum_{k \in \mathbb{Z}} |\langle f, D^j T_k \psi_0 \rangle|^2 \leq (1 + \epsilon) \|f\|^2.$$

Applying (18.17) j times shows that

$$\sum_{k \in \mathbb{Z}} |\langle f, T_k \psi_0 \rangle|^2 \leq \sum_{k \in \mathbb{Z}} |\langle f, D^j T_k \psi_0 \rangle|^2 \leq (1 + \epsilon) \|f\|^2.$$

Since $\epsilon > 0$ was arbitrary, it follows that $\sum_{k \in \mathbb{Z}} |\langle f, T_k \psi_0 \rangle|^2 \leq \|f\|^2$. Because this inequality holds on a dense subset of $L^2(\mathbb{R})$, it holds on $L^2(\mathbb{R})$ by Lemma 3.2.6. Thus, $\{T_k \psi_0\}_{k \in \mathbb{Z}}$ is a Bessel sequence, and the conclusion in (i) follows by Theorem 9.2.5.

For the proof of (ii), let $f \in L^2(\mathbb{R})$. By (i) and the fact that D^j is unitary, we know that $\{D^j T_k \psi_0\}_{k \in \mathbb{Z}}$ is a Bessel sequence with bound 1 for all $j \in \mathbb{Z}$. Let $I \subset \mathbb{R}$ be any bounded interval; then

$$\sum_{k \in \mathbb{Z}} |\langle f, D^j T_k \psi_0 \rangle|^2 \leq 2 \sum_{k \in \mathbb{Z}} |\langle f \chi_I, D^j T_k \psi_0 \rangle|^2$$

$$+ 2 \sum_{k \in \mathbb{Z}} |\langle f(1 - \chi_I), D^j T_k \psi_0 \rangle|^2$$

$$\leq 2 \sum_{k \in \mathbb{Z}} |\langle f \chi_I, D^j T_k \psi_0 \rangle|^2 + 2 \|f(1 - \chi_I)\|^2.$$

By choosing I sufficiently large, we can make $\|f(1 - \chi_I)\|^2$ arbitrarily small. Thus, it is enough to show that

$$\sum_{k \in \mathbb{Z}} |\langle f \chi_I, D^j T_k \psi_0 \rangle|^2 \to 0 \text{ as } j \to -\infty.$$

Now,

$$\sum_{k \in \mathbb{Z}} |\langle f \chi_I, D^j T_k \psi_0 \rangle|^2 = 2^j \sum_{k \in \mathbb{Z}} \left| \int_I f(x) \overline{\psi_0(2^j x - k)} dx \right|^2 ;$$

Thus,

$$\sum_{k\in\mathbb{Z}}|\langle f\chi_I, D^jT_k\psi_0\rangle|^2 \leq \|f\|^2 2^j \sum_{k\in\mathbb{Z}}\int_I |\psi_0(2^jx-k)|^2 dx$$

$$= \|f\|^2 \sum_{k\in\mathbb{Z}}\int_{2^jI-k} |\psi_0(x)|^2 dx.$$

An application of Lebesgue's dominated convergence theorem yields that the final expression goes to zero as $j \to -\infty$, which concludes the proof. □

We are now ready to formulate and prove the *unitary extension principle*.

Theorem 18.2.6 *Let $\{\psi_\ell, H_\ell\}_{\ell=0}^n$ be as in the general setup, and assume that $H(\gamma)^*H(\gamma) = I$ for a.e. $\gamma \in \mathbb{T}$. Then $\{D^jT_k\psi_\ell\}_{j,k\in\mathbb{Z},\ell=1,\dots,n}$ constitutes a tight frame for $L^2(\mathbb{R})$ with frame bound $A = 1$.*

Proof. Let $\epsilon > 0$ be given, and consider a function f for which $\hat{f} \in C_c(\mathbb{R})$. By Lemma 18.2.2, we can choose $J > 0$ such that for all $j > J$,

$$(1-\epsilon)\|f\|^2 \leq \sum_{k\in\mathbb{Z}}|\langle f, D^jT_k\psi_0\rangle|^2 \leq (1+\epsilon)\|f\|^2. \tag{18.18}$$

For *any* $j \in \mathbb{Z}$, Lemma 18.2.4 shows that

$$\sum_{k\in\mathbb{Z}}|\langle f, D^jT_k\psi_0\rangle|^2 = \sum_{\ell=0}^n\sum_{k\in\mathbb{Z}}|\langle f, D^{j-1}T_k\psi_\ell\rangle|^2$$

$$= \sum_{k\in\mathbb{Z}}|\langle f, D^{j-1}T_k\psi_0\rangle|^2 + \sum_{\ell=1}^n\sum_{k\in\mathbb{Z}}|\langle f, D^{j-1}T_k\psi_\ell\rangle|^2;$$

repeating the argument on $\sum_{k\in\mathbb{Z}}|\langle f, D^{j-1}T_k\psi_0\rangle|^2$, it follows that for all $m < j$,

$$\sum_{k\in\mathbb{Z}}|\langle f, D^jT_k\psi_0\rangle|^2 = \sum_{k\in\mathbb{Z}}|\langle f, D^mT_k\psi_0\rangle|^2 + \sum_{\ell=1}^n\sum_{p=m}^{j-1}\sum_{k\in\mathbb{Z}}|\langle f, D^pT_k\psi_\ell\rangle|^2.$$

Via (18.18), it follows that for all $j > J$ and $m < j$,

$$(1-\epsilon)\|f\|^2 \leq \sum_{k\in\mathbb{Z}}|\langle f, D^mT_k\psi_0\rangle|^2 + \sum_{\ell=1}^n\sum_{p=m}^{j-1}\sum_{k\in\mathbb{Z}}|\langle f, D^pT_k\psi_\ell\rangle|^2$$

$$\leq (1+\epsilon)\|f\|^2.$$

By Lemma 18.2.5, $\lim_{m\to-\infty}\sum_{k\in\mathbb{Z}}|\langle f, D^mT_k\psi_0\rangle|^2 = 0$. Therefore, letting $m \to -\infty$ above yields that for all $j > J$,

$$(1-\epsilon)\|f\|^2 \leq \sum_{\ell=1}^n\sum_{p=-\infty}^{j-1}\sum_{k\in\mathbb{Z}}|\langle f, D^pT_k\psi_\ell\rangle|^2 \leq (1+\epsilon)\|f\|^2.$$

Letting $j \to \infty$,

$$(1 - \epsilon)||f||^2 \le \sum_{\ell=1}^{n} \sum_{p=-\infty}^{\infty} \sum_{k\in\mathbb{Z}} |\langle f, D^p T_k \psi_\ell \rangle|^2 \le (1 + \epsilon)||f||^2.$$

Since $\epsilon > 0$ was arbitrary, we conclude that

$$\sum_{\ell=1}^{n} \sum_{p\in\mathbb{Z}} \sum_{k\in\mathbb{Z}} |\langle f, D^p T_k \psi_\ell \rangle|^2 = ||f||^2$$

for all the considered f; therefore, by Lemma 5.1.9, it holds for all $f \in L^2(\mathbb{R})$, which concludes the proof. □

As noted in [57] and [248], Theorem 18.2.6 holds slightly more general than explained here: it is enough to assume that $H(\gamma)^* H(\gamma)$ is the identity whenever $\mathcal{P}(|\widehat{\psi_0}|^2)(\gamma) > 0$.

The matrix $H(\gamma)^* H(\gamma)$ has four entries, so at a first glance, it seems that we have to solve four scalar equations in order to apply Theorem 18.2.6. However, it turns out that it is enough to verify two sets of equations (Exercise 18.4):

Corollary 18.2.7 *Let $\{\psi_\ell, H_\ell\}_{\ell=0}^{n}$ be as in the general setup on page 446, and assume that*

$$\begin{cases} \displaystyle\sum_{\ell=0}^{n} |H_\ell(\gamma)|^2 = 1, \\[2mm] \displaystyle\sum_{\ell=0}^{n} \overline{H_\ell(\gamma)} T_{1/2} H_\ell(\gamma) = 0, \end{cases} \tag{18.19}$$

for a.e. $\gamma \in \mathbb{T}$. Then the multiwavelet system $\{D^j T_k \psi_\ell\}_{j,k\in\mathbb{Z}, \ell=1,\dots,n}$ constitutes a tight frame for $L^2(\mathbb{R})$ with frame bound $A = 1$.

18.3 Applications to B-splines I

As an application of Theorem 18.2.6 we show how one can construct compactly supported tight multiwavelet frames based on splines. Note that a short survey on splines is in Section A.8. In contrast to the Battle–Lemarié wavelets discussed on page 100, the generators will be *finite* linear combinations of splines $B_m(2x - k), k \in \mathbb{Z}$, and thus have compact support. The price to pay is that we need multiple generators.

Example 18.3.1 For any $m = 1, 2, \dots$, we consider the B-spline

$$\psi_0 := B_{2m}$$

of order $2m$ as defined in (A.15). By Corollary A.8.2,

$$\widehat{\psi}_0(\gamma) = \left(\frac{\sin(\pi\gamma)}{\pi\gamma}\right)^{2m}.$$

It is clear that $\lim_{\gamma\to 0}\widehat{\psi}_0(\gamma) = 1$, and by direct calculation,

$$
\begin{aligned}
\widehat{\psi}_0(2\gamma) &= \left(\frac{\sin(2\gamma)}{2\pi\gamma}\right)^{2m} \\
&= \left(\frac{2\sin(\pi\gamma)\cos(\pi\gamma)}{2\pi\gamma}\right)^{2m} \\
&= \cos^{2m}(\pi\gamma)\widehat{\psi}_0(\gamma).
\end{aligned}
$$

Thus, ψ_0 satisfies a scaling equation with refinement mask

$$H_0(\gamma) = \cos^{2m}(\pi\gamma). \tag{18.20}$$

Now, let $\binom{2m}{\ell}$ denote the binomial coefficient $\frac{(2m)!}{(2m-\ell)!\ell!}$ and define the functions $H_1,\dots,H_{2m}\in L^\infty(\mathbb{T})$ by

$$H_\ell(\gamma) = \sqrt{\binom{2m}{\ell}}\,\sin^\ell(\pi\gamma)\cos^{2m-\ell}(\pi\gamma),\ \ell=1,\dots,2m. \tag{18.21}$$

Using that $\cos(\pi(\gamma-1/2)) = \sin(\pi\gamma)$ and $\sin(\pi(\gamma-1/2)) = -\cos(\pi\gamma)$, it follows that the matrix H in (18.6) is given by

$$H(\gamma) = \begin{pmatrix} H_0(\gamma) & T_{1/2}H_0(\gamma) \\ H_1(\gamma) & T_{1/2}H_1(\gamma) \\ \cdot & \cdot \\ \cdot & \cdot \\ H_{2m}(\gamma) & T_{1/2}H_{2m}(\gamma) \end{pmatrix} =$$

$$\begin{pmatrix} \cos^{2m}(\pi\gamma) & \sin^{2m}(\pi\gamma) \\ \sqrt{\binom{2m}{1}}\sin(\pi\gamma)\cos^{2m-1}(\pi\gamma) & -\sqrt{\binom{2m}{1}}\cos(\pi\gamma)\sin^{2m-1}(\pi\gamma) \\ \sqrt{\binom{2m}{2}}\sin^2(\pi\gamma)\cos^{2m-2}(\pi\gamma) & \sqrt{\binom{2m}{2}}\cos^2(\pi\gamma)\sin^{2m-2}(\pi\gamma) \\ \cdot & \cdot \\ \cdot & \cdot \\ \sqrt{\binom{2m}{2m}}\sin^{2m}(\pi\gamma) & \sqrt{\binom{2m}{2m}}\cos^{2m}(\pi\gamma) \end{pmatrix}.$$

Now consider the 2×2 matrix $M := H(\gamma)^* H(\gamma)$. Using the binomial formula

$$(x+y)^{2m} = \sum_{\ell=0}^{2m} \binom{2m}{\ell} x^\ell y^{2m-\ell}, \qquad (18.22)$$

we see that the first entry in the first row of M is

$$
\begin{aligned}
M_{1,1} &= \sum_{\ell=0}^{2m} \binom{2m}{\ell} \sin^{2\ell}(\pi\gamma) \cos^{2(2m-\ell)}(\pi\gamma) \\
&= \left(\sin^2(\pi\gamma) + \cos^2(\pi\gamma)\right)^{2m} = 1.
\end{aligned}
$$

A similar argument gives that $M_{2,2} = 1$. Also, using the binomial formula with $x = 1, y = -1$,

$$
\begin{aligned}
&M_{1,2} \\
&= \sin^{2m}(\pi\gamma) \cos^{2m}(\pi\gamma) \left(1 - \binom{2m}{1} + \binom{2m}{2} - \cdots + \binom{2m}{2m} \right) \\
&= \sin^{2m}(\pi\gamma) \cos^{2m}(\pi\gamma)(1-1)^{2m} = 0.
\end{aligned}
$$

Thus M is the identity on $\mathbb{C}^2$ for all γ; by Theorem 18.2.6 this implies that the $2m$ functions $\psi_1, \ldots, \psi_{2m}$ defined by

$$
\begin{aligned}
\widehat{\psi_\ell}(\gamma) &= H_\ell(\gamma/2) \widehat{\psi_0}(\gamma/2) \\
&= \sqrt{\binom{2m}{\ell}} \frac{\sin^{2m+\ell}(\pi\gamma/2) \cos^{2m-\ell}(\pi\gamma/2)}{(\pi\gamma/2)^{2m}}
\end{aligned}
$$

generate a tight multiwavelet frame $\{D^j T_k \psi_\ell\}_{j,k\in\mathbb{Z},\ell=1,\ldots,2m}$ for $L^2(\mathbb{R})$. □

We want to study the properties of the frame constructed in Example 18.3.1, but for a reason that will become clear soon we first change the definition slightly by multiplying each of the functions H_ℓ in (18.21) with a complex number of absolute value 1. This modification will not change the frame properties for the generated wavelet system.

Example 18.3.2 We continue Example 18.3.1, but now we define

$$H_\ell(\gamma) = i^\ell \sqrt{\binom{2m}{\ell}} \sin^\ell(\pi\gamma) \cos^{2m-\ell}(\pi\gamma), \quad \ell = 1, \ldots, 2m. \quad (18.23)$$

H_ℓ only differs from the choice in (18.21) by a constant of absolute value 1, so the functions $\psi_1, \ldots, \psi_{2m}$ given by

$$
\begin{aligned}
\widehat{\psi_\ell}(\gamma) &= H_\ell(\gamma/2) \widehat{\psi_0}(\gamma/2) \\
&= i^\ell \sqrt{\binom{2m}{\ell}} \sin^\ell(\pi\gamma/2) \cos^{2m-\ell}(\pi\gamma/2) \widehat{\psi_0}(\gamma/2) \quad (18.24)
\end{aligned}
$$

also generate a tight multiwavelet frame. Instead of inserting the expression for $\widehat{\psi_0}$ in (18.24), we now rewrite $H_\ell(\gamma/2)$ using Euler's formula:

$$
\begin{aligned}
H_\ell(\gamma/2) &= i^\ell \sqrt{\binom{2m}{\ell}} \left(\frac{e^{\pi i \gamma/2} - e^{-\pi i \gamma/2}}{2i}\right)^\ell \left(\frac{e^{\pi i \gamma/2} + e^{-\pi i \gamma/2}}{2}\right)^{2m-\ell} \\
&= 2^{-2m} \sqrt{\binom{2m}{\ell}} \left(e^{\pi i \gamma/2} - e^{-\pi i \gamma/2}\right)^\ell \left(e^{\pi i \gamma/2} + e^{-\pi i \gamma/2}\right)^{2m-\ell}.
\end{aligned}
$$
(18.25)

Via the binomial formula, we see that $H_\ell(\gamma/2)$ is a finite linear combination of terms

$$ e^{-\pi i m \gamma}, e^{-\pi i (m-1)\gamma}, \dots, e^{\pi i (m-1)\gamma}, e^{\pi i m \gamma}. $$

All coefficients in the linear combination are real. Writing $e^{\pi i k \gamma} = E_{k/2}(\gamma)$ and using that

$$ \widehat{\psi_\ell}(\gamma) = \sqrt{2} H_\ell(\gamma/2) D^{-1} \widehat{\psi_0}(\gamma), $$

we see that $\widehat{\psi_\ell}$ is a finite linear combination with real coefficients of terms

$$ E_{\frac{k}{2}} D^{-1} \widehat{\psi_0} = \mathcal{F} T_{-\frac{k}{2}} D \psi_0 = \mathcal{F} D T_k \psi_0, \quad k = -m, \dots, m. $$

Thus, ψ_ℓ is a finite linear combination with real coefficients of the functions

$$ D T_k \psi_0, \quad k = -m, \dots, m. $$
(18.26)

That is, ψ_ℓ is a real-valued spline. Since $D T_m \psi_0$ has support in $[0, m]$ and $D T_{-m} \psi_0$ has support in $[-m, 0]$, the spline ψ_ℓ has support in $[-m, m]$. Our arguments also show that the splines ψ_ℓ inherit other properties from ψ_0: they have degree $2m - 1$, belong to $C^{2m-2}(\mathbb{R})$, and have knots at $\mathbb{Z}/2$.

Let us be more concrete in the case $m = 1$. Here, we obtain two functions ψ_1 and ψ_2. First, via the expression (18.25) for H_1,

$$
\begin{aligned}
\widehat{\psi_1}(\gamma) &= H_1(\gamma/2) \widehat{\psi_0}(\gamma/2) \\
&= \frac{\sqrt{2}}{2^2} (e^{\pi i \gamma/2} - e^{-\pi i \gamma/2})(e^{\pi i \gamma/2} + e^{-\pi i \gamma/2}) \widehat{B_2}(\gamma/2) \\
&= \frac{1}{2}(e^{\pi i \gamma} - e^{-\pi i \gamma}) D^{-1} \widehat{B_2}(\gamma) \\
&= \frac{1}{2}\left(E_{\frac{1}{2}} D^{-1} \mathcal{F} B_2(\gamma) - E_{\frac{-1}{2}} D^{-1} \mathcal{F} B_2(\gamma)\right) \\
&= \frac{1}{2} \mathcal{F}\left(T_{-\frac{1}{2}} D B_2 - T_{\frac{1}{2}} D B_2\right)(\gamma).
\end{aligned}
$$

Thus,

$$
\begin{aligned}
\psi_1(x) &= \frac{1}{2}\left(T_{-\frac{1}{2}} D B_2(x) - T_{\frac{1}{2}} D B_2(x)\right) \\
&= \frac{1}{\sqrt{2}}(B_2(2x+1) - B_2(2x-1)).
\end{aligned}
$$
(18.27)

See Figure 18.1. Similarly one proves (Exercise 18.2) that

$$\psi_2(x) = \frac{1}{2}\left(B_2(2x+1) - 2B_2(2x) + B_2(2x-1)\right), \qquad (18.28)$$

which is shown in Figure 18.2.   □

We note that the computational effort in Example 18.3.1 and Example 18.3.2 increases with the order of the spline B_{2m}. For example, (18.26) shows that we need to calculate a larger number of coefficients in order to find ψ_ℓ when m increases. The number of generators $\psi_1, \ldots, \psi_{2m}$ also increases with the order of the spline B_{2m}. In particular, if we want high smoothness of the generators ψ_ℓ, we are forced to work with splines B_{2m} of high order and therefore a high number of generators. In contrast, for any $m \in \mathbb{N}$, the results in Section 18.5 will allow us to construct multiwavelets with two generators which are finite linear combinations of the spline B_{2m}. That is, any prescribed regularity can be obtained without increasing the number of generators.

Example 18.3.3 In continuation of Example 18.3.2, we can also construct spline frames with support on $[0, 2m]$. We ask the reader to provide the details in Exercise 18.3. Let ψ_0 be the translated B-spline of order $2m$ given by $\psi_0 = T_m \widehat{B_{2m}}$. Then

$$\widehat{\psi_0}(2\gamma) = H_0(\gamma)\widehat{\psi_0}(\gamma) \qquad (18.29)$$

with

$$H_0(\gamma) = \left(\frac{1 + e^{-2\pi i\gamma}}{2}\right)^{2m} = e^{-2\pi i m\gamma}\cos^{2m}(\pi\gamma).$$

Since H_0 appears from the corresponding function in (18.20) simply by multiplication with $e^{-2\pi i m\gamma}$, the functions

$$H_\ell(\gamma) = e^{-2\pi i m\gamma}\sqrt{\binom{2m}{\ell}}\sin^\ell(\pi\gamma)\cos^{2m-\ell}(\pi\gamma), \ \ell = 1, \ldots, 2m$$

satisfy the condition on H in the unitary extension principle. We prefer to multiply the functions with a complex number, i.e., to consider

$$H_\ell(\gamma) = i^\ell e^{-2\pi i m\gamma}\sqrt{\binom{2m}{\ell}}\sin^\ell(\pi\gamma)\cos^{2m-\ell}(\pi\gamma), \ \ell = 1, \ldots, 2m;$$

with this choice, we conclude that the functions $\psi_1, \ldots, \psi_{2m}$ defined by

$$\begin{aligned}
\widehat{\psi_\ell}(\gamma) &= H_\ell(\gamma/2)\widehat{\psi_0}(\gamma/2) \\
&= i^\ell e^{-2\pi i m\gamma}\sqrt{\binom{2m}{\ell}}\frac{\sin^{2m+\ell}(\pi\gamma/2)\cos^{2m-\ell}(\pi\gamma/2)}{(\pi\gamma/2)^{2m}}
\end{aligned}$$

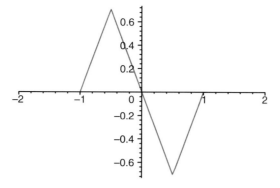

Figure 18.1. The function ψ_1 given by (18.27).

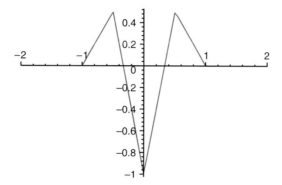

Figure 18.2. The function ψ_2 given by (18.28).

generate a tight multiwavelet frame for $L^2(\mathbb{R})$. Furthermore, the spline functions $\psi_1, \ldots, \psi_{2m}$ now have support on $[0, 2m]$. We return to the case $m = 1$ in Example 18.6.3. $\qquad\qquad\square$

18.4   The Oblique Extension Principle

We now return to the theoretical development. In the entire section, we keep the assumptions in the general setup in Section 18.1, and our purpose is to prove a more flexible version of the unitary extension principle. Let us first give some reasons why we want to do so.

Often, it is desirable that a multiwavelet frame is generated by functions $\{\psi_\ell\}_{\ell=1}^n$ having a large number of vanishing moments. If $\{\psi_\ell\}_{\ell=1}^n$ is constructed via the general setup and the unitary extension principle, we know that $\widehat{\psi_\ell}(\gamma) = H_\ell(\gamma/2)\widehat{\psi_0}(\gamma/2)$ and that $\widehat{\psi_0}(0) = 1$; it follows from here that the number of vanishing moments for ψ_ℓ is equal to the order of zero at $\gamma = 0$ of H_ℓ. This actually puts a restriction on the number of vanishing moments one can obtain for generators constructed via the unitary extension principle:

Example 18.4.1 We return to the B-spline B_{2m} considered in Example 18.3.1; it satisfies the scaling equation with refinement mask

$$H_0(\gamma) = \cos^{2m}(\pi\gamma).$$

If we want to construct a frame via the unitary extension principle, the condition $H(\gamma)^*H(\gamma) = I$ in particular implies that

$$1 = \sum_{\ell=0}^n |H_\ell(\gamma)|^2,$$

i.e., that

$$\sum_{\ell=1}^n |H_\ell(\gamma)|^2 = 1 - \cos^{2m}(\pi\gamma). \tag{18.30}$$

The order of the zero at $\gamma = 0$ for the function $1 - \cos^{2m}(\pi\gamma)$ is 2, so also on the left-hand side of (18.30) we can only factor γ^2 out; this implies that at least one of the functions $|H_\ell|^2$ can at most have a zero at $\gamma = 0$ of order 2, and therefore at least one of the functions ψ_ℓ can at most have one vanishing moment. $\qquad\square$

Another restriction on constructions via the unitary extension principle follows from Corollary 1.4.7: it shows that the assumption $H(\gamma)^*H(\gamma) = I$ implies that

$$|H_0(\gamma)|^2 + |H_0(\gamma + \frac{1}{2})|^2 \le 1.$$

Due to these restrictions, certain frame constructions seem impossible when working with a set of functions $\{\psi_i, H_i\}_{i=0}^n$ as in the general setup. However, sometimes another choice of these functions could lead to a surprising construction! Important reformulations of Theorem 18.2.6 were obtained simultaneously by Daubechies, Han, Ron, and Shen in [248] and Chui, He, and Stöckler in [209]. They give a more flexible recipe for construction of frames than Theorem 18.2.6; we state the version form [248], which is called the *oblique extension principle*:

Theorem 18.4.2 *Let* $\{\psi_\ell, II_\ell\}_{\ell=0}^n$ *be as in the general setup. Assume that there exists a strictly positive function* $\theta \in L^\infty(\mathbb{T})$ *for which*

$$\lim_{\gamma \to 0} \theta(\gamma) = 1$$

and such that for a.e. $\gamma \in \mathbb{T}$,

$$H_0(\gamma)\overline{H_0(\gamma + \nu)}\theta(2\gamma) \;+\; \sum_{\ell=1}^n H_\ell(\gamma)\overline{H_\ell(\gamma + \nu)}$$

$$= \begin{cases} \theta(\gamma) & \text{if } \nu = 0, \\ 0 & \text{if } \nu = \frac{1}{2}. \end{cases} \tag{18.31}$$

Then the functions $\{D^j T_k \psi_\ell\}_{j,k\in\mathbb{Z},\ell=1,\ldots,n}$ *constitute a tight frame for* $L^2(\mathbb{R})$ *with frame bound* $A = 1$.

Proof. Assume that the conditions in Theorem 18.4.2 are satisfied, and define the function $\widetilde{\psi_0} \in L^2(\mathbb{R})$ by

$$\widehat{\widetilde{\psi_0}}(\gamma) = \sqrt{\theta(\gamma)}\widehat{\psi_0}(\gamma). \tag{18.32}$$

Define the 1-periodic functions $\widetilde{H_0}, \ldots, \widetilde{H_n}$ by

$$\widetilde{H_0}(\gamma) = \sqrt{\frac{\theta(2\gamma)}{\theta(\gamma)}}\,H_0(\gamma), \quad \widetilde{H_\ell}(\gamma) = \sqrt{\frac{1}{\theta(\gamma)}}\,H_\ell(\gamma), \quad \ell = 1, \ldots, n. \tag{18.33}$$

The idea in the proof is to apply the unitary extension principle to $\widetilde{\psi_0}, \widetilde{H_0}, \ldots, \widetilde{H_n}$ and thereby obtain a tight frame $\{D^j T_k \widetilde{\psi_\ell}\}_{j,k\in\mathbb{Z},\ell=1,\ldots,n}$; finally, it turns out that $\widetilde{\psi_\ell} = \psi_\ell, \ell = 1, \ldots, n$.

We now prove that $\widetilde{\psi_0}, \widetilde{H_0}, \ldots, \widetilde{H_n}$ satisfy the conditions in the general setup. First,

$$\widehat{\widetilde{\psi_0}}(2\gamma) = \sqrt{\theta(2\gamma)}\widehat{\psi_0}(2\gamma) = \sqrt{\theta(2\gamma)}H_0(\gamma)\widehat{\psi_0}(\gamma)$$

$$= \sqrt{\frac{\theta(2\gamma)}{\theta(\gamma)}}\,H_0(\gamma)\widehat{\widetilde{\psi_0}}(\gamma) = \widetilde{H_0}(\gamma)\widehat{\widetilde{\psi_0}}(\gamma).$$

Also,

$$\lim_{\gamma \to 0} \widehat{\widetilde{\psi_0}}(\gamma) = \lim_{\gamma \to 0} \left(\sqrt{\theta(\gamma)}\widehat{\psi_0}(\gamma) \right) = 1.$$

Via the definition (18.33) and (18.31) with $\nu = 0$,

$$\sum_{\ell=0}^n |\widetilde{H_\ell}(\gamma)|^2 = \frac{\theta(2\gamma)}{\theta(\gamma)}|H_0(\gamma)|^2 + \sum_{\ell=1}^n \frac{|H_\ell(\gamma)|^2}{\theta(\gamma)} = 1,$$

so $\widetilde{H}_0, \ldots, \widetilde{H}_n \in L^\infty(\mathbb{T})$. Because $\theta(2(\gamma + \frac{1}{2})) = \theta(2\gamma)$, we also have

$$\sum_{\ell=0}^{n} \widetilde{H}_\ell(\gamma)\overline{\widetilde{H}_\ell(\gamma + \frac{1}{2})} = \frac{\theta(2\gamma)}{\sqrt{\theta(\gamma)\theta(\gamma + \frac{1}{2})}} H_0(\gamma)\overline{H_0(\gamma + \frac{1}{2})}$$

$$+ \frac{1}{\sqrt{\theta(\gamma)\theta(\gamma + \frac{1}{2})}} \sum_{\ell=1}^{n} H_\ell(\gamma)\overline{H_\ell(\gamma + \frac{1}{2})} = 0.$$

Defining the functions $\widetilde{\psi}_1, \ldots, \widetilde{\psi}_n$ by

$$\widehat{\widetilde{\psi}_\ell}(2\gamma) = \widetilde{H}_\ell(\gamma)\widehat{\widetilde{\psi}_0}(\gamma), \ \ell = 1, \ldots, n, \tag{18.34}$$

it follows from Theorem 18.2.6 that the functions $\{D^j T_k \widetilde{\psi}_\ell\}_{j,k \in \mathbb{Z}, \ell=1,\ldots,n}$ constitute a tight frame for $L^2(\mathbb{R})$ with frame bound $A = 1$. The proof is now completed by the observation that for $\ell = 1, \ldots, n$,

$$\widehat{\psi_\ell}(2\gamma) = H_\ell(\gamma)\widehat{\psi_0}(\gamma) = \sqrt{\theta(\gamma)}\widetilde{H}_\ell(\gamma)\frac{1}{\sqrt{\theta(\gamma)}}\widehat{\widetilde{\psi}_0}(\gamma) = \widehat{\widetilde{\psi}_\ell}(2\gamma),$$

which shows that $\psi_\ell = \widetilde{\psi}_\ell$. □

By taking $\theta = 1$ in Theorem 18.4.2, we obtain Theorem 18.2.6. From the extra freedom in Theorem 18.4.2 concerning the choice of θ, one could expect it to be a more general result than Theorem 18.2.6, but the proof shows that the class of frames which can be constructed is the same for the two theorems. However, in practice Theorem 18.4.2 gives more flexibility because it naturally leads to some constructions one would not expect from Theorem 18.2.6. Let us explain this in more detail.

First, it is clear that any construction via Theorem 18.2.6 can be performed in exactly the same way via Theorem 18.4.2, simply by taking $\theta = 1$. On the other hand, suppose that ψ_0 is a compactly supported function satisfying (18.4) for some function $H_0 \in L^\infty(\mathbb{T})$ and that the functions $\theta, H_\ell, \ell = 1, \ldots, n$ are trigonometric polynomials satisfying the conditions in Theorem 18.4.2. Writing $H_\ell(\gamma) = \sum_{k \in \mathbb{Z}} c_{k\ell} e^{2\pi i k\gamma}$ (a finite sum), the definition of ψ_ℓ yields that

$$\widehat{\psi_\ell}(2\gamma) = H_\ell(\gamma)\widehat{\psi_0}(\gamma) = \mathcal{F}\sum_{k \in \mathbb{Z}} c_{k\ell} T_{-k}\psi_0(\gamma).$$

This shows that the frame $\{D^j T_k \psi_\ell\}_{j,k \in \mathbb{Z}, \ell=1,\ldots,n}$ is generated by functions having compact support. Now, the proof of Theorem 18.4.2 shows that the same frame can be constructed via Theorem 18.2.6: If we define $\widetilde{\psi}_0$ by (18.32) and $\widetilde{\psi}_\ell$ by (18.34) and (18.33), then $\psi_\ell = \widetilde{\psi}_\ell$ and the functions $\widetilde{\psi}_\ell$ will satisfy the conditions in the unitary extension principle. However, $\widetilde{\psi}_0$ is in general not compactly supported, so the fact that the resulting frame $\{D^j T_k \psi_\ell\}_{j,k \in \mathbb{Z}, \ell=1,\ldots,n}$ is generated by compactly supported functions is somewhat miraculous and could certainly not be predicted in advance. In short, this shows that there are constructions which appear naturally via Theorem 18.4.2, but one would not even think about constructing them via Theorem 18.2.6.

The flexibility of the oblique extension principle is demonstrated in [248], where tight frames are obtained via some kind of interpolation between the B-splines considered in Example 18.3.1 and the functions that were used by Daubechies in her construction of orthonormal wavelet bases with compact support. We refer to [248] for details and examples.

18.5   Fewer Generators

The computational effort increases with the number of generators in a multiwavelet frame, so in general we wish to have as few generators as possible. Ignoring the issue of "good properties," the best would be to construct a pair of functions $\psi, \widetilde{\psi}$ such that the associated wavelet systems $\{D^j T_k \psi\}_{j,k \in \mathbb{Z}}, \{D^j T_k \widetilde{\psi}\}_{j,k \in \mathbb{Z}}$ form a pair of dual frames. However, as proved by Chui, He, and Stöckler [209], there are several natural cases where this is impossible:

Theorem 18.5.1 *Assume that $\{\psi_0, H_0\}$ are as in the general setup and that $\{T_k \psi_0\}_{k \in \mathbb{Z}}$ is a Riesz sequence. If $|H_0(-\frac{1}{4})| \neq \frac{1}{\sqrt{2}}$, then there does not exist a dual wavelet frame pair $\{D^j T_k \psi\}_{j,k \in \mathbb{Z}}, \{D^j T_k \widetilde{\psi}\}_{j,k \in \mathbb{Z}}$ for which $\psi, \widetilde{\psi}$ are compactly supported and $\psi \in V_1 = \overline{span}\{DT_k \psi_0\}_{k \in \mathbb{Z}}$.*

Although there seems to be several assumptions in Theorem 18.5.1, it excludes certain desirable constructions with B-splines. Consider, for example, a B-spline B_m of order $m > 1$. By Lemma 3.7.5, we know that $\{B_m(\cdot - k)\}_{k \in \mathbb{Z}}$ is a Riesz sequence, and B_m satisfies (by a calculation as in Example 18.3.1) a scaling equation with refinement mask $H_0(\gamma) = \cos^m(\pi\gamma)$. In particular, $|H_0(-\frac{1}{4})| = 2^{-m/2}$. Thus, for $m > 1$, there do not exist dual wavelet pairs $\{D^j T_k \psi\}_{j,k \in \mathbb{Z}}, \{D^j T_k \widetilde{\psi}\}_{j,k \in \mathbb{Z}}$, for which ψ is a finite linear combination of functions $DT_k B_m, k \in \mathbb{Z}$. However, multiwavelet frames with generators made up by finite linear combinations of $DT_k B_m$ exist, as we already saw in Example 18.3.2; other constructions will be given in Example 18.6.

In passing, we note that the case $m = 1$ actually has to be excluded in the above discussion. In fact, the Haar function in (3.43) can be written in terms of the B-spline $B_1 = \chi_{[-\frac{1}{2},\frac{1}{2}]}$, namely, as

$$\psi = \frac{1}{\sqrt{2}} \left(DT_{-\frac{1}{2}} B_1 - DT_{-\frac{3}{2}} B_1 \right) = \frac{1}{\sqrt{2}} T_{-\frac{1}{4}} \left(DB_1 - DT_{-1} B_1 \right);$$

since the Haar function generates an orthonormal basis for $L^2(\mathbb{R})$, also the function $\frac{1}{\sqrt{2}} \left(DB_1 - DT_{-1} B_1 \right)$ generates an orthonormal basis for $L^2(\mathbb{R})$.

Now we return to the oblique extension principle and show how to construct frames with two or three generators. We still follow the approach in [248]. Other constructions of multiwavelet frames with few generators were in fact previously given by Chui and He [207] and Petukhov [538]. They proved in particular that the general setup together with the assumption

$$|H_0(\gamma)|^2 + |H_0(\gamma + \tfrac{1}{2})|^2 \leq 1$$

always makes it possible to construct a frame with two generators; if H_0 is a polynomial of degree m, one can choose H_1 and H_2 as polynomials of degree at most m. Petukhov describes all solutions to the matrix equation $H(\gamma)^* H(\gamma) = I$ for $n = 2$ in [539].

In order to apply the oblique extension principle, one needs to choose the functions θ and $H_1, \ldots, H_n$ simultaneously such that (18.31) is satisfied. It is not clear how to do this in general, but we now prove that an extra condition on the choice of θ will make it easy to construct frames.

Corollary 18.5.2 *Let ψ_0 and H_0 be as in the general setup on page 446. Let $\theta \in L^\infty(\mathbb{T})$ be a strictly positive function for which $\lim_{\gamma \to 0} \theta(\gamma) = 1$, chosen such that the function*

$$\eta(\gamma) := \theta(\gamma) - \theta(2\gamma)\left(|H_0(\gamma)|^2 + |H_0(\gamma + \tfrac{1}{2})|^2\right) \qquad (18.35)$$

is positive as well. Fix an integer $n \geq 2$ and let $\{G_\ell\}_{\ell=2}^n$ be trigonometric polynomials for which

$$\sum_{\ell=2}^n |G_\ell(\gamma)|^2 = 1, \quad and \quad \sum_{\ell=2}^n G_\ell(\gamma)\overline{G_\ell(\gamma + \tfrac{1}{2})} = 0. \qquad (18.36)$$

Let ρ, σ be 1-periodic functions such that

$$|\rho(\gamma)|^2 = \theta(\gamma), \quad |\sigma(\gamma)|^2 = \eta(\gamma), \qquad (18.37)$$

and define $\{H_\ell\}_{\ell=1}^n$ by

$$H_1(\gamma) = e^{2\pi i \gamma}\rho(2\gamma)\overline{H_0(\gamma + \tfrac{1}{2})}, \quad H_\ell(\gamma) = G_\ell(\gamma)\sigma(\gamma), \quad \ell = 2, \ldots, n.$$

Then the functions $\{\psi_\ell\}_{\ell=1}^n$ given by (18.5) generate a tight frame $\{D^j T_k \psi_\ell\}_{j,k\in\mathbb{Z}, \ell=1,\ldots,n}$ for $L^2(\mathbb{R})$, with frame bound $A = 1$.

Proof. We check that the functions θ and H_ℓ satisfy (18.31). First,

$$|H_0(\gamma)|^2\theta(2\gamma) + \sum_{\ell=1}^n |H_\ell(\gamma)|^2$$

$$= |H_0(\gamma)|^2\theta(2\gamma) + |H_0(\gamma + \tfrac{1}{2})|^2|\rho(2\gamma)|^2 + |\sigma(\gamma)|^2 \sum_{\ell=2}^n |G_\ell(\gamma)|^2$$

$$= |H_0(\gamma)|^2\theta(2\gamma) + |H_0(\gamma + \tfrac{1}{2})|^2\theta(2\gamma) + \eta(\gamma) = \theta(\gamma).$$

Similarly,

$$\overline{H_0(\gamma)}H_0(\gamma + \frac{1}{2})\theta(2\gamma) + \sum_{\ell=1}^{n}\overline{H_\ell(\gamma)}H_\ell(\gamma + \frac{1}{2})$$

$$= \overline{H_0(\gamma)}H_0(\gamma + \frac{1}{2})\theta(2\gamma)$$

$$+\overline{\rho(2\gamma)}\rho(2(\gamma + \frac{1}{2}))e^{2\pi i\gamma}e^{-2\pi i(\gamma+1/2)}\overline{H_0(\gamma)}H_0(\gamma + \frac{1}{2})$$

$$+\overline{\sigma(\gamma)}\sigma(\gamma + \frac{1}{2})\sum_{\ell=2}^{n}\overline{G_\ell(\gamma)}G_\ell(\gamma + \frac{1}{2})$$

$$= \overline{H_0(\gamma)}H_0(\gamma + \frac{1}{2})\theta(2\gamma) - \theta(2\gamma)\overline{H_0(\gamma)}H_0(\gamma + \frac{1}{2}) = 0.$$

This concludes the proof. □

A necessary condition for application of the oblique extension principle is that

$$\theta(\gamma) - |H_0(\gamma)|^2\theta(2\gamma) = \sum_{\ell=1}^{n}|H_\ell(\gamma)|^2. \tag{18.38}$$

In particular, the expression on the left-hand side of (18.38) has to be positive; the condition (18.35) on η can naturally be considered as a strengthening of this. In the next section, we provide examples where condition (18.35) is satisfied; as soon as this is the case, Corollary 18.5.2 makes it relatively easy to obtain frames with, for example, three generators. In fact, (18.36) is satisfied with

$$G_2(\gamma) = \frac{1}{\sqrt{2}}, \quad G_3(\gamma) = \frac{1}{\sqrt{2}}e^{2\pi i\gamma}. \tag{18.39}$$

Thus, in order to apply Corollary 18.5.2, the remaining work consists in finding 1-periodic functions ρ, σ such that (18.37) is satisfied. This can be done via spectral factorization (cf. Lemma 3.8.5).

The assumption (18.35) even implies that we can construct a frame generated by two functions:

Corollary 18.5.3 *Let ψ_0 and H_0 be as in the general setup on page 446. Let $\theta \in L^\infty(\mathbb{T})$ be a strictly positive function for which $\lim_{\gamma\to 0}\theta(\gamma) = 1$, chosen such that the function η in (18.35) is positive as well. Define the functions ρ, σ as in (18.37) and let*

$$H_1(\gamma) = e^{2\pi i\gamma}\rho(2\gamma)\overline{H_0(\gamma + \frac{1}{2})}, \quad H_2(\gamma) = H_0(\gamma)\sigma(2\gamma). \tag{18.40}$$

Then the functions $\{\psi_\ell\}_{\ell=1}^{2}$ given by (18.5) generate a tight frame $\{D^jT_k\psi_\ell\}_{j,k\in\mathbb{Z},\ell=1,2}$ for $L^2(\mathbb{R})$, with frame bound $A = 1$.

The proof is similar to the proof of Corollary 18.5.2, except that one has to replace the function θ in the oblique extension principle by $\theta - \eta$.

Note that if θ and H_0 are trigonometric polynomials, then η defined in (18.35) is also a trigonometric polynomial. The assumption that θ and η are positive implies by Lemma 3.8.5 that we can choose ρ, σ in (18.37) to be trigonometric polynomials. In this case, the generators ψ_ℓ in the above corollaries are finite linear combinations of functions $DT_k\psi_0$.

18.6   Applications to B-splines II

The oblique extension principle turns out to be very useful in order to construct multiwavelet frames based on B-splines. Even the extra assumptions in Section 18.5 for reduction to two or three generators can be fulfilled:

Theorem 18.6.1 *Let B_{2m} denote the B-spline of order $2m$ with refinement mask $H_0(\gamma) = \cos^{2m}(\pi\gamma)$. Then, for each positive integer $M \leq 2m$, there exists a trigonometric polynomial θ of the form*

$$\theta(\gamma) = 1 + \sum_{j=1}^{M-1} c_j \sin^{2j}(\pi\gamma), \tag{18.41}$$

for which the following hold:

(i) $c_j \geq 0$ for all $j = 1, \ldots, M-1$,   i.e., $\theta(\gamma) > 0$ for all $\gamma \in \mathbb{R}$;

(ii) The function η in (18.35) is positive;

(iii) The generators in the tight wavelet frames constructed via the oblique extension principle and its corollaries have M vanishing moments.

The coefficients c_j, $j = 1, \ldots, M-1$ can be determined via the requirement that

$$\left(1 + \sum_{j=1}^{\infty} \frac{(2j-1)!!}{(2j)!!\,(2j+1)} y^j\right)^{4m} = 1 + \sum_{j=1}^{M-1} c_j y^j + O(|y|^M) \ \text{ as } y \to 0.$$

$$\tag{18.42}$$

Theorem 18.6.1 is proved in [248]. Thus, we can apply the results in Section 18.5 to construct multiwavelet frames with two or more generators based on *any* B-spline B_m. Let us for convenience consider Corollary 18.5.3; the same considerations will be valid for the other results in Section 18.5. If we choose the functions ρ, σ in (18.37) to be trigonometric polynomials, then the functions H_1, H_2 in (18.40) are trigonometric polynomials, which implies that the associated frame generators ψ_1, ψ_2 are finite linear combinations of functions $B_m(2x - k), k \in \mathbb{Z}$. By choosing m large enough, we can thus obtain generators belonging to any prescribed smoothness class

$C^N(\mathbb{R})$. In contrast with what we obtained for applications of the unitary extension principle, the number of generators is *not* forced to grow with the desired smoothness.

Example 18.6.2 Let us find the trigonometric polynomial associated with the B-spline B_{2m}, $m \in \mathbb{N}$, and $M = 2$. Note that

$$\left(1 + \sum_{j=1}^{\infty} \frac{(2j-1)!!}{(2j)!!\,(2j+1)} y^j \right)^{4m} = \left(1 + \frac{1}{6}y + \frac{1}{20}y^2 + \cdots \right)^{4m}$$

$$= 1 + \frac{2m}{3}y + O(|y|^2).$$

This proves that for $M = 2$, (18.42) is satisfied with $c_1 = 2m/3$. Thus, the desired trigonometric polynomial is

$$\theta(\gamma) = 1 + \frac{2m}{3}\sin^2(\pi\gamma) = 1 + \frac{2m}{3}\frac{1 - \cos(2\pi\gamma)}{2} = \frac{3+m}{3} - \frac{m}{3}\cos(2\pi\gamma). \ \square$$

We now give an example of frame constructions via Theorem 18.2.6 and Theorem 18.4.2.

Example 18.6.3 We return to the translated B-spline in Example 18.3.3 with $m = 1$; that is, we consider $\psi_0 = T_1 B_2$ and the refinement mask

$$H_0(\gamma) = \frac{(1 + e^{-2\pi i\gamma})^2}{4} = e^{-2\pi i\gamma}\cos^2(\pi\gamma).$$

We first revisit Example 18.3.3 and then give constructions via the oblique extension principle and its corollaries.

(i) Define H_1 and H_2 by

$$H_1(\gamma) = ie^{-2\pi i\gamma}\sqrt{2}\sin(\pi\gamma)\cos(\pi\gamma) = \frac{1}{\sqrt{2}}e^{-2\pi i\gamma}i\sin(2\pi\gamma)$$

$$= \frac{\sqrt{2}}{4}(1 - e^{-4\pi i\gamma}),$$

$$H_2(\gamma) = -e^{-2\pi i\gamma}\sin^2(\pi\gamma) = \frac{(1 - e^{-2\pi i\gamma})^2}{4}. \tag{18.43}$$

It follows from Example 18.3.3 that the functions $\psi_1^{(i)} := \psi_1$ and ψ_2 defined via (18.5) generate a tight frame for $L^2(\mathbb{R})$; they are obtained by translation by one of their counterparts in (18.27) and (18.28), i.e.,

$$\psi_1^{(i)}(x) = \frac{1}{\sqrt{2}}(B_2(2x - 1) - B_2(2x - 3)), \tag{18.44}$$

$$\psi_2(x) = \frac{1}{2}(B_2(2x - 1) - 2B_2(2x - 2) + B_2(2x - 3)). \tag{18.45}$$

See Figures 18.3 and 18.4.

(ii) An alternative construction can be obtained via the oblique extension principle. Let

$$\theta(\gamma) := \frac{4 - \cos(2\pi\gamma)}{3}. \qquad (18.46)$$

In this example, we keep the choice of H_2 in (18.43) and therefore the function ψ_2 in (18.45). Thus, if we want to use the oblique extension principle, we have to choose H_1 such that the two conditions in (18.31) are satisfied; that is, we require that

$$|H_1(\gamma)|^2 \;=\; \theta(\gamma) - |H_0(\gamma)|^2\theta(2\gamma) - |H_2(\gamma)|^2,$$

$$\overline{H_1(\gamma)}H_1(\gamma + \tfrac{1}{2}) \;=\; -\overline{H_0(\gamma)}H_0(\gamma + \tfrac{1}{2})\theta(2\gamma) - \overline{H_2(\gamma)}H_2(\gamma + \tfrac{1}{2}).$$

Inserting $\theta, H_0,$ and H_1 leads to the equations

$$|H_1(\gamma)|^2 \;=\; \frac{1}{6}(\cos(2\pi\gamma) + 2)^2(\cos(2\pi\gamma) - 1)^2,$$

$$\overline{H_1(\gamma)}H_1(\gamma + \tfrac{1}{2}) \;=\; \frac{1}{6}(\cos(2\pi\gamma) + 2)(\cos(2\pi\gamma) - 2)$$
$$\times(\cos(2\pi\gamma) - 1)(\cos(2\pi\gamma) + 1).$$

These equations are satisfied if we let

$$H_1(\gamma) \;=\; \frac{1}{\sqrt{6}}(\cos(2\pi\gamma) + 2)(\cos(2\pi\gamma) - 1)$$

$$=\; \frac{1}{\sqrt{6}}(\cos^2(2\pi\gamma) + \cos(2\pi\gamma) - 2)$$

$$=\; \frac{1}{4\sqrt{6}}(e^{4\pi i\gamma} + e^{-4\pi i\gamma} + 2e^{2\pi i\gamma} + 2e^{-2\pi i\gamma} - 6).$$

Via the choice of ψ_1 in the general setup,

$$\widehat{\psi_1}(\gamma) \;=\; H_1(\gamma/2)\widehat{\psi_0}(\gamma/2)$$

$$=\; \frac{1}{4\sqrt{6}}(e^{2\pi i\gamma} + e^{-2\pi i\gamma} + 2e^{\pi i\gamma} + 2e^{-\pi i\gamma} - 6)(\mathcal{F}T_1 B_2)(\gamma/2)$$

$$=\; \frac{1}{4\sqrt{3}}\left(E_1 + E_{-1} + 2E_{1/2} + 2E_{-1/2} - 6\right)D^{-1}(\mathcal{F}T_1 B_2)(\gamma)$$

$$=\; \frac{1}{4\sqrt{3}}\mathcal{F}\left(T_{3/2} + T_{-1/2} + 2T_1 + 2 - 6T_{1/2}\right)DB_2(\gamma).$$

Thus,

$$\psi_1(x) \;=\; \frac{1}{2\sqrt{6}}\left(B_2(2x - 3) + B_2(2x + 1) + 2B_2(2x - 2)\right)$$

$$+\frac{1}{2\sqrt{6}}\left(2B_2(2x) - 6B_2(2x - 1)\right).$$

This function has support on $[-1, 2]$. Instead of taking this generator, we take

$$\psi_1^{(ii)}(x) \; := \; \psi_1(x-1) \tag{18.47}$$

$$= \; \frac{1}{2\sqrt{6}}\,(B_2(2x-5) + 2B_2(2x-4) - 6B_2(2x-3))$$

$$+ \frac{1}{2\sqrt{6}}\,(2B_2(2x-2) + B_2(2x-1)),$$

which generate the same wavelet system and has support on $[0, 3]$. The function $\psi_1^{(ii)}$ is shown in Figure 18.5; by construction, the functions $\psi_1^{(ii)}$ and ψ_2 in (18.45) generate a tight frame.

(iii) Systematic constructions with two generators can be given via Corollary 18.5.3. We again define θ by (18.46). Then, the function η in (18.35) is

$$\eta(\gamma) \; = \; \theta(\gamma) - \theta(2\gamma)\left(|H_0(\gamma)|^2 + |H_0(\gamma + \tfrac{1}{2})|^2\right)$$

$$= \; \frac{4 - \cos(2\pi\gamma)}{3} - \frac{4 - \cos(4\pi\gamma)}{3}\left(\cos^4(\pi\gamma) + \cos^4(\pi(\gamma + 1/2))\right)$$

$$= \; \frac{2}{3}(8\cos^4(\pi\gamma) + 1)(\cos(\pi\gamma) - 1)^2(\cos(\pi\gamma) + 1)^2.$$

Since $\eta(\gamma) \geq 0$ for all γ, the conditions in Corollary 18.5.3 are satisfied. Thus, the remaining work consists in extracting a square root ρ of the function θ and a square root σ of the function η.

Writing $\theta(\gamma) = -\frac{1}{3}(\cos(2\pi\gamma) - 4)$, the general procedure for spectral factorization shows that we can take

$$\rho(\gamma) = \left(\frac{1}{2 \cdot 3(4 - \sqrt{15})}\right)^{1/2} (e^{2\pi i\gamma} - 4 - \sqrt{15}).$$

Concerning the square root of η, we first find a square root of $8\cos^4(\gamma)+1$. Note that

$$8y^2 + 1 \; = \; 8(y^4 + \frac{1}{8}) = 8(y^2 - 8^{1/2}i)(y^2 - 8^{1/2}i)$$

$$= \; 8(y - 8^{1/4}e^{i\frac{\pi}{4}})(y - 8^{1/4}e^{i\frac{5\pi}{4}})(y - 8^{1/4}e^{i\frac{3\pi}{4}})(y - 8^{1/4}e^{i\frac{7\pi}{4}}).$$

According to the general theory for spectral factorization, we let

$$z_1 \; = \; 8^{1/4}e^{i\frac{\pi}{4}} + \sqrt{8^{1/2}e^{i\frac{\pi}{2}} - 1} = 8^{1/4}e^{i\frac{\pi}{4}} + 1 + \sqrt{2}i,$$

$$z_2 \; = \; 8^{1/4}e^{i\frac{3\pi}{4}} + \sqrt{8^{1/2}e^{i\frac{3\pi}{2}} - 1} = 8^{1/4}e^{i\frac{3\pi}{4}} + 1 - \sqrt{2}i.$$

As a square root of $8\cos^4(\gamma) + 1$, we can take

$$\left(\frac{1}{2|z_1|\,|z_2|}\right)^{1/2} (e^{-2i\gamma} - 2e^{-i\gamma}Re(z_1) + |z_1|^2)(e^{-2i\gamma} - 2e^{-i\gamma}Re(z_2) + |z_2|^2);$$

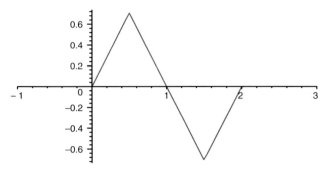

Figure 18.3. The function $\psi_1^{(i)}$ given by (18.44).

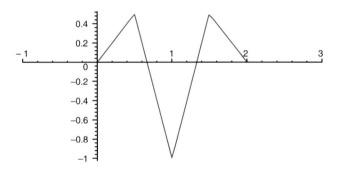

Figure 18.4. The function ψ_2 given by (18.45).

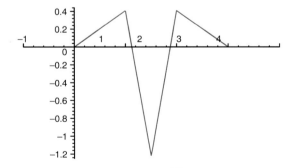

Figure 18.5. The function $\psi_1^{(ii)}$ given by (18.47).

evaluating this function in $\pi\gamma$ rather than γ and multiplying with the function $\sqrt{\frac{2}{3}}(\cos(\pi\gamma) - 1)(\cos(\pi\gamma) + 1)$, we obtain a square root σ of η. Inserting the expressions for ρ and η in Corollary 18.5.3, we again obtain a tight multiwavelet frame with two generators; we will not perform these calculations. The above computations are rather lengthy and cumbersome, and other choices of θ could be made such that the spectral factorization was easier. There is, however, a special reason for the choice of θ, which will be apparent from Section 18.7. We present a related construction in Example 18.9.1. □

18.7 Approximation Orders

In this section we give some more reasons for constructing frames via the oblique extension principle. More information can be found in [248]. We assume again that $\{H_\ell, \psi_\ell\}_{\ell=0}^n$ is as in the general setup and that $\{D^j T_k \psi_\ell\}_{j,k\in\mathbb{Z},\ell=1,\ldots,n}$ is a tight frame constructed via the oblique extension principle. Based on the refinable function ψ_0, we let

$$V_j = \overline{\mathrm{span}}\{D^j T_k \psi_0\}_{j,k\in\mathbb{Z}}.$$

The oblique extension principle and its corollaries give some freedom in the construction of tight frames, due to the different choices of θ one can start with. However, for practical purposes, the main point is which properties we can expect of the constructed frame, and it turns out that some desirable properties will restrict the class of usable functions θ considerably. Let H_s denote the Sobolev space defined in (A.5). We say that ψ_0 *provides approximation order s* if for all f in the Sobolev space $H^s(\mathbb{R})$,

$$\mathrm{dist}(f, V_j) = O(2^{-js}),$$

i.e., if there exists a constant $C > 0$ such that

$$\mathrm{dist}(f, V_j) \leq C 2^{-js}, \ \forall j \in \mathbb{Z}.$$

For the tight frame $\{D^j T_k \psi_\ell\}_{j,k\in\mathbb{Z},\ell=1,\ldots,n}$, we know from the frame decomposition (5.7) that for all $f \in L^2(\mathbb{R})$,

$$f = \sum_{\ell=1}^n \sum_{j,k\in\mathbb{Z}} \langle f, D^j T_k \psi_\ell\rangle D^j T_k \psi_\ell.$$

As an approximation of f, we can thus use

$$Q_J f := \sum_{\ell=1}^n \sum_{j<J} \sum_{k\in\mathbb{Z}} \langle f, D^j T_k \psi_\ell\rangle D^j T_k \psi_\ell$$

for a reasonably large value of $J \in \mathbb{Z}$. We say that *the frame* $\{D^j T_k \psi_\ell\}_{j,k \in \mathbb{Z}, \ell=1,\ldots,n}$ *provides approximation order* s if for all $f \in H^s(\mathbb{R})$,

$$\|f - Q_J f\| = O(2^{-sJ}).$$

When speaking about "the approximation order," it is in both cases understood that we mean the largest possible order.

By Lemma 17.3.1, we know that $\psi_1, \ldots, \psi_n \in V_1$, so $Q_J f \in V_J$ for all $J \in \mathbb{Z}$; thus, the approximation order of the frame $\{D^j T_k \psi_\ell\}_{j,k \in \mathbb{Z}, \ell=1,\ldots,n}$ cannot exceed the approximation order of the underlying refinable function ψ_0. Note that in the case of a classical multiresolution analysis, where a refinable function leads to the construction of an orthonormal basis $\{D^j T_k \psi\}_{j,k \in \mathbb{Z}}$ for $L^2(\mathbb{R})$, the operator Q_J is the orthogonal projection onto V_j and the two types of approximation orders coincide; in general they might be different.

Since every implementation has to be done with a finite collection of vectors, the approximation order of $\{D^j T_k \psi_\ell\}_{j,k \in \mathbb{Z}, \ell=1,\ldots,n}$ is clearly important in applications: we want it to be as large as possible. Assume that the refinable function ψ_0 provides approximation order s and consider the function

$$\Theta(\gamma) = \sum_{j=0}^{\infty} \sum_{\ell=1}^{n} |H_\ell(\gamma)|^2 \prod_{m=0}^{j-1} |H_0(2^m \gamma)|^2,$$

which appeared already in Theorem 18.1.1. One can prove that the approximation order of $\{D^j T_k \psi_\ell\}_{j,k \in \mathbb{Z}, \ell=1,\ldots,n}$ is $\min(s, p)$, where p is the order of zero of $1 - \Theta |\widehat{\psi_0}|^2$ at the origin. It is clear that the approximation order of $\{D^j T_k \psi_\ell\}_{j,k \in \mathbb{Z}, \ell=1,\ldots,n}$ only can be maximal (i.e., reach the value s) for some special functions Θ.

One can prove that for given functions $H_0, \ldots H_n$, the function Θ satisfies

$$\Theta(\gamma) = |H_0(\gamma)|^2 \Theta(\gamma) + \sum_{\ell=1}^{n} |H_\ell(\gamma)|^2,$$

which is one of the two conditions in the oblique extension principle if we take $\theta = \Theta$; thus, exactly this choice for θ is very natural.

Fortunately, in the important case where the general setup is based on B-splines, there exists an appropriate choice of θ, which also leads to fulfillment of the other conditions in the oblique extension principle. In fact, there exists a *unique* trigonometric polynomial θ of minimal order satisfying the conditions; for the B-spline B_2 of order 2, this is exactly the function $\theta(\gamma) = \frac{4 - \cos(\gamma)}{3}$ we used in Example 18.6.3(ii).

The approximation order also comes in if we want the functions $\{\psi_\ell\}_{\ell=1}^{n}$ to have a high number of vanishing moments. In fact, if the refinable function ψ_0 provides approximation order s and all the functions $\{\psi_\ell\}_{\ell=1}^{n}$ have at least m' vanishing moments, then $\{D^j T_k \psi_\ell\}_{j,k \in \mathbb{Z}, \ell=1,\ldots,n}$ has approximation order $\min(s, 2m')$. Thus, high approximation orders for ψ_0 as well as $\{D^j T_k \psi_\ell\}_{j,k\in\mathbb{Z},\ell=1,\dots,n}$ force a large number of vanishing moments.

18.8   Construction of Pairs of Dual Wavelet Frames

So far the constructions via the extension principles have concerned tight frames. However, the technique is more far-reaching, and one can actually extend the results and construct dual wavelet pairs. We cite a result from [248], which was (in a slightly different form) also obtained by Chui, He, and Stöckler [209]:

Theorem 18.8.1 *Let $\{H_\ell, \psi_\ell\}_{\ell=0}^n$ and $\{\widetilde{H_\ell}, \widetilde{\psi_\ell}\}_{\ell=0}^n$ be two sets of functions, satisfying the conditions in the general setup on page 446. Assume further that $\{D^j T_k \psi_\ell\}_{j,k\in\mathbb{Z},\ell=1,\dots,n}$ and $\{D^j T_k \widetilde{\psi_\ell}\}_{j,k\in\mathbb{Z},\ell=1,\dots,n}$ are Bessel sequences and that for some $C > 0$ and $\rho > \frac{1}{2}$,*

$$|\widehat{\psi_0}(\gamma)| \le \frac{C}{(1+|\gamma|)^\rho}, \quad |\widehat{\widetilde{\psi_0}}(\gamma)| \le \frac{C}{(1+|\gamma|)^\rho}, \quad a.e.\ \gamma \in \mathbb{R}. \qquad (18.48)$$

Assume that there exists a function $\theta \in L^\infty(\mathbb{T})$ such that $\lim_{\gamma\to 0} \theta(\gamma) = 1$ and

$$H_0(\gamma)\overline{\widetilde{H_0}(\gamma+\nu)}\theta(2\gamma) \quad + \quad \sum_{\ell=1}^n H_\ell(\gamma)\overline{\widetilde{H_\ell}(\gamma+\nu)}$$

$$= \begin{cases} \theta(\gamma) & \text{if } \nu = 0, \\ 0 & \text{if } \nu = \frac{1}{2}. \end{cases} \qquad (18.49)$$

Then $\{D^j T_k \psi_\ell\}_{j,k\in\mathbb{Z},\ell=1,\dots,n}$ and $\{D^j T_k \widetilde{\psi_\ell}\}_{j,k\in\mathbb{Z},\ell=1,\dots,n}$ are a pair of dual wavelet frames for $L^2(\mathbb{R})$.

Theorem 18.8.1 is a remarkable result. Assuming for the moment that we can construct the functions ψ_ℓ and $\widetilde{\psi_\ell}$ explicitly, it yields very satisfying answer to two fundamental problems for wavelet frames:

- Given a general wavelet frame, the canonical dual frame does not need to have wavelet structure. On the other hand, the assumptions in Theorem 18.8.1 automatically lead to a pair of dual frames, both having wavelet structure.

- For a general wavelet frame, application of the frame decomposition is complicated due to the necessary inversion of the frame operator. On the other hand, the assumptions in Theorem 18.8.1 lead to dual pairs of frames, and therefore an alternative to the frame decomposition.

The decay condition (18.48) is stronger than necessary, but on the other hand weak enough to be satisfied for almost all interesting constructions. To find a pair of dual frames via Theorem 18.8.1 is easier than to construct tight frames via the oblique extension principle: the main reason is that the function θ is not required to be positive.

We now return to the remaining question of explicit construction of the functions ψ_ℓ and $\widetilde{\psi_\ell}$. If the functions ψ_0 and $\widetilde{\psi_0}$ are known explicitly and the masks H_ℓ and $\widetilde{H_\ell}$ are trigonometric polynomials, then Theorem 18.8.1 yields explicitly given functions ψ_ℓ and $\widetilde{\psi_\ell}$.

Similar to what we saw for the oblique extension principle, we can use Theorem 18.8.1 to construct explicitly given dual pairs of frames with multiple generators. In the rest of this section, we will use the following:

Setup for construction of pairs of dual wavelet frames:

Let $\{\psi_0, H_0\}, \{\widetilde{H_0}, \widetilde{\psi_0}\}$ be as in the general setup on page 446. Assume further that $\{T_k \psi_0\}_{k \in \mathbb{Z}}$ and $\{T_k \widetilde{\psi_0}\}_{k \in \mathbb{Z}}$ are Bessel sequences and that (18.48) is satisfied. Let $\theta \in L^\infty(\mathbb{T})$ be a real-valued function for which $\lim_{\gamma \to 0} \theta(\gamma) = 1$, and assume that the function

$$\eta(\gamma) := \theta(\gamma) - \theta(2\gamma)\left(H_0(\gamma)\overline{\widetilde{H_0}(\gamma)} + H_0(\gamma + \frac{1}{2})\overline{\widetilde{H_0}(\gamma + \frac{1}{2})} \right) \quad (18.50)$$

is real-valued and has a zero of order at least 2 at the origin. Choose real-valued functions $\eta_1, \eta_2 \in L^\infty(\mathbb{T})$ such that

$$\eta(\gamma) = 2\eta_1(\gamma)\eta_2(\gamma), \text{ and } \eta_1(0) = \eta_2(0) = 0, \quad (18.51)$$

and choose two $\frac{1}{2}$-periodic and real-valued functions θ_1, θ_2 such that

$$\theta(2\gamma) = \theta_1(\gamma)\theta_2(\gamma). \quad (18.52)$$

$\square$

Let us comment on these assumptions and choices. First, the choice of $\frac{1}{2}$-periodic functions in (18.52) is possible because $\gamma \mapsto \theta(2\gamma)$ has period $\frac{1}{2}$. In the construction of tight multiwavelet frames in, e.g., Corollary 18.5.2, we had to perform a spectral factorization of the functions θ and η. The choices of the functions $\eta_1, \eta_2, \theta_1, \theta_2$ in (18.51) and (18.52) will replace the spectral factorization: in fact, we now prove how one can construct a multiwavelet frame based on these functions. We note that in general it is much easier to find functions satisfying (18.51) and (18.52) than to perform a spectral factorization.

Corollary 18.8.2 *Assume the setup on page 474 and define* $\{H_\ell\}_{\ell=1}^3$ *and* $\{\widetilde{H_\ell}\}_{\ell=1}^3$ *by*

$$H_1(\gamma) = e^{2\pi i\gamma}\theta_1(\gamma)\overline{\widetilde{H_0}(\gamma+\tfrac{1}{2})}, \qquad \widetilde{H_1}(\gamma) = e^{2\pi i\gamma}\theta_2(\gamma)\overline{H_0(\gamma+\tfrac{1}{2})},$$

$$H_2(\gamma) = \eta_1(\gamma), \qquad \widetilde{H_2}(\gamma) = \eta_2(\gamma),$$

$$H_3(\gamma) = e^{2\pi i\gamma}\eta_1(\gamma), \qquad \widetilde{H_3}(\gamma) = e^{2\pi i\gamma}\eta_2(\gamma).$$

Define the associated functions $\{\psi_\ell\}_{\ell=1}^3$ *and* $\{\widetilde{\psi_\ell}\}_{\ell=1}^3$ *as in the general setup on page 446. Then* $\{D^jT_k\psi_\ell\}_{j,k\in\mathbb{Z},\ell=1,2,3}$ *and* $\{D^jT_k\widetilde{\psi_\ell}\}_{j,k\in\mathbb{Z},\ell=1,2,3}$ *constitute a pair of dual wavelet frames for* $L^2(\mathbb{R})$.

Proof. For $\nu = 0$,

$$H_0(\gamma)\overline{\widetilde{H_0}(\gamma)}\theta(2\gamma) + \sum_{\ell=1}^3 H_\ell(\gamma)\overline{\widetilde{H_\ell}(\gamma)}$$

$$= H_0(\gamma)\overline{\widetilde{H_0}(\gamma)}\theta(2\gamma) + \theta_1(\gamma)\theta_2(\gamma)\overline{\widetilde{H_0}(\gamma+\tfrac{1}{2})}H_0(\gamma+\tfrac{1}{2}) + 2\eta_1(\gamma)\eta_2(\gamma)$$

$$= H_0(\gamma)\overline{\widetilde{H_0}(\gamma)}\theta(2\gamma) + \theta(2\gamma)\overline{\widetilde{H_0}(\gamma+\tfrac{1}{2})}H_0(\gamma+\tfrac{1}{2})$$

$$+\theta(\gamma) - \theta(2\gamma)\left(H_0(\gamma)\overline{\widetilde{H_0}(\gamma)} + H_0(\gamma+\tfrac{1}{2})\overline{\widetilde{H_0}(\gamma+\tfrac{1}{2})}\right)$$

$$= \theta(\gamma).$$

Similarly, for $\nu = \tfrac{1}{2}$,

$$H_0(\gamma)\overline{\widetilde{H_0}(\gamma+\tfrac{1}{2})}\theta(2\gamma) + \sum_{\ell=1}^3 H_\ell(\gamma)\overline{\widetilde{H_\ell}(\gamma+\tfrac{1}{2})}$$

$$= H_0(\gamma)\overline{\widetilde{H_0}(\gamma+\tfrac{1}{2})}\theta(2\gamma)$$

$$+e^{2\pi i\gamma}\theta_1(\gamma)\overline{\widetilde{H_0}(\gamma+\tfrac{1}{2})}\overline{e^{2\pi i(\gamma+1/2)}\theta_2(\gamma+\tfrac{1}{2})}H_0(\gamma)$$

$$+\eta_1(\gamma)\eta_2(\gamma+\tfrac{1}{2}) + e^{2\pi i\gamma}\overline{e^{2\pi i(\gamma+1/2)}}\eta_1(\gamma)\eta_2(\gamma+\tfrac{1}{2})$$

$$= 0.$$

$\square$

As for the oblique extension principle, the number of generators can be reduced to two:

Corollary 18.8.3 *Assume the setup on page 474 and let*

$$H_1(\gamma) = e^{2\pi i \gamma}\theta_1(\gamma)\overline{\widetilde{H}_0(\gamma + \tfrac{1}{2})}, \qquad \widetilde{H}_1(\gamma) = e^{2\pi i \gamma}\theta_2(\gamma)\overline{H_0(\gamma + \tfrac{1}{2})},$$

$$H_2(\gamma) = \eta_1(2\gamma)H_0(\gamma), \qquad \widetilde{H}_2(\gamma) = \eta_2(2\gamma)\widetilde{H}_0(\gamma).$$

Then $\{D^j T_k \psi_\ell\}_{j,k\in\mathbb{Z},\ell=1,2}$ and $\{D^j T_k \widetilde{\psi}_\ell\}_{j,k\in\mathbb{Z},\ell=1,2}$ constitute a pair of dual wavelet frames for $L^2(\mathbb{R})$.

We have assumed the factorizations of $\theta(2\cdot)$ and η to be real-valued. This is not strictly necessary. However, if θ, H_0 and $\widetilde{H}_0$ are trigonometric polynomials and η_1, η_2 and θ_1, θ_2 are real-valued trigonometric polynomials, then the frame generators $\{\psi_\ell\}_{\ell=1}^3$ and $\{\widetilde{\psi}_\ell\}_{\ell=1}^3$ are symmetric if the refinable functions ψ_0 and $\widetilde{\psi}_0$ are symmetric real-valued functions. Thus, the above process will lead to symmetric dual wavelet pairs when applied to even order B-splines.

Corollary 18.8.3 is related to a result by Daubechies and Han: they proved in [246] that based on any two refinable functions with compact support, one can construct a pair of dual wavelet frames having generators with compact support.

18.9   Applications to B-splines III

We now return to Example 18.6.3(iii), where construction of a tight frame turned out to be cumbersome.

Example 18.9.1 We give an example of a frame construction with two generators. We will base the choices of H_1, H_2, and $\widetilde{H}_1, \widetilde{H}_2$ on the same refinable function, namely, a translated B-spline of order 2. That is, we take $\psi_0 = \widetilde{\psi}_0 = T_1 B_2$; the associated refinement mask is

$$H_0(\gamma) = \frac{(1 + e^{-2\pi i \gamma})^2}{4} = e^{-2\pi i \gamma}\cos^2(\pi\gamma).$$

We again take

$$\theta(\gamma) = \frac{4 - \cos(2\pi\gamma)}{3};$$

as proved in Example 18.6.3, this leads to

$$\eta(\gamma) = \frac{2}{3}(8\cos^4(\pi\gamma) + 1)(\cos(\pi\gamma) - 1)^2(\cos(\pi\gamma) + 1)^2. \quad (18.53)$$

If we want to apply Corollary 18.8.3, we need to find functions $\eta_1, \eta_2, \theta_1, \theta_2$ satisfying (18.51) and (18.52). This is easy: the expression (18.53) immediately gives several choices for η_1, η_2, for example,

$$\eta_1(\gamma) = \frac{1}{3}(8\cos^4(\pi\gamma) + 1)(\cos(\pi\gamma) - 1)(\cos(\pi\gamma) + 1)^2,$$
$$\eta_2(\gamma) = (\cos(\pi\gamma) - 1).$$

Concerning θ_1, θ_2 we simply take

$$\theta_1(\gamma) = 1, \quad \theta_2(\gamma) = \theta(2\gamma) = \frac{4 - \cos(4\pi\gamma)}{3}.$$

The functions in Corollary 18.8.3 are now as follows:

$$H_1(\gamma) = e^{2\pi i\gamma}\theta_1(\gamma)\overline{\widetilde{H_0}}\left(\gamma + \frac{1}{2}\right) = e^{2\pi i\gamma}\frac{(1 - e^{2\pi i\gamma})^2}{4},$$

$$\widetilde{H_1}(\gamma) = e^{2\pi i\gamma}\theta_2(\gamma)H_0\left(\gamma + \frac{1}{2}\right)$$
$$= e^{2\pi i\gamma}\left(\frac{4}{3} - \frac{e^{4\pi i\gamma} + e^{-4\pi i\gamma}}{6}\right)\frac{(1 - e^{2\pi i\gamma})^2}{4},$$

$$H_2(\gamma) = \eta_1(2\gamma)H_0(\gamma)$$
$$= \frac{1}{3}\left(8\left(\frac{e^{2\pi i\gamma} + e^{-2\pi i\gamma}}{2}\right)^4 + 1\right)\left(\frac{e^{2\pi i\gamma} + e^{-2\pi i\gamma}}{2} - 1\right)$$
$$\times \left(\frac{e^{2\pi i\gamma} + e^{-2\pi i\gamma}}{2} + 1\right)^2\frac{(1 + e^{-2\pi i\gamma})^2}{4},$$

$$\widetilde{H_2}(\gamma) = \eta_2(2\gamma)\widetilde{H_0}(\gamma)$$
$$= \left(\frac{e^{2\pi i\gamma} + e^{-2\pi i\gamma}}{2} - 1\right)\frac{(1 + e^{-2\pi i\gamma})^2}{4}.$$

Define the functions $\{\psi_\ell\}_{\ell=1}^2$ and $\{\widetilde{\psi_\ell}\}_{\ell=1}^2$ associated with H_ℓ and $\widetilde{H_\ell}$ as in the general setup on page 446. Then $\{D^j T_k \psi_\ell\}_{j,k\in\mathbb{Z},\ell=1,2}$ and $\{D^j T_k \widetilde{\psi_\ell}\}_{j,k\in\mathbb{Z},\ell=1,2}$ constitute a pair of dual wavelet frames.   □

18.10   The MRA Literature and Applications

The development of MRA-based wavelets and the unitary extension principle has generated a huge number of explicit constructions and generalizations of the setup. For the case of the unitary extension principle, it is fair to say that the focus has been on obtaining a small number of generators, combined with attractive properties like symmetry and small support. Just to mention a few of such papers, see, e.g., the papers [539] by Petukhov, [440] by Jiang, and the papers [366, 367, 369] by Han and Mo. For example, in the paper [367], B-splines were used as scaling functions, while a more general approach, valid for real-valued, compactly supported, and symmetric scaling functions, was provided in [369].

Concerning more general approaches, Chui, He, and Stöckler have introduced and analyzed what they call *nonstationary wavelet frames* in a series of papers [210, 211]. The construction of such frames is based on a scale of spaces V_j, $j \in \mathbb{Z}$ as in the classical MRA setup, but without requiring that there is a scaling relation between the spaces; also, the spaces V_j are not necessarily defined in terms of translates of fixed function. The theory is applicable on bounded as well as unbounded intervals.

In this chapter, we have focused on MRA constructions in the one-dimensional case. For a discussion (and solution) of some of the problems that appear in the higher-dimensional case, we refer to, e.g., the papers [146, 147].

The wavelet frames constructed via the UEP and its variants have been applied to several problems in image analysis, e.g., in the papers by Cai, Osher, Shen, and their coauthors [105, 106]. The paper [105] gives a frame-based approach to image restoration that covers image denoising, deblurring, inpainting, and cartoon–texture image decomposition. In the paper [106] the authors provide a link between the spline-based wavelet frames and the variational methods that are traditionally used in image restoration. In fact, the authors show that the classical variational method can be viewed as a frame method.

18.11 Exercises

18.1 Prove (18.10) under the stated assumptions.

18.2 Derive the expression (18.28) for the function ψ_2.

18.3 Prove (18.29) and provide the missing details in Example 18.3.3.

18.4 Prove Corollary 18.2.7.

18.5 Prove Corollary 18.5.3.

18.6 Prove the reduction to two generators stated after Corollary 18.8.2.

18.7 Derive the expressions in (18.44) and (18.45).

18.8 Calculate the coefficients c_j in Theorem 18.6.1 for $m = 4$, $M = 2$.

19

Selected Topics on Wavelet Frames

Continuing the style from the chapters on Gabor frames, we will now present a few selected topics concerning wavelet frames. The sections deal with issues that appear in several places in the wavelet literature, and they can be read independently of each other. We begin in Section 19.1 with a discussion of irregular wavelet frames. Section 19.2 states a few results about oversampling of wavelet frames. We analyze the relationship between two wavelet systems with the same scaling parameter, but different translation parameters. In particular, we consider the case where one translation parameter is an integer multiple of the other; surprisingly, it turns out to play an important role whether this integer is even or odd. Section 19.3 returns to the extension problem considered for general frames in Section 6.4 and for Gabor frame in Section 12.7; while these sections contain complete and satisfying answers, the corresponding wavelet question is open and challenging. Section 19.4 gives a short description of wavelet theory from the signal processing perspective.

© Springer International Publishing Switzerland 2016
O. Christensen, *An Introduction to Frames and Riesz Bases*,
Applied and Numerical Harmonic Analysis,
DOI 10.1007/978-3-319-25613-9_19

19.1   Irregular Wavelet Frames

In Section 15.2 we exclusively considered translations with integer-multiples of the parameter b and dilations by $a^j, j \in \mathbb{Z}$. A more general and considerably more complicated question is:

Which conditions on a discrete sequence $\{(\lambda_j, \mu_j)\}_{j \in I}$ in $\mathbb{R}^+ \times \mathbb{R}$ and a function $\psi \in L^2(\mathbb{R})$ imply that

$$\{\lambda_j^{1/2} \psi(\lambda_j x - \mu_j)\}_{j \in I} \text{ is a frame for } L^2(\mathbb{R})?$$

A frame of this type is called an *irregular wavelet frame*. Only few results about irregular wavelet frames are known and, e.g., the proof of Theorem 15.2.3 does not extend to general sequences $\{(\lambda_j, \mu_j)\}_{j \in I}$. But if we assume that the translates are still along the set $b\mathbb{Z}$, the essence of Theorem 15.2.3 carries over. There are a few points where extra caution is needed, especially because $\{\lambda_j\}_{j \in \mathbb{Z}}$ is usually different from $\{\lambda_j^{-1}\}_{j \in \mathbb{Z}}$; in the proof of Theorem 15.2.3, we were frequently switching between $\{a^j\}_{j \in \mathbb{Z}}$ and $\{a^{-j}\}_{j \in \mathbb{Z}}$. Also, the function $\gamma \mapsto \sum_{j \in \mathbb{Z}} |\widehat{\psi}(\gamma/\lambda_j)|^2$ is in general not periodic, so in order to find its supremum or infimum, we have to investigate all $\gamma \in \mathbb{R}$. We encourage the reader to check that the proof of Theorem 15.2.3 works in the irregular case with these modifications taken into account (Exercise 19.1). A direct proof is in [171]; the result can also be obtained as a consequence of the theory for generalized shift-invariant systems; see Theorem 20.3.1.

Theorem 19.1.1 *Let $\{\lambda_j\}_{j \in \mathbb{Z}}$ be a sequence of positive real numbers, $b > 0$ and $\psi \in L^2(\mathbb{R})$. Suppose that*

$$A := \frac{1}{b} \inf_{\gamma \in \mathbb{R}} \left(\sum_{j \in \mathbb{Z}} |\widehat{\psi}(\frac{\gamma}{\lambda_j})|^2 - \sum_{k \neq 0} \sum_{j \in \mathbb{Z}} |\widehat{\psi}(\frac{\gamma}{\lambda_j})\widehat{\psi}(\frac{\gamma}{\lambda_j} + \frac{k}{b})| \right) > 0,$$

and

$$B := \frac{1}{b} \sup_{\gamma \in \mathbb{R}} \sum_{j,k \in \mathbb{Z}} |\widehat{\psi}(\frac{\gamma}{\lambda_j})\widehat{\psi}(\frac{\gamma}{\lambda_j} + \frac{k}{b})| < \infty.$$

Then $\{\lambda_j^{1/2}\psi(\lambda_j x - kb)\}_{j,k \in \mathbb{Z}}$ is a frame with frame bounds A and B.

One can check that parts of Chui and Shi's proof of Proposition 15.2.2 work for irregular wavelet frames of the type $\{\lambda_j^{1/2}\psi(\lambda_j \gamma - kb)\}_{j,k \in \mathbb{Z}}$:

Lemma 19.1.2 *Let $\psi \in L^2(\mathbb{R})$. If $\{\lambda_j\}_{j \in \mathbb{Z}}$ is a sequence in $\mathbb{R}^+$ and $\{\lambda_j^{1/2}\psi(\lambda_j \gamma - kb)\}_{j,k \in \mathbb{Z}}$ is a frame with upper bound B for some $b > 0$, then*

$$\frac{1}{b} \sum_{j \in \mathbb{Z}} |\widehat{\psi}(\frac{\gamma}{\lambda_j})|^2 \leq B, \text{ a.e. } \gamma \in \mathbb{R}.$$

Lemma 19.1.2 puts restrictions on the sequences $\{\lambda_j\}_{j\in\mathbb{Z}}$ for which $\{\lambda_j^{1/2}\psi(\lambda_j\gamma - kb)\}_{j,k\in\mathbb{Z}}$ can be a frame. Let us be more specific about this point. Following [194], we say that a sequence $\{\lambda_j\}_{j\in\mathbb{Z}}$ of positive numbers is *logarithmically separated by* $\lambda > 1$ if

$$|\log\lambda_j - \log\lambda_k| \geq \log\lambda, \quad \forall k \neq j.$$

If $\{\lambda_j\}_{j\in\mathbb{Z}}$ is ordered increasingly, this is equivalent to $\frac{\lambda_{j+1}}{\lambda_j} \geq \lambda, \ \forall j \in \mathbb{Z}$.

Proposition 19.1.3 *Let* $\psi \in L^1(\mathbb{R}) \cap L^2(\mathbb{R})$ *and assume that the sequence* $\{\lambda_j\}_{j\in\mathbb{Z}}$ *in* $\mathbb{R}^+$ *is chosen such that* $\{\lambda_j^{1/2}\psi(\lambda_j\gamma - kb)\}_{j,k\in\mathbb{Z}}$ *is a frame for* $L^2(\mathbb{R})$. *Then* $\{\lambda_j\}_{j\in\mathbb{Z}}$ *is a finite union of logarithmically separated sets.*

Proof. The assumption $\psi \in L^1(\mathbb{R})$ implies that $\widehat{\psi}$ is continuous. Let $s_j := \frac{1}{\lambda_j}$. Using Lemma 19.1.2, it follows that for each finite set $J \in \mathbb{Z}$,

$$\frac{1}{b}\sum_{j\in J}|\widehat{\psi}(\gamma s_j)|^2 \leq B \tag{19.1}$$

for all $\gamma \in \mathbb{R}$. Now, take $\gamma_0 \in \mathbb{R}$ such that $\widehat{\psi}(\gamma_0) \neq 0$; we can assume that $\gamma_0 > 0$. Let $c := |\widehat{\psi}(\gamma_0)|^2$ and choose $\delta > 0$ such that for all $\gamma \in I_0 := [\gamma_0, \gamma_0 + \delta[$,

$$|\widehat{\psi}(\gamma)|^2 \geq \frac{c}{2}.$$

By taking $\gamma = 1$ in (19.1), we see that the number N of elements from $\{s_j\}_{j\in\mathbb{Z}}$ belonging to the interval I_0 satisfies $N\frac{c}{2b} \leq B$, i.e., $N \leq \frac{2B}{c}b$. Now let $\sigma := \frac{\gamma_0 + \delta}{\gamma_0}$ and define the intervals

$$I_k := [\gamma_0\sigma^k, \gamma_0\sigma^{k+1}[.$$

Clearly $\{I_k\}_{k=-\infty}^{\infty}$ is a disjoint covering of $\mathbb{R}^+$ ($\mathbb{R}^+ = \cup_{k=-\infty}^{\infty} I_k$), and for given $k \in \mathbb{Z}$, the interval I_k contains at most N points from $\{s_j\}_{j\in\mathbb{Z}}$. Now observe that each point in I_0 is logarithmically separated from points in the intervals $I_{\pm 2}, I_{\pm 4}, \dots$. Similarly, a point from I_1 is logarithmically separated with points from the intervals $I_{-1}, I_{\pm 3}, I_{\pm 5}, \dots$. Thus $\{s_j\}_{j\in\mathbb{Z}}$ can be split into at most $2N$ logarithmically separated subsequences, from which the result follows. $\qquad\square$

Proposition 19.1.3 excludes the frame property for many types of sequences $\{\lambda_j\}_{j\in\mathbb{Z}}$. If, for example, $\lambda_j = j^\alpha$ for some $\alpha > 0$ and for j larger than a certain $J > 0$, then $\{\lambda_j^{1/2}\psi(\lambda_j\gamma - kb)\}_{j,k\in\mathbb{Z}}$ cannot even be a Bessel sequence if ψ satisfies the very weak condition in Proposition 19.1.3.

Sun and Zhou proved in [604] the following useful results concerning wavelet frames where both the dilation and the translation are allowed to be irregular:

Theorem 19.1.4 *Let $\psi \in L^2(\mathbb{R})$ be a real-valued function for which all the functions*

$$x \mapsto x\psi(x), \ x \mapsto \psi'(x), \ x \mapsto \psi''(x)$$

are in $L^2(\mathbb{R})$. Assume that $\widehat{\psi}(0) = 0$. Then there exist constants $a > 1, b > 0$ such that

$$\left\{ s_{j,k}^{-1/2} \psi \left(\frac{x - \mu_{j,k}}{s_{j,k}} \right) \right\}_{j,k \in \mathbb{Z}}$$

is a frame for $L^2(\mathbb{R})$ for all sequences $\{(s_{j,k}, \mu_{j,k})\}_{j,k \in \mathbb{Z}}$ for which

$$(s_{j,k}, \mu_{j,k}) \in [a^j, a^{j+1}] \times [a^j bk, a^j b(k+1)], \ j, k \in \mathbb{Z}. \qquad (19.2)$$

Theorem 19.1.4 can naturally be considered as a perturbation result; in fact, the conditions imply that $\{a^{-j/2}\psi(\frac{x - a^j bk}{a^j})\}_{j,k \in \mathbb{Z}}$ is a frame, and the condition (19.2) is "strong enough to guarantee that $\{s_{j,k}^{-1/2}\psi(\frac{x - \mu_{j,k}}{s_{j,k}})\}_{j,k \in \mathbb{Z}}$ is so close to $\{a^{-j/2}\psi(\frac{x - a^j bk}{a^j})\}_{j,k \in \mathbb{Z}}$ that it is itself a frame." We leave this as a rather intuitive statement, but we return to general perturbation theoretic methods in Chapter 22. For the full proof of Theorem 19.1.4 we refer to [604].

19.2   Oversampling of Wavelet Frames

If $\{a^{j/2}\psi(a^j x - kb)\}_{j,k \in \mathbb{Z}}$ is a frame for $L^2(\mathbb{R})$, the general frame theory tells us that the wavelet system contains enough elements to represent arbitrary functions in $L^2(\mathbb{R})$ as infinite linear combinations of the frame elements. It is clear from the definition of a frame that a wavelet system Ψ containing a frame $\{a^{j/2}\psi(a^j x - kb)\}_{j,k \in \mathbb{Z}}$ is itself a frame if and only if Ψ is a Bessel sequence. An example of a wavelet system containing $\{a^{j/2}\psi(a^j x - kb)\}_{j,k \in \mathbb{Z}}$ is

$$\{a^{j/2}\psi(a^j x - kb/n)\}_{j,k \in \mathbb{Z}}, \qquad (19.3)$$

where $n \in \mathbb{N}$. We say that the wavelet system in (19.3) is obtained via *oversampling* of $\{a^{j/2}\psi(a^j x - kb)\}_{j,k \in \mathbb{Z}}$.

Chui and Shi investigated the frame properties of an oversampled wavelet system in [212]:

Proposition 19.2.1 *Assume that $\{a^{j/2}\psi(a^j x - kb)\}_{j,k \in \mathbb{Z}}$ is a wavelet frame and that ψ satisfies the conditions in Proposition 15.2.8. Then the wavelet system in (19.3) is a wavelet frame for any $n \in \mathbb{N}$.*

Oversampling will in general change the frame bounds, and for a tight wavelet frame it might happen that the oversampled frame is no longer tight. A positive result was obtained in [212], where the given conditions imply that $\{a^{j/2}\psi(a^j x - kb/n)\}_{j,k\in\mathbb{Z}}$ is a tight frame if $\{a^{j/2}\psi(a^j x - kb)\}_{j,k\in\mathbb{Z}}$ is tight:

Theorem 19.2.2 *Let $a \geq 2$ be a positive integer and $b > 0$. Suppose that $\{a^{j/2}\psi(a^j x - kb)\}_{j,k\in\mathbb{Z}}$ is a frame for $L^2(\mathbb{R})$ with bounds A, B. Then, for any positive integer n which is relatively prime to a, the family in (19.3) is a frame for $L^2(\mathbb{R})$ with bounds nA, nB.*

In the special case $a = 2$, we see that tightness is preserved if n is odd. There exist examples, showing that tightness might not be preserved if n is even (cf. [215]).

If the wavelet frame $\{a^{j/2}\psi(a^j x - kb)\}_{j,k\in\mathbb{Z}}$ has a dual wavelet frame $\{a^{j/2}\widetilde{\psi}(a^j x - kb)\}_{j,k\in\mathbb{Z}}$ and n is a positive integer which is relatively prime to a, then the oversampled system (19.3) also has a dual with the wavelet structure, namely, $\{\frac{1}{n}a^{j/2}\widetilde{\psi}(a^j x - kb/n)\}_{j,k\in\mathbb{Z}}$. We refer to [212] for a proof.

We note that an alternative approach to oversampling was given by Ron and Shen in [562]. More recent results, especially concerning the case where the parameter a is allowed to be a rational number rather than just a positive integer, can be found in the papers [84, 83] by Bownik and Lemvig.

19.3   An Open Extension Problem

Extension problems have played a central role in our analysis of wavelet systems. So far, the results have been dealing with the unitary extension principle and its variants and are thus based on the assumption of an underlying refinable function. In this section we will take a more general viewpoint and consider extension problems that are not based on such an assumption. From this point of view the current section is a natural continuation of the sections about the extension problem for sequences in general Hilbert spaces (Section 6.4) and for Gabor systems (Section 12.7).

In the general setting of a separable Hilbert space $\mathcal{H}$, the extension problem concerns the question of how to extend a Bessel sequence $\{f_k\}_{k=1}^{\infty}$ in $\mathcal{H}$ to a (tight) frame $\{f_k\}_{k=1}^{\infty} \cup \{p_j\}_{j\in J}$ for $\mathcal{H}$; or how to extend a pair of Bessel sequences in $\mathcal{H}$ to a pair of dual frames for $\mathcal{H}$. In Theorem 6.2.1 and Theorem 6.4.1, we have seen that such extensions always exist. These results can of course be applied to wavelet systems in $L^2(\mathbb{R})$: in other words, if $\{D^j T_k \psi_1\}_{j,k\in\mathbb{Z}}$ is a Bessel sequence in $L^2(\mathbb{R})$, there exists a sequence $\{p_j\}_{j\in J}$ in $L^2(\mathbb{R})$ such that $\{D^j T_k \psi_1\}_{j,k\in\mathbb{Z}} \cup \{p_j\}_{j\in J}$ is a tight frame for $L^2(\mathbb{R})$. But this general result might not be the appropriate answer to the question! In fact, if a certain application asks for the wavelet structure of the sequence $\{D^j T_k \psi_1\}_{j,k\in\mathbb{Z}}$, it is probably essential that $\{p_j\}_{j\in J}$ has wavelet structure as well.

Based on this discussion, we will now formulate two key questions:

(i) Given a Bessel sequence $\{D^j T_k \psi_1\}_{j,k\in\mathbb{Z}}$ in $L^2(\mathbb{R})$, does there exist a wavelet system $\{D^j T_k \psi_2\}_{j,k\in\mathbb{Z}}$ such that

$$\{D^j T_k \psi_1\}_{j,k\in\mathbb{Z}} \cup \{D^j T_k \psi_2\}_{j,k\in\mathbb{Z}}$$

is a tight frame for $L^2(\mathbb{R})$?

(ii) Given Bessel sequences $\{D^j T_k \psi_1\}_{j,k\in\mathbb{Z}}$ and $\{D^j T_k \widetilde{\psi_1}\}_{j,k\in\mathbb{Z}}$ in $L^2(\mathbb{R})$, do there exist wavelet systems $\{D^j T_k \psi_2\}_{j,k\in\mathbb{Z}}$ and $\{D^j T_k \widetilde{\psi_2}\}_{j,k\in\mathbb{Z}}$ such that

$$\{D^j T_k \psi_1\}_{j,k\in\mathbb{Z}} \cup \{D^j T_k \psi_2\}_{j,k\in\mathbb{Z}} \text{ and } \{D^j T_k \widetilde{\psi_1}\}_{j,k\in\mathbb{Z}} \cup \{D^j T_k \widetilde{\psi_2}\}_{j,k\in\mathbb{Z}}$$

form dual frames for $L^2(\mathbb{R})$?

In the case of Gabor systems, we have seen in Section 12.7 that the answers to the analogue questions are affirmative (if $ab \leq 1$). Unfortunately, it turns out that the extension problem for wavelet systems is considerably more involved than for Gabor systems. In order to explain this, consider the proof of Theorem 6.4.1 and assume that the Bessel sequences $\{f_k\}_{k=1}^\infty$ and $\{g_k\}_{k=1}^\infty$ have wavelet structure, i.e., they have the form $\{D^j T_k \psi_1\}_{j,k\in\mathbb{Z}}$ and $\{D^j T_k \widetilde{\psi_1}\}_{j,k\in\mathbb{Z}}$ for some $\psi_1, \widetilde{\psi_1} \in L^2(\mathbb{R})$. Denote the synthesis operators by T and U, respectively. Then, letting $\{a_j\}_{j\in J}$ and $\{b_j\}_{j\in J}$ be a pair of dual frames for $L^2(\mathbb{R})$, the proof of Theorem 6.4.1 shows that

$$\{D^j T_k \psi_1\}_{j,k\in\mathbb{Z}} \cup \{(I - TU^*)a_j\}_{j\in J} \text{ and } \{D^j T_k \widetilde{\psi_1}\}_{j,k\in\mathbb{Z}} \cup \{b_j\}_{j\in J}$$

form a pair of dual frames for $L^2(\mathbb{R})$.

We can of course choose $\{a_j\}_{j\in J}$ and $\{b_j\}_{j\in J}$ to have wavelet structure; in this case, the remaining issue is whether the sequence $\{(I - TU^*)a_j\}_{j\in J}$ has wavelet structure. Unfortunately, as we have seen in Example 16.1.1, the operator TU^* in general does not commute with $D^j T_k$; thus, the system $\{(I - TU^*)a_j\}_{j\in J}$ will in general not be a wavelet system.

The conclusion of the above discussion is that the technique that worked very well in the Gabor case does not work in the wavelet case. In fact, the questions (i) and (ii) stated above are not answered in the literature.

The following partial result was obtained in [180]. It gives an affirmative answer to the question (ii) under the assumption that the Fourier transform of $\widetilde{\psi_1}$ has support within the interval $[-1, 1]$; furthermore, a slightly stronger condition implies that if also the Fourier transform of the function ψ_1 is compactly supported, then the extension can be performed with two functions ψ_2 and $\widetilde{\psi_2}$ enjoying the same property.

Theorem 19.3.1 *Let* $\{D^j T_k \psi_1\}_{j,k\in\mathbb{Z}}$ *and* $\{D^j T_k \widetilde{\psi_1}\}_{j,k\in\mathbb{Z}}$ *be Bessel sequences in* $L^2(\mathbb{R})$. *Assume that the Fourier transform of* $\widetilde{\psi_1}$ *satisfies that*

$$\text{supp } \widehat{\widetilde{\psi_1}} \subseteq [-1,1]. \tag{19.4}$$

Then there exist functions $\psi_2, \widetilde{\psi_2} \in L^2(\mathbb{R})$ *such that*

$$\{D^j T_k \psi_1\}_{j,k\in\mathbb{Z}} \cup \{D^j T_k \psi_2\}_{j,k\in\mathbb{Z}} \text{ and } \{D^j T_k \widetilde{\psi_1}\}_{j,k\in\mathbb{Z}} \cup \{D^j T_k \widetilde{\psi_2}\}_{j,k\in\mathbb{Z}}$$

form dual frames for $L^2(\mathbb{R})$. *If* $\widehat{\widetilde{\psi_1}}$ *is compactly supported and*

$$\text{supp } \widehat{\widetilde{\psi_1}} \subseteq [-1,1] \setminus [-\epsilon, \epsilon] \tag{19.5}$$

for some $\epsilon > 0$, *the functions* ψ_2 *and* $\widetilde{\psi_2}$ *can be chosen to have compactly supported Fourier transforms as well.*

The open questions in (i) and (ii) on page 484 are strongly connected to the following conjecture by Han [362]:

Conjecture by Deguang Han: *Let* $\{D^j T_k \psi_1\}_{j,k\in\mathbb{Z}}$ *be a wavelet frame with upper frame bound B. Then there exists $D > B$ such that for each $K \geq D$, there exists $\psi_2 \in L^2(\mathbb{R})$ such that $\{D^j T_k \psi_1\}_{j,k\in\mathbb{Z}} \cup \{D^j T_k \psi_2\}_{j,k\in\mathbb{Z}}$ is a tight frame for $L^2(\mathbb{R})$ with bound K.*

The paper [362] contains an example showing that (again in contrast with the Gabor setting) it might not be possible to extend the Bessel system $\{D^j T_k \psi_1\}_{j,k\in\mathbb{Z}}$ to a tight frame without enlarging the upper bound; hence it is essential that the conjecture includes the option that the extended wavelet system has a strictly larger frame bound K than the upper frame bound B for $\{D^j T_k \psi_1\}_{j,k\in\mathbb{Z}}$. We also note that Han's conjecture is based on an example, where supp $\widehat{\psi_1} \subseteq [-1,1]$, i.e., a case that is covered by Theorem 19.3.1.

Observe that a pair of wavelet Bessel sequences always can be extended to dual wavelet frame pairs by adding *two* pairs of wavelet systems. In fact, we can always add one pair of wavelet systems that cancels the action of the given wavelet system; and another one that yields a dual pair of wavelet frames by itself. Thus, the issue is really whether it is enough to add one pair of wavelet systems, as stated in the formulation of (ii) on page 484.

19.4   The Signal Processing Perspective

In Section 18.2, we gave a functional analytic presentation of the unitary extension principle. We will now look at this result once more and formulate it in signal processing terms.

We will first reformulate the equations in Corollary 18.2.7 in terms of the *Z-transform*. Formally, the Z-transform of a sequences $\{h_k\}_{k\in\mathbb{Z}}$ is defined as the infinite series (depending on a variable $z \in \mathbb{C}$)

$$\widetilde{H}(z) := \sum_{k\in\mathbb{Z}} h_k z^{-k}.$$

We will not worry too much about the exact domain of $z \in \mathbb{Z}$ for which the Z-transform of a given sequence $\{h_k\}_{k\in\mathbb{Z}}$ converges. The reason is that we mainly are interested in finite sequences $\{h_k\}_{k\in\mathbb{Z}}$, for which the Z-transform is defined for all $z \neq 0$. Besides such finite sequences, we will only consider the Z-transform of sequences $\{h_k\}_{k\in\mathbb{Z}}$, which are Fourier coefficients; for such sequences, the Z-transform converges for a.e. $z \in \mathbb{C}$ with $|z| = 1$, and this turns out to be sufficient for our purpose. In engineering language, the sequence $\{h_k\}_{k\in\mathbb{Z}}$ is often called a *filter*.

Consider the 1-periodic functions H_ℓ, $\ell = 0, \ldots, n$, in the general setup on page 446. We can write these functions in terms of their Fourier series, with Fourier coefficients $h_{k,\ell}$, $k \in \mathbb{Z}$:

$$H_\ell(\gamma) = \sum_{k\in\mathbb{Z}} h_{k,\ell} e^{2\pi i k\gamma}.$$

Note that in terms of the Z-transform, this means that

$$H_\ell(\gamma) = \sum_{k\in\mathbb{Z}} h_{k,\ell} \left(e^{-2\pi i\gamma}\right)^{-k} = \widetilde{H}_\ell(e^{-2\pi i\gamma}).$$

We can now formulate the main condition in the unitary extension principle in terms of the Z-transform:

Theorem 19.4.1 *Assume that the functions H_ℓ, $\ell = 0, \ldots, n$, have real Fourier coefficients $h_{k,\ell}$, $k \in \mathbb{Z}$. Then the conditions (18.19) hold if and only if the equations*

$$\begin{cases} \displaystyle\sum_{\ell=0}^{n} \widetilde{H}_\ell(z)\widetilde{H}_\ell(z^{-1}) = & 1, \\[2mm] \displaystyle\sum_{\ell=0}^{n} \widetilde{H}_\ell(z)\widetilde{H}_\ell(-z^{-1}) = & 0 \end{cases} \qquad (19.6)$$

hold for a.e. $z \in \mathbb{C}$ for which $|z| = 1$.

Proof. Let us rewrite the terms appearing in (18.19):

$$T_{1/2}H_\ell(\gamma) = \widetilde{H}_\ell(e^{-2\pi i(\gamma-1/2)}) = \widetilde{H}_\ell(-e^{-2\pi i\gamma}),$$

and, because the coefficients $h_{k,\ell}$ are assumed to be real,

$$\overline{H_\ell(\gamma)} = \sum_{k\in\mathbb{Z}} h_{k,\ell} e^{-2\pi i k\gamma} = \widetilde{H}_\ell(e^{2\pi i\gamma}).$$

Thus (18.19) is equivalent to the conditions

$$\begin{cases} \sum_{\ell=0}^{n} \widetilde{H_\ell}(e^{-2\pi i\gamma})\widetilde{H_\ell}(e^{2\pi i\gamma}) = 1, \\ \sum_{\ell=0}^{n} \widetilde{H_\ell}(e^{2\pi i\gamma})\widetilde{H_\ell}(-e^{-2\pi i\gamma}) = 0. \end{cases}$$

Putting $z = e^{2\pi i\gamma}$ now leads to the result.   □

Very often, conditions involving filters are formulated in terms of the so-called *polyphase decomposition* of the Z-transform. In order to introduce that, note that we can decompose a sequence $\{h_k\}_{k\in\mathbb{Z}}$ into "even" and "odd" parts:

$$\begin{aligned} (\ldots, h_{-2}, h_{-1}, h_0, h_1, h_2, \ldots) &= (\ldots, h_{-2}, 0, h_0, 0, h_2, \ldots) \\ &+ (\ldots, 0, h_{-1}, 0, h_1, 0, \ldots). \end{aligned}$$

By linearity, this decomposition implies that the Z-transformation of $\{h_k\}_{k\in\mathbb{Z}}$ can be written as

$$\begin{aligned} \widetilde{H}(z) &= \left[\cdots + h_{-2}z^2 + h_0 + h_2 z^{-2} + \cdots\right] \\ &+ \left[\cdots + h_{-1}z + h_1 z^{-1} + h_3 z^{-3} + \cdots\right] \\ &= \left[\cdots + h_{-2}z^2 + h_0 + h_2 z^{-2} + \cdots\right] \\ &+ z^{-1}\left[\cdots + h_{-1}z^2 + h_1 + h_3 z^{-2} + \cdots\right] \\ &= \sum_{k\in\mathbb{Z}} h_{2k} z^{-2k} + z^{-1} \sum_{k\in\mathbb{Z}} h_{2k+1} z^{-2k}. \end{aligned} \qquad (19.7)$$

The *polyphase components* of $\widetilde{H}(z)$ are now defined as the two functions

$$\widetilde{H_0}(z) := \sum_{k\in\mathbb{Z}} h_{2k} z^{-k}, \quad \widetilde{H_1}(z) = \sum_{k\in\mathbb{Z}} h_{2k+1} z^{-k};$$

thus, via (19.7), the Z-transformation has the *polyphase decomposition*

$$\widetilde{H}(z) = \widetilde{H_0}(z^2) + z^{-1}\widetilde{H_1}(z^2).$$

Consider now a given sequence of 1-periodic functions H_ℓ, $\ell = 0, \ldots, n$, or, equivalently, a sequence of filters $\{h_{k,\ell}\}_{k\in\mathbb{Z}}$, $\ell = 0, \ldots n$. Associated with the filter $\{h_{k,\ell}\}_{k\in\mathbb{Z}}$, we denote the polyphase components of $\widetilde{H_\ell}$ by $\widetilde{H_{\ell,0}}$ and $\widetilde{H_{\ell,1}}$. Define the $(n+1) \times 2$ matrix of polyphase components H_p by

$$H_p(z) = \begin{pmatrix} \widetilde{H_{0,0}}(z) & \widetilde{H_{0,1}}(z) \\ \widetilde{H_{1,0}}(z) & \widetilde{H_{1,1}}(z) \\ \cdot & \cdot \\ \cdot & \cdot \\ \widetilde{H_{n,0}}(z) & \widetilde{H_{n,1}}(z) \end{pmatrix}. \qquad (19.8)$$

We will now formulate Theorem 19.4.1 in terms of the matrix $H_p(z)$ and its transpose $H_p^T(z)$.

Theorem 19.4.2 *Assume that the functions H_ℓ, $\ell = 0, \ldots, n$ have real Fourier coefficients $h_{k,\ell}$, $k \in \mathbb{Z}$. Then the condition (19.6) is satisfied if and only if*

$$H_p^T(z^{-1})H_p(z) = \frac{1}{2}I \tag{19.9}$$

for almost all $z \in \mathbb{C}$ with $|z| = 1$.

Proof.   Note that $\widetilde{H_{\ell,k}}(z^{-1}) = \overline{\widetilde{H_{\ell,k}}(z)}$ for $\ell = 0, \ldots, n$, $k = 0, 1$; this implies that $H_p^T(z^{-1}) = \overline{H_p^T(z)}$. In terms of the entries of the matrix $H_p(z)$, the condition (19.9) means that for almost all $z \in \mathbb{C}$ with $|z| = 1$,

$$\begin{cases} \sum_{\ell=0}^{n} \left|\widetilde{H_{\ell,0}}(z)\right|^2 = \frac{1}{2}, \\ \sum_{\ell=0}^{n} \left|\widetilde{H_{\ell,1}}(z)\right|^2 = \frac{1}{2}, \\ \sum_{\ell=0}^{n} \widetilde{H_{\ell,0}}(z^{-1})\widetilde{H_{\ell,1}}(z) = 0, \\ \sum_{\ell=0}^{n} \widetilde{H_{\ell,1}}(z^{-1})\widetilde{H_{\ell,0}}(z) = 0. \end{cases} \tag{19.10}$$

On the other hand, the two terms in (19.6) can be rewritten using the polyphase decomposition. First,

$$\sum_{\ell=0}^{n} \widetilde{H_\ell}(z)\widetilde{H_\ell}(z^{-1})$$
$$= \sum_{\ell=0}^{n} \left(\widetilde{H_{\ell,0}}(z^2)) + z^{-1}\widetilde{H_{\ell,1}}(z^2)\right)\left(\widetilde{H_{\ell,0}}(z^{-2}) + z\widetilde{H_{\ell,1}}(z^{-2})\right),$$

i.e,

$$\sum_{\ell=0}^{n} \widetilde{H_\ell}(z)\widetilde{H_\ell}(z^{-1}) = \sum_{\ell=0}^{n} \left|\widetilde{H_{\ell,0}}(z^2)\right|^2 + \sum_{\ell=0}^{n} \left|\widetilde{H_{\ell,1}}(z^2)\right|^2 \tag{19.11}$$
$$+ z\sum_{\ell=0}^{n} \widetilde{H_{\ell,0}}(z^2)\widetilde{H_{\ell,1}}(z^{-2}) + z^{-1}\sum_{\ell=0}^{n} \widetilde{H_{\ell,0}}(z^{-2})\widetilde{H_{\ell,1}}(z^2).$$

For the second term in (19.6),

$$\sum_{\ell=0}^{n} \widetilde{H_\ell}(z)\widetilde{H_\ell}(-z^{-1}) \tag{19.12}$$

$$= \sum_{\ell=0}^{n} \left(\widetilde{H_{\ell,0}}(z^2) + z^{-1}\widetilde{H_{\ell,1}}(z^2)\right)\left(\widetilde{H_{\ell,0}}(z^{-2}) - z\widetilde{H_{\ell,1}}(z^{-2})\right)$$

$$= \sum_{\ell=0}^{n} \left|\widetilde{H_{\ell,0}}(z^2)\right|^2 - \sum_{\ell=0}^{n} \left|\widetilde{H_{\ell,1}}(z^2)\right|^2$$

$$- z\sum_{\ell=0}^{n} \widetilde{H_{\ell,0}}(z^2)\widetilde{H_{\ell,1}}(z^{-2}) + z^{-1}\sum_{\ell=0}^{n} \widetilde{H_{\ell,0}}(z^{-2})\widetilde{H_{\ell,1}}(z^2).$$

From here, it follows that if (19.10) is satisfied, then the conditions in (19.6) are satisfied as well.

Now assume that (19.6) holds. Adding and subtracting, respectively, the two equations in (19.6) and using the expressions derived in (19.11) and (19.12) lead to the equations

$$\begin{cases} 2\sum_{\ell=0}^{n} \left|\widetilde{H_{\ell,0}}(z^2)\right|^2 + 2z^{-1}\sum_{\ell=0}^{n} \widetilde{H_{\ell,0}}(z^{-2})\widetilde{H_{\ell,1}}(z^2) = 1, \\ 2\sum_{\ell=0}^{n} \left|\widetilde{H_{\ell,1}}(z^2)\right|^2 + 2z\sum_{\ell=0}^{n} \widetilde{H_{\ell,0}}(z^2)\widetilde{H_{\ell,1}}(z^{-2}) = 1. \end{cases} \tag{19.13}$$

The terms $z^{-1}\sum_{\ell=0}^{n} \widetilde{H_{\ell,0}}(z^{-2})\widetilde{H_{\ell,1}}(z^2)$ and $z\sum_{\ell=0}^{n} \widetilde{H_{\ell,0}}(z^2)\widetilde{H_{\ell,1}}(z^{-2})$ are the complex conjugated of each other, but by (19.13), they are also real; thus,

$$z^{-1}\sum_{\ell=0}^{n} \widetilde{H_{\ell,0}}(z^{-2})\widetilde{H_{\ell,1}}(z^2) = z\sum_{\ell=0}^{n} \widetilde{H_{\ell,0}}(z^2)\widetilde{H_{\ell,1}}(z^{-2}) \in \mathbb{R}. \tag{19.14}$$

Finally, applying the first equation in (19.6) with z replaced by $-z$ leads to

$$1 = \sum_{\ell=0}^{n} \widetilde{H_\ell}(-z)\widetilde{H_\ell}(-z^{-1})$$

$$= \sum_{\ell=0}^{n} \left(\widetilde{H_{\ell,0}}(z^2) - z^{-1}\widetilde{H_{\ell,1}}(z^2)\right)\left(\widetilde{H_{\ell,0}}(z^{-2}) - z\widetilde{H_{\ell,1}}(z^{-2})\right)$$

$$= \sum_{\ell=0}^{n} \left|\widetilde{H_{\ell,0}}(z^2)\right|^2 + \sum_{\ell=0}^{n} \left|\widetilde{H_{\ell,1}}(z^2)\right|^2$$

$$- z\sum_{\ell=0}^{n} \widetilde{H_{\ell,0}}(z^2)\widetilde{H_{\ell,1}}(z^{-2}) - z^{-1}\sum_{\ell=0}^{n} \widetilde{H_{\ell,0}}(z^{-2})\widetilde{H_{\ell,1}}(z^2).$$

Again by addition and subtraction with the equation in (19.12), this leads to

$$
\begin{cases}
2\sum_{\ell=0}^{n}\left|\widetilde{H_{\ell,0}}(z^2)\right|^2 - 2z\sum_{\ell=0}^{n}\widetilde{H_{\ell,0}}(z^2)\widetilde{H_{\ell,1}}(z^{-2})) = 1, \\[2mm]
2\sum_{\ell=0}^{n}\left|\widetilde{H_{\ell,1}}(z^2)\right|^2 - 2z^{-1}\sum_{\ell=0}^{n}\widetilde{H_{\ell,0}}(z^{-2})\widetilde{H_{\ell,1}}(z^2) = 1.
\end{cases}
\tag{19.15}
$$

Combining (19.15) with (19.13) and (19.14) finally leads to (19.10). □

It turns out that the condition (19.9) in Theorem 19.4.2 is well known in the context of *filter banks*. In the rest of this section, we discuss this connection.

Intuitively, a filter bank is some kind of "black box," which performs operations on an incoming signal (i.e., a sequence of numbers). Typically, a filter bank splits the incoming signal into certain subsignals, which contain particular information about the signal. For this reason, filter banks of that type are called *analysis filter banks*. After processing the subsequences coming out of the analysis filter bank, engineers usually wish to get back to the original input sequence. Therefore, it is essential that an analysis filter bank is followed by another filter bank, which reconstructs the original signal from the subsignals; such a filter bank is called a *synthesis filter bank*. In that case, the entire system consisting of the two filter banks is said to have the *perfect reconstruction* property.

The filter banks considered here will contain three operations on the incoming sequence $\{x_k\}_{k\in\mathbb{Z}}$:

- **Convolution with a sequence** $\{h_k\}_{k\in\mathbb{Z}}$**:** The outcome is a new sequence, whose kth coordinate is given by $\sum_{n\in\mathbb{Z}} h_n x_{k-n}$.

- **Downsampling:** The outcome is the sequence
$$
\downarrow \{x_k\}_{k\in\mathbb{Z}} := (\cdots x_{-2}, x_0, x_2, \cdots).
$$
Thus, downsampling removes each second element in the sequence.

- **Upsampling:** The outcome is the sequence
$$
\uparrow \{x_k\}_{k\in\mathbb{Z}} := (\cdots x_{-1}, 0, x_0, 0, x_1, \cdots).
$$
Thus, upsampling inserts zeroes between the elements in the sequence.

Note that downsampling is the left-inverse of upsampling but not the right-inverse.

We will now describe a particular filter bank. The analysis filter bank will split the incoming signal $\{x_k\}_{k\in\mathbb{Z}}$ into $n+1$ subsignals: each of these signals is obtained by convolving $\{x_k\}_{k\in\mathbb{Z}}$ with a sequence $h_{k,\ell}$, $\ell = 0,\ldots,n$, followed by a downsampling. The synthesis filter bank first upsamples each

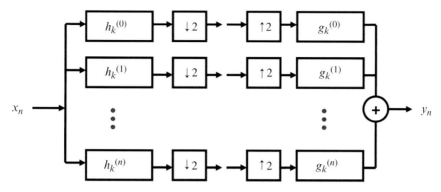

Figure 19.1. A filter bank consisting of an analysis filter bank composed with a synthesis filter bank.

of the incoming $n + 1$ subsignals, then convolves the resulting sequences with sequences $\{g_{k,\ell}\}_{k\in\mathbb{Z}}$, $\ell = 0, \ldots, n$, and finally adds the outcoming $n+1$ signals; see Figure 19.1. We will assume that the sequences $\{h_{k,\ell}\}_{k\in\mathbb{Z}}$ and $\{g_{k,\ell}\}_{k\in\mathbb{Z}}, \ell = 0, \ldots, n$ are related by

$$g_{k,\ell} = h_{-k,\ell}, \ k \in \mathbb{Z}, \ell = 0, \ldots, n.$$

For the above system consisting of the analysis filter bank followed by the synthesis filter bank, the perfect reconstruction property can be formulated in terms of the polyphase components associated with the filters $\{h_{k,\ell}\}_{k\in\mathbb{Z}}$:

Theorem 19.4.3 *For the considered filter bank, the perfect reconstruction property is equivalent to the condition*

$$H_p^T(z^{-1})H_p(z) = I \text{ for } z \in \mathbb{C} \text{ with } |z| = 1. \tag{19.16}$$

A proof of Theorem 19.4.3 can be found in [93]. Note that the conditions in (19.16) and (19.9) are really "identical": if one of these conditions is satisfied, the other will be satisfied if the filter sequences $\{h_{k,\ell}\}_{k\in\mathbb{Z}}$ are either multiplied or divided by $\sqrt{2}$. In other words, if the condition (19.9) (and the general setup for the unitary extension principle) is satisfied, then the functions

$$\psi_\ell = \sqrt{2} \sum_{k\in\mathbb{Z}} h_{k,\ell} DT_{-k}\psi_0, \ \ell = 1, \ldots, n,$$

generate a tight frame with frame bound 1; if (19.16) is satisfied, the functions

$$\psi_\ell = \sum_{k\in\mathbb{Z}} h_{k,\ell} DT_{-k}\psi_0, \ \ell = 1, \ldots, n,$$

generate a tight frame with frame bound 1.

Thus, the conditions in Theorem 19.4.2 for construction of a tight wavelet frame are equivalent to the perfect reconstruction property for the above filter bank.

19.5 Exercises

19.1 Prove Theorem 19.1.1.

19.2 Consider the B-spline B_2.

(i) Use the results in Example 18.3.1 and Example 18.6.3 to calculate the Z-transforms $\widetilde{H}_\ell$, $\ell = 0, 1, 2$, and verify that the conditions in Theorem 19.4.1 are satisfied.

(ii) Calculate the polyphase components for the Z-transforms $\widetilde{H}_\ell$, $\ell = 0, 1, 2$, and verify that the matrix $H_p^T(z)$ satisfies the conditions in Theorem 19.4.2.

20

Generalized Shift-Invariant Systems in $L^2(\mathbb{R}^d)$

We have already seen that a Gabor system $\{E_{mb}T_{na}g\}_{m,n\in\mathbb{Z}}$ in $L^2(\mathbb{R})$ is a special case of a shift-invariant system. In contrast, a wavelet system is not a shift-invariant system. Indeed, looking for example at a dyadic wavelet system $\{D^jT_k\psi\}_{j,k\in\mathbb{Z}}$, we can use the commutator relations for the operators D and T_k to rewrite the system as

$$\{D^jT_k\psi\}_{j,k\in\mathbb{Z}} = \{T_{k2^{-j}}D^j\psi\}_{j,k\in\mathbb{Z}};$$

thus, the system is in fact a collection of shifts of the functions $D^j\psi$, $j \in \mathbb{Z}$, but the translation parameters depend on $j \in \mathbb{Z}$. Therefore the system does not fall into the framework of shift-invariant systems in Section 10.

On the other hand, some of the results we have derived for Gabor systems and wavelet systems are very similar, with similar proofs. This is most evident from the sufficient conditions for such systems to form frames, derived in Theorem 11.4.2 and Theorem 15.2.3: except for the fact that the Gabor result is derived in the time domain and the wavelet result in the frequency domain, the results are clearly parallel.

This makes it natural to look for a general theory that covers these key results (and others) simultaneously for Gabor systems and wavelet systems. The way to do this is already apparent from the above description: we will consider systems of functions, generated by translates of a collection of "window functions," with translation parameters that depend on the window function.

The pioneers in the study of such systems were Hernandez, Labate, and Weiss [398] and Ron and Shen [564]. We will follow the approach in [398], but we will adopt the name *generalized shift-invariant system* coined by

© Springer International Publishing Switzerland 2016   493
O. Christensen, *An Introduction to Frames and Riesz Bases*,
Applied and Numerical Harmonic Analysis,
DOI 10.1007/978-3-319-25613-9_20

Ron and Shen. Note also the paper [5] by Aldroubi, Cabrelli, and Molter, which also provides a common framework for the analysis of Gabor systems and wavelet systems.

We will be more general than in the previous chapters and describe the theory in $L^2(\mathbb{R}^d)$. This implies that the relevant operators (translation, modulation, and scaling) will be defined in terms of vectors and matrices, but except for a slightly more involved notation, the proofs are similar to the one-dimensional case. Besides the convenience of having the result stated explicitly in $L^2(\mathbb{R}^d)$, this generalization can also be considered as an intermediate step to the analysis in Chapter 21, where we consider generalized shift-invariant systems on locally compact abelian groups.

We begin in Section 20.1 with a short discussion of analysis in $\mathbb{R}^d$, with focus on Fourier analysis and the operators that are necessary in order to define Gabor systems and wavelet systems. In Section 20.2 we consider systems generated by just one function; these systems are actually just shift-invariant systems like in Chapter 9 but now considered in $L^2(\mathbb{R}^d)$. The analysis of the generalized shift-invariant systems is split in two sections. First, we derive sufficient conditions for the Bessel property and the frame property in Section 20.3, and then, based on the results for one generator, we prove the results concerning dual pairs of frames in Section 20.4. In Section 20.5 we then return to the Gabor systems and state the corresponding special cases of the results in $L^2(\mathbb{R}^d)$; the same is done for wavelet systems in Section 20.6. The section also gives a short description of a very general class of systems that contain as well the Gabor structure as the wavelet structure, the so-called wave packet systems.

20.1   Analysis in $\mathbb{R}^d$ and Notation

Parallel to our development of frame theory in $L^2(\mathbb{R})$, we begin our analysis in $L^2(\mathbb{R}^d)$ with a formal definition of the relevant operators. We denote the standard inner product between two vectors $x, y \in \mathbb{R}^d$ by $x \cdot y$.

Definition 20.1.1 *Consider the following classes of linear operators on* $L^2(\mathbb{R}^d)$:

(i) *For* $a \in \mathbb{R}^d$, *the translation operator* T_a *is defined by*

$$(T_a f)(x) = f(x - a), \quad x \in \mathbb{R}^d. \tag{20.1}$$

(ii) *For* $b \in \mathbb{R}^d$, *the modulation operator* E_b *is*

$$(E_b f)(x) = e^{2\pi i b \cdot x} f(x), \quad x \in \mathbb{R}^d. \tag{20.2}$$

(iii) *The dilation operator associated with a real and invertible* $d \times d$ *matrix* C *is*

$$(D_C f)(x) = |\det C|^{1/2} f(Cx), \quad x \in \mathbb{R}^d. \tag{20.3}$$

Exactly as for the one-dimensional case considered in Section 2.9, one can prove that the operators in Definition 20.1.1 actually act as bounded operators from $L^2(\mathbb{R}^d)$ to $L^2(\mathbb{R}^d)$ and are unitary (Exercise 20.1).

Note that the definition of the scaling operator D_C differs slightly from the convention we used in the one-dimensional case, see (2.23): looking at (20.3) in the case $d = 1$, a scaling D_c would correspond to the definition in (2.23) with $a = c^{-1}$. This change will not cause any problems, but it implies a few minor changes in the commutator relations, as we will see in Lemma 20.1.3.

For $f \in L^1(\mathbb{R}^d) \cap L^2(\mathbb{R}^d)$, we denote the Fourier transform by

$$\mathcal{F}f(\gamma) = \hat{f}(\gamma) = \int_{\mathbb{R}^d} f(x)e^{-2\pi i \gamma \cdot x}dx, \ \gamma \in \mathbb{R}^d.$$

As usual, the Fourier transform is extended to a unitary operator on $L^2(\mathbb{R}^d)$.

Exactly as we have seen in the analysis of Gabor systems and wavelet systems in the one-dimensional case, the interaction between the operators T_a, E_b, D_C and $\mathcal{F}$ will be central for our analysis in $L^2(\mathbb{R}^d)$. The higher-dimensional setting will require slight modifications in the notation, which will be apparent already from the statement of the subsequent Lemma 20.1.3. In particular, we need the following notation:

- Given a matrix C, denote the transposed matrix by C^T;

- If C is an invertible $d \times d$ matrix with real entries, let

$$C^{\sharp} = (C^T)^{-1}. \tag{20.4}$$

- Let $\mathbb{T}^d$ denote the unit cube in $\mathbb{R}^d$, i.e.,

$$\mathbb{T}^d := [0, 1[^d. \tag{20.5}$$

We will often use the following change of variable (see [565]):

Lemma 20.1.2 *Let C be an invertible $d \times d$ matrix with real entries. Then, given a measurable set $X \subseteq \mathbb{R}^d$,*

$$\int_{C(X)} f(x)\,dx = |\det C| \int_X f(Cy)\,dy$$

whenever the function $f : \mathbb{R}^d \to \mathbb{C}$ is integrable over $C(X)$.

We will now state the commutator relations between the introduced operators; the proofs are left to the reader (Exercise 20.2). Note that the mentioned change in the definition of the scaling operator implies that the formulas in Lemma 20.1.3 differ slightly from the one-dimensional versions in (2.26) and (2.27).

Lemma 20.1.3 *Let C be an invertible $d \times d$ matrix with real entries, and let $a, b \in \mathbb{R}^d$. Then the following commutator relations hold:*

$$T_a E_b = e^{-2\pi i b \cdot a} E_b T_a; \quad D_C E_b = E_{C^T b} D_C; \quad D_C T_a = T_{C^{-1}a} D_C;$$

The operators are related to the Fourier transform via

$$\mathcal{F}T_a = E_{-a}\mathcal{F}; \quad \mathcal{F}E_b = T_b\mathcal{F}; \quad \mathcal{F}D_C = D_{C^\sharp}\mathcal{F}.$$

We will consider functions $f \in L^2(\mathbb{T}^d)$ as $\mathbb{Z}^d$-periodic functions on $\mathbb{R}^d$, i.e., $f(x+k) = f(x)$ for $x \in \mathbb{R}^d$ and $k \in \mathbb{Z}^d$. The *Fourier series* of a function $f \in L^2(\mathbb{T}^d)$ is given by

$$f(x) = \sum_{k \in \mathbb{Z}^d} c_k e^{2\pi i k \cdot x}, \tag{20.6}$$

where the Fourier coefficients c_k are

$$c_k = \int_{\mathbb{T}^d} f(x) e^{-2\pi i k \cdot x} \, dx, \ k \in \mathbb{Z}^d. \tag{20.7}$$

Parallel to the one-dimensional results described in Section 3.8, we have the following standard result about Fourier series in $L^2(\mathbb{T}^d)$:

Lemma 20.1.4

(i) *The functions $\{e^{2\pi i k \cdot x}\}_{k \in \mathbb{Z}^d}$ form an orthonormal basis for $L^2(\mathbb{T}^d)$.*

(ii) *If c_k and $\widetilde{c}_k$, $k \in \mathbb{Z}$, denote the Fourier coefficients for two functions F and $\widetilde{F}$ in $L^2(\mathbb{T}^d)$, then*

$$\langle F, \widetilde{F} \rangle = \sum_{k \in \mathbb{Z}^d} c_k \overline{\widetilde{c}_k}.$$

We will need some preparation before we can start the analysis of generalized shift-invariant systems, but let us state the central definition by Ron and Shen [564] already now.

Definition 20.1.5 *A generalized shift-invariant system (GSI system for short) is a system of functions $\{T_{C_j k}\phi_j\}_{j \in J, k \in \mathbb{Z}^d}$, where J is a countable index set, $\{\phi_j\}_{j \in J} \subset L^2(\mathbb{R}^d)$, and $\{C_j\}_{j \in J}$ is a collection of invertible $d \times d$ matrices with real entries.*

Our goal is to derive a characterization (under weak conditions) of dual frames of GSI systems. For technical reasons many of the results and proofs to follow will deal with functions in the space

$$\mathcal{D} := \left\{ f \in L^2(\mathbb{R}^d) \, \middle| \, \mathrm{supp}\widehat{f} \text{ is compact and } \widehat{f} \in L^\infty(\mathbb{R}^d) \right\}. \tag{20.8}$$

The space $\mathcal{D}$ is dense in $L^2(\mathbb{R}^d)$; see Exercise 20.3. We note that in certain applications it is necessary to consider a slightly more restrictive definition of $\mathcal{D}$. For example, in the wavelet case we must require that $\operatorname{supp}\widehat{f}$ is a compact set in $\mathbb{R}\setminus\{0\}$; this is still a dense subset in $L^2(\mathbb{R}^d)$, and the entire analysis will go through.

Note that variations of the generalized shift-invariant systems are known in the literature under various names. Considering a collection of invertible $d\times d$ matrices $\{C_j\}_{j\in\mathbb{Z}}$ with real entries and corresponding functions $\{g_j\}_{j\in\mathbb{Z}}\subset L^2(\mathbb{R}^d)$, systems on the form $\{E_{C_jk}g_j\}_{j\in\mathbb{Z},k\in\mathbb{Z}^d}$ were analyzed under the name *nonstationary Gabor systems* by Jaillet in [416] and further studied in, e.g., [257] by Dörfler and Matusiak and [409] by Holighaus; applying the Fourier transform immediately shows that such systems are equivalent with GSI systems. The special case of a nonstationary Gabor frame where the matrices C_j are independent of j was called a *Fourier-like system* in [197].

20.2   The Case of One Generator

In this section we will consider systems of functions on the form $\{T_{Ck}\phi\}_{k\in\mathbb{Z}^d}$, where $\phi\in L^2(\mathbb{R}^d)$ and C is an invertible $d\times d$ matrix with real entries. The main purpose is to provide the technical background for the analysis in Section 20.4, where the case of multiple generators is considered.

Lemma 20.2.1 *Let $\phi\in L^2(\mathbb{R}^d)$ and let C denote an invertible $d\times d$ matrix with real entries. If $f\in\mathcal{D}$, then the following hold:*

(i) *For any $k\in\mathbb{Z}^d$,*

$$\langle f,T_{Ck}\phi\rangle = \frac{1}{|\det C|}\int_{\mathbb{T}^d}\sum_{n\in\mathbb{Z}^d}\widehat{f}(C^\sharp(\mu+n))\overline{\widehat{\phi}(C^\sharp(\mu+n))}e^{2\pi ik\cdot\mu}d\mu. \quad (20.9)$$

(ii) *The function F defined by*

$$F(\mu) = \frac{1}{|\det C|}\sum_{n\in\mathbb{Z}^d}\widehat{f}(C^\sharp(\mu+n))\overline{\widehat{\phi}(C^\sharp(\mu+n))} \quad (20.10)$$

is $\mathbb{T}^d$-periodic and belongs to $L^2(\mathbb{T}^d)$.

(iii) *The $(-k)$th Fourier coefficient for the function F in (20.10) equals $\langle f,T_{Ck}\phi\rangle$.*

(iv) *We have*

$$\sum_{k\in\mathbb{Z}^d}|\langle f,T_{Ck}\phi\rangle|^2$$

$$= \frac{1}{|\det C|^2}\int_{\mathbb{T}^d}\left|\sum_{n\in\mathbb{Z}^d}\widehat{f}(C^\sharp(\mu+n))\overline{\widehat{\phi}(C^\sharp(\mu+n))}\right|^2 d\mu. \quad (20.11)$$

Proof. First, observe that for $f \in \mathcal{D}$ and $k \in \mathbb{Z}^d$,

$$\langle f, T_{Ck}\phi \rangle = \langle \mathcal{F}f, \mathcal{F}T_{Ck}\phi \rangle = \int_{\mathbb{R}^d} \widehat{f}(\mu)\overline{\widehat{\phi}(\mu)}e^{2\pi iCk\cdot\mu}d\mu$$

$$= \int_{\mathbb{R}^d} \widehat{f}(\mu)\overline{\widehat{\phi}(\mu)}e^{2\pi ik\cdot C^T\mu}d\mu.$$

Making the change of variable $\mu \to C^\sharp\mu$ and splitting the integral now yield that

$$\langle f, T_{Ck}\phi \rangle = |\det C^\sharp| \int_{\mathbb{R}^d} \widehat{f}(C^\sharp\mu)\overline{\widehat{\phi}(C^\sharp\mu)}e^{2\pi ik\cdot\mu}d\mu$$

$$= \frac{1}{|\det C|} \int_{\mathbb{T}^d} \sum_{n \in \mathbb{Z}^d} \widehat{f}(C^\sharp(\mu+n))\overline{\widehat{\phi}(C^\sharp(\mu+n))}e^{2\pi ik\cdot\mu}d\mu,$$

which proves (20.9). The assumption $f \in \mathcal{D}$ implies that the sum in (20.10) is finite for each $\mu \in \mathbb{R}^d$, with a uniform bound on the number of nonzero terms; thus the periodic function F belongs to $L^2(\mathbb{T}^d)$. Furthermore (20.9) shows that $\langle f, T_{Ck}\phi \rangle$ is the $(-k)$th Fourier coefficient for the function F. The result in (20.11) now follows from Lemma 20.1.4 (ii).   □

The expression in (20.11) is well suited to analyze the frame properties for a system of the form $\{T_{Ck}\phi\}_{k \in \mathbb{Z}^d}$. When we want to consider dual pairs of frames of this form, we need the following slight modification:

Lemma 20.2.2 *Let* $\phi, \widetilde{\phi} \in L^2(\mathbb{R}^d)$ *and let* C *denote an invertible* $d \times d$ *matrix with real entries. If* $f \in \mathcal{D}$, *then*

$$\sum_{k \in \mathbb{Z}^d} \langle f, T_{Ck}\phi \rangle\langle T_{Ck}\widetilde{\phi}, f \rangle = \tag{20.12}$$

$$\frac{1}{|\det C|^2} \int_{\mathbb{T}^d} \sum_{n \in \mathbb{Z}^d} \widehat{f}(C^\sharp(\mu+n))\overline{\widehat{\phi}(C^\sharp(\mu+n))} \sum_{\ell \in \mathbb{Z}^d} \overline{\widehat{f}(C^\sharp(\mu+\ell))}\widehat{\widetilde{\phi}}(C^\sharp(\mu+\ell))\, d\mu.$$

Proof. Note that the convergence of the series in (20.12) follows from Lemma 20.2.1. Using the function F in (20.10) and the function

$$\widetilde{F}(\mu) := \frac{1}{|\det C|} \sum_{\ell \in \mathbb{Z}^d} \widehat{f}(C^\sharp(\mu+\ell))\overline{\widehat{\widetilde{\phi}}(C^\sharp(\mu+\ell))}, \tag{20.13}$$

Lemma 20.1.4 and Lemma 20.2.1 yield that

$$\sum_{k \in \mathbb{Z}^d} \langle f, T_{Ck}\phi \rangle\langle T_{Ck}\widetilde{\phi}, f \rangle = \int_{\mathbb{T}^d} F(\mu)\overline{\widetilde{F}(\mu)}\, d\mu;$$

inserting the expressions for the functions F and $\widetilde{F}$ now leads to the desired conclusion.   □

The next lemma is a key step in the approach to frame properties for GSI systems. Basically it is an application of (20.12), with the function f replaced by its translates $T_y f$, $y \in \mathbb{R}^d$. Hereby we obtain a function $K : \mathbb{R}^d \to \mathbb{C}$, which turns out to equal the function in (20.14) below, with only a finite number of nonzero coefficients c_m, i.e., it is a trigonometric polynomial. The idea goes back to Janssen [425] (see Section 10.1) and Laugesen [476], and was again applied in [398].

Lemma 20.2.3 Let $\phi, \widetilde{\phi} \in L^2(\mathbb{R}^d)$ and let C denote an invertible $d \times d$ matrix with real entries. Fix $f \in \mathcal{D}$. Then the function

$$K : \mathbb{R}^d \to \mathbb{C}, \; K(y) := \sum_{k \in \mathbb{Z}^d} \langle T_y f, T_{Ck} \phi \rangle \langle T_{Ck} \widetilde{\phi}, T_y f \rangle \tag{20.14}$$

equals the trigonometric polynomial

$$K(y) = \sum_{m \in \mathbb{Z}^d} c_m e^{2\pi i C^{\sharp} m \cdot y},$$

where the Fourier coefficients are

$$c_m = \frac{1}{|\det C|} \int_{\mathbb{R}^d} \widehat{f}(\gamma) \overline{\widehat{f}(\gamma + C^{\sharp} m)} \overline{\widehat{\phi}(\gamma)} \widehat{\widetilde{\phi}}(\gamma + C^{\sharp} m) \, d\gamma.$$

Proof. The space $\mathcal{D}$ is clearly translation invariant, so the considerations in Lemma 20.2.2 show that the function K is well defined. Using the commutator relation $\widehat{T_y f}(\mu) = E_{-y} \widehat{f}(\mu) = e^{-2\pi i y \cdot \mu}$ and Lemma 20.2.2 with the function f replaced by $T_y f$, we see that

$$
\begin{aligned}
K(y) &= \frac{1}{|\det C|^2} \int_{\mathbb{T}^d} \sum_{n \in \mathbb{Z}^d} e^{-2\pi i y \cdot (C^{\sharp}(\mu+n))} \widehat{f}(C^{\sharp}(\mu+n)) \overline{\widehat{\phi}(C^{\sharp}(\mu+n))} \\
&\quad \times \sum_{\ell \in \mathbb{Z}^d} \overline{e^{-2\pi i y \cdot (C^{\sharp}(\mu+\ell))} \widehat{f}(C^{\sharp}(\mu+\ell))} \widehat{\widetilde{\phi}}(C^{\sharp}(\mu+\ell)) \, d\mu \\
&= \frac{1}{|\det C|^2} \int_{\mathbb{T}^d} \sum_{n \in \mathbb{Z}^d} \sum_{\ell \in \mathbb{Z}^d} e^{2\pi i y \cdot (C^{\sharp}(\ell-n))} \widehat{f}(C^{\sharp}(\mu+n)) \overline{\widehat{\phi}(C^{\sharp}(\mu+n))} \\
&\quad \times \overline{\widehat{f}(C^{\sharp}(\mu+\ell))} \widehat{\widetilde{\phi}}(C^{\sharp}(\mu+\ell)) \, d\mu.
\end{aligned}
$$

Letting $k = \ell - n$ and writing the sums in the above expression as sums over k and n, it follows that

$$
\begin{aligned}
K(y) &= \frac{1}{|\det C|^2} \int_{\mathbb{T}^d} \sum_{k \in \mathbb{Z}^d} \sum_{n \in \mathbb{Z}^d} e^{2\pi i y \cdot C^{\sharp} k} \widehat{f}(C^{\sharp}(\mu+n)) \overline{\widehat{\phi}(C^{\sharp}(\mu+n))} \\
&\quad \times \overline{\widehat{f}(C^{\sharp}(\mu+k+n))} \widehat{\widetilde{\phi}}(C^{\sharp}(\mu+k+n)) \, d\mu.
\end{aligned}
$$

Pulling out the sum over $n \in \mathbb{Z}^d$,

$$K(y) \quad = \quad \frac{1}{|\det C|^2} \sum_{n \in \mathbb{Z}^d} \int_{\mathbb{T}^d} \sum_{k \in \mathbb{Z}^d} e^{2\pi i y \cdot C^\sharp k} \widehat{f}(C^\sharp(\mu+n)) \overline{\widehat{\phi}(C^\sharp(\mu+n))}$$

$$\times \overline{\widehat{f}(C^\sharp(\mu+k+n))} \widehat{\widetilde{\phi}}(C^\sharp(\mu+k+n)) \, d\mu.$$

Using that the sum over $n \in \mathbb{Z}^d$ of the integrals over $\mathbb{T}^d$ yields an integral over $\mathbb{R}^d$,

$$K(y)$$

$$= \quad \frac{1}{|\det C|^2} \int_{\mathbb{R}^d} \sum_{k \in \mathbb{Z}^d} e^{2\pi i y \cdot C^\sharp k} \widehat{f}(C^\sharp \mu) \overline{\widehat{\phi}(C^\sharp \mu)} \overline{\widehat{f}(C^\sharp(\mu+k))} \widehat{\widetilde{\phi}}(C^\sharp(\mu+k)) \, d\mu$$

$$= \quad \frac{1}{|\det C|^2} \sum_{k \in \mathbb{Z}^d} \left(\int_{\mathbb{R}^d} \widehat{f}(C^\sharp \mu) \overline{\widehat{\phi}(C^\sharp \mu)} \overline{\widehat{f}(C^\sharp(\mu+k))} \widehat{\widetilde{\phi}}(C^\sharp(\mu+k)) \, d\mu \right) e^{2\pi i y \cdot C^\sharp k}.$$

Making the change of variable $\gamma = C^\sharp \mu$ in the expression for the coefficient in the above Fourier series and using that $y \cdot C^\sharp k = C^\sharp k \cdot y$ finally yield that

$$K(y) = \frac{1}{|\det C|} \sum_{k \in \mathbb{Z}^d} \left(\int_{\mathbb{R}^d} \widehat{f}(\gamma) \overline{\widehat{f}(\gamma + C^\sharp k)} \overline{\widehat{\phi}(\gamma)} \widehat{\widetilde{\phi}}(\gamma + C^\sharp k) \, d\gamma \right) e^{2\pi i C^\sharp k \cdot y},$$

as claimed. That the Fourier series actually is a trigonometric series follows from the assumption that $f \in \mathcal{D}$. □

The following is the d-dimensional pendant to a calculation that appeared in the proof of Theorem 9.5.1. See Exercise 20.4 for the corresponding extension of Theorem 9.5.1.

Lemma 20.2.4 *Let $\phi, \widetilde{\phi} \in L^2(\mathbb{R}^d)$ and let C denote an invertible $d \times d$ matrix with real entries. If $\{T_{Ck}\phi\}_{k \in \mathbb{Z}}$ and $\{T_{Ck}\widetilde{\phi}\}_{k \in \mathbb{Z}}$ are Bessel sequences, then for all $f \in \mathcal{D}$,*

$$\mathcal{F}\left(\sum_{k \in \mathbb{Z}^d} \langle f, T_{Ck}\widetilde{\phi} \rangle T_{Ck}\phi \right)(\gamma) = \frac{1}{|\det C|} \widehat{\phi}(\gamma) \sum_{n \in \mathbb{Z}^d} \widehat{f}(\gamma + C^\sharp n) \overline{\widehat{\widetilde{\phi}}(\gamma + C^\sharp n)}.$$

Proof. First,

$$\mathcal{F}\left(\sum_{k \in \mathbb{Z}^d} \langle f, T_{Ck}\widetilde{\phi} \rangle T_{Ck}\phi \right)(\gamma) = \sum_{k \in \mathbb{Z}^d} \langle f, T_{Ck}\widetilde{\phi} \rangle \mathcal{F} T_{Ck}\phi(\gamma);$$

using Lemma 20.2.1, it follows that

$$
\mathcal{F}\left(\sum_{k\in\mathbb{Z}^d}\langle f, T_{Ck}\widetilde{\phi}\rangle T_{Ck}\phi\right)(\gamma)
$$

$$
= \frac{\widehat{\phi}(\gamma)}{|\det C|}\sum_{k\in\mathbb{Z}^d}\left(\int_{\mathbb{T}^d}\sum_{n\in\mathbb{Z}^d}\widehat{f}(C^{\sharp}(\mu+n))\overline{\widehat{\widetilde{\phi}}(C^{\sharp}(\mu+n))}e^{2\pi ik\cdot\mu}d\mu\right)e^{-2\pi ik\cdot C^T\gamma}
$$

$$
= \frac{1}{|\det C|}\widehat{\phi}(\gamma)\sum_{n\in\mathbb{Z}^d}\widehat{f}(C^{\sharp}(C^T\gamma+n))\overline{\widehat{\widetilde{\phi}}(C^{\sharp}(C^T\gamma+n))}
$$

$$
= \frac{1}{|\det C|}\widehat{\phi}(\gamma)\sum_{n\in\mathbb{Z}^d}\widehat{f}(\gamma+C^{\sharp}n)\overline{\widehat{\widetilde{\phi}}(\gamma+C^{\sharp}n)}.
$$

This concludes the proof. □

20.3 Frames with Multiple Generators

The main results in this chapter describe how to construct dual pairs of frames with the GSI structure; see Theorem 20.4.3. In order to apply this result we need to know that the involved GSI systems form Bessel sequences. We now state a sufficient condition for the Bessel property and a corresponding condition for the frame property. The statements and proofs are similar to what we have seen for Gabor systems in Theorem 11.4.2 and for wavelet systems in Theorem 15.2.3. For this reason, we only sketch the proof; it originally appeared in [199] as a generalization of a result in [472].

Theorem 20.3.1 *Given a GSI system* $\{T_{C_jk}\phi_j\}_{j\in J, k\in\mathbb{Z}^d}$ *in* $L^2(\mathbb{R}^d)$*, the following hold:*

(i) *If*

$$
B := \sup_{\gamma\in\mathbb{R}^d}\sum_{j\in J}\sum_{k\in\mathbb{Z}^d}\frac{1}{|\det C_j|}|\widehat{\phi_j}(\gamma)\widehat{\phi_j}(\gamma - C_j^{\sharp}k)| < \infty, \qquad (20.15)
$$

then $\{T_{C_jk}\phi_j\}_{j\in J, k\in\mathbb{Z}^d}$ *is a Bessel sequence with bound* B.

(ii) *If also*

$$
A := \inf_{\gamma\in\mathbb{R}^d}\left(\sum_{j\in J}\frac{1}{|\det C_j|}|\widehat{\phi_j}(\gamma)|^2 - \sum_{j\in J}\sum_{k\neq 0}\frac{1}{|\det C_j|}|\widehat{\phi_j}(\gamma)\widehat{\phi_j}(\gamma - C_j^{\sharp}k)|\right) > 0,
$$

then $\{T_{C_jk}\phi_j\}_{j\in J, k\in\mathbb{Z}^d}$ *is a frame for* $L^2(\mathbb{R}^d)$ *with bounds* A *and* B.

Proof. It is sufficient to prove (i) and (ii) for functions f in the dense subspace $\mathcal{D}$ of $L^2(\mathbb{R}^d)$; see (20.8). Using arguments like in the proof of Theorem 11.4.2 or Theorem 15.2.3 (see the details in [472]), one can show that for $f \in \mathcal{D}$,

$$\sum_{j \in J} \sum_{k \in \mathbb{Z}^d} |\langle f, T_{C_j k} \phi_j \rangle|^2 = \int_{\mathbb{R}^d} |\widehat{f}(\gamma)|^2 \sum_{j \in J} \frac{1}{|\det C_j|} |\widehat{\phi_j}(\gamma)|^2 + R(f),$$

where

$$R(f) = \sum_{j \in J} \sum_{k \neq 0} \frac{1}{|\det C_j|} \int_{\mathbb{R}^d} \widehat{f}(\gamma) \overline{\widehat{\phi_j}(\gamma)} \widehat{f}(\gamma - C_j^\sharp k) \overline{\widehat{\phi_j}(\gamma - C_j^\sharp k)} d\gamma.$$

Furthermore

$$R(f) \leq \sum_{j \in J} \sum_{k \neq 0} \frac{1}{|\det C_j|} \int_{\mathbb{R}^d} |\widehat{f}(\gamma)|^2 |\widehat{\phi_j}(\gamma) \widehat{\phi_j}(\gamma - C_j^\sharp k)| d\gamma.$$

Thus

$$\sum_{j \in J} \sum_{k \in \mathbb{Z}^d} |\langle f, T_{C_j k} \phi_j \rangle|^2 \leq \int_{\mathbb{R}^d} |\widehat{f}(\gamma)|^2 \sum_{j \in J} \sum_{k \in \mathbb{Z}^d} \frac{1}{|\det C_j|} |\widehat{\phi_j}(\gamma) \widehat{\phi_j}(\gamma - C_j^\sharp k)| d\gamma.$$

Using (20.15), it now follows from the density of $\mathcal{D}$ in $L^2(\mathbb{R}^d)$ and Lemma 3.2.6 that $\{T_{C_j k} \phi_j\}_{j \in J, k \in \mathbb{Z}^d}$ is a Bessel sequence with bound B. Also, for $f \in \mathcal{D}$,

$$\sum_{j \in J} \sum_{k \in \mathbb{Z}^d} |\langle f, T_{C_j k} \phi_j \rangle|^2$$

$$\geq \int_{\mathbb{R}^d} |\widehat{f}(\gamma)|^2 \sum_{j \in J} \frac{1}{|\det C_j|} |\widehat{\phi_j}(\gamma)|^2 d\gamma$$

$$- |\sum_{j \in J} \sum_{k \neq 0} \frac{1}{|\det C_j|} \int_{\mathbb{R}^d} \widehat{f}(\gamma) \overline{\widehat{f}(\gamma - C_j^\sharp k)} \overline{\widehat{\phi_j}(\gamma)} \widehat{\phi_j}(\gamma - C_j^\sharp k) d\gamma|$$

$$\geq \int_{\mathbb{R}^d} |\widehat{f}(\gamma)|^2 \Big(\sum_{j \in J} \frac{1}{|\det C_j|} |\widehat{\phi_j}(\gamma)|^2$$

$$- \sum_{j \in J} \sum_{k \neq 0} \frac{1}{|\det C_j|} |\widehat{\phi_j}(\gamma) \widehat{\phi_j}(\gamma - C_j^\sharp k)| \Big) d\gamma.$$

Via Lemma 5.1.9, this concludes the proof of the lower bound in (ii). $\square$

We also note that a necessary condition for a GSI system to be a Bessel sequence was proved in [398]. We have already seen special cases of the result: see Proposition 11.3.4 for the Gabor case, Proposition 15.2.2 for dyadic wavelet systems, and Lemma 19.1.2 for irregular wavelet systems.

Proposition 20.3.2 *Assume that a GSI system* $\{T_{C_jk}\phi_j\}_{k\in\mathbb{Z}^d, j\in J}$ *is a Bessel sequence with bound B. Then*

$$\sum_{j\in J} \frac{1}{|\det C_j|} |\widehat{\phi_j}(\gamma)|^2 \leq B, \, a.e. \, \gamma \in \mathbb{R}^d. \tag{20.16}$$

Observe that the infinite sum in (20.16) equals one of the terms in the sufficient condition in (20.15), namely, the one corresponding to $k = 0$.

Even if $\{T_{C_jk}\phi_j\}_{k\in\mathbb{Z}^d, j\in J}$ is a frame for $L^2(\mathbb{R}^d)$, the sum in (20.16) might not be bounded below by a positive constant. This is in contrast to the special cases for Gabor systems in Proposition 11.3.4 for the wavelet systems in Proposition 15.2.2. We refer to [177] for a more detailed analysis.

20.4 Dual Pairs of Frames with Multiple Generators

In this section we will prove the main result about GSI systems, which yields a condition for two such systems to form dual frames for $L^2(\mathbb{R}^d)$. The results will be derived based on the calculations in Section 20.2. The original source is still the paper [398].

The basic idea is to consider a GSI system $\{T_{C_jk}\phi_j\}_{k\in\mathbb{Z}^d, j\in J}$ in $L^2(\mathbb{R}^d)$ as a countable union of the shift-invariant systems $\{T_{C_jk}\phi_j\}_{k\in\mathbb{Z}^d}$, i.e.,

$$\{T_{C_jk}\phi_j\}_{k\in\mathbb{Z}^d, j\in J} = \bigcup_{j\in J}\{T_{C_jk}\phi_j\}_{k\in\mathbb{Z}^d}.$$

Whenever we want to consider frame properties for $\{T_{C_jk}\phi_j\}_{k\in\mathbb{Z}^d, j\in J}$, we can then apply the result in (20.11) on each of the shift-invariant systems $\{T_{C_jk}\phi_j\}_{k\in\mathbb{Z}^d}$, simply by introducing an extra sum over $j \in J$. In a similar fashion, we can use Lemma 20.2.2 and Lemma 20.2.3 to examine the dual frame property for two GSI systems $\{T_{C_jk}\phi_j\}_{k\in\mathbb{Z}^d, j\in J}$ and $\{T_{C_jk}\widetilde{\phi}_j\}_{k\in\mathbb{Z}^d, j\in J}$.

Technically, an important step is to rewrite the sums that occur when we follow the sketched approach. Given a GSI system $\{T_{C_jk}\phi_j\}_{k\in\mathbb{Z}^d, j\in J}$, let

$$\Lambda = \left\{C_j^\sharp m \mid j \in J, m \in \mathbb{Z}^d\right\}, \tag{20.17}$$

where we remember that $C_j^\sharp = (C_j^T)^{-1}$. Given $\alpha \in \Lambda$, there might exist several pairs $(j, m) \in J \times \mathbb{Z}^d$ for which $\alpha = C_j^\sharp m$; let

$$J_\alpha = \left\{j \in J \mid \exists m \in \mathbb{Z}^d \text{ such that } \alpha = C_j^\sharp m\right\}. \tag{20.18}$$

We will need a technical condition, basically to guarantee that the function ω introduced in the subsequent Lemma 20.4.2 is continuous. Various versions of the condition can be found in the literature: the condition (i) in the following Definition 20.4.1 appeared in the paper [398], and the conditions (ii) & (iii) were introduced in [417] by Jakobsen and Lemvig.

Definition 20.4.1

(i) *A GSI system $\{T_{C_j k}\phi_j\}_{k \in \mathbb{Z}^d, j \in J}$ satisfies the local integrability condition (LIC) if*

$$\sum_{j \in J} \sum_{m \in \mathbb{Z}^d} \frac{1}{|\det C_j|} \int_{supp \, \widehat{f}} |\widehat{f}(\gamma + C_j^\sharp m)\widehat{\phi_j}(\gamma)|^2 \, d\gamma < \infty \qquad (20.19)$$

for all $f \in \mathcal{D}$.

(ii) *Two GSI systems $\{T_{C_j k}\phi_j\}_{k \in \mathbb{Z}^d, j \in J}$ and $\{T_{C_j k}\widetilde{\phi_j}\}_{k \in \mathbb{Z}^d, j \in J}$ satisfy the dual α-local integrability condition (dual α-LIC) if*

$$\sum_{j \in J} \sum_{m \in \mathbb{Z}^d} \frac{1}{|\det C_j|} \int_{supp \, \widehat{f}} |\widehat{f}(\gamma)\widehat{f}(\gamma + C_j^\sharp m)\widehat{\phi_j}(\gamma)\widetilde{\widehat{\phi_j}}(\gamma + C_j^\sharp m)| \, d\gamma < \infty$$

for all $f \in \mathcal{D}$.

(iii) *A GSI system $\{T_{C_j k}\phi_j\}_{k \in \mathbb{Z}^d, j \in J}$ satisfies the α-local integrability condition (α-LIC) if (ii) holds with $\phi_j = \widetilde{\phi_j}$.*

We leave it to the reader (Exercise 20.5) to prove that the α-LIC is weaker than the LIC: if $\{T_{C_j k}\phi_j\}_{k \in \mathbb{Z}^d, j \in J}$ satisfies the LIC, then $\{T_{C_j k}\phi_j\}_{k \in \mathbb{Z}^d, j \in J}$ also satisfies the α-LIC.

The following lemma is a key step toward the duality result for GSI systems.

Lemma 20.4.2 *Assume that the GSI systems $\{T_{C_j k}\phi_j\}_{k \in \mathbb{Z}^d, j \in J}$ and $\{T_{C_j k}\widetilde{\phi_j}\}_{k \in \mathbb{Z}^d, j \in J}$ satisfy the dual α-LIC, and let $f \in \mathcal{D}$. Then the function*

$$\omega(y) := \sum_{j \in J} \sum_{k \in \mathbb{Z}^d} \langle T_y f, T_{C_j k}\phi_j \rangle \langle T_{C_j k}\widetilde{\phi_j}, T_y f \rangle \qquad (20.20)$$

is continuous, and

$$\omega(y) = \sum_{\alpha \in \Lambda} \left(\int_{\mathbb{R}^d} \widehat{f}(\gamma)\overline{\widehat{f}(\gamma + \alpha)} \sum_{j \in J_\alpha} \frac{1}{|\det C_j|} \overline{\widehat{\phi_j}(\gamma)}\widetilde{\widehat{\phi_j}}(\gamma + \alpha) \right) e^{2\pi i \alpha \cdot y} d\gamma$$

pointwise for all $y \in \mathbb{R}^d$.

Proof. For $j \in J$, let

$$\omega_j(y) := \sum_{k \in \mathbb{Z}^d} \langle T_y f, T_{C_j k}\phi_j \rangle \langle T_{C_j k}\widetilde{\phi_j}, T_y f \rangle.$$

By Lemma 20.2.3 the function ω_j is continuous and equals a trigonometric polynomial, $\omega_j(y) = \sum_{m \in \mathbb{Z}^d} c_{m,j} e^{2\pi i C_j^\sharp m \cdot y}$, where the Fourier coefficients are

$$c_{m,j} = \frac{1}{|\det C_j|} \int_{\mathbb{R}^d} \widehat{f}(\gamma)\overline{\widehat{f}(\gamma + C_j^\sharp m)}\overline{\widehat{\phi_j}(\gamma)}\widetilde{\widehat{\phi_j}}(\gamma + C_j^\sharp m) \, d\gamma. \qquad (20.21)$$

The dual α-LIC immediately implies that

$$\sum_{j \in J} \sum_{m \in \mathbb{Z}^d} |c_{m,j}| < \infty. \tag{20.22}$$

Therefore the function ω in (20.20) is continuous, and

$$\omega(y) = \sum_{j \in J} \sum_{m \in \mathbb{Z}^d} c_{m,j} e^{2\pi i C_j^{\sharp} m \cdot y},$$

with absolute and uniform convergence of the infinite series. It follows that the terms can be reordered arbitrarily without affecting the convergence; using an indexing in terms of the sets Λ and J_α in (20.17) and (20.18),

$$
\begin{aligned}
\omega(y) &= \sum_{\alpha \in \Lambda} \sum_{j \in J_\alpha} \frac{1}{|\det C_j|} \left(\int_{\mathbb{R}^d} \widehat{f}(\gamma) \overline{\widehat{f}(\gamma + \alpha)} \widehat{\phi_j}(\gamma) \overline{\widehat{\widetilde{\phi}_j}(\gamma + \alpha)} \right) e^{2\pi i \alpha \cdot y} d\gamma \\
&= \sum_{\alpha \in \Lambda} \left(\int_{\mathbb{R}^d} \widehat{f}(\gamma) \overline{\widehat{f}(\gamma + \alpha)} \sum_{j \in J_\alpha} \frac{1}{|\det C_j|} \overline{\widehat{\phi_j}(\gamma)} \widehat{\widetilde{\phi}_j}(\gamma + \alpha) \right) e^{2\pi i \alpha \cdot y} d\gamma,
\end{aligned}
$$

as desired. $\qquad \square$

We are now ready to state a condition for two GSI systems being dual frames. Basically it appeared in the paper [398] by Hernandez, Labate, and Weiss, but following [417] we formulate it using the dual α-LIC rather than the stronger LIC.

Theorem 20.4.3 *Assume that the GSI systems $\{T_{C_j k} \phi_j\}_{k \in \mathbb{Z}^d, j \in J}$ and $\{T_{C_j k} \widetilde{\phi}_j\}_{k \in \mathbb{Z}^d, j \in J}$ are Bessel sequences and satisfy the dual α-LIC. Then $\{T_{C_j k} \phi_j\}_{k \in \mathbb{Z}^d, j \in J}$ and $\{T_{C_j k} \widetilde{\phi}_j\}_{k \in \mathbb{Z}^d, j \in J}$ are dual frames if and only if*

$$\sum_{j \in J_\alpha} \frac{1}{|\det C_j|} \overline{\widehat{\phi_j}(\gamma)} \widehat{\widetilde{\phi}_j}(\gamma + \alpha) = \delta_{\alpha,0}, \ a.e. \ \gamma \in \mathbb{R}^d \tag{20.23}$$

for all $\alpha \in \Lambda$.

Proof. Let us first assume that (20.23) is satisfied. Now, let $f \in \mathcal{D}$; then, by Lemma 20.4.2 the continuous function ω in (20.20) takes the form

$$
\begin{aligned}
\omega(y) &= \sum_{\alpha \in \Lambda} \left(\int_{\mathbb{R}^d} \widehat{f}(\gamma) \overline{\widehat{f}(\gamma + \alpha)} \sum_{j \in J_\alpha} \frac{1}{|\det C_j|} \overline{\widehat{\phi_j}(\gamma)} \widehat{\widetilde{\phi}_j}(\gamma + \alpha) \right) e^{2\pi i \alpha \cdot y} d\gamma \\
&= \int_{\mathbb{R}^d} \widehat{f}(\gamma) \overline{\widehat{f}(\gamma)} \, d\gamma = ||f||^2;
\end{aligned}
$$

in particular, for $y = 0$,

$$\sum_{j \in J} \sum_{k \in \mathbb{Z}^d} \langle f, T_{C_j k} \phi_j \rangle \langle T_{C_j k} \widetilde{\phi}, f \rangle = ||f||^2. \tag{20.24}$$

Since (20.24) holds on a dense subspace of $L^2(\mathbb{R}^d)$, Lemma 6.3.4 now shows that $\{T_{C_jk}\phi_j\}_{k\in\mathbb{Z}^d, j\in J}$ and $\{T_{C_jk}\widetilde{\phi}_j\}_{k\in\mathbb{Z}^d, j\in J}$ are dual frames.

In order to show the other implication, assume that $\{T_{C_jk}\phi_j\}_{k\in\mathbb{Z}^d, j\in J}$ and $\{T_{C_jk}\widetilde{\phi}_j\}_{k\in\mathbb{Z}^d, j\in J}$ are dual frames. Let $f\in\mathcal{D}$, and consider again the function ω in (20.20); the duality assumption implies that

$$\omega(y) = ||T_y f||^2 = ||f||^2, \forall y \in \mathbb{R}^d.$$

It now follows from the expression for the function ω in Lemma 20.4.2 and the independence of the exponentials (see [398] for details) that

$$\int_{\mathbb{R}^d} \widehat{f}(\gamma)\overline{\widehat{f}(\gamma+\alpha)} \sum_{j\in J_\alpha} \frac{1}{|\det C_j|} \overline{\widehat{\phi_j}(\gamma)}\widehat{\widetilde{\phi}_j}(\gamma+\alpha)d\gamma = \delta_{\alpha,0}||f||^2, \forall \alpha \in \Lambda.$$
(20.25)

In particular, taking $\alpha = 0$,

$$\int_{\mathbb{R}^d} |\widehat{f}(\gamma)|^2 \sum_{j\in J} \frac{1}{|\det C_j|} \overline{\widehat{\phi_j}(\gamma)}\widehat{\widetilde{\phi}_j}(\gamma)d\gamma = ||f||^2 = ||\widehat{f}||^2, \forall f \in \mathcal{D}.$$

By a standard argument this implies that

$$\sum_{j\in J} \frac{1}{|\det C_j|} \overline{\widehat{\phi_j}(\gamma)}\widehat{\widetilde{\phi}_j}(\gamma) = 1, \text{ a.e. } \gamma \in \mathbb{R}^d;$$

thus (20.23) holds for $\alpha = 0$. In order to prove that (20.23) holds for $\alpha \neq 0$, let

$$s_\alpha(\gamma) := \sum_{j\in J_\alpha} \frac{1}{|\det C_j|} \overline{\widehat{\phi_j}(\gamma)}\widehat{\widetilde{\phi}_j}(\gamma+\alpha), \ \gamma \in \mathbb{R}^d.$$

Applying polarization on (20.25) it follows that

$$\int_{\mathbb{R}^d} \widehat{f}(\gamma)\overline{\widehat{g}(\gamma+\alpha)}s_\alpha(\gamma)d\gamma = 0, \forall f, g \in \mathcal{D}.$$

Again, from here a standard argument implies that $s_\alpha(\gamma) = 0$ for a.e. $\gamma \in \mathbb{R}^d$, as desired. □

We note that the equivalence in Theorem 20.4.3 is wrong if the dual α-LIC is removed. In fact, an example by Bownik and Rzeszotnik in [86] shows that $\{T_{C_jk}\phi_j\}_{k\in\mathbb{Z}^d, j\in J}$ and $\{T_{C_jk}\widetilde{\phi}_j\}_{k\in\mathbb{Z}^d, j\in J}$ can be dual frames even if (20.23) does not hold.

In the subsequent sections, we will apply the GSI results to Gabor systems and wavelet systems. Here we will just note that Theorem 20.4.3 simplifies greatly for shift-invariant systems. In fact, if a shift-invariant system $\{T_{Ck}\phi_j\}_{j\in J, k\in\mathbb{Z}^d}$ in $L^2(\mathbb{R}^d)$ is a Bessel sequence, then the LIC is satisfied (Exercise 20.6), and Theorem 20.4.3 has the following consequence (Exercise 20.7):

Corollary 20.4.4 *Let C denote an invertible $d \times d$ matrix with real en-tries. Then two Bessel sequences $\{T_{Ck}\phi_j\}_{k\in\mathbb{Z}^d, j\in J}$ and $\{T_{Ck}\widetilde{\phi}_j\}_{k\in\mathbb{Z}^d, j\in J}$ in $L^2(\mathbb{R}^d)$ are dual frames if and only if for all $n \in \mathbb{Z}^d$,*

$$\sum_{j\in J} \overline{\widehat{\phi}_j(\gamma)}\widehat{\widetilde{\phi}}_j(\gamma + C^\sharp n) = |\det C|\,\delta_{n,0}, \ a.e.\ \gamma \in \mathbb{R}^d. \tag{20.26}$$

20.5 Gabor Systems in $L^2(\mathbb{R}^d)$

The results in Section 20.3 and Section 20.4 have immediate consequences for Gabor systems in $L^2(\mathbb{R}^d)$. In order to show the power of the approach in the previous sections, we will state very general results.

Given a function $g \in L^2(\mathbb{R}^d)$ and two invertible invertible $d \times d$ matrices $\mathcal{A}, \mathcal{B}$ with real entries, we will consider the Gabor system $\{E_{\mathcal{B}m}T_{\mathcal{A}n}g\}_{m,n\in\mathbb{Z}^d}$ in $L^2(\mathbb{R}^d)$. Explicitly, for $m, n \in \mathbb{Z}^d$,

$$E_{\mathcal{B}m}T_{\mathcal{A}n}g(x) = e^{2\pi i \mathcal{B}m \cdot x}g(x - \mathcal{A}n), \ x \in \mathbb{R}^d.$$

We will in fact be more general than that and consider a Gabor system generated by collection of windows $\{g_\ell\}_{\ell=1}^L \subset L^2(\mathbb{R}^d)$. Let us begin with sufficient conditions for such systems to form Bessel sequences or frames:

Theorem 20.5.1 *Let $\mathcal{A}$ and $\mathcal{B}$ denote invertible $d \times d$ matrices with real entries, and consider a finite collection of functions $\{g_\ell\}_{\ell=1}^L \subset L^2(\mathbb{R}^d)$. If*

$$B := \frac{1}{|\det \mathcal{A}|} \sup_{\gamma\in\mathbb{R}^d} \sum_{\ell=1}^L \sum_{j\in\mathbb{Z}^d} \sum_{m\in\mathbb{Z}^d} |\widehat{g}_\ell(\gamma - \mathcal{B}j)\widehat{g}_\ell(\gamma - \mathcal{B}j - \mathcal{A}^\sharp m)| < \infty,$$

then $\{E_{\mathcal{B}m}T_{\mathcal{A}n}g_\ell\}_{m,n\in\mathbb{Z}^d, \ell=1,\dots,L}$ is a Bessel sequence with bound B; if also

$$A := \frac{1}{|\det \mathcal{A}|} \inf_{\gamma\in\mathbb{R}^d} \sum_{\ell=1}^L \left(\sum_{j\in\mathbb{Z}^d} |\widehat{g}_\ell(\gamma - \mathcal{B}j)|^2 \right.$$

$$\left. - \sum_{j\in\mathbb{Z}^d} \sum_{m\neq 0} |\widehat{g}_\ell(\gamma - \mathcal{B}j)\widehat{g}_\ell(\gamma - \mathcal{B}j - \mathcal{A}^\sharp m)| \right) > 0,$$

then $\{E_{\mathcal{B}m}T_{\mathcal{A}n}g_\ell\}_{m,n\in\mathbb{Z}^d, \ell=1,\dots,L}$ is a frame for $L^2(\mathbb{R}^d)$ with bounds A, B.

Proof. First, using Lemma 20.1.3, we see that for $j, n \in \mathbb{Z}^d$ and $\ell = 1, \dots, L$,

$$E_{\mathcal{B}j}T_{\mathcal{A}n}g_\ell = e^{2\pi i \mathcal{B}j \cdot \mathcal{A}n}T_{\mathcal{A}n}E_{\mathcal{B}j}g_\ell; \tag{20.27}$$

this implies that $\{E_{\mathcal{B}j}T_{\mathcal{A}n}g_\ell\}_{j,n\in\mathbb{Z}^d, \ell=1,\dots,L}$ is a frame if and only if the (generalized) shift-invariant system $\{T_{\mathcal{A}n}E_{\mathcal{B}j}g_\ell\}_{j,n\in\mathbb{Z}^d, \ell=1,\dots,L}$ is a frame. The result now follows directly from Theorem 20.3.1. $\qquad\qquad\square$

Formulated in the time domain, the sufficient conditions for the upper and lower frame bounds read as follows (Exercise 20.8):

Theorem 20.5.2 *Let $\mathcal{A}$ and $\mathcal{B}$ denote invertible $d \times d$ matrices with real entries, and consider a finite collection of functions $\{g_\ell\}_{\ell=1}^L \subset L^2(\mathbb{R}^d)$. If*

$$B := \frac{1}{|\det \mathcal{B}|} \sup_{x \in \mathbb{R}^d} \sum_{\ell=1}^L \sum_{j \in \mathbb{Z}^d} \sum_{m \in \mathbb{Z}^d} |g_\ell(x - \mathcal{A}j)g_\ell(x - \mathcal{A}j - \mathcal{B}^\sharp m)| < \infty,$$

then $\{E_{\mathcal{B}m}T_{\mathcal{A}n}g_\ell\}_{m,n\in\mathbb{Z}^d, \ell=1,\dots,L}$ is a Bessel sequence with bound B; if also

$$A := \frac{1}{|\det \mathcal{B}|} \inf_{x \in \mathbb{R}^d} \sum_{\ell=1}^L \left(\sum_{j \in \mathbb{Z}^d} |g_\ell(x - \mathcal{A}j)|^2 \right.$$

$$\left. - \sum_{j \in \mathbb{Z}^d} \sum_{m \neq 0} |g_\ell(x - \mathcal{A}j)g_\ell(x - \mathcal{A}j - \mathcal{B}^\sharp m)| \right) > 0,$$

then $\{E_{\mathcal{B}m}T_{\mathcal{A}n}g_\ell\}_{m,n\in\mathbb{Z}^d, \ell=1,\dots,L}$ is a frame for $L^2(\mathbb{R}^d)$ with bounds A, B.

Corollary 20.4.4 leads to a number of equivalent conditions for two Gabor systems being dual frames for $L^2(\mathbb{R}^d)$. We collect them in the following theorem, which also appeared in our main source [398]. Note that the equivalence (i) $\Leftrightarrow$ (iv) is the *Wexler–Raz theorem*.

Theorem 20.5.3 *Let $\mathcal{A}$ and $\mathcal{B}$ denote invertible $d \times d$ matrices with real entries, and consider finite collections of functions $\{g_\ell\}_{\ell=1}^L, \{h_\ell\}_{\ell=1}^L \subset L^2(\mathbb{R}^d)$. Assuming that the Gabor systems $\{E_{\mathcal{B}m}T_{\mathcal{A}n}g_\ell\}_{m,n\in\mathbb{Z}^d, \ell=1,\dots,L}$ and $\{E_{\mathcal{B}m}T_{\mathcal{A}n}h_\ell\}_{m,n\in\mathbb{Z}^d, \ell=1,\dots,L}$ are Bessel sequences, the following are equivalent:*

(i) *$\{E_{\mathcal{B}m}T_{\mathcal{A}n}g_\ell\}_{m,n\in\mathbb{Z}^d, \ell=1,\dots,L}$ and $\{E_{\mathcal{B}m}T_{\mathcal{A}n}h_\ell\}_{m,n\in\mathbb{Z}^d, \ell=1,\dots,L}$ are dual frames for $L^2(\mathbb{R}^d)$.*

(ii) *For all $m \in \mathbb{Z}^d$ and a.e. $\gamma \in \mathbb{R}^d$,*

$$\sum_{\ell=1}^L \sum_{j \in \mathbb{Z}^d} \overline{\hat{g}_\ell(\gamma - \mathcal{B}j)}\hat{h}_\ell(\gamma - \mathcal{B}j + \mathcal{A}^\sharp m) = |\det \mathcal{A}|\, \delta_{m,0}. \quad (20.28)$$

(iii) *For all $m \in \mathbb{Z}^d$ and a.e. $x \in \mathbb{R}^d$,*

$$\sum_{\ell=1}^L \sum_{j \in \mathbb{Z}^d} \overline{g_\ell(x - \mathcal{A}j)}h_\ell(x - \mathcal{A}j + \mathcal{B}^\sharp m) = |\det \mathcal{B}|\, \delta_{m,0}. \quad (20.29)$$

(iv) *For any $m, n \in \mathbb{Z}^d$,*

$$\sum_{\ell=1}^L \langle h_\ell, E_{\mathcal{A}^\sharp m}T_{\mathcal{B}^\sharp n}g_\ell \rangle = |\det \mathcal{B}|\,|\det \mathcal{A}|\, \delta_{m,0}\delta_{n,0}. \quad (20.30)$$

Proof. We prove the result in the case of Gabor systems with a single window and leave the extension to several windows to the reader.

First, using the calculation (20.27) we see that two Gabor systems $\{E_{\mathcal{B}j}T_{\mathcal{A}n}g\}_{j,n\in\mathbb{Z}^d}$ and $\{E_{\mathcal{B}j}T_{\mathcal{A}n}h\}_{j,n\in\mathbb{Z}^d}$ are dual frames if and only if $\{T_{\mathcal{A}n}E_{\mathcal{B}j}g\}_{j,n\in\mathbb{Z}^d}$ and $\{T_{\mathcal{A}n}E_{\mathcal{B}j}h\}_{j,n\in\mathbb{Z}^d}$ are dual frames; these systems fall into the framework of shift-invariant systems in Corollary 20.4.4 by letting $C = \mathcal{A}$ and

$$\phi_j := E_{\mathcal{B}j}g, \ \widetilde{\phi}_j := E_{\mathcal{B}j}h, \ j \in \mathbb{Z}^d. \tag{20.31}$$

Now the equivalence (i) $\Leftrightarrow$ (ii) follows by direct calculation of the expressions in (20.23).

For the proof of the equivalence (i) $\Leftrightarrow$ (iii) we again apply Lemma 20.1.3, which shows that for $j, m \in \mathbb{Z}^d$,

$$\mathcal{F}^{-1}E_{\mathcal{B}m}T_{\mathcal{A}j}g = T_{-\mathcal{B}m}\mathcal{F}^{-1}T_{\mathcal{A}j}g;$$

thus $\{E_{\mathcal{B}m}T_{\mathcal{A}n}g\}_{m,n\in\mathbb{Z}^d}$ and $\{E_{\mathcal{B}m}T_{\mathcal{A}n}h\}_{m,n\in\mathbb{Z}^d}$ are dual frames if and only if $\{T_{-\mathcal{B}m}\mathcal{F}^{-1}T_{\mathcal{A}j}g\}_{m,n\in\mathbb{Z}^d}$ and $\{T_{-\mathcal{B}m}\mathcal{F}^{-1}T_{\mathcal{A}j}h\}_{m,n\in\mathbb{Z}^d}$ are dual frames. From here, the proof is completed in a similar way as the first part of the proof.

Finally, we will now prove the equivalence (ii) $\Leftrightarrow$ (iv). First, for any given $m \in \mathbb{Z}^d$, consider the $\mathcal{B}\mathbb{Z}^d$-periodic function

$$F_m(\gamma) := \sum_{j\in\mathbb{Z}^d} \overline{\hat{g}(\gamma - \mathcal{B}j)}\hat{h}(\gamma - \mathcal{B}j + \mathcal{A}^\sharp m), \ \gamma \in \mathbb{R}^d.$$

Recall that the functions $\{e^{2\pi ik\cdot x}\}_{k\in\mathbb{Z}^d}$ form an orthonormal basis for $L^2(\mathbb{T}^d)$. Using a change of variable (Lemma 20.1.2), it follows that $\{|\det\mathcal{B}|^{-1/2}e^{2\pi i\mathcal{B}^\sharp n\cdot\gamma}\}_{n\in\mathbb{Z}^d}$ is an orthonormal basis for $L^2(\mathcal{B}[0,1]^d)$. Thus, considering the Fourier expansion

$$F_m(\gamma) = \sum_{n\in\mathbb{Z}} c_{m,n}e^{2\pi i\mathcal{B}^\sharp n\cdot\gamma},$$

the coefficients can be calculated as

$$
\begin{aligned}
c_{m,n} &= |\det\mathcal{B}|^{-1}\int_{\mathcal{B}\mathbb{T}^d} F_m(\gamma)e^{-2\pi i\mathcal{B}^\sharp n\cdot\gamma}d\gamma \\
&= |\det\mathcal{B}|^{-1}\int_{\mathcal{B}\mathbb{T}^d} \sum_{j\in\mathbb{Z}^d} \overline{\hat{g}(\gamma - \mathcal{B}j)}\hat{h}(\gamma - \mathcal{B}j + \mathcal{A}^\sharp m)e^{-2\pi i\mathcal{B}^\sharp n\cdot\gamma}d\gamma \\
&= |\det\mathcal{B}|^{-1}\int_{\mathbb{R}^d} \overline{\hat{g}(\gamma)}\hat{h}(\gamma + \mathcal{A}^\sharp m)e^{-2\pi i\mathcal{B}^\sharp n\cdot\gamma}d\gamma \\
&= |\det\mathcal{B}|^{-1}\langle T_{-\mathcal{A}^\sharp m}\mathcal{F}h, E_{\mathcal{B}^\sharp n}\mathcal{F}g\rangle = |\det\mathcal{B}|^{-1}\langle\mathcal{F}E_{-\mathcal{A}^\sharp m}h, \mathcal{F}T_{-\mathcal{B}^\sharp n}g\rangle.
\end{aligned}
$$

Since the Fourier transform is unitary, we can continue with

$$c_{m,n} = |\det\mathcal{B}|^{-1}\langle E_{-\mathcal{A}^\sharp m}h, T_{-\mathcal{B}^\sharp n}g\rangle = |\det\mathcal{B}|^{-1}\langle h, E_{\mathcal{A}^\sharp m}T_{-\mathcal{B}^\sharp n}g\rangle.$$

The equivalence (ii) $\Leftrightarrow$ (iv) follows from this calculation. □

It is immediately clear that Theorem 20.5.3 generalizes Theorem 12.3.4 and Theorem 12.3.5 to the higher-dimensional case. One can use Theorem 20.5.3 to obtain explicit constructions of dual pairs of Gabor frames in $L^2(\mathbb{R}^d)$ along the same lines as what we saw in the one-dimensional setting in Section 12.5; however, for this particular approach the higher-dimensional case is more complicated [190]. Results that are easier to apply are given in [454] and [186]; the last-mentioned reference generalizes the results in Section 12.6 to the higher-dimensional case. We also refer to the paper [543] by Pfander, Rashkov, and Wang for constructions of tight Gabor frames in $L^2(\mathbb{R}^d)$ with compactly supported smooth windows.

Remember that a Gabor system $\{E_{mb}T_{na}g\}_{m,n\in\mathbb{Z}}$ in $L^2(\mathbb{R})$ only can be a frame for $L^2(\mathbb{R})$ if $ab \leq 1$. An elegant proof for this was given by Janssen in [425] and generalized to the higher-dimensional case by Labate in [471]:

Corollary 20.5.4 *Assume that $\mathcal{A}$ and $\mathcal{B}$ are invertible $d \times d$ matrices with real entries and that the Gabor system $\{E_{\mathcal{B}m}T_{\mathcal{A}n}g\}_{m,n\in\mathbb{Z}^d}$ is a frame for $L^2(\mathbb{R}^d)$. Then*

$$|\det\mathcal{A}|\,|\det\mathcal{B}| \leq 1. \qquad (20.32)$$

Proof. If $\{E_{\mathcal{B}m}T_{\mathcal{A}n}g\}_{m,n\in\mathbb{Z}^d}$ is a frame for $L^2(\mathbb{R}^d)$ with frame operator $S : L^2(\mathbb{R}^d) \to L^2(\mathbb{R}^d)$, calculations like in the one-dimensional case (Exercise 20.10) show that the frame decomposition takes the form

$$f = \sum_{m,n\in\mathbb{Z}^d} \langle f, E_{\mathcal{B}m}T_{\mathcal{A}n}S^{-1}g\rangle E_{\mathcal{B}m}T_{\mathcal{A}n}g, \ \forall f \in L^2(\mathbb{R}^d). \qquad (20.33)$$

We will now consider $f := g$, which has the trivial representation

$$g = \sum_{m,n\in\mathbb{Z}^d} \delta_{m,0}\delta_{n,0}E_{\mathcal{B}m}T_{\mathcal{A}n}g.$$

By Lemma 5.4.2, we know that the frame coefficients $\{\langle g, E_{\mathcal{B}m}T_{\mathcal{A}n}\,S^{-1}g\rangle\}_{m,n\in\mathbb{Z}^d}$ have minimal ℓ^2-norm among all coefficients representing g; thus,

$$\sum_{m,n\in\mathbb{Z}^d} |\langle g, E_{\mathcal{B}m}T_{\mathcal{A}n}S^{-1}g\rangle|^2 \leq 1,$$

which clearly implies that $|\langle g, S^{-1}g\rangle| \leq 1$. Now, by Theorem 20.5.3 (iv), we know that $\langle g, S^{-1}g\rangle = |\det\mathcal{B}|\,|\det\mathcal{A}|$, which completes the proof. $\square$

Parts of the proof of Theorem 20.5.3 generalize to the case where the modulations $\mathcal{B}m$, $m \in \mathbb{Z}^d$, are replaced by an irregular set and the translation matrix is allowed to depend on m. This no longer yields a shift-invariant system but a "genuine" GSI system. We will now state this more general version; note that it is not a "corollary" of Theorem 20.5.3 but rather a consequence of the insight gained from its proof. We leave the proof and a formulation of a sufficient condition for the Bessel property to the reader (Exercise 20.9).

Corollary 20.5.5 *Let $\{b_m\}_{m\in\mathbb{Z}}$ be a relatively separated sequence in $\mathbb{R}^d$ and $\{\mathcal{A}_m\}_{m\in\mathbb{Z}}$ a sequence of invertible $d\times d$ matrices with real entries. Furthermore, assume that two GSI Bessel sequences $\{T_{\mathcal{A}_m n}E_{b_m}g\}_{m\in\mathbb{Z},n\in\mathbb{Z}^d}$ and $\{E_{b_m}T_{\mathcal{A}_m n}h\}_{m\in\mathbb{Z},n\in\mathbb{Z}^d}$ satisfy the dual α-LIC. Then $\{T_{\mathcal{A}_m n}E_{b_m}g\}_{m\in\mathbb{Z},n\in\mathbb{Z}^d}$ and $\{E_{b_m}T_{\mathcal{A}_m n}h\}_{m\in\mathbb{Z},n\in\mathbb{Z}^d}$ are dual frames if and only if for all $\alpha\in\Lambda$ and a.e. $\gamma\in\mathbb{R}^d$,*

$$\sum_{j\in J_\alpha}\frac{1}{|\det\mathcal{A}_j|}\overline{\widehat{g}(\gamma-b_j)}\widehat{h}(\gamma-b_j+\alpha)=\delta_{\alpha,0}. \tag{20.34}$$

Here the sets Λ and J_α are, as usual, defined in (20.17) and (20.18).

20.6 Wavelet Systems in $L^2(\mathbb{R}^d)$

We will now turn our attention to wavelet systems. We will be very general and consider systems of the form $\{D_{\mathcal{A}_j}T_{\mathcal{B}k}\psi\}_{j\in J,k\in\mathbb{Z}^d}$, where $\mathcal{B}$ and $\{\mathcal{A}_j\}_{j\in J}$ are invertible $d\times d$ matrices with real entries. Explicitly, for $j\in J$ and $k\in\mathbb{Z}^d$,

$$D_{\mathcal{A}_j}T_{\mathcal{B}k}\psi(x)=|\det\mathcal{A}_j|\psi(\mathcal{A}_j x-\mathcal{B}k),\ x\in\mathbb{R}^d. \tag{20.35}$$

Note that $\{D_{\mathcal{A}_j}T_{\mathcal{B}k}\psi\}_{j\in J,k\in\mathbb{Z}^d}$ is the d-dimensional version of the system considered in Theorem 19.1.1.

We first generalize Theorem 15.2.3 and Theorem 19.1.1 to the d-dimensional case:

Theorem 20.6.1 *Let $\mathcal{B}$ and $\{\mathcal{A}_j\}_{j\in J}$ denote invertible $d\times d$ matrices with real entries, and let $\psi\in L^2(\mathbb{R}^d)$. If*

$$B:=\frac{1}{|\det\mathcal{B}|}\sup_{\gamma\in\mathbb{R}^d}\sum_{j\in J}\sum_{k\in\mathbb{Z}^d}|\widehat{\psi}(\mathcal{A}_j^\sharp\gamma)\widehat{\psi}(\mathcal{A}_j^\sharp\gamma-\mathcal{B}^\sharp k)|<\infty,$$

then $\{D_{\mathcal{A}_j}T_{\mathcal{B}k}\psi\}_{j\in J,k\in\mathbb{Z}^d}$ is a Bessel sequence in $L^2(\mathbb{R}^d)$. If furthermore

$$A:=\frac{1}{|\det\mathcal{B}|}\inf_{\gamma\in\mathbb{R}^d}\left(\sum_{j\in J}|\widehat{\psi}(\mathcal{A}_j^\sharp\gamma)|^2-\sum_{j\in J}\sum_{k\neq 0}|\widehat{\psi}(\mathcal{A}_j^\sharp\gamma)\widehat{\psi}(\mathcal{A}_j^\sharp\gamma-\mathcal{B}^\sharp k)|\right)>0,$$

then $\{D_{\mathcal{A}_j}T_{\mathcal{B}k}\psi\}_{j\in J,k\in\mathbb{Z}^d}$ is a frame for $L^2(\mathbb{R}^d)$ with bounds A,B.

Proof. Using Lemma 20.1.3, we see that

$$D_{\mathcal{A}_j}T_{\mathcal{B}k}\psi=T_{\mathcal{A}_j^{-1}\mathcal{B}k}D_{\mathcal{A}_j}\psi. \tag{20.36}$$

Let $C_j:=\mathcal{A}_j^{-1}\mathcal{B}$ and $\phi_j:=D_{\mathcal{A}_j}\psi$. Using again Lemma 20.1.3,

$$\widehat{\phi_j}(\gamma)=\mathcal{F}D_{\mathcal{A}_j}\psi(\gamma)=D_{\mathcal{A}_j^\sharp}\widehat{\psi}(\gamma)\ =\ |\det\mathcal{A}_j^\sharp|^{1/2}\widehat{\psi}(\mathcal{A}_j^\sharp\gamma)$$
$$=\ |\det\mathcal{A}_j|^{-1/2}\widehat{\psi}(\mathcal{A}_j^\sharp\gamma). \tag{20.37}$$

It is easy to see that $\mathcal{A}_j^\sharp(\mathcal{A}_j^{-1}\mathcal{B})^\sharp = \mathcal{B}^\sharp$; furthermore, by a direct calculation,

$$\frac{1}{|\det C_j|}|\widehat{\phi}_j(\gamma)\widehat{\phi}_j(\gamma - C_j^\sharp k)| = \frac{1}{|\det \mathcal{B}|}|\widehat{\psi}(\mathcal{A}_j^\sharp\gamma)\widehat{\psi}(\mathcal{A}_j^\sharp\gamma - \mathcal{B}^\sharp k)|.$$

Now the result follows from Theorem 20.3.1.   □

Observe that Theorem 20.6.1 has an immediate extension to wavelet systems $\{D_{\mathcal{A}_j}T_{\mathcal{B}_\ell k}\psi_\ell\}_{j\in J, k\in\mathbb{Z}^d, \ell=1,\ldots,L}$ generated by a finite collection of functions $\psi_\ell, \ell = 1,\ldots,L$ and matrices $\mathcal{B}_\ell$ depending on the generator ψ_ℓ.

We can also apply Theorem 20.4.3 to the very general wavelet systems considered in Theorem 20.6.1, but the condition (20.23) cannot be further simplified in this case. We will now consider the case where $J = \mathbb{Z}$ and $\mathcal{A}_j = \mathcal{A}^j$ for some matrix $\mathcal{A}$; in this case (20.23) can be simplified if we assume that the matrices $\mathcal{A}$ and $\mathcal{B}$ commute.

Thus, consider two wavelet systems $\{D_{\mathcal{A}^j}T_{\mathcal{B}k}\psi\}_{j\in\mathbb{Z}, k\in\mathbb{Z}^d}$, $\{D_{\mathcal{A}^j}T_{\mathcal{B}k}\widetilde{\psi}\}_{j\in\mathbb{Z}, k\in\mathbb{Z}^d}$; using the calculation (20.36) these wavelet systems can be considered as GSI systems $\{T_{C_j k}\phi_j\}_{k\in\mathbb{Z}^d, j\in\mathbb{Z}}, \{T_{C_j k}\widetilde{\phi}_j\}_{k\in\mathbb{Z}^d, j\in\mathbb{Z}}$, where

$$C_j = \mathcal{A}^{-j}\mathcal{B}, \quad \phi_j := D_{\mathcal{A}^j}\psi, \quad \widetilde{\phi}_j := D_{\mathcal{A}^j}\widetilde{\psi}. \tag{20.38}$$

We will state two versions of the characterizations of dual wavelet frames:

Corollary 20.6.2 *Let $\psi, \widetilde{\psi} \in L^2(\mathbb{R}^d)$, and let $\mathcal{A}$ and $\mathcal{B}$ denote invertible $d \times d$ matrices with real entries. Assume that*

- *$\mathcal{A}\mathcal{B} = \mathcal{B}\mathcal{A}$;*

- *The wavelet systems $\{D_{\mathcal{A}^j}T_{\mathcal{B}k}\psi\}_{j\in\mathbb{Z}, k\in\mathbb{Z}^d}, \{D_{\mathcal{A}^j}T_{\mathcal{B}k}\widetilde{\psi}\}_{j\in\mathbb{Z}, k\in\mathbb{Z}^d}$ are Bessel sequences;*

- *$\{T_{\mathcal{A}^{-j}\mathcal{B}k}D_{\mathcal{A}^j}\psi\}_{j\in\mathbb{Z}, k\in\mathbb{Z}^d}$ and $\{T_{\mathcal{A}^{-j}\mathcal{B}k}D_{\mathcal{A}^j}\widetilde{\psi}\}_{j\in\mathbb{Z}, k\in\mathbb{Z}^d}$ satisfy the LIC.*

Then the following are equivalent:

(i) $\{D_{\mathcal{A}^j}T_{\mathcal{B}k}\psi\}_{j\in\mathbb{Z}, k\in\mathbb{Z}^d}, \{D_{\mathcal{A}^j}T_{\mathcal{B}k}\widetilde{\psi}\}_{j\in\mathbb{Z}, k\in\mathbb{Z}^d}$ are dual frames for $L^2(\mathbb{R}^d)$;

(ii) For all $m \in \mathbb{Z}^d$ and a.e $\gamma \in \mathbb{R}^d$,

$$\sum_{\{j\in\mathbb{Z}\,|\,(\mathcal{A}^T)^{-j}m\in\mathbb{Z}^d\}} \overline{\widehat{\psi}((\mathcal{A}^T)^{-j}\gamma)}\widehat{\widetilde{\psi}}((\mathcal{A}^T)^{-j}(\gamma + \mathcal{B}^\sharp m)) = |\det\mathcal{B}|\,\delta_{m,0}; \tag{20.39}$$

(iii) For all $m' \in \mathbb{Z}^d, j' \in \mathbb{Z}$,

$$\sum_{\{(j,m)\in\mathbb{Z}\times\mathbb{Z}^d\,|\,(\mathcal{A}^T)^{j'}m'=(\mathcal{A}^T)^j m\}} \overline{\widehat{\psi}((\mathcal{A}^T)^{-j}\gamma)}\widehat{\widetilde{\psi}}((\mathcal{A}^T)^{-j}\gamma + \mathcal{B}^\sharp m) = |\det\mathcal{B}|\,\delta_{m,0}.$$

Proof. (i) $\Leftrightarrow$ (ii). We will show that the condition (20.39) is simply a reformulation of (20.23). Let $\alpha \in \Lambda$; then, for some $j' \in \mathbb{Z}, m \in \mathbb{Z}^d$,

$$\alpha = C^{\sharp}_{j'} m = ((A^{-j'} B)^T)^{-1} m = (A^T)^{j'} B^{\sharp} m. \tag{20.40}$$

We will now rewrite the condition (20.23) in terms of the parameters j', m. We split the argument into 4 steps:

Step 1: With α written as in (20.40), clearly $\alpha = 0 \Leftrightarrow m = 0$; thus, we can replace $\delta_{\alpha,0}$ in (20.23) by $\delta_{m,0}$.

Step 2: We now rewrite the term appearing in the sum in (20.23). With the definitions in (20.38), the calculation in (20.37) shows that

$$\widehat{\phi}(\gamma) = |\det \mathcal{A}^j|^{-1/2} \widehat{\psi}((\mathcal{A}^j)^{\sharp}\gamma) = |\det \mathcal{A}|^{-j/2} \widehat{\psi}((\mathcal{A}^T)^{-j}\gamma),$$

and similarly

$$\widehat{\widetilde{\phi}}(\gamma) = |\det \mathcal{A}|^{-j/2} \widehat{\widetilde{\psi}}((\mathcal{A}^T)^{-j}\gamma).$$

Thus, inserting the form of α stated in (20.40),

$$\frac{1}{|\det C_j|}\overline{\widehat{\widetilde{\phi}_j}(\gamma)}\widehat{\phi}_j(\gamma + \alpha) \tag{20.41}$$

$$= \frac{1}{|\det \mathcal{A}^{-j}\mathcal{B}|}|\det \mathcal{A}|^{-j}\overline{\widehat{\widetilde{\psi}}((\mathcal{A}^T)^{-j}\gamma)}\widehat{\psi}((\mathcal{A}^T)^{-j}[\gamma + (\mathcal{A}^T)^{j'} B^{\sharp} m])$$

$$= \frac{1}{|\det \mathcal{B}|}\overline{\widehat{\widetilde{\psi}}((\mathcal{A}^T)^{-(j-j')}(\mathcal{A}^T)^{-j'}\gamma)}\widehat{\psi}((\mathcal{A}^T)^{-(j-j')}[(\mathcal{A}^T)^{-j'}\gamma + B^{\sharp} m])$$

$$= \frac{1}{|\det \mathcal{B}|}\overline{\widehat{\widetilde{\psi}}((\mathcal{A}^T)^{-(j-j')}\eta)}\widehat{\psi}((\mathcal{A}^T)^{-(j-j')}[\eta + B^{\sharp} m]).$$

where we in the last step introduced a new variable $\eta := (\mathcal{A}^T)^{-j'}\gamma$. Since $(\mathcal{A}^T)^{-j'}$ is invertible, the variable γ runs through $\mathbb{R}^d$ if and only if η runs through $\mathbb{R}^d$.

Step 3: We now rewrite the index set J_α in (20.23) in terms of j', m. Note that, still with $\alpha = (\mathcal{A}^T)^{j'} B^{\sharp} m$ as in (20.40),

$$J_\alpha = \left\{ j \in \mathbb{Z} \,\middle|\, \exists n \in \mathbb{Z}^d : (C_j)^{\sharp}n = (\mathcal{A}^T)^{j'} B^{\sharp} m \right\}$$

$$= \left\{ j \in \mathbb{Z} \,\middle|\, \exists n \in \mathbb{Z}^d : (\mathcal{A}^T)^j B^{\sharp} n = (\mathcal{A}^T)^{j'} B^{\sharp} m \right\}$$

$$= \left\{ j \in \mathbb{Z} \,\middle|\, \exists n \in \mathbb{Z}^d : n = B^T (\mathcal{A}^T)^{-(j-j')}(B^T)^{-1} m \right\}.$$

The assumption that $\mathcal{A}$ and $\mathcal{B}$ commute implies that

$$B^T (\mathcal{A}^T)^{-(j-j')}(B^T)^{-1} = (\mathcal{A}^T)^{-(j-j')},$$

and the description of J_α reduces to

$$J_\alpha = \left\{ j \in \mathbb{Z} \,\middle|\, (A^T)^{-(j-j')} m \in \mathbb{Z}^d \right\}. \tag{20.42}$$

Step 4: We have now written all the ingredients in (20.23) in terms of j', m; thus, we can conclude that (20.23) holds for all $\alpha \in \Lambda$ if and only if

$$\sum_{\{j \in \mathbb{Z}\,|\,(A^T)^{-(j-j')}m \in \mathbb{Z}^d\}} \frac{1}{|\det \mathcal{B}|} \widehat{\psi}((A^T)^{-(j-j')}\eta)\widehat{\overline{\psi}}((A^T)^{-(j-j')}[\eta + B^\sharp m]) = \delta_{\alpha,0}$$

holds for a.e. $\eta \in \mathbb{R}^d$ and all $m \in \mathbb{Z}^d$; the change of summation index $j \to j + j'$ now yields the result.

(i) $\Leftrightarrow$ (iii). The proof is similar to the proof of (i) $\Leftrightarrow$ (ii). Let $\alpha \in \Lambda$; then, for some $j' \in \mathbb{Z}, m' \in \mathbb{Z}^d$, $\alpha = (A^T)^{j'} B^\sharp m'$. Now, using the first step in the calculation (20.41) [where m is now m'],

$$\frac{1}{|\det C_j|} \overline{\widehat{\phi_j}(\gamma)}\widehat{\phi_j}(\gamma + \alpha)$$

$$= \frac{1}{|\det \mathcal{B}|} \overline{\widehat{\psi}((A^T)^{-j}\gamma)}\widehat{\psi}((A^T)^{-j}\gamma + (A^T)^{-(j-j')}B^\sharp m')$$

$$= \frac{1}{|\det \mathcal{B}|} \overline{\widehat{\psi}((A^T)^{-j}\gamma)}\widehat{\psi}((A^T)^{-j}\gamma + B^\sharp(A^T)^{-(j-j')}m')$$

Using the form of J_α in (20.42), we can now rewrite the left-hand side of (20.23) as

$$\sum_{j \in J_\alpha} \frac{1}{|\det C_j|} \overline{\widehat{\phi_j}(\gamma)}\widehat{\phi_j}(\gamma + \alpha)$$

$$= \sum_{\{j\,|\,(A^T)^{-(j-j')}m' \in \mathbb{Z}^d\}} \frac{1}{|\det \mathcal{B}|} \overline{\widehat{\psi}((A^T)^{-j}\gamma)}\widehat{\psi}((A^T)^{-j}\gamma + B^\sharp(A^T)^{-(j-j')}m')$$

$$= \sum_{\{(j,m)\in\mathbb{Z}\times\mathbb{Z}^d\,|\,(A^T)^{-(j-j')}m'=m\}} \frac{1}{|\det \mathcal{B}|} \overline{\widehat{\psi}((A^T)^{-j}\gamma)}\widehat{\psi}((A^T)^{-j}\gamma + B^\sharp m)$$

$$= \sum_{\{(j,m)\in\mathbb{Z}\times\mathbb{Z}^d\,|\,(A^T)^{j'}m'=(A^T)^{j}m\}} \frac{1}{|\det \mathcal{B}|} \overline{\widehat{\psi}((A^T)^{-j}\gamma)}\widehat{\psi}((A^T)^{-j}\gamma + B^\sharp m).$$

Inserting this calculation in (20.23) now completes the proof. $\square$

We refer to the original paper [398] for more results about wavelet systems. In particular, [398] contains a discussion of the LIC and shows that it is satisfied if the matrix A is expanding on a subspace. Thus, for a wavelet system $\{D_{a^j}T_{kb}\psi\}_{j,k\in\mathbb{Z}}$ in $L^2(\mathbb{R})$ the LIC is automatically satisfied. The paper [398] also contains a number of wavelet constructions; see also the paper

[484] by Lemvig for explicit constructions of dual pairs of (band-limited) wavelet frames in $L^2(\mathbb{R}^d)$.

We will now derive some of the consequences of Corollary 20.6.2 in $L^2(\mathbb{R})$. First, the $L^2(\mathbb{R})$-version of the condition in Corollary 20.6.2 (iii) clearly yields precisely the conditions (i) and (ii) in Theorem 15.3.2; as explained above, the LIC is automatically satisfied, so the proof of Theorem 15.3.2 is completed.

Also Theorem 16.1.3 is a consequence of Corollary 20.6.2:

Proof of Theorem 16.1.3: We will apply the characterization of duality in Corollary 20.6.2 (ii). Thus, consider $m \in \mathbb{Z}$; in the case $m = 0$, the condition (20.39) clearly corresponds to the first condition in Theorem 16.1.3, so let us assume that $m \neq 0$. Write $m = 2^k q$, where $k \in \mathbb{N} \cup \{0\}$ and q is an odd integer. Then the index set in (20.39) is

$$\{j \in \mathbb{Z} \mid 2^{-j} m \in \mathbb{Z}\} = \{j \in \mathbb{Z} \mid 2^{k-j} q \in \mathbb{Z}\} = \{k, k-1, k-2, \dots\}.$$

Thus, the left-hand side in (20.39) takes the form

$$\sum_{j=-\infty}^{k} \overline{\widehat{\psi}(2^{-j}\eta)} \widehat{\widetilde{\psi}}(2^{-j}(\eta + m)) = \sum_{j=-\infty}^{k} \overline{\widehat{\psi}(2^{-j+k}[2^{-k}\eta])} \widehat{\widetilde{\psi}}(2^{-j+k}([2^{-k}\eta] + q))$$

$$= \sum_{j=0}^{\infty} \overline{\widehat{\psi}(2^{j}[2^{-k}\eta])} \widehat{\widetilde{\psi}}(2^{j}([2^{-k}\eta] + q)).$$

The variable $\gamma := 2^{-k}\eta$ runs through $\mathbb{R}$ whenever η runs through $\mathbb{R}$. Thus, the calculation shows that

$$\sum_{j=-\infty}^{k} \overline{\widehat{\psi}(2^{-j}\eta)} \widehat{\widetilde{\psi}}(2^{-j}(\eta + m)) = 0$$

for a.e. $\eta \in \mathbb{R}$ if and only if

$$\sum_{j=0}^{\infty} \overline{\widehat{\psi}(2^{j}\gamma)} \widehat{\widetilde{\psi}}(2^{j}(\gamma + q)) = 0$$

for a.e. $\gamma \in \mathbb{R}$. Thus, we arrive at the second condition in Theorem 16.1.3, and the proof is completed. □

Note that a variation of the wavelet setup, where the scalings are given in terms of products of certain (in general non-commuting) sets of matrices, was introduced and analyzed in [359]. The resulting sets, called *wavelet systems with composite dilations*, also fall within the framework of the GSI systems.

Both Gabor systems and wavelet systems have well-established applications in signal processing, and it is known when one of the systems is preferable compared to the other one. In order to obtain very flexible frame decompositions, it is natural to consider systems of functions that contain both the structures of Gabor systems and wavelet systems, i.e., systems of functions that are formed using all the operations' translation, modulation, and scaling. Let us state a definition of such a system, where we require a special structure of the translations.

Definition 20.6.3 *Let $\mathcal{B}$ and $\{\mathcal{A}_j\}_{j\in\mathbb{Z}}$ denote invertible $d \times d$ matrices with real entries, and let $\{c_m\}_{m\in\mathbb{Z}}$ be a sequence in $\mathbb{R}^d$. The associated wave packet system generated by a function $\psi \in L^2(\mathbb{R}^d)$ is the collection of functions*

$$\left\{ D_{\mathcal{A}_j} T_{\mathcal{B}k} E_{c_m} \psi \right\}_{j,m\in\mathbb{Z}, k\in\mathbb{Z}^d}. \qquad (20.43)$$

Wave packet systems first appeared in the frame context in the paper [399]. In [231], Czaja, Kutyniok, and Speegle proved that certain geometric conditions on the set of parameters in a wave packet systems are necessary in order for the system to form a frame, and also provided constructions of frames and orthonormal bases, based on characteristic functions.

As we have seen already in (20.36),

$$D_{\mathcal{A}_j} T_{\mathcal{B}k} E_{c_m} \psi = T_{\mathcal{A}_j^{-1}\mathcal{B}k} D_{\mathcal{A}_j} E_{c_m} \psi, \qquad (20.44)$$

so a wave packet system of the form in Definition 20.6.3 is a special case of a GSI system, with $C_j := \mathcal{A}_j^{-1}\mathcal{B}, j \in \mathbb{Z}$ and the functions $D_{\mathcal{A}_j} E_{c_m}\psi, j, m \in \mathbb{Z}$, playing the role as ϕ_j. Thus, we can easily state the formal conditions for such systems to form Bessel sequences or frames and for two such systems being dual frames. In practice, however, the situation is more complicated than that. In fact, since a wave packet system is formed by the action of three classes of operators, such a system might be heavily overcomplete, to an extent that involves a potential risk for the upper frame condition to be violated. In general, this forces us to choose in particular the matrices $\mathcal{A}_j$ and the points c_m with great care. We refer to the paper [199] for a more detailed discussion.

20.7   Exercises

20.1 Prove that the operators in Definition 20.1.1 actually map $L^2(\mathbb{R}^d)$ boundedly into $L^2(\mathbb{R}^d)$ and are unitary.

20.2 Prove Lemma 20.1.3.

20.3 Prove that the space $\mathcal{D}$ in (20.1.3) is dense in $L^2(\mathbb{R}^d)$.

20.4 Use Lemma 20.2.4 to formulate and prove a d-dimensional version of Theorem 9.5.1.

20.5 Prove the following:

(i) If $\{T_{C_j k}\phi_j\}_{k\in\mathbb{Z}^d, j\in J}$ satisfies the LIC, then it also satisfies the α-LIC.

(ii) If $\{T_{C_j k}\phi_j\}_{k\in\mathbb{Z}^d, j\in J}$ and $\{T_{C_j k}\widetilde{\phi_j}\}_{k\in\mathbb{Z}^d, j\in J}$ satisfy the LIC, then the dual α-LIC is satisfied.

20.6 Show that if a shift-invariant system $\{T_{Ck}\phi_j\}_{j\in J, k\in\mathbb{Z}^d}$ in $L^2(\mathbb{R}^d)$ is a Bessel sequence, then the LIC is satisfied.
Hint: Use Proposition 20.3.2.

20.7 Prove Corollary 20.4.4.

20.8 Prove Theorem 20.5.2.
Hint: Use that $\mathcal{F}^{-1}E_{\mathcal{B}m}T_{\mathcal{A}j} = T_{-\mathcal{B}m}\mathcal{F}^{-1}T_{\mathcal{A}j}$.

20.9 Prove Corollary 20.5.5, and state a sufficient condition for the Bessel property of $\{E_{b_m}T_{\mathcal{A}_m n}g\}_{m\in\mathbb{Z}, n\in\mathbb{Z}^d}$ along the lines of Theorem 20.5.1.

20.10 Assume that $\{E_{\mathcal{B}m}T_{\mathcal{A}n}g\}_{m,n\in\mathbb{Z}^d}$ is a frame for $L^2(\mathbb{R}^d)$ with frame operator $S : L^2(\mathbb{R}^d) \to L^2(\mathbb{R}^d)$. Show that the frame decomposition takes the form (20.33).

21

Frames on Locally Compact Abelian Groups

In this chapter we will consider frame theory from a broader viewpoint than before, namely, as a part of general harmonic analysis. A central part of harmonic analysis deals with functions on groups and ways to decompose such functions in terms of either series representations or integral representations of certain "basic functions." One of the strengths of harmonic analysis is that it allows very general results that cover several cases at once; for example, instead of developing parallel theories for various groups, we might obtain all of them as special manifestations of a single theory.

Even though the group aspect has not been explicitly discussed so far, it is evident that this is implicit in the treatment in the previous chapters. For example, Chapters 11–13 dealt with Gabor analysis in $L^2(\mathbb{R})$; in Section 20.5, we saw that completely similar results hold in $L^2(\mathbb{R}^d)$, and Chapter 14 presented parallel results for Gabor analysis in $\ell^2(\mathbb{Z})$. From the viewpoint of harmonic analysis, the sets $\mathbb{R}, \mathbb{R}^d$, and $\mathbb{Z}$ are just locally compact abelian groups, and the similarity between Gabor analysis in the three cases definitely calls for a general theory that covers all of them. The purpose of the current chapter is exactly to consider frame theory within the setup of harmonic analysis on locally compact abelian groups. On short form, the merits of the approach in the current chapter are as follows:

- It unifies harmonic analysis on the groups $\mathbb{R}, \mathbb{T}, \mathbb{Z}, \mathbb{Z}_L$, and $\mathbb{R}^d$;

- It unifies Gabor analysis and wavelet analysis;

- It unifies the continuous theory (integral representations) and the discrete theory (series expansions).

© Springer International Publishing Switzerland 2016
O. Christensen, *An Introduction to Frames and Riesz Bases*,
Applied and Numerical Harmonic Analysis,
DOI 10.1007/978-3-319-25613-9_21

The idea of connecting frame theory and harmonic analysis is not new. One of the first contributions to general frame theory, namely, the Feichtinger–Gröchenig theory developed around 1990 (see [280] or Section 24.2), was formulated in the language of square-integrable representations of locally compact groups. And the idea of considering Gabor analysis on locally compact abelian groups has appeared in several publications, including, e.g., [233] by Dahlke, [287] by Feichtinger and Kozek, and [339] by Gröchenig.

More recently Kutyniok and Labate [468] extended most of the results about generalized shift-invariant systems in [398] (i.e., Chapter 20 in the current book) to locally compact abelian groups. Cabrelli and Paternostro gave a detailed description of several aspects of the theory for shift-invariant systems in [101], and Christensen and Goh showed in [176] that also the explicit frame constructions (based, e.g., on B-splines) can be transferred to the setting of locally compact abelian groups. Also, a series of recent papers by Bownik and Ross [85] and Jakobsen and Lemvig [417, 418] presents a unified approach to Gabor analysis on groups that covers as well the discrete case (series expansions in terms of frames) as the integral case (continuous frames).

Our focus will be on some of the key results in the papers [468, 176, 417], and [418]. We do not aim at a detailed technical treatment but rather at an overview that connects the abstract theory with its concrete manifestations within Gabor analysis and wavelet theory. Thus, in many cases we will refer to the original sources for proofs and additional results.

The chapter is organized as follows. In Section 21.1, we present the basics about locally compact abelian groups, and Section 21.2 extends the central objects from classical Fourier analysis to this setting. Section 21.3 gives a formal introduction to the operators that will be used to generate the frames and the corresponding Gabor systems and GSI systems. Section 21.4 collects the basic frame calculations on locally compact abelian groups for the case of a single generator. Section 21.5 is based on [176] by Christensen and Goh; it shows that with the definition of B-splines on locally compact groups that was given independently by Dahlke [232, 233] and Tikhomirov [613] in 1994, not only the abstract theory for Gabor analysis but also the explicit constructions carry over to locally compact abelian groups. Section 21.6 extends the analysis of generalized shift-invariant systems to the setting of locally compact abelian groups; most of the results are taken from the paper [468] and follow the same pattern as our analysis in Chapter 20. Finally, Section 21.7 presents some of the recent results by Bownik & Ross and Jakobsen & Lemvig; the special case of Gabor systems is treated in Section 21.8.

21.1 LCA Groups

In this section we collect the main definitions and classical results related to locally compact abelian groups. Besides fixing the notation, the main purpose is to state the concrete forms of the abstract definitions whenever they are applied to the classical groups (see Example 21.1.2). Most results are presented without proofs; we refer to any of the standard references, i.e., the monographs [402] by Hewitt and Ross, [553] by Reiter and Stegeman, and [567] by Rudin for proofs and much more information.

Let us first give a precise definition of a locally compact abelian group, from now on called an LCA group; we refer to Definition A.3.1 for the definition of a locally compact group.

Definition 21.1.1 *An LCA group is a locally compact group for which the following hold:*

(i) *The group composition, denoted by the symbol "+," is abelian, i.e.,*

$$x + y = y + x, \ \forall x, y \in G;$$

(ii) *G is metrizable, i.e., G can be equipped with a metric.*

Note that we have included the condition of G being metrizable in the definition of an LCA group; remember also that it is already part of the definition of a locally compact group that the topology is a Hausdorff topology and that G is σ-compact, i.e., a countable union of compact sets. The assumption of G being metrizable and σ-compact implies (and is in fact equivalent with) that $\widehat{G}$ is metrizable and σ-compact; this is furthermore equivalent with $L^2(G)$ being separable.

Whenever G is an LCA group, we will always denote the group composition by the symbol "+" and the neutral element by 0. Let us introduce the standard cases of LCA groups.

Example 21.1.2 Let $d \in \mathbb{N}$.

(i) The set $\mathbb{R}^d$ equipped with the usual addition and topology is an LCA group.

(ii) The set $\mathbb{Z}^d$ equipped with the usual addition and the discrete topology is an LCA group.

(iii) The torus

$$\mathbb{T} = \{c \in \mathbb{C} \,|\, |c| = 1\}$$

is an LCA group with the composition defined by multiplication and the topology inherited from $\mathbb{C}$; similarly, $\mathbb{T}^d$ is an LCA group.

(iv) Given any $N \in \mathbb{N}$, let $\mathbb{Z}_N$ denote the set of integers modulo N. This group is finite and is usually identified with the set $\{0, 1, \ldots, N-1\}$;

it forms an LCA group with respect to addition and the discrete topology.

(v) Any direct product $\mathbb{R}^{d_1} \times \mathbb{Z}^{d_2} \times \mathbb{T}^{d_3} \times \mathbb{Z}_N$, where $d_1, d_2, d_3 \in \mathbb{N}$, forms an LCA group with respect to the natural composition and topology. In [552], Reiter called such groups *elementary LCA groups;* see also the paper [287] by Feichtinger and Kozek. □

Definition 21.1.3 *Let G denote an LCA group.*

(i) *A character on G is a function $\gamma : G \to \mathbb{T} := \{z \in \mathbb{C} \mid |z| = 1\}$, for which $\gamma(x + y) = \gamma(x)\gamma(y)$, $\forall x, y \in G$.*

(ii) *The set of continuous characters on G is denoted by $\widehat{G}$.*

The set $\widehat{G}$ forms an abelian group, the *dual group* of G, when equipped with the composition

$$(\gamma + \gamma')(x) := \gamma(x)\gamma'(x), \ \gamma, \gamma' \in \widehat{G}, x \in G.$$

We will now describe a topology on $\widehat{G}$ that even makes it an LCA group. Given a compact set $K \subset G$ and $\epsilon > 0$, let

$$U(K, \epsilon) := \left\{ \gamma \in \widehat{G} \,\big|\, |\gamma(x) - 1| < \epsilon, \forall x \in K. \right\}$$

Then the sets $U(K, \epsilon)$, with K ranging through all compact sets in G and $\epsilon > 0$ running through $\mathbb{R}_+$, form a basis of neighborhoods of the character $\gamma = 1$, for a topology that makes $\widehat{G}$ a locally compact space.

The *Pontryagin duality theorem* states that there exists a topological group isomorphism mapping the dual group of $\widehat{G}$, i.e., the group $\widehat{\widehat{G}}$, onto G. Usually we can identify $\widehat{\widehat{G}}$ and G and we will simply write

$$\widehat{\widehat{G}} = G. \tag{21.1}$$

Thus, $\gamma(x)$ can either be interpreted as the action of $\gamma \in \widehat{G}$ on $x \in G$, or as the action of $x \in \widehat{\widehat{G}} = G$ on $\gamma \in \widehat{G}$. For this reason, we will from now on use the notation

$$(x, \gamma) := \gamma(x), \ x \in G, \ \gamma \in \widehat{G}.$$

The following definition introduces one of the key concepts in an LCA group.

Definition 21.1.4 *Let G denote an LCA group and H a closed subgroup of G. The annihilator $H^\perp$ of H is defined by*

$$H^\perp := \left\{ \gamma \in \widehat{G} \mid (x, \gamma) = 1, \ \forall x \in H \right\}.$$

It follows from the definition of the topology on $\widehat{G}$ that the annihilator $H^\perp$ is a closed subgroup of $\widehat{G}$.

Now, let G denote an LCA group and H a closed subgroup of G. The sets $x + H$, $x \in G$, are called *cosets* of H. Two cosets $x + H$ and $y + H$ are identical if $x - y \in H$; if not, they are disjoint. Taking one representative of each coset yields the *quotient group* G/H; this set is in fact an abelian group under the composition

$$(x + H) + (y + H) := x + y + H.$$

Let us describe a topology on G/H, which actually makes it an LCA group. In order to do this, consider the *canonical quotient mapping*

$$\pi_H : G \to G/H, \ \pi_H(x) := x + H. \tag{21.2}$$

Now, a subset $E \subseteq G/H$ is said to be open if $\pi_H^{-1}(E)$ is an open set in G.

Let us collect some of the classical results about the introduced concepts. The proofs can be found, e.g., in [402].

Lemma 21.1.5 *Let G be an LCA group and H a closed subgroup in G. Then the following hold:*

(i) *If G is discrete, then $\widehat{G}$ is compact.*

(ii) *If G is compact, then $\widehat{G}$ is discrete.*

(iii) *There exists a topological group isomorphism mapping $\widehat{G/H}$ onto $H^\perp$, i.e., (in the sense of this isomorphism)*

$$\widehat{G/H} = H^\perp;$$

(iv) *There exists a topological group isomorphism mapping $\widehat{G}/H^\perp$ onto H, i.e., (in the sense of this isomorphism)*

$$\widehat{G}/H^\perp = H.$$

(v) $(H^\perp)^\perp = H.$

In Sections 21.2–21.6, we only need some special subgroups of G, the so-called *lattices*, to be defined next. Note that in the literature the name *uniform lattice* also appears.

Definition 21.1.6 *Let G denote an LCA group. A discrete subgroup Λ of G for which G/Λ is compact is called a (uniform) lattice.*

Lattices are known explicitly in most of the classical LCA groups, e.g., for the elementary LCA groups. However, there also exist LCA groups without (nontrivial) lattices; see, e.g., [446] and [457]. We return to such groups in Section 21.7.

A lattice in G leads to a splitting of the group G, as well as the dual group, into disjoint cosets:

Lemma 21.1.7 *Let G be an LCA group and Λ a lattice in G. Then the following hold:*

(i) *There exists a Borel measurable relatively compact set $Q \subset G$ such that*

$$G = \bigcup_{\lambda \in \Lambda} (\lambda + Q), \quad (\lambda + Q) \cap (\lambda' + Q) = \emptyset \text{ for } \lambda \neq \lambda', \ \lambda, \lambda' \in \Lambda. \quad (21.3)$$

(ii) *The set $\Lambda^\perp$ is a lattice in $\widehat{G}$, and there exists a Borel measurable relatively compact set $V \subset \widehat{G}$ such that*

$$\widehat{G} = \bigcup_{\omega \in \Lambda^\perp} (\omega + V), \quad (\omega + V) \cap (\omega' + V) = \emptyset \text{ for } \omega \neq \omega', \ \omega, \omega' \in \Lambda^\perp. \quad (21.4)$$

Proof. The result in (i) can be found, e.g., in [446]. Let us show how the results in Lemma 21.1.5 can be used to prove (ii). First, since G/Λ is compact by definition, Lemma 21.1.5(ii)+(iii) imply that $\Lambda^\perp$ is discrete. Also, since Λ is a discrete subgroup of G, its dual group is compact by Lemma 21.1.5(i). By (iv) in the same lemma, this implies that $\widehat{G}/\Lambda^\perp$ is compact (recall that the double dual of a group is the group itself). Thus, $\Lambda^\perp$ is a lattice in $\widehat{G}$, and the result in (ii) follows from (i). $\qquad\square$

A set Q as in Definition 21.1.7 (i) is called a *fundamental domain* associated with the lattice Λ; and the *density* (also called *lattice size*) of the lattice Λ is defined as

$$s(\Lambda) := \mu_G(Q), \quad (21.5)$$

where μ_G denotes the Haar measure on G (see the formal definition in Section 21.2). We note that in general the fundamental domain Q is not unique; however, the density is independent of the choice of the fundamental domain.

Let us determine the dual group, the lattices, and the corresponding annihilators for some of the standard LCA groups. Recall that, given an invertible matrix $\mathcal{A}$ with real entries, we use the notation

$$\mathcal{A}^\sharp := (\mathcal{A}^T)^{-1}.$$

Example 21.1.8 For the LCA group $G = \mathbb{R}^d$, the characters are the functions

$$\gamma_y : G \to \mathbb{C}, \ \gamma_y(x) = e^{2\pi i y \cdot x},$$

where $y \in \mathbb{R}^d$. That is, the dual group $\widehat{G}$ can be identified with $G = \mathbb{R}^d$, and we will simply write $(x, y) = e^{2\pi i y \cdot x}$, $x, y \in \mathbb{R}^d$. Any lattice has the form

$$\Lambda = \mathcal{A}\mathbb{Z}^d$$

for some invertible $d \times d$ matrix $\mathcal{A}$ with real entries; as a corresponding fundamental domain, we can take $Q = \mathcal{A}[0,1[^d$. The Haar measure on $\mathbb{R}^d$ is just the Lebesgue measure, so the density of the lattice is

$$s(\Lambda) = \mu(\mathcal{A}[0,1[^d) = |\det \mathcal{A}|.$$

The annihilator of the lattice Λ is

$$\Lambda^\perp = (\mathcal{A}^T)^{-1}\mathbb{Z}^d = \mathcal{A}^\sharp \mathbb{Z}^d,$$

and the set

$$V = \mathcal{A}^\sharp[0,1[^d$$

is a fundamental domain associated with the lattice $\Lambda^\perp$ in $\widehat{G}$.   □

Example 21.1.9 Consider the torus group $G = \mathbb{T}$, which we (as always) identify with $[0,1[$. As in Example 21.1.8, the characters have the form

$$\gamma_y : G \to \mathbb{C}, \ \gamma_y(x) = e^{2\pi i y x}, \ x \in [0,1[$$

for some $y \in \mathbb{R}$; but due to the periodicity, we have $\gamma_y(x+1) = \gamma_y(x)$ for all $x \in [0,1[$, and hence $y \in \mathbb{Z}$. The dual group of $\mathbb{T}$ is in fact $\widehat{G} = \mathbb{Z}$. There are only a countable number of lattices in G, namely, the sets of the form

$$\Lambda = \frac{1}{N}\mathbb{Z}_N = \left\{0, \frac{1}{N}, \ldots, \frac{N-1}{N}\right\}$$

for $N \in \mathbb{N}$. The set $Q = [0, 1/N[$ is a fundamental domain associated with the lattice Λ, and the annihilator of the lattice Λ is $\Lambda^\perp = N\mathbb{Z}$.   □

Example 21.1.10 For the group $G = \mathbb{Z}$, the dual group is $\widehat{G} = \mathbb{T}$; in fact, the characters are

$$\gamma_y : \mathbb{Z} \to \mathbb{C}, \ \gamma_y(x) = e^{2\pi i y x}, \ x \in \mathbb{Z},$$

for $y \in \mathbb{T}$. The lattices are of the form $\Lambda = N\mathbb{Z}$, $N \in \mathbb{N}$. The fundamental domain associated with Λ is $Q = \{0, \ldots, N-1\}$; the annihilator is $\Lambda^\perp = \frac{1}{N}\mathbb{Z}_N$, with corresponding fundamental domain $V = [0, 1/N[$.   □

Example 21.1.11 For $L \in \mathbb{N}$, let $G = \mathbb{Z}_L$; again, the characters have the form $\gamma_y : G \to \mathbb{C}, \ \gamma_y(x) = e^{2\pi i y x}, x \in \mathbb{Z}_L$, for some $y \in \mathbb{R}$. Due to the periodicity, $\gamma_y(x + L) = \gamma_y(x)$ for all $x \in \{0, 1, \ldots, L-1\}$, so $y \in L^{-1}\mathbb{Z}$. Let us change the notation slightly and write the characters as

$$\gamma_y : G \to \mathbb{C}, \ \gamma_y(x) = e^{2\pi i y x / L}$$

for $y \in \mathbb{Z}$. Note that since $x \in \{0, 1, \ldots, L-1\}$, the characters y and $y+L$ are identical; thus, the dual group of $\mathbb{Z}_L$ can be identified with $\mathbb{Z}_L$ itself.

Let $M \in \mathbb{N}$ be a divisor of L, i.e., $M\ell = L$ for some $\ell \in \mathbb{N}$. Then the set

$$\Lambda := \{0, M, \ldots, (\ell-1)M\} = M\mathbb{Z}_{L/M}$$

is a lattice in G. As corresponding fundamental domain, we can take $Q = \{0, 1, \ldots, M-1\} = \mathbb{Z}_M$. Also, by definition the annihilator $\Lambda^\perp$ consists of the $\gamma \in \hat{G} = \{0, 1, \ldots, L-1\}$ for which $e^{2\pi i \gamma x/L} = 1$ for all $x \in \Lambda$, i.e., such that $e^{2\pi i \gamma k M/L} = 1$ for all $k = 0, 1, \ldots, L/M - 1$; thus,

$$\Lambda^\perp = \{\ell, 2\ell, \ldots, (M-1)\ell\} = \frac{L}{M}\mathbb{Z}_M.$$

As fundamental domain associated with the lattice $\Lambda^\perp$, we can take $V = \{0, 1, \ldots, L/M - 1\} = \mathbb{Z}_{L/M}$. $\qquad\square$

We note that the dual group of any elementary LCA group can be found easily, based on the dual groups of $\mathbb{R}, \mathbb{T}, \mathbb{Z}$, and $\mathbb{Z}_N$:

Lemma 21.1.12 *Let $G_1, \ldots, G_n$ denote LCA groups, with corresponding dual groups $\widehat{G_1}, \ldots, \widehat{G_n}$. Then the product group*

$$G := G_1 \oplus \cdots \oplus G_n$$

is an LCA group with dual group

$$\hat{G} = \widehat{G_1} \oplus \cdots \oplus \widehat{G_n}.$$

21.2 Fourier Analysis on LCA Groups

We will now introduce the key elements in the Fourier analysis on an LCA group G.

Let $C_c(G)$ denote the set of continuous functions $f : G \to \mathbb{C}$ with compact support. As discussed in Section A.3, any LCA group G can be equipped with a positive measure μ_G which is translation invariant, i.e., such that for all $f \in C_c(G)$,

$$\int_G f(x+y)\, d\mu_G(x) = \int_G f(x)\, d\mu_G(x), \ \forall y \in G.$$

The measure is unique up to multiplication with a positive scalar and is called the *Haar measure*. Note that we have chosen to denote the measure by μ_G instead of just μ; the reason is that we will need to consider the Haar measure on several groups simultaneously.

With the Haar measure at hand, we can define the spaces $L^p(G)$, $1 \le p < \infty$, in the usual way; we will only need the spaces $L^1(G)$ and $L^2(G)$. The space $L^2(G)$ is a Hilbert space in the obvious way; furthermore, our assumption of G being a countable union of compact sets and metrizable implies (and is, in fact, equivalent to) that $L^2(G)$ is separable.

We will always consider the Haar measure μ_G as fixed. The *Fourier transform* is defined as the operator

$$\mathcal{F} : L^1(G) \to C_0(\widehat{G}), \ \mathcal{F}f(\gamma) := \int_G f(x)(-x,\gamma)\, d\mu_G(x), \ \gamma \in \widehat{G}; \quad (21.6)$$

here $C_0(\widehat{G})$ denotes the functions on $\widehat{G}$ vanishing at infinity. We will often use the notation $\widehat{f} := \mathcal{F}f$.

The standard results about the classical Fourier transform extend to LCA groups. In particular, we have the *inversion theorem* and *Plancherels theorem;* we refer to [567] for a proof of Theorem 21.2.1.

Theorem 21.2.1 *With appropriate normalization of the Haar measure $\mu_{\widehat{G}}$ on $\widehat{G}$, the following hold:*

(i) Whenever $f \in L^1(G)$ and $\widehat{f} \in L^1(\widehat{G})$, it holds that

$$f(x) = \int_{\widehat{G}} \widehat{f}(\gamma)\,(x,\gamma) d\mu_{\widehat{G}}(\gamma), \ x \in G. \quad (21.7)$$

(ii) The Fourier transform can be extended to a surjective isometry $\mathcal{F} : L^2(G) \to L^2(\widehat{G})$.

(iii) For any $f, g \in L^2(G)$,

$$\int_G f(x)\overline{g(x)}\, d\mu_G(x) = \int_{\widehat{G}} \widehat{f}(\gamma)\overline{\widehat{g}(\gamma)}\, d\mu_{\widehat{G}}(\gamma). \quad (21.8)$$

We will always choose the Haar measure on $\widehat{G}$ as in Theorem 21.2.1, i.e., such that the inversion formula (21.7) holds for the pairs G and $\widehat{G}$. Such Haar measures μ_G and $\mu_{\widehat{G}}$ are called *dual measures*. Putting $\phi := \widehat{f}$, we can write (21.7) as

$$\mathcal{F}^{-1}\phi(x) = \int_{\widehat{G}} \phi(\gamma)(x,\gamma) d\mu_{\widehat{G}}(\gamma). \quad (21.9)$$

Let us now again consider a closed subgroup H of G and the canonical quotient mapping $\pi_H : G \to G/H$ in (21.2). By definition, π_H is surjective, i.e., each $\dot{x} \in G/H$ has the form $\dot{x} = \pi_H(x)$ for some $x \in G$. Since G/H is an LCA group, we can equip it with a Haar measure $d\mu_{G/H}(\dot{x})$. Furthermore, letting $d\mu_H$ denote any Haar measure on H and taking any $f \in L^1(G)$, the integral $\int_H f(x+h)\, d\mu_H$ is constant on the coset $x + H$. Thus

$$\dot{x} \mapsto \int_H f(x+h)\, d\mu_H$$

defines a function on G/H. We can now state Weil's theorem.

Theorem 21.2.2 *Let H denote a closed subgroup of the LCA group G. Then the following hold:*

(i) *Taking any Haar measures on two of the LCA groups G, H, and G/H, the third Haar measure can be normalized such that for all $f \in L^1(G)$,*

$$\int_G f(x)\, d\mu_G(x) = \int_{G/H} \int_H f(x+h)\, d\mu_H(h)\, d\mu_{G/H}(\dot{x}). \quad (21.10)$$

(ii) *If the measures on G, H, and G/H are chosen such that (21.10) holds, then the corresponding dual measures on the dual groups $\widehat{G}, \widehat{H} = \widehat{G}/H^\perp$ and $\widehat{G/H} = H^\perp$ satisfy that for all $F \in L^1(\widehat{G})$,*

$$\int_{\widehat{G}} F(\gamma)\, d\gamma = \int_{\widehat{G}/H^\perp} \int_{H^\perp} F(\gamma+\omega)\, d\mu_{H^\perp}(\omega)\, d\mu_{\widehat{G}/H^\perp}(\dot{\gamma}). \quad (21.11)$$

The following example shows that in the case of the LCA group $\mathbb{R}$, Weil's theorem corresponds to the usual "periodization trick."

Example 21.2.3 Let $G = \mathbb{R}$, and consider the subgroup $H = a\mathbb{Z}$, for some $a > 0$. Then

$$\int_H f(x+h)\, d\mu_H(h) = \sum_{k \in \mathbb{Z}} f(x+ak).$$

Now let $d\mu_{G/H}(\dot{x})$ denote a Haar measure on G/H. Since the mapping

$$\Phi : C_c(G/H) \to \mathbb{C}, \ \Phi f := \int_0^a f(x+H)\, dx = \int_0^a f(\pi_H(x))\, dx$$

is a positive and translation-invariant linear functional on $C_c(G/H)$, the definition of the Haar measure on G/H implies that for some $c > 0$,

$$\Phi f = c \int_{G/H} f(\dot{x})\, d\mu_{G/H}(\dot{x})$$

for all $f \in C_c(G/H)$, or

$$c \int_{G/H} f(\dot{x})\, d\mu_{G/H}(\dot{x}) = \int_0^a f(\pi_H(x))\, dx.$$

Thus, Weil's theorem says that with appropriate normalization of the measures,

$$\int_0^a \sum_{k \in \mathbb{Z}} f(x+ak)\, dx = \int_{-\infty}^\infty f(x)\, dx;$$

this is the well-known "periodization trick."   □

We will need the following consequence of Weil's theorem.

Lemma 21.2.4 *Let Λ denote a lattice in the LCA group G, and let $V \subset \widehat{G}$ denote a fundamental domain associated with the annihilator $\Lambda^\perp$. Choose the Haar measure $\mu_{\widehat{G}}$ on $\widehat{G}$ such that the inversion formula (21.7) holds. Then, for all $F \in L^2(V)$,*

$$\mu_{\widehat{G}}(V) \int_V |F(\gamma)|^2 \, d\mu_{\widehat{G}}(\gamma) = \sum_{\lambda \in \Lambda} |\widehat{F\chi_V}(\lambda)|^2, \qquad (21.12)$$

where the hat denotes the Fourier transform on the group $\widehat{G}$.

Proof. Let $F \in L^2(V)$. Weil's formula implies that for an appropriate normalization of the measure on $\widehat{G}/\Lambda^\perp$ and with $\dot{g} = g + \Lambda^\perp$,

$$\int_V |F(\gamma)|^2 \, d\mu_{\widehat{G}}(\gamma)$$

$$= \int_{\widehat{G}} |F\chi_V(\gamma)|^2 \, d\mu_{\widehat{G}}(\gamma) = \mu_{\widehat{G}}(V) \int_{\widehat{G}/\Lambda^\perp} \sum_{h \in \Lambda^\perp} |F\chi_V(g+h)|^2 d\mu_{\widehat{G}/\Lambda^\perp}(\dot{g})$$

$$= \mu_{\widehat{G}}(V) \int_{\widehat{G}/\Lambda^\perp} \left| \sum_{h \in \Lambda^\perp} F\chi_V(g+h) \right|^2 d\mu_{\widehat{G}/\Lambda^\perp}(\dot{g}). \qquad (21.13)$$

Note that with this normalization of the measure on $\widehat{G}/\Lambda^\perp$, we have $\mu_{\widehat{G}/\Lambda^\perp}(\widehat{G}/\Lambda^\perp) = 1$. Thus, using the Plancherel theorem on $\widehat{G}/\Lambda^\perp$, as well as Lemma 21.1.5 (iv), followed by a new application of Weil's formula and the definition of the Fourier transform,

$$\int_{\widehat{G}/\Lambda^\perp} \left| \sum_{h \in \Lambda^\perp} F\chi_V(g+h) \right|^2 d\mu_{\widehat{G}/\Lambda^\perp}(\dot{g})$$

$$= \sum_{\lambda \in \Lambda} \left| \int_{\widehat{G}/\Lambda^\perp} \sum_{h \in \Lambda^\perp} (F\chi_V)(g+h)(-g, \lambda) \, d\mu_{\widehat{G}/\Lambda^\perp}(\dot{g}) \right|^2$$

$$= \sum_{\lambda \in \Lambda} \left| \frac{1}{\mu_{\widehat{G}}(V)} \int_{\widehat{G}} (F\chi_V)(\gamma)(-\gamma, \lambda) d\mu_{\widehat{G}}(\gamma) \right|^2$$

$$= \frac{1}{\mu_{\widehat{G}}(V)^2} \sum_{\lambda \in \Lambda} |\widehat{F\chi_V}(\lambda)|^2.$$

Inserting this in (21.13) yields the result. $\qquad\square$

We can consider Lemma 21.2.4 as a general version of Parseval's equation:

Example 21.2.5 Consider the group $G := \mathbb{R}$, with dual group $\widehat{G} = \mathbb{R}$. The functions $\{e^{2\pi ikx}\}_{k \in \mathbb{Z}}$ form an orthonormal basis for $L^2(0,1)$; thus, writing the Fourier coefficients for $f \in L^2(0,1)$ as $c_k = \int_0^1 f(x) \, e^{-2\pi ikx} \, dx$, $k \in \mathbb{Z}$,

Parseval's equation (see (3.35)) shows that

$$\int_0^1 |f(x)|^2 \, dx = \sum_{k \in \mathbb{Z}} |c_k|^2.$$

Via the Fourier transform on $L^2(\mathbb{R})$, this can be written as

$$\int_0^1 |f(x)|^2 \, dx = \sum_{k \in \mathbb{Z}} \left| \widehat{f \chi_{[0,1]}}(k) \right|^2. \qquad (21.14)$$

This is a special case of the result in Lemma 21.2.4. In fact, consider the set $V = [0, 1[$ as a subset of $\widehat{G} = \mathbb{R}$. Defining the lattice $\Lambda := \mathbb{Z}$ in $\mathbb{R}$, we have that $\Lambda^\perp = \mathbb{Z}$. Thus, the set $V = [0, 1[$ satisfies (21.4), and Lemma 21.2.4 tells us that (21.12) holds. Inserting the sets V and Λ in (21.12) shows that this is exactly the same as (21.14).   □

21.3 Gabor Systems on LCA Groups

As in the case of frame theory on $L^2(\mathbb{R})$, the concrete example of frames associated with LCA groups will be defined in terms of certain operators:

Definition 21.3.1 *Let G denote an LCA group with dual group $\widehat{G}$.*

(i) *For $y \in G$, the translation operator $T_y : L^2(G) \to L^2(G)$ is defined by*

$$T_y f(x) = f(x - y), \ x \in G. \qquad (21.15)$$

(ii) *For $\eta \in \widehat{G}$, the modulation operator $E_\eta : L^2(G) \to L^2(G)$ is*

$$E_\eta f(x) = (x, \eta) f(x), \ x \in G. \qquad (21.16)$$

(iii) *For $\eta \in \widehat{G}$, the translation operator $\mathcal{T}_\eta : L^2(\widehat{G}) \to L^2(\widehat{G})$ is*

$$\mathcal{T}_\eta f(\gamma) = f(\gamma - \eta), \ \gamma \in \widehat{G}. \qquad (21.17)$$

(iv) *For $y \in G$, the modulation operator $\mathcal{E}_y : L^2(\widehat{G}) \to L^2(\widehat{G})$ is*

$$\mathcal{E}_y f(\gamma) := (y, \gamma) \, f(\gamma), \ \gamma \in \widehat{G}. \qquad (21.18)$$

Note that the operators T_y and E_η act on $L^2(G)$, while $\mathcal{T}_\eta$ and $\mathcal{E}_y$ act on $L^2(\widehat{G})$. Like in the case of translation operators and modulation operators on $L^2(\mathbb{R})$, one can show (Exercise 21.1) that all four classes of operators are unitary and that they satisfy the commutator relations

$$T_y E_\eta = (-y, \eta) E_\eta T_y, \quad \mathcal{T}_\eta \mathcal{E}_y = (-y, \eta) \mathcal{E}_y \mathcal{T}_\eta, \ y \in G, \eta \in \widehat{G}, \qquad (21.19)$$

and

$$\mathcal{F} T_y = \mathcal{E}_{-y} \mathcal{F}, \quad \mathcal{F}^{-1} \mathcal{E}_y = T_{-y} \mathcal{F}^{-1}, \ y \in G. \qquad (21.20)$$

Note that in (21.19) and (21.20) the first stated relation is an operator equation in $L^2(G)$, while the second relation takes place in $L^2(\widehat{G})$. Based on the operators in Definition 21.3.1, we will consider various classes of systems of functions, in as well $L^2(G)$ as $L^2(\widehat{G})$. The results in Sections 21.4–21.5 will concern systems of functions arising by actions of a class of operators $\mathcal{E}_\lambda$ on a countable collection of functions $\{\Phi_k\}_{k\in I}$ in $L^2(\widehat{G})$, with λ belonging to lattices Λ_k in G that depend on $k \in I$. The resulting class of functions in $L^2(\widehat{G})$ is of the form $\{\mathcal{E}_\lambda\Phi_k\}_{\lambda\in\Lambda_k,k\in I}$; we refer to such a system as a *Fourier-like system*. We emphasize that $\{\mathcal{E}_\lambda\Phi_k\}_{\lambda\in\Lambda_k,k\in I}$ is a function system in $L^2(\widehat{G})$, not in $L^2(G)$.

In Section 21.6 the Fourier transform and the commutator relations in (21.20) are used to turn the Fourier-like systems $\{\mathcal{E}_\lambda\Phi_k\}_{\lambda\in\Lambda_k,k\in I}$ in $L^2(\widehat{G})$ into GSI systems $\{T_\lambda\phi_k\}_{\lambda\in\Lambda_k,k\in I}$ in $L^2(G)$.

In Section 21.8 we will consider systems in $L^2(G)$ of the form $\{E_\eta T_\lambda g\}_{\eta\in\Gamma,\lambda\in\Lambda}$, where translation and modulation of $g \in L^2(G)$ are along lattices $\Lambda \subset G$ and $\Gamma \subset \widehat{G}$ [in fact, the results will deal with a more general case where the lattices are replaced by certain subgroups]. We will call such a system a *Gabor system* in $L^2(G)$.

Note that we will encounter Gabor systems in as well $L^2(G)$ as in $L^2(\widehat{G})$. The Gabor systems in $L^2(\widehat{G})$ arise naturally as special cases of the Fourier-like systems $\{\mathcal{E}_\lambda\Phi_k\}_{\lambda\in\Lambda_k,k\in I}$ in $L^2(\widehat{G})$, simply by taking Φ_k on the form $\Phi_k = \mathcal{T}_k\Phi$ for a fixed function $\Phi \in L^2(\widehat{G})$. Thus, we are naturally led to Gabor systems in $L^2(\widehat{G})$ on the form $\{\mathcal{E}_\lambda \mathcal{T}_\eta\Phi\}_{\lambda\in\Lambda,\eta\in\Gamma}$, where $\Phi \in L^2(\widehat{G})$, Λ is a lattice in G, and Γ is a lattice in $\widehat{G}$.

In the case $G = \mathbb{R}$, the Gabor system in $L^2(G)$ is exactly the Gabor system considered in Chapters 11–13. We will now show that applying the above abstract approach to the elementary LCA groups leads to the classes of Gabor systems considered in Chapter 14.

Example 21.3.2 Let $G = \mathbb{Z}$; then $\widehat{G} = \mathbb{T}$, where we as usual identify $\mathbb{T}$ with $[0,1[$. Let $g = \Phi \in \ell^2(\mathbb{Z})$ and consider a lattice $\Lambda = N\mathbb{Z}$ in G and a lattice $\Gamma = \{0, \frac{1}{M}, \ldots, \frac{M-1}{M}\}$ in $\widehat{G}$; here $M, N \in \mathbb{N}$. Then $\{E_\eta T_\lambda g\}_{\eta\in\Gamma,\lambda\in\Lambda}$ takes the form

$$\{E_\eta T_\lambda g\}_{\eta\in\Gamma,\lambda\in\Lambda} = \{e^{2\pi im/Mx}\Phi(x-nN)\}_{n\in\mathbb{Z},m=0,\ldots,M-1}$$
$$= \{E_{m/M}T_{nN}\Phi\}_{n\in\mathbb{Z},m=0,\ldots,M-1} \qquad (21.21)$$

in $L^2(G) = \ell^2(\mathbb{Z})$; this is precisely the Gabor system in $\ell^2(\mathbb{Z})$ considered in Section 14.1. $\qquad\qquad\square$

Example 21.3.3 Let $G = \mathbb{T}$; then $\widehat{G} = \mathbb{Z}$. By Example 21.1.9, a lattice in G has the form

$$\Lambda = \left\{0, \frac{1}{N}, \ldots, \frac{N-1}{N}\right\}$$

for some $N \in \mathbb{N}$; and by Example 21.1.10, the lattices in $\widehat{G}$ have the form $\Gamma = M\mathbb{Z}$ for some $M \in \mathbb{Z}$. Thus, the Gabor system on $L^2(G) = L^2(0,1)$ with window $g = \Phi \in L^2(0,1)$ takes the form

$$
\begin{aligned}
\{E_\eta T_\lambda g\}_{\eta \in \Gamma, \lambda \in \Lambda} &= \{e^{2\pi i m M x}\Phi(x - n/N)\}_{m \in \mathbb{Z}, n=0,\ldots,N-1} \\
&= \{E_{mM}T_{n/N}\Phi\}_{m \in \mathbb{Z}, n=0,\ldots,N-1}. \qquad (21.22)
\end{aligned}
$$

If we choose to identify $\mathbb{T}$ with the interval $[0, L[$ for some $L \in \mathbb{N}$ instead of $[0,1[$, slight modifications of the calculations lead to the Gabor system $\{E_{mM/L}T_{nL/N}\Phi\}_{m \in \mathbb{Z}, n=0,\ldots,N-1}$ in $L^2(G) = L^2(0,L)$; this corresponds exactly to the form in (14.20). □

Example 21.3.4 Let $L \in \mathbb{N}$ and consider the finite group $G = \mathbb{Z}_L = \{0, 1, \ldots, L-1\}$. By Example 21.1.11, we know that the characters have the form

$$\gamma_\ell(x) = e^{2\pi i \ell x/L}, \ x \in \mathbb{Z}_L$$

for $\ell = 0, 1, \ldots, L-1$, i.e., $\widehat{G} = G = \mathbb{Z}_L$. Given $N \in \mathbb{N}$, such that $N' := L/N \in \mathbb{N}$, Example 21.1.11 shows that the set

$$\Lambda = N\mathbb{Z}_{L/N} = \{0, N, \ldots, (N'-1)N\}$$

is a lattice in G; similarly, given $M \in \mathbb{N}$ such that $M' := L/M \in \mathbb{N}$, we consider the lattice

$$\Gamma = M'\mathbb{Z}_{L/M'} = \{0, M', \ldots, (M-1)M'\}$$

in $\widehat{G}$. Then, given $g = \Phi \in \mathbb{C}^L$, we obtain the Gabor system

$$
\begin{aligned}
\{E_\eta T_\lambda g\}_{\eta \in \Gamma, \lambda \in \Lambda} &= \{e^{2\pi i m M' x/L}\Phi(x - nN)\}_{m=0,\ldots,M-1, n=0,\ldots,N'-1} \\
&= \{E_{m/M}T_{nN}\Phi\}_{m=0,\ldots,M-1, n=0,\ldots,N'-1} \qquad (21.23)
\end{aligned}
$$

in $\mathbb{C}^L$. This is precisely the Gabor system considered in Section 14.6; see (14.29). □

21.4 Basic Frame Calculations in $L^2(\widehat{G})$.

In the entire section we will consider an LCA group G, with dual group $\widehat{G}$. We will fix a Haar measure μ_G on G, and normalize the Haar measure $\mu_{\widehat{G}}$ on $\widehat{G}$ such that the inversion formula and Plancherel formula hold.

The material of the section is taken from the paper [176] by Christensen and Goh.

Our goal in this section is to analyze Fourier-like systems $\{\mathcal{E}_\lambda\Phi\}_{\lambda\in\Lambda}$, where Λ is a lattice in G and $\Phi \in L^2(\widehat{G})$; recall that the modulation operator is defined in (21.18). Compared to Section 21.2, we will simplify the notation and, for $f \in L^1(G)$, simply write

$$\int_G f(x)\, dx := \int_G f(x)\, d\mu_G(x).$$

Similarly, for $F \in L^1(\widehat{G})$, we write

$$\int_{\widehat{G}} F(\gamma)\, d\gamma := \int_{\widehat{G}} F(\gamma)\, d\mu_{\widehat{G}}(\gamma).$$

Also, for technical reasons, we will often need the dense subspace $\mathcal{D}(\widehat{G})$ of $L^2(\widehat{G})$ defined by

$$\mathcal{D}(\widehat{G}) := \{F \in L^2(\widehat{G}) \,|\, \operatorname{supp} F \text{ is compact and } F \in L^\infty(\widehat{G})\}. \quad (21.24)$$

This is analogue to our use of the space $\mathcal{D}$ in the context of generalized shift-invariant systems on $L^2(\mathbb{R}^d)$; see (20.8). We will use the notation from Section 21.2, and consequently let V denote a fundamental domain associated with the annihilator $\Lambda^\perp$.

We begin with a result, which will play a similar role as Lemma 20.2.1 in the analysis of GSI systems in $L^2(\mathbb{R}^d)$.

Lemma 21.4.1 *Let Λ be a lattice in G, and choose the relatively compact set $V \subset \widehat{G}$ as a fundamental domain for the lattice $\Lambda^\perp$. Let $F, \Phi \in L^2(\widehat{G})$. Then the following hold:*

(i) *The function*

$$\alpha : \widehat{G} \to \mathbb{C}, \ \ \alpha(\gamma) := \sum_{\omega\in\Lambda^\perp} F(\omega + \gamma)\overline{\Phi(\omega + \gamma)}, \quad (21.25)$$

is well-defined for a.e. $\gamma \in V$, belongs to $L^1(V)$, and satisfies that

$$\alpha(\gamma + \omega') = \alpha(\gamma), \ \forall \gamma \in \widehat{G}, \omega' \in \Lambda^\perp. \quad (21.26)$$

(ii) *For any $\lambda \in \Lambda$,*

$$\langle F, \mathcal{E}_\lambda\Phi \rangle = \int_V \alpha(\gamma)\overline{(\lambda, \gamma)}\, d\gamma = \widehat{\alpha\chi_V}(\lambda), \quad (21.27)$$

where the hat denotes the Fourier transform on the group $\widehat{G}$.

Proof. Using Lemma 21.1.7(ii),

$$\int_V \sum_{\omega \in \Lambda^\perp} |F(\omega + \gamma)\overline{\Phi(\omega + \gamma)}|\, d\gamma = \sum_{\omega \in \Lambda^\perp} \int_V |F(\omega + \gamma)\Phi(\omega + \gamma)|\, d\gamma$$

$$= \sum_{\omega \in \Lambda^\perp} \int_{\omega + V} |F(\gamma)\Phi(\gamma)|\, d\gamma = \int_{\widehat{G}} |F(\gamma)\Phi(\gamma)|\, d\gamma,$$

which is finite by the Cauchy–Schwarz inequality. This shows that $\alpha(\gamma)$ is well-defined pointwise for almost all $\gamma \in V$ and also implies that $\alpha \in L^1(V)$. Now, by Lemma 21.1.7 (ii), any $\gamma \in \widehat{G}$ can be written as $\gamma = \gamma' + \omega'$ for some $\gamma' \in V$, $\omega' \in \Lambda^\perp$. Then

$$\sum_{\omega \in \Lambda^\perp} |F(\omega + \gamma)\overline{\Phi(\omega + \gamma)}| = \sum_{\omega \in \Lambda^\perp} |F(\omega + \gamma' + \omega')\overline{\Phi(\omega + \gamma' + \omega')}|$$

$$= \sum_{\omega \in \Lambda^\perp} |F(\omega + \gamma')\overline{\Phi(\omega + \gamma')}|,$$

where we used the change of summation variable $\omega \to \omega - \omega'$ (which is allowed because $\Lambda^\perp$ is a group). Thus, the series defining $\alpha(\gamma)$ is absolutely convergent for a.e. $\gamma \in \widehat{G}$. The same argument, just without the absolute value, shows that $\alpha(\gamma + \omega') = \alpha(\gamma)$, which proves (i).

For the proof of (ii), using again Lemma 21.1.7(ii),

$$\langle F, \mathcal{E}_\lambda \Phi \rangle = \int_{\widehat{G}} F(\gamma)\overline{\Phi(\gamma)}\,\overline{(\lambda, \gamma)}\, d\gamma = \sum_{\omega \in \Lambda^\perp} \int_{\omega + V} F(\gamma)\overline{\Phi(\gamma)}\,\overline{(\lambda, \gamma)}\, d\gamma$$

$$= \sum_{\omega \in \Lambda^\perp} \int_V F(\omega + \gamma)\overline{\Phi(\omega + \gamma)}\,\overline{(\lambda, \omega + \gamma)}\, d\gamma.$$

Note that since $\lambda \in \Lambda$ and $\omega \in \Lambda^\perp$, we have that

$$(\lambda, \omega + \gamma) = (\lambda, \omega)(\lambda, \gamma) = (\lambda, \gamma).$$

Thus the calculation yields that

$$\langle F, \mathcal{E}_\lambda \Phi \rangle = \int_V \sum_{\omega \in \Lambda^\perp} F(\omega + \gamma)\overline{\Phi(\omega + \gamma)}\,\overline{(\lambda, \gamma)}\, d\gamma$$

$$= \int_V \alpha(\gamma)\overline{(\lambda, \gamma)}\, d\gamma = \int_V \alpha(\gamma)(-\lambda, \gamma)\, d\gamma = \widehat{\alpha\chi_V}(\lambda),$$

as desired. $\qquad\square$

The next result will be used to examine the Bessel condition for sequences on the form $\{\mathcal{E}_\lambda \Phi\}_{\lambda \in \Lambda}$.

Lemma 21.4.2 *Let Λ denote a lattice in the group G, and choose the relatively compact set $V \subset \widehat{G}$ as a fundamental domain for the lattice $\Lambda^\perp$. Let $\Phi \in L^2(\widehat{G})$. Then the following hold:*

(i) *If $F \in \mathcal{D}(\widehat{G})$, then*

$$
\begin{aligned}
\sum_{\lambda \in \Lambda} |\langle F, \mathcal{E}_\lambda \Phi \rangle|^2 &= \mu_{\widehat{G}}(V) \sum_{\omega \in \Lambda^\perp} \int_{\widehat{G}} F(\gamma) \overline{F(\omega + \gamma)} \, \overline{\Phi(\gamma)} \Phi(\omega + \gamma) \, d\gamma \\
&= \mu_{\widehat{G}}(V) \left(\int_{\widehat{G}} |F(\gamma) \Phi(\gamma)|^2 \, d\gamma + R(F) \right), \quad (21.28)
\end{aligned}
$$

where

$$
|R(F)| \leq \int_{\widehat{G}} |F(\gamma)|^2 \sum_{\omega \in \Lambda^\perp \setminus \{0\}} |\Phi(\gamma) \Phi(\gamma + \omega)| \, d\gamma.
$$

(ii) *Assume that*

$$
B := \mu_{\widehat{G}}(V) \sup_{\gamma \in \widehat{G}} \sum_{\omega \in \Lambda^\perp} |\Phi(\gamma) \Phi(\gamma + \omega)| < \infty.
$$

Then $\{\mathcal{E}_\lambda \Phi\}_{\lambda \in \Lambda}$ is a Bessel sequence in $L^2(\widehat{G})$ with bound B.

Proof. The assumption $F \in \mathcal{D}(\widehat{G})$ will justify all interchanges of summations and integrals in the following as $\widehat{G}$ is metrizable and $\Lambda^\perp$ is a discrete subgroup of $\widehat{G}$. In addition, applying the Cauchy–Schwarz inequality followed by Lemma 21.1.7(ii), it shows that $\alpha \in L^2(V)$. Thus, using Lemma 21.4.1(ii) and Lemma 21.2.4,

$$
\begin{aligned}
\sum_{\lambda \in \Lambda} |\langle F, \mathcal{E}_\lambda \Phi \rangle|^2 &= \sum_{\lambda \in \Lambda} |\widehat{\alpha \chi_V}(\lambda)|^2 = \mu_{\widehat{G}}(V) \int_V |\alpha(\gamma)|^2 \, d\gamma \\
&= \mu_{\widehat{G}}(V) \int_V \alpha(\gamma) \, \overline{\alpha(\gamma)} \, d\gamma. \quad (21.29)
\end{aligned}
$$

Inserting the expression for $\alpha(\gamma)$ (while keeping the term $\overline{\alpha(\gamma)}$) leads to

$$
\begin{aligned}
\sum_{\lambda \in \Lambda} |\langle F, \mathcal{E}_\lambda \Phi \rangle|^2 &= \mu_{\widehat{G}}(V) \int_V \sum_{\omega \in \Lambda^\perp} F(\omega + \gamma) \overline{\Phi(\omega + \gamma)} \, \overline{\alpha(\gamma)} \, d\gamma \\
&= \mu_{\widehat{G}}(V) \sum_{\omega \in \Lambda^\perp} \int_V F(\omega + \gamma) \overline{\Phi(\omega + \gamma)} \, \overline{\alpha(\gamma)} \, d\gamma \\
&= \mu_{\widehat{G}}(V) \sum_{\omega \in \Lambda^\perp} \int_{\omega + V} F(\gamma) \overline{\Phi(\gamma)} \, \overline{\alpha(\gamma - \omega)} \, d\gamma,
\end{aligned}
$$

where the last step used the translation invariance of the Haar measure. Now, by Lemma 21.4.1(i), we have that $\alpha(\gamma - \omega) = \alpha(\gamma)$ whenever $\omega \in \Lambda^\perp$. Thus, we arrive at

$$
\begin{aligned}
\sum_{\lambda \in \Lambda} |\langle F, \mathcal{E}_\lambda \Phi \rangle|^2 &= \mu_{\widehat{G}}(V) \sum_{\omega \subset \Lambda^\perp} \int_{\omega + V} F(\gamma) \overline{\Phi(\gamma)} \, \overline{\alpha(\gamma)} \, d\gamma \\
&= \mu_{\widehat{G}}(V) \int_{\widehat{G}} F(\gamma) \overline{\Phi(\gamma)} \, \overline{\alpha(\gamma)} \, d\gamma, \quad (21.30)
\end{aligned}
$$

where we used Lemma 21.1.7(ii) in the last step. Inserting again the expression for $\alpha(\gamma)$ now yields that

$$
\begin{aligned}
\sum_{\lambda \in \Lambda} |\langle F, \mathcal{E}_\lambda \Phi \rangle|^2 &= \mu_{\widehat{G}}(V) \int_{\widehat{G}} F(\gamma) \overline{\Phi(\gamma)} \sum_{\omega \in \Lambda^\perp} \overline{F(\omega + \gamma)} \Phi(\omega + \gamma) \, d\gamma \\
&= \mu_{\widehat{G}}(V) \sum_{\omega \in \Lambda^\perp} \int_{\widehat{G}} F(\gamma) \overline{F(\omega + \gamma)} \, \overline{\Phi(\gamma)} \Phi(\omega + \gamma) \, d\gamma.
\end{aligned}
$$

Pulling out the term corresponding to $\omega = 0$ gives that

$$
\sum_{\lambda \in \Lambda} |\langle F, \mathcal{E}_\lambda \Phi \rangle|^2 = \mu_{\widehat{G}}(V) \left(\int_{\widehat{G}} |F(\gamma)|^2 \, |\Phi(\gamma)|^2 \, d\gamma + R(F) \right),
$$

where

$$
R(F) := \sum_{\omega \in \Lambda^\perp \setminus \{0\}} \int_{\widehat{G}} F(\gamma) \overline{F(\omega + \gamma)} \, \overline{\Phi(\gamma)} \Phi(\omega + \gamma) \, d\gamma;
$$

from here, two applications of the Cauchy–Schwarz inequality and a use of the translation invariance of the measure prove (i) in the lemma (the proof is similar to the proof of Theorem 11.4.2).

For the proof of (ii), combining what we established in (i) shows that for $F \in \mathcal{D}(\widehat{G})$,

$$
\begin{aligned}
&\sum_{\lambda \in \Lambda} |\langle F, \mathcal{E}_\lambda \Phi \rangle|^2 \\
&\leq \mu_{\widehat{G}}(V) \left(\int_{\widehat{G}} |F(\gamma) \Phi(\gamma)|^2 \, d\gamma + \int_{\widehat{G}} |F(\gamma)|^2 \sum_{\omega \in \Lambda^\perp \setminus \{0\}} |\Phi(\gamma) \Phi(\gamma + \omega)| \, d\gamma \right) \\
&= \mu_{\widehat{G}}(V) \int_{\widehat{G}} |F(\gamma)|^2 \sum_{\omega \in \Lambda^\perp} |\Phi(\gamma) \Phi(\gamma + \omega)| \, d\gamma \\
&\leq \mu_{\widehat{G}}(V) \sup_{\gamma \in \widehat{G}} \sum_{\omega \in \Lambda^\perp} |\Phi(\gamma) \Phi(\gamma + \omega)| \int_{\widehat{G}} |F(\gamma)|^2.
\end{aligned}
$$

We conclude that the Bessel inequality holds on a dense subset of $L^2(\widehat{G})$. Therefore it holds on $L^2(\widehat{G})$, and we have now proved (ii).   □

We now state a consequence of the above results that will be of importance when we consider duality issues.

Lemma 21.4.3 *Let $\{\mathcal{E}_\lambda \Phi\}_{\lambda \in \Lambda}$ and $\{\mathcal{E}_\lambda \widetilde{\Phi}\}_{\lambda \in \Lambda}$ be Bessel sequences in $L^2(\widehat{G})$. Then for any $F, H \in \mathcal{D}(\widehat{G})$,*

$$
\sum_{\lambda \in \Lambda} \langle F, \mathcal{E}_\lambda \Phi \rangle \overline{\langle H, \mathcal{E}_\lambda \widetilde{\Phi} \rangle} = \mu_{\widehat{G}}(V) \sum_{\omega \in \Lambda^\perp} \int_{\widehat{G}} F(\gamma) \overline{H(\omega + \gamma)} \, \overline{\Phi(\gamma)} \widetilde{\Phi}(\omega + \gamma) \, d\gamma.
$$

$$
(21.31)
$$

Proof. The proof follows the lines of the proof of Lemma 21.4.2. First, the Cauchy–Schwarz inequality shows that the sum on the left-hand side of (21.31) is absolutely convergent. As in Lemma 21.4.1, letting

$$\beta(\gamma) := \sum_{\omega \in \Lambda^\perp} H(\omega + \gamma)\overline{\Phi}(\omega + \gamma),$$

we can write $\langle H, \mathcal{E}_\lambda \widetilde{\Phi} \rangle = \widehat{\beta \chi_V}(\lambda)$. Thus,

$$\sum_{\lambda \in \Lambda} \langle F, \mathcal{E}_\lambda \Phi \rangle \overline{\langle H, \mathcal{E}_\lambda \widetilde{\Phi} \rangle} = \sum_{\lambda \in \Lambda} \widehat{\alpha \chi_V}(\lambda)\overline{\widehat{\beta \chi_V}(\lambda)} = \mu_{\widehat{G}}(V) \int_V \alpha(\gamma)\overline{\beta(\gamma)} \, d\gamma,$$

where the last step used polarization of the identity in Lemma 21.2.4. Proceeding exactly as we did in the proof of Lemma 21.4.2 (see (21.29)), inserting the expression for $\alpha(\gamma)$ leads to

$$\sum_{\lambda \in \Lambda} \langle F, \mathcal{E}_\lambda \Phi \rangle \overline{\langle H, \mathcal{E}_\lambda \widetilde{\Phi} \rangle} = \mu_{\widehat{G}}(V) \int_{\widehat{G}} F(\gamma)\overline{\Phi(\gamma)}\, \overline{\beta(\gamma)} \, d\gamma,$$

corresponding to (21.30). Inserting the expression for $\beta(\gamma)$ now gives (21.31). □

21.5 Explicit Gabor Frame Constructions in $L^2(\widehat{G})$

In this section, we will consider Fourier-like systems $\{\mathcal{E}_\lambda \Phi_k\}_{\lambda \in \Lambda_k, k \in I}$, where $\{\Phi_k\}_{k \in I}$ is a countable collection of functions in $L^2(\widehat{G})$ and the sets Λ_k are lattices in G. Our purpose is to derive sufficient conditions for $\{\mathcal{E}_\lambda \Phi_k\}_{\lambda \in \Lambda_k, k \in I}$ being a frame for $L^2(\widehat{G})$; the results turn out to be parallel to what we have seen for frame constructions in $L^2(\mathbb{R}^d)$. Moreover, we show that not only the theoretical statements are similar; also on the level of concrete constructions, we are able to mimic constructions we have considered in $L^2(\mathbb{R}^d)$. The material of the section is taken from the paper [176] by Christensen and Goh.

In the entire section we will use the following

General setup: Let I denote a countable index set, and let $\{\Phi_k\}_{k \in I}$, $\{\widetilde{\Phi}_k\}_{k \in I}$ be two collections of functions in $L^2(\widehat{G})$. Furthermore, let $\{\Lambda_k\}_{k \in I}$ denote a family of lattices in G.

Letting $\Lambda_k^\perp$ denote the annihilator of Λ_k, Lemma 21.1.7(ii) shows that there exist relatively compact sets V_k in $\widehat{G}$ such that for each $k \in I$, and any $\omega, \omega' \in \Lambda_k^\perp$, $\omega \neq \omega'$,

$$\widehat{G} = \bigcup_{\omega \in \Lambda_k^\perp} (\omega + V_k), \quad (\omega + V_k) \cap (\omega' + V_k) = \emptyset. \tag{21.32}$$

We now state some sufficient conditions for $\{\mathcal{E}_\lambda \Phi_k\}_{\lambda \in \Lambda_k, k \in I}$ to be a Bessel sequence or frame for $L^2(\widehat{G})$; the reader will immediately recognize the similarity with the results in Theorem 11.4.2 (for Gabor systems in $L^2(\mathbb{R})$), Theorem 15.2.3 (for wavelet systems in $L^2(\mathbb{R})$), and Theorem 20.3.1 (for GSI systems in $L^2(\mathbb{R}^d)$).

Theorem 21.5.1 *Under the assumptions in the general setup, the following hold:*

(i) *$\{\mathcal{E}_\lambda \Phi_k\}_{\lambda \in \Lambda_k, k \in I}$ is a Bessel sequence in $L^2(\widehat{G})$ if*

$$B := \sup_{\gamma \in \widehat{G}} \sum_{k \in I} \mu_{\widehat{G}}(V_k) \sum_{\omega \in \Lambda_k^\perp} |\Phi_k(\gamma)\Phi_k(\gamma + \omega)| < \infty.$$

(ii) *If (i) holds, then $\{\mathcal{E}_\lambda \Phi_k\}_{\lambda \in \Lambda_k, k \in I}$ is a frame for $L^2(\widehat{G})$ if*

$$A := \inf_{\gamma \in \widehat{G}} \left(\sum_{k \in I} \mu_{\widehat{G}}(V_k) |\Phi_k(\gamma)|^2 - \right.$$

$$\left. - \sum_{k \in I} \mu_{\widehat{G}}(V_k) \sum_{\omega \in \Lambda_k^\perp \setminus \{0\}} |\Phi_k(\gamma)\Phi_k(\gamma + \omega)| \right) > 0.$$

Proof. Let $F \in \mathcal{D}(\widehat{G})$. For each $k \in I$, Lemma 21.4.2(i) implies that

$$\sum_{\lambda \in \Lambda_k} |\langle F, \mathcal{E}_\lambda \Phi_k \rangle|^2 \le \mu_{\widehat{G}}(V_k) \int_{\widehat{G}} |F(\gamma)|^2 \sum_{\omega \in \Lambda_k^\perp} |\Phi_k(\gamma)\Phi_k(\gamma + \omega)| \, d\gamma.$$

Thus,

$$\sum_{k \in I} \sum_{\lambda \in \Lambda_k} |\langle F, \mathcal{E}_\lambda \Phi_k \rangle|^2 \le \int_{\widehat{G}} |F(\gamma)|^2 \sum_{k \in I} \mu_{\widehat{G}}(V_k) \sum_{\omega \in \Lambda_k^\perp} |\Phi_k(\gamma)\Phi_k(\gamma + \omega)| \, d\gamma.$$

Under the assumption in (i), this implies that

$$\sum_{k \in I} \sum_{\lambda \in \Lambda_k} |\langle F, \mathcal{E}_\lambda \Phi_k \rangle|^2 \le B \int_{\widehat{G}} |F(\gamma)|^2 d\gamma = B \, \|F\|^2.$$

Since this holds on a dense set in $L^2(\widehat{G})$, we conclude that $\{\mathcal{E}_\lambda \Phi_k\}_{\lambda \in \Lambda_k, k \in I}$ is a Bessel sequence in $L^2(\widehat{G})$. The proof of (ii) is similar. $\square$

The following example shows that Theorem 21.5.1 generalizes Theorem 20.5.2.

Example 21.5.2 Consider the LCA group $G = \mathbb{R}^d$, and let $\mathcal{A}, \mathcal{B}$ denote invertible $d \times d$ matrices with real entries. Fixing a function $g \in L^2(\mathbb{R}^d)$, consider the Gabor system

$$\{E_{\underline{m}}T_{\mathcal{A}k}g\}_{m,k\in\mathbb{Z}^d} = \{e^{2\pi i \underline{m}\cdot x}g(x-\mathcal{A}k)\}_{m,k\in\mathbb{Z}^d}.$$

We immediately see from Example 21.1.8 that for any $\lambda \in G$ and $F \in L^2(\widehat{G})$,

$$\mathcal{E}_\lambda F(\gamma) = (\lambda,\gamma)F(\gamma) = e^{2\pi i \lambda \cdot \gamma}F(\gamma).$$

Thus, with $\Lambda := \mathcal{B}\mathbb{Z}^d$ and $\Phi_k(x) := g(x-\mathcal{A}k)$ for $k \in \mathbb{Z}^d$,

$$\{\mathcal{E}_\lambda \Phi_k\}_{\lambda\in\Lambda,k\in I} = \{E_{\mathcal{B}m}T_{\mathcal{A}k}g\}_{m,k\in\mathbb{Z}^d}.$$

Note that by Example 21.1.8, $\Lambda^\perp = \mathcal{B}^\sharp\mathbb{Z}$ and we can take $V = \mathcal{B}^\sharp[0,1[^d$ in (21.4). The measure of this set in $\mathbb{R}^d$ is

$$\mu(V) = |\det \mathcal{B}^\sharp| = |\det \mathcal{B}|^{-1}.$$

Thus, by Theorem 21.5.1 (i), the Gabor system $\{E_{\mathcal{B}m}T_{\mathcal{A}k}\phi\}_{m,k\in\mathbb{Z}^d}$ is a Bessel system with bound B if

$$B := \sup_{x\in\mathbb{R}^d}\sum_{k\in\mathbb{Z}^d}\mu(\mathcal{B}^\sharp[0,1]^s)\sum_{\omega\in\Lambda^\perp}|\phi(x-\mathcal{A}k)\phi(x-\mathcal{A}k+\omega)| < \infty,$$

or

$$B := \frac{1}{|\det\mathcal{B}|}\sup_{x\in\mathbb{R}^d}\sum_{k\in\mathbb{Z}^d}\sum_{n\in\mathbb{Z}^d}|\phi(x-\mathcal{A}k)\phi(x-\mathcal{A}k+\mathcal{B}^\sharp n)| < \infty.$$

This is precisely the Bessel condition in Theorem 20.5.2.   □

Under the assumption that the functions Φ_k have sufficiently small supports (in relation to the given lattices Λ_k), we obtain a characterization of the frame property for $\{\mathcal{E}_\lambda\Phi_k\}_{\lambda\in\Lambda_k,k\in I}$:

Corollary 21.5.3 *In addition to the general setup, assume that for each $k \in I$, the function Φ_k satisfies that*

$$\mathrm{supp}\,\Phi_k \cap \mathrm{supp}\,\Phi_k(\cdot+\omega) = \emptyset, \ \forall\omega\in\Lambda_k^\perp\setminus\{0\} \qquad (21.33)$$

(up to a set of measure zero in $\widehat{G}$). Then the following hold:

(i) *$\{\mathcal{E}_\lambda\Phi_k\}_{\lambda\in\Lambda_k,k\in I}$ is a Bessel sequence in $L^2(\widehat{G})$ if and only if*

$$B := \sup_{\gamma\in\widehat{G}}\sum_{k\in I}\mu_{\widehat{G}}(V_k)|\Phi_k(\gamma)|^2 < \infty.$$

(ii) *If (i) holds, then $\{\mathcal{E}_\lambda\Phi_k\}_{\lambda\in\Lambda_k,k\in I}$ is a frame for $L^2(\widehat{G})$ if and only if*

$$A := \inf_{\gamma\in\widehat{G}}\sum_{k\in I}\mu_{\widehat{G}}(V_k)|\Phi_k(\gamma)|^2 > 0.$$

Proof. The sufficiency of the conditions in (i) and (ii) follows directly from Theorem 21.5.1 and the assumption (21.33). Let us show that the condition in (i) is also necessary for $\{\mathcal{E}_\lambda \Phi_k\}_{\lambda \in \Lambda_k, k \in I}$ to be a Bessel sequence with bound B. First, by (21.28) and the assumption (21.33), for all $F \in \mathcal{D}(\widehat{G})$, we have

$$\sum_{k \in I} \sum_{\lambda \in \Lambda_k} |\langle F, \mathcal{E}_\lambda \Phi_k \rangle|^2 = \sum_{k \in I} \mu_{\widehat{G}}(V_k) \int_{\widehat{G}} |F(\gamma) \Phi_k(\gamma)|^2 \, d\gamma$$

$$= \int_{\widehat{G}} |F(\gamma)|^2 \sum_{k \in I} \mu_{\widehat{G}}(V_k) |\Phi_k(\gamma)|^2 \, d\gamma.$$

Thus, if $\{\mathcal{E}_\lambda \Phi_k\}_{\lambda \in \Lambda_k, k \in I}$ is a Bessel sequence with bound B,

$$\int_{\widehat{G}} |F(\gamma)|^2 \sum_{k \in I} \mu_{\widehat{G}}(V_k) |\Phi_k(\gamma)|^2 \, d\gamma \le B \, \|F\|^2$$

for all $F \in \mathcal{D}(\widehat{G})$. This implies that $\sum_{k \in I} \mu_{\widehat{G}}(V_k) |\Phi_k(\gamma)|^2 \le B$ almost everywhere, as desired: in fact, if $\sum_{k \in I} \mu_{\widehat{G}}(V_k) |\Phi_k(\gamma)|^2 > B$ on a set $\mathcal{S}$ of positive measure (we can assume that the measure is finite by switching to a subset, if necessary), taking $F := \chi_{\mathcal{S}}$ would lead to a contradiction. The necessity of the lower bound in (ii) is shown in a similar way. □

We will now provide simple and explicit frame constructions based on Theorem 21.5.1 and Corollary 21.5.3, in the full generality of LCA groups. The constructions will be based on a generalization of the classical B-splines to the setting of LCA groups; see Section A.10.

Using the general setup we will construct concrete Gabor-type frames for $L^2(\widehat{G})$ of the form $\{\mathcal{E}_\lambda \mathcal{T}_\eta \Phi\}_{\lambda \in \Lambda, \eta \in \Gamma}$, where $\Phi \in L^2(\widehat{G})$; furthermore, Λ is a lattice in G, and Γ is a lattice in $\widehat{G}$. The construction is based on splines of the type in Definition A.10.1, but defined on the group $\widehat{G}$.

Theorem 21.5.4 *Given a lattice Γ in $\widehat{G}$, let $\Omega \subset \widehat{G}$ denote a fundamental domain. For a fixed $r \in \mathbb{N}$, consider the function*

$$W_r := g_1 \chi_\Omega * g_2 \chi_\Omega * \cdots * g_r \chi_\Omega,$$

where $g_1, \ldots, g_r \in L^2(\Omega)$, with the assumption that $g_j > 0$ on Ω for $j = 1, \ldots, r$ and $g_j = C > 0$ for at least one index j. Given a lattice Λ in G, assume that the fundamental domain V associated with $\Lambda^\perp$ satisfies that $r\Omega \subseteq V$. Then $\{\mathcal{E}_\lambda \mathcal{T}_\eta W_r\}_{\lambda \in \Lambda, \eta \in \Gamma}$ is a frame for $L^2(\widehat{G})$.

Proof. Without loss of generality we can assume that $g_1 = C\chi_\Omega$. By Lemma A.10.2(iv), we have

$$\sum_{\eta \in \Gamma} W_r(\gamma - \eta) = C_r := \frac{1}{\mu_{\widehat{G}}(\Omega)} \prod_{j=1}^r \int_\Omega g_j(\eta) \, d\eta, \ \gamma \in \widehat{G}; \qquad (21.34)$$

it follows that

$$0 \leq W_r(\gamma) \leq C_r, \ \gamma \in \widehat{G}, \tag{21.35}$$

so the function W_r is bounded. Since $V \cap (V + \Lambda^\perp \setminus \{0\})) = \emptyset$ and $r\Omega \subseteq V$, we have $r\Omega \cap (r\Omega + (\Lambda^\perp \setminus \{0\})) = \emptyset$; thus, via Lemma A.10.2(ii),

$$\operatorname{supp} W_r \cap \operatorname{supp} W_r(\cdot + \omega) = \emptyset, \ \forall \omega \in \Lambda^\perp \setminus \{0\}. \tag{21.36}$$

We will now apply Corollary 21.5.3 with the functions Φ_k corresponding to $T_\eta W_r$, $\eta \in \Gamma$ i.e., we will estimate the supremum and infimum of

$$\sum_{\eta \in \Gamma} \mu_{\widehat{G}}(V)|W_r(\gamma - \eta)|^2 = \mu_{\widehat{G}}(V) \sum_{\eta \in \Gamma} |W_r(\gamma - \eta)|^2.$$

Note that (21.36) implies that (21.33) in Corollary 21.5.3 holds with $\Lambda_\eta = \Lambda$ for all $\eta \in \Gamma$. Now, by (21.35) and (21.34), we see that for any $\gamma \in \widehat{G}$,

$$\sum_{\eta \in \Gamma} |W_r(\gamma - \eta)|^2 \leq C_r \sum_{\eta \in \Gamma} |W_r(\gamma - \eta)| = C_r \sum_{\eta \in \Gamma} W_r(\gamma - \eta) = C_r^2.$$

We will now show that the term $\sum_{\eta \in \Gamma} |W_r(\gamma - \eta)|^2$ also has a strictly positive lower bound. To this end, we notice that

$$\inf_{\gamma \in \widehat{G}} \sum_{\eta \in \Gamma} |W_r(\gamma - \eta)|^2 = \inf_{\gamma \in \Omega} \sum_{\eta \in \Gamma} |W_r(\gamma - \eta)|^2. \tag{21.37}$$

The inequality $\leq$ is obvious. In order to show the opposite inequality, we use that any $\gamma \in \widehat{G}$ can be written in a unique way as $\gamma = \gamma' + k'$ with $k' \in \Gamma, \gamma' \in \Omega$. Thus

$$\sum_{\eta \in \Gamma} |W_r(\gamma - \eta)|^2 = \sum_{\eta \in \Gamma} |W_r(\gamma' + k' - \eta)|^2;$$

making the change of variable $\ell = \eta - k'$, this shows that

$$\sum_{\eta \in \Gamma} |W_r(\gamma - \eta)|^2 = \sum_{\ell \in \Gamma} |W_r(\gamma' - \ell)|^2 \geq \inf_{\zeta \in \Omega} \sum_{\eta \in \Gamma} |W_r(\zeta - \eta)|^2,$$

and (21.37) follows.

Now, for $r = 1$ the (strictly positive) lower bound of $\sum_{\eta \in \Gamma} |W_r(\gamma - \eta)|^2$ is obvious because $W_1 = C\chi_\Omega$ and Ω is the fundamental domain associated with Γ. Therefore we now assume that $r \geq 2$. Given any $\alpha \in \overline{\Omega}$, the partition of unity condition (21.34), with the nonnegative nature of W_r, shows that there is a lattice point $\eta_\alpha \in \Gamma$ such that $W_r(\alpha - \eta_\alpha) > 0$. Since W_r is continuous, for each $\alpha \in \overline{\Omega}$, there is a neighborhood $\mathcal{U}_\alpha$ around α such that $W_r(\gamma - \eta_\alpha) > 0$ for all $\gamma \in \mathcal{U}_\alpha$. The neighborhoods $\mathcal{U}_\alpha$, $\alpha \in \overline{\Omega}$, form an open cover of the compact set $\overline{\Omega}$, so we can select a finite collection of distinct points $\alpha_1, \ldots, \alpha_n \in \overline{\Omega}$ such that $\overline{\Omega} \subseteq \mathcal{U}_{\alpha_1} \cup \mathcal{U}_{\alpha_2} \cup \cdots \cup \mathcal{U}_{\alpha_n}$; thus, for any $\gamma \in \overline{\Omega}$, at least one of the terms $W_r(\gamma - \eta_{\alpha_j})$, $j = 1, \ldots, n$, is positive, and therefore $\sum_{j=1}^n |W_r(\gamma - \eta_{\alpha_j})|^2 > 0$. Since W_r is continuous and

$\overline{\Omega}$ is compact, this implies that $\inf_{\gamma \in \overline{\Omega}} \sum_{j=1}^{n} |W_r(\gamma - \eta_{\alpha_j})|^2 > 0$. Putting everything together, we conclude that

$$\inf_{\gamma \in \widehat{G}} \sum_{\eta \subset \Gamma} |W_r(\gamma - \eta)|^2 = \inf_{\gamma \in \Omega} \sum_{\eta \in I'} |W_r(\gamma - \eta)|^2 \geq \inf_{\gamma \in \overline{\Omega}} \sum_{j=1}^{n} |W_r(\gamma - \eta_{\alpha_j})|^2 > 0,$$

providing the promised lower bound. □

Example 21.5.5 Consider a Gabor system in $L^2(\mathbb{R})$, with translation parameter $a = 1$, i.e., $\{E_{mb}T_k g\}_{k,m \in \mathbb{Z}} = \{e^{2\pi imb \cdot} g(\cdot - k)\}_{k,m \in \mathbb{Z}}$. The technical condition $r\Omega \subseteq V$ in Theorem 21.5.4 means that $[0, r) \subseteq [0, 1/b)$. Thus, in this particular case we conclude that the (standard) B-spline B_r on $\mathbb{R}$ generates a Gabor frame $\{E_{mb}T_k B_r\}_{k,m \in \mathbb{Z}}$ for $L^2(\mathbb{R})$ if $b \leq 1/r$; this is a special case of Corollary 11.7.1. It is easy to follow the same approach and find explicit frame constructions for $L^2(\widehat{G})$ for any group of the form $G = \mathbb{R}^{d_1} \times \mathbb{Z}^{d_2} \times \mathbb{T}^{d_3} \times \mathbb{Z}_N$, as discussed in Example 21.1.2; we leave the concrete calculations to the reader. □

Let us now consider duality issues for two sequences $\{\mathcal{E}_\lambda \Phi_k\}_{\lambda \in \Lambda_k, k \in I}$ and $\{\mathcal{E}_\lambda \widetilde{\Phi}_k\}_{\lambda \in \Lambda_k, k \in I}$. First, as a direct consequence of Lemma 21.4.3, we have the following:

Proposition 21.5.6 If $\{\mathcal{E}_\lambda \Phi_k\}_{\lambda \in \Lambda_k, k \in I}$ and $\{\mathcal{E}_\lambda \widetilde{\Phi}_k\}_{\lambda \in \Lambda_k, k \in I}$ are Bessel sequences in $L^2(\widehat{G})$, then for all $F, H \in \mathcal{D}(\widehat{G})$,

$$\sum_{k \in I} \sum_{\lambda \in \Lambda_k} \langle F, \mathcal{E}_\lambda \Phi_k \rangle \overline{\langle H, \mathcal{E}_\lambda \widetilde{\Phi}_k \rangle}$$

$$= \sum_{k \in I} \mu_{\widehat{G}}(V_k) \sum_{\omega \in \Lambda_k^\perp} \int_{\widehat{G}} F(\gamma) \overline{H(\omega + \gamma)} \, \overline{\Phi_k(\gamma)} \widetilde{\Phi}_k(\omega + \gamma) \, d\gamma.$$

Proof. Note that the sum on the left-hand side is convergent by the Cauchy–Schwarz inequality and the Bessel assumption. Now the result follows immediately from Lemma 21.4.3. □

Theorem 21.5.7 In addition to the general setup, assume that for each $k \in I$,

$$\text{supp } \Phi_k \cap \text{supp } \widetilde{\Phi}_k(\cdot + \omega) = \emptyset, \ \forall \omega \in \Lambda_k^\perp \setminus \{0\} \qquad (21.38)$$

(up to a set of measure zero in $\widehat{G}$). If $\{\mathcal{E}_\lambda \Phi_k\}_{\lambda \in \Lambda_k, k \in I}$ and $\{\mathcal{E}_\lambda \widetilde{\Phi}_k\}_{\lambda \in \Lambda_k, k \in I}$ are Bessel sequences in $L^2(\widehat{G})$, they are dual frames for $L^2(\widehat{G})$ if and only if

$$\sum_{k \in I} \mu_{\widehat{G}}(V_k) \overline{\Phi_k(\gamma)} \widetilde{\Phi}_k(\gamma) = 1, \ a.e. \ \gamma \in \widehat{G}. \qquad (21.39)$$

Proof. If (21.39) holds, then Proposition 21.5.6 shows that for all $F, H \in \mathcal{D}(\widehat{G})$,

$$\sum_{k \in I} \sum_{\lambda \in \Lambda_k} \langle F, \mathcal{E}_\lambda \Phi_k \rangle \overline{\langle H, \mathcal{E}_\lambda \widetilde{\Phi}_k \rangle} = \langle F, H \rangle.$$

By continuity of the inner product, the above equation also holds for all $F, H \in L^2(\widehat{G})$. Combining with the assumption that $\{\mathcal{E}_\lambda \Phi_k\}_{\lambda \in \Lambda_k, k \in I}$ and $\{\mathcal{E}_\lambda \widetilde{\Phi}_k\}_{\lambda \in \Lambda_k, k \in I}$ are Bessel sequences, this proves that $\{\mathcal{E}_\lambda \Phi_k\}_{\lambda \in \Lambda_k, k \in I}$ and $\{\mathcal{E}_\lambda \widetilde{\Phi}_k\}_{\lambda \in \Lambda_k, k \in I}$ are dual frames for $L^2(\widehat{G})$; see Lemma 6.3.2.

Conversely, assume that $\{\mathcal{E}_\lambda \Phi_k\}_{\lambda \in \Lambda_k, k \in I}$ and $\{\mathcal{E}_\lambda \widetilde{\Phi}_k\}_{\lambda \in \Lambda_k, k \in I}$ are dual frames such that (21.38) holds. By Proposition 21.5.6, for $F = H \in \mathcal{D}(\widehat{G})$,

$$\int_{\widehat{G}} \sum_{k \in I} \mu_{\widehat{G}}(V_k) \, \overline{\Phi_k(\gamma)} \widetilde{\Phi}_k(\gamma) |F(\gamma)|^2 d\gamma = \int_{\widehat{G}} |F(\gamma)|^2 d\gamma.$$

Splitting $\sum_{k \in I} \mu_{\widehat{G}}(V_k) \, \overline{\Phi_k(\gamma)} \widetilde{\Phi}_k(\gamma)$ into real part and imaginary part, i.e. $a(\gamma) + ib(\gamma) = \sum_{k \in I} \mu_{\widehat{G}}(V_k) \, \overline{\Phi_k(\gamma)} \widetilde{\Phi}_k(\gamma)$, yields that

$$\int_{\widehat{G}} a(\gamma) |F(\gamma)|^2 d\gamma = \int_{\widehat{G}} |F(\gamma)|^2 d\gamma \text{ and } \int_{\widehat{G}} b(\gamma) |F(\gamma)|^2 d\gamma = 0$$

for all $F \in \mathcal{D}(\widehat{G})$, which implies that $a(\gamma) = 1$ and $b(\gamma) = 0$ for a.e. $\gamma \in \widehat{G}$, by exactly the same argument as in the proof of Corollary 21.5.3.   □

Let us return to the setup in Theorem 21.5.4 and consider a Gabor system in $L^2(\widehat{G})$ of the form $\{\mathcal{E}_\lambda \mathcal{T}_\eta W_r\}_{\lambda \in \Lambda, \eta \in \Gamma}$, where Γ is chosen as a lattice in $\widehat{G}$, Ω is a corresponding fundamental domain, and W_r is a weighted B-spline with $g_1 = C > 0$ and $g_j > 0, j = 2, \ldots, r$, on Ω. By Theorem 21.5.4, such a system is a frame for $L^2(\widehat{G})$ if $r\Omega \subseteq V$, but the proof shows that in fact it is sufficient that $r\Omega \cap (r\Omega + (\Lambda^\perp \setminus \{0\})) = \emptyset$. We will now impose a stronger assumption, which implies that we can find an explicitly given dual frame $\{\mathcal{E}_\lambda \mathcal{T}_\eta \widetilde{\Phi}\}_{\lambda \in \Lambda, \eta \in \Gamma}$.

Proposition 21.5.8 *In addition to the setup in Theorem 21.5.4, assume that the set*

$$\Delta := \{k \in \Gamma \mid r\Omega \cap (k + r\Omega) \neq \emptyset\} \tag{21.40}$$

satisfies that

$$r\Omega \cap (\Delta + r\Omega + (\Lambda^\perp \setminus \{0\})) = \emptyset. \tag{21.41}$$

Then, with the constant C_r defined as in (21.34), the function

$$\widetilde{\Phi}(\gamma) := \frac{1}{\mu_{\widehat{G}}(V) C_r^2} \sum_{k \in \Delta} W_r(\gamma - k), \quad \gamma \in \widehat{G}, \tag{21.42}$$

generates a dual frame $\{\mathcal{E}_\lambda \mathcal{T}_\eta \widetilde{\Phi}\}_{\lambda \in \Lambda, \eta \in \Gamma}$ of $\{\mathcal{E}_\lambda \mathcal{T}_\eta W_r\}_{\lambda \in \Lambda, \eta \in \Gamma}$ in $L^2(\widehat{G})$.

Proof. Note that in the described setup, the condition (21.39) takes the form

$$\sum_{\eta \in \Gamma} W_r(\gamma - \eta) \widetilde{\Phi}(\gamma - \eta) = \frac{1}{\mu_{\widehat{G}}(V)}, \quad \text{a.e. } \gamma \in \widehat{G},$$

or,

$$\sum_{\eta \in \Gamma} (W_r \widetilde{\Phi})(\gamma - \eta) = \frac{1}{\mu_{\widehat{G}}(V)}, \quad \text{a.e. } \gamma \in \widehat{G}, \tag{21.43}$$

Since $\sum_{\eta \in \Gamma} W_r(\gamma - \eta) = C_r$ as noted in (21.34), the condition (21.43) is obviously satisfied if we choose the function $\widetilde{\Phi}$ such that $W_r \widetilde{\Phi} = (\mu_{\widehat{G}}(V) C_r)^{-1} W_r$. Thus, it suffices to have that $\widetilde{\Phi}(\gamma) = (\mu_{\widehat{G}}(V) C_r)^{-1}$ for $\gamma \in r\Omega$, a condition that is satisfied if we take $\widetilde{\Phi}$ to be as in (21.42), with the index set Δ defined by (21.40). To see this, note that if $\gamma \in r\Omega$ and $k \in \Gamma \setminus \Delta$, then $\gamma \notin k + r\Omega$, which implies that $W_r(\gamma - k) = 0$. Therefore, for $\gamma \in r\Omega$,

$$
\begin{aligned}
\widetilde{\Phi}(\gamma) &= \frac{1}{\mu_{\widehat{G}}(V) C_r^2} \sum_{k \in \Delta} W_r(\gamma - k) \\
&= \frac{1}{\mu_{\widehat{G}}(V) C_r^2} \left(\sum_{k \in \Delta} W_r(\gamma - k) + \sum_{k \in \Gamma \setminus \Delta} W_r(\gamma - k) \right) \\
&= \frac{1}{\mu_{\widehat{G}}(V) C_r^2} \sum_{k \in \Gamma} W_r(\gamma - k) = \frac{1}{\mu_{\widehat{G}}(V) C_r},
\end{aligned}
$$

as desired. With the choice of $\widetilde{\Phi}$ in (21.42), the condition (21.41) ensures that (21.38) holds. Hence, the result follows from Theorem 21.5.7.   □

Example 21.5.9 Consider again a Gabor system $\{E_{mb} T_k g\}_{k,m \in \mathbb{Z}}$ in $L^2(\mathbb{R})$. Let $g := B_r$, we see that $\Delta = \{-r+1, \ldots, r-1\}$. Thus $\Delta + r\Omega = [-r+1, 2r-1[$; therefore (21.41) is satisfied if $1/b \geq 2r-1$, i.e., if $b \leq \frac{1}{2r-1}$. This is exactly the condition that was used in Theorem 12.5.1 in order to construct dual frame pairs. Similar to the case for Theorem 21.5.4, it is easy to apply Proposition 21.5.8 to construct explicit dual pairs of frames for $L^2(\widehat{G})$ for groups of the form $G = \mathbb{R}^s \times \mathbb{Z}^p \times \mathbb{T}^q \times \mathbb{Z}_m$, $d_1, d_2, d_3 \in \mathbb{N}$. We leave the calculations to the reader.   □

21.6   GSI Systems on LCA Groups

The results in Section 21.5 have immediate consequences for generalized shift-invariant systems. In fact, let Λ be a lattice in G. Then, for any $\phi \in$

$L^2(G)$ and $\lambda \in \Lambda$,

$$\mathcal{F}^{-1}\mathcal{E}_\lambda \mathcal{F}\phi(x) = T_{-\lambda}\mathcal{F}^{-1}\mathcal{F}\phi(x) = T_{-\lambda}\phi(x) = \phi(x + \lambda). \qquad (21.44)$$

Since the inverse Fourier transform is a unitary operator, it preserves the properties of Bessel sequences, frames, and dual frames from $L^2(\widehat{G})$ to $L^2(G)$. By setting $\Phi_k := \widehat{\phi_k}$ and $\widetilde{\Phi_k} := \widehat{\widetilde{\phi_k}}$, we obtain the following immediate consequences of Theorem 21.5.1 and Theorem 21.5.7:

Theorem 21.6.1 *Suppose that $\{\Lambda_k\}_{k \in I}$ is a countable family of lattices in G, and let $\{V_k\}_{k \in I}$ be fundamental domains for the annihilators $\{\Lambda_k^\perp\}_{k \in I}$. Consider two collections of elements $\{\phi_k\}_{k \in I}$, $\{\widetilde{\phi_k}\}_{k \in I}$ in $L^2(G)$. Then the following hold:*

(i) *$\{T_\lambda \phi_k\}_{\lambda \in \Lambda_k, k \in I}$ is a Bessel sequence in $L^2(G)$ if*

$$B := \sup_{\gamma \in \widehat{G}} \sum_{k \in I} \mu_{\widehat{G}}(V_k) \sum_{\omega \in \Lambda_k^\perp} \left| \widehat{\phi_k}(\gamma)\widehat{\phi_k}(\gamma + \omega) \right| < \infty.$$

(ii) *If (i) holds, then $\{T_\lambda \phi_k\}_{\lambda \in \Lambda_k, k \in I}$ is a frame for $L^2(G)$ if*

$$A := \inf_{\gamma \in \widehat{G}} \left(\sum_{k \in I} \mu_{\widehat{G}}(V_k) |\widehat{\phi_k}(\gamma)|^2 \right.$$

$$\left. - \sum_{k \in I} \mu_{\widehat{G}}(V_k) \sum_{\omega \in \Lambda_k^\perp \setminus \{0\}} \left| \widehat{\phi_k}(\gamma)\widehat{\phi_k}(\gamma + \omega) \right| \right) > 0.$$

(iii) *Assume that for each $k \in I$,*

$$\operatorname{supp} \widehat{\phi_k} \cap \operatorname{supp} \widehat{\widetilde{\phi_k}}(\cdot + \omega) = \emptyset, \ \forall \omega \in \Lambda_k^\perp \setminus \{0\},$$

up to a set of measure zero. If $\{T_\lambda \phi_k\}_{\lambda \in \Lambda_k, k \in I}$ and $\{T_\lambda \widetilde{\phi_k}\}_{\lambda \in \Lambda_k, k \in I}$ are Bessel sequences in $L^2(G)$, they are dual frames for $L^2(G)$ if and only if

$$\sum_{k \in I} \mu_{\widehat{G}}(V_k) \overline{\widehat{\phi_k}(\gamma)}\widehat{\widetilde{\phi_k}}(\gamma) = 1, \ a.e. \ \gamma \in \widehat{G}.$$

Let us apply Theorem 21.6.1 to derive a result about a matrix-generated wavelet system in $L^2(\mathbb{R}^d)$.

Example 21.6.2 Let $G = \mathbb{R}^d$, with dual group $\widehat{G} = \mathbb{R}^d$. Given real and invertible $d \times d$ matrices $\mathcal{A}_k$ and $\mathcal{B}_k$, $k \in I$, consider a wavelet system of the form

$$\{D_{\mathcal{A}_k} T_{\mathcal{B}_k j}\phi\}_{k \in I, j \in \mathbb{Z}^d} = \{|\det \mathcal{A}_k|^{1/2}\, \phi(\mathcal{A}_k \cdot - \mathcal{B}_k j)\}_{k \in I, j \in \mathbb{Z}^d},$$

where $\phi \in L^2(\mathbb{R}^d)$. Note that this general setup contains the classical wavelet systems as well as, e.g., the composite wavelets in [359] as special cases. Letting $\phi_k(x) := D_{A_k}\phi(x) = |\det A_k|^{1/2}\phi(A_k x)$, $k \in I, x \in \mathbb{R}^d$, we have that

$$T_\lambda \phi_k(x) = \phi_k(x - \lambda) = |\det A_k|^{1/2}\phi(A_k x - A_k \lambda).$$

Thus, taking $\Lambda_k := A_k^{-1}\mathcal{B}_k\mathbb{Z}^d$, the system $\{T_\lambda \phi_k\}_{k \in I, \lambda \in \Lambda_k}$ is exactly the wavelet system $\{D_{A_k}T_{\mathcal{B}_k j}\phi\}_{k \in I, j \in \mathbb{Z}^d}$. Since

$$\Lambda_k^\perp = ((A_k^{-1}\mathcal{B}_k)^T)^{-1}\mathbb{Z}^d = (A_k^{-1}\mathcal{B}_k)^\sharp \mathbb{Z}^d = A_k^T \mathcal{B}_k^\sharp \mathbb{Z}^d$$

and

$$\mathbb{R}^d = \bigcup_{n \in \mathbb{Z}^d} (n + [0, 1[^d),$$

we can take $V_k = A_k^T \mathcal{B}_k^\sharp [0, 1)^d$ in (21.32). Now,

$$\widehat{\phi_k}(\gamma) = \mathcal{F}D_{A_k}\phi(\gamma) = D_{A_k^\sharp}\widehat{\phi}(\gamma),$$

so the condition in Theorem 21.6.1(i) amounts to

$$B := \sup_{\gamma \in \mathbb{R}^d} \sum_{k \in I} |\det(A_k^T \mathcal{B}_k^\sharp)| \sum_{\omega \in \Lambda_k^\perp} |\det A_k^\sharp| \, |\widehat{\phi}(A_k^\sharp \gamma)\widehat{\phi}(A_k^\sharp \gamma + A_k^\sharp \omega)| < \infty,$$

or

$$B = \sup_{\gamma \in \mathbb{R}^d} \sum_{k \in I} \frac{1}{|\det \mathcal{B}_k|} \sum_{n \in \mathbb{Z}^d} |\widehat{\phi}(A_k^\sharp \gamma)\widehat{\phi}(A_k^\sharp \gamma + \mathcal{B}_k^\sharp n)| < \infty.$$

This generalizes the Bessel condition in Theorem 20.6.1. □

Theorem 21.6.1 shows that the sufficient conditions for a GSI system $\{T_\lambda \phi_k\}_{\lambda \in \Lambda_k, k \in I}$ to be a frame for $L^2(\mathbb{R}^d)$ generalize to the setting of LCA groups. In [468], Kutyniok and Labate showed that also the characterization of tight frames of GSI systems in $L^2(\mathbb{R}^d)$ (which is a special case of Theorem 20.4.3) generalizes to LCA groups. Their proof follows closely the proof for GSI systems in $L^2(\mathbb{R}^d)$ given in [398], and thus it is not a surprise that also the characterization of dual frames in Theorem 20.4.3 generalizes to LCA groups. A detailed proof of this is given in [417], and we will state the exact result in Theorem 21.6.4.

Given a countable collection $\{\Lambda_k\}_{k \in I}$ of lattices in G, let

$$\Gamma := \bigcup_{k \in I} \Lambda_k^\perp; \qquad (21.45)$$

and, given $\alpha \in \Gamma$, let

$$J_\alpha := \{k \in I \,|\, \alpha \in \Lambda_k^\perp\}. \qquad (21.46)$$

Parallel with the definition in (20.8) for GSI systems on $\mathbb{R}^d$, let

$$\mathcal{D} := \left\{ f \in L^2(G) \,\middle|\, \widehat{f} \in L^\infty(\widehat{G}) \text{ and } \mathrm{supp}\widehat{f} \text{ is compact} \right\}. \qquad (21.47)$$

The space $\mathcal{D}$ is dense in $L^2(G)$. As for the analysis of GSI systems in $L^2(\mathbb{R}^d)$, we will assume that a local integrability condition holds:

Definition 21.6.3 *Two GSI systems* $\{T_\lambda \phi_k\}_{\lambda \in \Lambda_k, k \in I}$ *and* $\{T_\lambda \widetilde{\phi}_k\}_{\lambda \in \Lambda_k, k \in I}$ *are said to satisfy the dual α-LIC if*

$$\sum_{k \in I} \sum_{\omega \in \Lambda_k^\perp} \frac{1}{s(\Lambda_k)} \int_{\widehat{G}} |\widehat{f}(\gamma)\widehat{f}(\gamma+\omega)\widehat{\phi_k}(\gamma)\widetilde{\widehat{\phi}_k}(\gamma+\omega)| \, d\gamma < \infty \qquad (21.48)$$

for all $f \in \mathcal{D}$.

We can now state the announced characterization of GSI systems in $L^2(G)$.

Theorem 21.6.4 *Assume that the GSI systems* $\{T_\lambda \phi_k\}_{\lambda \in \Lambda_k, k \in I}$ *and* $\{T_\lambda \widetilde{\phi}_k\}_{\lambda \in \Lambda_k, k \in I}$ *are Bessel sequences in $L^2(G)$ and satisfy the dual α-LIC. Then* $\{T_\lambda \phi_k\}_{\lambda \in \Lambda_k, k \in I}$ *and* $\{T_\lambda \widetilde{\phi}_k\}_{\lambda \in \Lambda_k, k \in I}$ *are dual frames if and only if*

$$\sum_{k \in J_\alpha} \frac{1}{s(\Lambda_k)} \overline{\widehat{\phi_k}(\gamma)} \widetilde{\widehat{\phi}_k}(\gamma+\alpha) = \delta_{\alpha,0}, \ a.e. \ \gamma \in \widehat{G} \qquad (21.49)$$

for all $\alpha \in \Gamma$.

As an application of Theorem 21.6.4 we will consider a shift-invariant system in $L^2(\mathbb{R}^d)$.

Example 21.6.5 Consider a countable collection of functions $\{\phi_k\}_{k \in I}$ in $L^2(\mathbb{R}^d)$ and an invertible $d \times d$ matrix $\mathcal{A}$. Consider two Bessel sequences $\{T_{\mathcal{A}n}\phi_k\}_{k \in I, n \in \mathbb{Z}^d}$ and $\{T_{\mathcal{A}n}\widetilde{\phi}_k\}_{k \in I, n \in \mathbb{Z}^d}$; it is easy to see that they automatically satisfy the dual α-LIC. Letting $\Lambda_k := \mathcal{A}\mathbb{Z}^d$, (21.45) and the calculation in Example 21.1.8 yield the set $\Gamma = (\mathcal{A}\mathbb{Z}^d)^\perp = \mathcal{A}^\sharp \mathbb{Z}^d$. It follows that $J_\alpha = I$ for all $\alpha \in \Gamma$. Finally, again according to Example 21.1.8, the density of the lattice Λ_k is $s(\Lambda_k) = |\det \mathcal{A}|$. By Theorem 21.6.4, we now conclude that $\{T_{\mathcal{A}n}\phi_k\}_{k \in I, n \in \mathbb{Z}^d}$ and $\{T_{\mathcal{A}n}\widetilde{\phi}_k\}_{k \in I, n \in \mathbb{Z}^d}$ are dual frames if and only if

$$\sum_{k \in I} \overline{\widehat{\phi_k}(\gamma)} \widetilde{\widehat{\phi}_k}(\omega + \mathcal{A}^\sharp n) = |\det \mathcal{A}| \, \delta_{n,0}, \ \forall n \in \mathbb{Z}^d. \qquad (21.50)$$

This result was also proved in Corollary 20.4.4 and, in the one-dimensional case, in Theorem 10.1.7. Remember that for notational convenience we focussed on the one-dimensional case in the presentation of shift-invariant systems in Chapter 10. The group-theoretical approach shows that there actually is no difference between the one-dimensional setting and the higher-dimensional case at all – even the matrix case follows the same way.   □

Let us end this section with one more result from [468], which generalizes Proposition 20.3.2.

Proposition 21.6.6 *If the GSI system* $\{T_\lambda \phi_k\}_{\lambda \in \Lambda_k, k \in I}$ *is a Bessel sequence in* $L^2(G)$ *with bound* B, *then*

$$\sum_{k \in I} \frac{1}{s(\Lambda_k)} |\widehat{\phi_k}(\gamma)|^2 \leq B, \, a.e. \, \gamma \in \widehat{G}. \tag{21.51}$$

21.7 Generalized Translation-Invariant Systems

The purpose of this section is to derive integral/series expansions in terms of classes of functions that are much more general than the GSI systems in Section 21.6. The high degree of generality yields a platform on which a unifying theory for the continuous case and the discrete case can be derived. Thus, the results about the short-time Fourier transform in Section 11.1 and the Gabor systems $\{E_{mb}T_{na}g\}_{m,n \in \mathbb{Z}}$ in Section 11.2 can now be considered as special cases of general results; the same remark applies to the results about the continuous wavelet transform and wavelet systems $\{D^j T_k \psi\}_{j,k \in \mathbb{Z}}$.

The Fourier-like systems $\{\mathcal{E}_\lambda \Phi_k\}_{\lambda \in \Lambda_k, k \in I}$ in Section 21.5 and the GSI systems $\{T_\lambda \phi_k\}_{\lambda \in \Lambda_k, k \in I}$ in Section 21.6 are based on a countable set of lattices $\{\Lambda_k\}_{k \in I}$ in the group G. However, the reader might have observed that in the proofs the main role is not played by the lattices Λ_k themselves: the key point is that the annihilators $\Lambda_k^\perp$ are discrete and that we have their corresponding fundamental domains at our disposal. To illustrate this, look at the formulation of, e.g., the general setup in Section 21.5, or Theorem 21.6.1.

In 2014, it was observed, independently by Bownik and Ross [85] and Jakobsen and Lemvig [417], that a more general perspective is possible. In fact, assume that G is an LCA group and that H is a closed subgroup for which the quotient group G/H is compact. Then $\widehat{G/H}$ is discrete by Lemma 21.1.5 (ii); furthermore, since G/H is metrizable and compact, the dual group $\widehat{G/H}$ is countable; see (24.15) in [402]. By Lemma 21.1.5 (iii) we now conclude that the annihilator $H^\perp$ is discrete and countable. This observation turns out to be the key to a generalization of several of the results considered in the previous sections, with subgroups replacing the lattices Λ_k. This allows to obtain frame decompositions based on LCA groups that do not have (nontrivial) lattices, e.g., the group of p-adic numbers [417] and the Prüfer group [418].

The work by Bownik and Ross focusses on what they call *translation invariant systems;* see the following Definition 21.7.5. Such systems correspond to shift-invariant systems, except that the shifts now can take place along a possible uncountable subgroup instead of a discrete lattice. The work by Jakobsen and Lemvig deals with *generalized translation-invariant systems;* they are related to translation-invariant systems in a similar fashion as GSI systems are related to shift-invariant systems. Generalized translation-invariant systems turn out to give a unified approach to (discrete) frames and continuous frames, with general formulations of, e.g., admissibility conditions. We will give a presentation of some of the key results by Jakobsen and Lemvig, and we refer to their papers [417, 418] for much more information.

Let us first state the central definitions for this section.

Definition 21.7.1 *Let G denote an LCA group. A subgroup H of G is said to be co-compact if G/H is compact.*

In the language of Definition 21.7.1, we can now say that if H is a closed co-compact subgroup of an LCA group G, then the annihilator $H^\perp$ is discrete and countable.

Even if lattices exist in a given LCA group G, the class of subgroups is often larger than the class of lattices. That is, even in this case the work by Bownik & Ross and Jakobsen & Lemvig adds new information, as illustrated by the following example.

Example 21.7.2 In the LCA group $\mathbb{R}^d$, the subgroups and lattices are characterized as follows:

- The closed subgroups of $\mathbb{R}^d$ have the form

$$H = \mathcal{A}(\{0\}^\ell \times \mathbb{R}^s \times \mathbb{Z}^k),$$

  where $\mathcal{A}$ is an invertible $d \times d$ matrix with real entries, and the parameters $\ell, k, s \in \{0, 1, \dots, d\}$ satisfy that $\ell + k + s = d$. The subgroup H is discrete if and only if $s = 0$.

- The closed co-compact subgroups of $\mathbb{R}^d$ have the form

$$H = \mathcal{A}(\mathbb{R}^s \times \mathbb{Z}^{d-s}),$$

  where $\mathcal{A}$ is an invertible $d \times d$ matrix with real entries and the parameter $s \in \{0, 1, \dots, d\}$. The annihilator is $H^\perp = \mathcal{A}^\sharp(\{0\}^s \times \mathbb{Z}^{d-s})$.

- The lattices in $\mathbb{R}^d$ have the form $\Lambda = \mathcal{A}\mathbb{Z}^d$ for some invertible $d \times d$ matrix $\mathcal{A}$ with real entries. Recall that the annihilator is $\Lambda^\perp = \mathcal{A}^\sharp\mathbb{Z}^d$.

$\square$

Now, consider an LCA group G and a closed subgroup H. Assume that we have fixed Haar measures on the group G and on the closed subgroup H. As always, we normalize the Haar measure on the quotient group G/H such that Weil's formula (21.11) holds. Generalizing the density of a lattice in (21.5), the *size* of the subgroup H is defined by

$$s(H) = \int_{G/H} d\mu_{G/H}(\dot{x}). \qquad (21.52)$$

Note that $s(H)$ is finite if and only if G/H is compact.

Before we state the definition of the systems of functions to be analyzed in this section, let us recall a few facts about Gabor systems in $L^2(\mathbb{R})$. Given $g \in L^2(\mathbb{R})$ and two parameters $a, b > 0$, it follows from the commutator relation (2.25) that $\{E_{mb}T_{na}g\}_{m,n\in\mathbb{Z}}$ is a frame for $L^2(\mathbb{R})$ if and only if the shift-invariant system $\{T_{na}T_{mb}g\}_{m,n\in\mathbb{Z}}$ is a frame for $L^2(\mathbb{R})$. Similarly, the set of all time–frequency shifts of g, i.e., the set $\{E_aT_bg\}_{a,b\in\mathbb{R}}$, is a continuous frame for $L^2(\mathbb{R})$ if and only if $\{T_bE_ag\}_{a,b\in\mathbb{R}}$ is a continuous frame for $L^2(\mathbb{R})$.

Following Bownik and Ross, we will now define the translation-invariant systems in $L^2(G)$.

Definition 21.7.3 *Let G denote an LCA group and Λ a closed, co-compact subgroup of G. Let P be a countable or uncountable index set and let $g_p, p \in P$, denote a collection of functions in $L^2(G)$. Then the system of functions*

$$\{T_\lambda g_p\}_{\lambda\in\Lambda,p\in P} \qquad (21.53)$$

is called a translation-invariant system (TI system).

Definition 21.7.3 is broad enough to cover a large part of the systems considered in this book, e.g., the continuous Gabor systems, discrete Gabor systems, shift-invariant systems, and continuous wavelets. However, it does not cover the discrete wavelet systems:

Example 21.7.4 In (i) and (ii), we let G denote an LCA group with dual group $\widehat{G}$.

(i) For any given $g \in L^2(G)$, the Gabor system $\{E_\eta T_\lambda g\}_{\eta\in\widehat{G},\lambda\in G}$ is a continuous frame for $L^2(G)$ if and only if $\{T_\lambda E_\eta g\}_{\eta\in\widehat{G},\lambda\in G}$ is a continuous frame for $L^2(G)$; the later system is a TI system with $\Lambda = G$, $P = \widehat{G}$, and the functions g_p corresponding to $E_\eta g, \eta \in \widehat{G}$.

(ii) Let $\Gamma \subset \widehat{G}$ and $\Lambda \subset G$ be lattices and let $g \in L^2(G)$. Then the (discrete) Gabor system $\{E_\eta T_\lambda g\}_{\eta\in\Gamma,\lambda\in\Lambda}$ is a frame for $L^2(G)$ if and only if $\{T_\lambda E_\eta g\}_{\eta\in\Gamma,\lambda\in G}$ is a frame for $L^2(G)$; the later system is a TI system in $L^2(G)$ with $P = \Gamma$ and the functions g_p corresponding to $E_\eta g, \eta \in \Gamma$.

(iii) A continuous wavelet system in $L^2(\mathbb{R})$ has the form $\{T_bD_a\psi\}_{b\in\mathbb{R},a\in\mathbb{R}\setminus\{0\}}$ (see (15.1)): it is a TI system in $L^2(\mathbb{R})$ with $\Lambda=\mathbb{R}, P=\mathbb{R}\setminus\{0\}$, and the functions g_p corresponding to $D_a\psi$, $a\in\mathbb{R}\setminus\{0\}$.

(iv) A discrete dyadic wavelet system in $L^2(\mathbb{R})$ can be rewritten as

$$\{D^jT_k\psi\}_{j,k\in\mathbb{Z}}=\{T_{2^{-j}k}D^j\psi\}_{j,k\in\mathbb{Z}}=\bigcup_{j\in\mathbb{Z}}\{T_{2^{-j}k}D^j\psi\}_{k\in\mathbb{Z}}.$$

This is clearly not a TI system. □

In order also to cover the discrete wavelet systems, an extension of the setup in Definition 21.7.3 was proposed by Jakobsen and Lemvig [417]. The extension is similar to the step from the shift-invariant systems in Chapter 10 to the generalized shift-invariant systems in Chapter 20:

Definition 21.7.5 *Let G denote an LCA group and let $J\subseteq\mathbb{Z}$ be a countable index set. Furthermore, for each $j\in J$,*

(i) Let Λ_j denote a closed, co-compact subgroup of G.

(ii) Let P_j be a countable or uncountable index set; further, for $j\in J$, let $\{g_{j,p}\}_{p\in P_j}$ be a collection of functions in $L^2(G)$.

Then the family of functions

$$\bigcup_{j\in J}\{T_\lambda g_{j,p}\}_{\lambda\in\Lambda_j,p\in P_j}\tag{21.54}$$

is called a generalized translation-invariant system (GTI system for short).

Example 21.7.6 The dyadic wavelet system $\{D^jT_k\psi\}_{j,k\in\mathbb{Z}}$ in Example 21.7.4 is a GTI system with $J=\mathbb{Z}, \Lambda_j=2^{-j}\mathbb{Z}$ and P_j being a singleton for each $j\in\mathbb{Z}$. For each $j\in\mathbb{Z}$ there is only one function $g_{j,p}$, namely, $D^j\psi$. □

The analysis of GTI systems in full generality needs some weak technical conditions that are automatically satisfied for all the systems appearing in the current book. First, each index set P_j must be equipped with a σ-algebra of subsets and a corresponding measure satisfying a few very mild but technical conditions; see [417, 419]. In the cases of interest, P_j will either be a countable set equipped with the counting measure, or P_J will itself be an LCA group with a corresponding Haar measure; the technical conditions are always satisfied in these cases. Along the same line, we need that the map $p\mapsto g_{j,p}$ is continuous from P_j to $\mathbb{C}$ for all $j\in J$. This is clearly the case in Example 21.7.6, which corresponds to Example 21.7.4 (iv); it also holds in Example 21.7.4 (i)–(iii), due to the fact that the modulation operators and the scaling operators depend continuously on their parameters (see Lemma 2.9.2 for the $L^2(\mathbb{R})$-version of this statement). We will not go into these details.

We are now ready to consider frame properties for GTI systems $\bigcup_{j\in J}\{T_\lambda g_{j,p}\}_{\lambda\in\Lambda_j,p\in P_j}$. The following definition is really just a special case of Definition 5.6.1:

Definition 21.7.7 *Under the setup in Definition 21.7.5, assume that we have fixed the Haar measures on the groups G and Λ_j, $j \in J$. Then $\bigcup_{j\in J}\{T_\lambda g_{j,p}\}_{\lambda\in\Lambda_j,p\in P_j}$ is a GTI frame for $L^2(G)$ if there exist constants $A, B > 0$ such that*

$$A\,||f||^2 \leq \sum_{j\in J}\int_{\Lambda_j}\int_{P_j}|\langle f, T_\lambda g_{j,p}\rangle|^2 d\mu_{\Lambda_j}(\lambda)d\mu_{P_j}(p) \leq B\,||f||^2, \quad (21.55)$$

for all $f \in L^2(G)$. The GTI system is a Bessel family if at least the upper condition in (21.55) is satisfied.

Note that the formulation of Definition 21.7.7 allows us to choose the normalization of the Haar measure on the groups Λ_j freely. In contrast, the definition in [417] required the normalization to be chosen such that the measure on annihilator $\Lambda_j^\perp$ becomes the counting measure by Weil's formula. Due to the freedom in Definition 21.7.7, we can obtain the "standard form" for, e.g., the frame decomposition whenever we apply the definition to Gabor systems $\{E_{mb}T_{na}g\}_{m,n\in\mathbb{Z}}$ in $L^2(\mathbb{R})$. On the other hand, certain normalization factors appear in [417]. The freedom in Definition 21.7.7 agrees with the setup in [419].

We note that having two dual GTI frames $\bigcup_{j\in J}\{T_\lambda g_{j,p}\}_{\lambda\in\Lambda_j,p\in P_j}$ and $\bigcup_{j\in J}\{T_\lambda h_{j,p}\}_{\lambda\in\Lambda_j,p\in P_j}$, the frame decomposition (5.26) takes the form

$$f = \sum_{j\in J}\int_{\Lambda_j}\int_{P_j}\langle f, T_\lambda h_{j,p}\rangle T_\lambda g_{j,p}\,d\mu_{\Lambda_j}(\gamma)d\mu_{P_j}(p). \quad (21.56)$$

We need a version of the LIC for GTI systems.

Definition 21.7.8 *Two GTI systems $\bigcup_{j\in J}\{T_\lambda g_{j,p}\}_{\lambda\in\Lambda_j,p\in P_j}$ and $\bigcup_{j\in J}\{T_\lambda h_{j,p}\}_{\lambda\in\Lambda_j,p\in P_j}$ satisfy the dual α local integrability condition (dual α-LIC) if*

$$\sum_{j\in J}\frac{1}{s(\Lambda_j)}\int_{P_j}\sum_{\alpha\in\Lambda_j^\perp}\int_{\widehat{G}}|\widehat{f}(\gamma)\widehat{f}(\gamma+\alpha)\widehat{g_{j,p}}(\gamma)\widehat{h_{j,p}}(\gamma+\alpha)|\,d\mu_{\widehat{G}}(\gamma)\,d\mu_{P_j}(p) < \infty$$

$$(21.57)$$

for all $f \in \mathcal{D}$; see (21.47). In the case $g_{j,p} = h_{j,p}$, the condition (21.57) is called the α local integrability condition (α-LIC).

It is proved in [417] that:

- The dual α-LIC is automatically satisfied for any pair of TI systems;

- An analogue of the CC-condition for GTI systems implies that the α-LIC is satisfied.

The following result from [417] generalizes Proposition 21.6.6.

Proposition 21.7.9 *If a GTI system* $\bigcup_{j \in J} \{T_\lambda g_{j,p}\}_{\lambda \in \Lambda_j, p \in P_j}$ *is a Bessel family with bound* B, *then*

$$\sum_{j \in J} \frac{1}{s(\Lambda_j)} \int_{P_j} |\widehat{g_{j,p}}(\eta)|^2 d\mu_{P_j}(p) \le B \quad \text{for a.e. } \eta \in \widehat{G}. \quad (21.58)$$

We will now state one of the key results in [417], namely, a characterization of dual frames with the GTI structure.

Theorem 21.7.10 *Suppose that the GTI systems* $\bigcup_{j \in J} \{T_\lambda g_{j,p}\}_{\lambda \in \Lambda_j, p \in P_j}$ *and* $\bigcup_{j \in J} \{T_\lambda h_{j,p}\}_{\lambda \in \Lambda_j, p \in P_j}$ *are Bessel families satisfying the dual* α-LIC. *Then the following statements are equivalent:*

(i) $\bigcup_{j \in J} \{T_\lambda g_{j,p}\}_{\lambda \in \Lambda_j, p \in P_j}$ *and* $\bigcup_{j \in J} \{T_\lambda h_{j,p}\}_{\lambda \in \Lambda_j, p \in P_j}$ *are dual frames for* $L^2(G)$;

(ii) *For each* $\alpha \in \bigcup_{j \in J} \Lambda_j^\perp$, *the equation*

$$\sum_{\{j \in J \,|\, \alpha \in \Lambda_j^\perp\}} \frac{1}{s(\Lambda_j)} \int_{P_j} \overline{\widehat{g_{j,p}}(\eta)} \widehat{h_{j,p}}(\eta + \alpha) \, d\mu_{P_j}(p) = \delta_{\alpha,0} \quad (21.59)$$

holds for a.e. $\eta \in \widehat{G}$.

The generality in Theorem 21.7.10 is remarkable. As we have seen in Example 21.7.4 and Example 21.7.6, the GTI systems contain the Gabor systems and the wavelet systems in as well the discrete case as the continuous case: thus, the duality condition in Theorem 21.7.10 covers all these cases at once. It also contains several other systems of interest, e.g., the shearlet systems [469]. For the purpose of applications to continuous Gabor and wavelet systems, we note that the condition (21.59) simplifies significantly for TI systems with $\Lambda = G$: in fact, since $G^\perp = \{0\}$, there is only one equation in this case.

Lemma 21.7.11 *Suppose that* $\Lambda_j = G$ *for all* $j \in J$. *Then* (21.59) *reduces to*

$$\sum_{j \in J} \int_{P_j} \overline{\widehat{g_{j,p}}(\eta)} \widehat{h_{j,p}}(\eta) \, d\mu_{P_j}(p) = 1, \quad a.e. \ \eta \in \widehat{G}.$$

Let us show how to derive the admissibility condition for the continuous wavelet transform based on Theorem 21.7.10.

Example 21.7.12 As discussed in Example 21.7.4, a continuous wavelet system $\{T_b D_a \psi\}_{b \in \mathbb{R}, a \in \mathbb{R} \setminus \{0\}}$ in $L^2(\mathbb{R})$ is a TI system with $\Lambda = \mathbb{R}, P = \mathbb{R} \setminus \{0\}$, and the functions g_p corresponding to $D_a \psi$, $a \in \mathbb{R} \setminus \{0\}$. We will equip P with the measure $|a|^{-2} da$. Recall that the α-LIC is satisfied for TI systems, thus, Theorem 21.7.10 and Lemma 21.7.11 imply that $\{T_b D_a \psi\}_{b \in \mathbb{R}, a \in \mathbb{R} \setminus \{0\}}$ is a continuous Parseval frame for $L^2(\mathbb{R})$ (w. r. t. the set $\mathbb{R} \times (\mathbb{R} \setminus \{0\})$ equipped with the measure $\frac{1}{a^2} da\, db$) if and only if

$$\int_{\mathbb{R} \setminus \{0\}} \frac{1}{|a|} |\widehat{\psi}(a\gamma)|^2 \, da = 1, \ a.e.\, \gamma \in \mathbb{R}. \tag{21.60}$$

Via a change of variable the condition (21.60) is equivalent with the condition

$$\int_{\mathbb{R} \setminus \{0\}} \frac{1}{|a|} |\widehat{\psi}(a)|^2 \, da = 1;$$

thus, we have recovered the admissibility condition in Corollary 15.1.2. $\square$

The generality of Theorem 21.7.10 also allows to derive a characterization of Parseval wavelet systems $\{D^j T_k \psi\}_{j,k \in \mathbb{Z}}$, written on the form in Example 21.7.6 (Exercise 21.3). Applications to the Gabor case will be considered in Section 21.8.

21.8 Co-compact Gabor Systems

The value of the LCA approach and the unification in Section 21.7 is perhaps most evident by considering Gabor analysis. In the current book, we have considered continuous Gabor systems $\{E_b T_a g\}_{a,b \in \mathbb{R}}$ in $L^2(\mathbb{R})$, discrete Gabor systems $\{E_{mb} T_{na} g\}_{m,n \in \mathbb{Z}}$ in $L^2(\mathbb{R})$, as well as higher-dimensional versions and Gabor systems in $\ell^2(\mathbb{Z}), L^2(0, L)$, and $\mathbb{C}^L$; all of these cases are covered by the general approach in Section 21.7. In this section, we will derive a few of the concrete manifestations of the theory and leave the other cases as exercises.

Let G denote an LCA group, and let $g \in L^2(G)$. We will consider Gabor systems of the form $\{E_\eta T_\lambda g\}_{\eta \in \Gamma, \lambda \in \Lambda}$, where translation and modulation of $g \in L^2(G)$ are along closed co-compact subgroups $\Lambda \subset G$ and $\Gamma \subset \widehat{G}$, respectively. Such a system is called a *co-compact Gabor system*. The first systematic treatment of co-compact Gabor systems appeared in the paper [418] by Jakobsen and Lemvig.

In $L^2(\mathbb{R}^d)$, co-compact Gabor systems are of the form (see Example 21.7.2)

$$\left\{ e^{2\pi i \gamma \cdot x} g(x - \lambda) \right\}_{\lambda \in \mathcal{A}(\mathbb{R}^s \times \mathbb{Z}^{d-s}), \gamma \in \mathcal{B}(\mathbb{R}^r \times \mathbb{Z}^{d-r})} \tag{21.61}$$

for some choice of invertible $d \times d$ matrices $\mathcal{A}, \mathcal{B}$ with real entries and parameters $r, s \in \{0, 1, \ldots, d\}$. Depending on the parameters r and s, these Gabor systems range from discrete over semicontinuous to continuous families. Thus, the setup unifies discrete and continuous Gabor theory.

Theorem 21.7.10 has the following immediate consequence for Gabor systems.

Corollary 21.8.1 *Assume that the co-compact Gabor systems $\{E_\eta T_\lambda g\}_{\eta \in \Gamma, \lambda \in \Lambda}$ and $\{E_\eta T_\lambda h\}_{\eta \in \Gamma, \lambda \in \Lambda}$ are Bessel families in $L^2(G)$. Then the following statements are equivalent:*

(i) $\{E_\eta T_\lambda g\}_{\eta \in \Gamma, \lambda \in \Lambda}$ and $\{E_\eta T_\lambda h\}_{\eta \in \Gamma, \lambda \in \Lambda}$ are dual frames for $L^2(G)$;

(ii) For each $\alpha \in \Gamma^\perp$,

$$\frac{1}{s(\Gamma)} \int_\Lambda \overline{g(x+\lambda)} h(x + \lambda + \alpha) \, d\mu_\Lambda(\lambda) = \delta_{\alpha, 0}, \quad a.e. \ x \in G;$$

(iii) For each $\beta \in \Lambda^\perp$,

$$\frac{1}{s(\Lambda)} \int_\Gamma \overline{\widehat{g}(\eta + \gamma)} \widehat{h}(\eta + \gamma + \beta) \, d\mu_\Gamma(\gamma) = \delta_{\beta, 0}, \quad a.e. \ \eta \in \widehat{G}.$$

Corollary 21.8.1 covers the explicit characterizations we have obtained for Gabor systems in $L^2(\mathbb{R})$ (as well the discrete cases as the continuous case), $\ell^2(\mathbb{Z}), L^2(0, L)$, and $\mathbb{C}^L$. We leave the concrete calculations to the reader (Exercises 14.7, 21.4, 21.5). Here we will only consider the extreme case, where $\Lambda = G$ and $\Gamma = \widehat{G}$:

Example 21.8.2 Let $g, h \in L^2(G)$ and consider the co-compact Gabor systems $\{E_\eta T_\lambda g\}_{\eta \in \widehat{G}, \lambda \in G}$ and $\{E_\eta T_\lambda h\}_{\eta \in \widehat{G}, \lambda \in G}$. We equip G and $\widehat{G}$ with their respective Haar measures μ_G and $\mu_{\widehat{G}}$. For $f \in L^2(G)$, a standard calculation (see the proof of Proposition 11.1.2 for the case $G = \mathbb{R}$) yields that

$$\int_G \int_{\widehat{G}} |\langle f, E_\gamma T_\lambda g \rangle|^2 \, d\mu_{\widehat{G}}(\gamma) \, d\mu_G(\lambda) = \|f\|^2 \|g\|^2;$$

clearly, the result also holds with the window g replaced by the window h. We conclude that both Gabor systems are Bessel families. Since $\Gamma^\perp = \widehat{G}^\perp = \{0\}$, Corollary 21.8.1 implies that $\{E_\eta T_\lambda g\}_{\eta \in \widehat{G}, \lambda \in G}$ and $\{E_\eta T_\lambda h\}_{\eta \in \widehat{G}, \lambda \in G}$ are dual frames for $L^2(G)$ if and only if for a.e. $x \in G$

$$\int_G \overline{g(x - \lambda)} h(x - \lambda) \, d\mu_G(\lambda) = 1.$$

For the case $G = \mathbb{R}$, this is clearly equivalent with the condition $\langle g, h \rangle = 1$. Thus, we have recovered the result in Corollary 11.1.3 but now in the general setting of LCA groups. $\square$

In [418], Jakobsen and Lemvig extend two of the key results in classical Gabor analysis to the setting of co-compact Gabor systems. The *Wexler–Raz theorem* takes the following form:

Theorem 21.8.3 *Two co-compact Gabor systems* $\{E_\eta T_\lambda g\}_{\eta\in\Gamma,\lambda\in\Lambda}$ *and* $\{E_\eta T_\lambda h\}_{\eta\in\Gamma,\lambda\in\Lambda}$ *that form Bessel families are dual frames if and only if*

$$\langle h, E_\beta T_\alpha g\rangle = s(\Lambda)s(\Gamma)\delta_{\beta,0}\delta_{\alpha,0} \qquad \forall \alpha \in \Gamma^\perp, \beta \in \Lambda^\perp. \qquad (21.62)$$

Also the *duality principle* extends to co-compact Gabor systems:

Theorem 21.8.4 *A co-compact Gabor system* $\{E_\eta T_\lambda g\}_{\eta\in\Gamma,\lambda\in\Lambda}$ *is a frame for* $L^2(G)$ *with bounds* A *and* B *if and only if* $\{E_\beta T_\alpha g\}_{\alpha\in\Gamma^\perp,\beta\in\Lambda^\perp}$ *is a Riesz sequence with bounds* $s(\Lambda)s(\Gamma)A$ *and* $s(\Lambda)s(\Gamma)B$.

Note that while the co-compact Gabor system $\{E_\eta T_\lambda g\}_{\eta\in\Gamma,\lambda\in\Lambda}$ might be continuous or discrete depending on the choice of the subgroups Λ and Γ, the system $\{E_\beta T_\alpha g\}_{\alpha\in\Gamma^\perp,\beta\in\Lambda^\perp}$ is always discrete.

21.9   Exercises

21.1 Show that the operators in Definition 21.3.1 are unitary and that the commutator relations (21.19) and (21.18) hold.

21.2 Suppose that $\Gamma \subset \hat{G}$ and $\Lambda \subset G$ are closed subgroups. Let $g, h \in L^2(G)$ and assume that $\{E_\eta T_\lambda g\}_{\eta\in\Gamma,\lambda\in\Lambda}, \{E_\eta T_\lambda h\}_{\eta\in\Gamma,\lambda\in\Lambda}$ are Bessel families, with mixed frame operator

$$S_{g,h} : L^2(G) \to L^2(G), \quad S_{g,h}f = \int_\Gamma \int_\Lambda \langle f, E_\eta T_\lambda g\rangle E_\eta T_\lambda g\, d\lambda\, d\eta,$$

as usual understood in the weak sense. Prove the following:

(i) $S_{g,h}E_\eta T_\lambda = E_\eta T_\lambda S_{g,h}$ for all $\eta \in \Gamma$ and $\lambda \in \Lambda$,

(ii) If $\{E_\eta T_\lambda g\}_{\eta\in\Gamma,\lambda\in\Lambda}$ is a frame, then

$$S^{-1}E_\eta T_\lambda = E_\eta T_\lambda S^{-1}, \forall \eta \in \Gamma, \lambda \in \Lambda.$$

21.3 Derive a characterization of Parseval wavelet systems $\{D^j T_k \psi\}_{j,k\in\mathbb{Z}}$, based on Theorem 21.7.10.

21.4 Derive Theorem 12.3.4 based on Theorem 21.7.10.

21.5 Derive Theorem 14.2.1 based on Theorem 21.7.10.

22
Perturbation of Frames

The question of *stability* plays an important role in connection with bases. That is, if $\{f_k\}_{k=1}^{\infty}$ is a basis and $\{g_k\}_{k=1}^{\infty}$ is in some sense "close" to $\{f_k\}_{k=1}^{\infty}$, does it follow that $\{g_k\}_{k=1}^{\infty}$ is also a basis? A classical result states that if $\{f_k\}_{k=1}^{\infty}$ is a basis for a Banach space X, then a sequence $\{g_k\}_{k=1}^{\infty}$ in X is also a basis if there exists a constant $\lambda \in]0,1[$ such that

$$\left\| \sum_{k=1}^{\infty} c_k (f_k - g_k) \right\| \leq \lambda \left\| \sum_{k=1}^{\infty} c_k f_k \right\| \tag{22.1}$$

for all finite sequences of scalars $\{c_k\}_{k=1}^{\infty}$. The result is usually attributed to Paley and Wiener [533], but it can be traced back to Neumann [524]: in fact, it is an almost immediate consequence of Theorem 2.2.3 with $Uf_k := g_k$.

In this chapter we concentrate on frames, so the perturbation theory takes place in a Hilbert space $\mathcal{H}$. Note that if $\{f_k\}_{k=1}^{\infty}$ is a Riesz basis for $\mathcal{H}$ and (22.1) holds for all finite sequences, then (22.1) automatically holds for all $\{c_k\}_{k=1}^{\infty} \in \ell^2(\mathbb{N})$; thus, we can consider (22.1) as a condition on the operator

$$K : \ell^2(\mathbb{N}) \to \mathcal{H}, \quad K\{c_k\}_{k=1}^{\infty} = \sum_{k=1}^{\infty} c_k (f_k - g_k). \tag{22.2}$$

For this reason K is called the *perturbation operator*. The same philosophy applies to the results in this chapter: all theoretical results will be obtained by putting appropriate conditions on the operator K. We begin by stating the general results, and in later sections they are applied to Gabor frames and wavelet frames.

© Springer International Publishing Switzerland 2016
O. Christensen, *An Introduction to Frames and Riesz Bases*,
Applied and Numerical Harmonic Analysis,
DOI 10.1007/978-3-319-25613-9_22

22.1   A Paley–Wiener Theorem for Frames

In the entire section, we assume that $\{f_k\}_{k=1}^{\infty}$ is a frame for a Hilbert space $\mathcal{H}$. We wish to find conditions on a perturbed family $\{g_k\}_{k=1}^{\infty}$ which implies that it is a frame. Let us denote the synthesis operators for $\{f_k\}_{k=1}^{\infty}$ and $\{g_k\}_{k=1}^{\infty}$ by T and U, respectively, i.e.,

$$T, U : \ell^2(\mathbb{N}) \to \mathcal{H}, \; T\{c_k\}_{k=1}^{\infty} = \sum_{k=1}^{\infty} c_k f_k, \; U\{c_k\}_{k=1}^{\infty} = \sum_{k=1}^{\infty} c_k g_k.$$

Note that T is well-defined and bounded by assumption; the synthesis operator U is at least well-defined on finite sequences, but we have to *prove* that $\{g_k\}_{k=1}^{\infty}$ is a Bessel sequence before we know that U is well-defined on $\ell^2(\mathbb{N})$. See Theorem 3.2.3.

We first note that the condition (22.1) with $\lambda < 1$ is too restrictive if $\{f_k\}_{k=1}^{\infty}$ is an overcomplete frame. In fact, if (22.1) holds for all finite sequences $\{c_k\}_{k=1}^{\infty}$ and some $\lambda \in]0,1[$, then for all such sequences it holds that

$$\sum_{k=1}^{\infty} c_k f_k = 0 \Leftrightarrow \sum_{k=1}^{\infty} c_k g_k = 0;$$

thus, the condition can only handle perturbations $\{g_k\}_{k=1}^{\infty}$ that have the "same linear dependence" as $\{f_k\}_{k=1}^{\infty}$. A much more flexible result can be obtained by adding an extra term in the perturbation condition as in the following Theorem 22.1.1, first proved by Christensen in [154].

Theorem 22.1.1 *Let $\{f_k\}_{k=1}^{\infty}$ be a frame for $\mathcal{H}$ with bounds A, B. Let $\{g_k\}_{k=1}^{\infty}$ be a sequence in $\mathcal{H}$ and assume that there exist constants $\lambda, \mu \geq 0$ such that $\lambda + \frac{\mu}{\sqrt{A}} < 1$ and*

$$\left\| \sum_{k=1}^{\infty} c_k(f_k - g_k) \right\| \leq \lambda \left\| \sum_{k=1}^{\infty} c_k f_k \right\| + \mu \left(\sum_{k=1}^{\infty} |c_k|^2 \right)^{1/2} \quad (22.3)$$

for all finite scalar sequences $\{c_k\}_{k=1}^{\infty}$. Then $\{g_k\}_{k=1}^{\infty}$ is a frame for $\mathcal{H}$ with bounds

$$A\left(1 - \left(\lambda + \frac{\mu}{\sqrt{A}}\right)\right)^2, \quad B\left(1 + \lambda + \frac{\mu}{\sqrt{B}}\right)^2.$$

Moreover, if $\{f_k\}_{k=1}^{\infty}$ is a Riesz basis, then $\{g_k\}_{k=1}^{\infty}$ is a Riesz basis.

Proof. $\{f_k\}_{k=1}^{\infty}$ is assumed to be a frame, so by Theorem 3.2.3, the synthesis operator T is bounded and $\|T\| \leq \sqrt{B}$. The condition (22.3) implies that for all finite sequences $\{c_k\}_{k=1}^{\infty}$,

$$
\begin{aligned}
\left\| \sum_{k=1}^{\infty} c_k g_k \right\| &= \left\| -\sum_{k=1}^{\infty} c_k(f_k - g_k) + \sum_{k=1}^{\infty} c_k f_k \right\| \\
&\leq \left\| -\sum_{k=1}^{\infty} c_k(f_k - g_k) \right\| + \left\| \sum_{k=1}^{\infty} c_k f_k \right\| \\
&\leq (1+\lambda) \left\| \sum_{k=1}^{\infty} c_k f_k \right\| + \mu \left(\sum_{k=1}^{\infty} |c_k|^2 \right)^{1/2}.
\end{aligned}
$$

This calculation even holds for all $\{c_k\}_{k=1}^{\infty} \in \ell^2(\mathbb{N})$. To see this, we first have to prove that $\sum_{k=1}^{\infty} c_k g_k$ is convergent for any given $\{c_k\}_{k=1}^{\infty} \in \ell^2(\mathbb{N})$. Given $n, m \in \mathbb{N}$ with $n > m$,

$$
\begin{aligned}
\left\| \sum_{k=1}^{n} c_k g_k - \sum_{k=1}^{m} c_k g_k \right\| &= \left\| \sum_{k=m+1}^{n} c_k g_k \right\| \\
&\leq (1+\lambda) \left\| \sum_{k=m+1}^{n} c_k f_k \right\| + \mu \left(\sum_{k=m+1}^{n} |c_k|^2 \right)^{1/2}.
\end{aligned}
$$

Since $\{c_k\}_{k=1}^{\infty} \in \ell^2(\mathbb{N})$ and $\sum_{k=1}^{\infty} c_k f_k$ is convergent, this implies that $\{\sum_{k=1}^{n} c_k g_k\}_{n=1}^{\infty}$ is a Cauchy sequence in $\mathcal{H}$ and therefore convergent. Thus, the analysis operator U is well defined on $\ell^2(\mathbb{N})$; it follows that for all $\{c_k\}_{k=1}^{\infty} \in \ell^2(\mathbb{N})$,

$$
\left\| \sum_{k=1}^{\infty} c_k g_k \right\| \leq (1+\lambda) \left\| \sum_{k=1}^{\infty} c_k f_k \right\| + \mu \left(\sum_{k=1}^{\infty} |c_k|^2 \right)^{1/2}. \tag{22.4}
$$

In terms of the operators T, U, (22.4) states that

$$
\begin{aligned}
\|U\{c_k\}_{k=1}^{\infty}\| &\leq (1+\lambda) \, \|T\{c_k\}_{k=1}^{\infty}\| + \mu \left(\sum_{k=1}^{\infty} |c_k|^2 \right)^{1/2} \\
&\leq \left((1+\lambda)\sqrt{B} + \mu \right) \left(\sum_{k=1}^{\infty} |c_k|^2 \right)^{1/2}, \quad \forall \{c_k\}_{k=1}^{\infty} \in \ell^2(\mathbb{N}).
\end{aligned}
$$

Via Theorem 3.2.3 this estimate shows that $\{g_k\}_{k=1}^{\infty}$ is a Bessel sequence with bound

$$
\left((1+\lambda)\sqrt{B} + \mu \right)^2 = B \left(1 + \lambda + \frac{\mu}{\sqrt{B}} \right)^2.
$$

Now we prove that $\{g_k\}_{k=1}^{\infty}$ has a lower frame bound. Since $\{f_k\}_{k=1}^{\infty}$ is a frame, the frame operator $S = TT^*$ is invertible by Lemma 5.1.5, and we can define an operator

$$
T^{\dagger} : \mathcal{H} \to \ell^2(\mathbb{N}), \quad T^{\dagger} f := T^*(TT^*)^{-1} f = \{ \langle f, (TT^*)^{-1} f_k \rangle \}_{k=1}^{\infty}. \tag{22.5}
$$

Note that $\{(TT^*)^{-1}f_k\}_{k=1}^\infty$ is the dual frame of $\{f_k\}_{k=1}^\infty$, so by Lemma 5.1.5,

$$\|T^\dagger f\|^2 = \sum_{k=1}^\infty |\langle f, (TT^*)^{-1}f_k\rangle|^2$$

$$\leq \frac{1}{A}\|f\|^2, \ \forall f \in \mathcal{H}.$$

Since $\sum_{k=1}^\infty c_k f_k$ and $\sum_{k=1}^\infty c_k g_k$ are convergent for all $\{c_k\}_{k=1}^\infty \in \ell^2(\mathbb{N})$ and the synthesis operators T and U are bounded, the inequality (22.3) holds for all $\{c_k\}_{k=1}^\infty \in \ell^2(\mathbb{N})$. In terms of the operators T and U,

$$\|T\{c_k\}_{k=1}^\infty - U\{c_k\}_{k=1}^\infty\| \leq \lambda \, \|T\{c_k\}_{k=1}^\infty\| + \mu \left(\sum_{k=1}^\infty |c_k|^2\right)^{1/2}, \quad (22.6)$$

for all $\{c_k\}_{k=1}^\infty \in \ell^2(\mathbb{N})$. Note that for $f \in \mathcal{H}$,

$$TT^\dagger f = TT^*(TT^*)^{-1}f = f,$$

$$UT^\dagger f = \sum_{k=1}^\infty (T^\dagger f)_k g_k = \sum_{k=1}^\infty \langle f, (TT^*)^{-1}f_k\rangle g_k.$$

Using (22.6) on the sequence $\{c_k\}_{k=1}^\infty = T^\dagger f$ yields

$$\|f - UT^\dagger f\| \leq \lambda \, \|f\| + \mu \, \|T^\dagger f\|$$

$$\leq \left(\lambda + \frac{\mu}{\sqrt{A}}\right) \|f\|, \ \forall f \in \mathcal{H}.$$

Since we have assumed that $\lambda + \frac{\mu}{\sqrt{A}} < 1$, this implies that the operator $UT^\dagger$ is invertible, and (Exercise 22.1)

$$\|UT^\dagger\| \leq 1 + \lambda + \frac{\mu}{\sqrt{A}}, \quad \|(UT^\dagger)^{-1}\| \leq \frac{1}{1 - \left(\lambda + \frac{\mu}{\sqrt{A}}\right)}. \quad (22.7)$$

Now, $f \in \mathcal{H}$ can be written as

$$f = UT^\dagger(UT^\dagger)^{-1}f = \sum_{k=1}^\infty \langle (UT^\dagger)^{-1}f, (TT^*)^{-1}f_k\rangle g_k.$$

Inserting this in the first entry of $\langle f, f \rangle$ leads to

$$
\begin{aligned}
||f||^4 &= |\langle f, f \rangle|^2 \\
&= \left| \sum_{k=1}^{\infty} \langle (UT^\dagger)^{-1}f, (TT^*)^{-1}f_k \rangle \langle g_k, f \rangle \right|^2 \\
&\leq \sum_{k=1}^{\infty} |\langle (UT^\dagger)^{-1}f, (TT^*)^{-1}f_k \rangle|^2 \sum_{k=1}^{\infty} |\langle g_k, f \rangle|^2 \\
&\leq \frac{1}{A} ||(UT^\dagger)^{-1}f||^2 \sum_{k=1}^{\infty} |\langle g_k, f \rangle|^2 \\
&\leq \frac{1}{A} \left(\frac{1}{1 - \left(\lambda + \frac{\mu}{\sqrt{A}}\right)} \right)^2 ||f||^2 \sum_{k=1}^{\infty} |\langle g_k, f \rangle|^2, \ \forall f \in \mathcal{H}.
\end{aligned}
$$

So

$$
\sum_{k=1}^{\infty} |\langle g_k, f \rangle|^2 \geq A \left(1 - \left(\lambda + \frac{\mu}{\sqrt{A}} \right) \right)^2 ||f||^2,
$$

i.e., $\{g_k\}_{k=1}^{\infty}$ is a frame for $\mathcal{H}$.

For the rest of the proof we now assume that $\{f_k\}_{k=1}^{\infty}$ is a Riesz basis. To prove that $\{g_k\}_{k=1}^{\infty}$ is a Riesz basis, we use Theorem 7.1.1 and assume that $\sum_{k=1}^{\infty} c_k g_k = 0$ for some coefficients $\{c_k\}_{k=1}^{\infty} \in \ell^2(\mathbb{N})$. By Theorem 5.4.1, the lower frame bound for $\{f_k\}_{k=1}^{\infty}$ is also a lower Riesz basis bound, so (22.6) implies that

$$
\begin{aligned}
\left\| \sum_{k=1}^{\infty} c_k f_k \right\| &\leq \lambda \left\| \sum_{k=1}^{\infty} c_k f_k \right\| + \mu \left(\sum_{k=1}^{\infty} |c_k|^2 \right)^{1/2} \\
&\leq \left(\lambda + \frac{\mu}{\sqrt{A}} \right) \left\| \sum_{k=1}^{\infty} c_k f_k \right\|.
\end{aligned}
$$

Since $\lambda + \frac{\mu}{\sqrt{A}} < 1$, it follows that $\sum_{k=1}^{\infty} c_k f_k = 0$. Using Theorem 7.1.1 on the Riesz basis $\{f_k\}_{k=1}^{\infty}$, we conclude that $c_k = 0$ for all $k \in \mathbb{N}$; therefore $\{g_k\}_{k=1}^{\infty}$ is a Riesz basis. $\qquad \square$

We already argued for the role of the μ-term in the condition (22.3). Most applications of Theorem 22.1.1 actually take place with $\lambda = 0$, so a natural question is whether the appearance of the λ-term improves the result. In fact, it does: in Exercise 22.7, we consider an example where the λ-term guarantees the frame property for a larger class of sequences than the corresponding result without the λ-term.

We now illustrate Theorem 22.1.1 by an example in a general Hilbert space. In particular, the example shows that the conclusion in

Theorem 22.1.1 might fail if the condition $\lambda + \frac{\mu}{\sqrt{A}} < 1$ is replaced by $\lambda + \frac{\mu}{\sqrt{A}} = 1$. In that sense, Theorem 22.1.1 is the best possible perturbation result.

Example 22.1.2 Let $\{e_k\}_{k=1}^{\infty}$ be an orthonormal basis for $\mathcal{H}$. Given a sequence $\{a_k\}_{k=1}^{\infty}$ of complex numbers, we consider the family of vectors $\{g_k\}_{k=1}^{\infty}$ defined by

$$g_k = e_k + a_k e_{k+1}, \ k \in \mathbb{N}.$$

Then, for all finite scalar sequences $\{c_k\}_{k=1}^{\infty}$,

$$\left\|\sum_{k=1}^{\infty} c_k (g_k - e_k)\right\| = \left\|\sum_{k=1}^{\infty} c_k a_k e_{k+1}\right\| = \left(\sum_{k=1}^{\infty} |c_k a_k|^2\right)^{1/2}$$

$$\leq \sup_k |a_k| \left(\sum_{k=1}^{\infty} |c_k|^2\right)^{1/2}.$$

Thus, if $a := \sup_k |a_k| < 1$, Theorem 22.1.1 shows that $\{g_k\}_{k=1}^{\infty}$ is a frame (in fact, a Riesz basis) with bounds $(1-a)^2, (1+a)^2$.

By taking $a_k = 1$ for all $k \in \mathbb{N}$, we obtain the family

$$g_k = e_k + e_{k+1} \ k \in \mathbb{N},$$

which was considered in Example 5.4.6. In particular, we know that $\{g_k\}_{k=1}^{\infty}$ is not a frame. For any sequence $\{c_k\}_{k=1}^{\infty} \in \ell^2(\mathbb{N})$,

$$\left\|\sum_{k=1}^{\infty} c_k (g_k - e_k)\right\| = \left\|\sum_{k=1}^{\infty} c_k e_{k+1}\right\| = \left\|\sum_{k=1}^{\infty} c_k e_k\right\| = \left(\sum_{k=1}^{\infty} |c_k|^2\right)^{1/2}.$$

Thus, the condition (22.3) is satisfied with $(\lambda, \mu) = (1, 0)$, or $(\lambda, \mu) = (0, 1)$; in either case, it shows that the condition $\lambda + \frac{\mu}{\sqrt{A}} < 1$ is necessary for Theorem 22.1.1 to hold in this particular case.   □

The operator $T^{\dagger}$ defined in (22.5) is the pseudo-inverse of T; see Theorem 5.4.3. By stressing this point it is possible to prove a more general result than Theorem 22.1.1, where the condition (22.3) is replaced by a more "symmetric" version which also involves $\sum c_k g_k$ on the right-hand side; the exact condition is

$$\left\|\sum_{k=1}^{\infty} c_k (f_k - g_k)\right\| \leq \lambda \left\|\sum_{k=1}^{\infty} c_k f_k\right\| + \gamma \left\|\sum_{k=1}^{\infty} c_k g_k\right\| + \mu \left(\sum_{k=1}^{\infty} |c_k|^2\right)^{1/2},$$

where $\lambda + \frac{\mu}{\sqrt{A}} < 1$ and $\gamma \in [0, 1[$. The conclusion is again that $\{g_k\}_{k=1}^{\infty}$ is a frame, but now the bounds also involve the parameter γ. One can actually construct examples where this condition is satisfied, but where the condition (22.3) is not satisfied. This extension of Theorem 22.1.1 is remarkable in light of Example 22.1.2, which showed that one cannot extend the range of the parameters λ, μ in the condition (22.3). We refer to [119], where Casazza and Christensen derive the extension as a consequence of the following interesting generalization of Neumann's Theorem, due to Hilding [404]:

Lemma 22.1.3 *Let $U : \mathcal{H} \to \mathcal{H}$ be a bounded operator, and assume that there exist constants $\lambda, \mu \in [0, 1[$ for which*

$$\|Ux - x\| \leq \lambda \|Ux\| + \mu \|x\|, \ \forall x \in \mathcal{H}.$$

Then U is invertible.

The fact that a perturbation (in the sense of Theorem 22.1.1) of a Riesz basis is again a Riesz basis makes it plausible that if $\{f_k\}_{k=1}^{\infty}$ is a near-Riesz basis, then a family $\{g_k\}_{k=1}^{\infty}$ satisfying (22.3) is a near-Riesz basis having the same excess. A proof of this fact can be found in [121]. Based on this result, one could easily believe that a perturbation of *any* frame containing a Riesz basis would again contain a Riesz basis, but this turns out to be wrong. Since this is a surprising result and forces us to deal with perturbations with great care, we present an example from [121].

Example 22.1.4 Let $\{e_k\}_{k=1}^{\infty}$ be an orthonormal basis for a Hilbert space $\mathcal{K}$, and consider the Hilbert space $\mathcal{H}$ constructed in the proof of Theorem 7.5.2, together with the frame $\{f_k^n\}_{k=1,n=1}^{n+1,\infty}$. Recall that

$$\begin{cases} f_k^n = e_{\frac{(n-1)n}{2}+k} - \dfrac{1}{n}\sum_{j=1}^{n} e_{\frac{(n-1)n}{2}+j}, \ 1 \leq k \leq n; \\[4mm] f_{n+1}^n = \dfrac{1}{\sqrt{n}}\sum_{j=1}^{n} e_{\frac{(n-1)n}{2}+j}. \end{cases}$$

Given $\epsilon > 0$, define the sequence $\{g_k^n\}_{k=1,n=1}^{n+1,\infty}$ by

$$\begin{cases} g_k^n = e_{\frac{(n-1)n}{2}+k} - \dfrac{1-\epsilon}{n}\sum_{j=1}^{n} e_{\frac{(n-1)n}{2}+j}, \ 1 \leq k \leq n \\[4mm] g_{n+1}^n = \dfrac{1}{\sqrt{n}}\sum_{j=1}^{n} e_{\frac{(n-1)n}{2}+j}. \end{cases}$$

Now, given a finite scalar sequence $\{c_k^n\}_{k=1,n=1}^{n+1,\infty}$, we have

$$\left\|\sum_{n=1}^{\infty}\sum_{k=1}^{n+1} c_k^n(f_k^n - g_k^n)\right\| = \epsilon \left\|\sum_{n=1}^{\infty}\left(\sum_{k=1}^{n} c_k^n\right)\frac{1}{n}\sum_{j=1}^{n} e_{\frac{(n-1)n}{2}+j}\right\|$$

$$\leq \epsilon\sqrt{\sum_{n=1}^{\infty}\left|\sum_{k=1}^{n} c_k^n \frac{1}{\sqrt{n}}\right|^2} \leq \epsilon\sqrt{\sum_{n=1}^{\infty}\sum_{k=1}^{n}|c_k^n|^2}. \tag{22.8}$$

By the proof of Theorem 7.5.2 we know that $\{f_k^n\}_{k=1,n=1}^{n+1,\infty}$ is a tight frame for $\mathcal{H}$ with frame bound 1. If we choose $\epsilon < 1$, then the perturbation condition in Theorem 22.1.1 is satisfied with $\lambda = 0, \mu = \epsilon$, implying that $\{g_k^n\}_{k=1,n=1}^{n+1,\infty}$ is a frame for $\mathcal{H}$ with bounds $(1-\epsilon)^2, (1+\epsilon)^2$. It contains the subfamily $\{g_k^n\}_{k=1,n=1}^{n,\infty}$, which is a Riesz basis. To see this, note that (Exercise 22.2)

$$\overline{\text{span}}\{g_k^n\}_{k=1,n=1}^{n,\infty} = \mathcal{H}. \tag{22.9}$$

Furthermore, consider an arbitrary finite sequence $\{c_k^n\}$, and observe that via the opposite triangle inequality and the calculation leading to (22.8),

$$\left\| \sum_{n=1}^{\infty} \sum_{k=1}^{n} c_k^n g_k^n \right\| \geq \left\| \sum_{n=1}^{\infty} \sum_{k=1}^{n} c_k^n e_{\frac{(n-1)n}{2}+k} \right\|$$

$$-(1-\epsilon) \left\| \sum_{n=1}^{\infty} \left(\sum_{k=1}^{n} c_k^n \right) \frac{1}{n} \sum_{j=1}^{n} e_{\frac{(n-1)n}{2}+j} \right\|$$

$$\geq \sqrt{\sum_{n=1}^{\infty} \sum_{k=1}^{n} |c_k^n|^2} - (1-\epsilon) \sqrt{\sum_{n=1}^{\infty} \sum_{k=1}^{n} |c_k^n|^2}$$

$$\geq \epsilon \sqrt{\sum_{n=1}^{\infty} \sum_{k=1}^{n} |c_k^n|^2}.$$

Thus $\{g_k^n\}_{k=1,n=1}^{n,\infty}$ is a Riesz basis by Theorem 3.6.6. So actually we have an example where $\{f_k^n\}_{k=1,n=1}^{n+1,\infty}$ does not contain a Riesz basis, but the perturbed family does. The opposite situation is also possible. In fact, since $\{g_k^n\}_{k=1,n=1}^{n+1,\infty}$ has the lower frame bound $(1-\epsilon)^2$, we can by (22.8) consider $\{f_k^n\}_{k=1,n=1}^{n+1,\infty}$ as a perturbation of $\{g_k^n\}_{k=1,n=1}^{n+1,\infty}$ if $\frac{\epsilon}{1-\epsilon} < 1$, i.e., if $\epsilon < \frac{1}{2}$. So we get our example by choosing $\epsilon < 1/2$ and switching the roles of $\{f_k^n\}_{k=1,n=1}^{n+1,\infty}$ and $\{g_k^n\}_{k=1,n=1}^{n+1,\infty}$. $\qquad\square$

An important special case of Theorem 22.1.1 is given by

Corollary 22.1.5 *Let* $\{f_k\}_{k=1}^{\infty}$ *be a frame for* $\mathcal{H}$ *with bounds* A, B, *and let* $\{g_k\}_{k=1}^{\infty}$ *be a sequence in* $\mathcal{H}$. *If there exists a constant* $R < A$ *such that*

$$\sum_{k=1}^{\infty} |\langle f, f_k - g_k \rangle|^2 \leq R \, ||f||^2, \ \forall f \in \mathcal{H}, \tag{22.10}$$

then $\{g_k\}_{k=1}^{\infty}$ *is a frame for* $\mathcal{H}$ *with bounds*

$$A\left(1 - \sqrt{\frac{R}{A}}\right)^2, \ B\left(1 + \sqrt{\frac{R}{B}}\right)^2. \tag{22.11}$$

If $\{f_k\}_{k=1}^{\infty}$ *is a Riesz basis, then* $\{g_k\}_{k=1}^{\infty}$ *is a Riesz basis.*

Proof. The condition (22.10) corresponds to the condition in Theorem 22.1.1 with $\lambda = 0$, $\mu = \sqrt{R}$, just formulated in terms of the adjoint of the synthesis operator (i.e., the analysis operator) instead of the synthesis operator itself. However, an easier way to prove the frame part is to apply the triangle inequality in $\ell^2(\mathbb{N})$ to the sequence

$$\{\langle f, g_k\rangle\}_{k=1}^{\infty} = \{\langle f, f_k\rangle\}_{k=1}^{\infty} - \{\langle f, f_k - g_k\rangle\}_{k=1}^{\infty}. \qquad \square$$

Corollary 22.1.5 implies that if $\{f_k\}_{k=1}^{\infty}$ is a frame for $\mathcal{H}$ with lower frame bound A and $\{g_k\}_{k=1}^{\infty}$ is a sequence such that

$$\sum_{k=1}^{\infty} ||f_k - g_k||^2 < A,$$

then $\{g_k\}_{k=1}^{\infty}$ is a frame for $\mathcal{H}$ (Exercise 22.4). A related result was recently obtained by Chen, Li, and Zheng [150]:

Proposition 22.1.6 *Let $\{f_k\}_{k=1}^{\infty}$ be a frame for $\mathcal{H}$ with bounds A, B, and let $\{h_k\}_{k=1}^{\infty}$ denote a dual frame with Bessel bound D. Consider any sequence $\{g_k\}_{k=1}^{\infty}$ in $\mathcal{H}$ such that*

$$\lambda := \sum_{k=1}^{\infty} ||f_k - g_k||^2 < \infty$$

and

$$\mu := \sum_{k=1}^{\infty} ||f_k - g_k||\, ||h_k|| < 1.$$

Then $\{h_k\}_{k=1}^{\infty}$ is a frame for $\mathcal{H}$ with bounds $D^{-1}(1-\mu)^2, B(1 + \sqrt{\frac{\lambda}{B}})^2$.

22.2 Compact Perturbation

Another type of condition on the perturbation operator appeared in the paper [179] by Christensen and Heil:

Theorem 22.2.1 *Let $\{f_k\}_{k=1}^{\infty}$ be a frame for $\mathcal{H}$, and let $\{g_k\}_{k=1}^{\infty}$ be a sequence in $\mathcal{H}$. If*

$$K : \ell^2(\mathbb{N}) \to \mathcal{H}, \ K\{c_k\}_{k=1}^{\infty} := \sum_{k=1}^{\infty} c_k(f_k - g_k)$$

is a well-defined compact operator, then $\{g_k\}_{k=1}^{\infty}$ is a frame sequence.

Proof. Since $\{f_k\}_{k=1}^{\infty}$ is a frame and the perturbation operator K is bounded, the synthesis operator U for $\{g_k\}_{k=1}^{\infty}$ is well-defined and bounded, and

$$\|U\| = \|T - K\| \le \|T\| + \|K\|.$$

By Theorem 3.2.3 this implies that $\{g_k\}_{k=1}^{\infty}$ is a Bessel sequence. The frame operator for $\{g_k\}_{k=1}^{\infty}$ is given by

$$UU^* = (T - K)(T - K)^* = S - TK^* - KT^* + KK^*,$$

where $S = TT^*$ is the frame operator for $\{f_k\}_{k=1}^{\infty}$. Since S is invertible, we can write

$$UU^* = S\left(I + S^{-1}(-TK^* - KT^* + KK^*)\right). \tag{22.12}$$

Using Lemma 2.4.2, we see that $S^{-1}(TK^* - KT^* + KK^*)$ is a compact operator and that $I + S^{-1}(TK^* - KT^* + KK^*)$ has closed range. By (22.12) also UU^* has closed range. Since $\mathcal{R}_U = \mathcal{R}_{UU^*}$ (Exercise 22.3), we conclude by Corollary 5.5.2 that $\{g_k\}_{k=1}^{\infty}$ is a frame sequence. □

Note that Theorem 22.2.1 only states that $\{g_k\}_{k=1}^{\infty}$ is a frame sequence, i.e., it might not span the entire Hilbert space. An example where K is compact and $\{g_k\}_{k=1}^{\infty}$ only spans a subspace is obtained by letting $\{f_k\}_{k=1}^{\infty}$ be an orthonormal basis for $\mathcal{H}$ and taking

$$\{g_k\}_{k=1}^{\infty} := \{0, f_2, f_3, f_4, \dots\}.$$

Perturbation via a compact operator as in Theorem 22.2.1 preserves the excess: if $\{f_k\}_{k=1}^{\infty}$ contains a Riesz basis, then a *total* family $\{g_k\}_{k=1}^{\infty}$ satisfying the compactness condition also contains a Riesz basis, and the two frames have the same excess (finite or not). This is proved by Casazza and Christensen in [121].

An extreme case of "perturbing" an element f_ℓ in a frame $\{f_k\}_{k=1}^{\infty}$ is to replace f_ℓ by zero. We have already in Theorem 5.4.7 seen that either $\{f_k\}_{k \ne \ell}$ is still a frame for $\mathcal{H}$ or $\{f_k\}_{k \ne \ell}$ is no longer complete. As a consequence of Theorem 22.2.1 we now prove that in the latter case, we still have a frame for the closed span of the remaining elements:

Corollary 22.2.2 *Let $\{f_k\}_{k=1}^{\infty}$ be a frame for $\mathcal{H}$ and $\{g_k\}_{k=1}^{\infty}$ a sequence in $\mathcal{H}$. If $g_k = f_k$ except for a finite set of $k \in \mathbb{N}$, then $\{g_k\}_{k=1}^{\infty}$ is a frame sequence.*

Proof. Suppose that $g_k = f_k$ except for $k \in I$, where I is a finite subset of $\mathbb{N}$. Then the operator

$$K\{c_k\}_{k=1}^{\infty} = \sum_{k=1}^{\infty} c_k(f_k - g_k) = \sum_{k \in I} c_k(f_k - g_k)$$

has a finite-dimensional range and is thus compact. We conclude by Theorem 22.2.1 that $\{g_k\}_{k=1}^\infty$ is a frame sequence. $\qquad\square$

Corollary 22.2.2 connects to the theme of erasure frame elements, discussed in a more applied context in Section 1.9 and Section 1.10. It can also be applied the other way around to conclude that certain sequences do not have "simple" extensions to frames:

Example 22.2.3 Let $\{e_k\}_{k=1}^\infty$ denote an orthonormal basis for $\mathcal{H}$, and consider again the family $\{e_k + e_{k+1}\}_{k=1}^\infty$ from Example 5.4.6. The family $\{e_k + e_{k+1}\}_{k=1}^\infty$ is not a frame for $\overline{\mathrm{span}}\{e_k + e_{k+1}\}_{k=1}^\infty = \mathcal{H}$, and Corollary 22.2.2 shows that $\{e_k + e_{k+1}\}_{k=1}^\infty$ cannot be extended to a frame for $\mathcal{H}$ by adding a finite number of elements. $\qquad\square$

22.3 Perturbation of Frame Sequences

In Theorem 22.1.1 we assumed that $\{f_k\}_{k=1}^\infty$ was a frame for the entire Hilbert space $\mathcal{H}$, and this is actually an essential assumption. If $\{f_k\}_{k=1}^\infty$ only spans a subspace of $\mathcal{H}$, a perturbation $\{g_k\}_{k=1}^\infty$ might not belong to this subspace, and we can not conclude anything based on the inequality (22.3):

Example 22.3.1 Let $\{e_k\}_{k=1}^\infty$ be an orthonormal basis for $\mathcal{H}$ and define the sequence

$$\{f_k\}_{k=1}^\infty = \{e_1, e_2, 0, 0, \ldots, 0, \ldots\}.$$

Then $\{f_k\}_{k=1}^\infty$ is a frame sequence with bounds $A = B = 1$. Now let $\epsilon > 0$ be given, and consider the sequence

$$\{g_k\}_{k=1}^\infty = \left\{e_1, e_2, \frac{\epsilon}{3}e_3, \frac{\epsilon}{4}e_4, \frac{\epsilon}{5}e_5, \ldots, \frac{\epsilon}{k}e_k, \cdots\right\}.$$

For any sequence $\{c_k\}_{k=1}^\infty \in \ell^2(\mathbb{N})$,

$$\left\|\sum_{k=1}^\infty c_k(f_k - g_k)\right\| = \left\|\sum_{k=3}^\infty c_k \frac{\epsilon}{k} e_k\right\| \le \frac{\epsilon}{3}\left(\sum_{k=1}^\infty |c_k|^2\right)^{1/2}.$$

Thus, we can satisfy (22.3) with $\lambda = 0$ and an arbitrarily small value of μ. However, $\{g_k\}_{k=1}^\infty$ is not a frame sequence for any $\epsilon > 0$. $\qquad\square$

If $\{f_k\}_{k=1}^\infty$ is a Riesz sequence the situation in Example 22.3.1 does not occur, and the perturbation condition in (22.3) is enough to guarantee that the perturbed sequence $\{g_k\}_{k=1}^\infty$ is also a Riesz sequence:

Theorem 22.3.2 *Let $\{f_k\}_{k=1}^\infty$ be a Riesz sequence in a Hilbert space $\mathcal{H}$, with bounds A, B. Let $\{g_k\}_{k=1}^\infty$ be a sequence in $\mathcal{H}$ and assume that there exist constants $\lambda, \mu \geq 0$ such that $\lambda + \frac{\mu}{\sqrt{A}} < 1$ and*

$$\left\| \sum_{k=1}^\infty c_k(f_k - g_k) \right\| \leq \lambda \left\| \sum_{k=1}^\infty c_k f_k \right\| + \mu \left(\sum_{k=1}^\infty |c_k|^2 \right)^{1/2} \tag{22.13}$$

for all finite scalar sequences $\{c_k\}_{k=1}^\infty$. Then $\{g_k\}_{k=1}^\infty$ is a Riesz sequence with bounds

$$A\left(1 - \left(\lambda + \frac{\mu}{\sqrt{A}}\right)\right)^2, \quad B\left(1 + \lambda + \frac{\mu}{\sqrt{B}}\right)^2.$$

Proof. We ask the reader to prove that $\{g_k\}_{k=1}^\infty$ is a Bessel sequence (check the proof of Theorem 22.1.1). Now let $\{c_k\}_{k=1}^\infty$ be an arbitrary finite scalar sequence. Then the opposite triangle inequality together with the assumption (22.13) implies that

$$\begin{aligned}
\left\| \sum_{k=1}^\infty c_k g_k \right\| &\geq \left\| \sum_{k=1}^\infty c_k f_k \right\| - \left\| \sum_{k=1}^\infty c_k(f_k - g_k) \right\| \\
&\geq (1-\lambda) \left\| \sum_{k=1}^\infty c_k f_k \right\| - \mu \left(\sum_{k=1}^\infty |c_k|^2 \right)^{1/2} \\
&\geq \left((1-\lambda)\sqrt{A} - \mu\right) \left(\sum |c_k|^2 \right)^{1/2} \\
&= \sqrt{A}\left(1 - \left(\lambda + \frac{\mu}{\sqrt{A}}\right)\right) \left(\sum |c_k|^2 \right)^{1/2}.
\end{aligned}$$

$\square$

By involving the *gap* (a notion introduced by Kato [447]) between certain subspaces of $\mathcal{H}$ one can obtain versions of Theorem 22.1.1 which apply to frame sequences. Given two arbitrary non-empty subspaces V, W of $\mathcal{H}$, the gap from V to W is defined by

$$\delta(V, W) = \sup_{x \in V, ||x||=1} \mathrm{dist}(x, W) = \sup_{x \in V, ||x||=1} \inf_{y \in W} ||x - y||.$$

Let $\{f_k\}_{k=1}^\infty$ be a frame sequence. As before, let T and U denote the synthesis operators corresponding to $\{f_k\}_{k=1}^\infty$ and $\{g_k\}_{k=1}^\infty$; furthermore, denote their kernels by $\mathcal{N}_T$ and $\mathcal{N}_U$, respectively. Involving the gap between the kernels of T and U, it turns out that (22.13) is sufficient for $\{g_k\}_{k=1}^\infty$ being a frame sequence if

$$\delta(\mathcal{N}_T, \mathcal{N}_U) < 1 \quad \text{and} \quad \lambda + \frac{\mu}{\sqrt{A}(1 - \delta(\mathcal{N}_T, \mathcal{N}_U)^2)^{1/2}} < 1.$$

We refer to [159] for the proof. Another sufficient condition for $\{g_k\}_{k=1}^\infty$ being a frame sequence, now in terms of the gap between $\overline{\mathrm{span}}\{g_k\}_{k=1}^\infty$ and

$\overline{\text{span}}\{f_k\}_{k=1}^\infty = \mathcal{R}_T$, is given by

$$\lambda + \frac{\mu}{\sqrt{A}} < \sqrt{1 - \delta(\overline{\text{span}}\,\{g_k\}_{k=1}^\infty, \overline{\text{span}}\{f_k\}_{k=1}^\infty)^2}.$$

This version is proved in [167]. A further analysis of the case where $\lambda = 0$ was performed by Bishop, Heil, Koo, and Lim in [61]; in that case the authors prove that perturbation in the sense of (22.13) preserves the dimension of the space (i.e., $\dim(\overline{\text{span}}\{f_k\}_{k=1}^\infty) = \dim(\overline{\text{span}}\,\{g_k\}_{k=1}^\infty)$), the excess, and the deficit.

It is known that it generally is very hard to calculate the gap. This is the reason that we do not go into more detail with these results. It would be interesting to have general results that were easier to apply.

We mention a special case where the condition (22.13) applies without a bound on (λ, μ) involving the gap. Suppose that $\{f_k\}_{k=1}^\infty$ is a frame sequence for which the analysis operator T has an *index*, i.e.,

$$\text{either} \quad \dim(\mathcal{N}_T) < \infty \quad \text{or} \quad \text{codim}(\mathcal{R}_T) := \dim(\mathcal{R}_T^\perp) < \infty. \qquad (22.14)$$

Recall that $\dim(\mathcal{N}_T) < \infty$ means that $\{f_k\}_{k=1}^\infty$ is a near-Riesz basis for its closed span, and that $\dim(\mathcal{N}_T)$ measures the excess. In the case (22.14), the index of T is defined as

$$\text{ind(T)} := \dim(\mathcal{N}_T) - \text{codim}(\mathcal{R}_T).$$

Under the stated assumptions it is proved in [159] that a sequence $\{g_k\}_{k=1}^\infty$ satisfying (22.13) with $\lambda + \frac{\mu}{\sqrt{A}} < 1$ also is a frame sequence, and that the corresponding synthesis operator U has an index; in fact,

$$\dim(\mathcal{N}_U) \le \dim(\mathcal{N}_T), \ \text{codim}(\mathcal{R}_U) \le \text{codim}(\mathcal{R}_T), \ \text{and ind}(U) = \text{ind}(T).$$

The relation between the various dimensions is particularly interesting in the case where T is a *Fredholm operator*, meaning that both $\dim(\mathcal{N}_T)$ and $\text{codim}(\mathcal{R}_T)$ are finite. In this case we see that a perturbation can increase the dimension of the spanned space, but the excess will decrease with the same amount. This can be illustrated by an example in $\mathbb{R}^3$:

Example 22.3.3 Let $\{e_i\}_{i=1}^3$ be an orthonormal basis for $\mathbb{R}^3$ and let

$$\{f_i\}_{i=1}^3 = \{e_1, 0, 0\}, \quad \{g_i\}_{i=1}^3 = \left\{e_1, \frac{1}{2}e_2, 0\right\}.$$

$\{f_i\}_{i=1}^3$ spans a one-dimensional subspace, and the excess is 2. $\{g_i\}_{i=1}^3$ is a perturbation of $\{f_i\}_{i=1}^3$ in the sense that (22.13) is satisfied with $(\lambda, \mu) = (0, 1/2)$; however, $\{g_i\}_{i=1}^3$ spans a 2-dimensional subspace, and the excess is 1. $\qquad \square$

22.4   Perturbation of Gabor frames

In this section we return to Gabor frames $\{E_{mb}T_{na}g\}_{m,n\in\mathbb{Z}}$ for $L^2(\mathbb{R})$. There are several important perturbation questions related to a Gabor frame. We will deal with three of them, namely:

(i) If $\{E_{mb}T_{na}g\}_{m,n\in\mathbb{Z}}$ is a Gabor frame and $h \in L^2(\mathbb{R})$ is "close" to g, does it follows that $\{E_{mb}T_{na}h\}_{m,n\in\mathbb{Z}}$ is a frame?

(ii) If $\{E_{mb}T_{na}g\}_{m,n\in\mathbb{Z}}$ is a Gabor frame and the points $\{(\mu_{m,n}, \lambda_{m,n})\}_{m,n\in\mathbb{Z}}$ are "close" to $\{(na, mb)\}_{m,n\in\mathbb{Z}}$, does it follows that $\{E_{\lambda_{m,n}}T_{\mu_{m,n}}g\}_{m,n\in\mathbb{Z}}$ is a frame?

(iii) If $\{E_{mb}T_{na}g\}_{m,n\in\mathbb{Z}}$ is a Gabor frame and (a', b') is "close" to (a, b), does it follow that $\{E_{mb'}T_{na'}g\}_{m,n\in\mathbb{Z}}$ is a frame?

In all three cases we have to specify what "close" should mean. We begin with (i). If $\{E_{mb}T_{na}g\}_{m,n\in\mathbb{Z}}$ is a Gabor frame, one could expect $\{E_{mb}T_{na}h\}_{m,n\in\mathbb{Z}}$ to be a frame if $||g - h||$ is sufficiently small, but a result of this type turns out not to hold. Consider, for example, the orthonormal basis $\{E_m T_n \chi_{[0,1]}\}_{m,n\in\mathbb{Z}}$ from Example 3.8.3; no matter how small we choose $\epsilon > 0$, the functions $\{E_m T_n \chi_{[0,1-\epsilon]}\}_{m,n\in\mathbb{Z}}$ are not complete in $L^2(\mathbb{R})$ and therefore cannot form a frame for $L^2(\mathbb{R})$, despite the fact that the norm difference

$$||\chi_{[0,1]} - \chi_{[0,1-\epsilon]}|| = \epsilon$$

can be arbitrarily small. This shows that for the perturbation problem (i), it is not appropriate just to use $||g - h||$ as a measure for how close g and h are.

A positive result can be obtained directly via Theorem 11.4.2 combined with Corollary 22.1.5:

Theorem 22.4.1 *Let $g, h \in L^2(\mathbb{R})$ and $a, b > 0$ be given, and suppose that $\{E_{mb}T_{na}g\}_{m,n\in\mathbb{Z}}$ is a frame for $L^2(\mathbb{R})$ with frame bounds A, B. If*

$$R := \frac{1}{b} \sup_{x\in[0,a]} \sum_{k\in\mathbb{Z}} \left| \sum_{n\in\mathbb{Z}} (g - h)(x - na)\overline{(g - h)(x - na - k/b)} \right| < A, \quad (22.15)$$

then $\{E_{mb}T_{na}h\}_{m,n\in\mathbb{Z}}$ is a frame for $L^2(\mathbb{R})$ with bounds

$$A\left(1 - \sqrt{\frac{R}{A}}\right)^2, \quad B\left(1 + \sqrt{\frac{R}{B}}\right)^2.$$

If $\{E_{mb}T_{na}g\}_{m,n\in\mathbb{Z}}$ is a Riesz basis for $L^2(\mathbb{R})$, then $\{E_{mb}T_{na}h\}_{m,n\in\mathbb{Z}}$ is also a Riesz basis for $L^2(\mathbb{R})$.

As a consequence of Theorem 22.4.1, the frame property is preserved under small perturbations measured in the Wiener space norm $||\cdot||_{W,a}$; see the definition in (11.30):

Corollary 22.4.2 *Let* $g, h \in L^2(\mathbb{R})$ *and* $a, b > 0$ *be given, and suppose that* $\{E_{mb}T_{na}g\}_{m,n\in\mathbb{Z}}$ *is a frame for* $L^2(\mathbb{R})$ *with frame bounds* A, B. *If* $||g-h||_{W,a} < \sqrt{\frac{bA}{2}}$, *then* $\{E_{mb}T_{na}h\}_{m,n\in\mathbb{Z}}$ *is a frame for* $L^2(\mathbb{R})$ *with bounds*

$$A\left(1 - \sqrt{\frac{2}{bA}}||g-h||_{W,a}\right)^2, \quad B\left(1 + \sqrt{\frac{2}{bB}}||g-h||_{W,a}\right)^2.$$

Proof. Define again R by (22.15). By Lemma 11.5.1,

$$R \le \frac{2}{b}||g-h||_{W,a}^2;$$

from here, the result now follows from Corollary 22.1.5 with R replaced by $\frac{2}{b}||g-h||_{W,a}^2$. $\qquad\square$

We now consider the problem (ii) of perturbing the lattice points $\{(na, mb)\}_{m,n\in\mathbb{Z}}$. This problem was first considered by Favier and Za-lik [273] in 1995. Since then, several authors have studied the problem. Common for most of the results is that only the translations na or the modulations mb were perturbed. Finally, in 2001 Sun and Zhou [604] gave conditions such that both could be perturbed simultaneously. To be more precise, they proved that reasonable conditions on g imply that if $\{E_{mb}T_{na}g\}_{m,n\in\mathbb{Z}}$ is a frame and the Euclidean distance between (na, mb) and $(\mu_{m,n}, \lambda_{m,n})$ is sufficiently small for all $m, n \in \mathbb{Z}$, then the irregular Gabor system $\{E_{\lambda_{m,n}}T_{\mu_{m,n}}g\}_{m,n\in\mathbb{Z}}$ is a frame as well. We state their result, which is formulated in terms of a function H depending on the choice of $g \in L^2(\mathbb{R})$,

$$H(g) := \left(\frac{1}{b}\sup_{x\in[0,a]}\sum_{n,k\in\mathbb{Z}}|g(x-na)g(x-na-k/b)|\right)^{1/2}.$$

Theorem 22.4.3 *Let* $g \in L^2(\mathbb{R})$ *be continuously differentiable and assume that there exist constants* $C > 0, \alpha > 2$ *such that*

$$|g(x)|, |g'(x)| \le \frac{C}{(1+|x|)^\alpha}, \quad \forall x \in \mathbb{R}.$$

Define $\tilde{g}(x) = xg(x)$. *Let* $a, b > 0$ *be given, and assume that* $\{E_{mb}T_{na}g\}_{m,n\in\mathbb{Z}}$ *is a frame with bounds* A, B. *Let* δ, η *be any positive numbers for which*

$$R := (8\eta H(g') + 8\sigma H(\tilde{g}) + 64\sigma\eta H(\tilde{g}'))^2 < A.$$

Then, for any sequence $\{(\mu_{m,n}, \lambda_{m,n})\}_{m,n \in \mathbb{Z}} \subset \mathbb{R}^2$ *for which*

$$|\mu_{m,n} - na| \leq \eta, \ |\lambda_{m,n} - mb| \leq \sigma, \ \forall m, n \in \mathbb{Z},$$

the Gabor system $\{E_{\lambda_{m,n}} T_{\mu_{m,n}} g\}_{m,n \in \mathbb{Z}}$ *is a frame with frame bounds*

$$A \left(1 - \sqrt{\frac{R}{A}} \right)^2, \ B \left(1 + \sqrt{\frac{R}{B}} \right)^2.$$

The conditions in Theorem 22.4.3 imply that there exists an open ball $B(0, \epsilon)$ in $\mathbb{R}^2$ centered at the origin and with radius ϵ, such that any choice of points $\{(\mu_{m,n}, \lambda_{m,n})\}_{m,n \in \mathbb{Z}}$ with

$$(\mu_{m,n}, \lambda_{m,n}) \in (na, mb) + B(0, \epsilon), \ \forall m, n \in \mathbb{Z}$$

will lead to a frame $\{E_{\lambda_{m,n}} T_{\mu_{m,n}} g\}_{m,n \in \mathbb{Z}}$. It is remarkable that all points (na, mb) are allowed to be perturbed equally. A significantly weaker conclusion can be obtained directly via Exercise 22.4, without any decay condition on g. In fact, assume that $\{E_{mb} T_{na} g\}_{m,n \in \mathbb{Z}}$ is a frame. Then, since the mapping $(x, y) \mapsto ||E_x T_y g||$ is continuous by Lemma 2.9.2, we can choose a sequence $\{(\mu_{m,n}, \lambda_{m,n})\}_{m,n \in \mathbb{Z}} \neq \{(na, mb)\}_{m,n \in \mathbb{Z}}$ such that

$$\sum_{m,n \in \mathbb{Z}} ||E_{mb} T_{na} g - E_{\lambda_{m,n}} T_{\mu_{m,n}} g||^2 < A; \tag{22.16}$$

then $\{E_{\lambda_{m,n}} T_{\mu_{m,n}} g\}_{m,n \in \mathbb{Z}}$ is a frame. However, the condition (22.16) will force that

$$|(na - \mu_{m,n}, mb - \lambda_{m,n})| \to 0 \text{ as } m, n \to \infty.$$

The statement of Theorem 22.4.3 indicates that the key to obtain reasonable perturbation results is to put the right assumptions on g. Feichtinger and Kaiblinger [286] have provided strong support to this statement by proving the following important result, where g is assumed to belong to the Feichtinger algebra S_0:

Theorem 22.4.4 *Assume that* $g, h \in S_0$ *and let* $a, b > 0$ *be given. If* $\{E_{mb} T_{na} g\}_{m,n \in \mathbb{Z}}$ *is a frame, then there exists* $\epsilon > 0$ *such that* $\{E_{mb'} T_{na'} h\}_{m,n \in \mathbb{Z}}$ *is a frame if*

$$|a - a'| < \epsilon, \ |b - b'| < \epsilon, \ ||g - h||_{S_0} < \epsilon.$$

Theorem 22.4.4 is in a sense very surprising, even when we let $h = g$. In fact, when $(a', b') \neq (a, b)$, the functions

$$x \mapsto E_{mb} T_{na} g(x) = e^{2\pi i mbx} g(x - na)$$

and

$$x \mapsto E_{mb'} T_{na'} g(x) = e^{2\pi i mb'x} g(x - na')$$

are moving far apart from each other for large values of m, n, so in a pointwise sense one cannot consider $\{E_{mb'}T_{na'}g\}_{m,n\in\mathbb{Z}}$ as a perturbation of $\{E_{mb}T_{na}g\}_{m,n\in\mathbb{Z}}$. The assumption $g \in \mathcal{S}_0$ is important in order to obtain that $\{E_{mb'}T_{na'}h\}_{m,n\in\mathbb{Z}}$ is nevertheless a frame when (a', b') and (a, b) are sufficiently close. To illustrate this, we can look at the function $g = \chi_{[0,1]}$; then $\{E_m T_n g\}_{m,n\in\mathbb{Z}}$ is a frame, but $\{E_m T_{n(1+\epsilon)}g\}_{m,n\in\mathbb{Z}}$ is not a frame for any $\epsilon > 0$; it is not even complete, because the shifts $T_{n(1+\epsilon)}g$ do not cover the entire real axis.

Theorem 22.4.4 was later generalized to the irregular case, i.e., the case of a Gabor frame $\{E_{\lambda_n}T_{\mu_n}g\}_{n\in I}$ where $\{(\mu_n, \lambda_n)\}_{n\in I}$ is an arbitrary sequence of points in $\mathbb{R}^2$. In this case it was proved in [18] that $\{E_{\rho\lambda_n}T_{\rho\mu_n}h\}_{n\in I}$ is also a frame, provided that ρ is sufficiently close to 1 and that the window h sufficiently close to g, measured in the $\mathcal{S}_0$-norm. Clearly, the assumption $g \in \mathcal{S}_0$ is essential here as well.

More general nonlinear deformations that keep the frame property have recently been considered by Gröchenig, Ortega-Cerda, and Romero [353].

22.5 Perturbation of Wavelet Frames

The perturbation theory for wavelet frames is less developed than its Gabor counterpart: some results about perturbation of the generator are known, but the literature for perturbation of the translation/scaling parameters is sparse.

We leave the proof of the following result concerning perturbation of the generator to the reader (Exercise 22.5).

Theorem 22.5.1 *Let $\psi, \varphi \in L^2(\mathbb{R})$ and $a > 1, b > 0$ be given, and assume that $\{a^{j/2}\psi(a^j x - kb)\}_{j,k\in\mathbb{Z}}$ is a frame with bounds A, B. If $\varphi \in L^2(\mathbb{R})$ and*

$$R := \frac{1}{b} \sup_{|\gamma|\in[1,a]} \sum_{j,k\in\mathbb{Z}} \left|(\widehat{\psi} - \widehat{\varphi})(a^j\gamma)(\widehat{\psi} - \widehat{\varphi})(a^j\gamma + k/b)\right| < A,$$

then $\{a^{j/2}\varphi(a^j x - kb)\}_{j,k\in\mathbb{Z}}$ is a frame for $L^2(\mathbb{R})$ with frame bounds

$$A\left(1 - \sqrt{\frac{R}{A}}\right)^2, \quad B\left(1 + \sqrt{\frac{R}{A}}\right)^2.$$

Favier and Zalik [273] have proved that if $\psi \in L^2(\mathbb{R})$ satisfies some mild conditions and $\{a^{j/2}\psi(a^j x - kb)\}_{j,k\in\mathbb{Z}}$ is a frame for some $a > 1, b > 0$, then $\{a^{j/2}\psi(a^j x - kb')\}_{j,k\in\mathbb{Z}}$ is also a frame if b' is sufficiently close to b. A result where both a and b are perturbed has apparently not been proved yet.

Note that the result by Sun and Zhou stated in Theorem 19.1.4 can also be considered as a perturbation result: in fact, the conditions imply that $\{a^{j/2}\varphi(a^j x - kb)\}_{j,k\in\mathbb{Z}}$ is a frame for $L^2(\mathbb{R})$ and that

$\left\{ s_{j,k}^{-1/2} \psi\left(\frac{x - \mu_{j,k}}{s_{j,k}}\right) \right\}_{j,k\in\mathbb{Z}}$ is also a frame for $L^2(\mathbb{R})$ whenever the points $\{(s_{j,k}, \mu_{j,k})\}_{j,k\in\mathbb{Z}}$ are "sufficiently close" to $\{(a^j, a^j kb)\}_{j,k\in\mathbb{Z}}$. However, this result concerns the possibility of constructing irregular wavelet frames, which is another question than that of perturbing the parameters a and b.

22.6   Perturbation of the Haar Wavelet

Let us return to the wavelet orthonormal basis $\{D^j T_k \psi\}_{j,k\in\mathbb{Z}}$ generated by the Haar wavelet,

$$\psi = \chi_{[0,1/2[} - \chi_{[1/2,1[}.$$

One of the main problems with this wavelet is the missing regularity, which, e.g., leads to bad localization of its Fourier transform $\hat{\psi}$. A very natural idea is to consider perturbations $\tilde{\psi}$ of ψ, and ask for $\tilde{\psi}$ to belong to a certain smoothness class $C^m(\mathbb{R})$ and $\{D^j T_k \tilde{\psi}\}_{j,k\in\mathbb{Z}}$ to be a Riesz basis for $L^2(\mathbb{R})$. Govil and Zalik did that in [326]. To be more precise, they modified the Haar wavelet pointwise by adding linear combinations of mth order splines with support in small neighborhoods of the discontinuity points $0, \frac{1}{2}, 1$. Hereby they obtained, for any integer $m \geq 2$ and any $\epsilon > 0$, a function $\tilde{\psi} \in L^2(\mathbb{R})$ (we suppress the dependence on the parameters in the notation) such that

(i) $\tilde{\psi} \in C^m(\mathbb{R})$,

(ii) supp $\tilde{\psi} \subseteq [-\epsilon, 1 + \epsilon]$,

(iii) $\{D^j T_k \tilde{\psi}\}_{j,k\in\mathbb{Z}}$ is a Riesz basis for $L^2(\mathbb{R})$.

The proof is based on Theorem 22.1.1. By letting $\epsilon \to 0$, the function $\tilde{\psi}$ will approach the Haar wavelet in $L^p(\mathbb{R}), 0 < p < \infty$, and the frame bounds converge to 1. The constructed functions $\tilde{\psi}$ cannot be generated by a multiresolution analysis: Zalik proved in [638] that there does not exist a multiresolution analysis with associated scaling function ϕ such that

$$\tilde{\psi}(x) = \sum_{k\in\mathbb{Z}} c_k \phi(2x - k) \text{ for some } \{c_k\}_{k\in\mathbb{Z}} \in \ell^2(\mathbb{Z}).$$

22.7   Exercises

22.1 Prove the estimates (22.7).

22.2 Prove (22.9).

22.3 Let U be a bounded operator between Hilbert spaces. Prove that if at least one of the spaces $\mathcal{R}_U$ and $\mathcal{R}_{UU^*}$ is closed, then

$$\mathcal{R}_U = \mathcal{R}_{UU^*}.$$

22.4 Let $\{f_k\}_{k=1}^{\infty}$ be a frame for $\mathcal{H}$ with frame bounds A, B, and let $\{g_k\}_{k=1}^{\infty}$ be a sequence in $\mathcal{H}$. Prove that if

$$\sum_{k=1}^{\infty} ||f_k - g_k||^2 < A,$$

then $\{g_k\}_{k=1}^{\infty}$ is a frame for $\mathcal{H}$, with bounds as in (22.11)

22.5 Prove Theorem 22.5.1.

22.6 Extend Corollary 22.1.5 to Riesz sequences.

22.7 Let $\{e_1, e_2\}$ be an orthonormal basis for $\mathbb{C}^2$, and consider the frame $\{f_1, f_2\}$ given by

$$f_1 = e_1, f_2 = 2e_2.$$

Given a number $c \in \mathbb{C}$, let

$$g_1 = e_1, g_2 = ce_2.$$

Based on the frame $\{f_1, f_2\}$, we want to find the range of parameter c for which $\{g_1, g_2\}$ is also a frame.

(i) Apply Theorem 22.1.1 with $\lambda = 0$ – for which $c \in \mathbb{C}$ does the result guarantee that $\{g_1, g_2\}$ is a frame?

(ii) Apply Theorem 22.1.1 with $\mu = 0$ – for which $c \in \mathbb{C}$ does the result guarantee that $\{g_1, g_2\}$ is a frame?

(iii) What is the exact range of $c \in \mathbb{C}$ for which $\{g_1, g_2\}$ is a frame?

23

Approximation of the Inverse Frame Operator

Consider a frame $\{f_k\}_{k=1}^{\infty}$ for a Hilbert space $\mathcal{H}$ and the associated frame operator,

$$S : \mathcal{H} \to \mathcal{H}, \quad Sf = \sum_{k=1}^{\infty} \langle f, f_k \rangle f_k.$$

One of the main results in frame theory, the frame decomposition (5.7), states that each $f \in \mathcal{H}$ has the representation

$$f = \sum_{k=1}^{\infty} \langle f, S^{-1} f_k \rangle f_k. \tag{23.1}$$

In practice it can be very difficult (or impossible) to apply the frame decomposition directly: the reason is that $\mathcal{H}$ usually is an infinite-dimensional Hilbert space, which makes it hard to invert the frame operator. In case we cannot find S^{-1} explicitly, we need to approximate S^{-1} (or at least approximate the frame coefficients $\{\langle f, S^{-1} f_k \rangle\}_{k=1}^{\infty}$). In this chapter we present some methods for approximation that only use vectors in finite-dimensional vector spaces. This has the consequence that all calculations in principle can be done using linear algebra.

The first method for approximation of the inverse frame operator will be discussed in Section 23.1. It does not work for all frames, but it leads in a natural way to the Casazza–Christensen method in Section 23.2 – a method that works for all frames. The analysis of the method is continued in Section 23.3, where convergence estimates due to Song and Gelb are presented. In Section 23.4 the method is applied to Gabor frames; the special

© Springer International Publishing Switzerland 2016
O. Christensen, *An Introduction to Frames and Riesz Bases*,
Applied and Numerical Harmonic Analysis,
DOI 10.1007/978-3-319-25613-9_23

case of integer-oversampled Gabor frames is considered in Section 23.5. Finally, Section 23.6 deals with the finite section method, with applications to Gabor frames given in Section 23.7.

23.1   The First Approach

In the entire chapter we let $\mathcal{H}$ denote a Hilbert space. Given a frame $\{f_k\}_{k=1}^{\infty}$ with frame operator $S : \mathcal{H} \to \mathcal{H}$, we know from general frame theory that S is invertible. However, explicit calculation of the inverse S^{-1} is usually not possible, which clearly makes it impossible to apply the frame decomposition (23.1) directly. It is natural to try to circumvent the problem by suitable approximations of S^{-1}, e.g., using finite subsets of $\{f_k\}_{k=1}^{\infty}$. Given $n \in \mathbb{N}$, the family $\{f_k\}_{k=1}^{n}$ is a frame for $\mathcal{H}_n := \mathrm{span}\{f_k\}_{k=1}^{n}$ by Proposition 1.1.2; denote its frame operator by

$$S_n : \mathcal{H}_n \to \mathcal{H}_n, \quad S_n f = \sum_{k=1}^{n} \langle f, f_k \rangle f_k. \tag{23.2}$$

From Lemma 5.2.3 we know that the orthogonal projection P_n of $\mathcal{H}$ onto $\mathcal{H}_n$ is given by

$$P_n f = \sum_{k=1}^{n} \langle f, S_n^{-1} f_k \rangle f_k, \quad f \in \mathcal{H}. \tag{23.3}$$

Note that $\mathcal{H}_n$ is finite-dimensional; thus, at least in principle, we can find S_n^{-1} using linear algebra. Now, since

$$P_n f = \sum_{k=1}^{n} \langle f, S_n^{-1} f_k \rangle f_k \to f = \sum_{k=1}^{\infty} \langle f, S^{-1} f_k \rangle f_k \text{ for } n \to \infty,$$

it is natural to ask whether S_n^{-1} approximates S^{-1} in the sense that

$$\langle f, S_n^{-1} f_k \rangle \to \langle f, S^{-1} f_k \rangle \text{ as } n \to \infty, \ \forall f \in \mathcal{H}, \forall k \in \mathbb{N}. \tag{23.4}$$

The question makes sense: for a given value of $k \in \mathbb{N}$, f_k is in the domain $\mathcal{H}_n$ for S_n^{-1} as soon as $n \geq k$. The following result was proved in [152].

Theorem 23.1.1 *Let $\{f_k\}_{k=1}^{\infty}$ be a frame for $\mathcal{H}$. Then (23.4) holds if and only if*

$$\forall j \in \mathbb{N} \ \exists c_j \in \mathbb{R} : ||S_n^{-1} f_j|| \leq c_j, \ \forall n \geq j. \tag{23.5}$$

Proof. First, suppose that (23.5) is satisfied. Fix $j \in \mathbb{N}$, and define

$$\phi_n := S_n^{-1} f_j - S^{-1} f_j, \ n \geq j.$$

We need to prove that for all $f \in \mathcal{H}$, $\langle f, \phi_n \rangle \to 0$ as $n \to \infty$. Observe that

$$Sf = \sum_{k=1}^{\infty} \langle f, f_k \rangle f_k = S_n f + \sum_{k=n+1}^{\infty} \langle f, f_k \rangle f_k. \qquad (23.6)$$

We will use this to obtain an alternative formula for ϕ_n. First, since

$$S\phi_n = SS_n^{-1} f_j - f_j,$$

an application of (23.6) on $S_n^{-1} f_j$ yields

$$\begin{aligned}
S\phi_n &= S_n S_n^{-1} f_j + \sum_{k=n+1}^{\infty} \langle S_n^{-1} f_j, f_k \rangle f_k - f_j \\
&= \sum_{k=n+1}^{\infty} \langle S_n^{-1} f_j, f_k \rangle f_k.
\end{aligned}$$

It follows that

$$\phi_n = \sum_{k=n+1}^{\infty} \langle S_n^{-1} f_j, f_k \rangle S^{-1} f_k, \ n \geq j.$$

Therefore, for $f \in \mathcal{H}$,

$$\begin{aligned}
|\langle f, \phi_n \rangle|^2 &= \left| \sum_{k=n+1}^{\infty} \langle f_k, S_n^{-1} f_j \rangle \langle f, S^{-1} f_k \rangle \right|^2 \\
&\leq \sum_{k=n+1}^{\infty} |\langle S_n^{-1} f_j, f_k \rangle|^2 \sum_{k=n+1}^{\infty} |\langle f, S^{-1} f_k \rangle|^2 \\
&\leq B \left\| S_n^{-1} f_j \right\|^2 \sum_{k=n+1}^{\infty} |\langle f, S^{-1} f_k \rangle|^2 \\
&\leq B c_j^2 \sum_{k=n+1}^{\infty} |\langle S^{-1} f, f_k \rangle|^2.
\end{aligned}$$

Since $\{f_k\}_{k=1}^{\infty}$ is a frame, $\sum_{k=n+1}^{\infty} |\langle S^{-1} f, f_k \rangle|^2 \to 0$ as $n \to \infty$. Therefore, our estimate proves that $\langle f, \phi_n \rangle \to 0$ as $n \to \infty$, as desired. On the other hand, if we assume that (23.4) is satisfied, we can fix an arbitrary $j \in \mathbb{N}$ and consider the functionals

$$A_n : \mathcal{H} \to \mathbb{C}, \quad A_n f = \langle f, S_n^{-1} f_j \rangle, \ n \geq j.$$

Each A_n is bounded, and by (23.4) the family of operators $\{A_n\}_{n \geq j}$ is pointwise convergent; by Theorem 2.2.1 the family of norms $\{\|A_n\|\}_{n \geq j}$ is therefore bounded, i.e., there is a constant $c_j > 0$ such that

$$\|A_n\| = \|S_n^{-1} f_j\| \leq c_j, \ \forall n \geq j. \qquad \square$$

Via Proposition 5.4.4 we obtain the following immediate consequence of Theorem 23.1.1:

Corollary 23.1.2 *Assume that $\{f_k\}_{k=1}^{\infty}$ is a Riesz frame. Then* (23.4) *holds.*

In particular, (23.4) holds if $\{f_k\}_{k=1}^{\infty}$ is a Riesz basis. Intuitively, one could expect the same to be true if $\{f_k\}_{k=1}^{\infty}$ is "close to be a Riesz basis," but this turns out not to be true. After adding a single element to a Riesz basis, the property (23.4) might no longer hold:

Example 23.1.3 Let $\{e_k\}_{k=1}^{\infty}$ be an orthonormal basis for $\mathcal{H}$, and define

$$f_1 = e_1, \quad f_k = e_{k-1} + \frac{1}{k}e_k, \quad k \geq 2.$$

By Example 22.1.2 we know that $\{f_k\}_{k=2}^{\infty}$ is a Riesz basis with bounds $\frac{1}{4}, \frac{9}{4}$; so $\{f_k\}_{k=1}^{\infty}$ is a frame with excess equal to 1. For $n \in \mathbb{N}$ we want to find $f := S_n^{-1}f_1$, i.e., to solve the equation

$$\sum_{k=1}^{n}\langle f, f_k\rangle f_k = f_1, \quad f \in \mathcal{H}_n.$$

In terms of the orthonormal basis $\{e_k\}_{k=1}^{\infty}$, the equation can be written as

$$\sum_{k=1}^{n-1}\left(\frac{1}{k}\langle f, f_k\rangle + \langle f, f_{k+1}\rangle\right)e_k + \frac{1}{n}\langle f, f_n\rangle e_n = e_1. \tag{23.7}$$

It follows from here that

$$\langle f, f_n\rangle = 0 \text{ and that } \langle f, f_k\rangle = -k\langle f, f_{k+1}\rangle, \quad k = 2, \ldots, n-1,$$

so $\langle f, f_k\rangle = 0$ for all $k = 2, \ldots, n$. Again by (23.7) we have $\langle f, f_1\rangle = 1$; expressing the last two conclusions in terms of $\{e_k\}_{k=1}^{\infty}$, we have

$$\langle f, e_1\rangle = 1, \quad \langle f, e_2\rangle = -2\langle f, e_1\rangle = -2,$$

and in general

$$\langle f, e_k\rangle = -k\langle f, e_{k-1}\rangle = (-1)^{k-1}k!, \quad k = 2, \ldots, n.$$

Since $f \in \mathcal{H}_n = \text{span}\{e_k\}_{k=1}^{n}$, this implies that

$$f = \sum_{k=1}^{n}\langle f, e_k\rangle e_k = \sum_{k=1}^{n}(-1)^{k-1}k!e_k. \tag{23.8}$$

In particular,

$$\|S_n^{-1}f_1\| = \left(\sum_{k=1}^{n}(k!)^2\right)^{1/2} \to \infty \text{ as } n \to \infty.$$

Therefore (23.5) does not hold. We can actually be more concrete and exhibit a vector $g \in \mathcal{H}$ for which the desired convergence in (23.4) fails. In fact, with

$$g := \sum_{k=1}^{\infty} \frac{(-1)^{k-1}}{k!} e_k,$$

the expression for $S_n^{-1} f_1$ in (23.8) shows that

$$\langle g, S_n^{-1} f_1 \rangle = n.$$

Thus, $\{\langle g, S_n^{-1} f_1 \rangle\}_{k=1}^{\infty}$ is divergent, and (23.4) does not hold. $\qquad\square$

The question of convergence in (23.4) can in fact be used to give yet another characterization of a frame being a Riesz basis:

Proposition 23.1.4 *A frame $\{f_k\}_{k=1}^{\infty}$ is a Riesz basis if and only if $\{f_k\}_{k=1}^{\infty}$ is linearly independent and (23.4) holds.*

Proof. A Riesz basis is linearly independent and satisfies (23.4) by Corollary 23.1.2. Now assume that $\{f_k\}_{k=1}^{\infty}$ is linearly independent and that (23.4) holds. Let $n \in \mathbb{N}$. The linear independence of $\{f_k\}_{k=1}^{\infty}$ implies that $\{f_k\}_{k=1}^{n}$ is a (Riesz) basis for $\mathcal{H}_n$. By Corollary 1.1.7, the dual basis is $\{S_n^{-1} f_k\}_{k=1}^{n}$, so

$$\langle f_k, S_n^{-1} f_j \rangle = \delta_{k,j}, \quad k, j = 1, 2, \ldots, n.$$

By letting $n \to \infty$ and using (23.4), we obtain that

$$\langle f_k, S^{-1} f_j \rangle = \delta_{k,j}, \quad \forall k, j \in \mathbb{N}.$$

By Theorem 7.1.1 we conclude that $\{f_k\}_{k=1}^{\infty}$ is a Riesz basis. $\qquad\square$

Example 23.1.3 is of course disappointing because it provides an example where the approximation method does not work. We will now prove that the method does not apply for overcomplete Gabor frames either:

Corollary 23.1.5 *Let $\{E_{mb} T_{na} g\}_{m,n \in \mathbb{Z}}$ be a Gabor frame with frame operator S, and let $\{f_k\}_{k=1}^{\infty}$ denote an arbitrary re-indexing of the frame elements. Then (23.4) holds if and only if $\{E_{mb} T_{na} g\}_{m,n \in \mathbb{Z}}$ is a Riesz basis, i.e., if and only if $ab = 1$.*

Proof. As discussed on page 36 and page 343, the elements in $\{E_{mb} T_{na} g\}_{m,n \in \mathbb{Z}}$ are linearly independent. Thus, by Proposition 23.1.4 we know that (23.4) holds if and only if $\{E_{mb} T_{na} g\}_{m,n \in \mathbb{Z}}$ is a Riesz basis. $\square$

It is not known whether there exist overcomplete wavelet frames for which (23.4) hold. But the negative outcome for Gabor frames prompts us to develop a more general theory.

23.2   The Casazza–Christensen Method

In this section we derive a method for approximation of the inverse frame operator which works for all frames. It can be considered as an improvement of the method from the last section. The initial results (up to Theorem 23.2.3) were proved by Casazza and Christensen in [124], and the rest are from [160]. We keep our previous notation, and let $\{f_k\}_{k=1}^\infty$ denote a frame with frame operator S; we again consider finite subfamilies $\{f_k\}_{k=1}^n$ and the associated finite-dimensional vector space $\mathcal{H}_n = \mathrm{span}\{f_k\}_{k=1}^n$. Let P_n denote the orthogonal projection of $\mathcal{H}$ onto $\mathcal{H}_n$, and let

$$I_n := \{1, 2, \ldots, n\}, \ n \in \mathbb{N}. \tag{23.9}$$

With this notation, the purpose is to approximate S^{-1} via sets of the form $\{f_k\}_{k\in I_n}, n \in \mathbb{N}$. It is only for notational convenience that the following results are formulated for a frame indexed by $\mathbb{N}$ and for this choice of I_n; given a frame indexed by a countable set I, similar results with identical proofs hold for *any* family $\{I_n\}_{n=1}^\infty$ of finite subsets of I for which

$$I_1 \subset I_2 \subset \cdots \subset I_n \uparrow I.$$

In order to make the general result clear from our presentation, we will denote the number of elements in I_n by $|I_n|$, despite the fact that with our choice (23.9), we simply have $|I_n| = n$.

We begin with a lemma.

Lemma 23.2.1 *Let $\{f_k\}_{k=1}^\infty$ be a frame for $\mathcal{H}$ with lower bound A. Given $n \in \mathbb{N}$, there exists a positive integer $m(n)$ such that*

$$\frac{A}{2}\|f\|^2 \le \sum_{k=1}^{n+m(n)} |\langle f, f_k\rangle|^2, \ \ \forall f \in \mathcal{H}_n. \tag{23.10}$$

Proof. Let $n \in \mathbb{N}$. Given $\epsilon > 0$, choose a finite set of elements $\{g_j\}_{j=1}^J$ in $\mathcal{H}_n$ such that $\|g_j\| = 1$ for all $j = 1, \ldots, J$, and such that the balls

$$B(g_j, \epsilon) := \{f \in \mathcal{H}_n \ : \ \|f - g_j\| \le \epsilon\}$$

cover the compact set $\{f \in \mathcal{H}_n \mid \|f\| = 1\}$. Since

$$A \le \sum_{k=1}^\infty |\langle g_j, f_k\rangle|^2, \ \forall j = 1, \ldots, J,$$

we can choose $m(n)$ such that

$$A\frac{2}{3} \le \sum_{k=1}^{n+m(n)} |\langle g_j, f_k\rangle|^2, \ \ \forall j = 1, \ldots, J.$$

Let B denote an upper frame bound for $\{f_k\}_{k=1}^\infty$, and consider $f \in \mathcal{H}_n, \|f\| = 1$. Choose j such that $f \in B(g_j, \epsilon)$. By the opposite triangle inequality applied to

$$\{\langle f, f_k\rangle\}_{k=1}^{n+m(n)} = \{\langle g_j, f_k\rangle - \langle g_j - f, f_k\rangle\}_{k=1}^{n+m(n)},$$

we have

$$\left(\sum_{k=1}^{n+m(n)} |\langle f, f_k\rangle|^2\right)^{1/2} \geq \left(\sum_{k=1}^{n+m(n)} |\langle g_j, f_k\rangle|^2\right)^{1/2}$$
$$-\left(\sum_{k=1}^{n+m(n)} |\langle g_j - f, f_k\rangle|^2\right)^{1/2}$$
$$\geq \sqrt{A\frac{2}{3}} - \sqrt{B}\,\|g_j - f\| \geq \sqrt{A\frac{2}{3}} - \sqrt{B}\epsilon.$$

By choosing ϵ small enough, $\sqrt{A\frac{2}{3}} - \sqrt{B}\epsilon \geq \sqrt{\frac{A}{2}}$, from which the result follows. $\qquad\square$

The next lemma shows that for any frame $\{f_k\}_{k=1}^\infty$, we can construct a family of frames "approaching $\{f_k\}_{k=1}^\infty$," which have common frame bounds. Remember that (23.4) holds for every Riesz frame; the lemma below turns out to be the key to an improved method that works for every frame.

Lemma 23.2.2 *Let $\{f_k\}_{k=1}^\infty$ be a frame with bounds A, B. For any $n \in \mathbb{N}$, choose a positive integer $m(n)$ such that (23.10) is satisfied. Then $\{P_n f_k\}_{k=1}^{n+m(n)}$ is a frame for $\mathcal{H}_n$ with bounds $\frac{A}{2}, B$; the associated frame operator is*

$$P_n S_{n+m(n)} : \mathcal{H}_n \to \mathcal{H}_n,$$

and

$$\|P_n S_{n+m(n)}\| \leq B, \quad \|(P_n S_{n+m(n)})^{-1}\| \leq \frac{2}{A}.$$

Proof. Fix $n \in \mathbb{N}$ and let $f \in \mathcal{H}_n$. Then, with our choice of $m(n)$,

$$\sum_{k=1}^{n+m(n)} |\langle f, P_n f_k\rangle|^2 = \sum_{k=1}^{n+m(n)} |\langle f, f_k\rangle|^2 \geq \frac{A}{2}\,\|f\|^2.$$

Also,

$$\sum_{k=1}^{n+m(n)} |\langle f, P_n f_k\rangle|^2 = \sum_{k=1}^{n+m(n)} |\langle f, f_k\rangle|^2 \leq \sum_{k=1}^\infty |\langle f, f_k\rangle|^2 \leq B\,\|f\|^2.$$

So $\{P_n f_k\}_{k=1}^{n+m(n)}$ is a frame for $\mathcal{H}_n$ with the claimed bounds. The frame operator is given by the mapping

$$f \longmapsto \sum_{k=1}^{n+m(n)} \langle f, P_n f_k \rangle P_n f_k = P_n \sum_{k=1}^{n+m(n)} \langle f, f_k \rangle f_k = P_n S_{n+m(n)} f, \quad f \in \mathcal{H}_n.$$

Now where $P_n S_{n+m(n)}$ is identified as the frame operator for a frame with bounds $\frac{A}{2}, B$, the norm estimates for $P_n S_{n+m(n)}$ and $(P_n S_{n+m(n)})^{-1}$ follow from Proposition 5.4.4. □

Technically, the merit of the frames constructed in Lemma 23.2.2 is that the frame operators are well-conditioned: the condition number for $P_n S_{n+m(n)}$ is at most $2B/A$, regardless of $n \in \mathbb{N}$. In contrast, the frame operators for the finite subfamilies $\{f_k\}_{k=1}^n$ of $\{f_k\}_{k=1}^\infty$ might be badly conditioned, as we have discussed in Section 7.2.

We are now ready to prove that S^{-1} can be approximated arbitrarily well in the strong operator topology using the operators

$$(P_n S_{n+m(n)})^{-1} P_n : \mathcal{H}_n \to \mathcal{H}_n, \quad n \in \mathbb{N}.$$

Note that $P_n S_{n+m(n)}$ is an operator on a finite-dimensional vector space. This implies that its inverse, and therefore $(P_n S_{n+m(n)})^{-1} P_n$, in principle can be found using finite-dimensional linear algebra. In practice, of course, large values of n will complicate the calculations. The method is called the *Casazza–Christensen method*.

Theorem 23.2.3 *Let $\{f_k\}_{k=1}^\infty$ be a frame with bounds A, B. For $n \in \mathbb{N}$, choose a positive integer $m(n)$ such that (23.10) is satisfied. Then*

$$(P_n S_{n+m(n)})^{-1} P_n f \to S^{-1} f \text{ for } n \to \infty, \quad \forall f \in \mathcal{H}. \qquad (23.11)$$

Proof. Let $f \in \mathcal{H}$. Then

$$S^{-1} f - (P_n S_{n+m(n)})^{-1} P_n f = P_n S^{-1} f - (P_n S_{n+m(n)})^{-1} P_n f$$
$$+ (I - P_n) S^{-1} f.$$

Since $(I - P_n) S^{-1} f \to 0$ as $n \to \infty$, it is enough to show that

$$\psi_n := P_n S^{-1} f - (P_n S_{n+m(n)})^{-1} P_n f \to 0 \text{ as } n \to \infty.$$

Since $\psi_n \in \mathcal{H}_n$ we can apply the operator $P_n S_{n+m(n)}$ to get

$$\psi_n = (P_n S_{n+m(n)})^{-1} (P_n S_{n+m(n)} P_n S^{-1} f - P_n f).$$

Consequently, via Lemma 23.2.2,

$$\|\psi_n\| \leq \|(P_n S_{n+m(n)})^{-1}\| \, \|P_n S_{n+m(n)} P_n S^{-1} f - P_n f\|$$
$$\leq \frac{2}{A} \|S_{n+m(n)} P_n S^{-1} f - f\| \to 0 \text{ for } n \to \infty,$$

as desired. □

At the moment the approximation method in Theorem 23.2.3 is purely theoretical: it depends on the choice of the positive number $m(n)$, which has not been estimated yet. We now want to obtain a more explicit result, giving more information about how to choose $m(n)$ such that (23.10) is satisfied. Recall that according to our chosen notation, $I_n = \{1, 2, \ldots, n\}$. In the rest of the section we will, for each given $n \in \mathbb{N}$, consider an index set J_n containing I_n. The reader can simply consider

$$J_n := I_{n+m(n)}; \tag{23.12}$$

this is somehow the "natural choice," but (23.12) is not necessary for the following results to hold.

We will need the following estimate.

Lemma 23.2.4 *Let $\{f_k\}_{k=1}^{\infty}$ be a sequence in $\mathcal{H}$ and let $n \in \mathbb{N}$. Let A_n denote a lower frame bound for the frame sequence $\{f_k\}_{k=1}^{n}$. Then for any set J_n containing I_n,*

$$\sum_{k \notin J_n} |\langle f, f_k \rangle|^2 \leq \frac{|I_n|}{A_n} \, max_{j \in I_n} \sum_{k \notin J_n} |\langle f_k, f_j \rangle|^2 \, ||f||^2, \; \forall f \in \mathcal{H}_n.$$

Proof. Let $f \in \mathcal{H}_n$. Since $\{f_k\}_{k=1}^{n}$ is a frame for $\mathcal{H}_n$, we can use the frame decomposition $f = \sum_{j \in I_n} \langle f, S_n^{-1} f_j \rangle f_j$ to get

$$|\langle f, f_k \rangle|^2 = \left| \langle \sum_{j \in I_n} \langle f, S_n^{-1} f_j \rangle f_j, f_k \rangle \right|^2 = \left| \sum_{j \in I_n} \langle f, S_n^{-1} f_j \rangle \langle f_j, f_k \rangle \right|^2.$$

Now, by Cauchy–Schwarz' inequality and the fact that $\{S_n^{-1} f_j\}_{j \in I_n}$ is a frame for $\mathcal{H}_n$ with upper bound $\frac{1}{A_n}$, we have

$$|\langle f, f_k \rangle|^2 \leq \sum_{j \in I_n} |\langle f, S_n^{-1} f_j \rangle|^2 \sum_{j \in I_n} |\langle f_j, f_k \rangle|^2$$

$$\leq \frac{1}{A_n} ||f||^2 \sum_{j \in I_n} |\langle f_j, f_k \rangle|^2.$$

Thus

$$\sum_{k \notin J_n} |\langle f, f_k \rangle|^2 \leq \sum_{k \notin J_n} \frac{1}{A_n} ||f||^2 \sum_{j \in I_n} |\langle f_j, f_k \rangle|^2$$

$$= \frac{1}{A_n} ||f||^2 \sum_{j \in I_n} \sum_{k \notin J_n} |\langle f_j, f_k \rangle|^2$$

$$\leq \frac{|I_n|}{A_n} ||f||^2 \, max_{j \in I_n} \sum_{k \notin J_n} |\langle f_j, f_k \rangle|^2,$$

as claimed.   □

We will now state a more explicit version of Theorem 23.2.3, formulated in terms of the frame operator V_n for the finite family $\{P_n f_k\}_{k \in J_n}$; by Lemma 23.2.2 this operator is precisely the one that appears in Theorem 23.2.3. Compared with Theorem 23.2.3, the result gives more information about how to choose the set J_n and also provides some insight concerning the speed of convergence in (23.11). More direct convergence estimates will be considered in Section 23.3.

Theorem 23.2.5 *Let $\{f_k\}_{k=1}^{\infty}$ be a frame for $\mathcal{H}$ with bounds A, B. Let $\{\epsilon_n\}_{n=1}^{\infty} \subseteq\,]0, A[$ be a decreasing sequence of numbers converging to zero. For $n \in \mathbb{N}$, choose a finite set J_n containing I_n such that*

$$\sum_{k \notin J_n} |\langle f, f_k \rangle|^2 \leq \epsilon_n \|f\|^2, \quad \forall f \in \mathcal{H}_n. \tag{23.13}$$

Let $V_n : \mathcal{H}_n \to \mathcal{H}_n$ denote the frame operator for the finite family $\{P_n f_k\}_{k \in J_n}$. Then, for all $f \in \mathcal{H}$,

$$\|S^{-1}f - V_n^{-1}P_n f\| \leq \frac{\epsilon_n}{A(A - \epsilon_n)} \|f\| + \left(\frac{B}{A - \epsilon_n} + 1 \right) \|(I - P_n)S^{-1}f\|.$$

Proof. Let $n \in \mathbb{N}$. Denote the restriction of $P_n S - V_n$ to $\mathcal{H}_n$ by $(P_n S - V_n)_{|\mathcal{H}_n}$; the reader can check that $(P_n S - V_n)_{|\mathcal{H}_n}$ is self-adjoint. Furthermore, for $f \in \mathcal{H}_n$, we have

$$
\begin{aligned}
\langle (P_n S - V_n)_{|\mathcal{H}_n} f, f \rangle &= \langle (P_n S - V_n)f, f \rangle \\
&= \langle P_n S f, f \rangle - \langle V_n f, f \rangle \\
&= \left\langle \sum_{k=1}^{\infty} \langle f, f_k \rangle P_n f_k, f \right\rangle - \left\langle \sum_{k \in J_n} \langle f, P_n f_k \rangle P_n f_k, f \right\rangle \\
&= \sum_{k=1}^{\infty} |\langle f, f_k \rangle|^2 - \sum_{k \in J_n} |\langle f, f_k \rangle|^2 \\
&= \sum_{k \notin J_n} |\langle f, f_k \rangle|^2 \geq 0.
\end{aligned}
$$

It follows from (2.8) and the condition (23.13) that

$$
\begin{aligned}
\|(P_n S - V_n)_{|\mathcal{H}_n}\| &= \sup_{f \in \mathcal{H}_n, \|f\|=1} |\langle (P_n S - V_n)f, f \rangle| \\
&= \sup_{f \in \mathcal{H}_n, \|f\|=1} \sum_{k \notin J_n} |\langle f, f_k \rangle|^2 \leq \epsilon_n.
\end{aligned}
$$

Also, for $f \in \mathcal{H}_n$,

$$
\begin{aligned}
\sum_{k \in J_n} |\langle f, P_n f_k \rangle|^2 &= \sum_{k \in J_n} |\langle f, f_k \rangle|^2 \\
&= \sum_{k=1}^{\infty} |\langle f, f_k \rangle|^2 - \sum_{k \notin J_n} |\langle f, f_k \rangle|^2 \geq (A - \epsilon_n)\|f\|^2.
\end{aligned}
$$

So $A - \epsilon_n$ is a lower frame bound for $\{P_n f_k\}_{k \in J_n}$; by Proposition 5.4.4, this implies that $||V_n^{-1}|| \le \frac{1}{A-\epsilon_n}$. Now let $f \in \mathcal{H}$. We have

$$
\begin{aligned}
||S^{-1}f - V_n^{-1}P_n f|| &\le ||(I-P_n)S^{-1}f|| + ||P_n S^{-1}f - V_n^{-1}P_n f|| \\
&\le ||(I-P_n)S^{-1}f|| + ||V_n^{-1}|| \, ||V_n P_n S^{-1}f - P_n f|| \\
&\le ||(I-P_n)S^{-1}f|| + \frac{1}{A-\epsilon_n} \, ||V_n P_n S^{-1}f - P_n f||.
\end{aligned}
$$

Now,

$$
\begin{aligned}
||V_n P_n S^{-1}f - P_n f|| &\le ||V_n P_n S^{-1}f - P_n S P_n S^{-1}f|| \\
&\quad + ||P_n S P_n S^{-1}f - P_n f|| \\
&\le ||(V_n - P_n S)P_n S^{-1}f|| + ||S P_n S^{-1}f - f|| \\
&\le \epsilon_n ||P_n S^{-1}f|| + ||S|| \, ||P_n S^{-1}f - S^{-1}f|| \\
&\le \frac{\epsilon_n}{A}||f|| + B \, ||(I-P_n)S^{-1}f||.
\end{aligned}
$$

Altogether,

$$
||S^{-1}f - V_n^{-1}P_n f|| \le \frac{\epsilon_n}{A(A-\epsilon_n)} \, ||f|| + \left(\frac{B}{A-\epsilon_n}+1\right) ||(I-P_n)S^{-1}f||.
$$

This completes the proof. $\qquad\square$

It is always possible to chose a set J_n such that (23.13) is satisfied (Exercise 23.2). By Theorem 23.2.5, this choice of J_n implies that

$$
V_n^{-1}P_n f \to S^{-1}f \text{ for } n \to \infty, \quad \forall f \in \mathcal{H}.
$$

That is, the operators $\{V_n^{-1}P_n\}_{n=1}^\infty$ converge to S^{-1} in the strong operator topology. In particular, the frame coefficients can be approximated:

$$
\langle f, V_n^{-1}P_n f_k\rangle \to \langle f, S^{-1}f_k\rangle \text{ for } n \to \infty, \quad \forall f \in \mathcal{H}, \ k \in \mathbb{N}. \quad (23.14)
$$

Let us compare the conclusion in (23.14) with the initial question (23.4). While the convergence in (23.4) was proved only to hold for some frames, the convergence in (23.14) holds for all frames $\{f_k\}_{k=1}^\infty$. Since $\{P_n f_k\}_{k \in J_n}$ is a finite set, the frame operator V_n and its inverse can be computed using finite-dimensional linear algebra, exactly as in the case of the projection method in Section 23.1. This does not make it trivial to apply the results, but calculation of $V_n^{-1}P_n$ is a drastic simplification compared to inversion of the frame operator S.

Under the conditions in Theorem 23.2.5, it even holds that the *sequence* of coefficients $\{\langle f, V_n^{-1}P_n f_k\rangle\}_{k \in J_n}$, $n \in \mathbb{N}$, converges to $\{\langle f, S^{-1}f_k\rangle\}_{k=1}^\infty$ in ℓ^2-sense as $n \to \infty$:

Theorem 23.2.6 *For* $n \in \mathbb{N}$, *choose* J_n *as in Theorem 23.2.5. Then*

$$\sum_{k \in J_n} |\langle f, V_n^{-1} P_n f_k \rangle - \langle f, S^{-1} f_k \rangle|^2 + \sum_{k \notin J_n} |\langle f, S^{-1} f_k \rangle|^2$$

$$\to 0 \text{ for } n \to \infty, \quad \forall f \in \mathcal{H}.$$

Proof. Let $f \in \mathcal{H}$. It is clear that $\sum_{k \notin J_n} |\langle f, S^{-1} f_k \rangle|^2 \to 0$ for $n \to \infty$. Concerning the first term, we have

$$\sum_{k \in J_n} |\langle f, V_n^{-1} P_n f_k \rangle - \langle f, S^{-1} f_k \rangle|^2$$

$$= \sum_{k \in J_n} |\langle V_n^{-1} P_n f, f_k \rangle - \langle S^{-1} f, f_k \rangle|^2$$

$$\leq B \, ||(V_n^{-1} P_n - S^{-1}) f||^2 \to 0 \text{ as } n \to \infty,$$

as desired. $\qquad\qquad\square$

For applications of Theorem 23.2.5 the pure existence of sets J_n satisfying (23.13) is not enough: we need to be able to *find* J_n. The condition (23.13) is quite complicated because it has to be satisfied for all $f \in \mathcal{H}_n$. Combining Theorem 23.2.5 and Lemma 23.2.4, we will now show that it can be replaced by a condition only involving the finite set of vectors $f_j, j \in I_n$:

Theorem 23.2.7 *Let* $\{f_k\}_{k=1}^{\infty}$ *be a frame for* $\mathcal{H}$ *with bounds* A, B. *Let* $\{\epsilon_n\}_{n=1}^{\infty} \subseteq]0, A[$ *be a decreasing sequence of numbers converging to zero. For* $n \in \mathbb{N}$, *let* A_n *denote a lower frame bound for the frame sequence* $\{f_k\}_{k=1}^{n}$ *and choose a finite set* J_n *containing* I_n *such that*

$$\sum_{k \notin J_n} |\langle f_j, f_k \rangle|^2 \leq \frac{\epsilon_n A_n}{|I_n|}, \quad \forall j \in I_n. \qquad (23.15)$$

Let $V_n : \mathcal{H}_n \to \mathcal{H}_n$ *denote the frame operator for the finite family* $\{P_n f_k\}_{k \in J_n}$. *Then, for all* $f \in \mathcal{H}$,

$$||S^{-1} f - V_n^{-1} P_n f|| \leq \frac{\epsilon_n}{A(A - \epsilon_n)} \, ||f|| + \left(\frac{B}{A - \epsilon_n} + 1 \right) ||(I - P_n) S^{-1} f||.$$

Observe that for $n \in \mathbb{N}$, (23.15) consists of $|I_n|$ conditions on the set J_n. In Section 23.4 we apply this result to Gabor frames, where the number of conditions can be reduced further. Applications to wavelet frames are given in [160]; also in this case the number of conditions can be reduced.

One important issue remains. In fact, Theorem 23.2.5 and Theorem 23.2.7 do not provide concrete information about how fast the operators $V_n^{-1} P_n$ converge to S^{-1}; the main problem is that the estimates in these results contain the term $||(I - P_n) S^{-1} f||$. In Section 23.3 we will address the question of how to obtain more explicit estimates for the speed of convergence.

23.3 Convergence Estimates for Localized Frames

In this section, we continue to consider a frame $\{f_k\}_{k=1}^{\infty}$ for a Hilbert space $\mathcal{H}$, with frame operator denoted by S. In [578] Song and Gelb studied the convergence rate in Theorem 23.2.5 under the extra assumption that the frame $\{f_k\}_{k=1}^{\infty}$ is self-localized with decay rate $s > 1$; see Definition 8.2.3. The paper [578] also marks an interesting shift in the viewpoint: while Theorem 23.2.5 aims at approximation of $S^{-1}f$ for arbitrary $f \in \mathcal{H}$, we will now restrict the attention to the vectors f satisfying the condition

$$|\langle f, f_k \rangle| \le C_0 \, k^{-s}, \, k \in \mathbb{N} \qquad (23.16)$$

for some $C_0 > 0$. The results in [578] show that the value of the parameter s plays an important role for the speed of convergence of the approximation method.

As in Section 23.2 we let $\mathcal{H}_n = \mathrm{span}\{f_k\}_{k=1}^{n}$; P_n denotes the orthogonal projection of $\mathcal{H}$ onto $\mathcal{H}_n$, and $I_n := \{1, 2, \ldots, n\}$. The results in Theorem 23.2.5 and Theorem 23.2.7 show that we for any given $f \in \mathcal{H}$ can approximate $S^{-1}f$ using calculations taking place in the finite-dimensional spaces $\mathcal{H}_n$. However, in order to obtain a more quantitative information, we need an estimate for $||(I - P_n)S^{-1}f||$ for the given element $f \in \mathcal{H}$; this is exactly what the following result will give us.

Lemma 23.3.1 *Assume that the frame $\{f_k\}_{k=1}^{\infty}$ is self-localized with decay rate $s > 1$ and consider an element $f \in \mathcal{H}$ satisfying the condition (23.16) for some $C_0 > 0$. Then there exists a constant $C > 0$ such that the following estimates hold:*

$$|\langle f, S^{-1}f_k \rangle| \; \le \; Ck^{-s}, \forall k \in \mathbb{N}; \qquad (23.17)$$
$$||(I - P_n)f|| \; \le \; Cn^{-(s-1/2)}, \forall n \in \mathbb{N}; \qquad (23.18)$$
$$||(I - P_n)S^{-1}f|| \; \le \; Cn^{-(s-1/2)}, \forall n \in \mathbb{N}. \qquad (23.19)$$

Proof. We first prove (23.17). Via the frame decomposition on the form

$$f = \sum_{j=1}^{\infty} \langle f, f_j \rangle S^{-1}f_j,$$

we have that

$$|\langle f, S^{-1}f_k \rangle| \;\; = \;\; \left| \langle \sum_{j=1}^{\infty} \langle f, f_j \rangle S^{-1}f_j, S^{-1}f_k \rangle \right|$$

$$\le \;\; \sum_{j=1}^{\infty} |\langle f, f_j \rangle| \, |\langle S^{-1}f_j, S^{-1}f_k \rangle|. \qquad (23.20)$$

We have assumed that $\{f_k\}_{k=1}^{\infty}$ is self-localized, so by Lemma 8.2.4 (ii) the canonical dual frame $\{S^{-1}f_k\}_{k=1}^{\infty}$ is also self-localized, with the same decay rate. Using the assumption (23.16) and (23.20), we conclude that there exists a constant $C_1 > 0$ such that

$$|\langle f, S^{-1}f_k \rangle| \leq C_1 \sum_{j=1}^{\infty} j^{-s}(1 + |k - j|)^{-s}. \tag{23.21}$$

Standard arguments now imply that the sum in (23.21) is bounded by a constant times k^{-s}, which leads to the estimate in (23.17).

In order to prove (23.18), we use the frame decomposition on the form

$$f = \sum_{j=1}^{\infty} \langle f, S^{-1}f_j \rangle f_j.$$

Let us consider the partial sum $\sum_{j=1}^{n} \langle f, S^{-1}f_j \rangle f_j$, which belongs to $\mathcal{H}_n$. Since the orthogonal projection $P_n f$ is the element in $\mathcal{H}_n$ which is closest to f, we have that

$$\begin{aligned}
\|(I - P_n)f\|^2 &= \|f - P_n f\|^2 \\
&\leq \left\| f - \sum_{j=1}^{n} \langle f, S^{-1}f_j \rangle f_j \right\|^2 \\
&= \left\| \sum_{j=n+1}^{\infty} \langle f, S^{-1}f_j \rangle f_j \right\|^2 \\
&= \sum_{j,k=n+1}^{\infty} \langle f, S^{-1}f_j \rangle \langle S^{-1}f_k, f \rangle \langle f_j, f_k \rangle.
\end{aligned}$$

Using (23.17) and that $\{f_k\}_{k=1}^{\infty}$ is self-localized, this shows that there is a constant $C_2 > 0$, which is independent of n and such that

$$\|(I - P_n)f\|^2 \leq C_2 \sum_{j,k=n+1}^{\infty} j^{-s}k^{-s}(1 + |k - j|)^{-s}. \tag{23.22}$$

The sum in (23.22) is bounded by a constant (depending on s) times $n^{-(2s-1)}$. This proves (23.18).

Finally, we need to prove the estimate (23.19); note that this is "the same" as (23.18), just with the element f replaced by $S^{-1}f$. Thus, we can prove (23.19) simply by proving that the condition (23.16) holds with f replaced by $S^{-1}f$. Now, using that the inverse of the frame operator is self-adjoint followed by an application of (23.17),

$$|\langle S^{-1}f, f_k \rangle| = |\langle f, S^{-1}f_k \rangle| \leq Ck^{-s}, \forall k \in \mathbb{N};$$

this is exactly what we need in order to complete the proof of (23.19). □

Let us now return to the Casazza–Christensen method. Under the assumptions in Lemma 23.3.1, Song and Gelb [578] found a suitable set J_n in Theorem 23.2.5 and showed that for some $C > 0$,

$$\|S^{-1}f \quad V_n^{-1}P_nf\| \le Cn^{-(s-1/2)}. \tag{23.23}$$

We will not repeat the proof here. Instead we show how such an estimate can be obtained by an appropriate choice of the numbers ϵ_n in (23.15) and the corresponding sets J_n. Recall that $I_n = \{1,\dots,n\}$.

Theorem 23.3.2 *Assume that*

(i) $\{f_k\}_{k=1}^\infty$ *is a frame with bounds A, B;*

(ii) $\{f_k\}_{k=1}^\infty$ *is self-localized with decay rate $s > 1$;*

(iii) *For $n \in \mathbb{N}$, the finite set J_n containing I_n is chosen such that*

$$\sum_{k \notin J_n} |\langle f_j, f_k\rangle|^2 \le \frac{A_n}{|I_n|} n^{-(s-1/2)}, \; \forall j \in I_n, \tag{23.24}$$

where A_n denote a lower frame bound for the frame sequence $\{f_k\}_{k=1}^n$;

(iv) *f is a given element in $\mathcal{H}$ and (23.16) holds for some $C_0 > 0$.*

Let $V_n : \mathcal{H}_n \to \mathcal{H}_n$ denote the frame operator for the finite family $\{P_nf_k\}_{k \in J_n}$. Then there is a constant $C > 0$ such that

$$\|S^{-1}f - V_n^{-1}P_nf\| \le Cn^{-(s-1/2)}, \; \forall n \in \mathbb{N}.$$

Proof. Take $\epsilon_n := n^{-(s-1/2)}$. Then, for n sufficiently large, we have $A - \epsilon_n \ge A/2$, and therefore

$$\frac{\epsilon_n}{A(A - \epsilon_n)} \le \frac{2}{A^2}\epsilon_n \le \frac{2}{A^2}n^{-(s-1/2)}.$$

The result now follows by combining Theorem 23.2.7 with the estimate (23.19). □

23.4   Applications to Gabor Frames

As noted at the beginning of Section 23.2, the methods for approximation of the inverse frame operator can also be applied to frames indexed by $\mathbb{Z}^2$: we only have to replace the index sets $\{I_n\}_{n=1}^\infty$ in (23.9) by finite subsets of $\mathbb{Z}^2$ for which

$$I_1 \subset I_2 \subset \cdots \subset I_n \uparrow \mathbb{Z}^2.$$

In this chapter, we denote Gabor frames by $\{E_{kb}T_{la}g\}_{k,l\in\mathbb{Z}}$; recall that here $a, b > 0$ and $g \in L^2(\mathbb{R})$. For a Gabor frame it is natural to choose the sets I_n as "finite lattices,"

$$I_n := \{(k,l) \in \mathbb{Z}^2 : |kb| \le Dn, |la| \le Cn\}, \tag{23.25}$$

where C, D are positive constants; the freedom in the "lattice size" gained by introducing these constants will prove useful. Our purpose is to show that the condition on the finite set J_n in Theorem 23.2.7 can be simplified in the case of a Gabor frame. By taking J_n of the form $J_n = I_{n+m(n)}$, the question is how to find a value for the positive integer $m(n)$ such that (23.15) is satisfied. We begin with a technical lemma.

Lemma 23.4.1 *Let $n \in \mathbb{N}$ and $m(n)$ be an arbitrary nonnegative integer. Then, for all $(k', l') \in I_n$, we have*

$$\sum_{(k,l) \notin I_{n+m(n)}} |\langle E_{k'b} T_{l'a} g, E_{kb} T_{la} g\rangle|^2 \leq \sum_{(k,l) \notin I_{m(n)}} |\langle E_{kb} T_{la} g, g\rangle|^2.$$

Proof. Let $(k', l') \in I_n$. Then

$$\sum_{(k,l) \notin I_{n+m(n)}} |\langle E_{k'b} T_{l'a} g, E_{kb} T_{la} g\rangle|^2$$

$$= \sum_{(k,l) \notin I_{n+m(n)}} |\langle E_{(k-k')b} T_{(l-l')a} g, g\rangle|^2$$

$$= \sum_{k,l \in \mathbb{Z}} |\langle E_{(k-k')b} T_{(l-l')a} g, g\rangle|^2 - \sum_{(k,l) \in I_{n+m(n)}} |\langle E_{(k-k')b} T_{(l-l')a} g, g\rangle|^2$$

$$= \sum_{k,l \in \mathbb{Z}} |\langle E_{kb} T_{la} g, g\rangle|^2 - \sum_{(k,l) \in I_{n+m(n)}} |\langle E_{(k-k')b} T_{(l-l')a} g, g\rangle|^2.$$

Let $C, D > 0$ be the constants in (23.25); then

$$\sum_{(k,l) \in I_{n+m(n)}} |\langle E_{(k-k')b} T_{(l-l')a} g, g\rangle|^2$$

$$= \sum_{|kb| \leq (n+m(n))D} \sum_{|la| \leq (n+m(n))C} |\langle E_{(k-k')b} T_{(l-l')a} g, g\rangle|^2$$

$$\geq \sum_{|kb| \leq (m(n))D} \sum_{|la| \leq (m(n))C} |\langle E_{kb} T_{la} g, g\rangle|^2 = \sum_{(k,l) \in I_{m(n)}} |\langle E_{kb} T_{la} g, g\rangle|^2.$$

It follows that for all $(k', l') \in I_n$,

$$\sum_{(k,l) \notin I_{n+m(n)}} |\langle E_{k'b} T_{l'a} g, E_{kb} T_{la} g\rangle|^2$$

$$\leq \sum_{k,l \in \mathbb{Z}} |\langle E_{kb} T_{la} g, g\rangle|^2 - \sum_{(k,l) \in I_{m(n)}} |\langle E_{kb} T_{la} g, g\rangle|^2$$

$$= \sum_{(k,l) \notin I_{m(n)}} |\langle E_{kb} T_{la} g, g\rangle|^2.$$

This concludes the proof.   □

Combining Theorem 23.2.7 and Lemma 23.4.1, we get

Theorem 23.4.2 *Let* $\{E_{kb}T_{la}g\}_{k,l\in\mathbb{Z}}$ *be a Gabor frame with bounds* A, B, *and let* $\{\epsilon_n\}_{n=1}^{\infty} \subseteq]0, A[$ *be a decreasing sequence of numbers converging to zero. For* $n \in \mathbb{N}$, *let* A_n *be a lower frame bound for the frame sequence* $\{E_{kb}T_{la}g\}_{(k,l)\in I_n}$ *and choose a positive integer* $m(n)$ *such that*

$$\sum_{(k,l)\notin I_{m(n)}} |\langle E_{kb}T_{la}g, g\rangle|^2 \le \frac{A_n\epsilon_n}{|I_n|}. \qquad (23.26)$$

Let $V_n : \mathcal{H}_n \to \mathcal{H}_n$ *be the frame operator for* $\{P_n E_{kb}T_{la}g\}_{(k,l)\in I_{n+m(n)}}$. *Then, for all* $f \in L^2(\mathbb{R})$,

$$\|S^{-1}f - V_n^{-1}P_nf\| \le \frac{\epsilon_n}{A(A - \epsilon_n)}\|f\| + \left(\frac{B}{A - \epsilon_n} + 1\right)\|(I - P_n)S^{-1}f\|.$$

Thus, in the case of a Gabor frame the single condition (23.26) is enough to determine the choice of J_n. Observe that by the frame condition $\sum_{k,l\in\mathbb{Z}} |\langle E_{kb}T_{la}g, g\rangle|^2$ is finite; thus, to satisfy (23.26) is "only" a question of choosing $m(n)$ sufficiently big.

For the Gaussian, a direct estimate for $\sum_{(k,l)\notin I_{m(n)}} |\langle E_{kb}T_{la}g, g\rangle|^2$ can be given:

Example 23.4.3 Let $g(x) = 2^{1/4}e^{-\pi x^2}$. It is well known, cf. [300], that

$$|\langle E_{kb}T_{la}g, g\rangle| = e^{-(k^2a^2+l^2b^2)\pi/2}.$$

Thus, taking $C = D = 1$ in (23.25),

$$\sum_{(k,l)\notin I_{m(n)}} |\langle E_{kb}T_{la}g, g\rangle|^2$$

$$\le \sum_{|k|>m(n)}\sum_{l\in\mathbb{Z}} e^{-\pi(k^2a^2+l^2b^2)} + \sum_{|l|>m(n)}\sum_{k\in\mathbb{Z}} e^{-\pi(k^2a^2+l^2b^2)}$$

$$\le 4\left(\sum_{k=(m(n)+1)^2}^{\infty} e^{-\pi ka^2}\sum_{l=0}^{\infty} e^{-\pi lb^2} + \sum_{l=(m(n)+1)^2}^{\infty} e^{-\pi la^2}\sum_{k=0}^{\infty} e^{-\pi kb^2}\right)$$

$$= 4\left(\frac{e^{-\pi a^2(m(n)+1)^2} + e^{-\pi b^2(m(n)+1)^2}}{(1 - e^{-\pi b^2})(1 - e^{-\pi a^2})}\right).$$

If a lower frame bound A_n for $\{E_{kb}T_{la}g\}_{(k,l)\in I_n}$ is known, we can now use (23.26) to find a suitable value for the positive integer $m(n)$. Note that by Theorem 13.4.4 (see also Lemma 7.2.1), we know that $A_n \to 0$ when $n \to \infty$. □

23.5   Integer Oversampled Gabor Frames

In this section we consider a Gabor frame $\{E_{kb}T_{la}g\}_{k,l\in\mathbb{Z}}$ which is *integer oversampled*, i.e., we assume that

$$ab = \frac{1}{N}, \text{ where } N \in \mathbb{N}.$$

In this case we choose the index sets I_n in (23.25) as

$$I_n := \left\{(k,l) \in \mathbb{Z}^2 \mid |k|, |l| \leq nN\right\}.$$

With this choice of the index set I_n, Theorem 23.4.2 applies with

$$|I_n| = (2nN+1)^2.$$

We will show how to obtain estimates for the approximation rate for the dual window $S^{-1}g$ in the case of integer oversampling. As before P_n will denote the projection of $L^2(\mathbb{R})$ onto

$$\mathcal{H}_n = \text{span}\{E_{kb}T_{la}g\}_{(k,l)\in I_n} = \text{span}\{E_{kb}T_{la}g\}_{|k|,|l|\leq nN}.$$

Let HH^* be the Gram matrix for $\{E_{m/a}T_{n/b}g\}_{m,n\in\mathbb{Z}}$, as defined in (12.32). In case g and the Fourier transform $\hat{g}$ decay exponentially, Strohmer proved in [588] that there exist constants $C, \lambda > 0$ such that

$$|(HH^*)_{l,k;l',k'}| \leq Ce^{-\lambda(|k-k'|+|l-l'|)}.$$

Using Lemma 12.4.3, it follows that for some C', λ',

$$|[(HH^*)^{-1}]_{l,k;0,0}| \leq C'e^{-\lambda'(|k|+|l|)}. \tag{23.27}$$

In the context of the finite section method (see Section 23.7), Strohmer proved that the duality principle in Gabor analysis leads to an estimate for the speed of convergence of the approximation method. The same principle can be applied in the setup discussed here:

Theorem 23.5.1 *Suppose that $\{E_{kb}T_{la}g\}_{k,l\in\mathbb{Z}}$ is an integer oversampled frame and that g and its Fourier transform $\hat{g}$ decay exponentially. Under the assumptions in Theorem 23.4.2, there exist constants $\lambda, C > 0$ such that*

$$||S^{-1}g - V_n^{-1}P_ng|| \leq \frac{\epsilon_n}{A(A-\epsilon_n)}||f|| + Ce^{-\lambda n}, \ \forall n \in \mathbb{N}.$$

Proof.   By the Janssen representation (12.33) of the inverse frame operator, we have

$$\begin{aligned} S^{-1}g &= ab \sum_{k,l\in\mathbb{Z}} [(HH^*)^{-1}]_{l,k;00} E_{k/a}T_{l/b}g \\ &= ab \sum_{k,l\in\mathbb{Z}} [(HH^*)^{-1}]_{l,k;00} E_{kNb}T_{lNa}g. \end{aligned}$$

For $|k|, |l| \leq n$, we have that $P_n E_{kNb} T_{lNa} g = E_{kNb} T_{lNa} g$. Thus,

$$(I - P_n) S^{-1} g = ab(I - P_n) \sum_{|k|>n \text{ or } |l|>n} [(HH^*)^{-1}]_{l,k;00} E_{kNb} T_{lNa} g.$$

By Theorem 13.1.1 we know that $\{E_{k/a} T_{l/b} g\}_{k,l \in \mathbb{Z}}$ is a Riesz sequence with upper bound Bab. Therefore the subfamily

$$\{E_{kNb} T_{lNa} g\}_{|k|>n \text{ or} \|l\|>n}$$

is also a Riesz sequence with upper bound Bab. Using the estimate (23.27) for $[(HH^*)^{-1}]_{l,k;00}$, we get

$$
\begin{aligned}
& \|(I - P_n) S^{-1} g\|^2 \\
\leq\ & Bab(ab)^2 \sum_{|k|>n \text{ or } |l|>n} |[(HH^*)^{-1}]_{l,k;0,0}|^2 \\
\leq\ & B(C')^2 (ab)^3 \sum_{|k|>n \text{ or } |l|>n} e^{-2\lambda'(|k|+|l|)} \\
\leq\ & B(C')^2 (ab)^3 \left(\sum_{|k|>n} e^{-2\lambda'|k|} \sum_{l \in \mathbb{Z}} e^{-2\lambda'|l|} + \sum_{|l|>n} e^{-2\lambda'|l|} \sum_{k \in \mathbb{Z}} e^{-2\lambda'|k|} \right) \\
\leq\ & 8B(C')^2 (ab)^3 \frac{e^{-2\lambda'}}{(1 - e^{-2\lambda'})^2} e^{-2\lambda' n}.
\end{aligned}
$$

Now the result follows from Theorem 23.4.2. $\qquad\square$

23.6   The Finite Section Method

The finite section method is a standard tool to approximate solutions to an infinite system of linear equations,

$$Kx = y, \qquad\qquad (23.28)$$

where $K : \ell^2(\mathbb{Z}^2) \to \ell^2(\mathbb{Z}^2)$ is a given invertible operator and $y \in \ell^2(\mathbb{Z}^2)$ is a given vector. Depending on the properties of the given operator K, several variants of the method can be found in the literature. The problem of inverting the frame operator can easily be turned into the form (23.28):

Example 23.6.1 Consider a frame $\{f_k\}_{k=1}^\infty$ for a Hilbert space $\mathcal{H}$, with frame operator $S : \mathcal{H} \to \mathcal{H}$. Then, to find $S^{-1} f$ for a given $f \in \mathcal{H}$ amounts to solve the equation

$$Sg = f. \qquad\qquad (23.29)$$

Letting now $\{e_k\}_{k=1}^{\infty}$ denote any orthonormal basis for $\mathcal{H}$, we can write any $g \in \mathcal{H}$ as

$$g = \sum_{k=1}^{\infty} c_k e_k \qquad (23.30)$$

for some coefficients $\{c_k\}_{k=1}^{\infty} \in \ell^2(\mathbb{N})$. If g is a solution to (23.29), then $f = Sg = \sum_{k=1}^{\infty} c_k Se_k$; thus

$$\langle f, e_j \rangle = \sum_{k=1}^{\infty} c_k \langle Se_k, e_j \rangle, \ j \in \mathbb{N}. \qquad (23.31)$$

This is an infinite set of linear equations to determine the coefficients $\{c_k\}_{k=1}^{\infty}$ in the solution g in (23.30). Letting K denote the bi-infinite matrix where the jkth entry is $\langle Se_k, e_j \rangle$, the system of equations takes the form (23.28) with $y := \{\langle f, e_j \rangle\}_{j=1}^{\infty}$ and $x = \{c_k\}_{k=1}^{\infty}$. We note that K simply is the matrix representation of the operator S and hence is a positive operator. $\qquad \square$

In the rest of this section, we give a short presentation of the finite section methods for positive operators. In Section 23.7 we apply the results to Gabor frames.

For $y \in \ell^2(\mathbb{Z}^2)$ and $n \in \mathbb{N}$, we define the orthogonal projections P_n by

$$P_n : \ell^2(\mathbb{Z}^2) \to \ell^2(\mathbb{Z}^2), \ (P_n y)_{k,l} = \begin{cases} y_{k,l} & \text{if } \max\{|k|, |l|\} \leq n, \\ 0 & \text{otherwise.} \end{cases} \qquad (23.32)$$

The first version of the method reads as follows:

Lemma 23.6.2 *Let $K : \ell^2(\mathbb{Z}^2) \to \ell^2(\mathbb{Z}^2)$ be a bounded and invertible operator and $\{K_n\}_{n=1}^{\infty}$ a sequence of positive bounded operators on $\ell^2(\mathbb{Z}^2)$ which converge strongly to K. If each operator K_n maps $P_n \ell^2(\mathbb{Z}^2)$ onto itself (thus, K_n restricted to this space is invertible) and there exists a constant $C > 0$ such that*

$$CI \leq K_n \text{ on } P_n \ell^2(\mathbb{Z}^2), \ \forall n \in \mathbb{N}, \qquad (23.33)$$

then

$$K_n^{-1} P_n y \to K^{-1} y, \ \forall y \in \ell^2(\mathbb{Z}^2). \qquad (23.34)$$

Proof. Let $y \in \ell^2(\mathbb{Z}^2)$. Then

$$\left\| K^{-1} y - K_n^{-1} P_n y \right\| \leq \left\| K^{-1} y - P_n K^{-1} y \right\| + \left\| P_n K^{-1} y - K_n^{-1} P_n y \right\|.$$

We see immediately that $\left\| K^{-1} y - P_n K^{-1} y \right\| \to 0$ as $n \to \infty$. Thus we have to show that also the second term converges to zero. Now, for any

$n \in \mathbb{N}$, Theorem 2.4.3 implies that $K_n^{-1} \leq \frac{1}{C}I$ on $P_n \ell^2(\mathbb{Z}^2)$, so

$$\left\| \left(K_n \mid_{P_n \ell^2(\mathbb{Z}^2)} \right)^{-1} \right\| \leq \frac{1}{C}.$$

Thus

$$\begin{aligned}
\left\| P_n K^{-1}y - K_n^{-1} P_n y \right\| &\leq \left\| \left(K_n \mid_{P_n \ell^2(\mathbb{Z}^2)} \right)^{-1} \right\| \left\| K_n P_n K^{-1}y - P_n y \right\| \\
&\leq \frac{1}{C} \left\| K_n P_n K^{-1}y - P_n y \right\| \\
&\to 0 \text{ as } n \to \infty.
\end{aligned}$$

$\square$

Note that the proof of Lemma 23.6.2 does not give much information about the speed of convergence in (23.34). A natural candidate for the operator K_n in Lemma 23.6.2 is $K_n := P_n K P_n$, a so-called *finite section* of the operator K; however, we cannot be sure that this operator satisfies the technical condition (23.33). We will now formulate a slightly revised version of Lemma 23.6.2 for this particular choice of K_n, which also yields an error estimate. The result is taken from [99] and [356].

Lemma 23.6.3 *Let* $K : \ell^2(\mathbb{Z}^2) \to \ell^2(\mathbb{Z}^2)$ *be a bounded and invertible operator, and assume that the finite section* $K_n := P_n K P_n$ *is invertible on* $P_n \ell^2(\mathbb{Z}^2)$ *for some* $n \in \mathbb{N}$. *Then*

$$\|K^{-1}y - K_n^{-1} P_n y\| \leq \left(1 + \|K_n^{-1} P_n\| \, \|K\| \right) \|(I - P_n)K^{-1}y\|, \ \forall y \in \ell^2(\mathbb{Z}^2).$$

Proof. We immediately get that

$$\|K^{-1}y - K_n^{-1} P_n y\| \leq \|(I - P_n)K^{-1}y\| + \|P_n K^{-1}y - K_n^{-1} P_n y\|. \quad (23.35)$$

Now observe that

$$\begin{aligned}
K_n^{-1} P_n K (1 - P_n) K^{-1}y &= K_n^{-1} P_n y - K_n^{-1} P_n K P_n K^{-1}y \\
&= K_n^{-1} P_n y - P_n K^{-1}y;
\end{aligned}$$

inserting this in (23.35) yields that

$$\begin{aligned}
\|K^{-1}y &- K_n^{-1} P_n y\| \\
&\leq \|(I - P_n)K^{-1}y\| + \|K_n^{-1} P_n K (1 - P_n) K^{-1}y\| \\
&\leq \|(I - P_n)K^{-1}y\| + \|K_n^{-1} P_n\| \, \|K\| \, \|(1 - P_n) K^{-1}y\|,
\end{aligned}$$

which leads to the desired result. $\square$

If there exists an $n_0 \in \mathbb{N}$ such that the operator $K_n = P_n K P_n$ is invertible on $P_n \ell^2(\mathbb{Z}^2)$ for all $n \geq n_0$ and $\sup_{n \geq n_0} \|K_n^{-1} P_n\| < \infty$, then Lemma 23.6.3 yields that

$$\|K^{-1}y - K_n^{-1} P_n y\| \leq C \, \|(I - P_n)K^{-1}y\|, \ \forall n \geq n_0,$$

for some constant $C > 0$. Note that a similar term as $||(I - P_n)K^{-1}y||$ also appeared in the estimates for the methods in Section 23.2; see, e.g., Theorem 23.2.5. A more general version of Lemma 23.6.3 on weighted ℓ^p−spaces and quantitative error estimates are given in the paper [356] by Gröchenig, Rzeszotnik, and Strohmer.

23.7   The Finite Section Method for Gabor Frames

In this section we present a direct method for approximation of the dual frame for a Gabor frame $\{E_{kb}T_{la}g\}_{k,l\in\mathbb{Z}}$. It was developed by Strohmer [588] and is based on the finite section method in Section 23.6.

Formula (12.33) and Theorem 13.1.1 are the main ingredients for this approach. The key point is that the approximation problem for Gabor frames can be translated into an approximation problem for Gabor Riesz sequences via the duality principle.

The starting point is to return to the discussion in Section 12.4 and consider the analysis operator H associated with the frame $\{E_{kb}T_{la}g\}_{k,l\in\mathbb{Z}}$, i.e.,

$$H : L^2(\mathbb{R}) \to \ell^2(\mathbb{Z}^2), \ \ Hf = \{\langle f, E_{m/a}T_{n/b}g\rangle\}_{m,n\in\mathbb{Z}}. \quad (23.36)$$

Letting again P_n denote the orthogonal projection defined in (23.32), we define truncated versions of the operator H in (23.36) by

$$H_n : L^2(\mathbb{R}) \to \ell^2(\mathbb{Z}^2), \ \ H_nf = P_nHf,$$

which we identify with

$$H_n : L^2(\mathbb{R}) \to \mathbb{C}^{(2n+1)^2}, \ \ H_nf = \{\langle f, E_{k/a}T_{l/b}g\rangle\}_{|k|,|l|\leq n}. \quad (23.37)$$

The matrix

$$H_nH_n^* = P_nHH^*P_n = \{\langle E_{k'/a}T_{l'/b}g, E_{k/a}T_{l/b}g\rangle\}_{|k|,|l|,|k'|,|l'|\leq n}, \quad (23.38)$$

is a finite section of the infinite-dimensional matrix HH^*. Motivated by (12.30), we let

$$\gamma^{(n)} := abH_n^*(H_nH_n^*)^{-1}P_n\{\delta_{k,0}\delta_{l,0}\}_{k,l\in\mathbb{Z}} \quad (23.39)$$
$$= abH_n^*(H_nH_n^*)^{-1}\{\delta_{k,0}\delta_{l,0}\}_{|k|,|l|\leq n} \quad \text{for } n \in \mathbb{N}.$$

We now prove that the functions $\gamma^{(n)}$ in (23.39) indeed converge to $S^{-1}g$ for $n \to \infty$.

Theorem 23.7.1 *Let $g \in L^2(\mathbb{R})$ and $a, b > 0$ be given, and assume that $\{E_{kb}T_{la}g\}_{k,l\in\mathbb{Z}}$ is a frame for $L^2(\mathbb{R})$. Then*

$$\gamma^{(n)} \to S^{-1}g \quad \text{for } n \to \infty.$$

Proof. Letting A, B be frame bounds for $\{E_{kb}T_{la}g\}_{k,l\in\mathbb{Z}}$, we know by Theorem 13.1.1 that $\{E_{k/a}T_{l/b}g\}_{k,l\in\mathbb{Z}}$ is a Riesz sequence with bounds abA, abB. In particular, for each finite scalar sequence $\{c_{k,l}\}$,

$$abA \sum_{|k|,|l|\leq n} |c_{k,l}|^2 \leq \left\| \sum_{|k|,|l|\leq n} c_{k,l} E_{k/a}T_{l/b}g \right\|^2 \leq abB \sum_{|k|,|l|\leq n} |c_{k,l}|^2.$$

In terms of the operator H_n, this means that

$$\|H_n H_n^*\| = \|H_n^*\|^2 \geq abA.$$

Since $H_n H_n^* \to HH^*$ strongly for $n \to \infty$, we can now apply Lemma 23.6.2 to conclude that

$$(H_n H_n^*)^{-1} P_n \to (HH^*)^{-1} \text{ for } n \to \infty.$$

Now the definition of $\gamma^{(n)}$ in (23.39) combined with (12.30) yields that

$$\begin{aligned}
\gamma^{(n)} &= abH_n^*(H_n H_n^*)^{-1} P_n \{\delta_{k,0}\delta_{l,0}\}_{k,l\in\mathbb{Z}} \\
&\to abH^*(HH^*)^{-1}\{\delta_{k,0}\delta_{l,0}\}_{k,l\in\mathbb{Z}} = S^{-1}g.
\end{aligned}$$

$\square$

In [588] Strohmer also proves that the above method converges exponentially if g as well as the Fourier transform $\widehat{g}$ decay exponentially. That is, the assumptions imply that for some constants $C', \lambda' > 0$,

$$\left\| S^{-1}g - \gamma^{(n)} \right\| \leq C' e^{-\lambda' n}$$

The proof uses Lemma 12.4.3 and is not constructive. It would be very useful to have knowledge of concrete values of C', λ'.

Prior to the paper [588], Strohmer proved similar results for approximation of the inverse frame operator associated to shift-invariant systems in $\ell^2(\mathbb{Z})$. We refer to [587] for details.

23.8 Exercises

23.1 Prove that the near-Riesz basis in Example 23.1.3 is not a Riesz frame.

23.2 Prove that for an arbitrary frame $\{f_k\}_{k=1}^\infty$ and $n \in \mathbb{N}$, one can choose a set J_n such that (23.13) is satisfied.

24

Expansions in Banach Spaces

The material presented in this book naturally splits in two parts: a functional analytic treatment of frames in general Hilbert spaces, and a more direct approach to structured frames like Gabor frames and wavelet frames. For the second part the most general results were presented in Chapter 21, in the setting of generalized shift-invariant systems on an LCA group.

The current chapter is in a certain sense a natural continuation of both tracks. We consider connections between frame theory and abstract harmonic analysis and show how we can construct frames in Hilbert spaces via the theory for group representations. In special cases the general approach will bring us back to the Gabor systems and wavelet systems. The abstract framework adds another new aspect to the theory: we will not only obtain expansions in Hilbert spaces but also in a class of Banach spaces.

In Section 24.1 we show how the orthogonality relations for square-integrable group representations lead to integral representations in terms of continuous frames for the underlying Hilbert space; on a concrete level, this gives an alternative approach to Gabor systems and wavelet systems. Section 24.2 presents that basics of Feichtinger–Gröchenig theory, showing that the group-theoretic setup allows us to obtain frames, as well as series expansions in a large scale of Banach spaces. This naturally led Gröchenig to define frames in Banach spaces, which are discussed in Section 24.3. In Section 24.4 we consider p-frames. They are defined by removing some of Banach frame conditions and were first studied separately by Aldroubi, Sun, and Tang in the context of the L^p-spaces. Later they were generalized to other classes of Banach spaces. Finally, Section 24.5 discusses a few aspect of the theory for Gabor systems and wavelet systems in L^p-spaces.

© Springer International Publishing Switzerland 2016   601
O. Christensen, *An Introduction to Frames and Riesz Bases*,
Applied and Numerical Harmonic Analysis,
DOI 10.1007/978-3-319-25613-9_24

This chapter is more advanced than the previous chapters. It is less detailed and also states open problems for future research.

24.1   Representations of Locally Compact Groups

The elements in a group can be quite abstract objects, so it is desirable to transfer questions on a group into a more familiar setting. This is done by the concept of a *group representation*, which identifies (see the comment after the definition) the group elements with certain operators on a Hilbert space.

Definition 24.1.1 *Let $\mathcal{G}$ be a locally compact group with left Haar measure μ, and let $\mathcal{H}$ be a Hilbert space. A representation of $\mathcal{G}$ on $\mathcal{H}$ is a family of bounded invertible operators $\{\pi(x)\}_{x\in\mathcal{G}}$ on $\mathcal{H}$ for which*

(i) $\pi(xy) = \pi(x)\pi(y), \ \forall x, y \in \mathcal{G}$.

(ii) for all $f \in \mathcal{H}$, the mapping $x \mapsto \pi(x)f$ is continuous from $\mathcal{G}$ into $\mathcal{H}$.

We further say that

(iii) π is unitary if all the operators $\{\pi(x)\}_{x\in\mathcal{G}}$ are unitary.

(iv) π is irreducible if the only closed subspaces of $\mathcal{H}$ which are invariant under all the operators $\{\pi(x)\}_{x\in\mathcal{G}}$ are $\{0\}$ and $\mathcal{H}$.

(v) A unitary irreducible representation π is integrable if

$$\mathcal{A} := \left\{ f \in \mathcal{H} \mid \int_{\mathcal{G}} |\langle \pi(x)f, f \rangle| d\mu(x) < \infty \right\} \neq \{0\}. \qquad (24.1)$$

A square-integrable representation is defined similarly.

Condition (ii) (called *strong continuity* of π) is not always part of the definition of a group representation. As said before, the idea behind a group representation is to identify elements in $\mathcal{G}$ with operators. For this to hold, we also need the mapping $x \mapsto \pi(x)$ to be injective; a representation with this property is said to be *proper*.

A representation π is irreducible if and only if (Exercise 24.2)

$$\overline{\text{span}}\{\pi(x)g\}_{x\in\mathcal{G}} = \mathcal{H}, \ \forall g \in \mathcal{H} \setminus \{0\}.$$

Assuming that π is irreducible and fixing an arbitrary $g \in \mathcal{H}\setminus\{0\}$, we can thus approximate any $f \in \mathcal{H}$ arbitrarily well by finite linear combinations of vectors $\pi(x)g, x \in \mathcal{G}$. It is therefore very natural to ask if we can find $g \in \mathcal{H}$ and a sequence $\{x_k\}_{k=1}^{\infty}$ in $\mathcal{G}$ such that $\{\pi(x_k)g\}_{k=1}^{\infty}$ is a frame. The answer turns out to be yes in a very general case if π is an integrable representation. Before we present results in that direction, we give some concrete examples of groups and their representations.

Example 24.1.2 The *Heisenberg group* is the set $G := \mathbb{R} \times \mathbb{R} \times \mathbb{T}$ equipped with the product topology and the group composition

$$(a_1, b_1, t_1) \cdot (a_2, b_2, t_2) = (a_1 + a_2, b_1 + b_2, t_1 t_2 e^{2\pi i b_1 a_2}).$$

The Heisenberg group is not abelian, but it is unimodular and the Haar measure is the product measure of the three involved Lebesgue measures; see, e.g., [395]. The definition of the group composition implies that we can define a representation of G on $L^2(\mathbb{R})$ by

$$[\pi(a, b, t)g](y) = te^{2\pi i b(y-a)}g(y-a), \quad g \in L^2(\mathbb{R}), (a, b, t) \in G, y \in \mathbb{R}. \quad (24.2)$$

This is the *Schrödinger representation*. To see that π actually defines a representation, note that in terms of the operators E_a and T_b from Section 2.9,

$$[\pi(a, b, t)g](y) = te^{-2\pi i ab}E_b T_a g(y).$$

Using the commutator relations for the operators E_a and T_b, one can now prove that (i) of Definition 24.1.1 is satisfied (Exercise 24.3) and (ii) follows by Lemma 2.9.2, which also shows that π is unitary. To see that π is irreducible, let $g \in \mathcal{H} \setminus \{0\}$ and assume that $f \perp \pi(a, b, t)g$ for all $(a, b, t) \in G$. Then, by Proposition 11.1.2,

$$
\begin{aligned}
0 &= \int_{-\infty}^{\infty} \int_{-\infty}^{\infty} \int_0^1 |\langle f, \pi(a, b, t)g \rangle|^2 \, dt \, da \, db \\
&= \int_{-\infty}^{\infty} \int_{-\infty}^{\infty} |\langle f, E_a T_b g \rangle|^2 \, da \, db \\
&= \|f\|^2 \|g\|^2.
\end{aligned}
$$

Therefore $f = 0$ and π is irreducible. We also observe that for the Schrödinger representation, the set in (24.1) is $\mathcal{A} = S_0$, the Feichtinger algebra, which is dense in $L^2(\mathbb{R})$; thus π is integrable. Finally,

$$[\pi(na, mb, 1)g](y) = e^{-2\pi i mnab}E_{mb}T_{na}g(y), \quad m, n \in \mathbb{Z},$$

i.e., up to a (usually irrelevant) factor of absolute value 1, the Schrödinger representation sampled on the set $\{(na, mb, 1)\}_{m,n \in \mathbb{Z}}$ and applied to $g \in L^2(\mathbb{R})$ corresponds to the regular Gabor system $\{E_{mb}T_{na}g\}_{m,n \in \mathbb{Z}}$.

A technical detail: the torus component in the Heisenberg group will never play any practical role in this context. It is only introduced in order to obtain a group representation involving the operators $E_a T_b$; in fact, the operators defined by $\rho(a, b) = E_a T_b$ do not form a representation of $\mathbb{R}^2$ on $L^2(\mathbb{R})$; they form a so-called *projective group representation*; see Exercise 24.3, page 611, and [156]. □

Example 24.1.3 The $ax + b$ *group* is the set $\mathcal{G} = \mathbb{R} \times (\mathbb{R} \setminus \{0\})$ equipped with the product topology and the composition

$$(b, a) \cdot (x, s) = (ax + b, as).$$

The left Haar measure is $\frac{1}{a^2} dadb$ and the right Haar measure is $\frac{1}{|a|} dadb$, where $dadb$ is the Lebesgue measure on $\mathbb{R}^2$. In particular, the group is not unimodular. The $ax + b$ group is also called the *affine group* in the literature; we refer to [335] for a more detailed discussion of its properties. One can define a unitary representation on $L^2(\mathbb{R})$ by

$$[\pi(b, a)f](y) = T_b D_a f(y) = \frac{1}{\sqrt{|a|}} f(\frac{y - b}{a}), \ (b, a) \in \mathcal{G}, \ f \in L^2(\mathbb{R}), \ y \in \mathbb{R}.$$

Note that

$$[\pi(bka^j, a^j)g](y) = a^{-j/2} g(a^{-j}y - kb), \ j, k \in \mathbb{Z},$$

i.e., the wavelet systems appear by appropriate samplings of the representation. The representation satisfies the integrability condition (24.1), but is not irreducible. This is not a problem in practice: it is possible to extend the $ax + b$ group to a larger group such that an appropriate extension of π is a unitary irreducible representation satisfying the integrability condition. We shall not go into the technical details, but this is the reason that we still speak about this representation in the context of integrable representations (see, e.g., the discussion on page 605). We refer to [311] for a description of the role played by irreducibility in the general case. $\square$

Given a representation π of $\mathcal{G}$ on $\mathcal{H}$, we choose $g \in \mathcal{H}$ and consider the transformation

$$V_g : \mathcal{H} \to C(\mathcal{G}), \ V_g(f)(x) = \langle f, \pi(x)g \rangle. \tag{24.3}$$

Here $C(\mathcal{G})$ denotes the set of continuous complex-valued functions on $\mathcal{G}$. With our convention for the inner product, V_g is a linear operator – but it depends conjugated linear on g. Note that if π is the representation of the $ax + b$ group considered in Example 24.1.3, then $V_g(f)$ is the continuous wavelet transform of f with respect to g. For this reason, the misleading word "wavelet transform" has also been associated to the transform in the general case. The correct terminology used in abstract harmonic analysis is that $V_g(f)$ is a *representation coefficient* for the representation π. Similarly, g has frequently been called the "mother wavelet," while "analyzing atom" or "generator" is more appropriate.

Our purpose is to show how an integrable group representation π leads to expansions of the elements in the Hilbert space associated with π, as well as in a class of related Banach spaces. The starting point is the *orthogonality relations*, first proved in [261] (see also [334]). They give an expression for the inner product between two representation coefficients in $L^2(\mathcal{G})$ in terms of an (in general unbounded) operator U on a domain $\mathcal{D}(U) \subseteq \mathcal{H}$:

Theorem 24.1.4 *Let $\mathcal{G}$ be a locally compact group with left Haar measure μ, and assume that π is a square-integrable representation of $\mathcal{G}$ on $\mathcal{H}$. Then there exists a unique positive self-adjoint operator*

$$U : \mathcal{D}(U) \subseteq \mathcal{H} \to \mathcal{H},$$

such that

(i) $V_g(g) \in L^2(\mathcal{G}) \Leftrightarrow g \in \mathcal{D}(U)$.

(ii) *For all $g_1, g_2 \in \mathcal{D}(U)$ and $f_1, f_2 \in \mathcal{H}$,*

$$\int_{\mathcal{G}} \overline{\langle f_1, \pi(x)g_1 \rangle} \langle f_2, \pi(x)g_2 \rangle d\mu(x) = \langle Ug_1, Ug_2 \rangle \langle f_2, f_1 \rangle. \qquad (24.4)$$

The domain $\mathcal{D}(U)$ is dense in $\mathcal{H}$. If $\mathcal{G}$ is unimodular, then $\mathcal{D}(U) = \mathcal{H}$ and U is a multiple of the identity on $\mathcal{H}$.

Theorem 24.1.4 immediately leads to continuous frames as discussed in Section 5.6:

Corollary 24.1.5 *Let π be a square-integrable representation of $\mathcal{G}$ on $\mathcal{H}$. Then, for all $g \in \mathcal{D}(U) \setminus \{0\}$, $\{\pi(x)g\}_{x \in \mathcal{G}}$ is a tight continuous frame for $\mathcal{H}$ (with respect to $\mathcal{G}$ equipped with the left Haar measure). In particular, this holds for all $g \in \mathcal{H} \setminus \{0\}$ if $\mathcal{G}$ is unimodular.*

Corollary 24.1.5 gives an abstract explanation of the differences we have observed between Gabor analysis and wavelet analysis. The Weyl–Heisenberg group is unimodular and the Schrödinger representation is square-integrable, so all $g \in L^2(\mathbb{R}) \setminus \{0\}$ leads to continuous frames, in accordance with our direct proof in Corollary 11.1.4. On the other hand the $ax+b$-group is not unimodular, so there might be $g \in L^2(\mathbb{R}) \setminus \{0\}$ which does not generate a continuous frame $\{\pi(x)g\}_{x \in \mathcal{G}}$. This fact is expressed by the admissibility condition in Corollary 15.1.2.

It is worth noting that the two representation coefficients appearing in the orthogonality relations (24.4) might be with respect to different analyzing atoms g_1, g_2 in the transforms. In Proposition 15.1.1 we did not use this freedom: the same function ψ was used for both of the appearing wavelet transforms. Additional freedom is actually obtained if we choose different functions for the two wavelet transforms: this allows to obtain dual continuous frame pairs in $L^2(\mathbb{R})$ rather than just tight frames.

24.2 Feichtinger–Gröchenig Theory

The Feichtinger–Gröchenig theory was presented in a series of papers appearing around 1990. The purpose of the papers was to obtain series expansions in a large class of Banach spaces based on the theory for integrable group representations. We will give a short introduction to the theory and refer to [280], [281], and [336] for more details and further results.

Let $\mathcal{G}$ be a locally compact group with left Haar measure μ. Define the translation operator T_x, $x \in \mathcal{G}$, acting on functions $f : \mathcal{G} \to \mathbb{C}$, by

$$(T_x f)(y) = f(x^{-1}y), \ y \in \mathcal{G}.$$

In the generality discussed here, T_x is called the *left regular representation*; in the special case $\mathcal{G} = \mathbb{R}$, it equals our translation operator in Section 2.9.

For functions $F, G \in L^1(\mathcal{G})$, the *convolution* $F * G : \mathcal{G} \to \mathbb{C}$ is defined by

$$F * G(y) \quad := \quad \int_{\mathcal{G}} F(x)G(x^{-1}y)d\mu(x) = \int_{\mathcal{G}} F(x)T_x G(y)d\mu(x), \ y \in \mathcal{G}.$$

The assumption $F, G \in L^1(\mathcal{G})$ implies that $F * G$ is well-defined and belongs to $L^1(\mathcal{G})$. However, the convolution is well-defined under many other conditions on F, G, and we will use the convolution symbol for any pair of functions F, G for which $F * G$ is a well-defined function.

If the function F does not oscillate too much, the convolution $F * G$ can be considered as the limit of a sequence of linear combinations of translates of the function G, with weights determined by F.

Lemma 24.2.1 below relates convolution and the orthogonality relations. It gives a reformulation of the orthogonality relations, which is the starting point for Feichtinger–Gröchenig theory.

Lemma 24.2.1 *Let π be a square-integrable representation of $\mathcal{G}$ on $\mathcal{H}$. Then the following hold:*

(i) $V_g(\pi(y)f) = T_y V_g(f), \ \forall f \in \mathcal{H}, \ y \in \mathcal{G}.$

(ii) *The operator U introduced in Theorem 24.1.4 is injective.*

(iii) *Choosing $g \in \mathcal{D}(U)$ such that $||Ug|| = 1$, we have*

$$V_g(f) = V_g(f) * V_g(g), \ \forall f \in \mathcal{H}, \tag{24.5}$$

and the orthogonal projection of $L^2(\mathcal{G})$ onto the range $\mathcal{R}_{V_g}$ of V_g is

$$F \mapsto F * V_g(g), \ F \in L^2(\mathcal{G}). \tag{24.6}$$

Proof. (i) follows by computation. To prove (ii) we let $g \in \mathcal{D}(U)$, and assume that $Ug = 0$. Via the orthogonality relations, we see that for $f \neq 0$,

$$0 = ||Ug||^2 = \frac{1}{||f||^2} \int_{\mathcal{G}} |\langle f, \pi(x)g \rangle|^2 d\mu(x);$$

by the continuity of the representation coefficients we conclude that

$$\langle f, \pi(x)g \rangle = 0 \text{ for all } x \in \mathcal{G}.$$

Since this holds for all $f \neq 0$, we have $\pi(x)g = 0$ for all $x \in \mathcal{G}$, and therefore $g = 0$.

For the proof of (iii) we first show that $F * V_g(g)$ is well defined for any $F \in L^2(\mathcal{G})$. Note that by (i) and the left invariance of the Haar measure,

$$\int_{\mathcal{G}} |V_g(\pi(y)g)(x)|^2 d\mu(x) = \int_{\mathcal{G}} |V_g(g)(y^{-1}x)|^2 d\mu(x)$$
$$= \int_{\mathcal{G}} |V_g(g)(x)|^2 d\mu(x) < \infty,$$

i.e., $V_g(\pi(y)g) \in L^2(\mathcal{G})$. Since

$$F(x)V_g(g)(x^{-1}y) = F(x)\langle g, \pi(x^{-1}y)g \rangle = F(x)\overline{\langle \pi(y)g, \pi(x)g \rangle}$$
$$= F(x)\overline{V_g(\pi(y)g)(x)}, \tag{24.7}$$

it follows that the function

$$x \mapsto F(x)V_g(g)(x^{-1}y)$$

is integrable for $y \in \mathcal{G}$, i.e., that $F * V_g(g)$ is well defined.

As a consequence of (ii), an arbitrary $g \in \mathcal{D}(U) \setminus \{0\}$ can be normalized such that $\|Ug\| = 1$. Doing so, and applying the orthogonality relations with $g_1 = g_2 = g$, $f_1 = \pi(y)g$, and $f_2 = f$ for an arbitrary $f \in \mathcal{H}$,

$$V_g(f)(y) = \langle f, \pi(y)g \rangle$$
$$= \int_{\mathcal{G}} \overline{\langle \pi(y)g, \pi(x)g \rangle}\langle f, \pi(x)g \rangle d\mu(x)$$
$$= \int_{\mathcal{G}} \langle g, \pi(x^{-1}y)g \rangle\langle f, \pi(x)g \rangle d\mu(x)$$
$$= V_g(f) * V_g(g)(y).$$

This proves in particular that the mapping $F \mapsto F * V_g(g)$ is the identity on $\mathcal{R}_{V_g}$. For the second part of (iii) we only need to prove that $F * V_g(g) = 0$ for all F belonging to the orthogonal complement of $\mathcal{R}_{V_g}$ in $L^2(\mathcal{G})$. But for these F, (24.7) shows that

$$F * V_g(g)(y) = \int_{\mathcal{G}} F(x)V_g(g)(x^{-1}y)d\mu(x) = \langle F, V_g(\pi(y)g) \rangle = 0,$$

as desired. $\qquad\square$

The result in (i) is expressed by saying that V_g is an *intertwining operator* for the representations $\pi(x)$ and T_x. Formula 24.5 gives an *integral representation* of all functions $V_g(f)$,

$$V_g(f)(y) = V_g(f) * V_g(g)(y) = \int_{\mathcal{G}} V_g(f)(x)T_x V_g(g)(y)d\mu(x), \quad f \in \mathcal{H}, y \in \mathcal{G}.$$

With our interpretation of the convolution on page 606, it is natural to search for a representation of $F = V_g(f)$ via an infinite superposition of translates of $V_g(g)$; formulated in short, to search for expansions

$$F = \sum_{k=1}^{\infty} c_k(F) T_{x_k} V_g(g), \ F \in \mathcal{R}_{V_g}. \tag{24.8}$$

The question is how to choose the points $\{x_k\}_{k=1}^{\infty}$ in $\mathcal{G}$ and the coefficient functionals c_k. We will base the choice of $\{x_k\}_{k=1}^{\infty}$ on our knowledge from series expansions via a Gabor frame.

Recall from Theorem 11.3.1 that in order for a Gabor system $\{E_{mb}T_{na}g\}_{m,n\in\mathbb{Z}}$ to be a frame, it is necessary that $ab \leq 1$. We can interpret this in terms of the Schrödinger representation π of the Weyl–Heisenberg group, discussed in Example 24.1.2: $\{E_{mb}T_{na}g\}_{m,n\in\mathbb{Z}}$ corresponds (up to a constant) to $\{\pi(na, mb, 1)g\}_{m,n\in\mathbb{Z}}$, so the condition $ab \leq 1$ means that the points $\{(na, mb)g\}_{m,n\in\mathbb{Z}}$ have to be "sufficiently dense" in $\mathbb{R}^2$. On the other hand, for an irregular Gabor family $\{\pi(x_k, y_k, 1)g\}_{k=1}^{\infty}$ to be a frame, the points $\{(x_k, y_k)\}_{k=1}^{\infty}$ are not allowed to be "too dense" (Exercise 24.4). Motivated by these considerations, we introduce some definitions related to general locally compact groups.

Definition 24.2.2 *Let $\{x_k\}_{k=1}^{\infty}$ be a sequence in $\mathcal{G}$.*

(i) *Let $V \in \mathcal{O}(e)$ be relatively compact. If*

$$\bigcup_{k=1}^{\infty} x_k V = \mathcal{G},$$

then $\{x_k\}_{k=1}^{\infty}$ is said to be V-dense.

(ii) *If there exists a relatively compact neighborhood $V \in \mathcal{O}(e)$ such that $x_k V \cap x_j V = \emptyset$ for $k \neq j$, then $\{x_k\}_{k=1}^{\infty}$ is said to be separated. $\{x_k\}_{k=1}^{\infty}$ is relatively separated if it is a finite union of separated sets.*

In order to derive appropriate coefficient functionals such that (24.8) holds, we need to introduce a new version of the *partition of unity condition*:

Definition 24.2.3 *Let $V \in \mathcal{O}(e)$ be compact. A family $\Psi = \{\psi_k\}_{k=1}^{\infty}$ of continuous functions on $\mathcal{G}$ is a partition of unity of size V if*

(i) $0 \leq \psi_k(x) \leq 1$ *and* $\sum_{k=1}^{\infty} \psi_k(x) = 1, \ \forall x \in \mathcal{G}.$

(ii) *There exists a relatively separated and V-dense set $\{x_k\}_{k=1}^{\infty}$ in $\mathcal{G}$ for which* $\text{supp } \psi_k \subseteq x_k V, \ \forall k \in \mathbb{N}.$

It is important to notice that such partitions of unity can be constructed for arbitrarily small neighborhoods $V \in \mathcal{O}(e)$ (see [275]).

We will now state a special case of the main result in [280], formulated within the frame work of Hilbert spaces. Recall that the set $\mathcal{A}$ was introduced in Definition 24.1.1 (v).

Theorem 24.2.4 *Let $\mathcal{G}$ be a unimodular locally compact group and π an integrable representation of $\mathcal{G}$ on $\mathcal{H}$. Given $g \in \mathcal{A} \setminus \{0\}$, there exists a neighborhood $V \in \mathcal{O}(e)$ with the following property: for every V-dense and relatively separated family $\{x_k\}_{k=1}^{\infty}$, there exists a bounded operator*

$$\Lambda : \mathcal{H} \to \ell^2(\mathbb{N}), \quad \Lambda f = \{\lambda_k(f)\}_{k=1}^{\infty},$$

such that

$$f = \sum_{k=1}^{\infty} \lambda_k(f)\pi(x_k)g, \ \forall f \in \mathcal{H}. \tag{24.9}$$

Proof. We will not give a full proof but only sketch the main points. The basic idea is to approximate the convolution operator $F \mapsto F * V_g(g)$ by operators of the type

$$C_\Psi : \mathcal{R}_{V_g} \to \mathcal{R}_{V_g}, \quad C_\Psi(F) = \sum_{k=1}^{\infty} \langle F, \psi_k \rangle T_{x_k} V_g(g),$$

where $\Psi = \{\psi_k\}_{k=1}^{\infty}$ is a partition of unity of size V. One can prove that if V is chosen small enough and $\{x_k\}_{k=1}^{\infty}$ denotes a set of points in $\mathcal{G}$ as in Definition 24.2.3, then $C_\Psi(F)$ is well defined and for some constant $C < 1$,

$$\|F * V_g(g) - C_\Psi(F)\| \leq C\|F\|, \ \forall F \in \mathcal{R}_{V_g}.$$

Since the operator $F \mapsto F * V_g(g)$ is the identity on the Banach space $\mathcal{R}_{V_g}$ according to Lemma 24.2.1, this implies that the operator C_Ψ is invertible on $\mathcal{R}_{V_g}$; thus each $F \in \mathcal{R}_{V_g}$ has a representation

$$F = C_\Psi C_\Psi^{-1}(F) = \sum_{k=1}^{\infty} \langle C_\Psi^{-1}(F), \psi_k \rangle T_{x_k} V_g(g).$$

That is, for $f \in \mathcal{H}$,

$$V_g(f) = \sum_{k=1}^{\infty} \langle C_\Psi^{-1}(V_g(f)), \psi_k \rangle T_{x_k} V_g(g).$$

Applying the intertwining property in Lemma 24.2.1(i), we obtain a representation of $f \in \mathcal{H}$:

$$\begin{aligned}
f = V_g^{-1} V_g(f) &= \sum_{k=1}^{\infty} \langle C_\Psi^{-1}(V_g(f)), \psi_k \rangle V_g^{-1} T_{x_k} V_g(g) \\
&= \sum_{k=1}^{\infty} \langle C_\Psi^{-1}(V_g(f)), \psi_k \rangle \pi(x_k)g.
\end{aligned}$$

The proof that the operator $f \mapsto \{\langle C_\Psi^{-1}(V_g(f)), \psi_k\rangle\}_{k=1}^\infty$ is bounded from $\mathcal{H}$ into $\ell^2(\mathbb{N})$ can be found in [280]. $\qquad\square$

Corollary 24.2.5 *The conditions in Theorem 24.2.4 imply that* $\{\pi(x_k)g\}_{k=1}^\infty$ *is a frame for* $\mathcal{H}$.

Proof. We only verify the lower frame condition and refer to [281] for the proof that $\{\pi(x_k)g\}_{k=1}^\infty$ is a Bessel sequence. Let $f \in \mathcal{H}$. Then, putting the expression for f from Theorem 24.2.4 into the first entry of $\langle f, f\rangle$, we obtain that

$$
\begin{aligned}
\|f\|^4 &= \left[\sum_{k=1}^\infty \lambda_k(f)\langle\pi(x_k)g, f\rangle\right]^2 \\
&\leq \sum_{k=1}^\infty |\lambda_k(f)|^2 \sum_{k=1}^\infty |\langle\pi(x_k)g, f\rangle|^2 \\
&\leq \|\Lambda\|^2 \|f\|^2 \sum_{k=1}^\infty |\langle\pi(x_k)g, f\rangle|^2,
\end{aligned}
$$

from which the result follows. $\qquad\square$

When we apply Theorem 24.2.4 to the Heisenberg group and the Schrödinger representation, we obtain a result about irregular Gabor frames. For the proof we only need to recall from Example 24.1.2 that in this case the set $\mathcal{A}$ in Definition 24.1.1 (v) equals the Feichtinger algebra $\mathcal{S}_0$.

Corollary 24.2.6 *Let* $g \in \mathcal{S}_0 \setminus \{0\}$. *Then there exists an open set* $V \subset \mathbb{R}^2$ *such that* $\{E_{\lambda_k} T_{\mu_k} g\}_{k=1}^\infty$ *is a frame for* $L^2(\mathbb{R})$ *for every separated sequence* $\{(\mu_k, \lambda_k)\}_{k=1}^\infty$ *in* $\mathbb{R}^2$ *for which*

$$
\bigcup_{k=1}^\infty [(\mu_k, \lambda_k) + V] = \mathbb{R}^2.
$$

This short description is far from giving full justice to the work by Feichtinger and Gröchenig; we will now mention a few central points where the theory is more general than described here.

From the sketch of the proof of Theorem 24.2.4 it is not clear why $\mathcal{G}$ needs to be unimodular, and it is in fact an unnecessary assumption. However, without this assumption, we need to be slightly more restrictive with the choice of $g \in \mathcal{H}$. The class of usable $g \in \mathcal{H}$ is still dense in $\mathcal{H}$, but its definition is slightly more involved; see [280].

Of even more importance is the fact that Feichtinger–Gröchenig theory extends to series expansions in a scale of Banach spaces. The key point in [280] is the observation that the convolution identity (24.5) can be extended to hold for a large class of distributions f; here the notation

$$
V_g(f) = \langle f, \pi(x)g\rangle
$$

has to be reinterpreted as the action of the distribution f on $\pi(x)g$. To a large class of Banach function spaces Y (including weighted L^p-spaces), one can associate a sequence space Y_d and a Banach space CoY such that Theorem 24.2.4 holds with $\mathcal{H}$ replaced by CoY and $\ell^2(I)$ replaced by Y_d. The spaces CoY are called *coorbit spaces*, and many classical function spaces are found among these spaces. For example, it is proved in [336] that Besov spaces and Triebel–Lizorkin spaces appear as coorbit spaces via the representation in Example 24.1.3 and certain choices of the function space Y. Via the Schrödinger representation in Example 24.1.2, one obtains the modulation spaces originally introduced by Feichtinger and discussed in detail in [278] and [340]; see also the short introduction in Section A.5.

Note that Feichtinger–Gröchenig theory has been generalized in many different directions over the years. An extension to projective group representations appeared in [156]; this allows to apply the theory directly to the operators $\rho(a,b) = E_a T_b$ instead of extending the group by an irrelevant torus component. Extensions to homogeneous spaces are treated in the papers [234, 235, 236] by Dahlke et al. Also, a generalization to continuous frames not arising from a square-integrable representation was given by Fornasier and Rauhut in [303].

24.3   Banach Frames

In the entire book we have focused on series expansions in Hilbert spaces, with most concrete constructions taking place in $L^2(\mathbb{R})$ and $L^2(\mathbb{R}^d)$. We have already mentioned that a more general viewpoint is possible: the original theory by Feichtinger and Gröchenig takes place in a scale of Banach spaces, and we have also noticed that localized frames lead to series expansions in certain Banach spaces.

Extensions of frame theory to Banach spaces are an important issue, not only from the theoretically point of view: several applications involve signals in Banach spaces and call for associated expansions in terms of well-chosen "building blocks." Mathematically, the extension of frame theory to Banach spaces is highly nontrivial and deeply connected with certain well-known problems in Banach space theory.

The purpose of this section is to introduce some of the natural generalizations of frame theory to Banach spaces. We will highlight some points where the extended theory behaves radically different from the frame theory in Hilbert spaces; and we will also discuss how localization assumptions on a Gabor frame in $L^2(\mathbb{R})$ imply that the associated frame decomposition automatically extends to a class of Banach spaces.

In the entire section we let X denote a Banach space; the *dual Banach space*, i.e., the set of continuous linear functionals $g : X \to \mathbb{C}$, will be denoted by X^*. Natural examples will be to take $X = L^p(\mathbb{R})$ for some

$p \in]1, \infty[$; then the dual space X^* can be identified with $L^q(\mathbb{R})$ for q chosen such that $p^{-1} + q^{-1} = 1$. Another central ingredient will be a Banach space consisting of sequences $\{c_k\}_{k=1}^\infty$, where $c_k \in \mathbb{C}$ for each $k \in \mathbb{N}$; such a space will be called a *Banach sequence space* and will be denoted by X_d.

There are several ways to extend frame theory to Banach spaces. We first state the definition of an *atomic decomposition*:

Definition 24.3.1 *Let X be a Banach space and X_d a Banach sequence space indexed by $\mathbb{N}$. Let $\{f_k\}_{k=1}^\infty$ be a sequence in X and $\{g_k\}_{k=1}^\infty$ a sequence in X^*. Then the pair $(\{g_k\}_{k=1}^\infty, \{f_k\}_{k=1}^\infty)$ is an atomic decomposition of X with respect to X_d if*

  (i) $\{g_k(f)\}_{k=1}^\infty \in X_d$ for all $f \in X$;

  (ii) there exist constants $A, B > 0$ such that
  $$A \, \|f\|_X \leq \|\{g_k(f)\}_{k=1}^\infty\|_{X_d} \leq B \, \|f\|_X, \ \forall f \in X;$$

  (iii) $f = \sum_{k=1}^\infty g_k(f) f_k, \ \forall f \in X.$

Note that the analog to frames in Hilbert spaces would consist only of parts (i) and (ii) in Definition 24.3.1. In a Hilbert space $\mathcal{H}$ the assumptions (i) and (ii) with $X = \mathcal{H}$ and $X_d = \ell^2(\mathbb{N})$ are enough to obtain the frame decomposition, but in the current Banach space setting they do not imply the existence of a sequence $\{g_k\}_{k=1}^\infty$ in X^* such that (iii) holds. We come back to this point on page 618.

In [336] Gröchenig defined *Banach frames* as follows.

Definition 24.3.2 *Let X be a Banach space and X_d a Banach sequence space indexed by $\mathbb{N}$. Let $\{g_k\}_{k=1}^\infty$ be a sequence in X^* and $S : X_d \to X$ be a bounded operator. Then $(\{g_k\}_{k=1}^\infty, S)$ is a Banach frame for X with respect to X_d if*

  (i) $\{g_k(f)\}_{k=1}^\infty \in X_d$ for all $f \in X$;

  (ii) there exist constants $A, B > 0$ such that
  $$A \, \|f\|_X \leq \|\{g_k(f)\}_{k=1}^\infty\|_{X_d} \leq B \, \|f\|_X, \ \forall f \in X;$$

  (iii) $S\{g_k(f)\} = f, \ \forall f \in X.$

We note that Definition 24.3.2 is closely related to the development of Feichtinger–Gröchenig: in fact, in the "general version" of Theorem 24.2.4 from [336] Gröchenig shows that $\{\pi(x_k)g\}_{k=1}^\infty$ is a Banach frame for a coorbit space (for an appropriate operator S and with respect to a certain sequence space X_d).

Let us compare atomic decompositions and Banach frames. First, the definition of an atomic decomposition expresses the desire to obtain a *series expansion* of $f \in X$ as for frames in Hilbert spaces. On the other hand, the definition of a Banach frame opens up for the possibility of a *reconstruction formula* not necessarily given by an infinite series, allowing one to come back to $f \in X$ from the coefficients $\{g_k(f)\}_{k=1}^{\infty}$. In [136], Casazza, Han, and Larson show that in case the canonical unit vectors $\{\delta_k\}_{k=1}^{\infty}$ belong to the sequence space X_d and constitute a basis for X_d, there is a simple relationship between Banach frames and atomic decompositions:

Proposition 24.3.3 *Let X be a Banach space and X_d a Banach sequence space indexed by $\mathbb{N}$. Assume that the canonical unit vectors $\{\delta_k\}_{k=1}^{\infty}$ constitute a basis for X_d; finally, let $\{g_k\}_{k=1}^{\infty}$ be a sequence in X^* and consider a bounded operator $S : X_d \to X$. Then the following are equivalent:*

(i) *$(\{g_k\}_{k=1}^{\infty}, S)$ is a Banach frame for X with respect to X_d.*

(ii) *$(\{g_k\}_{k=1}^{\infty}, \{S(\delta_k)\}_{k=1}^{\infty})$ is an atomic decomposition of X with respect to X_d.*

In [136] it is also proved that every separable Banach space possesses a Banach frame:

Proposition 24.3.4 *Every separable Banach space X can be equipped with a Banach frame with respect to an appropriately chosen sequence space X_d.*

Proof. Since X is assumed to be separable, we can choose a dense sequence $\{x_j\}_{j=1}^{\infty}$ in $X \backslash \{0\}$. Given $j \in \mathbb{N}$, there exists (see, e.g., [401], Theorem 28.3) an element $g_j \in X^*$ such that

$$g_j(x_j) = \|x_j\|, \text{ and } \|g_j\| = 1.$$

Given $f \in X$, we can choose a subsequence of $\{x_j\}_{j=1}^{\infty}$, say, $\{x_{k_j}\}_{j=1}^{\infty}$, which converges to f as $j \to \infty$. Since

$$\|x_{k_j}\| = \|g_{k_j}(x_{k_j})\| \le \|g_{k_j}(f)\| + \|g_{k_j}(x_{k_j} - f)\| \le \|g_{k_j}(f)\| + \|f - x_{k_j}\|,$$

it follows that

$$\|f\| \le \sup_{j \in \mathbb{N}} \|g_j(f)\|.$$

Since we also have $\|f\| \ge \sup_{j \in \mathbb{N}} \|g_j(f)\|$, we have proved that $\{g_k\}_{k=1}^{\infty} \subset X^*$ satisfies

$$\|f\| = \sup_{k \in \mathbb{N}} |g_k(f)|, \ \forall f \in X.$$

Let X_d be the subspace of $\ell^{\infty}(\mathbb{N})$ consisting of all sequences $\{g_k(f)\}_{k=1}^{\infty}$, where $f \in X$. Defining the operator $S : X_d \to X$ by $S\{g_k(f)\}_{k=1}^{\infty} = f$, we obtain that $(\{g_k\}_{k=1}^{\infty}, S)$ is a Banach frame for X with respect to X_d. $\square$

Already in Chapter 3 we mentioned that there exist separable Banach spaces having no basis. From this point of view one could say that the concept of Banach frames is very satisfying because they always exist. However, one could also be suspicious and ask if the pure existence of Banach frames is interesting. In order for a Banach frame to be practically useful, it has to be defined with respect to a convenient and easily identifiable sequence space X_d; this is not the case with the Banach frame constructed in the proof of Proposition 24.3.4. One should rather ask for the existence of Banach frames with respect to a nice class of sequence spaces, which would make Banach frames share more of the properties we know from frames in Hilbert spaces. This point of view is supported by a result by Stoeva, published in [131]: it says that every total sequence in X^* is a Banach frame for X with respect to *some* sequence space X_d. For example, let $\{e_k\}_{k=1}^\infty$ be an orthonormal basis for a Hilbert space $\mathcal{H}$, and consider the family $\{e_k + e_{k+1}\}_{k=1}^\infty$ in Example 5.4.6: then

- $\{e_k + e_{k+1}\}_{k=1}^\infty$ is a Banach frame for $\mathcal{H}$ with respect to a certain sequence space X_d;

- $\{e_k + e_{k+1}\}_{k=1}^\infty$ is not a frame for $\mathcal{H}$.

The existence of such examples shows that the definition of Banach frames does not match the definition of frames in Hilbert spaces, and it gives a strong argument for restricting the class of Banach frames to more useful ones. Several attempts to the "right definition" of a Banach frame can be found in the literature. Feichtinger and Gröchenig have advocated to use some special frames in Hilbert spaces as the starting point: in fact, there exist Hilbert space frames which are at the same time frames for a scale of Banach spaces. A general framework for this was developed by Gröchenig [341], who introduced localized frames in Hilbert spaces; see Section 8.2. To such a frame one can associate a class of Banach spaces, and all the central frame objects carry over from the Hilbert space to these spaces. In particular, the frame operator extends to a bounded bijection on each space, which leads to frame decompositions exactly as in Theorem 5.1.6. Following Fornasier and Gröchenig [304], we state the definition of the relevant class of Banach spaces and some of the key results concerning the associated series expansions. Note that weight functions are discussed in Section A.5.

Definition 24.3.5 *Let* $\{f_k\}_{k=1}^\infty$ *denote a frame for a Hilbert space* $\mathcal{H}$ *and let* $\{g_k\}_{k=1}^\infty$ *be any dual frame. Let*

$$\mathcal{H}_0 = \left\{ \sum_{k=1}^\infty c_k f_k \,\middle|\, \{c_k\}_{k=1}^\infty \text{ is finite} \right\}.$$

For $p \in [1, \infty[$ *and any weight function* m *on* $\mathbb{N}$, *define a norm* $\|\cdot\|_{\mathcal{H}_m^p}$ *on* $\mathcal{H}_0$ *by*

$$\|f\|_{\mathcal{H}_m^p} := \|\{\langle f, g_k \rangle\}_{k=1}^\infty\|_{\ell_m^p}, \quad f \in \mathcal{H}_0.$$

Furthermore, let $\mathcal{H}_m^p$ *denote the completion of the space* $\mathcal{H}_0$ *with respect to this norm.*

The following results from [304] show that the frame decomposition (5.7) extends to the Banach spaces $\mathcal{H}_m^p$:

Theorem 24.3.6 *Assume that the frame $\{f_k\}_{k=1}^\infty$ for $\mathcal{H}$ is self-localized with decay rate $s > 1$. Denote the frame operator by S. Given a weight function m on $\mathbb{N}$, the canonical dual frame $\{S^{-1}f_k\}_{k=1}^\infty$ is a Banach frame for $\mathcal{H}_m^p$ for all $1 \le p < \infty$, and the reconstruction formula*

$$f = \sum_{k=1}^\infty \langle f, S^{-1}g_k \rangle f_k$$

holds for all $f \in \mathcal{H}_m^p$, with unconditional convergence.

Concrete manifestations of Theorem 24.3.6 appear by considering a Gabor frame for $L^2(\mathbb{R})$. Under certain conditions on the weight functions m and v, it was proved in [304] that the assumptions in Proposition 13.5.1 imply that the space $\mathcal{H}_m^p$ equals the modulation space M_m^p; see Section A.5; furthermore, $\{\pi(\gamma)g\}_{\gamma \in \Gamma}$ is a Banach frame for M_m^p, and the frame expansion

$$f = \sum_{\gamma \in \Gamma} \langle f, \pi(\gamma)S^{-1}g \rangle \pi(\gamma)g \tag{24.10}$$

holds for all $f \in M_m^p$. Note that the expansion (24.10) also can be obtained without reference to localized frames and the theory for the space $\mathcal{H}_m^p$; see Corollary 12.2.6 in [340].

Let us now assume that that the conditions (i) and (ii) in Definition 24.3.2 are satisfied. Then the *analysis operator*

$$U : X \to X_d, \ Uf := \{g_k(f)\}_{k=1}^\infty \tag{24.11}$$

is injective and thus has an inverse on its range $\mathcal{R}_U$,

$$U^{-1} : \mathcal{R}_U \subseteq X_d \to X, \ U^{-1}\{g_k(f)\}_{k=1}^\infty = f, \ f \in X.$$

The only condition that is missing in order for $(\{g_k\}_{k=1}^\infty, U^{-1})$ to be a Banach frame is that the operator U^{-1} can be extended to an operator on X_d. If the range $\mathcal{R}_U$ is complemented in X_d, the operator U^{-1} can be extended by zero on a complement; however, no easily verifiable condition for a subspace to be complemented exists in general, even in the case where X_d is an ℓ^p-space. Thus, in general, it is not very fruitful to try to obtain Banach frames this way.

Note that [302] by Fornasier and [75, 76] by Borup and Nielsen construct atomic decompositions and Banach frames in the context of Besov spaces and α-modulation spaces.

24.4   p-frames

A different approach to series expansions in a class of Banach spaces was given in the paper [9] by Aldroubi, Sun, and Tang. They considered the first part of the definition of a Banach frame separately in the case where X is a general Banach space and X_d is an ℓ^p-space:

Definition 24.4.1 *Let $p \in]1, \infty[$ be given. A sequence $\{g_k\}_{k=1}^\infty$ in X^* is a p-frame for X if there exist constants $A, B > 0$ such that*

$$A \, ||f||_X \leq \left(\sum_{k=1}^\infty |g_k(f)|^p \right)^{1/p} \leq B \, ||f||_X, \ \forall f \in X. \qquad (24.12)$$

$\{g_k\}_{k=1}^\infty$ *is a p-Bessel sequence if at least the upper p-frame condition is satisfied.*

In [9] p-frames are used to obtain series expansions in shift-invariant subspaces of $L^p(\mathbb{R})$. Let

$$\widetilde{W} = \left\{ f : \mathbb{R} \to \mathbb{C} \ \Big| \ \sup_{x \in \mathbb{R}} \sum_{k \in \mathbb{Z}} |T_k f(x)| < \infty \right\}.$$

If $\phi \in \widetilde{W}$ and $p \in]1, \infty[$ is given, then $\sum_{k \in \mathbb{Z}} c_k T_k \phi$ converges in $L^p(\mathbb{R})$ for all $\{c_k\}_{k \in \mathbb{Z}} \in \ell^p(\mathbb{Z})$, and we can consider the space

$$S_p := \left\{ \sum_{k \in \mathbb{Z}} c_k T_k \phi \ \Big| \ \{c_k\}_{k \in \mathbb{Z}} \in \ell^p(\mathbb{Z}) \right\}.$$

Note that for $p = 2$, the space S_p appeared in Section 9.7.

The main result in [9] yields expansions in the spaces S_p in terms of p-frames:

Theorem 24.4.2 *Let $\phi \in \widetilde{W}$ and $p \in]1, \infty[$. Then the following statements are equivalent:*

(i) S_p is closed in $L^p(\mathbb{R})$.

(ii) $\{T_k \phi\}_{k \in \mathbb{Z}}$ is a p-frame for S_p.

(iii) There exists a function $\psi \in \widetilde{W}$ such that each $f \in S_p$ has unconditionally convergent expansions

$$f = \sum_{k \in \mathbb{Z}} \langle f, T_k \phi \rangle T_k \psi = \sum_{k \in \mathbb{Z}} \langle f, T_k \psi \rangle T_k \phi. \qquad (24.13)$$

Note that in the coefficients in (24.13) the function f might not belong to $L^2(\mathbb{R})$, so the notation $\langle f, g \rangle$ should be interpreted as $\int f \bar{g}$ and not as an inner product. The original article [9] is more general than stated above, and it applies to a space S_p generated by a finite collection of functions rather than just the single function ϕ. The cases $p = 1$ and $p = \infty$ are covered by requiring that ϕ belongs to the Wiener space W, which is a stronger condition than membership of $\widetilde{W}$. If $\phi \in W$, it is also proved that the conditions in Theorem 24.4.2 are independent of the choice of p; in particular, if $\{T_k\phi\}_{k\in\mathbb{Z}}$ is a p_0-frame for one value of p_0, it is a p-frame for all $p \in [1, \infty]$. This implies that if $\{T_k\phi\}_{k\in\mathbb{Z}}$ is a frame sequence in $L^2(\mathbb{R})$, we automatically obtain series expansions like (24.13) in a scale of Banach spaces. In other words: we can focus on the simpler task of designing appropriate frame sequences $\{T_k\phi\}_{k\in\mathbb{Z}}$ in $L^2(\mathbb{R})$ and obtain expansions in a class of Banach spaces as a consequence.

The definition of Riesz bases can also be extended to Banach spaces. The definition will often be applied in the dual Banach space of the given Banach space X, so in order to avoid confusion we state it in a Banach space to be denoted by Y. Furthermore, as a standard convention we let q denote the conjugated exponent of $p \in]1, \infty[$, i.e.,

$$\frac{1}{p} + \frac{1}{q} = 1.$$

Definition 24.4.3 *Let $q \in]1, \infty[$ be given, and let Y be a Banach space. A sequence $\{g_k\}_{k=1}^\infty$ in Y is a q-Riesz basis for Y if $\overline{span}\{g_k\}_{k=1}^\infty = Y$ and there exist constants $A, B > 0$ such that for all finite scalar sequences $\{d_k\}_{k=1}^\infty$,*

$$A \left(\sum_{k=1}^\infty |d_k|^q \right)^{1/q} \leq \left\| \sum_{k=1}^\infty d_k g_k \right\|_Y \leq B \left(\sum_{k=1}^\infty |d_k|^q \right)^{1/q}. \qquad (24.14)$$

Note that completeness is part of our definition of a q-Riesz basis (in contrast to the definition in [9]). Standard arguments show that the assumptions in Definition 24.4.3 imply that $\sum_{k=1}^\infty d_k g_k$ converges unconditionally for all $\{d_k\}_{k=1}^\infty \in \ell^q(\mathbb{N})$ and that (24.14) holds also for these sequences (Exercise 24.7).

For windows in the Feichtinger algebra $\mathcal{S}_0$, it was recently proved by Gröchenig, Ortega-Cerda, and Romero [353] that the p-frame property for a Gabor system in the modulation space M^p is independent of the choice of $p \in [1, \infty]$:

Theorem 24.4.4 *Assume that the set $\{(\mu_k, \lambda_k)\}_{k\in I} \subset \mathbb{R}^2$ is relatively separated and that $g \in \mathcal{S}_0$. Then the following hold:*

(i) *If $\{E_{\lambda_k} T_{\mu_k} g\}_{k\in I}$ is a p-frame for the modulation space M^p, then it is a p-frame for M^p for all $p \in [1, \infty]$.*

(ii) *If $\{E_{\lambda_k} T_{\mu_k} g\}_{k\in I}$ is a q-Riesz sequence in the modulation space M^q, then it is a q-Riesz sequence for M^q for all $q \in [1, \infty]$.*

In the rest of this section we discuss results by Christensen and Stoeva [200]. As standing assumption, we will consider a Banach space X. First, note that if X can be equipped with a p-frame, then X is isomorphic to a closed subspace of ℓ^p and therefore reflexive. A characterization of the p-frame property in terms of the *synthesis operator* is given by the following result, which generalizes Theorem 3.2.3:

Theorem 24.4.5 *Let X be a reflexive Banach space and $\{g_k\}_{k=1}^{\infty}$ a sequence in X^*. Then $\{g_k\}_{k=1}^{\infty}$ is a p-frame for X if and only if*

$$T : \{d_k\}_{k=1}^{\infty} \to \sum_{k=1}^{\infty} d_k g_k$$

is a well-defined mapping of ℓ^q onto X^*.

Note that Theorem 24.4.5 does not mean that the p-frame property is enough to obtain frame-like expansions in X^*. The result only says that if $\{g_k\}_{k=1}^{\infty}$ is a p-frame, then each $g \in X^*$ has a representation $g = \sum_{k=1}^{\infty} d_k g_k$ for some $\{d_k\}_{k=1}^{\infty} \in \ell^q(\mathbb{N})$, but nothing guarantees that the coefficients $\{d_k\}_{k=1}^{\infty}$ can be chosen as continuous linear functionals on X^*.

Theorem 24.4.5 sheds some light on the reason for adding the condition (iii) to the definition of a Banach frame, see Definition 24.3.2: it shows that in the special case of $X_d = \ell^p(\mathbb{N})$, the norm-equivalence in Definition 24.3.2 (ii) alone is equivalent to some kind of "expansion property" in the dual Banach space X^*. This is clearly different from obtaining expansions in the Banach space X itself.

In Proposition 3.6.4 we saw that a Riesz basis for a Hilbert space $\mathcal{H}$ also is a frame for $\mathcal{H}$. This result has a natural extension to q-Riesz bases and p-frames:

Corollary 24.4.6 *Let $\{g_k\}_{k=1}^{\infty}$ be a q-Riesz basis for X^* with q-Riesz basis bounds A, B. Then $\{g_k\}_{k=1}^{\infty}$ is a p-frame for X with p-frame bounds A and B.*

For q-Riesz bases, the desired expansions exist without further assumptions, in X as well as in X^*:

Theorem 24.4.7 *If $\{g_k\}_{k=1}^{\infty}$ is a q-Riesz basis for X^* with bounds A, B, there exists a unique p-Riesz basis $\{f_k\}_{k=1}^{\infty}$ for X for which*

$$f = \sum_{k=1}^{\infty} g_k(f) f_k, \ \forall f \in X, \tag{24.15}$$

$$g = \sum_{k=1}^{\infty} g(f_k) g_k, \ \forall g \in X^*. \tag{24.16}$$

Furthermore, $\{f_k\}_{k=1}^{\infty}$ has the bounds $\frac{1}{B}, \frac{1}{A}$.

In [131] it is proved that for some Banach spaces X and $p \neq 2$, there exist p-frames $\{g_k\}_{k=1}^{\infty}$ for X, for which no sequence $\{f_k\}_{k=1}^{\infty}$ in X satisfies that

$$f = \sum_{k=1}^{\infty} g_k(f) f_k, \ \forall f \in X. \tag{24.17}$$

Since a q-Riesz basis for X^* is a special case of a p-frame for X, Theorem 24.4.7 suggests the following question: given a p-frame $\{g_k\}_{k=1}^{\infty} \subset X^*$ for X, under what conditions can we find a q-frame $\{f_k\}_{k=1}^{\infty}$ for X^* such that (24.17) is satisfied? A theoretical answer is contained in the following theorem; it is formulated on terms of the analysis operator U in (24.11) and its range $\mathcal{R}_U$.

Theorem 24.4.8 *Suppose that the sequence $\{g_k\}_{k=1}^{\infty}$ in X^* is a p-frame for X. Then the following are equivalent:*

(i) *$\mathcal{R}_U$ is complemented in ℓ^p.*

(ii) *The operator $U^{-1} : \mathcal{R}_U \to X$ can be extended to a bounded linear operator $V : \ell^p \to X$.*

(iii) *There exists a q-Bessel sequence $\{f_k\}_{k=1}^{\infty} \subset X$ for X^* such that*

$$f = \sum_{k=1}^{\infty} g_k(f) f_k, \ \forall f \in X.$$

(iv) *There exists a q-Bessel sequence $\{f_k\}_{k=1}^{\infty} \subset X$ for X^* such that*

$$g = \sum_{k=1}^{\infty} g(f_k) g_k, \ \forall g \in X^*.$$

(v) *$\{g_k\}_{k=1}^{\infty}$ is a Banach frame for X with respect to ℓ^p.*

If (one of) the conditions are satisfied, the sequence $\{f_k\}_{k=1}^{\infty}$ in (iii) is a q-frame.

24.5   Gabor Systems and Wavelets in $L^p(\mathbb{R})$ and Related Spaces

For $p \neq 2$ the L^p-spaces are not the right spaces to search for unconditionally convergent Gabor expansions. Feichtinger–Gröchenig theory leads to unconditionally convergent expansions in coorbit spaces, but in [284] it is proved that $L^p(\mathbb{R})$ is not a coorbit space under the Schrödinger representation for $p \in [1, \infty[\backslash\{2\}$. The "right spaces" in connection with Gabor analysis are the modulation spaces as described on page 611 and explained in detail in [340].

If one is satisfied with *conditionally* convergent Gabor expansions, one can obtain convergence for functions in $L^p(\mathbb{R})$ by requiring that the window for the Gabor frame as well as the window for the canonical dual frame belong to the Wiener space W. The following result was proved by Gröchenig and Hcil in [345].

Theorem 24.5.1 *Let* $p \in]1, \infty[$, $g \in W$, *and* $a, b > 0$ *be given. Assume that the Gabor system* $\{E_{mb}T_{na}g\}_{m,n\in\mathbb{Z}}$ *is a frame for* $L^2(\mathbb{R})$. *Furthermore, denote the frame operator by* S *and assume that* $S^{-1}g \in W$. *Then, for an arbitrary* $f \in L^p(\mathbb{R})$,

$$\sum_{|m|\leq N} \sum_{|n|\leq N} \langle f, E_{mb}T_{na}S^{-1}g\rangle E_{mb}T_{na}g \to f \ in \ L^p(\mathbb{R}) \ as \ N \to \infty.$$

Similar results using pairs of dual Gabor frames generated by Schwartz functions were obtained by Grafakos and Lennard [329]; that paper also covers the case $p = 1$ in terms of a certain Cesaro-type sum.

For wavelet systems it is well known that a large class of wavelet orthonormal bases in $L^2(\mathbb{R})$ are unconditional bases for $L^p(\mathbb{R})$ for all $p \in]1, \infty[$ (most wavelet books contain versions of this statement). Chui and Shi [213] have stated sufficient conditions for the frame operator for a Bessel sequence $\{2^{j/2}\psi(2^j x - k)\}_{j,k\in\mathbb{Z}}$ in $L^2(\mathbb{R})$ to extend to a bounded operator on $L^p(\mathbb{R})$, $p \in]1, \infty[$; sufficient conditions for the mixed frame operator associated with two wavelet systems to extend to a bijection on $L^p(\mathbb{R})$ were found by Bui and Laugesen [92]. The completeness problem, e.g., for the Mexican hat wavelet, is considered by the same authors in [90] and [91]. Atomic decompositions in $L^p(\mathbb{R})$ and Sobolev spaces based on the oblique extension principle were obtained by Borup, Gribonval, and Nielsen in [74].

24.6 Exercises

24.1 Prove that every abelian locally compact group is unimodular.

24.2 Prove that a representation π is irreducible if and only if $\overline{\text{span}}\{\pi(x)g\}_{x\in\mathcal{G}} = \mathcal{H}, \ \forall g \in \mathcal{H} \setminus \{0\}$.

24.3 Prove that π defined by (24.2) satisfies condition (i) in Definition 24.1.1. Prove also that the operators $\rho(a,b) = E_a T_b$ do not form a representation of $(\mathbb{R}^2, +)$ on $L^2(\mathbb{R})$.

24.4 Prove that if $\{(x_k, y_k)\}_{k=1}^{\infty} \subset \mathbb{R}^2$ has an accumulation point, then the samples of the Schrödinger representation $\{\pi(x_k, y_k, 1)g\}_{k=1}^{\infty}$ cannot be a frame for any $g \in L^2(\mathbb{R})$.

24.5 Prove that an integrable representation is also square-integrable.

24.6 Let π be an integrable representation. Show that the set $\mathcal{A}$ defined in (24.1) is dense in $\mathcal{H}$.

24.7 Show that the assumptions in Definition 24.4.3 imply that $\sum_{k=1}^{\infty} d_k g_k$ converges unconditionally for all $\{d_k\}_{k=1}^{\infty} \in \ell^q(\mathbb{N})$ and that (24.14) holds for $\{d_k\}_{k=1}^{\infty} \in \ell^q(\mathbb{N})$.

Appendix A
Appendix

A.1 Linear Algebra

Let V, W be finite-dimensional vector spaces, equipped with inner products $\langle \cdot, \cdot \rangle_V$ and $\langle \cdot, \cdot \rangle_W$, respectively (when it is clear from the context in which space the inner product is taken we will skip the subscript). Assume that

$$\dim V = n, \quad \dim W = m.$$

Assume that $T : V \rightarrow W$ is a linear map and that we have fixed an orthonormal basis $\{e_k\}_{k=1}^n$ in V and an orthonormal basis $\{\widetilde{e}_j\}_{j=1}^m$ in W. The *matrix* of T with respect to the chosen bases is the $m \times n$ matrix, where the kth column consists of the coordinates of the image under T of the kth basis vector in V, in terms of the given basis in W. The jkth entry in the matrix representation is $\langle Te_k, \widetilde{e}_j \rangle$, $k = 1, \ldots, n; j = 1, \ldots, m$.

The matrix representation gives a convenient way to find the action of the linear map T on a given $v \in V$: by writing $v = \sum_{k=1}^n c_k e_k$, the result of multiplying the matrix representation of T with $\{c_k\}_{k=1}^n$ is the sequence of coordinates representing Tv in the basis for W. We will always identify the linear map and its matrix representation.

Given a linear operator $T : V \rightarrow W$, the *adjoint* operator $T^* : W \rightarrow V$ is characterized by

$$\langle Tx, y \rangle = \langle x, T^* y \rangle, \quad x \in V, y \in W.$$

© Springer International Publishing Switzerland 2016
O. Christensen, *An Introduction to Frames and Riesz Bases*,
Applied and Numerical Harmonic Analysis,
DOI 10.1007/978-3-319-25613-9

In matrix language, T^* is represented by the Hermitian transpose of the matrix for T, i.e., the matrix we obtain by complex conjugation and transposing.

The *kernel* for T is

$$\mathcal{N}_T = \{x \in V \mid Tx = 0\},$$

and the *range* is

$$\mathcal{R}_T = \{Tx \mid x \in V\}.$$

The vector spaces $\mathcal{N}_T$ and $\mathcal{R}_{T^*}$ are subspaces of V, and

$$\mathcal{N}_T = \mathcal{R}_{T^*}^{\perp};$$

in particular, the linear map T induces orthogonal decompositions of V and (via T^*) W given by

$$V = \mathcal{N}_T \oplus \mathcal{R}_{T^*}, \tag{A.1}$$
$$W = \mathcal{N}_{T^*} \oplus \mathcal{R}_T. \tag{A.2}$$

In case $T = T^*$ (this can only happen when $V = W$), we say that T is *self-adjoint.* The finite-dimensional version of the *spectral theorem* says that a self-adjoint operator has enough eigenvectors to span the entire space:

Theorem A.1.1 *If $T : V \to V$ is self-adjoint, then all eigenvalues are real, and V has an orthonormal basis consisting of eigenvectors for T.*

A.2   Integration

Here we state some basic facts from the theory of integration. The proofs and further results can be found in any standard book on the subject, e.g., [565].

Let X be a set and $\mathcal{M}$ a σ-algebra of subsets of X, in which there is defined a measure μ. We will exclusively consider *positive measures,* which means that $\mu(A) \in [0, \infty]$ for all $A \in \mathcal{M}$. An example is the real numbers $\mathbb{R}$ with the Borel subsets as σ-algebra, and the Lebesgue measure. Another example is the natural numbers $\mathbb{N}$ equipped with the σ-algebra consisting of all subsets, and the counting measure. See [565, 567] or any other standard text on integration for more information.

A *null-set* is a measurable set with measure zero. A condition holds *almost everywhere* (abbreviated a.e.) if it holds except on a null set.

We now state *Fatou's Lemma*:

Lemma A.2.1 *Let $f_n : X \to [0, \infty]$, $n \in \mathbb{N}$ be a sequence of measurable functions. Then*

$$\int_X \liminf_{n \to \infty} f_n d\mu \le \liminf_{n \to \infty} \int_X f_n d\mu.$$

Lebesgue's dominated convergence theorem is the main tool to interchange sums and integrals:

Theorem A.2.2 *Suppose that* $f_n : X \to \mathbb{C}$, $n \in \mathbb{N}$ *is a sequence of measurable functions, that* $f_n(x) \to f(x)$ *pointwise, and that there exists a positive, measurable function* g *such that* $|f_n| \leq g$ *for all* $n \in \mathbb{N}$ *and* $\int_X g d\mu < \infty$. *Then*

$$\lim_{n \to \infty} \int_X f_n d\mu = \int_X f d\mu.$$

We will frequently need the following standard result.

Lemma A.2.3 *Let* μ *be a positive measure on a* σ-*algebra* $\mathcal{M}$. *Assume that* $\{A_n\}_{n=1}^\infty \subset \mathcal{M}$ *and*

$$A_1 \supseteq A_2 \supseteq \cdots \supseteq A_n \supseteq \ldots.$$

If $\mu(A_1) < \infty$, *then*

$$\mu\left(\bigcap_{n=1}^\infty A_n\right) = \lim_{n \to \infty} \mu(A_n).$$

A.3   Locally Compact Groups

Let $\mathcal{G}$ denote a group with neutral element e. The group composition of two elements $x, y \in \mathcal{G}$ will be written $x \cdot y$ or simply xy; the inverse of $x \in \mathcal{G}$ is denoted by x^{-1}. Let us give the formal definition of a locally compact group.

Definition A.3.1 *Let* $\mathcal{G}$ *denote a group which is equipped with a Hausdorff topology, i.e., every pair of distinct points in* $\mathcal{G}$ *has disjoint neighborhoods. We say that* $\mathcal{G}$ *is a locally compact group if the following conditions are satisfied:*

(i) *The neutral element* e *has a neighborhood whose closure is compact;*

(ii) $\mathcal{G}$ *can be covered by a countable union of compact sets, i.e.,* $\mathcal{G}$ *is* σ-*compact.*

Let $\mathcal{O}(e)$ denote the family of neighborhoods of e, i.e., the sets $V \subseteq \mathcal{G}$ containing e in the interior.

Every locally compact group $\mathcal{G}$ can be equipped with a unique (up to scalar multiplication) positive measure μ which is regular and *left-invariant* in the sense that for all continuous functions $F : \mathcal{G} \to \mathbb{C}$ with compact support

$$\int_\mathcal{G} F(yx) d\mu(x) = \int_\mathcal{G} F(x) d\mu(x), \quad \forall y \in \mathcal{G}. \tag{A.3}$$

The measure μ is called the *left Haar measure*. The *right Haar measure* is defined similarly, simply by replacing the composition yx in (A.3) by xy. If the right and left Haar measures coincide (after appropriate normalizations), we simply speak about the *Haar measure*, and $\mathcal{G}$ is said to be *unimodular*. In particular, a locally compact abelian group is obviously unimodular.

The simplest example of a locally compact group is $\mathbb{R}^n$ equipped with the composition "+" and the Euclidean topology. Another example is the *torus*

$$\mathbb{T} = \{z \in \mathbb{C} \mid |z| = 1\};$$

here the composition is complex multiplication and the topology is inherited from $\mathbb{C}$.

A.4   Some Infinite-Dimensional Vector Spaces

1) Given a family of Hilbert spaces $\{\mathcal{H}_n\}_{n=1}^{\infty}$, their *direct sum* is denoted by

$$\mathcal{H} = \left(\sum_{n=1}^{\infty} \oplus \, \mathcal{H}_n\right)_{\ell^2}; \qquad (A.4)$$

by definition, $\mathcal{H}$ consists of all sequences $g = (g_1, g_2, \dots)$ for which $g_n \in \mathcal{H}_n$ for all $n \in \mathbb{N}$, and $\sum_{n=1}^{\infty} ||g_n||^2 < \infty$. $\mathcal{H}$ is a Hilbert space with respect to the inner product

$$\langle f, g \rangle = \sum_{n=1}^{\infty} \langle f_n, g_n \rangle_{\mathcal{H}_n}, \ f, g \in \mathcal{H};$$

the associated norm is

$$||g||^2 = \sum_{n=1}^{\infty} ||g_n||^2.$$

2) Given a parameter $s > 0$, we define the *Sobolev space*

$$H_s(\mathbb{R}) = \left\{ f \in L^2(\mathbb{R}) \,\middle|\, \int_{-\infty}^{\infty} |\hat{f}(\gamma)|^2 (1 + |\gamma|^2)^s d\gamma < \infty \right\}. \qquad (A.5)$$

$H_s(\mathbb{R})$ is a Banach space with respect to the natural norm,

$$||f||_{H_s} = \left(\int_{-\infty}^{\infty} |\hat{f}(\gamma)|^2 (1 + |\gamma|^2)^s d\gamma\right)^{1/2}.$$

3) The *Schwartz space* $\mathcal{S}$ consists of all $f \in C^{\infty}(\mathbb{R})$ which decay faster than any inverse polynomial; that is, for any $\alpha, k \in \mathbb{N} \cup \{0\}$,

$$\sup_{x \in \mathbb{R}} \left| x^{\alpha} \frac{d^k f}{dx^k}(x) \right| < \infty.$$

A.5   Modulation Spaces

Modulation spaces were introduced by Feichtinger in [274] and play a key role in Gabor analysis. They are well described in the literature; see, e.g., Feichtinger's survey article [278] and the book [340] by Gröchenig. We will just state the definition and some of the general results.

Before we define the modulation spaces we need to consider various *weight functions.* We state the general definition for weight functions on $\mathbb{R}^d$, but we actually only need the case $d = 2$.

Definition A.5.1

(i) *A continuous function $v : \mathbb{R}^d \to [0, \infty[$ is called a weight function.*

(ii) *A weight function v is said to be submultiplicative if*

$$v(z_1 + z_2) \le v(z_1)\, v(z_2), \forall z_1, z_2 \in \mathbb{R}^d.$$

(iii) *Given weight functions m and v, we say that m is v-moderate if there exists a constant $C > 0$ such that*

$$m(z_1 + z_2) \le Cv(z_1)\, m(z_2), \forall z_1, z_2 \in \mathbb{R}^d.$$

Letting $|\cdot|$ denote the Euclidean norm on $\mathbb{R}^d$, some standard examples of weight functions are:

(i) The *polynomial weights,* which, for a given $s \ge 0$, have the form

$$v(z) := (1 + |z|)^s, \ z \in \mathbb{R}^d; \tag{A.6}$$

these weights are submultiplicative.

(i) The *exponential weights,* which, for a given $a > 0$, have the form

$$v(z) := e^{a|z|}, \ z \in \mathbb{R}^d. \tag{A.7}$$

(iii) The *sub-exponential weights,* which, for given $a > 0$ and $b \in]0, 1[$, have the form

$$v(z) := e^{a|z|^b}, \ z \in \mathbb{R}^d. \tag{A.8}$$

The modulation spaces are defined in terms of the behavior of the short-time Fourier transform, given in (11.3). We will formulate the definition directly in terms of the involved modulation and translation operators. First, given any function g belonging to the Schwartz space $\mathcal{S}$, we will denote the action of a tempered distribution $f \in \mathcal{S}'$ on $g \in \mathcal{S}$ by $\langle f, g \rangle$. We note that the Schwartz space is invariant under the action of the translation operators and modulation operators. In the definition of the modulation spaces we will first consider a particular choice of a Schwartz function, namely, the Gaussian:

Definition A.5.2 *Let m and v denote weight functions on $\mathbb{R}^2$, and assume that m is v-moderate. Furthermore, let $g(x) := e^{-x^2}$, $x \in \mathbb{R}$.*

(i) *For any $1 \le p,q < \infty$, the modulation space $M_m^{p,q}$ consists of all tempered distributions $f \in \mathcal{S}'$ such that*

$$\int_{-\infty}^{\infty} \left(\int_{-\infty}^{\infty} |\langle f, E_x T_y g \rangle|^p \, m(x,y)^p \, dx \right)^{q/p} dy < \infty. \qquad (A.9)$$

(ii) *For $p = q$, we write $M_m^p := M_m^{p,p}$.*

(iii) *If $m = 1$, we write $M^{p,q} := M_m^{p,q}$.*

(iv) *The above definitions extend to the case where $p = \infty$ or $q = \infty$ (or both) via standard modifications; in particular the space M_m^∞ consists of the tempered distributions $f \in \mathcal{S}'$ for which*

$$\sup_{(x,y)\in\mathbb{R}^2} |\langle f, E_x T_y g \rangle| \, m(x,y) < \infty. \qquad (A.10)$$

Note that Proposition 11.1.2 and (11.3) with $f_1 = f_2, g_1 = g_2$ immediately shows that $M^2 = M^{2,2} = L^2(\mathbb{R})$. The modulation space M^1 is also known as the *Feichtinger algebra* and will be treated separately in Section A.6.

The following result collects some of the key properties of the modulation spaces. Note in particular that in the definition of the modulation spaces, the Gaussian can be replaced by any other function $g \in \mathcal{S} \setminus \{0\}$; this yields the same space and an equivalent norm.

Lemma A.5.3 *Let m and v denote weight functions on $\mathbb{R}^2$, and assume that m is v-moderate. Let $g(x) = e^{-x^2}$, $x \in \mathbb{R}$. Then the following hold:*

(i) *For any $1 \le p,q \le \infty$, the space $M_m^{p,q}$ is a Banach space with respect to the natural norm, for the case $1 \le p,q < \infty$ given by*

$$\|f\|_{M_m^{p,q}} = \left(\int_{-\infty}^{\infty} \left(\int_{-\infty}^{\infty} |\langle f, E_x T_y g \rangle|^p \, m(x,y)^p \, dx \right)^{q/p} dy \right)^{1/q}$$

and for the space M_m^∞ given by

$$\|f\|_{M_m^\infty} = \sup_{(x,y)\in\mathbb{R}^2} |\langle f, E_x T_y g \rangle| \, m(x,y).$$

(ii) *For any $1 \le p,q \le \infty$, the space $M_m^{p,q}$ is dense in $L^2(\mathbb{R})$.*

(iii) *For any $1 \le p,q \le \infty$, the function $g(x) = e^{-x^2}$ in the definition of $M_m^{p,q}$ can be replaced by any function $g \in \mathcal{S} \setminus \{0\}$; this yields the same space and an equivalent norm.*

(iv) *For any $1 \le p,q \le \infty$, the space $M_m^{p,q}$ is invariant under time–frequency shifts, and there is a constant $C > 0$ such that*

$$\|T_x E_y f\|_{M_m^{p,q}} \le C \, v(x,y) \, \|f\|_{M_m^{p,q}}, \ \forall f \in M_m^{p,q}, \forall x,y \in \mathbb{R}.$$

The modulation spaces with $p - q - 1$ play a particular role and can for submultiplicative weights be characterized as follows:

Lemma A.5.4 *Assume that the weight function* $v : \mathbb{R}^2 \to [0, \infty[$ *is submultiplicative. Then a function* f *belongs to* M_v^1 *if and only if*

$$\sum_{(k_1, k_2) \in \mathbb{Z}^2} \sup_{(x,y) \in [0,1]^2} \left(|\langle f, E_{x+k_1} T_{y+k_2} f \rangle| \, v(x, y) \right) < \infty. \tag{A.11}$$

Note that for the case $v = 1$, the condition in (A.11) means that the function $(x, y) \mapsto |\langle f, E_x T_y f \rangle|$ belongs to a two-dimensional variant of the Wiener amalgam space defined in (11.29). Besides their applications within frame theory, modulation spaces have found applications in many topics that are not treated in this book, e.g., in the analysis of pseudodifferential operators and as symbol classes.

A.6   Feichtinger's algebra $\mathcal{S}_0$

The *Feichtinger algebra* $\mathcal{S}_0$ is a special case of the modulation spaces M^p discussed in Section A.5. Letting $g(x) := e^{-x^2}$, $x \in \mathbb{R}$, the Feichtinger algebra is defined as the vector space consisting of all $f \in L^2(\mathbb{R})$ for which the short-time Fourier transform $\Psi_g(f)$ introduced in Definition 11.1.1 belongs to $L^1(\mathbb{R}^2)$, i.e.,

$$\int_{-\infty}^{\infty} \int_{-\infty}^{\infty} |\langle f, E_x T_y g \rangle| dx dy < \infty. \tag{A.12}$$

Thus $\mathcal{S}_0 = M^1$, and we have the general results about modulation spaces in Section A.5 at our disposal. In particular, $\mathcal{S}_0$ is a Banach space with respect to the norm

$$\|f\|_{\mathcal{S}_0} = \|\Psi_g(f)\|_{L^1(\mathbb{R}^2)} = \int_{-\infty}^{\infty} \int_{-\infty}^{\infty} |\langle f, E_x T_y g \rangle| dx dy,$$

and it is dense in $L^2(\mathbb{R})$. Also, Lemma A.5.3 shows that in the definition of $\mathcal{S}_0$, we can replace the Gaussian by any nonzero function in $\mathcal{S}_0$; this yields the same space and an equivalent norm.

We also see that $\mathcal{S}_0$ corresponds to the set $\mathcal{A}$ in (24.1) whenever π is chosen as the Schrödinger representation, see Example 24.1.2.

Several characterizations of $\mathcal{S}_0$ can be found in the literature; see, e.g., Feichtinger's paper [274], and [340]; in particular, $\mathcal{S}_0$ consists of all *countable* superpositions of time–frequency shifts of the Gaussian with ℓ^1-coefficients:

$$\mathcal{S}_0 = \left\{ f = \sum_{k=1}^{\infty} c_k E_{y_k} T_{x_k} g \;\middle|\; \{(x_k, y_k)\}_{k=1}^{\infty} \subset \mathbb{R}^2, \{c_k\}_{k=1}^{\infty} \in \ell^1(\mathbb{N}) \right\}.$$

The infimum of all ℓ^1-norms $\sum |c_k|$, taken over coefficients representing a given f, gives an equivalent norm on $\mathcal{S}_0$.

It is generally accepted by the scientific community that $\mathcal{S}_0$ is the "correct window class" for Gabor analysis, considered as a branch of time–frequency analysis. In the rest of the section we will collect some of the important results about functions in $\mathcal{S}_0$.

We first state a result from [274], giving a convenient criterion for a function to belong to $\mathcal{S}_0$.

Lemma A.6.1 *The set of functions*

$$\left\{ f \in C_c(\mathbb{R}) \,\middle|\, \widehat{f} \in L^1(\mathbb{R}) \right\}$$

is contained in $\mathcal{S}_0$.

In particular, Lemma A.6.1 implies that the B-splines B_n in Section A.8 belong to $\mathcal{S}_0$ whenever $n \geq 2$; see Theorem A.8.1 and Corollary A.8.2.

The following result relates the Schwartz space $\mathcal{S}$, the Wiener space W, and the Feichtinger algebra $\mathcal{S}_0$.

Lemma A.6.2 $\mathcal{S} \subset \mathcal{S}_0 \subset W$.

Proof. We will only show the inclusion between the Feichtinger algebra and the Wiener space. Thus, consider a function $f \in \mathcal{S}_0$. By Lemma A.5.3 we can replace the Gaussian by any nonzero function $g \in \mathcal{S}_0$ in the definition of $\mathcal{S}_0$; we will take a compactly supported function g such that $0 \leq g(x) \leq 1$ for all $x \in \mathbb{R}$ and $g(x) = 1$ for $x \in [-1, 1]$. An example of such a function would be $g(x) = \sum_{k=-1}^{1} \widetilde{B}_2(x + k)$, where $\widetilde{B}_2$ is the B-spline defined in (A.18); up to the factor 3, this function equals the function h_2 in (12.41), shown in Figure 12.2 (a). Now, $\chi_{[0,1]}(x) \leq T_t g(x)$, $\forall t \in [0,1]$, $x \in \mathbb{R}$, so via Fourier's inversion formula

$$||f T_k \chi_{[0,1]}||_\infty \leq ||f T_{k+t} g||_\infty \quad \leq \quad ||\mathcal{F}(f T_{k+t} g)||_1$$
$$= \int_{-\infty}^{\infty} |\langle f, E_\omega T_{k+t} g \rangle| \, d\omega, \ \forall t \in [0,1].$$

Integrating over $t \in [0,1]$ yields that

$$||f T_k \chi_{[0,1]}||_\infty \leq \int_0^1 \left(\int_{-\infty}^{\infty} |\langle f, E_\omega T_{k+t} g \rangle| \, d\omega \right) dt;$$

thus,

$$\sum_{k \in \mathbb{Z}} ||f T_k \chi_{[0,1]}||_\infty \quad \leq \quad \int_{-\infty}^{\infty} \left(\int_{-\infty}^{\infty} |\langle f, E_\omega T_x g \rangle| \, d\omega \right) dt < \infty,$$

i.e., $f \in W$. $\qquad\square$

The following versions of the *Poisson summation formula* can also be found in [274]. The stated condition implies that the left-hand side of (A.13) converges absolutely and defines an α-periodic function in the variable x; now the proof follows by expanding this function in a Fourier series.

Lemma A.6.3 *Assume that either $f \in \mathcal{S}_0$ or the decay conditions*

$$|f(x)| \leq C(1+|x|)^{-1-\epsilon}, \ |\widehat{f}(\gamma)| \leq C(1+|\gamma|)^{-1-\epsilon}$$

hold for some $\epsilon > 0$. Let $\alpha > 0$ be given. Then for all $x \in \mathbb{R}$,

$$\sum_{k\in\mathbb{Z}} f(x+k\alpha) = \frac{1}{\alpha}\sum_{k\in\mathbb{Z}} \widehat{f}(k/\alpha)e^{2\pi ikx/\alpha}, \tag{A.13}$$

with absolute convergence on both sides.

Let us collect some of the key properties of $\mathcal{S}_0$:

- $\mathcal{S}_0$ is invariant under translation, modulation, scaling, and the Fourier transform;

- If $f, g \in L^2(\mathbb{R})$ and the short-time Fourier transform $\Psi_g(f)$ (see Definition 11.1.1) belongs to $L^1(\mathbb{R})$, then $f, g \in \mathcal{S}_0$; see [340].

- The Zak transform Zf (see Section 13.2) is continuous if $f \in \mathcal{S}_0$;

- If a Gabor system $\{E_{mb}T_{na}g\}_{m,n\in\mathbb{Z}}$ is a frame with frame operator S and $g \in \mathcal{S}_0$, then $S^{-1}g \in \mathcal{S}_0$; see [349];

- If $g \in \mathcal{S}_0$, the synthesis operator for a Gabor system $\{E_{mb}T_{na}g\}_{m,n\in\mathbb{Z}}$ is bounded from $\ell^p(\mathbb{Z}^2)$ into the modulation space M^p, for all parameters $a, b > 0$ and $p \in [1, \infty]$; the analysis operator is bounded from M^p into $\ell^p(\mathbb{Z}^2)$, and the frame operator is bounded on M^p. See [340].

- The Walnut representation of the Gabor frame operator is available whenever the window belongs to $\mathcal{S}_0$; see Theorem 12.2.1;

- Condition A (see Section 12.1) is satisfied whenever $f \in \mathcal{S}_0$; in particular, the Janssen representation of the Gabor frame operator is available; see Theorem 12.2.5.

- Condition R (see Section 12.1) is satisfied whenever $f \in \mathcal{S}_0$.

- The linear mapping $f \mapsto \{f(k)\}_{k\in\mathbb{Z}}$ is bounded from $\mathcal{S}_0$ into $\ell^1(\mathbb{Z})$; see [282].

- $\mathcal{S}_0$ is continuously embedded in $L^1(\mathbb{R})$, i.e., there exists a constant $C > 0$ such that

$$||f||_{L^1(\mathbb{R})} \leq C\,||f||_{\mathcal{S}_0}, \ \forall f \in \mathcal{S}_0.$$

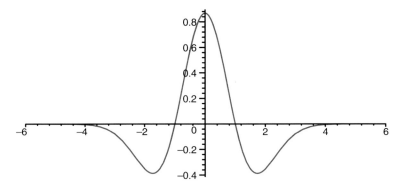

Figure A.1. The Mexican hat

A.7   Some Special Functions

The *Gaussian* with parameter $a > 0$ is the function

$$g_a(x) := e^{-ax^2}, \ x \in \mathbb{R}.$$

It plays a special role in Fourier analysis, partly because it is (up to constants) invariant under the Fourier transform:

Lemma A.7.1 *For $a > 0$,*

$$\mathcal{F}g_a(\gamma) = \sqrt{\frac{\pi}{a}} e^{-\frac{\pi^2}{a}\gamma^2}.$$

Another important function is the *Mexican hat* (Figure A.1), which can be derived from the Gaussian with $a = \frac{1}{2}$:

Example A.7.2 We consider the Gaussian $g(x) = e^{-\frac{1}{2}x^2}$. Its first derivatives are

$$g'(x) = -xe^{-\frac{1}{2}x^2}, \ g''(x) = -(1 - x^2)e^{-\frac{1}{2}x^2}.$$

One can show that $\int_{-\infty}^{\infty} |g''(x)|^2 dx = \frac{3}{4}\pi^{1/2}$; normalizing g'' in $L^2(\mathbb{R})$ gives the *Mexican hat*

$$\psi(x) := \frac{2}{\sqrt{3}}\pi^{-1/4}(1 - x^2)e^{-\frac{1}{2}x^2}.$$

Direct calculation yields that

$$\widehat{\psi}(\gamma) = 8\sqrt{\frac{2}{3}}\pi^{9/4}\gamma^2 e^{-2\pi^2\gamma^2}.$$

□

A.8   B-Splines

In short, *splines* are functions which are piecewise polynomials; in the one-dimensional case, this means that one can split the domain of a spline into intervals in such a way that the function is a polynomial on each interval. The points where the function changes from one polynomial to another polynomial are called *knots*. In the general setting no assumption on the knots are made, and one can also consider splines in more variables.

For our purpose, however, the most elementary splines, namely, *B*-splines, will suffice. They are defined inductively: the first is simply

$$B_1(x) = \chi_{[-\frac{1}{2},\frac{1}{2}]}(x), \tag{A.14}$$

and, assuming that we have defined B_n for some $n \in \mathbb{N}$, the next is defined by a convolution:

$$
\begin{aligned}
B_{n+1}(x) = B_n * B_1(x) &= \int_{-\infty}^{\infty} B_n(x-t)B_1(t)dt \\
&= \int_{-\frac{1}{2}}^{\frac{1}{2}} B_n(x-t)dt. \tag{A.15}
\end{aligned}
$$

The functions B_n defined by (A.14) and (A.15) are called *B-splines*, and n is the *order*. See Figure A.2 for graphs of B-splines B_2 and B_3. We collect some of their fundamental properties; they can be proved by induction.

Theorem A.8.1 *Given $n \in \mathbb{N}$, B_n has the following properties:*

(i) *If $n \geq 2$, then $B_n \in C^{n-2}(\mathbb{R})$.*

(ii) *supp $B_n = [-\frac{n}{2}, \frac{n}{2}]$ and $B_n > 0$ on $]-\frac{n}{2}, \frac{n}{2}[$.*

(iii) *$\int_{-\infty}^{\infty} B_n(x)dx = 1$.*

(iv) *$\sum_{k \in \mathbb{Z}} B_n(x-k) = 1$ for all $x \in \mathbb{R}$ (for $n = 1$, except for $x \in \mathbb{Z}$).*

(v) *For any continuous function $f : \mathbb{R} \to \mathbb{C}$,*

$$\int_{-\infty}^{\infty} B_n(x)f(x)dx = \int_{[-\frac{1}{2},\frac{1}{2}]^n} f(x_1 + \cdots + x_n)dx_1 \cdots dx_n. \tag{A.16}$$

If n is even, the restriction of B_n to each interval $[k, k+1]$, $k \in \mathbb{Z}$, is a polynomial of degree at most $n-1$; if n is odd, the restriction of B_n to each interval $[k - \frac{1}{2}, k + \frac{1}{2}]$, $k \in \mathbb{Z}$, is a polynomial of degree at most $n-1$.

Explicit expressions for the B-splines B_2 and B_3 are given by (Exercise 12.10)

$$
B_2(x) = \begin{cases} 1+x & \text{if } x \in [-1,0], \\ 1-x & \text{if } x \in [0,1], \\ 0 & \text{otherwise}, \end{cases}
$$

and

$$B_3(x) = \begin{cases} \frac{1}{2}x^2 + \frac{3}{2}x + \frac{9}{8} & \text{if } x \in [-\frac{3}{2}, -\frac{1}{2}], \\ -x^2 + \frac{3}{4} & \text{if } x \in [-\frac{1}{2}, \frac{1}{2}], \\ \frac{1}{2}x^2 - \frac{3}{2}x + \frac{9}{8} & \text{if } x \in [\frac{1}{2}, \frac{3}{2}], \\ 0 & \text{otherwise.} \end{cases}$$

Note in particular that the integer-translates of any B_n form a partition of unity for all $n \in \mathbb{N}$, and that the regularity and the support size of B_n increase with n. Via (A.16) we can find the Fourier transform of B_n:

Corollary A.8.2 *For $n \in \mathbb{N}$,*

$$\widehat{B_n}(\gamma) = \left(\frac{e^{\pi i \gamma} - e^{-\pi i \gamma}}{2\pi i \gamma}\right)^n = \left(\frac{\sin(\pi\gamma)}{\pi\gamma}\right)^n. \tag{A.17}$$

Proof. Using that $\widehat{f * g} = \widehat{f} * \widehat{g}$ for $f, g \in L^1(\mathbb{R})$, the definition of the B-spline B_n immediately gives that

$$\widehat{B_n}(\gamma) = \left(\widehat{B_1}(\gamma)\right)^n = \left(\int_{-\frac{1}{2}}^{\frac{1}{2}} e^{-2\pi i x \gamma} dx\right)^n = \left(\frac{e^{\pi i \gamma} - e^{-\pi i \gamma}}{2\pi i \gamma}\right)^n,$$

as desired. □

With our definition of the B-splines, all the functions B_n have support on a symmetric interval around zero. By Theorem A.8.1, the *translated spline*

$$\widetilde{B_n}(x) := T_{\frac{n}{2}} B_n(x) = B_n(x - \frac{n}{2}) \tag{A.18}$$

has support on the interval $[0, n]$. Alternatively, the splines $\widetilde{B_n}$ can be defined inductively exactly as the B-splines, starting with the function $\widetilde{B_1} = \chi_{[0,1]}$. Explicit expressions for the B-splines $\widetilde{B_2}$ and $\widetilde{B_3}$ are given by

$$\widetilde{B_2}(x) = \begin{cases} x, & x \in [0, 1[, \\ 2 - x, & x \in [1, 2[, \\ 0, & x \notin [0, 2[, \end{cases}$$

$$\widetilde{B_3}(x) = \begin{cases} 1/2\, x^2, & x \in [0, 1[, \\ -3/2 + 3x - x^2, & x \in [1, 2[, \\ 9/2 - 3x + 1/2\, x^2, & x \in [2, 3[, \\ 0, & x \notin [0, 3[. \end{cases}$$

Using Corollary A.8.2 we can find the Fourier transform of the translated splines $\widetilde{B_n}$:

Corollary A.8.3 *For $n \in \mathbb{N}$,*

$$\widehat{\widetilde{B_n}}(\gamma) = \left(\frac{1 - e^{-2\pi i \gamma}}{2\pi i \gamma}\right)^n. \tag{A.19}$$

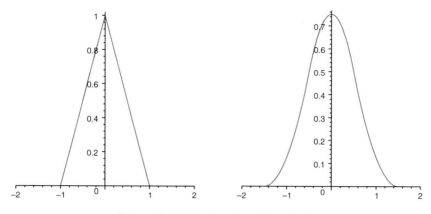

Figure A.2. The B-splines B_2 and B_3

We will now derive an alternative expression for the B-splines $\widetilde{B}_n$. For a real-valued function f, let

$$f(x)_+ := \max\{0, f(x)\}.$$

Also, for any nonnegative integer n, let

$$f(x)_+^n := (f(x)_+)^n.$$

Finally, for $n \in \mathbb{N}$ and $j = 0, 1, \ldots, n$, let

$$\binom{n}{j} := \frac{n!}{j!(n-j)!}.$$

Theorem A.8.4 *For each $n = 2, 3, \ldots,$ the B-spline $\widetilde{B}_n$ can be written*

$$\widetilde{B}_n(x) = \frac{1}{(n-1)!} \sum_{j=0}^{n} (-1)^j \binom{n}{j} (x-j)_+^{n-1}, \quad x \in \mathbb{R}. \qquad \text{(A.20)}$$

Proof. We prove (A.20) by induction. For $n = 2$, the result can be proved by a direct calculation. Now, assume that (A.20) holds for the B-spline $\widetilde{B}_n$ for some $n \in \mathbb{N}$, and consider the B-spline $\widetilde{B}_{n+1}$; we want to show that

$$\widetilde{B}_{n+1}(x) = \frac{1}{n!} \sum_{j=0}^{n+1} (-1)^j \binom{n+1}{j} (x-j)_+^n, \quad x \in \mathbb{R}. \qquad \text{(A.21)}$$

First we notice that for $x < 0$, we have $\widetilde{B}_{n+1}(x) = 0$ and $(x-j)_+ = 0$ for all $j = 0, \ldots, n+1$; thus, the equation in (A.21) holds. Let us now consider

$x \in [0, n+1]$. Via the induction hypothesis, we derive that

$$
\begin{aligned}
\widetilde{B_{n+1}}(x) &= \int_0^1 \widetilde{B_n}(x-t)\, dt \\
&= \frac{1}{(n-1)!} \sum_{j=0}^{n} (-1)^j \binom{n}{j} \int_0^1 (x-t-j)_+^{n-1}\, dt. \quad (\text{A.22})
\end{aligned}
$$

For technical reasons, we will now split the interval $[0, n+1]$ into subintervals and consider $x \in [J, J+1]$ for some arbitrary but fixed $J \in \{0, 1, \ldots, n\}$; if we can prove (A.21) for such x, the result holds for all $x \in [0, n+1]$. In order to calculate the integrals in (A.22), we split the index set $j = 0, 1, \ldots, n$ into three groups:

- For $j = J+1, J+2, \ldots, n$,

$$
\int_0^1 (x-t-j)_+^{n-1}\, dt = 0.
$$

- For $j = J$,

$$
\begin{aligned}
\int_0^1 (x-t-J)_+^{n-1}\, dt &= \int_0^{x-J} (x-t-J)^{n-1}\, dt \\
&= \frac{1}{n}(x-J)^n.
\end{aligned}
$$

- For $j = 0, 1, \ldots, J-1$,

$$
\begin{aligned}
\int_0^1 (x-t-j)_+^{n-1}\, dt &= \int_0^1 (x-t-j)^{n-1}\, dt \\
&= \frac{1}{n}\left((x-j)^n - (x-1-j)^n\right).
\end{aligned}
$$

We now have all the information needed to calculate the sum in (A.22). Let us first consider the partial sum corresponding to $j = 0, \ldots, J-1$:

$$
\begin{aligned}
&\sum_{j=0}^{J-1} (-1)^j \binom{n}{j} \int_0^1 (x-t-j)_+^{n-1}\, dt \\
=\ & \frac{1}{n} \sum_{j=0}^{J-1} (-1)^j \binom{n}{j} \left((x-j)^n - (x-1-j)^n\right) \\
=\ & \frac{1}{n} \sum_{j=0}^{J-1} (-1)^j \binom{n}{j} (x-j)^n - \frac{1}{n} \sum_{j=0}^{J-1} (-1)^j \binom{n}{j} (x-1-j)^n = (*).
\end{aligned}
$$

Splitting of the sum into two and reordering of the terms lead to

$$
(*) = \frac{1}{n}\sum_{j=0}^{J-1}(-1)^j\binom{n}{j}(x-j)^n + \frac{1}{n}\sum_{j=1}^{J}(-1)^j\binom{n}{j-1}(x-j)^n
$$

$$
= \frac{1}{n}x^n + \frac{1}{n}\sum_{j=1}^{J-1}(-1)^j\left(\binom{n}{j}+\binom{n}{j-1}\right)(x-j)^n
$$

$$
+\frac{1}{n}(-1)^J\binom{n}{J-1}(x-J)^n.
$$

Using that

$$
\binom{n}{j}+\binom{n}{j-1}=\binom{n+1}{j}, \tag{A.23}
$$

this implies that

$$
\sum_{j=0}^{J-1}(-1)^j\binom{n}{j}\int_0^1(x-t-j)_+^{n-1}\,dt
$$

$$
= \frac{1}{n}x^n + \frac{1}{n}\sum_{j=1}^{J-1}(-1)^j\binom{n+1}{j}(x-j)^n + \frac{1}{n}(-1)^J\binom{n}{J-1}(x-J)^n.
$$

We can now find $\widetilde{B_{n+1}}$ using (A.22):

$$
\widetilde{B_{n+1}}(x)
$$

$$
= \frac{1}{(n-1)!}\sum_{j=0}^{n}(-1)^j\binom{n}{j}\int_0^1(x-t-j)_+^{n-1}\,dt
$$

$$
= \frac{1}{(n-1)!}\sum_{j=0}^{J}(-1)^j\binom{n}{j}\int_0^1(x-t-j)_+^{n-1}\,dt
$$

$$
= \frac{1}{(n-1)!}\left(\frac{1}{n}x^n + \frac{1}{n}\sum_{j=1}^{J-1}(-1)^j\binom{n+1}{j}(x-j)^n\right)
$$

$$
+\frac{1}{(n-1)!}\frac{1}{n}(-1)^J\binom{n}{J-1}(x-J)^n
$$

$$
+\frac{1}{(n-1)!}\frac{1}{n}(-1)^J\binom{n}{J}(x-J)^n
$$

$$
= \frac{1}{n!}x^n + \frac{1}{n!}\sum_{j=1}^{J-1}(-1)^j\binom{n+1}{j}(x-j)^n
$$

$$
+\frac{1}{n!}(-1)^J\left(\binom{n}{J-1}+\binom{n}{J}\right)(x-J)^n.
$$

Using (A.23) again, this leads to

$$
\begin{aligned}
\widetilde{B_{n+1}}(x) &= \frac{1}{n!} \sum_{j=0}^{J} (-1)^j \binom{n+1}{j} (x-j)^n \\
&= \frac{1}{n!} \sum_{j=0}^{n+1} (-1)^j \binom{n+1}{j} (x-j)_+^n.
\end{aligned}
$$

This proves (A.21) for $x \in [0, n+1]$. The proof that (A.21) holds for $x > n+1$ is similar and is left to the reader. $\qquad\square$

Theorem A.8.4 has some direct consequences.

Corollary A.8.5 *For $n = 2, 3, \ldots$, the B-spline $\widetilde{B_n}$ has the following properties:*

(i) $\widetilde{B_n} \in C^{n-2}(\mathbb{R})$.

(ii) The restriction of $\widetilde{B_n}$ to each interval $[k, k+1]$, $k \in \mathbb{Z}$, is a polynomial of degree at most $n - 1$.

We now state a lemma concerning linear independence of translated versions of a B-spline. We will only need the lemma in Section 9.5.

Lemma A.8.6 *Let $n \in \mathbb{N}$. Then the functions $\widetilde{B_n}(\cdot + k)$, $k = 0, \ldots, n-1$, are linearly independent on $[0, 1]$.*

Proof. For $0 \leq x \leq 1$ and $k = 0, \ldots, n-1$, it follows from (A.20) that

$$
\begin{aligned}
\widetilde{B_n}(x+k) &= \frac{1}{(n-1)!} \sum_{j=0}^{n} (-1)^j \binom{n}{j} (x+k-j)_+^{n-1} \\
&= \frac{1}{(n-1)!} \sum_{j=0}^{k} (-1)^j \binom{n}{j} (x+k-j)^{n-1} \\
&= \frac{1}{(n-1)!} \sum_{\ell=0}^{k} (-1)^{k-\ell} \binom{n}{k-\ell} (x+\ell)^{n-1} \\
&= \frac{(-1)^k}{(n-1)!} \sum_{\ell=0}^{k} (-1)^\ell \binom{n}{k-\ell} (x+\ell)^{n-1}.
\end{aligned}
$$

This calculation shows that the linear operator that maps the functions

$$
(\cdot + \ell)^{n-1}, \ell = 0, 1, \ldots, n-1,
$$

onto the functions

$$
\widetilde{B_n}(\cdot + k), \ k = 0, 1, \ldots, n-1
$$

is lower triangular, with nonzero diagonal entries; thus, the transformation is invertible. Consider the operator

$$\nabla f(x) := f(x+1) - f(x);$$

in the literature, this is often called the *forward difference operator*. For $k = 0, \ldots, n-1$, $\nabla^k(x^{n-1})$ is a polynomial of exact degree $n-1-k$, which is a linear combination of $x^{n-1}, (x+1)^{n-1}, \cdots, (x+k)^{n-1}$. It follows that

$$\mathrm{span}\{(\cdot + k)^{n-1} : k = 0, \ldots, n-1\}$$

equals the space of all polynomials of degree less than n. Therefore, the polynomials $(\cdot + k)^{n-1}, k = 0, \ldots, n-1$ are linearly independent on $[0,1]$. As we have seen, $(\cdot + k)^{n-1}, k = 0, \ldots, n-1$ and $\widetilde{B}_n(\cdot + k), \ k = 0, 1, \ldots, n-1$ are related by an invertible operator; as a consequence, we infer that the functions $\widetilde{B}_n(\cdot + k), k = 0, 1, \ldots, n-1$ are linearly independent on $[0,1]$. $\square$

We refer to the books by Chui [205] and de Boor [70] for information about general splines.

A.9 Exponential B-Splines

Exponential B-splines are defined in a similar way as the B-splines, but with the characteristic functions multiplied by exponential functions. For the purpose in this book it is enough to consider the following types:

Definition A.9.1 *Consider a finite sequence of scalars* $\beta_1, \beta_2, \ldots, \beta_n \in \mathbb{R}$, *for some* $n \in \mathbb{N}$, *and let*

$$e_k(x) := e^{\beta_k x} \chi_{[0,1]}(x), \ k = 1, \ldots, n.$$

An exponential B-spline is a function $\mathcal{E}_n : \mathbb{R} \to \mathbb{C}$ *of the form*

$$\mathcal{E}_n := e_1 * e_2 * \cdots * e_n. \tag{A.24}$$

The exponential B-splines share many properties with the classical B-splines given by the choice $\beta_k = 0, k = 1, \ldots, n$. In particular, the exponential B-spline $\mathcal{E}_n$ is $n-2$ times differentiable (for $n \geq 2$) and its support is $[0, n]$. For more general information about exponential B-splines, we refer to the papers [240], [558], and [624].

Some of the exponential B-splines have a property that is close to the partition of unity property for B-splines. In fact, Theorem 3.1 in [107] shows that the function $\sum_{k \in \mathbb{Z}} \mathcal{E}_n(x - k)$ is constant if and only if $\beta_k = 0$ for at least one value of $k \in \{1, \ldots, n\}$.

The stated references contain several formulas that are appropriate for calculation of the exponential B-splines. We will only need the following results from [107]:

Lemma A.9.2 *Consider an exponential B-spline of the form* (A.24), *$n \geq 2$, and assume that*

$$\beta_k = (k-1)\beta, k = 1, \ldots, n \tag{A.25}$$

for some $\beta > 0$. Then

$$\sum_{k \in \mathbb{Z}} \mathcal{E}_n(x-k) = \frac{\displaystyle\prod_{m=1}^{n-1} (e^{\beta m} - 1)}{\beta^{n-1}(n-1)!}, \tag{A.26}$$

and

$$\mathcal{E}_n(x) = \begin{cases} \dfrac{1}{\beta^{N-1}} \displaystyle\sum_{k=0}^{n-1} \dfrac{1}{\displaystyle\prod_{\substack{j=1 \\ j \neq k+1}}^{n} (k+1-j)} e^{\beta kx}, & x \in [0,1], \\[2em] \dfrac{(-1)^{\ell-1}}{\beta^{n-1}} \displaystyle\sum_{k=0}^{n-1} \left(\dfrac{\displaystyle\sum_{\substack{0 \leq j_1 < \cdots < j_{\ell-1} \leq n-1 \\ j_1, \ldots, j_{\ell-1} \neq k}} e^{\beta j_1 + \cdots + \beta j_{\ell-1}}}{\displaystyle\prod_{\substack{j=1 \\ j \neq k+1}}^{n} (k+1-j)} \right) e^{\beta k(x-\ell+1)}, & \begin{array}{l} x \in [\ell-1, \ell], \\ \ell = 2, \ldots, n, \end{array} \\[2em] 0, & x \notin [0, n]. \end{cases}$$

Note that Lemma A.9.2 corrects a typo in the expression for $\mathcal{E}_n(x)$ for $x \in [k-1, k]$ on page 304 of [107]: in the result in [107] and using the notation from [107], the term $e^{a_{j_1}} + \cdots + e^{a_{j_{k-1}}}$ should be $e^{a_{j_1} + \cdots + a_{j_{k-1}}}$.

Example A.9.3 Consider the exponential B-spline $\mathcal{E}_2$ in (A.24) with the choice $\beta_1 = 0, \beta_2 = 1$, i.e., (A.25) holds with $n = 2$ and $\beta = 1$. Then

$$\mathcal{E}_2(x) = \begin{cases} e^x - 1, & \text{if } x \in [0, 1], \\ e - e^{-1}e^x, & \text{if } x \in [1, 2], \\ 0, & \text{if } x \notin [0, 2]. \end{cases}$$

By (A.26) we have

$$\sum_{k \in \mathbb{Z}} \mathcal{E}_2(x-k) = e - 1, \ x \in \mathbb{R}.$$

This shows that the function $g_2(x) := (e-1)^{-1} \mathcal{E}_2(x)$ satisfies the partition of unity condition.

Similarly, taking $n - 3$ and letting $\beta_k - k - 1$, $k = 1, 2, 3$,

$$\mathcal{E}_3(x) = \begin{cases} \frac{1}{2}\left(1 - 2e^x + e^{2x}\right), & x \in [0,1], \\ \frac{1}{2}\left(-(e + e^2) + 2(e^{-1} + e)e^x - (e^{-2} + e^{-1})e^{2x}\right), & x \in [1,2], \\ \frac{1}{2}\left(e^3 - 2e^x + e^{-3}e^{2x}\right), & x \in [2,3], \\ 0, & x \notin [0,3]. \end{cases}$$

By (A.26) we have

$$\sum_{k \in \mathbb{Z}} \mathcal{E}_3(x - k) = \frac{1}{2}(e - 1)(e^2 - 1), \ x \in \mathbb{R}.$$

Thus $g_3(x) := 2(e - 1)^{-1}(e^2 - 1)^{-1}\mathcal{E}_3(x)$ satisfies the partition of unity condition.   □

A.10   Splines on Locally Compact Abelian Groups

The classical B-splines were generalized to the setting of locally compact abelian (LCA) groups in 1994, independently by Dahlke [232] and Tikhomirov [613]. We will now introduce a slight extension of these splines from [176]; in fact, we will allow certain weight functions to appear, see (A.27) below. This enlarges the class of obtained splines in the same sense as the exponential splines generalize the classical B-splines.

We will use the notation and terminology from Chapter 21 without further comments. Given an LCA group G and functions $f, g : G \to \mathbb{C}$, the *convolution* is the function $f * g$ given by

$$(f * g)(y) := \int_G f(y - x)g(x)\,dx, \ y \in G,$$

whenever the integral exists; as usual, the integral is with respect to the Haar measure.

Definition A.10.1 *Let Λ denote a lattice in the LCA group G, with associated fundamental domain Q. Let $n \in \mathbb{N}$. Given functions $g_1, \ldots, g_n \in L^2(Q)$, the function on G defined by the n-fold convolution*

$$W_n := g_1 \chi_Q * g_2 \chi_Q * \cdots * g_n \chi_Q \tag{A.27}$$

is called a weighted B-spline of order n.

Note that since Q is relatively compact, the assumption $g_j \in L^2(Q)$ implies that $g_j \in L^1(Q)$. Therefore the convolution in (A.27) is well defined, and the terms in the convolution can be reordered without changing the function W_n.

Lemma A.10.2 *Let Λ denote a lattice in the LCA group G, with associated fundamental domain Q. Given functions $g_1, \ldots, g_n \in L^2(Q)$, the weighted B-spline W_n has the following properties:*

(i) *$\{T_\lambda W_n\}_{\lambda \in \Lambda}$ is a Bessel sequence with bound $\prod_{j=1}^n \|g_j\|_{L^2(Q)}^2$.*

(ii) *For $x \in G$, $W_n(x) \neq 0$ only if $x \in nQ := Q + Q + \cdots + Q$; therefore $\operatorname{supp} W_n \subseteq \overline{nQ}$.*

(iii) *If $n \geq 2$, then $W_n \in C_c(G)$; in particular, $W_n \in L^p(G)$ for all $p \geq 1$.*

(iv) *If $g_j > 0$ on Q for $j = 1, \ldots, n$ and $g_j = C$ for at least one index j, then W_n is nonnegative on G and satisfies the partition of unity condition up to a constant, i.e.,*

$$\sum_{\lambda \in \Lambda} W_n(x - \lambda) = \frac{1}{\mu_G(Q)} \prod_{j=1}^n \int_Q g_j(y)\, dy, \quad x \in G.$$

Proof. (i) Given just one function $g \in L^2(Q)$, the system $\{T_\lambda(g\chi_Q)\}_{\lambda \in \Lambda}$ is an orthogonal system (the orthogonality follows from Lemma 21.1.7) and therefore a Bessel sequence, with Bessel bound $\|g\|_{L^2(Q)}^2$. We will now use a result by Cabrelli and Paternostro [101], which states that a system $\{T_\lambda \phi\}_{\lambda \in \Lambda}$ is a Bessel sequence with bound B if and only if

$$\sum_{\omega \in \Lambda^\perp} |\hat{\phi}(\gamma + \omega)|^2 \leq B, \quad a.e. \ \gamma \in V,$$

where V is a fundamental domain in $\hat{G}$ associated with the lattice $\Lambda^\perp$. Applied to $W_1 := g\chi_Q$, this shows that

$$\sum_{\omega \in \Lambda^\perp} |\widehat{g\chi_Q}(\gamma + \omega)|^2 \leq \|g\|_{L^2(Q)}^2. \tag{A.28}$$

Consider now any weighted B-spline W_n. It follows from (A.28) that all function values of $\widehat{g_j\chi_Q}, j = 1, \ldots, n-1$, are bounded by $\|g_j\|_{L^2(Q)}$. Then (A.28) applied to g_n implies that

$$\sum_{\omega \in \Lambda^\perp} |\widehat{W_r}(\gamma + \omega)|^2 = \sum_{\omega \in \Lambda^\perp} \prod_{j=1}^n |\widehat{g_j\chi_Q}(\gamma + \omega)|^2$$

$$\leq \prod_{j=1}^{n-1} \|g_j\|_{L^2(Q)}^2 \sum_{\omega \in \Lambda^\perp} |\widehat{g_r\chi_Q}(\gamma + \omega)|^2 \leq \prod_{j=1}^n \|g_j\|_{L^2(Q)}^2.$$

This shows that $\{T_\lambda W_n\}_{\lambda \in \Lambda}$ is a Bessel sequence with the claimed bound.

(ii) This is an immediate consequence of the definition of the convolution.

(iii) Since $g_1, g_2 \in L^2(Q)$, it follows from standard results for convolution (see [567, pp. 4–5]) that

$$W_2 := g_1\chi_Q * g_2\chi_Q \in C_c(G),$$

implying that $W_2 \in L^p(G)$ for all $p \geq 1$. Iterating the argument leads to the result.

(iv) First, let f_1 be any nonnegative compactly supported function on G for which there is a constant C_1 such that $\sum_{\lambda \in \Lambda} f_1(x - \lambda) = C_1$, $x \in G$. Then, if $f_2 \in L^1(G)$, we have

$$
\begin{aligned}
\sum_{\lambda \in \Lambda} f_1 * f_2(x - \lambda) &= \sum_{\lambda \in \Lambda} \int_G f_1(x - y - \lambda) f_2(y) \, dy \\
&= \int_G \sum_{\lambda \in \Lambda} f_1(x - y - \lambda) f_2(y) \, dy \\
&= C_1 \int_G f_2(y) \, dy. \quad (A.29)
\end{aligned}
$$

This eventually shows that a convolution has the partition of unity property (up to a constant) if at least one of the factors has the property. Indeed, if $g_j = C$ for at least one $j \in \{1, \ldots, n\}$, let us reorder the terms and assume that $g_1 = C$. Then $W_1 := g_1 \chi_Q = C \chi_Q$ satisfies that

$$
\sum_{\lambda \in \Lambda} W_1(x - \lambda) = C = \frac{1}{\mu_G(Q)} \int_Q g_1(y) \, dy, \, x \in G.
$$

The general result now follows by induction based on (A.29). $\square$

A.11   Notes

In this section we provide references for further reading.

Chapter 1: Frames in finite-dimensional spaces have recently attracted more attention because of their use in signal processing. See [46] by Benedetto and Fickus, as well as [138], [143], and [116] by Casazza et al. Frames $\{f_k\}_{k=1}^\infty$ where the frame condition even holds if some of the terms $|\langle f, f_k \rangle|^2$ are replaced by $-|\langle f, f_k \rangle|^2$ are studied by Peng and Waldron in [537], with special emphasis on finite frames. Results on tight frames and applications to coding and communication are given by Strohmer and Heath [591].

Chapter 3: Donoho proved in [255] that the use of unconditional bases leads to optimal sparsity in signal representations. A survey on local trigonometric bases is in [62]. He and Volkmer use Riesz bases in the context of Sturm–Liouville equations in [384]. Local trigonometric bases were introduced by Malvar [512] and Coifman & Meyer [224].

Chapter 5: Frames and operator algebras are studied by Han and Larson [377]. An algorithm (the matching pursuit algorithm) to represent elements in a Hilbert space via an overcomplete system was proposed by Mallat and Zhang in [511]; it applies to highly overcomplete systems which are not Bessel sequences, for example, combined wavelet and Gabor systems with arbitrary parameters. See also [151]. There is a large literature on greedy algorithms and m-term approximation; see e.g., [612] by Temlyakov, where overcomplete systems are used. A survey on frame theory which discusses several open problems is given by Casazza [115]. Frames in Bargmann spaces are considered by Daubechies and Grossmann [243], by Gröchenig and Walnut [358], and by Lyubarskii [505]. Frames in Hilbert C^*-modules were introduced in [306] by Frank and Larson; see also the papers [449] and [448].

Chapter 7: Casazza and Christensen proved in [118] that a frame is unconditional if and only if it is a near-Riesz basis.

Chapter 9: Results on Riesz–Fischer sequences of exponentials are given by Reid in [551].

Chapter 11-14: Gabardo and Han [317] considered Gabor frames for subspaces of $L^2(\mathbb{R})$ and operator algebras.

Chapters 15–18: An early approach to wavelet frames in $L^2(\mathbb{R})$ is given by Frazier and Jawerth [310]. Libraries of frames, i.e., wavelet packet frames, are studied by Long and Chen in [500] and by Chen in [148]; a less advanced approach is in [162]. Wavelet frames in Sobolev are considered by Oswald [532]. Gribonval and Nielsen [331] use spline-generated wavelet frames in approximation theory. Among the alternatives to wavelet bases we mention brushlets, which were introduced by Laeng [474] and Meyer and Coifman [518]; they are based on a local trigonometric basis multiplied with a bell function. Ridgelets were introduced by Candes [108] as a tool to obtain better performance in image processing with images having edges. Around 2005 shearlets were introduced by Guo, Kutyniok, and Labate, also with the purpose of obtaining efficient representations of high-dimensional signals; a comprehensive collection of papers concerning shearlets appear in [469].

Chapter 24: Atomic decompositions of Hardy spaces appeared already in [225] by Coifman and Weiss. Wilson bases in coorbit spaces are discussed by Feichtinger, Gröchenig, and Walnut in [284].

List of Symbols

$\mathbb{R}$: The real numbers

$\mathbb{R}^+$: The strictly positive real numbers

$\mathbb{N}$: The natural numbers: 1,2,3,...

$\mathbb{Z}$: The integers

$\mathbb{Z}_M$: The integers modulo M, i.e., the (cyclic) set $\{0, 1, \ldots, M-1\}$

$\mathbb{Q}$: The rational numbers

$\mathbb{C}$: The complex numbers

$gcd(p, q)$: The largest common divisor for $p, q \in \mathbb{N}$

$\lfloor x \rfloor$: The integer part of $x \in \mathbb{R}$, i.e., the largest integer not exceeding x

$\overline{x}$: The complex conjugated of $x \in \mathbb{C}$

X, Y : Banach spaces

X^* : dual of the Banach space X

X_d : Banach sequence space

$\mathcal{H}, \mathcal{K}$: Hilbert spaces

$L^p(\mathbb{R})$: The space of measurable functions $f : \mathbb{R} \mapsto \mathbb{C}$ for which $\int_{\mathbb{R}} |f(x)|^p dx < \infty$

$C^k(\mathbb{R})$: The space of k times differentiable functions with a continuous k-th derivative

© Springer International Publishing Switzerland 2016
O. Christensen, *An Introduction to Frames and Riesz Bases*,
Applied and Numerical Harmonic Analysis,
DOI 10.1007/978-3-319-25613-9

$C_0(\mathbb{R})$:   The space of continuous functions vanishing at infinity

$\mathcal{F}f(\gamma) = \hat{f}(\gamma)$:   The Fourier transform, for $f \in L^1(\mathbb{R})$ given by $\hat{f}(\gamma) = \int_{\mathbb{R}} f(x)e^{-2\pi i x \gamma}dx$

$\ell^2(I)$:   The space of square summable sequences on I

$|I|$:   The Lebesgue measure of a Borel set I, or when I is discrete, the number of elements in I

χ_A :   The characteristic function for a set A, $\chi_A(x) = 1$ if $x \in A$, otherwise 0

$\overline{A}$:   The closure of a set A

$A^{\perp}$:   The orthogonal complement of a subset A in a Hilbert space

$\operatorname{supp}f$:   The support of the function f: $\operatorname{supp}f = \overline{\{x \in \mathbb{R} : f(x) \neq 0\}}$

$\delta_{k,j}$:   The Kronecker delta: $\delta_{k,j} = 1$ if $k = j$, $\delta_{k,j} = 0$ if $k \neq j$

T_a :   The translation operator $(T_a f)(x) = f(x - a)$

E_b :   The modulation operator $(E_b f)(x) = e^{2\pi i b x} f(x)$

D_a :   The dilation operator $(D_a f)(x) = \frac{1}{\sqrt{a}}f(\frac{x}{a})$, $a > 0$

D :   The dilation operator $(Df)(x) = 2^{1/2}f(2x)$

S :   The frame operator

T :   The pre-frame operator

$U^{\dagger}$:   The pseudo-inverse of the operator U

$\mathcal{N}_U$:   The kernel of the operator U

$\mathcal{R}_U$:   The range of the operator U

$\widehat{G}$:   The dual group of an LCA group G

$H^{\perp}$:   The annihilator of a subgroup H of an LCA group G

π_H :   The canonical quotient map $\pi_H(x) = x + H$ from an LCA group G into G/H

$\mathcal{A}^{\sharp}$:   The matrix $(\mathcal{A}^{-1})^T$

$\mathcal{S}_0$:   The Feichtinger algebra

References

[1] Aldroubi, A.: Portraits of frames. Proc. Am. Math. Soc. **123**, 1661–1668 (1995)

[2] Aldroubi, A.: Non-uniform weighted average sampling and reconstruction in shift-invariant and wavelet spaces. Appl. Comput. Harmon. Anal. **13**, 151–161 (2002)

[3] Aldroubi, A., Baskakov, A., Krishtal, I.: Slanted matrices, Banach frames, and sampling. J. Funct. Anal. **255**(7), 1667–1691 (2008)

[4] Aldroubi, A., Cabrelli, C., Hardin, D., Molter, U.: Optimal shift invariant spaces and their Parseval generators. Appl. Comput. Harmon. Anal. **23**, 273–283 (2007)

[5] Aldroubi, A., Cabrelli, C., Molter, U.: Wavelets on irregular grids with arbitrary dilation matrices and frame atoms for $L^2(\mathbb{R}^d)$. Appl. Comp. Harmon. Anal. **17**, 119–140 (2004)

[6] Aldroubi, A., Gröchenig, K.: Beurling-Landau type theorems for non-uniform sampling in shift invariant spaces. J. Fourier Anal. Appl. **6**(1), 93–103 (2000)

[7] Aldroubi, A., Gröchenig, K.: Nonuniform sampling and reconstruction in shift-invariant spaces. SIAM Rev. **43**(4), 585–620 (2001)

© Springer International Publishing Switzerland 2016   647
O. Christensen, *An Introduction to Frames and Riesz Bases*,
Applied and Numerical Harmonic Analysis,
DOI 10.1007/978-3-319-25613-9

[8] Aldroubi, A., Larson, D., Tang, W.S., Weber, E.: Geometric aspects of frame representations of abelian groups. Trans. Am. Math. Soc. **356**, 4767–4786 (2004)

[9] Aldroubi, A., Sun, Q., Tang, W.: p-Frames and shift invariant subspaces of L^p. J. Fourier Anal. Appl. **7**(1), 1–22 (2001)

[10] Alexeev, B., Cahill, J., Mixon, D.G.: Full spark frames. J. Fourier Anal. Appl. **18**(6), 1167–1194 (2012)

[11] Ali, S.T., Antoine, J.-P., Gazeau, J.-P.:Continuous frames in Hilbert space. Ann. Phys. **222**(1), 1–37 (1993)

[12] Ali, S.T., Antoine, J.-P., Gazeau, J.-P.: Relativistic quantum frames. Ann. Phys. **222**(1), 38–88 (1993)

[13] Ali, S.T., Antoine, J.-P., Gazeau, J.-P.: Coherent States, Wavelets and Their Generalizations, 2nd edn. Springer, New York (2014)

[14] Aniello, P., Cassinelli, G., De Vito, E., Levrero, A.: On discrete frames associated with semidirect products. J. Fourier Anal. Appl. **7**(2), 199–206 (2001)

[15] Antezana, J., Corach, G., Ruiz, M., Stojanoff, D.: Oblique projections and frames. Proc. Am. Math. Soc. **134**(4), 1031–1037 (2006)

[16] Antezana, J., Massey, P., Ruiz, M., Stojanoff, D.: The Schur-Horn theorem for operators and frames with prescribed norms and frame operator. Ill. J. Math. **51**(2), 537–560 (2007)

[17] Ascensi, G., Lyubarskii, Y., Seip, K.: Phase space distribution of Gabor expansions. Appl. Comput. Harmon. Anal. **26**, 277–282 (2009)

[18] Ascensi, G., Feichtinger, H.G., Kaiblinger, N.: Dilation of the Weyl symbol and the Balian-Low theorem. Trans. Am. Math. Soc. **366**(7), 3865–3880 (2004)

[19] Asgari, M.S., Khosravi, A.: Frames and bases of subspaces in Hilbert spaces. J. Math. Anal. Appl. **308**, 541–553 (2005)

[20] Au-Yeung, E., Datta, S.: Tight frames, partial isometries, and signal reconstruction. Appl. Anal. **94**(4), 653–671 (2015)

[21] Auscher, P.: Remarks on the local Fourier bases. In: Benedetto, J., Frazier, M. (eds.) Wavelets: Mathematics and Applications, pp. 203–218. CRC, Boca Raton (1994)

[22] Bachman, G., Narici, L., Beckenstein, E.: Fourier and Wavelet Analysis. Springer, New York (2000)

[23] Baggett, L.W., Merrill, K.D.: Abstract harmonic analysis and wavelets in $\mathbb{R}^n$. Contemp. Math. **247**, 17–27 (1999)

[24] Baggett, L.W.; Medina, H.A., Merrill, K.D.: Generalized multi-resolution analyses and a construction procedure for all wavelet sets in $\mathbb{R}^n$. J. Fourier Anal. Appl. **5**(6), 563–573 (1999)

[25] Bakić, D., Berić, T.: On excesses of frames (2014, preprint)

[26] Balan, R.: Stability theorems for Fourier frames and wavelet Riesz bases. J. Fourier Anal. Appl. **3**, 499–504 (1997)

[27] Balan, R.: Equivalence relations and distances between Hilbert frames. Proc. Am. Math. Soc. **127**(8), 2353–2366 (1999)

[28] Balan, R.: Extensions of no-go theorems to many signal systems. Contemp. Math. **216**, 3–14 (1997)

[29] Balan, R., Bodmann, B., Casazza, P., Eddin, D.: Painless reconstruction from magnitudes of frame coefficients. J. Fourier Anal. Appl. **15**(4), 488–501 (2009)

[30] Balan, R., Casazza, P., Edidin, D.: On signal reconstruction without phase. Appl. Comput. Harmon. Anal. **20**, 345–356 (2006)

[31] Balan, R., Casazza, P., Kutyniok, G., Edidin, D.: A new identity for Parseval frames. Proc. Am. Math. Soc. **135**(4), 1007–1015 (2007)

[32] Balan, R., Casazza, P., Heil, C., Landau, Z.: Deficits and excesses of frames. Adv. Comput. Math. **18**, 93–116 (2002)

[33] Balan, R., Casazza, P.G., Heil, C., Landau, Z.: Density, overcompleteness, and localization of frames I. Theory. J. Fourier Anal. Appl. **12**, 105–143 (2006)

[34] Balan, R., Casazza, P.G., Heil, C., Landau, Z.: Density, overcompleteness, and localization of frames II. Gabor systems. J. Fourier Anal. Appl. **12**, 309–344 (2006)

[35] Balan, R., Casazza, P.G., Heil, C., Landau, Z.: Excesses of Gabor frames. Appl. Comput. Harmon. Anal. **14**(2), 87–106 (2003)

[36] Balan, R., Casazza, P.G., Landau, Z.: Redundancy for localized frames. Isr. J. Math. **185**, 445–476 (2011)

[37] Balan, R., Landau, Z.: Measure functions for frames. J. Funct. Anal. **252**(2), 630–676 (2007)

[38] Balan, R., Wang, Y.: Invertibility and robustness of phaseless reconstruction. Appl. Comput. Harmon. Anal. **38**, 469–488 (2015)

[39] Balazs, P.: Basic definition and properties of Bessel multipliers. J. Math. Anal. Appl. **325**(1), 571–585 (2007)

[40] Balazs, P., Dörfler, M., Jaillet, F., Holighaus, N., Velasco, G.: Theory, implementation and applications of nonstationary Gabor frames. J. Comput. Appl. Math. **236**(6), 1481–1496 (2011)

[41] Bannert, S., Gröchenig, K., Stöckler, J.: Discretized Gabor frames of totally positive functions. IEEE Trans. Inf. Theory **60**, 159–169 (2014)

[42] Bastiaans, M.J.: Gabor's expansion of a signal into Gaussian elementary signals. Proc. IEEE **68**(4), 538–539 (1980)

[43] Ben-Israel, A., Greville, T.N.E.: Generalized Inverses. Theory and Applications. Canadian Mathematical Society. Springer, New York (2002)

[44] Benedetto, J.: Noise reduction in terms of the theory of frames. In: Wavelet Analysis and Applications, vol. 7, pp. 259–284. Academic, San Diego (1998)

[45] Benedetto, J., Czaja, W., Gadzinski, P., Powell, A.: Balian-Low theorem and regularity of Gabor systems. J. Geom. Anal. **13**(2), 239–254 (2003)

[46] Benedetto, J., and Fickus, M.: Finite normalized tight frames. Adv. Comput. Math. **18**(2–4), 357–385 (2003)

[47] Benedetto J., Frazier, M.: Wavelets: Mathematics and Applications. CRC, Boca Raton (1993)

[48] Benedetto, J., Heil, C., Walnut, D.: Differentiation and the Balian-Low theorem. J. Fourier Anal. Appl. **1**(4), 355–402 (1995)

[49] Benedetto, J., Heller, W.: Irregular sampling and the theory of frames. Note Mat. **10**, 103–125 (1990)

[50] Benedetto, J., King, E.J.: Smooth functions associated with wavelet sets on $\mathbb{R}^d, d \geq 1$, and frame bound gaps. Acta Appl. Math. **107**(1–3), 121–142 (2009)

[51] Benedetto, J., Li, S.: The theory of multiresolution analysis frames and applications to filter banks. Appl. Comput. Harmon. Anal., **5**, 389–427 (1998)

[52] Benedetto, J., Li, S.: Subband coding and noise reduction in frame multiresolution analysis. In: Proceedings of SPIE Conference on Mathematical Imaging, San Diego (1994)

[53] Benedetto, J., Pfander, P.: Frame expansions for Gabor multipliers. Appl. Comput. Harmon. Anal. **20**, 26–40 (2006)

[54] Benedetto, J., Powell A., Yilmaz, Ö.: Sigma-Delta quantization and finite frames. IEEE. Trans. Inf. Theory, **52**, 1990–2005 (2006)

[55] Benedetto, J., Powell A., Yilmaz, Ö.: Second order sigma-delta quantization of finite frame expansions. Appl. Comput. Harmon. Anal. **20**, 126–148 (2006)

[56] Benedetto, J., Teolis, Λ.: Local frames. In: Wavelet Applications in Signal and Image Processing. SPIE, Bellingham (1993)

[57] Benedetto, J., Treiber, O.: Wavelet frames: multiresolution analysis and extension principles. In: Debnath, L. (ed.) Wavelet Transforms and Time-Frequency Signal Analysis, pp. 1–36. Birkhäuser, Boston (2001)

[58] Benedetto, J., Walnut, D.: Gabor frames for L^2 and related spaces. In: Benedetto, J., Frazier, M. (eds.) Wavelets: Mathematics and Applications, pp. 97–162. CRC, Boca Raton (1993)

[59] Beutler, F.J., Root, W.L.: The operator pseudo-inverse in control and systems identifications. In: Zuhair Nashed, M. (ed.) Generalized Inverses and Applications. Academic, New York (1976)

[60] Beylkin, G., Coifman, R., Rokhlin, V.: Fast wavelet transforms and numerical algorithms. Commun. Pure Appl. Math. **44**, 141–183 (1991)

[61] Bishop, S., Heil, C., Koo, Y.Y., Lim, J.K.: Invariances of frame sequences under perturbations. Linear Algebra Appl. **432**, 1501–1574 (2010)

[62] Bittner, K.: Biorthogonal local trigonometric bases. In: Anastassiou, G. (ed.) Handbook of Analytic-Computational Methods in Applied Mathematics. Chapman & Hall/CRC, Boca Raton (2000)

[63] Bittner, K.: Folding operators, Wilson bases, and Zak transforms. In: Feichtinger, H.G., Strohmer, T. (eds.) Advances in Gabor Analysis. (Birkhäuser, Boston, 2002)

[64] Blum, J., Lammers, M., Powell, A.M., Yilmaz, Ö.: Sobolev duals in frame theory and sigma-delta quantization. J. Fourier Anal. Appl. **16**, 365–381 (2010)

[65] Blum, J., Lammers, M., Powell, A.M., Yilmaz, Ö.: Errata to: Sobolev duals in frame theory and sigma-delta quantization. J. Fourier Anal. Appl. **16**, 382–382 (2010)

[66] Bodmann, B.G.: Frames as codes. In: Kutyniok, G., Casazza, P.G. (eds.) Finite Frames. Applied and Numerical Harmonic Analysis, pp. 241–266. Birkhäuser, New York (2013)

[67] Bodmann, B.G., Casazza, P.G.: The road to equal-norm Parseval frames. J. Funct. Anal. **258**(2), 397–420 (2010)

[68] Bodmann, B.G., Casazza, P.G.; Kutyniok, G.: A quantitative notion of redundancy for finite frames. Appl. Comput. Harmon. Anal. **30**(3), 348–362 (2011)

[69] Bodmann, B.G., Casazza, P.G.; Paulsen, V.I., Speegle, D.: Spanning and independence properties of frame partitions. Proc. Am. Math. Soc. **140**(7), 2193–2207 (2012)

[70] de Boor, C.: A Practical Guide to Splines. Springer, New York (2001)

[71] de Boor, C., DeVore, R., Ron, A.: On the construction of multivariate (pre)wavelets. Constr. Approx. **9**, 123–166 (1993)

[72] Borichev, A., Gröchenig, K., Lyubarskii, Yu.: Frame constants of Gabor frames near the critical density. J. Math. Pures Appl. **94**, 170–182 (2010)

[73] Borup, L., Gribonval, R., Nielsen, M.: Bi-framelet systems with few vanishing moments characterize Besov spaces. Appl. Comput. Harmon. Anal. **17**, 3–28 (2004)

[74] Borup, L., Gribonval, R., Nielsen, M.: Tight wavelet frames in Lebesgue and Sobolev spaces. J. Funct. Spaces Appl. **2**(3), 227–252 (2004)

[75] Borup, L., Nielsen, M.: Banach frames for multivariate α-modulation spaces. J. Math. Anal. Appl. **321**, 880–895 (2006)

[76] Borup, L., Nielsen, M.: Frame decomposition of decomposition spaces. J. Fourier Anal. Appl. **13**(1), 39–70 (2007)

[77] Bownik, M.: The structure of shift-invariant subspaces of $L^2(\mathbb{R}^n)$. J. Funct. Anal. **177**, 282–309 (2000)

[78] Bownik, M.: A characterization of affine dual frames in $L^2(\mathbb{R}^n)$. Appl. Comput. Harmon. Anal. **8**, 203–221 (2000)

[79] Bownik, M.: Tight frames of multidimensional wavelets. J. Fourier Anal. Appl. **3**(5), 525–542 (1997)

[80] Bownik, M., Christensen, O., Huang, X., Yu, B.: Extensions of shift-invariant systems to frames. Numer. Funct. Anal. Optim. **33**(7–9), 833–846 (2012)

[81] Bownik, M., Jasper, J., Speegle, D.: Orthonormal dilations of non-tight frames. Proc. Am. Math. Soc. **139**(9), 3247–3256 (2011)

[82] Bownik, M., Lemvig, J.: The canonical and alternate duals of a wavelet frame. Appl. Comput. Harmon. Anal. **23**, 263–272 (2007)

[83] Bownik, M., Lemvig, J.: Affine and quasi-affine frames for rational dilations. Trans. Am. Math. Soc. **363**, 1887–1924 (2011)

[84] Bownik, M., Lemvig, J.: Oversampling of wavelet frames for real dilations. J. Lond. Math. Soc. **85**, 765–788 (2012)

[85] Bownik, M., Ross, K.: The structure of translation-invariant spaces on locally compact abelian groups. J. Fourier Anal. Appl. **21**, 849–884 (2015)

[86] Bownik, M., Rzeszotnik, Z.: The spectral function of shift-invariant spaces on general lattices. Contemp. Math. **345**, 49–59 (2004)

[87] Bownik, M., Speegle, D.: Linear independence of Parseval wavelets. Ill. J. Math. **54**, 771–785 (2010)

[88] Bownik, M., Speegle, D.: Linear independence of time-frequency translates of functions with faster than exponential decay. Bull. Lond. Math. Soc. **45**, 554–566 (2013)

[89] Bownik, M., Weber, E.: Affine frames, GMRA's, and the canonical dual. Stud. Math. **159**, 453–479 (2003)

[90] Bui, H.-Q., Laugesen, R.: Frequency-scale frames and the solution of the Mexican hat problem. Constr. Approx. **33**, 163–189 (2011)

[91] Bui, H.-Q., Laugesen, R.: Wavelets in Littlewood-Paley space, and Mexican hat completeness. Appl. Comput. Harmon. Anal. **30**(2), 204–213 (2011)

[92] Bui, H.-Q., Laugesen, R.: Wavelet frame bijectivity on Lebesgue and Hardy spaces. J. Fourier Anal. Appl. **19**, 376–409 (2013)

[93] Burrus, C.S., Gopinath, R.A., Guo, H.: Wavelets and Wavelet Transforms. Prentice Hall, Englewood Cliffs (1998)

[94] Bölcskei, H.: A necessary and sufficient condition for dual Weyl-Heisenberg frames to be compactly supported. J. Fourier Anal. Appl. **5**(5), 409–419 (1999)

[95] Bölcskei, H., Gröchenig, K., Hlawatsch, F., Feichtinger, H.G.: Oversampled Wilson expansions. IEEE Signal Process Lett. **4**(4), 106–108 (1997)

[96] Bölcskei, H., Hlawatsch, F.: Oversampled modulated filter banks. In: Feichtinger, H.G., Strohmer, T. (eds.) Gabor Analysis: Theory and Application. Birkhäuser, Boston (1998)

[97] Bölcskei, H., Hlawatsch, F., Feichtinger, H. G.: Frame-theoretic analysis of oversampled filter banks. IEEE Trans. Signal Process. **46**(12), 3256–3268 (1998)

[98] Bölcskei, H., Janssen, A.J.E.M.: Gabor frames, unimodularity, and window decay. J. Fourier Anal. Appl. **6**(3), 255–276 (2000)

[99] Böttcher, A., Silbermann, B.: Analysis of Toeplitz Operators. Springer, Berlin (1990)

654   References

[100] Cabrelli, C., Mosquera, C.A., Paternostro, V.: Linear combinations of frame generators in systems of translates. J. Math. Anal. Appl. **413**(2), 776–788 (2014)

[101] Cabrelli, C., Paternostro, V.: Shift-invariant spaces on LCA groups. J. Funct. Anal. **258**, 2034–2059 (2010)

[102] Cabrelli, C., Molter, U.: Density of the set of generators of wavelet systems. Constr. Approx. **26**(1), 65–81 (2007)

[103] Cahill, J., Casazza, P., Li, S.: Non-orthogonal fusion frames and the sparsity of fusion frame operators. J. Fourier Anal. Appl. **18**, 287–308 (2012)

[104] Cahill, J., Fickus, M., Mixon, D.G., Poteet, M.J., Strawn, N.: Constructing finite frames of a given spectrum and set of lengths. Appl. Comput. Harmon. Anal. **35**(1), 52–73 (2013)

[105] Cai, J.F., Osher, S., Shen, Z.: Split Bregman methods and frame based image restoration. Multiscale Model. Simul. **8**, 337–369 (2009)

[106] Cai, J.F., Dong, B., Osher, S., Shen, Z.: Image restoration: total variation, wavelet frames, and beyond. J. Am. Math. Soc. **25**, 1033–1089 (2012)

[107] Campbell, S.L., Meyer, C.D.: Generalized inverses of linear transformations. Corrected reprint of the 1979 original. Dover, New York (1991)

[108] Candes, E.: Ridgelets and the representation of multilated Sobolev functions. SIAM J. Math. Anal. **33**, 347–368 (2001)

[109] Candes, E., Donoho, D.L.: New tight frames of curvelets and optimal representations of objects with piecewise C^2-singularities. Commun. Pure Appl. Math. **57**(2), 219–266 (2004)

[110] Carrizo, I., Favier, S.: Perturbation of wavelet and Gabor frames. Anal. Theory Appl. **19**(3), 238–254 (2003)

[111] Carrizo, I., Heineken, S.: Critical pairs of sequences of a mixed frame potential. Numer. Funct. Anal. Optim. **35**(6), 665–684 (2014)

[112] Casazza, P.G.: Characterizing Hilbert space frames with the subframe property. Ill. J. Math. **41**(4), 648–666 (1997)

[113] Casazza, P.G.: Every frame is a sum of three (but not two) orthonormal bases - and other frame representations J. Fourier Anal. Appl. **4**(6), 727–732 (1998)

[114] Casazza, P.G.: Modern tools for Weyl-Heisenberg (Gabor) frame theory. Adv. Imaging Electron Phys. **115**, 1–127 (2000)

[115] Casazza, P.G.: The art of frame theory. Taiwan. J. Math. **4**(2), 129–201 (2000)

[116] Casazza, P.G.: Custom building finite frames. In: Wavelets, Frames and Operator Theory. Contemporary Mathematics, vol. 345, pp. 61–86. American Mathematical Society, Providence (2004)

[117] Casazza, P.G.: The Kadison–Singer problem and Paulsen problems in finite frames theory. In: Casazza, P., Kutyniok, G. (eds.) Finite Frames, Theory and Applications. Birkhäuser, Boston (2012)

[118] Casazza, P.G., Christensen, O.: Hilbert space frames containing a Riesz basis and Banach spaces which have no subspace isomorphic to c_0. J. Math. Anal. Appl. **202**, 940–950 (1996)

[119] Casazza, P.G., Christensen, O.: Perturbation of operators and applications to frame theory. J. Fourier Anal. Appl. **3**, 543–557 (1997)

[120] Casazza, P.G., Christensen, O.: Frames and Schauder bases. In: Govil, N.K., Mohapatra, R.N., Nashed, Z., Sharma, A., Szabados, J. (eds.) Approximation Theory: In Memory of A.K. Varna, pp. 133–139. Marcel Dekker, New York (1998)

[121] Casazza, P.G., Christensen, O.: Frames containing a Riesz basis and preservation of this property under perturbation. SIAM J. Math. Anal. **29**(1), 266–278 (1998)

[122] Casazza, P.G., Christensen, O.: Riesz frames and approximation of the frame coefficients. Appraisals Theory Appl. **14**(2), 1–11 (1998)

[123] Casazza, P.G., Christensen, O., Janssen, A.J.E.M.: Classifying tight Weyl-Heisenberg frames. Contemp. Math. **247**, 131–148 (1999)

[124] Casazza, P.G., Christensen, O.: Approximation of the inverse frame operator and applications to Weyl-Heisenberg frames. J. Approx. Theory **103**(2), 338–356 (2000)

[125] Casazza, P.G., Christensen, O.: Weyl-Heisenberg frames for subspaces of $L^2(\mathbb{R})$. Proc. Am. Math. Soc. **129**, 145–154 (2001)

[126] Casazza, P.G., Christensen, O.: A perturbation theorem for Banach spaces. Can. Bull. Math. **51**, 348–358 (2008)

[127] Casazza, P.G., Christensen, O., Janssen, A.J.E.M.: Weyl-Heisenberg frames, translation invariant systems, and the Walnut representation. J. Funct. Anal. **180**, 85–147 (2001)

[128] Casazza, P.G., Christensen, O., Li, S., Lindner, A.: On Riesz-Fischer sequences and lower frame bounds. Z. Anal. Anwend. **21**(2), 305–314 (2002)

[129] Casazza, P.G., Christensen, O., Kalton, N: Frames of translates. Collect. Math. **52**(1), 35–54 (2001)

[130] Casazza, P.G., Christensen, O., Lindner, A., Vershynin, R.: Frames and the Feichtinger conjecture. Proc. Am. Math. Soc. **133**, 1025–1033 (2005)

[131] Casazza, P.G., Christensen, O., Stoeva, D.T.: Frame expansions in separable Banach spaces. J. Math. Anal. Appl. **307**(2), 710–723 (2005)

[132] Casazza, P.G., Fickus, M., Tremain, J.C., Weber, E.:: The Kadison-Singer problem in mathematics and engineering – a detailed account. Contemp. Math. **414**, 297–356 (2006)

[133] Casazza, P. G., Fickus, M.,, Heinecke, A., Wang, Y., Zhou, Z.: Spectral tetris fusion frame constructions. J. Fourier Anal. Appl. **18**(4), 828–851 (2012)

[134] Casazza, P.G., Fickus, M., Kovačević, J., Leon, M.T., Tremain, J.C.: A physical interpretation of tight frames. In: Heil C. (ed.) Harmonic Analysis and Applications. Applied and Numerical Harmonic Analysis, pp. 51–76. Birkhäuser, Boston (2006)

[135] Casazza, P.G., Fickus, M., Mixon, D., Wang, Y., Zhou, Z.: Constructing tight fusion frames. Appl. Comput. Harmon. Anal. **30**, 175–187 (2011)

[136] Casazza, P.G., Han, D., Larson, D.: Frames for Banach spaces. Contemp. Math. **247**, 149–182 (1999)

[137] Casazza, P.G., Kalton, N.J.: Roots of complex polynomials and Weyl-Heisenberg frame sets. Proc. Am. Math. Soc. **130**(8), 2313–2318 (2002)

[138] Casazza, P.G., Kovačević, J.: Equal-norm tight frames with erasures. Adv. Comput. Math. **18**, 387–430 (2003)

[139] Casazza, P.G., Kutyniok, G. (eds): Finite Frames: Theory and Applications. Birkhäuser, Boston (2012)

[140] Casazza, P.G., Kutyniok, G.: Frames and subspaces. In: Wavelets, Frames, and Operator Theory. Contemporary Mathematics, vol. 345, pp. 87–113. American Mathematical Society, Providence (2004)

[141] Casazza, P.G., Kutyniok, G., Li, S.: Fusion frames and distributed processing. Appl. Comput. Harmon. Anal. **25**, 114–132 (2008)

[142] Casazza, P.G., Kutyniok, G., Lammers, M.: Duality principles in abstract frame theory. J. Fourier Anal. Appl. **10**(4), 383–408 (2004)

[143] Casazza, P.G., Leon, M.: Existence and construction of finite tight frames. J. Concr. Appl. Math. **4**(3), 277–289 (2006)

[144] Casazza, P.G., Leon, M.: Existence and construction of finite frames with a given frame operator. Int. J. Pure Appl. Math. **63**(2), 149–157 (2010)

[145] Casazza, P.G., Leonhard, N.: Classes of finite equal norm Parseval frames. Contemp. Math. **451**, 11–31 (2008)

[146] Charina, M., Putinar, M., Scheiderer, C., Stöckler, J.: An algebraic perspective on multivariate tight wavelet frames. Constr. Approx. **38**, 253–276 (2013)

[147] Charina, M., Putinar, M., Scheiderer, C., Stöckler, J.: An algebraic perspective on multivariate tight wavelet frames II. Appl. Comput. Harmon. Anal. **39**(2), 185–213 (2015)

[148] Chen, D.: On the splitting trick and wavelet frame packets. SIAM J. Math. Anal. **31**(4), 726–739 (2000)

[149] Chen, D.: Frames of periodic shift-invariant spaces. J. Appr. Theory **107**, 204–211 (2000)

[150] Chen, D.Y., Li, L., Zheng, B.T.: Perturbations of frames. Acta Math. Sin. Engl. Ser. **30**(7), 1089–1108 (2014)

[151] Chen, S., Donoho, D., Saunders, M.A.: Atomic decomposition by basis pursuit. SIAM Rev. **43**(1), 129–157 (2001)

[152] Christensen, O.: Frames and the projection method. Appl. Comp. Harm Anal. **1** 50–53 (1993)

[153] Christensen, O.: Frames and pseudo-inverse operators. J. Math. Anal. Appl. **195**, 401–414 (1995)

[154] Christensen, O.: A Paley–Wiener theorem for frames. Proc. Am. Math. Soc. **123**, 2199–2202 (1995)

[155] Christensen, O.: Frame perturbations. Proc. Am. Math. Soc. **123**, 1217–1220 (1995)

[156] Christensen, O.: Atomic decomposition via projective group representations. Rocky Mt. J. Math. **26**(4), 1289–1312 (1996)

[157] Christensen, O.: Frames containing a Riesz basis and approximation of the frame coefficients using finite dimensional methods. J. Math. Anal. Appl. **199**, 256–270 (1996)

[158] Christensen, O.: Perturbations of frames and applications to Gabor frames. In: Feichtinger, H.G., Strohmer, T. (eds.) Gabor Analysis and Algorithms: Theory and Applications, pp. 193–209. Birkhäuser, Boston (1998)

[159] Christensen, O.: Operators with closed range and perturbation of frames for a subspace. Can. Math. Bull **42**(1), 37–45 (1999)

[160] Christensen, O.: Finite-dimensional approximation of the inverse frame operator and applications to Weyl-Heisenberg frames and wavelet frames. J. Fourier Anal. Appl. **6**(1), 79–91 (2000)

[161] Christensen, O.: Frames, bases, and discrete Gabor/wavelet expansions. Bull. Am. Math. Soc. **38**(3), 273–291 (2001)

[162] Christensen, O.: Linear combinations of frames and frame packets. Z. Anal. Anwend. **20**(4), 805–815 (2001)

[163] Christensen, O.: Frames and generalized shift-invariant systems. In: Operator Theory: Advances and Applications, vol. 164, pp. 193–209. Birkhäuser, Boston (2006)

[164] Christensen, O.: Pairs of dual Gabor frames with compact support and desired frequency localization. Appl. Comput. Harmon. Anal. **20**, 403–410 (2006)

[165] Christensen, O.: Functions, Spaces and Expansions: Mathematical Tools in Physics and Engineering. Birkhäuser, Boston (2010)

[166] Christensen, O.: Six problems in frame theory. In: Schmeisser, G., Zayed, A. (eds.) New Perspectives on Approximation and Sampling Theory (Festschrift in honor of Paul Butzer.). Springer, New York (2014)

[167] Christensen, O., deFlicht, C., Lennard, C.: Perturbation of frames for a subspace of a Hilbert space. Rocky Mt. J. Math. **30**(4), 1237–1249 (2000)

[168] Christensen, O., Deng, B., Heil, C.: Density of Gabor frames. Appl. Comput. Harmon. Anal. **7**, 292–304 (1999)

[169] Christensen, O., Eldar, Y.: Oblique dual frames and shift-invariant spaces. Appl. Comput. Harmon. Anal. **17**(1), 48–68 (2004)

[170] Christensen, O. and Eldar, Y.: Frames for subspaces and generalized duals. J. Fourier Anal. Appl. **11**(3), 299–313 (2005)

[171] Christensen, O., Favier, S., Zó, F.: Irregular Wavelet Frames and Gabor Frames. Appraisal Theory Appl. **17**(3), 90–101 (2001)

[172] Christensen, O., Forster, B., Massopust, P.: Directional time-frequency analysis via continuous frames. Bull. Aust. Math. Soc. **92**(2), 268–281 (2015)

[173] Christensen, O., Goh, S.S.: Pairs of oblique dual frames in spaces of periodic functions. Adv. Comput. Math. **32**(3), 353–379 (2010)

[174] Christensen, O., Goh, S.S.: Pairs of dual periodic frames. Appl. Comput. Harmon. Anal. **33**, 315–329 (2012)

[175] Christensen, O., Goh, S.S.: From dual pairs of Gabor frames to dual pairs of wavelet frames and vice versa. Appl. Comput. Harmon. Anal. **46**, 198–214 (2014)

[176] Christensen, O., Goh, S.S.: Fourier-like frames on locally compact abelian groups. J. Approx. Theory **192**, 82–101 (2015)

[177] Christensen, O., Hasannasab, M., Lemvig, J.: Explicit constructions and properties of generalized shift-invariant systems in $L^2(\mathbb{R})$ (2015, preprint)

[178] Christensen, O., Jensen, T.K.: An introduction to the theory of bases, frames, and wavelets. Lecture Notes. Technical University of Denmark (2000)

[179] Christensen, O., Heil, C.: Perturbations of Banach frames and atomic decompositions. Math. Nachr. **185**, 33–47 (1997)

[180] Christensen, O., Kim, H.O., Kim, R.Y.: Gabor windows supported on $[-1,1]$ and compactly supported dual windows. Appl. Comput. Harmon. Anal. **28**, 89–103 (2010)

[181] Christensen, O., Kim, H.O., Kim, R.Y.: On the duality principle by Casazza, Kutyniok, and Lammers. J. Fourier Anal. Appl. **17**, 640–655 (2011)

[182] Christensen, O., Kim, H.O., Kim, R.Y.: Gabor windows supported on $[-1,1]$ and dual windows with small support. Adv. Comput. Math. **36**, 525–545 (2012)

[183] Christensen, O., Kim, H.O., Kim, R.Y.: Extensions of Bessel sequences to dual pairs of frames. Appl. Comput. Harmon. Anal. **34**, 224–233 (2013)

[184] Christensen, O., Kim, H.O., Kim, R.Y.: On Parseval wavelet frame with two or three generators via the unitary extension principle. Can. Math. Bull. **57**, 254–263 (2014)

[185] Christensen, O., Kim, H.O., Kim, R.Y.: On entire functions restricted to intervals, partition of unities, and dual Gabor frames. Appl. Comput. Harmon. Anal. **38**, 72–86 (2015)

[186] Christensen, O., Kim, H.O., Kim, R.Y.: On partition of unities generated by entire functions and construction of Gabor frames in $L^2(\mathbb{R}^d)$ and $\ell^2(\mathbb{Z}^d)$. J. Fourier Anal. Appl. (2016, to appear)

[187] Christensen, O., Kim, H.O., Kim, R.Y.: On extensions of wavelet systems to dual pairs of frames. Adv. Comp. Math. (2016, to appear)

[188] Christensen, O., Kim, H.O., Kim, R.Y.: On Gabor frames generated by sign-changing windows and B-splines. Appl. Comput. Harmon. Anal. **39**(3), 535–544 (2015)

[189] Christensen, O., Kim, H.O., Kim, R.Y., Lim, J.K.: Riesz sequences of translates and their generalized duals. J. Geom. Anal. **16**(4), 585–596 (2006)

[190] Christensen, O., Kim, R.Y.: Pairs of explicitly given dual Gabor frames in $L^2(R^d)$. J. Fourier Anal. Appl. **12**(3), 243–255 (2006)

[191] Christensen, O., Kim, R.Y.: On dual Gabor frame pairs generated by polynomials. J. Fourier Anal. Appl. **16**, 1–16 (2010)

[192] Christensen, O., Laugesen, R.: Approximately dual frames in Hilbert spaces and applications to Gabor frames. Sampling Theory Signal Image Process. **9**, 77–90 (2011)

[193] Christensen, O., Lindner, A.: Frames of exponentials: lower frame bounds for finite subfamilies, and approximation of the inverse frame operator. Linear Algebra Appl. **323**(1–3), 117–130 (2001)

[194] Christensen, O., Lindner, A.: Lower bounds for finite Gabor and wavelet systems. Appraisal Theory Appl. **17**(1), 18–29 (2001)

[195] Christensen, O., Lindner, A.: Decompositions of wavelets and Riesz frames into a finite number of linearly independent sets. Linear Algebra Appl. **355**, 147–159 (2002)

[196] Christensen, O., Massopust, P.: Exponential B-splines and the partition of unity property. Adv. Comput. Math. **37**(3), 301–318 (2012)

[197] Christensen, O., Osgooei, E.: On frame-properties for Fourier-like systems. J. Approx. Theory **172**, 47–57 (2013)

[198] Christensen, O., Powell, A.M., Xiao, X.C.: A note on finite dual frame pairs. Proc. Am. Math. Soc. **140**(11), 3921–3930 (2012)

[199] Christensen, O., Rahimi, A.: Frame properties of wave packet systems in $L^2(R^d)$. Adv. Comput. Math. **29**(2), 101–111 (2008)

[200] Christensen, O., Stoeva, D.T.: p-Frames in separable Banach spaces. Adv. Comput. Math. **18**(2–4), 117–126 (2003)

[201] Christensen, O., Strohmer, T.: Methods for approximation of the inverse (Gabor) frame operator. In: Feichtinger, H.G., Strohmer, T. (eds.) Advances in Gabor Analysis. Birkhäuser, Boston (2002)

[202] Christensen, O., Strohmer, T.: The finite section method and problems in frame theory. J. Approx. Theory **133**, 221–237 (2005)

[203] Christensen, O., Sun, W.: Explicitly given pairs of dual frames with compactly supported generators and applications to irregular B-splines. J. Approx. Theory **151**, 155–163 (2008)

[204] Chui, C.: Wavelets - A Tutorial in Theory and Practice. Academic, San Diego (1992)

[205] Chui, C.: Multivariate Splines. SIAM, Philadelphia (1988)

[206] Chui, C., Czaja, C., Maggioni, M., Weiss, G.: Characterization of general tight wavelet frames with matrix dilations and tightness preserving oversampling. J. Fourier Anal. Appl. **8**(2), 173–200 (2002)

[207] Chui, C., He, W.: Compactly supported tight frames associated with refinable functions. Appl. Comput. Harmon. Anal. **8**, 293–319 (2000)

[208] Chui, C., He, W.: Construction of multivariate tight frames via Kronecker products. Appl. Comput. Harmon. Anal. **11**, 305–312 (2001)

[209] Chui, C., He, W., Stöckler, J.: Compactly supported tight and sibling frames with maximum vanishing moments. Appl. Comput. Harmon. Anal. **13**(3), 226–262 (2002)

[210] Chui, C., He, W., Stöckler, J.: Nonstationary tight wavelet frames I. Bounded intervals. Appl. Comput. Harmon. Anal. **17**(2), 141–197 (2004)

[211] Chui, C., He, W., Stöckler, J.: Nonstationary tight wavelet frames II. Unbounded intervals. Appl. Comput. Harmon. Anal. **18**(1), 25–66 (2005)

[212] Chui, C., Shi, X.: Bessel sequences and affine frames. Appl. Comput. Harmon. Anal. **1**, 29–49 (1993)

[213] Chui, C., Shi, X.: On L^p-boundedness of affine frame operators. Indag. Math. **4**(4), 431–438 (1993)

[214] Chui, C., Shi, X.: Inequalities of Littlewood-Paley type for frames and wavelets. SIAM J. Math. Anal. **24**(1), 263–277 (1993)

[215] Chui, C., Shi, X.: $N\times$ oversampling preserves any tight affine frame for odd N. Proc. Am. Math. Soc. **121**(2), 511–517 (1994)

[216] Chui, C., Shi, X.: Wavelets of Wilson type with arbitrary shapes. Appl. Comput. Harmon. Anal. **8**, 1–23 (2000)

[217] Chui, C., Shi, X.: Orthonormal wavelets and tight frames with arbitrary real dilations. Appl. Comput. Harmon. Anal. **9**(3), 243–264 (2000)

[218] Chui, C., Shi, X., Stöckler, J.: Affine frames, quasi-affine frames, and their duals. Adv. Comput. Math. **8**, 1–17 (1998)

[219] Chui, C., Sun, Q.: Tight frame oversampling and its equivalence to shift-invariance of affine frame operators. Proc. Am. Math. Soc. **131**(5), 1527–1538 (2003)

[220] Chui, C., Sun, Q.: Affine frame decompositions and shift-invariant spaces. Appl. Comput. Harmon. Anal. **20**(1), 74–107 (2006)

[221] Chui, C., Sun, Q.: Characterizations of tight over-sampled affine frame systems and over-sampling rates. Appl. Comput. Harmon. Anal. **22**(1), 1–15 (2007)

[222] Chui, C., Wang, J.: On compactly supported spline wavelets and a duality principle. Trans. Am. Math. Soc. **330**, 903–915 (1992)

[223] Cohen, A., Daubechies, I., Feauveau, J.-C.: Biorthogonal bases of compactly supported wavelets. Commun. Pure Appl. Math. **45**, 485–560 (1993)

[224] Coifman, R., Meyer, Y.: Remarques sur l'analyse de Fourier à fenêtre. C. R. Acad. Sci. Paris Sér. I Math. **312**(3), 259–261 (1991)

[225] Coifman, R., Weiss, G.: Extensions of Hardy space and their use in analysis. Bull. Am. Math. Soc. **83**, 569–645 (1977)

[226] Cordero, E., Gröchenig, K.: Localization of frames II. Appl. Comput. Harmon. Anal. **17**, 29–47 (2004)

[227] Cordero, E., Gröchenig, K., Nicola, F.: Approximation of Fourier integral operators by Gabor multipliers. J. Fourier Anal. Appl. **18**(4), 661–684 (2012)

[228] Currey, B., Mayeli, A.: The orthonormal dilation property for abstract Parseval wavelet frames. Can. Math. Bull. **56**(4), 729–736 (2013)

[229] Cvetković, Z., Vetterli, M.: Tight Weyl-Heisenberg frames. IEEE Trans. Signal Process. **46**(5), 1256–1259 (1998)

[230] Cvetković, Z., Vetterli, M.: Oversampled filter banks. IEEE Trans. Signal Process. **46**(5), 1245–1255 (1998)

[231] Czaja, W., Kutyniok, G., Speegle, D.: The Geometry of sets of parameters of wave packets. Appl. Comput. Harmon. Anal. **20**(1), 108–125 (2006)

[232] Dahlke, S.: Multiresolution analysis and wavelets on locally compact abelian groups. In: Wavelets, Images, and Surface Fittings, pp. 141–156. AK Peters, Wellesley (1994)

[233] Dahlke, S.: A note on generalized Weyl-Heisenberg frames. Appl. Math. Lett. **7**, 79–82 (1994)

[234] Dahlke, S., Fornasier, M., Rauhut, H. Steidl, G., Teschke, G.: Generalized coorbit theory, Banach frames, and the relation to a-modulation spaces. Proc. Lond. Math. Soc. **96**(2), 464–506 (2008)

[235] Dahlke, S., Kutyniok, G., Steidl, G., Teschke, G.: Shearlet coorbit spaces and associated Banach frames. Appl. Comput. Harmon. Anal. **27**(2), 195–214 (2009)

[236] Dahlke, S. Steidl, G., Teschke, G.: Frames and coorbit theory on homogeneous spaces with a special guidance on the sphere. J. Fourier Anal. Appl. **13**(4), 387–404 (2007)

[237] Dai, X., Diao, Y., Gu, Q.: Frame wavelet sets in $\mathbb{R}$. Proc. Am. Math. Soc. **129**(7), 2045–2055 (2000)

[238] Dai, X., Diao, Y., Gu, Q., Han, D.: Frame wavelets in subspaces of $\mathbb{R}^d$. Proc. Am. Math. Soc. **130**(11), 3259–3267 (2002)

[239] Dai, X.R., Sun, Q.: The abc-problem for Gabor systems. Memoirs of the American Mathematical Society (2015, to appear)

[240] Dahmen, W., Micchelli, C.A.: On multivariate E-splines. Adv. Math. **76**, 33–93 (1989)

[241] Daubechies, I.: The wavelet transformation, time-frequency localization and signal analysis. IEEE Trans. Inf. Theory **36**, 961–1005 (1990)

[242] Daubechies, I.: Ten Lectures on Wavelets. SIAM, Philadelphia (1992)

[243] Daubechies, I., Grossmann, A.: Frames in the Bargmann space of entire functions. Commun. Pure and Appl. Math. **41**, 151–164 (1988)

[244] Daubechies, I., Grossmann, A., Meyer, Y.: Painless nonorthogonal expansions. J. Math. Phys. **27**, 1271–1283 (1986)

[245] Daubechies, I., Jaffard, S., Journé, J.L.: A simple Wilson orthonormal basis with exponential decay. SIAM J. Math. Anal. **22**, 554–572 (1991)

[246] Daubechies, I., Han, B.: The canonical dual of a wavelet frame. Appl. Comput. Harmon. Anal. **12**(3), 269–285 (2002)

[247] Daubechies, I., Han, B.: Pairs of dual wavelet frames from any two refinable functions. Constr. Appr. **20**, 325–352 (2004)

[248] Daubechies, I., Han, B., Ron, A., Shen, Z.: Framelets: MRA-based constructions of wavelet frames. Appl. Comp. Harm. Anal. **14**(1), 1–46 (2003)

[249] Daubechies, I., Landau, H.J., Landau, Z.: Gabor time-frequency lattices and the Wexler-Raz identity. J. Fourier Anal. Appl. **1**, 437–478 (1995)

[250] Davis, M.J., Heller, E.J.: Semiclassical Gaussian basis set method for molecular vibrational wave functions. J. Chem. Phys **71**, 3383–3395 (1979)

[251] Del Prete, V.: Estimates, decay properties, and computation of the dual function for Gabor frames. J. Fourier Anal. Appl. **5**, 545–562 (1999)

[252] Del Prete, V.: On a necessary condition fo B-spline Gabor frames. Ricerche Mat. **59**, 161–164 (2010)

[253] Deng, B., Heil, C.: Density of Gabor Schauder bases. In: Aldroubi, A., Laine, A., Unser, M. (eds.) Wavelet Applications in Signal and Image Processing, VIII, pp. 153–164. SPIE, Bellingham (2000)

[254] Ding, J., Huang, L.J.: Perturbation of generalized inverses of linear operators in Hilbert spaces. J. Math. Anal. Appl. **198**, 505–516 (1996)

[255] Donoho, D.: Unconditional bases are optimal bases for data compression and for statistical estimation. Appl. Comput. Harmon. Anal. **1**, 100–115 (1993)

[256] Donovan, G., Geronimo, J.S., Hardin, D.P.: Intertwining multiresolution analyses and the construction of piecewise-polynomial splines. SIAM J. Math. Anal. **27**(6), 1791–1815 (1996)

[257] Dörfler, M., Matusiak, E.: Nonstationary Gabor frames — existence and construction. Int. J. Wavelets Multiresolution Inf. Process. **12** (2014)

[258] Dörfler, M., Matusiak, E.: Nonstationary Gabor frames - approximately dual frame and reconstruction errors. Adv. Comput. Math. **41**, 293–316 (2015)

[259] Dörfler, M., Romero, J.L.: Frames adapted to a phase-space cover. Constr. Approx. **39**(3), 445–484 (2014)

[260] Dörfler, M., Torresani, B.: Representation of operators in the time-frequency domain and generalized Gabor multipliers. J. Fourier Anal. Appl. **16**(2), 261–293 (2010)

[261] Duflo, M., Moore, C.C.: On the regular representation of a non unimodular locally compact group. J. Funct. Anal. **21**, 209–243 (1976)

[262] Duffin, R.J., Schaeffer, A.C.: A class of nonharmonic Fourier series. Trans. Am. Math. Soc. **72**, 341–366 (1952)

[263] Dutkay, D., Han, D., Larson, D.: A duality principle for groups. J. Funct. Anal. **257**, 1133–1143 (2009)

[264] Dykcman, K., Freeman, D., Kernelson, K., Larson, D., Ordower, M., Weber, E.: Elliposidal tight frames and projection decompositions of operators. Ill. J. Math. **48**(2), 477–489 (2004)

[265] Easwaran Nambudiri, T.C.; Parthasarathy, K.: A characterisation of Weyl-Heisenberg frame operators. Bull. Sci. Math. **137**(3), 322–324 (2013)

[266] Ehler, M.: On multivariate compactly supported bi-frames. J. Fourier Anal. Appl. 13, 511–532 (2007)

[267] Ehler, M.: Nonlinear approximation schemes associated with non-separable wavelet bi-frames. J. Approx. Theory **161**(1), 292–313 (2009)

[268] Eldar, Y.: Sampling Theory: Beyond Bandlimited Systems. Cambridge University Press, Cambridge (2015)

[269] Enflo, P.: A counterexample to the approximation property in Banach spaces. Acta Math. **130**, 309–317 (1973)

[270] Fan, Z., Shen, Z.: Dual Gramian analysis: duality principle and unitary extension principle. Math. Comput. (to appear)

[271] Fan, Z., Heinecke, A., Shen, Z.: Duality for Frames. J. Fourier Anal. Appl. (2016, to appear)

[272] Farrell, B., Strohmer, T.: Eigenvalue estimates and mutual information for the linear time-varying channel. IEEE Trans. Inf. Theory **57**(9), 5710–5719 (2011)

[273] Favier, S.J., Zalik, R.A.: On the stability of frames and Riesz bases. Appl. Comput. Harmon. Anal. **2**, 160–173 (1995)

[274] Feichtinger, H.G.: On a new Segal algebra. Monatsh. Math. **92**(4), 269–289 (1981)

[275] Feichtinger, H.G.: Minimal Banach spaces and atomic decomposition. Publ. Math. **34**(3–4), 231–240 (1987)

[276] Feichtinger, H.G.: Atomic characterizations of modulation spaces through Gabor-type representations. Rocky Mt. J. Math. **19**(1), 113–125 (1989)

[277] Feichtinger, H.G.: Pseudo-inverse matrix methods for signal reconstruction from partial data. In: Visual Communications and Image Processing, pp. 766–772. SPIE, Boston (1991)

[278] Feichtinger, H.G.: Modulation spaces: looking back and ahead. Sampling Theory Signal Image Process. **5**(2), 109–140 (2006)

[279] Feichtinger, H.G., Grybos, A., Onchis, D.M.: Approximate dual Gabor atoms via the adjoint lattice method. Adv. Comput. Math. **40**, 651–665 (2014)

[280] Feichtinger, H.G., Gröchenig, K.: Banach spaces related to integrable group representations and their atomic decomposition I. J. Funct. Anal. **86**, 307–340 (1989)

[281] Feichtinger, H.G., Gröchenig, K.: Banach spaces related to integrable group representations and their atomic decomposition II. Monatsh. Math. **108**, 129–148 (1989)

[282] Feichtinger, H.G., Gröchenig, K.: Gabor frames and time-frequency analysis of distributions. J. Funct. Anal. **146**, 464–495 (1997)

[283] Feichtinger, H.G., Gröchenig, K.: Irregular sampling theorems and series expansions of band-limited functions. J. Math. Anal. Appl. **167**(2), 530–556 (1992)

[284] Feichtinger, H.G., Gröchenig, K., Walnut, D.: Wilson bases and modulation spaces. Math. Nachr. **155**, 7–17 (1992)

[285] Feichtinger, H.G., Janssen, A.J.E.M.: Validity of WH-frame conditions depends on lattice parameters. Appl. Comput. Harmon. Anal. **8**, 104–112 (2000)

[286] Feichtinger, H.G., Kaiblinger, N.: Varying the time-frequency lattice of Gabor frames. Trans. Am. Math. Soc. **356**, 2001–2023 (2004)

[287] Feichtinger, H.G., Kozek, W.: Quantization of TF lattice-invariant operators on elementary LCA groups. In: Feichtinger, H.G., Strohmer, T. (eds.) Gabor Analysis and Algorithms: Theory and Applications, pp. 233–266. Birkhäuser, Boston (1998)

[288] Feichtinger, H.G., Kozek, W., Luef, F.: Gabor analysis over finite abelian groups. Appl. Comput. Harmon. Anal. **26**(2), 230–248 (2009)

[289] Feichtinger, H.G., Onchis, D.M., Wiesmeyr, C.: Construction of approximate dual wavelet frames. Adv. Comput. Math. **40**, 273–282 (2014)

[290] Feichtinger, H.G., Qiu, S.: Discrete Gabor structures and optimal representations. IEEE Signal Process. **43**(10), 2258–2268 (1995)

[291] Feichtinger, H.G., Strohmer, T. (eds.): Gabor Analysis and Algorithms: Theory and Applications. Birkhäuser, Boston (1998)

[292] Feichtinger, H.G., Strohmer, T. (eds.): Advances in Gabor Analysis. Birkhäuser, Boston (2002)

[293] Feichtinger, H.G., Sun, W.: Two Banach spaces of atoms for stable wavelet frame expansions. J. Approx. Theory **146**, 28–70 (2007)

[294] Feichtinger, H.G., Sun, W.: Sufficient conditions for irregular Gabor frames. Adv. Comput. Math. **26**, 403–430 (2007)

[295] Feichtinger, H.G., Zimmermann, G.: A Banach space of test functions for Gabor analysis. In: Feichtinger, H.G., Strohmer, T. (eds.) Gabor Analysis and Algorithms: Theory and Applications, pp. 123–170. Birkhäuser, Boston (1998)

[296] Fernandez-Morales, H.R., Garcia, A.G.; Hernandez-Medina, M.A., Munoz-Bouzo, M.J.: On some sampling-related frames in U-invariant spaces. Abstr. Appl. Anal. (2013)

[297] Fickus, M.: Maximally equiangular frames and Gauss sums. J. Fourier Anal. Appl. **15**(3), 413–427 (2009)

[298] Fickus, M., Johnson, B.D., Kornelson, K., Okoudjou, K.A.: Convolutional frames and the frame potential. Appl. Comput. Harmon. Anal. **19**(1), 77–91 (2005)

[299] Fickus, M., Mixon, D.G, Tremain, J.C.: Steiner equiangular tight frames. Linear Algebra Appl. **436**(5), 1014–1027 (2012)

[300] Folland, G.B.: Harmonic Analysis in Phase Space. Annals of Mathematics Studies. Princeton University Press, Princeton (1989)

[301] Folland, G.B.: A Course in Abstract Harmonic Analysis. CRC, Boca Raton (1995)

[302] Fornasier, M.: Banach frames for α-modulation spaces. Appl. Comput. Harmon. Anal. **22**(2), 157–175 (2007)

[303] Fornasier, M., Rauhut, H.: Continuous frames, function spaces, and the discretization problem. J. Fourier Anal. Appl. **11**(3), 245–287 (2005)

[304] Fornasier, M., Gröchenig, K.: Intrinsic localization of frames. Constr. Approx. **22**, 395–415 (2005)

[305] Forster, B., Blu, T., Unser, M.: Complex B-splines. Appl. Comput. Harmon. Anal. **20**(2), 261–282 (2006)

[306] Frank, M., Larson, D.R.: Frames in Hilbert C^*-modules and C^*-algebras. J. Oper. Theory **48**(2), 273–314 (2002)

[307] Frank, M., Paulsen, V.I., Tiballi, T.R.: Symmetric approximation of frames and bases in Hilbert spaces. Trans. Am. Math. Soc. **354**(2), 777–793 (2002)

[308] Frazier, M.: An Introduction to Wavelets Through Linear Algebra. Springer, New York (2001)

[309] Frazier, M., Garrigos, G., Wang, K., Weiss, G.: A characterization of functions that generate wavelet and related expansion. J. Fourier Anal. Appl. **3**, 883–906 (1997)

[310] Frazier, M., Jawerth, B.: A discrete transform and decomposition of distribution spaces. J. Funct. Anal. **93**(1), 34–170 (1990)

[311] Führ, H.: Admissible vectors for the regular representation. Proc. Am. Math. Soc. **130**(10), 2959–2970 (2002)

[312] Führ, H.: Abstract harmonic analysis of continuous wavelet transforms. Lecture Notes in Mathematics, vol. 1863. Springer, Berlin (2004)

[313] Garcia, A.G., Kim, J.M., Kwon, K.H., Yoon, G.J. Multi-channel sampling on shift-invariant spaces with frame generators. Int. J. Wavelets Multiresolution Inf. Process. **10**(1), (2012)

[314] Gabor, D.: Theory of communications. J. IEE (Lond.) **93**(3), 429–457 (1946)

[315] Gabardo, J.-P.: Tight Gabor frames associated with non-separable lattices and the hyperbolic secant. Acta Appl. Math. **107**(1–3), 49–73 (2009)

[316] Gabardo, J.-P.: Weighted irregular Gabor tight frames and dual systems using windows in the Schwartz class. J. Funct. Anal. **256**(3), 635–672 (2009)

[317] Gabardo, J.-P., Han, D.: Subspace Weyl-Heisenberg frames. J. Fourier Anal. Appl. **7**(4), 419–433 (2001)

[318] Gabardo, J.-P., Han, D., Li, Y.Z.: Lattice tiling and density conditions for subspace Gabor frames. J. Funct. Anal. **265**(7), 1170–1189 (2013)

[319] Gibson, P.C., Lamoureux, M.P., Margrave, G.F.: Representation of linear operators by Gabor multipliers. In: Excursions in Harmonic Analysis. Applied and Numerical Harmonic Analysis, vol. 2, pp. 229–250. Birkhäuser, Boston (2013)

[320] Goh, S.S., Goodman, T.N.T., Lee, S.L. Constructing tight frames of multivariate functions. J. Approx. Theory **158**(1), 49–68 (2009)

[321] Goh, S.S., Han, B., Shen, Z.: Tight periodic wavelet frames and approximation orders. Appl. Comput. Harmon. Anal. **31**, 228–248 (2011)

[322] Goh, S.S., Lim, Z.Y., Shen, Z.: Symmetric and antisymmetric tight wavelet frames. Appl. Comput. Harmon. Anal. **20**(3), 411–421 (2006)

[323] Goh, S.S., Ron, A., Shen, Z. (eds.) : Gabor and Wavelet Frames. Lecture Notes Series. Institute for Mathematical Sciences. National University of Singapore/World Scientific Publishing, Singapore (2007)

[324] Goh, S.S., Teo, K.M.: Extension principles for tight wavelet frames of periodic functions. Appl. Comput. Harmon. Anal. **25**, 168–186 (2008)

[325] Gohberg, I., Krein, M.: Introduction to the Theory of Linear Non-selfadjoint Operators. American Mathematical Society, Providence (1969)

[326] Govil, N.K., Zalik, R.A.: Perturbation of the Haar wavelet. Proc. Am. Math. Soc. **125**(11), 3363–3370 (1997)

[327] Goyal, V.K., Vetterli, M., Thao, N.T.: Quantized expansions in $\mathbb{R}^n$; analysis, synthesis, and algorithms. IEEE Trans. Inf. Theory **44**, 16–31 (1998)

[328] Goyal, V.K., Kovačević, J., Kelner, A.J.: Quantized frame expansions with erasures. Appl. Comput. Harmon. Anal. **10**(3), 203–233 (2000)

[329] Grafakos, L., Lennard, C.: Characterization of $L^p(\mathbb{R}^n)$ using Gabor frames. J. Fourier Anal. Appl. **7**, 101–126 (2001)

[330] Grafakos, L., Sansing, C.: Gabor frames and directional time-frequency analysis. Appl. Comput. Harmon. Anal. **25**, 47–67 (2008)

[331] Gribonval, R., Nielsen, M.: On approximation with spline generated framelets. Constr. Approx. **20**, 207–232 (2004)

[332] Grip, N., Sun, W.: Remarks on the article: on the stability of wavelet and Gabor frames (Riesz bases) by J. Zhang. J. Fourier Anal. Appl. **9**(1), 97–100 (2003)

[333] Grossmann, A., Morlet, J.: Decomposition of Hardy functions into square integrable wavelets of constant shape. SIAM J. Math. Anal. **15**, 723–736 (1984)

[334] Grossmann, A., Morlet, J., Paul, T.: Transforms associated to square integrable group representations I. J. Math. Phys. **26**(10), 2473–2479 (1985)

[335] Grossmann, A., Morlet, J., Paul, T.: Transforms associated to square integrable group representations II. Ann. Inst. H. Poincaré Phys. Théor. **45**(10), 293–309 (1986)

[336] Gröchenig, K.: Describing functions: frames versus atomic decompositions. Monatsh. Math. **112**, 1–41 (1991)

[337] Gröchenig, K.: Irregular sampling of wavelet and short time Fourier transforms. Constr. Approx. **9**, 283–297 (1993)

[338] Gröchenig, K.: Acceleration of the frame algorithm. IEEE Trans. Signal Process. **41**(12), 3331–3340 (1993)

[339] Gröchenig, K.: Aspects of Gabor analysis on locally compact abelian groups. In: Feichtinger, H.G., Strohmer, T. (eds.) Gabor Analysis and Algorithms: Theory and Applications, pp 211–231. Birkhäuser, Boston (1998)

[340] Gröchenig, K.: Foundations of Time-Frequency Analysis. Birkhäuser, Boston (2000)

[341] Gröchenig, K.: Localization of frames, Banach frames, and the invertibility of the frame operator. J. Fourier Anal. Appl. **10**, 105–132 (2004)

[342] Gröchenig, K.: Localized frames are finite unions of Riesz sequences. Adv. Comput. Math. **18**, 149–157 (2003)

[343] Gröchenig, K.: The mystery of Gabor frames. J. Fourier Anal. Appl. **20**, 865–895 (2014)

[344] Gröchenig, K.: Linear independence of time-shifts? Monatsh. Math. **177**(1), 67–77 (2015)

[345] Gröchenig, K., Heil, C.: Gabor meets Littlewood-Paley: Gabor expansions in $L^p(\mathbb{R}^d)$. Stud. Math.**146**(1), 15–33 (2001)

[346] Gröchenig, K., Janssen, A.J.E.M.: A new criterion for Gabor frames. J. Fourier Anal. Appl. **8**(5), 507–512 (2002)

[347] Gröchenig, K., Janssen, A.J.E.M., Kaiblinger, N., Pfander, G.: Note on B-splines, wavelet scaling functions, and Gabor frames. IEEE Trans. Inf. Theory **49**(12), 3318–3320 (2003)

[348] Gröchenig, K., Kutyniok, G., Seip, K.: Landau's necessary density conditions for LCA groups. J. Funct. Anal. **255**, 1831–1850 (2008)

[349] Gröchenig, K., Leinert, M.: Wiener's lemma for twisted convolution and Gabor frames. J. Am. Math. Soc. **17**, 1–18 (2004)

[350] Gröchenig, K., Lyubarskii, Y.: Gabor frames with Hermite functions. C. R. Math. Acad. Sci. Paris **344**(3), 157–162 (2007)

[351] Gröchenig, K., Lyubarskii, Y.: Gabor (super)frames with Hermite functions. Math. Ann. **345**(2), 267–286 (2009)

[352] Gröchenig, K., Malinnikova, E.: Phase space localization of Riesz bases for $L^2(\mathbb{R}^d)$. Rev. Mat. Iberoam. **29**(1), 115–134 (2013)

[353] Gröchenig, K., Ortega-Cerda, J., Romero, J.L.: Deformation of Gabor systems. Adv. Math. **277**, 388–425 (2015)

[354] Gröchenig, K. and Ron, A.: Tight compactly supported wavelet frames of arbitrarily high smoothness. Proc. Am. Math. Soc. **126**(4), 1101–1107 (1998)

[355] Gröchenig, K., Rzeszotnik, Z.: Banach algebras of pseudodifferential operators and their almost diagonalization. Ann. Inst. Fourier (Grenoble) **58**(7), 2279–2314 (1998)

[356] Gröchenig, K., Rzeszotnik, Z., Strohmer, T.: Convergence analysis of the finite section method and Banach algebras of matrices. Integr. Equ. Oper. Theory **67**, 183–202 (2010)

[357] Gröchenig, K., Stöckler, J.: Gabor frames and totally positive functions. Duke Math. J. **162**, 1003–1031 (2013)

[358] Gröchenig, K., Walnut, D.: A Riesz basis for Bargmann-Fock space related to sampling and interpolation. Ark. Mat. **30**(2), 283–295 (1992)

[359] Guo, K., Labate, D., Lim, W., Weiss, G., Wilson, E.: Wavelets with composite dilations and their MRA-properties. Appl. Comput. Harmon. Anal. **20**, 202–236 (2006)

[360] Ha, Y.H., Ryu, H.Y., Shin, I.S.: Angle criteria for frame sequences and frames containing a Riesz basis. J. Math. Anal. Appl. **347**(1), 90–95 (2008)

[361] Haar, A.: Zur Theorie der Orthogonalen Funktionen-Systeme. Math. Ann. **69**, 331–371 (1910)

[362] Han, B.: On dual wavelet tight frames. Appl. Comput. Harmon. Anal. **4**(4), 380–413 (1997)

[363] Han, B.: Compactly supported tight wavelet frames and orthonormal wavelets of exponential decay with a general dilation matrix. J. Comput. Appl. Math. **155**, 43–67 (2003)

[364] Han, B.: Matrix splitting with symmetry and symmetric tight framelet filter banks with two high-pass filters. Appl. Comput. Harmon. Anal. **35**(2), 200–227 (2013)

[365] Han, B.: Symmetric tight framelet filter banks with three high-pass filters. Appl. Comput. Harmon. Anal. **37**(1), 140–161 (2014)

[366] Han, B., Mo, Q.: Tight wavelet frames generated by three symmetric B-spline functions with high vanishing moments. Proc. Am. Math. Soc. **132**(1), 77–86 (2003)

[367] Han, B., Mo, Q.: Multiwavelet frames from refinable function vectors. Adv. Comput. Math. **18**, 211–245 (2003)

[368] Han, B., Mo, Q.: Splitting a matrix of Laurent polynomials with symmetry and its applications to symmetric framelet filter banks. SIAM J. Matrix Anal. Appl. **26**, 97–124 (2004)

[369] Han, B., Mo, Q.: Symmetric MRA tight wavelet frames with three generators and high vanishing moments. Appl. Comput. Harmon. Anal. **18**, 67–93 (2005)

[370] Han, B., Shen, Z.: Dual wavelet frames and Riesz bases in Sobolev spaces. Constr. Approx. **29**, 369–406 (2009)

[371] Han, B., Shen, Z.: Characterization of Sobolev spaces of arbitrary smoothness using nonstationary tight wavelet frames. Isr. J. Math. **172**, 371–398 (2009)

[372] Han, D., Approximations for Gabor and wavelet frames. Trans. Am. Math. Soc. **355**(8), 3329–3342 (2003)

[373] Han, D., Frame representations and Parseval duals with applications to Gabor frames. Trans. Am. Math. Soc. **360**(6), 3307–3326 (2008)

[374] Han, D., Dilations and completions for Gabor systems, J. Fourier Anal. Appl. **15**, 201–217 (2009)

[375] Han, D.: The existence of tight Gabor duals for Gabor frames and subspace Gabor frames. J. Funct. Anal. **256**(1), 129–148 (2009)

[376] Han, D., Kornelson, K., Larson, D., Weber, E.: Frames for undergraduates. Student Mathematical Library, vol. 40. American Mathematical Society, Providence (2007)

[377] Han, D., Larson, D.: Frames, bases and group representations. Mem. Am. Math. Soc. **147**(697) (2000)

[378] Han, D., Larson, D., Papadakis, M., Stavropoulos, Th.: Multiresolution analyses of abstract Hilbert spaces and wandering subspaces. Contemp. Math. **247**, 259–284 (1999)

[379] Han, D., Li. P., Tang, W.S.: Frames and their associated H_F^p-subspaces. Adv. Comp. Math. **34**(2), 185–200 (2011)

[380] Han, D., Wang, Y.: Lattice tiling and the Weyl-Heisenberg frames. Geom. Funct. Anal. **11**(4), 742–758 (2001)

[381] Hayashi, E., Li, S., Sorrells, T.: Gabor duality characterizations. In: Heil, C. (ed.) Harmonic Analysis and Applications. Applied and Numerical Harmonic Analysis, pp. 127–137. Birkhäuser, Boston (2006)

[382] Heckel, R., Bölcskei, H.: Identification of sparse linear operators. IEEE Trans. Inf. Theory **59**(12), 7985–8000 (2013)

[383] He, X., Key, E., Volkmer, H.: Perturbation of orthonormal bases in L^2-spaces. Integr. Equ. Oper. Theory **41**, 396–409 (2001)

[384] He, X., Volkmer, H.: Riesz bases of solutions of Sturm-Liouville equations. J. Fourier Anal. Appl. **7**(3), 297–308 (2001)

[385] Heil, C.: Integral operators, pseudodifferential operators, and Gabor frames. In: Feichtinger, H.G., Strohmer, T. (eds.) Advances in Gabor Analysis. Birkhäuser, Boston (2002)

[386] Heil, C. (Ed.): Harmonic Analysis and Applications. Applied and Numerical Harmonic Analysis. Birkhäuser, Boston (2006)

[387] Heil, C.: Linear independence of finite Gabor systems. In: Harmonic Analysis and Applications. Applied and Numerical Harmonic Analysis, pp. 171–206. Birkhäuser, Boston (2006)

[388] Heil, C.: History and evolution of the density theorem for Gabor frames. J. Fourier Anal. Appl. **13**, 113–166 (2007)

[389] Heil, C.: A basis theory primer. Applied and Numerical Harmonic Analysis, Expanded edn. Birkhäuser/Springer, New York (2011)

[390] Heil, C., Kutyniok, G.: Density of weighted wavelet frames. J. Geom. Anal. **13**, 479–493 (2003)

[391] Heil, C., Kutyniok, G.: The homogeneous approximation property for wavelet frames. J. Approx. Theory **147**, 28–46 (2007)

[392] Heil, C., Kutyniok, G.: Density of frames and Schauder bases of windowed exponentials. Houst. J. Math. **34**, 565–600 (2008)

[393] Heil, C., Ramanathan, J., Topiwala, P.: Linear independence of time-frequency translates. Proc. Am. Math. Soc. **124**, 2787–2795 (1996)

[394] Heil, C., Ramanathan, J., Topiwala, P.: Singular values of compact pseudodifferential operators. J. Funct. Anal. **150**, 426–452 (1997)

[395] Heil, C., Walnut, D.: Continuous and discrete wavelet transforms. SIAM Rev. **31**, 628–666 (1989)

[396] Heil, C., Walnut, D. (eds.): Fundamental Papers in Wavelet Theory. Princeton University Press, Princeton, (2006)

[397] Heineken, S.B., Morillas, P.M., Benavente, A.M., Zakowicz, M.I.: Dual fusion frames. Arch. Math. **103**(4), 355–365 (2014)

[398] Hernandez, E., Labate, D., Weiss, G.: A unified characterization of reproducing systems generated by a finite family II. J. Geom. Anal. **12**(4), 615–662 (2002)

[399] Hernandez, E. Labate, D. Weiss, G., Wilson, E.: Oversampling, quasi affine frames and wave packets. Appl. Comput. Harmon. Anal **16**, 111–147 (2003)

[400] Hernandez, E., Weiss, G.: A First Course on Wavelets. CRC, Boca Raton (1996)

[401] Heuser, H.: Functional Analysis. Wiley, New York (1982)

[402] Hewitt, E., Ross, K.: Abstract Harmonic Analysis, vols. 1 and 2. Springer, Berlin (1963)

[403] Higgins, J. R.: Completeness and Basis Properties of Sets of Special Functions. Cambridge University Press, Cambridge (1977)

[404] Hilding, S.: Note on completeness theorems of Paley-Wiener type. Ann. Math. **49**(4), 953–955 (1948)

[405] Hirsch, M., Smale, S.: Differential Equations, Dynamical Systems, and Linear Algebra. Academic, New York (1970)

[406] Hogan, J.A., Lakey, J.D. Time-frequency and time-scale methods. Adaptive decompositions, uncertainty principles, and sampling. Applied and Numerical Harmonic Analysis. Birkhäuser, Boston (2005)

[407] Hogan, J.A., Lakey, J.D. BMO, boundedness of affine operators, and frames. Appl. Comput. Harmon. Anal. **18**(1), 3–24 (2005)

[408] Hogan, J.A., Lakey, J.D. Frame properties of shifts of prolate spheroidal wave functions. Appl. Comput. Harmon. Anal. **39**(1), 21–32 (2015)

[409] Holighaus, N.: Structure of nonstationary Gabor frames and their dual systems. Appl. Comput. Harmon. Anal. **37**, 442–463 (2014)

[410] Holighaus, N., Wiesmeyr, C.: Construction of warped time-frequency representations on nonuniform frequency scales, Part I: Frames (2015, preprint)

[411] Holmes, R.B., Paulsen, V.I.: Optimal frames for erasures. Linear Algebra Appl. **377**, 31–51 (2004)

[412] Holschneider, M., Tchamitchiam, P.: Régularité locale de la fonction "nondifférentiable" de Riemann. In: Lemarié, P.G. (ed.) Les ondelettes en 1989. Lecture Notes in Mathematics, vol. 1438. Springer, New York (1989)

[413] Holub, J.: Pre-frame operators, Besselian frames and near-Riesz bases. Proc. Am. Math. Soc. **122**, 779–785 (1994)

[414] Jaffard, S.: A density criterion for frames of complex exponentials. Michigan J. Math. **38**, 339–348 (1991)

[415] Jaffard, S., Young, R.: A representation theorem for Schauder bases in Hilbert spaces. Proc. Am. Math. Soc. **126**(2), 553–560 (1998)

[416] Jaillet, F.: Représentation et traitement temps-fréquence des signaux audionumérique pour des applications de design sonore. Ph.D Thesis, Université de la Méditerranée – Aix-Marseille II (2005)

[417] Jakobsen, M.S., Lemvig, J.: Reproducing formulas for generalized translation invariant systems on locally compact groups. Trans. Am. Math. Soc. (2015, to appear)

[418] Jakobsen, M.S., Lemvig, J.: Co-compact Gabor systems on locally compact groups. J. Fourier Anal. Appl. (2016, to appear)

[419] Jakobsen, M.S., Lemvig, J.: A characterization of tight and dual generalized translation invariant frames. In: Proceeding of the SAMPTA Conference, Washington (2015)

[420] Jang, S., Jeong, B., Kim, H.O.: Techniques for smoothing and splitting in the construction of tight frame Gabor windows. Int. J. Wavelets Multiresolution Inf. Process. **11**(1) (2013)

[421] Jang, S., Jeong, B., Kim, H.O.: Compactly supported multiwindow dual Gabor frames of rational sampling density. Adv. Comput. Math. **38**(1), 159–186 (2013)

[422] Janssen, A.J.E.M.: Gabor representation of generalized functions. J. Math. Anal. Appl. **83**, 377–394 (1981)

[423] Janssen, A.J.E.M.: Bargmann transform, Zak transform, and coherent states. J. Math. Phys. **23**, 720–731 (1982)

[424] Janssen, A.J.E.M.: The Zak transform: a signal transform for sampled time-continuous signals. Philips J. Res. **43**, 23–69 (1988)

[425] Janssen, A.J.E.M.: Signal analytic proofs of two basic results on lattice expansions. Appl. Comput. Harmon. Anal. **1**, 350–354 (1994)

[426] Janssen, A.J.E.M.: On rationally oversampled Weyl-Heisenberg frames. Signal Process. **47**, 239–245 (1995)

[427] Janssen, A.J.E.M.: Duality and biorthogonality for Weyl-Heisenberg frames. J. Fourier Anal. Appl. **1**(4), 403–436 (1995)

[428] Janssen, A.J.E.M.: Some Weyl-Heisenberg frame bound calculations. Indag. Math **7**, 165–183 (1996)

[429] Janssen, A.J.E.M.: From continuous to discrete Weyl-Heisenberg frames through sampling. J. Fourier Anal. Appl. **3**(5), 583–596 (1997)

[430] Janssen, A.J.E.M.: The duality condition for Weyl-Heisenberg frames. In: Feichtinger, H.G., Strohmer, T. (eds.) Gabor Analysis: Theory and Application. Birkhäuser, Boston (1998)

676   References

[431] Janssen, A.J.E.M.: Representations of Gabor frame operators. In: Twentieth Century Harmonic Analysis–a Celebration. Nato Science Series II: Mathematics, Physics and Chemistry, vol. 33, pp. 73–101. Kluwer Academic, Dordrecht (2001)

[432] Janssen, A.J.E.M.: Zak transforms with few zeros and the tie. In: Feichtinger, H.G., Strohmer, T. (eds.) Advances in Gabor Analysis. Birkhäuser, Boston (2002)

[433] Janssen, A.J.E.M.: On generating tight Gabor frames at critical density. J. Fourier Anal. Appl. **9**(2), 175–214 (2003)

[434] Janssen, A.J.E.M., Strohmer, T.: Characterization and computation of canonical tight windows for Gabor frames. J. Fourier Anal. Appl. **8**(1), 1–28 (2002)

[435] Janssen, A.J.E.M., Strohmer, T.: Hyperbolic secants yield Gabor frames. Appl. Comput. Harmon. Anal. **12**(3), 259–267 (2002)

[436] Janssen, A.J.E.M., Søndergaard, P.: Iterative algorithms to approximate canonical Gabor windows: computational aspects. J. Fourier Anal. Appl. **13**(2), 211–241 (2007)

[437] Jasper, J., Mixon, D.G., Fickus, M.: Kirkman equiangular tight frames and codes. IEEE Trans. Inf. Theory **60**(1), 170–181 (2014)

[438] Jensen, H.E., Høholdt, T., Justesen, J.: Double series representations of bounded signals. IEEE Trans. Inf. Theory **34**(4), 613–624 (1988)

[439] Jetter, K., Stöckler, J.: Riesz bases and regularized splines with multiple knots. J. Appr. Theory **87**(3), 338–359 (1996)

[440] Jiang, Q.T.: Parametrizations of masks for tight affine frames with two symmetric/antisymmetric generators. Adv. Comput. Math. **18**, 247–268 (2003)

[441] Jing, Z.: On the stability of wavelet and Gabor frames (Riesz bases) J. Fourier Anal. Appl. **5**(1), 105–125 (1999)

[442] Johnson, B.D.: On the oversampling of affine wavelet frames. SIAM J. Math. Anal. **35**(3), 623–638 (2003)

[443] Kaiblinger, N.: Approximation of the Fourier transform and the dual Gabor window. J. Fourier Anal. Appl. **11**(1), 25–42 (2005)

[444] Kaiser, G.: A Friendly Guide to Wavelets. Birkhäuser, Boston (1994)

[445] Kaiser, G.: Quantum Physics, Relativity, and Complex Spacetime: Towards a New Synthesis. North-Holland, Amsterdam (1990)

[446] Kaniuth, E., Kutyniok, G.: Zeroes of the Zak transform on locally compact abelian groups. Proc. Am. Math. Soc. **126**, 3561–3569 (1998)

[447] Kato, T.: Perturbation Theory for Linear Operators. Springer, New York (1976)

[448] Khosravi, A., Hasannasab, M.: Modular Riesz bases and modular g-Riesz bases in Hilbert C^*-modules (2015, preprint)

[449] Khosravi, A., Khosravi, B.: g-Frames and modular Riesz bases in Hilbert C^*-modules. Int. J. Wavelets Multiresolution Inf. Process. **10**(2) (2012)

[450] Kim, H.O., Lim, J.K.: New characterizations of Riesz bases. Appl. Comput. Harmon. Anal. **4**, 222–229 (1997)

[451] Kim, H.O., Lim, J.K.: On frame wavelets associated with frame multiresolution analysis. Appl. Comput. Harmon. Anal. **10**(1), 61–70 (2001)

[452] Kim, H.O., Lim, J.K.: Frame multiresolution analysis. Commun. Korean Math. Soc. **15**, 285–308 (2000)

[453] Kim, I.: Gabor Frames in one dimension with Trigonometric Spline Dual Windows. Asian-European J. Math. **8**(4) 1550072 (2015)

[454] Kim, I.: Gabor frames with trigonometric spline dual windows. Ph.D. Dissertation, University of Illinois at Urbana-Champaign (2011). https://www.ideals.illinois.edu/handle/2142/26039

[455] Kim, J.M., Kwon, K.H.: Frames by integer translates. J. Korean Soc. Ind. Appl. Math. (2007, to appear)

[456] King, E.J.: Smooth Parseval frames for $L^2(\mathbb{R})$ and generalizations to $L^2(\mathbb{R}^d)$. Int. J. Wavelets Multiresolution Inf. Process. **11**(6) (2013)

[457] King, E.J., Skopina, M.A.: Quincunx multiresolution analysis for $L^2(\mathbb{Q}_2^2)$. p-Adic Numbers Ultrametric Anal. Appl. **2**, 222–231 (2010)

[458] Kloos, T.: Zeros of the Zak transform of totally positive functions. J. Fourier Anal. Appl. **21**(5), 1130–1145 (2015)

[459] Kloos, T., Stöckler, J.: Zak transforms and Gabor frames of totally positive functions and exponential B-splines. J. Approx. Theory **184**, 209–237 (2014)

[460] Koo, Y.Y., Lim, J.K.: Perturbation of frame sequences and its applications to shift-invariant spaces. Linear Algebra Appl. **420**, 295–309 (2007)

[461] Koo, Y.Y., Lim, J.K.: Existence of Parseval oblique duals of a frame sequence. J. Math. Anal. Appl. **404**(2), 470–476 (2013)

[462] Kozek, W., Molisch, A.: Nonorthogonal pulseshapes for multicarrier communications in doubly dispersive channels. IEEE Trans. Sel. Area. Commun. **16**(8), 1579–1589 (1998)

[463] Krahmer, F., Kutyniok, G., Lemvig, J.: Sparsity and spectral properties of dual frames. Linear Algebra Appl. **439**(4), 982–998 (2013)

[464] Kreyzig, E.: Introductory Functional Analysis with Applications. Wiley, New York (1989)

[465] Körner, T.W.: Fourier Analysis. Cambridge University Press, Cambridge (1989)

[466] Kutyniok, G.: Linear independence of time-frequency shifts under a generalized Schrödinger representation. Arch. Math. **78**(2), 135–144 (2002)

[467] Kutyniok, G.: Affine density, frame bounds, and the admissibility condition for wavelet frames. Constr. Approx. **25**(3), 239–253 (2007)

[468] Kutyniok, G., Labate, D.: Theory of reproducing systems on locally compact abelian group. Colloq. Math. **106**, 197–220 (2006)

[469] Kutyniok, G., Labate, D. (eds.) : Shearlets: Multiscale Analysis for Multivariate Data. Birkhäuser, Boston (2012)

[470] Kutyniok, G., Strohmer, T.: Wilson bases for general time-frequency lattices. SIAM J. Math. Anal. **37**(3), 685–711 (2005)

[471] Labate, D.: A unified characterization of reproducing systems generated by a finite family. J. Geom. Anal.**12**(3), 469–491 (2002)

[472] Labate, D. Weiss, G., Wilson, E.: An Approach to the Study of Wave Packet systems. Contemp. Math. Wavelets Frames Oper. Theory, **345**, 215–235 (2004)

[473] Lammers, M., Powell, A., Yilmaz, "O.: Alternative dual frames for digital-to-analog conversion in sigma-delta quantization. Adv. Comput. Math. **32**, 73–102 (2010)

[474] Laeng, E.: Une base orthonormale de $L^2(\mathbb{R})$ dont les éléments sont bien localisés dans l'espace de phase et leurs supports adapté à toute partition symétrique de l'espace des fréquences. C. R. Acad. Sci. Paris Sér. I. Math. **311**(11), 677–680 (1990)

[475] Landau, H.J.: A sparse regular sequence of exponentials closed on large sets. Bull. Am. Math. Soc. **70**, 566–569 (1964)

[476] Laugesen, R.S.: Completeness of orthonormal wavelet systems for arbitrary real dilations. Appl. Comput. Harmon. Anal. **11**, 455–473 (2001)

[477] Laugesen, R.S.: Translational averaging for completeness, characterization and oversampling of wavelets. Collect. Math. **53**(3), 211–249 (2002)

[478] Laugesen, R.S.: On affine frames with transcendental dilations. Proc. Am. Math. Soc. **135**, 211–216 (2007)

[479] Laugesen, R.S.: Gabor dual spline windows. Appl. Comput. Harmon. Anal. **27**, 180–194 (2009)

[480] Lawrence, J., Pfander, G.E., Walnut, D.: Linear independence of Gabor systems in finite dimensional vector spaces. J. Fourier Anal. Appl. **11**(6), 715–726 (2005)

[481] Lemarié, P.G.: Une nouvelle base d'ondelettes de $L^2(\mathbb{R}^n)$. J. Math. Pures Appl. **67**, 227–236 (1988)

[482] Lemarié, P.G., Meyer, Y.: Ondelettes et bases hilbertiennes. Rev. Math. Iberoam. **2**, 1–18 (1986)

[483] Lemvig, J.: Constructing pairs of dual bandlimited framelets with desired time localization. Adv. Comput. Math. **30**, 231–247 (2009)

[484] Lemvig, J.: Constructing pairs of dual bandlimited frame wavelets in $L^2(\mathbb{R}^n)$. Appl. Comput. Harmon. Anal. **32**, 313–328 (2012)

[485] Lemvig, J., Miller, C., Okoudjou, K.: Prime tight frames. Adv. Comput. Math. **40**(2), 315–334 (2014)

[486] Lemvig, J., Nielsen, K.H.: A counterexample to the B-spline conjecture for Gabor frames (2015, preprint)

[487] Leng, J., Han, D.: Optimal dual frames for erasures II. Linear Algebra Appl. **435**(6), 1464–1472 (2011)

[488] Li, D.F., Sun, W.: Expansion of frames to tight frames. Acta Math. Sin. (Engl. Ser.) **25**, 287–292 (2009)

[489] Li, S.: On general frame decompositions. Numer. Funct. Anal. Optim. **16**(9 & 10), 1181–1191 (1995)

[490] Li, S.: A theory of generalized multiresolution structure and pseudo-frames of translates. J. Fourier Anal. Appl. **7**(1), 23–40 (2001)

[491] Li, S. Liu, Y., Mi, T.: Sparse dual frames and dual Gabor functions of minimal time and frequency supports. J. Fourier Anal. Appl. **19**(1), 48–76 (2013)

[492] Li, S., Ogawa, H.: Pseudo-duals of frames with applications. Appl. Comput. Harmon. Anal. **11**, 289–304 (2001)

[493] Li, S., Ogawa, H.: Pseudoframes for subspaces with applications. J. Fourier Anal. Appl. **10**(4), 409–431 (2004)

[494] Liu, Y.M., Walter, G.: Irregular sampling in wavelet subspaces. J. Fourier Anal. Appl. **2**(2), 181–189 (1995)

[495] Lindenstrauss, J., Tzafriri, L.: Classical Banach Spaces 1. Springer, New York (1977)

[496] Lindner, A.: On lower bounds for exponential frames. J. Fourier. Anal. Appl. **5**, 187–194 (1999)

[497] Lindner, A.: A universal constant for exponential Riesz sequences. Z. Anal. Anwend. **19**(2), 553–559 (2000)

[498] Lindner, A.: Growth estimates for sine-type functions and applications to Riesz bases of exponentials. Appr. Theory Appl. **18**(3), 26–41 (2002)

[499] Linnell, P.: Von Neumann algebras and linear independence of translates. Proc. Am. Math. Soc. **127**(11), 3269–3277 (1999)

[500] Long, R., Chen, W.: Wavelet basis packets and wavelet frame packets. J. Fourier Anal. Appl. **3**(3), 239–256 (1997)

[501] Lopez, J., Han, D.: Optimal dual frames for erasures. Linear Algebra Appl. **432**(1), 471–482 (2010)

[502] Lopez, J., Han, D.: Discrete Gabor frames in $\ell^2(\mathbb{Z}^d)$. Proc. Am. Math. Soc. **141**(11), 3839–3851 (2013)

[503] Luef, F.: Projective modules over noncommutative tori are multi-window Gabor frames for modulation spaces. J. Funct. Anal. **257**(6), 1921–1946 (2009)

[504] Luef, F.: Projections in noncommutative tori and Gabor frames. Proc. Am. Math. Soc. **139**(2), 571–582 (2011)

[505] Lyubarskii, Y.: Frames in the Bargmann space of entire functions. Adv. Sov. Math. **11**, 167–180 (1992)

[506] Lyubarskii, Y., Nes, P.G.: Gabor frames with rational density. Appl. Comput. Harmon. Anal. **34**, 488–494 (2013)

[507] Lyubarskii, Y., Seip, K.: Convergence and summability of Gabor expansions at the Nyquist density. J. Fourier. Anal. Appl. **5**(2/3), 127–157 (1999)

[508] Malikiosis, R.-D.: A note on Gabor frames in finite dimensions. Appl. Comput. Harmon. Anal. **38**, 318–330 (2015)

[509] Mallat, S.: A Wavelet Tour of Signal Processing. Academic, San Diego (1999)

[510] Mallat, S.: Multiresolution approximations and wavelet orthonormal bases of $L^2(\mathbb{R})$. Trans. Am. Math. Soc. **315**(1), 69–87 (1989)

[511] Mallat, S., Zhang, Z.: Matching pursuit with time-frequency dictionaries. IEEE Trans. Signal Process. **41**(12), 3397–3415 (1993)

[512] Malvar, H.: Lapped transforms for efficient transform/subband coding. IEEE Trans. Acoust. Speech Signal Process. **38**, 969–978 (1990)

[513] Massey, P.G., Ruiz, M.A., Stojanoff, D.: Optimal dual frames and frame completions for majorization. Appl. Comput. Harmon. Anal. **34**(2), 201–223 (2013)

[514] Massey, P.G., Ruiz, M.A., Stojanoff, D.: Optimal frame completions. Adv. Comput. Math. **40**(5–6), 1011–1042 (2014)

[515] Massey, P.G., Ruiz, M.A., Stojanoff, D.: Optimal frame completions with prescribed norms for majorization. J. Fourier Anal. Appl. **20**(5), 1111–1140 (2014)

[516] Massopust, P.: Interpolation and Approximation with Splines and Fractals. Oxford University Press, Oxford (2010)

[517] Marcus, A., Spielman, D.A., Srivastava: Interlacing families II: Mixed characteristic polynomials and the Kadison–Singer problem (2013, preprint)

[518] Meyer, F.G., Coifman, R.: Brushlets: a tool for directional image analysis and image compression. Appl. Comput. Harm.Anal. **4**(2), 147–187 (1997)

[519] Meyer, Y.: Principe d'incertitude, bases hilbertiennes et algebres d'operateurs. Seminaire Bourbaki nr. 662 (1985–1986)

[520] Meyer, Y.: Wavelets and Operators. Herman, Paris (1990)

[521] Mixon, D.G., Quinn, C.J., Kiyavash, N., Fickus, M.: Fingerprinting with equiangular tight frames. IEEE Trans. Inf. Theory **59**(3), 1855–1865 (2013)

[522] Munch, N.J.: Noise reduction in tight Weyl-Heisenberg frames. IEEE Trans. Inf. Theory **38**(2), 608–616 (1992)

[523] Naimark, M.A.: Normed rings. Translated from the first russian version by L. Boron. P. Noordhoff N.V., Groningen (1964)

[524] Neumann, C.: Untersuchungen über das Logarithmische und Newtonsche Potential. Teubner, Leipzig (1877)

[525] von Neumann, J.: Mathematische Grundlagen der Quantenmechanik. Springer, Berlin (1932). English translation: Mathematical Foundations of Quantum Mechanics. Princeton University Press, Princeton (1955)

[526] Nielsen, M.: Frames for decomposition spaces generated by a single function. Collect. Math. **65**(2), 183–201 (2014)

[527] Nielsen, M., Rasmussen, K.N.: Compactly supported frames for decomposition spaces. J. Fourier Anal. Appl. **18**(1), 87–117 (2012)

[528] Nitzan, S., Olevskii, A.: Quasi-frames of translates. C. R. Math. Acad. Sci. Paris **347**(13–14), 739–742 (2009)

[529] Olsen, P.A., Seip, K.: A note on irregular discrete wavelet transforms. IEEE Trans. Inf. Theory **38**(2), 861–863 (1992)

[530] Olson, T.E., Zalik, R.A.: Nonexistence of a Riesz basis of translates. In: Anastassiou, G.A. (ed.) Approximation Theory. Lecture Notes in Pure and Applied Mathematics, vol. 138, pp. 401–418. Marcel Dekker, New York (1992)

[531] Ortega-Cerda, J., Seip, K.: Fourier frames. Ann. Math. **155**(3), 789–806 (2002)

[532] Oswald, P.: Frames in Sobolev spaces. Notes (1997)

[533] Paley, R.E.A.C., Wiener, N.: Fourier transforms in complex domains. AMS Colloq. Publ. **19** (1934)

[534] Paluszynski, M., Sikic, H., Weiss, G., Xiao, S.: Generalized low pass filters and MRA frame wavelets. J. Geom. Anal. **11**(2), 311–342 (2001)

[535] Papadakis, M.: Generalized frame multiresolution analysis of abstract Hilbert spaces. In: Sampling, Wavelets, and Tomography. Applied and Numerical Harmonic Analysis, pp. 179-223. Birkhäuser, Boston (2004)

[536] Pelczynski, A., Singer, I.: On non-equivalent bases and conditional bases in Banach spaces. Stud. Math. **25**, 5–25 (1964)

[537] Peng, I., Waldron, S.: Signed frames and Hadamard products of Gram matrices. Linear Algebra Appl. **347**, 131–157 (2002)

[538] Petukhov, A.: Explicit construction of framelets. Appl. Comput. Harmon. Anal. **11**, 313–327 (2001)

[539] Petukhov, A.: Symmetric framelets. Constr. Approx. **19**, 309–328 (2003)

[540] Pfander, G:. Sampling of operators. J. Fourier Anal. Appl. **19**(3), 612–650 (2013)

[541] Pfander, G.: Gabor frames in finite dimensions. In: Casazza, P., Kutyniok, G. (eds.) Finite frames, chap. 6, pp. 193–240. Birkhäuser, Boston (2013)

[542] Pfander, G., Rashkov, P.: Remarks on multivariate Gaussian Gabor frames. Monatsh. Math. **172**, 179–187 (2013)

[543] Pfander, G., Rashkov, P., Wang, Y.: A geometric construction of tight multivariate Gabor frames with compactly supported smooth windows. J. Fourier Anal. Appl. **18**(2), 223–239 (2012)

[544] Pfander, G., Zheltov, P.: Sampling of stochastic operators. IEEE Trans. Inf. Theory **60**(4), 2359–2372 (2014)

[545] Pfander, G., Walnut, D.: Measurement of time-variant linear channels. IEEE Trans. Inf. Theory **52**(11), 4808–4820 (2006)

[546] Pilipovic, S., Stoeva, D.T.: Series expansions in Frechet spaces and their duals, construction of Frechet frames. J. Approx. Theory **163**(11), 1729–1747 (2011)

[547] Powell, A.M., Saab, R., Yilmaz, Ö: Quantization and finite frames. In: Casazza, P., Kutyniok, G. (eds.) Finite Frames, chap. 8. Birkhäuser, Boston (2012)

[548] Qiu, S.: The undersampled discrete Gabor transform. IEEE Trans. Signal. Proc. **46**(5), 1221–1228 (1998)

[549] Qiu, S.: Block-circulant Gabor matrix structure and Gabor transforms. Opt. Eng. **34**, 2872–2878 (1995)

[550] Ramanathan, R., Steger, T.: Incompleteness of spares coherent states. Appl. Comput. Harmon. Anal. **2**, 148–153 (1995)

[551] Reid, R.M.: A class of Riesz-Fischer sequences. Proc. Am. Math. Soc. **123**(3), 827–829 (1995)

[552] Reiter, H.: Classical Harmonic Analysis and Locally Compact Groups. Oxford Mathematical Monographs (1968)

[553] Reiter, H., Stegeman, J.D.: Classical Harmonic Analysis and Locally Compact Groups, 2nd edn. Oxford University Press, Oxford (2000)

[554] Rieffel, M.A.: Von Neumann algebras associated with pairs of lattices in Lie groups. Math. Anal. **257**, 403–418 (1981)

[555] Rieffel, M.A.: Projective modules over higher-dimensional noncommutative tori. Can. J. Math. **40**(2), 257–338 (1988)

[556] Rochberg, R., Tachizawa, K.: Pseudodifferential operators, Gabor frames, and local trigonometric bases. In: Feichtinger, H.G., Strohmer, T. (eds.) Gabor Analysis: Theory and Application. Birkhäuser, Boston (1998)

[557] Romero, J.L.: Surgery of spline-type and molecular frames. J. Fourier Anal. Appl. **17**(1), 135–174 (2011)

[558] Ron, A.: Exponential B-splines. Const.Appr. **4**, 357–378 (1998)

[559] Ron, A., Shen, Z.: Frames and stable bases for shift-invariant subspaces of $L^2(\mathbb{R}^d)$. Can. J. Math. **47**(5), 1051–1094 (1995)

[560] Ron, A., Shen, Z.: Weyl-Heisenberg systems and Riesz bases in $L^2(\mathbb{R}^d)$. Duke Math. J. **89**, 237–282 (1997)

[561] Ron, A., Shen, Z.: Affine systems in $L_2(\mathbb{R}^d)$: the analysis of the analysis operator. J. Funct. Anal. **148**, 408–447 (1997)

[562] Ron, A., Shen, Z.: Affine systems in $L_2(R^d)$ II: dual systems. J. Fourier Anal. Appl. **3**, 617–637 (1997)

[563] Ron, A., Shen, Z.: Compactly supported tight affine spline frames in $L_2(R^d)$. Math. Comput. **67**, 191–207 (1998)

[564] Ron, A., Shen, Z.: Generalized shift-invariant systems. Const. Appr. **22**(1), 1–45 (2005)

[565] Rudin, W.: Real and Complex Analysis. McGraw-Hill, New York (1986)

[566] Rudin, W.: Functional Analysis. McGraw-Hill, New York (1973)

[567] Rudin, W.: Fourier Analysis on Groups. Interscience Publishers, New York (1962)

[568] San Antolin, A., Zalik, R.: Some smooth compactly supported tight framelets. Commun. Math. Appl. **3**, 343–353 (2012)

[569] Schauder, J.: Zur theorie stetiger Abbildingen in Funktionalräumen. Math. Z. **26**, 47–65, 417–431 (1927)

[570] Seip, K.: On the connection between exponential bases and certain related sequences in $L^2(-\pi, \pi)$. J. Funct. Anal. **130**, 131–160 (1995)

[571] Seip, K.: Density theorems for sampling and interpolation in the Bargmann-Fock space I. J. Reine Angew. Math. **429**, 91–106 (1992)

[572] Seip, K.: A simple construction of exponential bases in L^2 of the union of several intervals. Proc. Edinb. Math. Soc. **38**, 171–177 (1995)

[573] Seip, K.: Wavelets in $H^2(\mathbb{R})$: sampling, interpolation, and phase space density. In: Chui, C. (ed.) Wavelets: A Tutorial in Theory and Practice, pp. 529–540. Academic, San Diego (1992)

[574] Seip, K., Ortega, J.: Fourier frames. Ann. Math. **155**, 789–806 (2002)

[575] Seip, K., Wallsten, R.: Density theorems for sampling and interpolation in the Bargmann-Fock space II. J. Reine Angew. Math. **429**, 107–113 (1992)

[576] Selesnick, I.: Smooth wavelet tight frames with zero moments. Appl. Comput. Harmon. Anal. **10**(2), 163–181 (2001)

[577] Singer, I.: Bases in Banach Spaces 1. Springer, New York (1970)

[578] Song, G., Gelb, A.: Approximating the inverse frame operator from localized frames. Appl. Comput. Harmon. Anal. **35**, 94–110 (2013)

[579] Stoeva, D.T.: Generalization of the frame operator and the canonical dual frame to Banach spaces. Asian-Eur. J. Math. **1**(4), 631–643 (2008)

[580] Stoeva, D.T., Christensen, O.: On various R-duals and the duality principle. J. Fourier Anal. Appl. **21**(2), 383–400 (2015)

[581] Stöckler, J.: A Laurent operator technique for multivariate frames and wavelet bases. In: Fontanella, F., Jetter, K., Laurent, P.J. (eds.) Advanced Topics in Multivariate Approximation, pp. 339–354. World Scientific Publishing Co., River Edge (1996)

[582] Stöckler, J.: The general structure of multivariate affine frames. In: Haußmann, W., Jetter, K., Reimer, M. (eds.) Multivariate Approximation: Recent Trends and Results, pp. 287–302. Akademie, Berlin (1997)

[583] Stöckler, J.: Affine frames and multiresolution. In: Nürnberger, G., Schmidt, J.W., Walz, G. (eds.) Multivariate Approximation and Splines, pp. 307–320. Birkhäuser, Boston (1997)

[584] Strang, G.: Linear Algebra and Its Applications. Academic, New York (1980)

[585] Strang, G., Nguyen, T.: Wavelets and Filter Banks. Wellesley-Cambridge Press, Cambridge (1997)

[586] Strohmer, T.: Numerical algorithms for discrete Gabor expansions. In: Feichtinger, H.G., Strohmer, T. (eds.) Gabor Analysis: Theory and Application. Birkhäuser, Boston (1998)

[587] Strohmer, T.: Rates of convergence for the approximation of dual shift-invariant systems. J. Fourier Anal. Appl. **5**(6), 599–615 (1999)

[588] Strohmer, T.: Approximation of dual Gabor frames, window decay and wireless communications. Appl. Comput. Harmon. Anal. **11**(2), 243–262 (2001)

[589] Strohmer, T.: Pseudodifferential operators and Banach algebras in mobile communications. Appl. Comput. Harmon. Anal. **20**(2), 237–249 (2006)

[590] Strohmer, T.: A note on equiangular tight frames. Linear Algebra Appl. **429**(1), 326–330 (2008)

[591] Strohmer, T., Heath, R.: Grassmannian frames with applications to coding and communication. Appl. Comput. Harmon. Anal. **14**, 257–275 (2003)

[592] Strömberg, J.O.:A modified Franklin system and higher order spline systems on $\mathbb{R}^n$ as unconditional bases for Hardy spaces. In: Beckner, W. (ed.) Conference on Harmonic Analysis in Honor of A. Zygmund. The Wadsworth Mathematics Series, vol. II, pp. 475–493. Wadsworth, Belmont (1983)

[593] Sun, W.: G-frames and g-Riesz bases. J. Math. Anal. Appl. **322**, 437–452 (2006)

[594] Sun, W.: Density of wavelet frames. Appl. Comput. Harmon. Anal. **22**, 264–272 (2007)

[595] Sun, W.: Stability of g-frames. J. Math. Anal. Appl. **326**, 858–868 (2007)

[596] Sun, W.: Homogeneous approximation property for wavelet frames with matrix dilations. Math. Nachr. **283**, 1488–1505 (2010)

[597] Sun, W.: Homogeneous approximation property for wavelet frames. Monatsh. Math. **159**, 289–324 (2010)

[598] Sun, W.: Asymptotic properties of Gabor frame operators as sampling density tends to infinity. J. Funct. Anal. **258**, 913–932 (2010)

[599] Sun, W., Zhou, X.: On Kadec's 1/4-theorem and the stability of Gabor frames. Appl. Comput. Harmon. Anal. **7**(2), 239–242 (1999)

[600] Sun, W., Zhou, X.: On the stability of multivariate trigonometric systems. J. Math. Anal. Appl. **235**, 159–167 (1999)

[601] Sun, W., Zhou, X.: A sharper stability bound of Fourier frames. J. Fourier Anal. Appl. **5**(1), 67–71 (1999)

[602] Sun, W., Zhou, X.: On the stability of Gabor frames. Adv. Appl. Math. **26**, 181–191 (2001)

[603] Sun, W., Zhou, X.: Reconstruction of band-limited functions from local averages. Constr. Approx. **18**, 205–222 (2002)

[604] Sun, W., Zhou, X.: Irregular wavelet/Gabor frames. Appl. Comput. Harmon. Anal. **13**(1), 63–76 (2002)

[605] Sun, W., Zhou, X.: Density and stability of wavelet frames. Appl. Comput. Harmon. Anal. **15**, 117–133 (2003)

[606] Sustik, M.A., Tropp, J.A., Dhillon, I.S., Heath, R.W., Jr.: On the existence of equiangular tight frames. Linear Algebra Appl. **426**(2–3), 619–635 (2007)

[607] Søndergaard, P.: Gabor frames by sampling and periodization. Adv. Comput. Math. **27**(4), 355–373 (2007)

[608] Søndergaard, P., Torresani, B., Balazs, P.: The linear time frequency analysis toolbox. Int. J. Wavelets Multiresolution Inf. Process. **10**(4) (2012)

[609] Tachizawa, K.: The pseudodifferential operators and Wilson bases. J. Math. Pures appl. **75**, 509–529 (1996)

[610] Tang, W.S.: Oblique projections, biorthogonal Riesz bases and multiwavelets in Hilbert spaces. Proc. Am. Math. Soc. **128**, 463–473 (1999)

[611] Tang, W.S., Weber, E.: Frame vectors for representations of abelian groups. Appl. Comput. Harmon. Anal. **20**, 283–297 (2006)

[612] Temlyakov, V.N.: Greedy algorihms and m-term approximation with regard to overcomplete dictionaries. J. Approx. Theory **98**(1), 117–145 (1999)

[613] Tikhomirov, V.M.: Harmonic tools for approximation and splines on locally compact abelian groups. Usp. Mat. Nauk **49** (1994), 193–194. Translated in Russian Math. Surveys **49**, 200–201 (1994)

[614] Toft, J.: Continuity properties for modulation spaces with applications to pseudo-differential operators I. J. Funct. Anal. **207**, 399–429 (2004)

[615] Tolimieri, R., Orr, R.: Poisson summation, the ambiguity function, and the theory of Weyl-Heisenberg frames. J. Fourier Anal. Appl. **1**(3), 233–247 (1995)

[616] Trebels, B., Steidl, G.: Riesz bounds of Wilson bases generated by B-splines. J. Fourier Anal. Appl. **6**(2), 171–184 (2000)

[617] Tropp, J.A., Dhillon, I.S.; Heath, R.W., Jr., Strohmer, T.: Designing structured tight frames via an alternating projection method. IEEE Trans. Inf. Theory **51**(1), 188–209 (2005)

[618] Xia, P., Zhou, S., Giannakis, S.B.: Achieving the Welch Bound with difference sets. IEEE Trans. Inf. Theory **51**(5), 1900–1907 (2005)

[619] Yang, D.Y., Zhou, X.W., Yuan, Z.Z.: Frame wavelets with compact supports for $L^2(\mathbb{R}^n)$. Acta Math. Sin. (Engl. Ser.) **23**(2), 349–356 (2007)

[620] Yoon, G.J., Heil, C.: Duals of weighted exponential systems. Acta. Appl. Math. **119**, 97–112 (2012)

[621] Yosida, K.: Functional Analysis. Springer, Berlin (1980)

[622] Young, R.: An Introduction to Nonharmonic Fourier Series. Academic, New York (1980) (revised first edition 2001)

[623] Young, R.: Interpolation in a classical Hilbert space of entire functions. Trans. Am. Math. Soc. **192**, 97–114 (1974)

[624] Unser, M., Blu, T.: Cardinal exponential B-splines: Part I – Theory and filtering algorithms. IEEE Trans. Signal Process. **53**, 1425–1438 (2005)

[625] Vershynin, R.: Subsequences of frames. Stud. Math. **145**(3), 185–197 (2001)

[626] Vetterli, M., Kovačević, J.: Wavelets and Subband Coding. Prentice-Hall, Englewood Cliffs (1995)

[627] Vetterli, M., Kovačević, J., Goyal, V.K.: Foundations of Signal Processing. Cambridge University Press, Cambridge (2015)

[628] Voss, J.: On discrete and continuous norms in Paley-Wiener spaces and consequences for exponential frames. J. Fourier Anal. Appl. **5**(2/3), 193–201 (1999)

[629] Vretblad, A.: Fourier Analysis and Applications. Springer, New York (2003)

[630] Walnut, D.: Weyl-Heisenberg wavelet expansions: existence and stability in weighted spaces. Ph.D. Thesis, University of Maryland, College Park (1989)

[631] Walnut, D.: Continuity properties of the Gabor frame operator. J. Math. Anal. Appl. **165**(2), 479–504 (1992)

[632] Walnut, D.: Lattice size estimates for Gabor decompositions. Monathsh. Math. **115**(3), 245–256 (1993)

[633] Walnut, D.: An Introduction to Wavelet Analysis. Birkhäuser, Boston (2001)

[634] Walter, G.: A sampling theorem for wavelet subspaces. IEEE Trans. Inf. Theory **38**(2), 881–884 (1992)

[635] Weber, E.: Orthogonal frames of translates. Appl. Comput. Harmon. Anal. **17**, 69–90 (2004)

[636] Wexler, J., Raz, S.: Discrete Gabor expansions. Signal Process. **21**, 207–220 (1990)

[637] Wojtaszczyk, P.: A Mathematical Introduction to Wavelets. Cambridge University Press, Cambridge (1999)

[638] Zalik, R.A.: Riesz bases and multiresolution analysis. Appl. Comput. Harmon. Anal. **7**, 315–331 (1999)

[639] Zalik, R.A.: On MRA Riesz wavelets. Proc. Am. Math. Soc. **135**, 777–785 (2007)

[640] Zalik, R.A.: Bases of translates and multiresolution analyses. Appl. Comput. Harmon. Anal. **24**, 41–57 (2008). Corrigendum in **29**, 121 (2010)

[641] Zang, L., Sun, W.: Inequalities for irregular Gabor frames. Monatsh. Math. **154**(1), 71–81 (2008)

[642] Zimmermann, G.: Normalized tight frames in finite dimensions. In: Jetter, K., Haußmann, W., Reimer, M. (eds.) Recent Progress in Multivariate Approximation, pp. 249–252. Birkhäuser, Boston (2001)

[643] Zibulski, M., Zeevi, Y.Y.: Oversampling in the Gabor scheme. IEEE Trans. SP **41**(8), 2679–2687 (1993)

[644] Zwaan, M.: Error estimates for nonuniform sampling. Numer. Funct. Anal. Optim. **11**(5 & 6), 589–599 (1990)

[645] Zwaan, M.: Approximation of the solution to the moment problem. Numer. Funct. Anal. Optimi. **11**(5 & 6), 601–608 (1990)

Index

© Springer International Publishing Switzerland 2016

O. Christensen, *An Introduction to Frames and Riesz Bases*,

Applied and Numerical Harmonic Analysis,

DOI 10.1007/978-3-319-25613-9

Applied and Numerical Harmonic Analysis (72 volumes)

A. Saichev and W.A. Woyczyski: *Distributions in the Physical and Engineering Sciences* (ISBN 978-0-8176-3924-2)

C.E. D'Attellis and E.M. Fernandez-Berdaguer: *Wavelet Theory and Harmonic Analysis in Applied Sciences* (ISBN 978-0-8176-3953-2)

H.G. Feichtinger and T. Strohmer: *Gabor Analysis and Algorithms* (ISBN 978-0-8176-3959-4)

R. Tolimieri and M. An: *Time-Frequency Representations* (ISBN 978-0-8176-3918-1)

T.M. Peters and J.C. Williams: *The Fourier Transform in Biomedical Engineering* (ISBN 978-0-8176-3941-9)

G.T. Herman: *Geometry of Digital Spaces* (ISBN 978-0-8176-3897-9)

A. Teolis: *Computational Signal Processing with Wavelets* (ISBN 978-0-8176-3909-9)

J. Ramanathan: *Methods of Applied Fourier Analysis* (ISBN 978-0-8176-3963-1)

J.M. Cooper: *Introduction to Partial Differential Equations with MATLAB* (ISBN 978-0-8176-3967-9)

A. Procházka, N.G. Kingsbury, P.J. Payner, and J. Uhlir: *Signal Analysis and Prediction* (ISBN 978-0-8176-4042-2)

W. Bray and C. Stanojevic: *Analysis of Divergence* (ISBN 978-1-4612-7467-4)

© Springer International Publishing Switzerland 2016   701
O. Christensen, *An Introduction to Frames and Riesz Bases*,
Applied and Numerical Harmonic Analysis,
DOI 10.1007/978-3-319-25613-9

G.T. Herman and A. Kuba: *Discrete Tomography*
(ISBN 978-0-8176-4101-6)
K. Gröchenig: *Foundations of Time-Frequency Analysis*
(ISBN 978-0-8176-4022-4)
L. Debnath: *Wavelet Transforms and Time-Frequency Signal Analysis*
(ISBN 978-0-8176-4104-7)
J.J. Benedetto and P.J.S.G. Ferreira: *Modern Sampling Theory*
(ISBN 978-0-8176-4023-1)
D.F. Walnut: *An Introduction to Wavelet Analysis*
(ISBN 978-0-8176-3962-4)
A. Abbate, C. DeCusatis, and P.K. Das: *Wavelets and Subbands*
(ISBN 978-0-8176-4136-8)
O. Bratteli, P. Jorgensen, and B. Treadway: *Wavelets Through a Looking
Glass* (ISBN 978-0-8176-4280-80)
H.G. Feichtinger and T. Strohmer: *Advances in Gabor Analysis*
(ISBN 978-0-8176-4239-6)
O. Christensen: *An Introduction to Frames and Riesz Bases*
(ISBN 978-0-8176-4295-2)
L. Debnath: *Wavelets and Signal Processing* (ISBN 978-0-8176-4235-8)
G. Bi and Y. Zeng: *Transforms and Fast Algorithms for Signal Analysis
and Representations* (ISBN 978-0-8176-4279-2)
J.H. Davis: *Methods of Applied Mathematics with a MATLAB Overview*
(ISBN 978-0-8176-4331-7)
J.J. Benedetto and A.I. Zayed: *Modern Sampling Theory*
(ISBN 978-0-8176-4023-1)
E. Prestini: *The Evolution of Applied Harmonic Analysis*
(ISBN 978-0-8176-4125-2)
L. Brandolini, L. Colzani, A. Iosevich, and G. Travaglini: *Fourier Analysis
and Convexity* (ISBN 978-0-8176-3263-2)
W. Freeden and V. Michel: *Multiscale Potential Theory*
(ISBN 978-0-8176-4105-4)
O. Christensen and K.L. Christensen: *Approximation Theory*
(ISBN 978-0-8176-3600-5)
O. Calin and D.-C. Chang: *Geometric Mechanics on Riemannian Manifolds*
(ISBN 978-0-8176-4354-6)
J.A. Hogan: *Time?Frequency and Time?Scale Methods*
(ISBN 978-0-8176-4276-1)
C. Heil: *Harmonic Analysis and Applications* (ISBN 978-0-8176-3778-1)
K. Borre, D.M. Akos, N. Bertelsen, P. Rinder, and S.H. Jensen: *A Software-
Defined GPS and Galileo Receiver* (ISBN 978-0-8176-4390-4)
T. Qian, M.I. Vai, and Y. Xu: *Wavelet Analysis and Applications* (ISBN
978-3-7643-7777-9)
G.T. Herman and A. Kuba: *Advances in Discrete Tomography and Its
Applications* (ISBN 978-0-8176-3614-2)

M.C Fu, R.A. Jarrow, J.-Y. Yen, and R.J. Elliott: *Advances in Mathematical Finance* (ISBN 978-0-8176-4544-1)

O. Christensen: *Frames and Bases* (ISBN 978-0-8176-4677-6)

P.E.T. Jorgensen, J.D. Merrill, and J.A. Packer: *Representations, Wavelets, and Frames* (ISBN 978-0-8176-4682-0)

M. An, A.K. Brodzik, and R. Tolimieri: *Ideal Sequence Design in Time-Frequency Space* (ISBN 978-0-8176-4737-7)

S.G. Krantz: *Explorations in Harmonic Analysis* (ISBN 978-0-8176-4668-4)

B. Luong: *Fourier Analysis on Finite Abelian Groups*
(ISBN 978-0-8176-4915-9)

G.S. Chirikjian: *Stochastic Models, Information Theory, and Lie Groups, Volume 1* (ISBN 978-0-8176-4802-2)

C. Cabrelli and J.L. Torrea: *Recent Developments in Real and Harmonic Analysis* (ISBN 978-0-8176-4531-1)

M.V. Wickerhauser: *Mathematics for Multimedia*
(ISBN 978-0-8176-4879-4)

B. Forster, P. Massopust, O. Christensen, K. Gröchenig, D. Labate, P. Vandergheynst, G. Weiss, and Y. Wiaux: *Four Short Courses on Harmonic Analysis* (ISBN 978-0-8176-4890-9)

O. Christensen: *Functions, Spaces, and Expansions*
(ISBN 978-0-8176-4979-1)

J. Barral and S. Seuret: *Recent Developments in Fractals and Related Fields* (ISBN 978-0-8176-4887-9)

O. Calin, D.-C. Chang, and K. Furutani, and C. Iwasaki: *Heat Kernels for Elliptic and Sub-elliptic Operators* (ISBN 978-0-8176-4994-4)

C. Heil: *A Basis Theory Primer* (ISBN 978-0-8176-4686-8)

J.R. Klauder: *A Modern Approach to Functional Integration*
(ISBN 978-0-8176-4790-2)

J. Cohen and A.I. Zayed: *Wavelets and Multiscale Analysis*
(ISBN 978-0-8176-8094-7)

D. Joyner and J.-L. Kim: *Selected Unsolved Problems in Coding Theory*
(ISBN 978-0-8176-8255-2)

G.S. Chirikjian: *Stochastic Models, Information Theory, and Lie Groups, Volume 2* (ISBN 978-0-8176-4943-2)

J.A. Hogan and J.D. Lakey: *Duration and Bandwidth Limiting*
(ISBN 978-0-8176-8306-1)

G. Kutyniok and D. Labate: *Shearlets* (ISBN 978-0-8176-8315-3)

P.G. Casazza and P. Kutyniok: *Finite Frames* (ISBN 978-0-8176-8372-6)

V. Michel: *Lectures on Constructive Approximation*
(ISBN 978-0-8176-8402-0)

D. Mitrea, I. Mitrea, M. Mitrea, and S. Monniaux: *Groupoid Metrization Theory* (ISBN 978-0-8176-8396-2)

T.D. Andrews, R. Balan, J.J. Benedetto, W. Czaja, and K.A. Okoudjou: *Excursions in Harmonic Analysis, Volume 1* (ISBN 978-0-8176-8375-7)

T.D. Andrews, R. Balan, J.J. Benedetto, W. Czaja, and K.A. Okoudjou: *Excursions in Harmonic Analysis, Volume 2* (ISBN 978-0-8176-8378-8)

D.V. Cruz-Uribe and A. Fiorenza: *Variable Lebesgue Spaces* (ISBN 978-3-0348-0547-6)

W. Freeden and M. Gutting: *Special Functions of Mathematical (Geo-)Physics* (ISBN 978-3-0348-0562-9)

A. Saichev and W.A. Woyczyski: *Distributions in the Physical and Engineering Sciences, Volume 2: Linear and Nonlinear Dynamics of Continuous Media* (ISBN 978-0-8176-3942-6)

S. Foucart and H. Rauhut: *A Mathematical Introduction to Compressive Sensing* (ISBN 978-0-8176-4947-0)

G. Herman and J. Frank: *Computational Methods for Three-Dimensional Microscopy Reconstruction* (ISBN 978-1-4614-9520-8)

A. Paprotny and M. Thess: *Realtime Data Mining: Self-Learning Techniques for Recommendation Engines* (ISBN 978-3-319-01320-6)

A. Zayed and G. Schmeisser: *New Perspectives on Approximation and Sampling Theory: Festschrift in Honor of Paul Butzer's 85th Birthday* (978-3-319-08800-6)

R. Balan, M. Begue, J. Benedetto, W. Czaja, and K.A Okoudjou: *Excursions in Harmonic Analysis, Volume 3* (ISBN 978-3-319-13229-7)

H. Boche, R. Calderbank, G. Kutyniok, J. Vybiral: *Compressed Sensing and its Applications* (ISBN 978-3-319-16041-2)

S. Dahlke, F. De Mari, P. Grohs, and D. Labate: *Harmonic and Applied Analysis: From Groups to Signals* (ISBN 978-3-319-18862-1)

G. Pfander: *Sampling Theory, a Renaissance* (ISBN 978-3-319-19748-7)

R. Balan, M. Begue, J. Benedetto, W. Czaja, and K.A Okoudjou: *Excursions in Harmonic Analysis, Volume 4* (ISBN 978-3-319-20187-0)

O. Christensen: *An Introduction to Frames and Riesz Bases, Second Edition* (ISBN 978-3-319-25611-5)

For an up-to-date list of ANHA titles, please visit http://www.springer.com/series/4968

Printed in the United States
By Bookmasters